Lecture Notes in Mathematics

2034

Editors:
J.-M. Morel, Cachan
B. Teissier, Paris

For further volumes:
http://www.springer.com/series/304

Andrea Bonfiglioli • Roberta Fulci

Topics in Noncommutative Algebra

The Theorem of Campbell, Baker, Hausdorff and Dynkin

 Springer

Andrea Bonfiglioli
Roberta Fulci
University of Bologna
Department of Mathematics
Piazza di Porta San Donato 5
40126 Bologna
Italy
bonfigli@dm.unibo.it
fulci@dm.unibo.it

ISBN 978-3-642-22596-3   e-ISBN 978-3-642-22597-0
DOI 10.1007/978-3-642-22597-0
Springer Heidelberg Dordrecht London New York

Lecture Notes in Mathematics ISSN print edition: 0075-8434
                              ISSN electronic edition: 1617-9692

Library of Congress Control Number: 2011937714

Mathematics Subject Classification (2010): 22-XX, 01-XX, 17-XX, 53-XX

Printed on acid-free paper

Springer is part of Springer Science+Business Media (www.springer.com)

In memory of Raffaele Bonfiglioli

Preface

SUPPOSE you were asked if the so-called Fundamental Theorem of Algebra is a result of Algebra or of Analysis. What would you answer? And again, among the arguments proving that the complex field is algebraically closed, which would you choose? A proof making use of the concept of continuity? Or of analytic functions, even? Or of Galois theory, instead?

The central topic of this book, the mathematical result named after the mathematicians *Henry Frederick Baker, John Edward Campbell, Eugene Borisovich Dynkin* and *Felix Hausdorff*, shares – together with the Fundamental Theorem of Algebra – the remarkable property to cross, by its nature, the realms of many mathematical disciplines: Algebra, Analysis, Geometry.

As for the proofs of this theorem (named henceforth CBHD, in chronological order of the contributions), it will be evident in due course of the book that the intertwining of Algebra and Analysis is especially discernible: We shall present arguments making use – all at the same time – of topological algebras, the theory of power series, ordinary differential equations techniques, the theory of Lie algebras, metric spaces; and more.

If we glance at the fields of application of the CBHD Theorem, it is no surprise that so many different areas are touched upon: the theory of Lie groups and Lie algebras; linear partial differential equations; quantum and statistical mechanics; numerical analysis; theoretical physics; control theory; sub-Riemannian geometry; and more.

Curiously, the CBHD Theorem crosses our path already at an early stage of our secondary school education ($e^x e^y = e^{x+y}$, $x, y \in \mathbb{R}$); then it reappears as a nontrivial problem at the very beginning of our university studies, yet in the simplest of non-Abelian contexts (is there a formula for $e^A e^B$ when A, B are square matrices?); then, as our mathematical background progresses, we become acquainted with deeper and more natural settings where this theorem plays a rôle (for example if we face the problem of writing $\mathrm{Exp}(X) \cdot \mathrm{Exp}(Y)$ in logarithmic coordinates, when X, Y belong to the Lie algebra of a Lie group); finally, when our undergraduate studies are complete, we may happen – in all likelihood – to meet the CBHD Theorem again if we are researchers into one of the mathematical fields mentioned a few lines above.

Since the early 1897-98 studies by Campbell on this problem, more than 110 years have passed. Still, the problem of the "multiplication of two exponentials" (whatever the context) has not ceased to provide sources for new questions. Take, for instance, the problem of finding the optimal domain of convergence for the series naturally attached to $\log(e^x e^y)$ (when x, y belong to an arbitrary non-Abelian Banach algebra), a problem which is still not solved in complete generality. Or consider the question of finding more natural and more fitting proofs of the CBHD Theorem, a question which has been renewed – at repeated intervals – in the literature. Indeed, mathematicians have gone on feeling the need for new and simpler proofs of the CBHD Theorem throughout the last century: See for example the papers [183, *1937*], [33, *1956*], [59, *1968*], [48, *1975*], [174, *1980*], [169, *2004*]; and many books may be cited: [99, *1962*], [159, *1964*], [85, *1965*], [79, *1968*], [27, *1972*], [151, *1973*], [171, *1974*], [70, *1982*], [84, *1991*], [144, *1993*], [72, *1997*], [91, *1998*], [52, *2000*], [149, *2002*], [77, *2003*], [1, *2007*], [158, *2007*] – and this is just a small sample; an exhaustive list would be very much longer.

The interest in the CBHD Theorem straddling the decades, the very nature of this result ranging over Algebra, Analysis and Geometry, its fields of application stretching across so many branches of Mathematics and Physics, the proofs so variegated and rich in ideas, the engrossing history of the early contributions: all these facts have seemed to us a sufficient incentive and stimulus in devoting this monograph to such a fascinating Theorem.

This book is intended to present in a unified and *self-contained* way the natural context in which the CBHD Theorem can be formulated and proved. This context is purely algebraic, but the proofs – as mentioned – are very rich and diversified. Also, in order to understand and appreciate the varied arguments attacking the proof of the CBHD Theorem, a historical overview of its early proofs is also needed, without forgetting – in due course later in the monograph – to catch a glimpse of more modern studies related to it, the current state-of-the-art and some open problems.

Most importantly, our aim is to look ahead to applications of the CBHD Theorem. In order to arrive at these applications, it is first necessary to deal with the statements and proofs of the CBHD Theorem in the domain of *Algebra*. Then the applications in *Geometry* and in *Analysis* will eventually branch off from this algebraic setting.

Since this book may be used by a non-specialist in Algebra (as he may be, for example, a researcher in PDEs or a quantum physicist who has felt the need for a deeper understanding of a theorem which has been a cornerstone for some part of his studies), our biggest effort here is to furnish an exposition complete with all details and prerequisites. Any Reader, more or less acquainted with the algebraic background, will be free to skip those details he feels fully conversant with.

Now, before revealing the detailed contents of this book (and to avoid singing further praises of the CBHD Theorem), we shall pass on to some brief historical notes about this theorem.

§1. A Brief Historical *Résumé*. A more exhaustive historical overview is provided in Chapter 1. Here we confine ourselves to disclosing some forgotten facts about the history of the theorem we are concerned with.

Though the exponential nature of the composition of two 'exponential transformations' is somehow implicit in Lie's original theory of finite continuous groups (tracing back to the late nineteenth century), the need for an autonomous study of the symbolic identity "$e^x e^y = e^z$" becomes prominent at the very beginning of the twentieth century.

In this direction, F.H. Schur's papers [154]–[157] present some explicit formulas containing – in a very "quantitative" fashion – the core of the Second and Third Fundamental Theorems of Lie: given a Lie algebra, he exhibits suitable multivariate series expansions, only depending on the structure constants of the algebra and on some universal numbers (Bernoulli's), reconstructing a (local) Lie group with prescribed structure. *This is a precursor of the CBHD Theorem.*

Meanwhile, in 1897 [28] – motivated by group theory – Campbell takes up the study of the existence of an element z such that the composition of two finite transformations e^x, e^y of a continuous transformation group satisfies the identity $e^x \circ e^y = e^z$. By means of a not completely transparent series expansion [30], Campbell solves this problem with little reference (if any) to group theory, showing that z can be expressed as a *Lie series* in x, y.

Using arguments not so far from those of Schur, and inspired by the same search for direct proofs of the Fundamental Theorems of Lie, J.H. Poincaré and E. Pascal attack "Campbell's problem" by the use of suitable algebraic manipulations of polynomials (around 1900-1903).

For instance, in studying the identity $e^x e^y = e^z$ from a more symbolic point of view [142], Poincaré brilliantly shapes a tool (he invents the universal enveloping algebra!) allowing him to manage both algebraic and analytic aspects of the problem. Aiming to give full analytical meaning to his formulas, Poincaré introduces a successful *ODE technique*: he derives an ordinary differential equation for $z(t)$ equivalent to the identity $e^{z(t)} = e^x e^{ty}$, whose solution (at $t = 1$) solves Campbell's problem and, at the same time, the Second and Third Fundamental Theorems of Lie. In fact, $z(t)$ can be expressed, by means of the residue calculus, in a suitable (integral) form exhibiting its Lie-series nature. Although Poincaré's contribution is decisive for the late history of the CBHD Theorem, his latent appeal to group theory and the lack of a formula expressing z as a *universal* Lie series in the symbols x, y probably allowed this contribution to die out amid the twists and turns of mathematical history.

Much in the spirit of Poincaré, but with the use of more direct – and more onerous – computations, Pascal pushes forward Poincaré's "symmetrization" of polynomials, in such a way that he is able to rebuild the formal power series $\sum_{m,n\geq 0} \frac{x^m y^n}{m!\, n!}$ as a pure exponential $\sum_{k\geq 0} \frac{z(x,y)^k}{k!}$, where $z(x, y)$ is a Lie series in x, y involving the Bernoulli numbers. Furthermore, Pascal sketches the way the commutator series for $z(x, y)$ can be produced: after the in-embryo formula by Campbell, Pascal's fully fledged results point out (for the first time) the universal Lie series expansibility of z in terms of x, y, a fact which had escaped Poincaré's notice. Though Pascal's papers [135]–[140] will leave Hausdorff and Bourbaki unsatisfied (mostly for the failure to treat the convergence issue and for the massive computations), the fact that in modern times Pascal's contribution to the CBHD Theorem has been almost completely forgotten seems to us to be highly unwarranted.

The final impulse towards a completely *symbolical* version of the Exponential Theorem $e^x e^y = e^z$ is given by Baker [8] and by Hausdorff [78]. The two papers (reasonably independent of each other, for Hausdorff's 1906 paper does not mention Baker's of 1905) use the same technique of 'polar differentiation' to derive for $z(x, y)$ suitable recursion formulas, exhibiting its Lie-series nature. Both authors obtain the same expansion $z = \exp(\delta)(y)$ in terms of a "PDE operator" $\delta = \omega(x, y) \frac{\partial}{\partial y}$, where

$$\omega(x, y) = x + \frac{\operatorname{ad} y}{e^{\operatorname{ad} y} - 1}(x).$$

The Lie series $\omega(x, y)$, besides containing the Bernoulli numbers[1] (reappearing, after Schur, in every proof mentioned above), is nothing but the subseries obtained by collecting – from the expansion of $\log(e^x e^y)$ – the summands containing x precisely once. The same series appeared clearly in Pascal, implicitly in Campbell and in an integral form in Poincaré: it can be rightly considered as the *fil rouge* joining all the early proofs cited so far.

In proving the representation $z = \exp(\delta)(y)$, Baker makes use of quite a puzzling formalism on Lie polynomials, but he is able to draw out of his machinery such abundance of formulas, that it is evident that this is much more than a pure formalism.

However, Hausdorff's approach in proving the formula $z = \exp(\delta)(y)$ is so efficacious and authoritative that it became the main source for future work on the exponential formula, to such an extent that Baker's contribution went partly – but undeservedly – forgotten. (As a proof of this fact we must

---

[1]Indeed, $\frac{z}{e^z - 1} = \sum_{n=0}^{\infty} \frac{B_n}{n!} z^n$.

recall that, in a significant part of the related literature, the Exponential Theorem is named just "Campbell-Hausdorff".) This is particularly true of the commentary of Bourbaki on the early history of this formula, citing Hausdorff as the only "perfectly precise" and reliable source. Admittedly, Hausdorff must be indeed credited (together with fruitful recursion formulas for the coefficients of z) for providing the long awaited *convergence* argument in the set of Lie's transformation groups.

After Hausdorff's 1906 paper, about 40 years elapsed before Dynkin [54] proved another long awaited result: *an explicit presentation for the commutator series of* $\log(e^x e^y)$. Dynkin's paper indeed contains much more: thanks to the explicitness of his representation, Dynkin provides a direct estimate – the first in the history of the CBHD Theorem – for the convergence domain, more general than Hausdorff's. Moreover, the results can be generalized to the infinite-dimensional case of the so-called *Banach-Lie algebras* (to which Dynkin extensively returns in [56]). Finally, Dynkin's series allows us to prove Lie's Third Theorem (a concern for Schur, Poincaré, Pascal and Hausdorff) in an incredibly simple way.

Two years later [55], Dynkin will give another proof of the Lie-series nature of $\log(e^x e^y)$, independently of all his predecessors, a proof disclosing all the *combinatorial* aspects behind the exponential formula. As for the history of the CBHD Theorem, Dynkin's papers [54]–[56] paved the way, by happenstance, for the study of other possible presentations of $\log(e^x e^y)$ and consequently for the problem of further *improved domains of convergence*, dominating the "modern era" of the CBHD Theorem (from 1950 to present days).

This modern age of the Exponential Theorem begins with the first applications to Physics (dating back to the 1960-70s) and especially to Quantum and Statistical Mechanics (see e.g., [51, 64, 65, 104, 119, 128, 178, 180, 181]).

In parallel, a rigorous mathematical formalization of the CBHD Theorem became possible thanks to the Bourbakist refoundation of Mathematics, in particular of Algebra. Consequently, the new proofs of the CBHD Theorem (rather than exploiting *ad hoc* arguments as in all the contributions we cited so far) are based on very general algebraic tools. For example, they can be based on characterizations of Lie elements (as given e.g. by Friedrichs's criterion for primitive elements) as in Bourbaki [27], Hochschild [85], Jacobson [99], Serre [159]. As a consequence, the CBHD Theorem should be regarded (mathematically and historically) as a result from noncommutative algebra, rather than a result of Lie group theory, as it is often popularized in undergraduate university courses.

As a matter of fact, this popularization is caused by the remarkable application of the Exponential Theorem to the structure theory of Lie groups. Indeed, as is well known, in this context the CBHD Theorem allows us to prove a great variety of results: the effective analytic regularity of all smooth Lie groups (an old result of Schur's!), the local "reconstruction" of the group law via the bracket in the Lie algebra, many interesting results on the duality group/algebra homomorphisms, the classifying of the simply connected Lie groups by their Lie algebras, the local version of Lie's Third Theorem, and many others.

For this reason, all major books in Lie group theory starting from the 1960s comprise the CBHD Theorem (mainly named after Campbell, Hausdorff or Baker, Campbell, Hausdorff): See e.g., the classical books (ranging over the years sixties–eighties) Bourbaki [27], Godement [70], Hausner, Schwartz [79], Hochschild [85], Jacobson [99], Sagle, Walde [151], Serre [159], Varadarajan [171]; or the more recent books Abbaspour, Moskowitz [1], Duistermaat, Kolk [52], Gorbatsevich, Onishchik, Vinberg [72], Hall [77], Hilgert, Neeb [84], Hofmann, Morris [91], Rossmann [149], Sepanski [158]. (Exceptions are Chevalley [38], which is dated 1946, and Helgason [81], where only expansions up to the second order are used.)

A remarkable turning point in the history of the CBHD Theorem is provided by Magnus's 1954 paper [112]. In studying the exponential form $\exp(\Omega(t))$ under which the solution $Y(t)$ to the nonautonomous linear ODE system $Y'(t) = A(t)Y(t)$ can be represented, Magnus introduced a formula – destined for a great success – for expanding $\Omega(t)$, later referred to also as the *continuous Campbell-Baker-Hausdorff Formula*. (See also [10, 37, 119, 160, 175].) In fact, in proper contexts and when $A(t)$ has a suitable form, a certain evaluation of the expanded Ω gives back the CBHD series. For a comprehensive treatise on the Magnus expansion, the Reader is referred to Blanes, Casas, Oteo, Ros, 2009 [16] (and to the detailed list of references therein).

Here we confine ourselves to pointing out that the modern literature (mainly starting from the 1980s) regarding the CBHD Theorem has mostly concentrated on the problem of *improved domains of convergence* for the possible different presentations of $\log(e^x e^y)$, both in commutator or non-commutator series expansions (for the latter, see the pioneering paper by Goldberg [71]). Also, the problem of efficient algorithms for computing the terms of this series (in suitable bases for free Lie algebras and with minimal numbers of commutators) has played a major rôle. For this and the above topics, see [9, 14–16, 23, 35, 36, 46, 53, 61, 103, 106, 107, 109, 118, 122, 123, 131, 134, 145, 146, 166, 177].

In the study of convergence domains, the use of the Magnus expansion has proved to be a very useful tool. However, the problem of the best domain of convergence of the CBHD series in the setting of general *Banach algebras* and of general *Banach-Lie algebras* is still open, though many optimal results exist for matrix algebras and in the setting of Hilbert spaces (see the references in Section 5.7 on page 359).

In parallel, starting from the mid seventies, the CBHD Theorem has been crucially employed in the study of wide classes of PDEs, especially of subelliptic type, for example those involving the so-called *Hörmander operators* (see Folland [62], Folland, Stein [63], Hörmander [94], Rothschild,

Stein [150], Nagel, Stein, Wainger [129], Varopoulos, Saloff-Coste, Coulhon [172]).

The rôle of the CBHD Theorem is not only prominent for usual (finite-dimensional) Lie groups, but also within the context of *infinite dimensional* Lie groups (for a detailed survey, see Neeb [130]). For example, among infinite dimensional Lie groups, the so-called BCH-groups (Baker-Campbell-Hausdorff groups) are particularly significant. For some related topics (a comprehensive list of references on infinite-dimensional Lie groups being out of our scope), see e.g., [12, 13, 24, 42, 43, 56, 66–69, 73, 83, 86, 87, 92, 93, 130, 133, 147, 148, 152, 170, 173, 182].

Finally, the early years of the twenty-first century have seen a renewed interest in CBHD-type theorems (both continuous and discrete) within the field of numerical analysis (specifically, in *geometric integration*) see e.g. [76, 97, 98, 101, 114].

§ **2. The Main Contents of This book.** We furnish a very brief digest of the contents of this book. After a historical preamble given in Chapter 1 (also containing reference to modern applications of the CBHD Theorem), the book is divided into two parts.

Part I (Chapters 2–6) begins with an introduction of the background algebra (comprehensive of all the involved notations) which is a prerequisite to the rest of the book. Immediately after such preliminaries, we jump into the heart of the subject. Indeed, Part I treats widely all the qualitative properties and the problems arising from the statement of the CBHD Theorem and from its various proofs, such as the well-posedness of the 'CBHD operation', its associativity and convergence, or the relationship between the CBHD Theorem, the Theorem of Poincaré, Birkhoff and Witt and the existence of free Lie algebras. The results given in Chapter 2, although essential to the stream of the book, would take us a long distance away if accompanied by their proofs. For this reason they are simply stated in Part I, while all the missing proofs can be found in **Part II** (Chapters 7-10).

Let us now have a closer look at the contents of each chapter.

Chapter 2 is entirely devoted to recalling algebraic prerequisites and to introduce the required notations. Many essential objects are introduced, such as tensor algebras, completions of graded algebras, formal power series, free Lie algebras, universal enveloping algebras. Some of the results are demonstrated in Chapter 2 itself, but most of the proofs are deferred to Chapter 7. Chapter 2 is meant to provide the necessary algebraic background to non-specialist Readers and may be skipped by those trained in Algebra. Section 2.3 also contains some needed results (on metric spaces) from Analysis.

Chapter 3 illustrates a complete proof of the CBHD Theorem, mainly relying on the book by Hochschild [85, Chapter X]. The proof is obtained from general results of Algebra, such as Friedrichs's characterization of Lie elements and the use of the Hausdorff group. Afterwards, Dynkin's Formula is produced. This result is conceptually subordinate to the so called Campbell-Baker-Hausdorff Theorem, and it is based, as usual, on the application of the Dynkin-Specht-Wever Lemma.

Our inquiry into the meaningful reasons why the CBHD Theorem holds widens in **Chapter 4**, where several shorter (but more specialized) proofs of the Theorem, differing from each other and from the one given in Chapter 3, are presented. We deal here with the works by M. Eichler [59], D. Ž. Djoković [48], V. S. Varadarajan [171], C. Reutenauer [144], P. Cartier [33].

Eichler's argument is the one most devoid of prerequisites, though crucially *ad hoc* and tricky; Djoković's proof (based on an "ODE technique", partially tracing back to Hausdorff) has the merit to rediscover early arguments in a very concise way; Varadarajan's proof (originally conceived for a Lie group context) completes, in a very effective fashion, Djoković's proof and allows us to obtain recursion formulas perfectly suited for convergence questions; Reutenauer's argument fully formalizes the early approach by Baker and Hausdorff (based on so-called polar differentiation); Cartier's proof, instead, differs from the preceding ones, based as it is on a suitable characterization of Lie elements, in line with the approach of Chapter 3. Each of the strategies presented in Chapter 4 has its advantages, so that the Reader has the occasion to compare them thoroughly and to go into the details for every different approach (and, possibly, to choose the one more suited for his taste or requirements).

In **Chapter 5** the convergence of the Dynkin series is studied, in the context of finite-dimensional Lie algebras first, and then in the more general setting of normed Banach-Lie algebras. Besides, the "associativity" of the operation defined by the Dynkin series is afforded. Throughout this chapter, we shall be exploiting identities implicitly contained (and hidden) in the CBHD Theorem. As a very first taste of the possible (geometrical) applications of the results presented here, we shall have the chance to prove – in a direct and natural fashion – the Third Fundamental Theorem of Lie for finite-dimensional nilpotent Lie algebras (and more). Finally, the chapter closes with a long list (briefly commented, item by item) of modern bibliography on the convergence problem and on related matters.

Chapter 6 clarifies the deep and – in some ways – surprising intertwinement occurring between the CBHD and PBW Theorems (PBW is short for Poincaré-Birkhoff-Witt). As it arises from Chapter 3, CBHD is classically derived by PBW, although other strategies are possible. In Chapter 6 we will show how the opposite path may be followed, thus proving the PBW Theorem by means of CBHD. This less usual approach was first provided by Cartier, whose work [33] is at the basis of the chapter. An essential tool is represented by free Lie algebras, whose rôle – in proving CBHD and PBW – is completely clarified here.

Chapter 7 consists of a collection of the missing proofs from Chapter 2.

Chapter 8 is intended to complete those results of Chapter 2 which deal with the existence of the free Lie algebra Lie(X) related to a set X. The characterization of Lie(X) as the Lie subalgebra (contained in the algebra of the polynomials in the elements of X) consisting of Lie-polynomials is also given in detail (without requiring the PBW Theorem or any of its corollaries). Furthermore, some results about free nilpotent Lie algebras are presented here, helpful e.g., in constructing free Carnot groups (as in [21, Chapters 14, 17]).

An algebraic approach to formal power series can be found in **Chapter 9**.

Finally, **Chapter 10** contains all the machinery about symmetric algebras which is needed in Chapter 6.

§ **3. How to Read This Book.** Since this book is intended for a readership potentially not acquainted with graduate level Algebra, the main effort is to make the presentation *completely self-contained*. The only prerequisites are a basic knowledge of Linear Algebra and undergraduate courses in Analysis and in Algebra. The book opens with a historical overview, Chapter 1, of the early proofs of the CBHD Theorem and a glimpse into more modern results: it is designed not only for historical scholars, but also for the Reader who asked himself the question "Campbell, Baker, Hausdorff, Dynkin: who proved what?".

The algebraic prerequisites are collected in Chapter 2, where the notations used throughout are also collected. The Reader interested in the corresponding proofs will find them in Part II (Chapters from 7 to 10). Chapter 2 and Part II can be skipped by the Reader fully conversant with the algebraic prerequisites. In any case, Chapter 2 must be used as a complete reference for the notations. The Reader interested in the proofs of the CBHD Theorem can directly refer to Chapter 3 (for a more elaborate proof, making use of general algebraic results, in the spirit of the Bourbaki exposition of the subject) or to Chapter 4 (where shorter proofs are presented, but with more *ad hoc* arguments). These chapters require the background results of Chapter 2.

Once the main CBHD Theorem has been established, Chapter 5 presents a primer on the convergence question. The Reader will also find an extended list of references on related topics. Chapter 5 can also be read independently of the preceding chapters, once any proof of the CBHD Theorem is taken for granted. Analogously, Chapter 6 can be read on its own, only requiring some theory of free Lie algebras and of symmetric algebras (coming from Chapters 8 and 10, respectively). A synopsis of the book structure together with the interdependence of the different chapters is given in Figure 1 below.

Acknowledgements. We would like to thank Jean Michel for making available to us the papers [115,118]. We are grateful to Luca Migliorini for his encouragement in the collaboration which has led to this book. We would also like to thank Ian D. Marshall for his help with editing the manuscript. We finally express our sincere thanks to Rüdiger Achilles for enabling us to read the original papers by Hausdorff, Dynkin and Schur.

The present version of the book benefits substantially from several remarks and suggestions – both in form and contents – by the three Referees of the original manuscript, which we most gratefully acknowledge.

It is also a pleasure to thank the Springer-Verlag staff for the kind collaboration, in particular Catriona M. Byrne and Ute McCrory.

Some bibliographical researches, especially those preceding the early 1900s, have been possible only with the help of the staffs of the Libraries of the Department of Mathematics (Rosella Biavati) and of the Department of Physics of the University of Bologna, which we gratefully thank.

Bologna   *Andrea Bonfiglioli*
                                                                                             Roberta Fulci

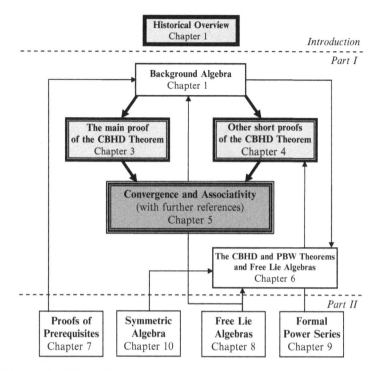

Fig. 1 Synopsis of the book structure

Contents

1   Historical Overview 1
     1.1   The Early Proofs of the CBHD Theorem 1
          1.1.1   The Origin of the Problem 2
          1.1.2   Schur, Poincaré and Pascal 9
          1.1.3   Campbell, Baker, Hausdorff, Dynkin 19
     1.2   The "Modern Era" of the CBHD Theorem 37
     1.3   The "Name of the Game" 44

Part I   Algebraic Proofs of the Theorem of Campbell,
           Baker, Hausdorff and Dynkin

2   Background Algebra 49
     2.1   Free Vector Spaces, Algebras and Tensor Products 49
          2.1.1   Vector Spaces and Free Vector Spaces 49
          2.1.2   Magmas, Algebras and (Unital)
                    Associative Algebras 56
          2.1.3   Tensor Product and Tensor Algebra 72
     2.2   Free Lie Algebras 87
     2.3   Completions of Graded Topological Algebras 93
          2.3.1   Topology on Some Classes of Algebras 94
          2.3.2   Completions of Graded Topological Algebras 98
          2.3.3   Formal Power Series 101
          2.3.4   Some More Notation on Formal Power Series 106
     2.4   The Universal Enveloping Algebra 108

3   The Main Proof of the CBHD Theorem 115
     3.1   Exponential and Logarithm 117
          3.1.1   Exponentials and Logarithms 119
          3.1.2   The Statement of Our Main CBHD Theorem 124
          3.1.3   The Operation $\blacklozenge$ on $\widehat{\mathscr{T}_+}(V)$ 126
          3.1.4   The Operation $\diamond$ on $\widehat{\mathscr{T}_+}(V)$ 128

3.2   The Campbell, Baker, Hausdorff Theorem 132
       3.2.1   Friedrichs's Characterization of Lie Elements......... 133
       3.2.2   The Campbell, Baker, Hausdorff Theorem............ 139
       3.2.3   The Hausdorff Group 142
3.3   Dynkin's Formula 145
       3.3.1   The Lemma of Dynkin, Specht, Wever 145
       3.3.2   Dynkin's Formula 151
       3.3.3   The Final Proof of the CBHD Theorem 154
       3.3.4   Some "Finite" Identities Arising
               from the Equality Between ♦ and ◇ 156
3.4   Résumé: The "Spine" of the Proof of the CBHD Theorem ... 158
3.5   A Few Summands of the Dynkin Series 159
3.6   Further Reading: Hopf Algebras................................. 162

4 Some "Short" Proofs of the CBHD Theorem 173
4.1   Statement of the CBHD Theorem for Formal
       Power Series in Two Indeterminates 178
4.2   Eichler's Proof 187
       4.2.1   Eichler's Inductive Argument 189
4.3   Djokovic's Proof 199
       4.3.1   Polynomials and Series in t over a UA Algebra...... 199
       4.3.2   Background of Djoković's Proof........................ 205
       4.3.3   Djoković's Argument 210
4.4   The "Spine" of the Proof 216
       4.4.1   Yet Another Proof with Formal Power Series 220
4.5   Varadarajan's Proof 223
       4.5.1   A Recursion Formula for the CBHD Series 227
       4.5.2   Another Recursion Formula 229
4.6   Reutenauer's Proof... 231
4.7   Cartier's Proof 253
       4.7.1   Some Important Maps 254
       4.7.2   A New Characterization of Lie Elements 258

5 Convergence of the CBHD Series and Associativity
   of the CBHD Operation 265
5.1   "Finite" Identities Obtained from the CBHD Theorem....... 268
5.2   Convergence of the CBHD Series 277
       5.2.1   The Case of Finite Dimensional Lie Algebras 278
       5.2.2   The Case of Banach-Lie Algebras 292
       5.2.3   An Improved Domain of Convergence............... 301
5.3   Associativity of the CBHD Operation.......................... 305
       5.3.1   "Finite" Identities from the Associativity
               of ◇ 306
       5.3.2   Associativity for Banach-Lie Algebras 312

5.4 Nilpotent Lie Algebras and the Third Theorem of Lie........ 320
    5.4.1 Associativity for Nilpotent Lie Algebras.............. 320
    5.4.2 The Global Third Theorem of Lie
         for Nilpotent Lie Algebras 328
5.5 The CBHD Operation and Series in Banach Algebras 337
    5.5.1 An Alternative Approach Using
         Analytic Functions 347
5.6 An Example of Non-convergence of the CBHD Series 354
5.7 Further References 359

6 Relationship Between the CBHD Theorem,
the PBW Theorem and the Free Lie Algebras 371
6.1 Proving PBW by Means of CBHD 375
    6.1.1 Some Preliminaries 375
    6.1.2 Cartier's Proof of PBW via CBHD 383

Part II Proofs of the Algebraic Prerequisites

7 Proofs of the Algebraic Prerequisites................................. 393
7.1 Proofs of Sect. 2.1.1 393
7.2 Proofs of Sect. 2.1.2 396
7.3 Proofs of Sect. 2.1.3 396
7.4 Proofs of Sect. 2.3.1 407
7.5 Proofs of Sect. 2.3.2 417
7.6 Proofs of Sect. 2.3.3 428
7.7 Proofs of Sect. 2.4 435
7.8 Miscellanea of Proofs 445

8 Construction of Free Lie Algebras 459
8.1 Construction of Free Lie Algebras Continued 459
    8.1.1 Free Lie Algebras over a Set............................ 463
8.2 Free Nilpotent Lie Algebra Generated by a Set.............. 469

9 Formal Power Series in One Indeterminate 479
9.1 Operations on Formal Power Series
in One Indeterminate 480
    9.1.1 The Cauchy Product of Formal Power Series 480
    9.1.2 Substitution of Formal Power Series.................. 481
    9.1.3 The Derivation Operator on Formal
         Power Series... 486
    9.1.4 The Relation Between the $\exp$ and the $\log$ Series 488
9.2 Bernoulli Numbers.. 494

10 Symmetric Algebra 501

    10.1 The Symmetric Algebra and the Symmetric Tensor Space ... 501

        10.1.1 Basis for the Symmetric Algebra 512

    10.2 Proofs of Sect. 10.1 514

A List of the Basic Notation 523

Reference 529

Index 537

Chapter 1
Historical Overview

The period since the CBHD Theorem first came to life, about 120 years ago, can be divided into two distinct phases. The range 1890–1950, beginning with Schur's paper [154] and ending with Dynkin's [56], and the remaining 60-year range, up to the present day. The first range comprises, besides the contributions by the authors whose names are recalled in our acronym CBHD, other significant papers (often left unmentioned) by *Ernesto Pascal, Jules Henri Poincaré, Friedrich Heinrich Schur*.

The second range covers what we will call the "modern era" of the CBHD Theorem: this includes a rigorous formalization of the theorem, along with several new proofs and plenty of applications – from Mathematics and Physics – in the fields we already mentioned in the Preface. The beginning of this renewed age for the CBHD Theorem begins – not surprisingly – with the Bourbakist school in the middle fifties, whose rigorous and selfcontained refoundation of all Mathematics involved also the CBHD Theorem and emblazoned it in a well-established algebraic context.

The aim of the next few paragraphs is to give the essential coordinates to orient oneself in the history of this theorem. *An exhaustive treatise concerned with the early proofs of the CBHD Theorem (during the years 1897–1950) can be found in [3] and in [18].*

1.1 The Early Proofs of the CBHD Theorem

To begin with, we hope that a chronological prospect of the early original contributions (range 1890–1950) to the CBHD Theorem may be illuminating: this is furnished in Table 1.1. We split our exposition into three subparagraphs:

A. Bonfiglioli and R. Fulci, *Topics in Noncommutative Algebra*, Lecture Notes in Mathematics 2034, DOI 10.1007/978-3-642-22597-0_1,
© Springer-Verlag Berlin Heidelberg 2012

- *The origin of the problem* (we describe how the problem came about of finding an intrinsic formula for the composition of two exponentials generated by Lie's theory of finite continuous groups of transformations).
- *Schur, Poincaré and Pascal* (we outline the – long forgotten – contributions of these three precursors of the CBHD formula, showing that their papers contain – in fact – much more than forerunning results).
- *Campbell, Baker, Hausdorff, Dynkin* (we describe and compare the results of the mathematicians whose names we have chosen for the acronym of the theorem to which this whole book is devoted).

Table 1.1 Comprehensive references for the early proofs of the CBHD Theorem. *Dark-gray* lines refer to the main references in the acronym CBHD; *light-gray* lines refer to very significant papers on the subject which – in our opinion – cannot be ignored to picture the full history of the theorem

Year	Author	Paper
1890	Schur	[154, 155]
1891	Schur	[156]
1893	Schur	[157]
1897	Campbell	[28, 29]
1898	Campbell	[30]
1899	Poincaré	[141]
1900	Poincaré	[142]
1901	Baker	[5]
1901	Pascal	[135, 136]
1901	Poincaré	[143]
1902	Baker	[6]
1902	Pascal	[137–139]
1903	Baker	[7]
1903	Campbell	[31]
1903	Pascal	[140]
1905	Baker	[8]
1906	Hausdorff	[78]
1937	Yosida	[183]
1947	Dynkin	[54]
1949	Dynkin	[55]
1950	Dynkin	[56]

1.1.1 The Origin of the Problem

The main theorem we are concerned with in this book finds its roots in Sophus Lie's theory of the *finite continuous groups of transformations*. What we now know as "Lie groups" are quite different from what Lie himself had occasion to call "his groups" and which he studied during the second half of the 19th century. Since we shall have no occasion to insist elsewhere on Lie's original theory, we confine ourselves to recalling a simplified definition of a group of transformations.[1]

A *finite continuous group of transformations* is a family of maps $x' = f(x, a)$, indexed with a (belonging to some open neighborhood of 0 in the Euclidean r-dimensional space), where x, x' vary on some domain of the Euclidean n-dimensional space, and such that the following properties hold:

(1) The maps $(x, a) \mapsto f(x, a)$ are analytic (real or complex) and to different a, a' there correspond different functions $f(\cdot, a)$ and $f(\cdot, a')$.
(2) $f(x, 0) = x$ for every x; moreover for every parameter a there exists a' such that $f(\cdot, a)$ and $f(\cdot, a')$ are inverse to each other.
(3) (The main axiom) *for every pair of parameters* a, b (sufficiently close to 0) *there exists* $\varphi(a, b)$ *in the domain of the parameters – which is unique by* (i) – *such that*

$$f(f(x, a), b) = f(x, \varphi(a, b)), \quad \text{for every } x. \tag{1.1}$$

Throughout this chapter, we shall also, for short, use the terminology *transformation group*, with the above meaning. We warn the Reader that here the adjective "finite" and the substantive "group" are not meant in their (modern) algebraic sense[2]: in modern words, one should rather speak of a *local* group, or a group *chunk*. Indeed, it is not difficult to see that the above map $(a, b) \mapsto \varphi(a, b)$ defines a *local* analytic group. For example, the "associativity"[3] of the local group law follows from the associativity of the composition of functions:

$$f\big(x, \varphi(\varphi(a, b), c)\big) = f\big(f(x, \varphi(a, b)), c\big) = f\big(f(f(x, a), b), c\big)$$
$$= f\big(f(x, a), \varphi(b, c)\big) = f\big(x, \varphi(a, \varphi(b, c))\big),$$

so that (see the second part of (i) above) $\varphi(\varphi(a, b), c) = \varphi(a, \varphi(b, c))$. Despite this underlying (local) group, as a matter of fact *the abstract group structure* never played such a central rôle in the original theory of Lie.

---

[1]For instance, point (2) below was not present in the first definitions of transformation groups given by Lie. More details can be found e.g., in [126, 171]. Books on ODE's from the first decades of 1900 are also of interest: see e.g., [31, 41, 60, 96].

[2]Finiteness is not referred here to the cardinality of the group, but it is used in contrast with the so-called infinitesimal transformations.

[3]Note that the existence of $\varphi(a, b)$ is not assumed to hold throughout, but only for a, b in some neighborhood, say $\mathcal{U}$, of the origin; more precisely, the word "associativity" should not be even allowed, since it is not required that $\varphi(a, b)$ belongs to $\mathcal{U}$, for every $a, b \in \mathcal{U}$.

What, instead, played a key rôle from the beginning were the so-called *infinitesimal transformations* of the transformation group (roughly speaking, the vector fields related to the columns of the Jacobian matrix of $u \mapsto f(x, u)$ at the origin): something very similar to what we would now call the associated *Lie algebra*. Let us see how these were thought of at the time of Lie. Let a be fixed in the domain of the parameters and let ε be a small scalar. The point x is transformed into x' by the "infinitesimal" transformation $x' = f(x, \varepsilon\, a)$. If F is any function sufficiently regular around x, its increment $F(x') - F(x)$ can be approximated by the Maclaurin expansion with respect to ε in the following obvious way (recall that $f(x, 0) = x$ by axiom (2)):

$$ F(x') - F(x) = F(f(x, \varepsilon\, a)) - F(f(x, 0)) = \varepsilon \cdot \sum_{j=1}^{r} a_j\, X_j F(x) + \mathcal{O}(\varepsilon^2), $$

as $\varepsilon \to 0$, where for every $j = 1, \ldots, r$ we have set

$$ X_j F(x) = \sum_{i=1}^{n} \frac{\partial}{\partial u_j}\Big|_{u=0} f_i(x, u) \cdot \frac{\partial F(x)}{\partial x_i}. \tag{1.2} $$

The differential operator $\sum_{j=1}^{r} a_j\, X_j$ (whose action on F gives, as shown above, the limit of the incremental ratio $(F(x') - F(x))/\varepsilon$ is then the appropriate "weight" of the infinitesimal transformation in the "direction" of the parameter a.

The Second Fundamental Theorem of Lie ensures that the span of the operators $X_1, \ldots, X_r$ is closed under the bracket operation, that is, there exist constants c_{ijs} (called the *structure constants* of the transformation group) such that

$$ [X_i, X_j] = \sum_{s=1}^{r} c_{ijs}\, X_s, \quad \text{for every } i, j = 1, \ldots, r. \tag{1.3} $$

The verification of this fact is a simple exercise of Calculus: by differentiating the identity (1.1) with respect to a_i, b_j at $(a, b) = (0, 0)$ one gets (setting $f = f(x, u)$ and $\varphi = \varphi(u, v)$)

$$ \sum_{h=1}^{n} \frac{\partial^2 f_k(x, 0)}{\partial x_h\, \partial u_j} \cdot \frac{\partial f_h(x, 0)}{\partial u_i} = \frac{\partial^2 f_k(x, 0)}{\partial u_i\, \partial u_j} + \sum_{s=1}^{r} \frac{\partial f_k(x, 0)}{\partial u_s} \cdot \frac{\partial^2 \varphi_s(0, 0)}{\partial u_i\, \partial v_j}. $$

If X_j is as in (1.2), the above identity gives at once (1.3) with the choice

$$ c_{ijs} = \frac{\partial^2 \varphi_s(0, 0)}{\partial u_i\, \partial v_j} - \frac{\partial^2 \varphi_s(0, 0)}{\partial u_j\, \partial v_i}. \tag{1.4} $$

The converse of this fact (part of the Second Fundamental Theorem too) is less straightforward, and it states that if a set of independent analytic vector fields $\{X_1, \ldots, X_r\}$ on some domain $\Omega \subseteq \mathbb{R}^n$ spans a finite-dimensional Lie algebra, then these vector fields are the infinitesimal transformations of some transformation group. The arguments to prove this fact make use of the "one parameter groups" generated by the family of vector fields, namely the solutions to the ODE system

$$\begin{cases} \dot{\gamma}(t) = (a \cdot X)(\gamma(t)) \\ \gamma(0) = x \end{cases} \quad \text{where} \quad a \cdot X := a_1 X_1 + \cdots + a_r X_r, \quad (1.5)$$

$a_1, \ldots, a_r$ being scalars and $x \in \Omega$. If the scalars a_i are small enough, the formula

$$f(x, a) := \gamma(1) \quad (1.6)$$

makes sense and it defines the desired transformation group. The proof of the existence of the corresponding local-group law $\varphi(a, b)$ is not so direct: φ is the solution of a suitable system of (first order) PDE problems whose integrability derives from the existence of constants c_{ijs} satisfying (1.3) (see e.g. [60, 154]).

Despite the concealed nature of φ, *the above facts contain the core of the CBHD Theorem*. Indeed, as can be trivially shown, the Maclaurin expansion of the above γ is (setting $A := a \cdot X$ for short)

$$\gamma(t) = \sum_{k=0}^{\infty} \frac{t^k A^k(x)}{k!} =: e^{tA}(x), \quad (1.7)$$

justifying the exponential notation for $\gamma(t)$. From (1.6) and from the group property (1.1), it follows that the composition of two exponentials must obey the following law (here $\circ$ is the composition of functions)

$$e^{a \cdot X} \circ e^{b \cdot X} = e^{\varphi(a,b) \cdot X}, \quad (1.8)$$

that is, $e^A \circ e^B = e^C$, where $A = a \cdot X, B = b \cdot X, C = c \cdot X$ with $c = \varphi(a, b)$. Moreover, given two vector fields A, B in the span of $\{X_1, \ldots, X_r\}$, another Calculus exercise shows that the Taylor expansion of *the composition of two exponentials* $(\alpha, \beta) \mapsto (e^{\alpha A} \circ e^{\beta B})(x)$ at the point $(0, 0)$ is equal to

$$\sum_{i,j \geq 0} \frac{B^i A^j(x)}{i! j!} \beta^i \alpha^j. \quad (1.9)$$

Note the reversed order for A, B. As a consequence, (1.8) can also be suggestively rewritten as follows (here a, b are sufficiently small so that we can take $\alpha = \beta = 1$ in (1.9))

$$\sum_{i,j\geq 0} \frac{B^i A^j}{i!j!} = \sum_{k\geq 0} \frac{C^k}{k!}, \qquad (1.10)$$

for some C belonging to span$\{X_1, \ldots, X_r\}$. Once again this naturally led to an "exponential formula" of the type $e^X e^Y = e^Z$, where this time e^X, e^Y, e^Z are formal power series.

Followed by Baker, it was Campbell who first proposed to investigate the quantity $e^A e^B$ from a purely symbolic point of view. Later, this was to be named (by Poincaré and by Hausdorff) *Campbell's problem*.

It then became clear that the study of the "product" of two exponentials, originating from Lie's theory of transformation groups, deserved autonomous attention and to be studied in its more general form. Lie had not paid much attention to the problem, for the equation $e^A e^B = e^C$ was more or less contained in the axiomatic identity (1.1) of his theory. What Lie failed to prove is that the rule describing the form of C as a series of commutators of A and B is in fact "universal"

$$C = A + B + \tfrac{1}{2}[A, B] + \tfrac{1}{12}[A[A, B]] - \tfrac{1}{12}[B[A, B]] + \tfrac{1}{24}[A[B[B, A]]] + \cdots,$$

and that the structure constants c_{ijs} of the specific group intervene only if one wants to write C in some particular basis $\{X_1, \ldots, X_r\}$. The universal symbolic nature of C is the subject to which Campbell, Baker, Hausdorff and Dynkin in due course addressed, thus untying the exponential formula from the original group context and lending it an interest in its own right.

Before closing this digression on the foundations of Lie's theory, we should not neglect to recall the Third Fundamental Theorem of Lie (in its original form), which is even more closely related to the CBHD Formula than the Second Theorem. Originally, Lie's Third Theorem was stated as follows:

Lie's Third Theorem, original version. *Given r^3 real constants $\{c_{ijk}\}_{i,j,k\leq r}$ satisfying the conditions*

$$c_{ijk} = -c_{jik}, \qquad \sum_{s=1}^r (c_{ijs}c_{skh} + c_{jks}c_{sih} + c_{kis}c_{sjh}) = 0, \qquad (1.11)$$

for every $i, j, h, k = 1, \ldots, r$, there exists a transformation group whose structure constants are c_{ijk}, that is, such that the infinitesimal transformations $X_1, \ldots, X_r$ of this transformation group satisfy

$$[X_i, X_j] = \sum_{k=1}^r c_{ijk} X_k. \qquad (1.12)$$

Throughout this chapter, when we speak of the Third Theorem (of Lie), we mean this original version of the theorem, which is very different from the modern (more profound) formulation of this result, which we now recall.

Indeed, it is nowadays commonly understood that "the Third (Fundamental) Theorem of Lie" is the following result:

A. Lie's Third Theorem, global version. *Given a finite-dimensional (real or complex) Lie algebra* $\mathfrak{g}$, *there exists a Lie group whose Lie algebra is isomorphic to* $\mathfrak{g}$.

Along with this global result, the following statements are also invoked as local versions of Lie's Third Theorem:

B. Lie's Third Theorem, local version (for finite-dimensional Lie algebras). *Given a finite-dimensional (real or complex) Lie algebra* $\mathfrak{g}$, *there exists a local Lie group whose Lie algebra is isomorphic to* $\mathfrak{g}$.

C. Lie's Third Theorem, local version (for Banach-Lie algebras). *Given a real or complex Banach-Lie algebra* $\mathfrak{g}$ *(see Definition 5.22, page 292), there exists a local Lie group whose Lie algebra is isomorphic to* $\mathfrak{g}$.

Obviously, Theorem B is a particular case of Theorem C, whilst Theorem A implies Theorem B. It is also easily seen that the original version of the theorem proves – in particular – the local version in the (real) finite-dimensional case (indeed, recall that a transformation group is a special class of local Lie group). It is also of some interest to observe that, as a consequence of the original version of Lie's Third Theorem, one obtains that every finite-dimensional (real) Lie algebra is isomorphic to one in a special class of Lie algebras, namely the Lie algebras spanned by analytic vector fields on some open domain of $\mathbb{R}^n$.

In this book, we shall sketch some proofs of the original version of Lie's Third Theorem, by recalling the arguments of its original contributors (Schur, Poincaré, Campbell, Pascal, ...), whereas Theorem C (hence Theorem B) easily derives from the results in Chap. 5 (see Theorem 5.52). We shall not attempt to prove (the very profound) Theorem A in this book, apart from a (very!) particular case: that of the real *nilpotent* finite-dimensional Lie algebras (see Sect. 5.4.2, page 328). The different caliber distinguishing global and local forms is also evident from the corresponding proofs, the global form requiring highly nontrivial results, such as, e.g., Ado's Theorem or a large amount of structure theory of Lie groups (see e.g., Varadarajan [171, Section 3.15]; see also Serre's proof in [159]). Instead, we shall see that Theorems B and C are simple corollaries of the convergence and associativity of the CBHD operation on a Banach-Lie algebra, a fundamental result for this type of noncommutative algebras.

In this book, as a convention, *when we speak of Lie's Third Theorem we mean the original version*, whereas we shall have no occasion – in general – to invoke the statements of Theorems A, B, C. However, due to the significant distinction between global and local results, we shall occasionally add the adjective *local* when speaking of the original version. Roughly speaking, the conflict "global vs local" depends on the fact that, according to its modern definition, a Lie group is a "global" object (a smooth manifold equipped with a smooth group law...), whereas a transformation group – in its original meaning – has to be seen as a local object, as we remarked at the beginning of Sect. 1.1.1.

For a modern functorial version of Lie's Third Theorem, see Hofmann and Morris [90]. For the different meanings attached to it, see also the headword "Lie's Theorem" in [80].

We now return to the historical background. Besides precursory studies by Schur, the link between the Third Theorem and the exponential formula was first understood by Poincaré.

With modern words, we would like to anticipate how the identity $e^A e^B = e^C$ intervenes in the proof of the Third Theorem. Indeed, suppose the constants c_{ijk} are as in (1.11) and let $X_1, \ldots, X_r$ be r independent letters. Formula (1.12) then uniquely defines (by bilinearity) a structure on

$$\mathfrak{g} := \mathrm{span}\{X_1, \ldots, X_r\}$$

of a real Lie algebra (obviously conditions (1.11) are equivalent to skew-symmetry and the Jacobi identity). Suppose we have proved the formal power series identity $e^A e^B = e^C$, where $\psi(A, B) := C$ is a uniquely determined formal Lie series in A, B. The associativity of the product of formal power series ensures that

$$\psi(A, \psi(B, C)) = \psi(\psi(A, B), C). \tag{1.13}$$

Suppose further that, for $U = u_1 X_1 + \cdots + u_r X_r$, $V = v_1 X_1 + \cdots + v_r X_r$, the $\mathfrak{g}$-valued series $\psi(U, V)$ converges in the finite dimensional vector space $\mathfrak{g}$, provided that $u = (u_1, \ldots, u_r)$ and $v = (v_1, \ldots, v_r)$ belong to a sufficiently small neighborhood of the origin, say Ω.

Let us set $\varphi(u, v) = (\varphi_1(u, v), \ldots, \varphi_r(u, v))$ where

$$\psi(U, V) = \varphi_1(u, v)\, X_1 + \cdots + \varphi_r(u, v)\, X_r. \tag{1.14}$$

and U, V are as above. Then (1.13) implies that $u' := \varphi(u, v)$ defines a finite transformation group (parametrized by v in Ω) on the open domain $\Omega \subseteq \mathbb{R}^r$. Note that the formal identity $e^A e^B = e^{\psi(A,B)}$ forces the condition

$$\psi(A, B) = A + B + \tfrac{1}{2}[A, B] + \{\text{higher orders in } A, B\}.$$

Consequently, (1.12) and (1.14) produce the Maclaurin expansion

$$\varphi_s(u, v) = u_s + v_s + \tfrac{1}{2} \sum_{i,j=1}^{r} c_{ijs}\, u_i v_j + \{\text{higher orders in } u, v\}. \tag{1.15}$$

Now, since the structure constants of a transformation group are given by formula (1.4), we recognize by a direct differentiation of (1.15) that these constants are precisely c_{ijk}. The (local) Third Theorem then follows.

This is more or less Poincaré's argument in [142], dated 1900. Despite its being over 100 years old, this argument hits the mark for its intertwinement of algebraic objects, group theory and calculus and it exemplifies how the CBHD Theorem can be used for algebraic and geometric purposes alike.

1.1.2 Schur, Poincaré and Pascal

Along with the fact that there is still no conventional agreement for the "parenthood" of the CBHD Theorem,[4] the most surprising fact concerning the early investigations is that the contributions of Schur, Poincaré and Pascal have nowadays almost become forgotten.[5] In our opinion, there are three good (or perhaps bad?) reasons for this:

1. Firstly, they focussed on Lie's setting of the theory of groups of transformations, instead of taking up a purely symbolical approach.
2. Secondly, the very cold judgements towards the contributions of Schur, Poincaré and Pascal given by Hausdorff himself and then also by Bourbaki have certainly played a rôle.
3. Thirdly, the fact that none of their original papers on the subject was written in English has contributed to their becoming obsolete. (Hausdorff's paper though, written in German, did not share this lot.)

To be more specific, we recall a few of the comments of Hausdorff and Bourbaki on the works by Schur, Poincaré and Pascal.

- In Hausdorff's paper [78, page 20] it is stated that Poincaré had *not* proved the exponential-theorem in its general and symbolic form, but only in the form of the group theory. Moreover, according to Hausdorff, the massive computations in Pascal's papers make it hard to check the validity of the arguments and Pascal is faulted for omitting the convergence matter. Finally, a comment on Schur's contribution is missing, apart for an acknowledgement (see [78, page 34]) concerning some expansions involving Bernoulli numbers.
- In the book of Bourbaki [27, Chapter III, Historical Note, V] Poincaré is faulted (together with Campbell and Baker) for a lack of clarity about the

---

[4]Baker's name is frequently unfairly omitted and Dynkin, unfairly as well, is acknowledged only for his explicit series, but not for his proof.

[5]To the best of our knowledge, the only modern sources quoting Schur, Poincaré or Pascal regarding the CBHD Theorem are: Czyż [43] (who quotes Pascal [139] and Poincaré [142]) and Duistermaat, Kolk (who quote Poincaré [141] and Schur [156, 157]).

question of whether the brackets in his exponential formulas are symbolic expressions or elements of some fixed Lie algebra. Pascal is named with those who returned to the question of $e^A e^B = e^C$, but nothing is said about his results and his papers are not mentioned in the list of references. Hausdorff – instead – is acknowledged as the only reliable source. Finally, Schur's results on the product of exponentials are not mentioned.[6]

We would like to express some opinions on the above comments.

Hausdorff insists on reasserting over and over again that his own theorem implies Poincaré's and not vice versa, but eventually his words become more prudent[7]: this legitimately leads us to suppose that Hausdorff recognized that Poincaré proved more in his paper [142] than it seems on the surface.[8] Obviously, it is not possible to fail to recognize Hausdorff's merits in shifting all the attention on purely symbolic arguments. Nonetheless, our opinion is that Poincaré's ingenious approach (take a finite dimensional Lie algebra $\mathfrak{g}$, obtain the exponential theorem on $\mathscr{U}(\mathfrak{g})$, then go back to $\mathfrak{g}$ and prove the convergence with the residue calculus) is definitely more modern than a typical publication from 1900.

Moreover, Hausdorff's complaint on the over-abundance of computations in Pascal's papers is not enough to doubt the correctness of Pascal's proof. Indeed, one may praise a large part of Pascal's arguments in the series of papers [135]–[139] for being more transparent than part of the symbolic computations by Campbell or Baker.

Also, even though the historical notes in Bourbaki do not pay attention to their contributions, Pascal and Schur should be regarded as noteworthy precursors of what later came to be known only as the Campbell-Hausdorff formula (or Campbell-Baker-Hausdorff formula). Following Schmid [153, page 177], this formula is already contained "in disguise" in Schur's papers; and the same should be said for Pascal's works, as we shall see below.

In order to let the Reader form his own opinions, we describe more closely the contributions by Schur, Poincaré and Pascal.

---

[6]Schur is recalled only for his proof of the fact that C^2 assumptions on transformation groups are (up to isomorphism) sufficient to get analytic regularity.

[7]After repeated reassertions that his theorem implies Poincaré's and not vice versa, Hausdorff finally recognizes that Poincaré's theorem *seems not* to be without relevance for his own result.

[8]In [142] Poincaré introduced for the very first time the universal enveloping algebra of a Lie group, but the relevance of this invention was not caught for several decades (see Ton-That, Tran [168] and Grivel [74] for related topics). More importantly, as Schmid neatly points out (see [153, page 184]), Poincaré's identity $e^U . e^V = e^T$ makes no reference at all to the Lie algebra of the infinitesimal transformations of a group.

1.1.2.1 Schur

Schur's[9] papers [154]–[157] are mainly devoted to providing new derivations of Lie's Theorems by means of a shrewd way of rewriting the group condition (1.1) via systems of PDEs. Actually, in dealing with a new proof of Lie's Third Theorem (see [156, §1] and [154, §3]), he constructs a (local) transformation group only by means of suitable *power series* depending on the structure constants and on the *Bernoulli numbers*. [Almost all of his successors will attribute to Schur the discovery of the rôle of these numbers in the exponential formula.] *This is clearly a forerunner of the Campbell-Hausdorff series.*

Like Lie, Schur did not capture the universal law of bracket-formation which underlies the analytic expression of $\varphi(a, b)$. However, he perfectly captured the idea that an explicit formula for φ depending only on the constants c_{ijk} was the right tool to prove Lie's Third Fundamental Theorem, a theorem which is – as we now understand it – deeply related to the CBHD Theorem.

As an example of Schur's notable computations with power series, we would like to recall a remarkable formula [155, eq.(2) page 2], which catches one's eyes for its explicitness: If $\omega = (\omega_{i,j}(u))_{i,j}$ is the Jacobian matrix of $v \mapsto \varphi(u, v)$ at $v = 0$, then

$$\omega_{i,j}(u) = \delta_{i,j} + \sum_{m=1}^{\infty} \lambda_m U_{i,j}^{(m)}(u), \tag{1.16}$$

where $\delta_{i,j}$ is Kronecker's symbol, the constants λ_m are a variant of the Bernoulli numbers,[10] and the functions U are explicitly defined by

$$U_{i,j}^{(m)}(u) = \sum_{\substack{1 \leqslant h_1,\ldots,h_m \leqslant r \\ 1 \leqslant k_1,\ldots,k_{m-1} \leqslant r}} c_{j,h_1}^{k_1} c_{k_1,h_2}^{k_2} \cdots c_{k_{m-2},h_{m-1}}^{k_{m-1}} c_{k_{m-1},h_m}^{i} u_{h_1} u_{h_2} \cdots u_{h_m}.$$

$$\tag{1.17}$$

---

[9] Friedrich Heinrich Schur; Maciejewo, near Krotoschin, Prussia (now Krotoszyn, Poland), 1856 – Breslau, Prussia (now Wrocław, Poland), 1932. Friedrich Heinrich Schur should not be confused with the (presumably more famous and more influential) coeval mathematician Issai Schur (Mogilev, 1875; Tel Aviv, 1941). Whereas the former was a follower of the school of Sophus Lie, the latter was a student of Frobenius and Fuchs, Berlin. For a source on Berlin mathematicians of that time, see [11].

[10] Schur uses the following definition for the Bernoulli numbers B_n:

$$\frac{x}{e^x - 1} = 1 - \frac{x}{2} + \sum_{q=1}^{\infty} (-1)^{q-1} \frac{B_{2q-1}}{(2q)!} x^{2q}.$$

Then λ_m is defined by $\lambda_1 = -\frac{1}{2}$, $\lambda_{2q+1} = 0$, $\lambda_{2q} = (-1)^{q+1} \frac{B_{2q-1}}{(2q)!}$.

Here $u = (u_1, \ldots, u_r)$ is a point of some neighborhood of the origin of $\mathbb{R}^r$ and $c_{i,j}^s$ are the structure constants of the group. Conversely (see [156, Satz 2, page 271]), Schur uses this formula as an *Ansatz* to give a very explicit proof of Lie's (local) Third Theorem.

Later, Schur proves in [157] the remarkable result that C^2 assumptions on the functions f and φ in (1.1) actually guarantee that they can be transformed (by a change of variables) into analytic functions. This notable result (closely related to Hilbert's Fifth Problem) has become a central topic of modern Lie group theory: The fact that this result is usually proved – by sheer chance – by means of the CBHD formula highlights the deep link between Schur's studies and the theorem we are concerned with in this book.

1.1.2.2 Poincaré

It was Poincaré[11] who first realized the link between Schur's proof of the Third Theorem and what he defined "le problème de Campbell".

Indeed, given a transformation group and its infinitesimal transformations $\{X_1, \ldots, X_r\}$ as in (1.2), the one-parameter subgroups (see (1.7))

$$t \mapsto e^{tX}, \qquad (X \in \mathfrak{g} := \mathrm{span}\{X_1, \ldots, X_r\})$$

generate the transformation group. Hence, if Campbell's identity $e^V e^T = e^W$ has a precise analytical meaning, it follows that the whole group can be reconstructed by $\mathfrak{g}$. *Poincaré anticipated the modern way of thinking about the CBHD Formula as the tool permitting the reconstruction of the group law by means of the Lie algebra.*

We refer the Reader to Schmid [153] and to Ton-That, Tran [168] for, respectively, a treatise on Poincaré's contribution to the theory of continuous groups and for Poincaré's contribution to the Poincaré-Birkhoff-Witt Theorem (see also Grivel [74] for this last topic); here we confine ourselves in pointing out his contribution to the development of the CBHD Theorem.

The most impressive fact about Poincaré's contribution is that, in solving an important problem, he creates an even more important device: *the universal enveloping algebra* of a Lie algebra. Indeed, suppose there is given a Lie algebra[12] (over a field of characteristic zero) $\mathfrak{g} := \mathrm{span}\{X_1, \ldots, X_r\}$ (the X_i are called generators), equipped with the bracket defined by

$$[X_i, X_j] = \sum_{s=1}^{r} c_{ijs} X_s. \tag{1.18}$$

---

[11]Jules Henri Poicaré; Nancy (France), 1854 – Paris (France), 1912.

[12]Lacking, as he did, the general definition of Lie algebra – not introduced until the 1930s – Poincaré pictured his Lie algebras as spanned, typically, by analytic vector fields.

Then Poincaré introduces an equivalence relation on the set of *formal polynomials* in the generators, by calling *equivalent* any two polynomials whose difference is a linear combination of two-sided products of

$$X_i X_j - X_j X_i - [X_i, X_j] \qquad (i, j \le r). \tag{1.19}$$

As claimed by Hausdorff-Bourbaki, Poincaré's construction is not devoid of some ambiguity.[13] Subsequently, Poincaré shows that any homogeneous polynomial of degree p is equivalent to a uniquely determined span of polynomials (called *regular*) of the form[14]

$$(\alpha_1 X_1 + \alpha_2 X_2 + \cdots)^p.$$

Then, if $V, T \in \mathfrak{g}$, Poincaré considers a formal product of the type

$$e^V e^T = \sum_{m,n \ge 0} \frac{V^m T^n}{m! \, n!},$$

motivated by the theory of transformation groups. He aims to prove – by direct methods – the existence of W such that $e^V e^T = e^W$ (a fact implicitly ensured by the Second Fundamental Theorem).

Poincaré's insightful idea is to use the universal enveloping algebra to do that. Indeed, the monomial $V^m T^n$ can be made *equivalent* to sums of regular polynomials, say

$$\frac{V^m T^n}{m! \, n!} = \sum_{p=0}^{m+n} W(p, m, n), \tag{1.20}$$

where $W(p, m, n)$ is symmetric and homogeneous of degree p. Consequently,

$$e^V e^T = \sum_{m,n \ge 0} \frac{V^m T^n}{m! \, n!} = \sum_{p=0}^{\infty} \left(\sum_{m+n \ge p} W(p, m, n) \right) =: \sum_{p=0}^{\infty} W_p. \tag{1.21}$$

Poincaré observes that, if we seek W – homogeneous of degree 1 – such that $e^V e^T = \sum_{p=0}^{\infty} W^p$, then W^p is regular and homogeneous of degree p. Hence, by the *uniqueness* of the regularization process, we must have

---

[13]For example, identity (1) on page 224 in [142] writes $XY - YX = [X, Y]$ and identity (2) contains $XY - YX - [X, Y]$, claiming that it is not null, since $XY - YX$ is of 2nd degree whereas $[X, Y]$ is of 1st degree. Actual identities and congruences repeatedly overlap with no clear distinctions. See [3] for an explanation of this fact.

[14]As a matter of fact, this is what we now call the *(linear) isomorphism of $\mathscr{U}(\mathfrak{g})$ with the symmetric algebra of the vector space $\mathfrak{g}$.*

$$W_p = \frac{(W_1)^p}{p!}, \quad p \geq 0. \tag{1.22}$$

As a matter of fact, W_1 *must be exactly the CBHD series associated to* V, T.

Let us check how W_1 can be computed by repeated applications of the following identity (here the "=" sign can be interpreted as a congruence or as a true equality, if V, T are smooth vector fields)

$$VT = TV + [V, T]. \tag{1.23}$$

For example, consider the following computation:

$$VT = \tfrac{1}{2} VT + \tfrac{1}{2} VT = \tfrac{1}{2} (VT + TV) + \tfrac{1}{2} [V, T]$$
$$= \tfrac{1}{2} (V + T)^2 - \tfrac{1}{2} V^2 - \tfrac{1}{2} T^2 + \tfrac{1}{2} [V, T].$$

With the notation in (1.20), this gives

$$W(0, 1, 1) = 0, \quad W(1, 1, 1) = \tfrac{1}{2} [V, T], \quad W(2, 1, 1) = \tfrac{1}{2} \left((V + T)^2 - V^2 - T^2 \right).$$

Again, consider the identity (which we derived by means of (1.23) only)

$$\frac{VT^2}{2} = \frac{1}{6} (VTT + TVT + TTV) + \frac{1}{4} ([V, T]T + T[V, T]) + \frac{1}{12} [T, [T, V]],$$

so that $W(1, 1, 2) = \tfrac{1}{12} [T, [T, V]]$. An analogous computation ensures that we have $W(1, 2, 1) = \tfrac{1}{12} [V, [V, T]]$, whence

$$W_1 = W(1, 1, 0) + W(1, 0, 1) + W(1, 1, 1) + W(1, 2, 1) + W(1, 1, 2) + \cdots$$
$$= V + T + \tfrac{1}{2} [V, T] + \tfrac{1}{12} [V, [V, T]] + \tfrac{1}{12} [T, [T, V]] + \cdots$$

We recognize the first few terms of the CBHD series, and the original contribution by Poincaré to the CBHD Theorem is crystal clear.

As anticipated, Poincaré derives from (1.22) a new proof of the (local) Third Fundamental Theorem, with an argument similar to that given at the end of §1. It is also of interest to exhibit some of Poincaré's formulas (returning in the subsequent literature devoted to the CBHD Formula), aimed at giving a precise analytical meaning to (1.22).

Once more Poincaré's intuitions will leave their mark: *He is the first to characterize the solution* W *of* $e^W = e^V e^T$ *as solving a suitable ODE.* For instance, he proves that $e^V e^{\beta T} = e^{W(\beta)}$ if and only if $W(\beta)$ solves

$$\begin{cases} \frac{dW(\beta)}{d\beta} = \phi(\operatorname{ad} W(\beta))(T) \\ W(0) = V \end{cases} \quad \text{where } \phi(z) = \frac{z}{1 - e^{-z}}. \tag{1.24}$$

If V, T are sufficiently near the origin in $\mathfrak{g}$, then $W := W(1)$ is the solution to $e^W = e^V e^T$, an identity in the enveloping algebra of $\mathfrak{g}$. This identity can be transformed into an identity between the operators e^V, e^T, e^W, which solves Campbell's problem and the Second and Third Theorems of Lie.

Instead of using formal power series to define $\phi(\mathrm{ad}\, W)$, Poincaré invokes the residue calculus: thus the solution $W(\beta)$ of (1.24) can be found in the form $\sum_{i=1}^r w_i(\beta) X_i \in \mathfrak{g}$, where

$$\frac{\mathrm{d} w_i(\beta)}{\mathrm{d}\beta} = \frac{1}{2\pi\iota} \int \frac{\sum_{j=1}^r t_j P_{i,j}(\beta,\xi)}{\det\left(\mathrm{ad}\,(W(\beta)) - \xi\right)} \frac{\xi}{1-e^{-\xi}}\, \mathrm{d}\xi. \tag{1.25}$$

The crucial fact is that these ODEs can be integrated.[15]

Poincaré obtains the ODE (1.24) by combining his algebraic symmetrization process, differential calculus, explicit computations of noncommutative algebra, etc. As we shall repeatedly observe, the compound of Algebra and Analysis is typical of the arguments needed to prove the CBHD Theorem. For example, a key rôle is played by the symbolic identity

$$e^{\alpha V + \beta W} = e^{\alpha V} e^{\beta Y}, \quad \text{where } Y = \frac{1-e^{-\mathrm{ad}\,(\alpha V)}}{\mathrm{ad}\,(\alpha V)}(W).$$

It is interesting to observe that this identity is derived both with algebraic arguments, and with a direct approach based on an expansion modulo $\mathcal{O}(\beta^2)$:

$$e^{\alpha V + \beta W} = \sum_{n=0}^{\infty} \frac{(\alpha V + \beta W)^n}{n!}$$

$$= e^{\alpha V} + \beta \sum_{n=1}^{\infty} \frac{\alpha^{n-1}}{n!}\left(\sum_{k=0}^{n-1} V^{n-1-k} W V^k\right) + \mathcal{O}(\beta^2)$$

$$= e^{\alpha V} + \beta \sum_{n=1}^{\infty} \frac{\alpha^{n-1}}{n!}\left(\sum_{k=0}^{n-1}(-1)^k \binom{n}{k+1} V^{n-1-k}(\mathrm{ad}\,V)^k(W)\right) + \mathcal{O}(\beta^2)$$

$$= e^{\alpha V} + \beta \sum_{k=0}^{\infty}\sum_{j=0}^{\infty} \frac{(\alpha V)^j}{j!}\frac{(-\alpha\,\mathrm{ad}\,V)^k}{(k+1)!}(W)\right) + \mathcal{O}(\beta^2)$$

$$= e^{\alpha V}\left(1 + \beta \frac{1-e^{-\mathrm{ad}\,(\alpha V)}}{\mathrm{ad}\,(\alpha V)}(W)\right) + \mathcal{O}(\beta^2). \tag{1.26}$$

---

[15]Here ι is the imaginary unit, the coefficients t_j are the coefficients of T with respect to $\{X_1, \ldots, X_r\}$, whilst $(P_{i,j})$ is the adjugate (i.e., transpose of the cofactor) matrix of $\mathrm{ad}\,(W(\beta)) - \xi$. Also, the integral is taken over a contour around $0 \in \mathbb{C}$ which does not contain the poles of $\phi(z)$, for example a circle about 0 of radius $R < 2\pi$.

As we shall see, similar computations will reappear in Pascal, Baker, Hausdorff and in various modern proofs of the CBHD Theorem (see Chap. 4).

1.1.2.3 Pascal

In a series of five papers dated 1901–1902, Pascal[16] undertook the study of the composition of two exponentials. He also collected an abridged version of these papers, together with a didactic exposition of the main results of those years about groups of transformations, in the book [140], 1903. We here give a brief overview of these papers.[17]

In [135], Pascal announces a new – as direct as possible – proof of the Second Theorem of Lie, the crucial part being played by a formula for the product of two finite transformations, each presented in canonical exponential form $x \mapsto e^{tX}(x)$. To this aim, he provides suitable identities valid in any associative, noncommutative algebra: for example

$$k\, X_2 X_1^{k-1} = \sum_{j=0}^{k-1} j! \binom{k}{j} \gamma^{(j)} \left(\sum_{i=1}^{k-j} X_1^{i-1} (\operatorname{ad} X_1)^j (X_2)\, X_1^{k-j-2} \right), \qquad (1.27)$$

where $k \in \mathbb{N}$ and the constants $\gamma^{(j)}$ are variants of Bernoulli numbers:

$$\gamma^{(0)} = 1, \quad \gamma^{(j)} = -\left(\tfrac{1}{2!}\, \gamma^{(j-1)} + \tfrac{1}{3!}\, \gamma^{(j-2)} + \cdots + \tfrac{1}{j!}\, \gamma^{(1)} + \tfrac{1}{(j+1)!}\, \gamma^{(0)} \right).$$
$$(1.28)$$

A certain analogy with Campbell's algebraic computations, with Poincaré's symmetrized polynomials, and with the use of Bernoulli numbers as in Schur are evident; but Pascal's methods are quite different from those of his predecessors.

As an application, in [136] Pascal furnishes a new proof of the Second Fundamental Theorem. He claims that the explicit formula $e^{X_2} \circ e^{X_1} = e^{X_3}$ is "a uniquely comprehensive source" for many Lie group results. The analogy with Poincaré's point of view (recognizing the exponential formula as a unifying tool) is evident, but there's no way of knowing if Pascal knew, at the time, Poincaré's paper [142] (which is not mentioned in [136]).

In order to obtain the Second Theorem, Pascal generalizes (1.27) by decomposing the product (obviously coming from $e^{X_2} e^{X_1}$)

---

[16]Ernesto Pascal; Naples (Italy), 1865–1940.

[17]We plan to return to this subject (with more mathematical contents) in a forthcoming study [18].

$$\frac{X_2^r}{r!} \frac{X_1^{k-r}}{(k-r)!} \qquad (0 \le r \le k, \ k \in \mathbb{N}) \tag{1.29}$$

as a linear combination of *symmetric* sums (called elementary) based on X_1, X_2 and nested commutators of the form

$$[X_{i_1} X_{i_2} \cdots X_{i_s} X_1 X_2] := [X_{i_1}, [X_{i_2} \cdots [X_{i_s}, [X_1, X_2]] \cdots]], \tag{1.30}$$

where $i_1, \ldots, i_s \in \{1, 2\}$. The law of composition of the coefficient of such an elementary sum is described very closely, though an explicit formula is not given. *With a delicate – yet very direct – analysis of the coefficients of the monomials decomposing* (1.29), he obtains the following representation

$$e^{t' X_2} e^{t X_1} = \sum_{k \ge 0} \left(\sum_{r=0}^{k} t'^r t^{k-r} \frac{X_2^r}{r!} \frac{X_1^{k-r}}{(k-r)!} \right) \tag{1.31}$$

$$= \sum_{k=0}^{\infty} \frac{1}{k!} \Big\{ tX_1 + t'X_2 + \gamma^{(1)} tt' [X_1, X_2] + \gamma^{(2)} t^2 t' [X_1 X_1 X_2]$$

$$+ -\gamma^{(2)} tt'^2 [X_2 X_1 X_2] - \gamma^{(2)} \gamma^{(1)} t^2 t'^2 [X_1 X_2 X_1 X_2] + \cdots \Big\}^k.$$

Note the similarity with Poincaré's decomposition in (1.21) and (1.22).

Nonetheless, whereas Poincaré uses an *a posteriori* argument to prove $e^V e^T = \sum_{p=0}^{\infty} \frac{(W_1)^p}{p!}$ (by showing that W_1 satisfies a suitable system of ODEs), Pascal's construction is much more direct: as he announced, "at the cost of longer computations" (which will unfortunately provoke a complaint from Hausdorff) he provides a way to reconstruct the expansion of $e^{t' X_2} e^{t X_1}$ as a pure exponential e^{X_3}, this reconstruction being uniquely based on a *direct unraveling* of $\sum_{i,j} \frac{(t' X_2)^j}{j!} \frac{(t X_1)^i}{i!}$.

Besides the partial expansion in (1.31), Pascal proves that the series in curly braces is a series of terms as in (1.30). As we have remarked, *Poincaré did not succeed to prove this result explicitly*. Though Pascal's derivation of (1.31) is computationally onerous (and lacking in some arguments – especially in inductive proofs – which Pascal probably considered redundant), this is clearly an argument towards a symbolic version of the later-to-be-known-as Campbell-Baker-Hausdorff Theorem, second to the one proposed a few years before by Campbell [30].

Unfortunately, what is really missing in Pascal's papers [135]–[138] is the study of the convergence of the series in the far right-hand side of (1.31).[18]

---

[18]In his review of [136], Engel faults Pascal for this omission with hard words, but he also acknowledges Pascal's formula (1.31) for being very remarkable.

The convergence matter is missing also in the proof of the Third Theorem which he gives in his last paper [139], which must be considered the *summa* of Pascal's work on group theory, for the following reasons:

- He only uses his original algebraic identities in [135] (in neat dissimilarity with Schur's methods for proving the Third Theorem).
- With these identities he provides a series of explicit coefficients in the expansion (1.31) (those being really crucial, as Baker and Hausdorff will later rediscover with different techniques, but leaving Pascal un-acknowledged), thus improving the results in [136].
- He constructs, with a more natural method, if compared to his previous paper [138], the infinitesimal transformations of the parameter group (this time using some prerequisites from group theory).

Let us briefly run over the key results in [139]. Pascal shows that, in the expansion (1.31), all the summands containing t' with degree 1 or those containing t with degree 1 are respectively given by

$$\sum_{n=1}^{\infty} \gamma^{(n)} t' t^n \,(\operatorname{ad} X_1)^n (X_2), \qquad \sum_{n=2}^{\infty} \gamma^{(n)} t'^n t \,(\operatorname{ad} X_2)^n (X_1), \qquad (1.32)$$

where the constants $\gamma^{(n)}$ are as in (1.28).

Let now $Z_1, \ldots, Z_r$ be the infinitesimal transformations generating a group with r parameters. Let us set

$$X_1 = v_1 Z_1 + \cdots + v_r Z_r, \quad X_2 = u_1 Z_1 + \cdots + u_r Z_r.$$

Then, by general group theory, it is known that the composition $e^{X_1} \circ e^{X_2}$ is given by e^{X_3}, where X_3 is a linear combination of the transformations Z_i, say

$$X_3 = u_1' Z_1 + \cdots + u_r' Z_r.$$

The u_i are functions of u and v (and of the structure constants), say

$$u_h' = \varphi_h(u, v), \quad h = 1, \ldots, r. \qquad (1.33)$$

Again from the group theory, it is known that (1.33) defines a transformation group (the parameter group).

Pascal's crucial remark is that, if we know that (1.33) defines a group, *it is sufficient to consider only the terms containing v with first degree* (which de facto furnish the infinitesimal transformations): these correspond to the summands in (1.31) with t of degree one. By formula (1.32), we thus get

$$X_3 = \sum_i v_i Z_i + \sum_i u_i Z_i + \gamma^{(1)} \left[\sum_i v_i Z_i, \sum_j u_j Z_j \right]$$

$$+ \sum_{n=2}^{\infty} \gamma^{(n)} \left(\operatorname{ad} \sum_i u_i Z_i \right)^n \left(\sum_j v_j Z_j \right) + \mathcal{O}(|v|^2).$$

By Lie's Second Theorem (viz. $[Z_i, Z_j] = \sum_k c_{ijk} Z_k$) we derive

$$\varphi_h(u,v) = u_h + v_h + \frac{1}{2}\sum_{jk} c_{jkh}\, u_j v_k + \sum_{n=1}^{\infty} \sum_{t_1,\dots,t_{2n}} \sum_{s_1,\dots,s_{2n-1}} \sum_{k}$$

$$\times\, \gamma^{(2n)}\, c_{t_1 k s_1}\cdots c_{t_{2n} s_{2n-1} h}\, u_{t_1}\cdots u_{t_{2n}} v_k + \mathcal{O}(|v|^2).$$

Differentiating this with respect to the coordinates v, Pascal derives at once

$$U_k = \sum_{h=1}^{r} \frac{\partial \varphi_h}{\partial v_k}(u,0)\, \frac{\partial}{\partial u_h} = \sum_{h=1}^{r}\Bigg(\delta_{hk} + \frac{1}{2}\sum_{j} c_{jkh}\, u_j +$$

$$+\sum_{n=1}^{\infty} \gamma^{(2n)} \sum_{t_1,t_2,\dots} \sum_{s_1,s_2,\dots} c_{t_1 k s_1}\cdots c_{t_{2n} s_{2n-1} h}\, u_{t_1}\cdots u_{t_{2n}}\Bigg)\frac{\partial}{\partial u_h}.$$

Finally, φ can be recovered by exponentiation:

$$\varphi(u,v) = e^{\sum_k v_k U_k}(u) = u + \Big(\sum_k v_k U_k\Big)(u) + \frac{1}{2!}\Big(\sum_k v_k U_k\Big)^2(u) + \cdots$$

In view of the above explicit formula for the vector fields U_k, this identity (already appearing in Schur's studies) contains a "quantitative" version of the Third Theorem: it shows that an explicit local group can be constructed by the use of the Bernoulli numbers $\gamma^{(2n)}$ and by a set of constants c_{ijk} satisfying the structure relations (1.11).

Finally, it is not a rash judgement to say that Pascal's contribution to the CBHD Theorem is of prime importance: he was the first to construct explicitly a local group by using the *commutator series* $X \diamond Y$ for $\log(e^X e^Y)$, or more precisely, by using the subseries derived from $X \diamond Y$ of *the summands of degree one* in one of the indeterminates X, Y. Analogous results were to be reobtained by Baker and, mostly, by Hausdorff, and reappear in more modern proofs of the CBHD Theorem (see e.g., Reutenauer [144, Section 3.4]).

1.1.3 Campbell, Baker, Hausdorff, Dynkin

In this section we take up a brief overview of the results, concerning the exponential formula, by those authors whose names are recalled in our acronym "CBHD". Since the parentage of the formula is not well established – neither nowadays, nor immediately after the original papers were published – we hope that a résumé and a comparison of the contributions by each of the four authors might help.

1.1.3.1   Campbell

Campbell's[19] 1897 paper [28] is the very first in the history of the exponential formula. In a very readable and concise style, it contains plenty of formulas which will reappear in subsequent papers on the composition of exponentials. To begin, he establishes the following purely algebraic identity[20]

$$\frac{y\,x^r}{r!} = \sum_{j=0}^{r} \frac{(-1)^j a_j}{(r+1-j)!} \sum_{i=0}^{r-j} x^i \,(\mathrm{ad}\,x)^j (y)\, x^{r-j-i}, \quad r \geq 0, \tag{1.34}$$

where the constants a_j are defined by the recursion formula

$$\begin{cases} a_0 = 1, \quad a_1 = 1/2 \\ a_j = \frac{1}{j+1}\left(a_{j-1} - \sum_{i=1}^{j-1} a_i\, a_{j-i}\right), \quad j \geq 2. \end{cases} \tag{1.35}$$

He acknowledges Schur for the discovery of the constants a_j (which are indeed related to the Bernoulli numbers).

   Though Campbell will eventually be faulted by Bourbaki for a lack of clarity on the context of his calculations (what do x, y mean? formal series, infinitesimal transformations of a group, non-associative indeterminates?), his results on the exponential formula are undoubtedly important. For example, we exhibit one of his most fruitful computations. Let us set

$$z = z(x,y) := \sum_{j=0}^{\infty} (-1)^j a_j\, (\mathrm{ad}\,x)^j (y). \tag{1.36}$$

(This will turn out to be the subseries of $\mathrm{Log}(\mathrm{Exp}\,y\,\mathrm{Exp}\,x)$ containing y precisely once, a crucial series in the CBHD Formula!) Then one has

$$y\,e^x = \sum_{r=0}^{\infty} \frac{y\,x^r}{r!} \quad \left(\text{use (1.34) and interchange sums}\right)$$

$$= \sum_{j=0}^{\infty}\sum_{r=j}^{\infty} \frac{(-1)^j a_j}{(r+1-j)!} \sum_{i=0}^{r-j} x^i\,(\mathrm{ad}\,x)^j(y)\,x^{r-j-i} \quad \left(\text{rename } s = r - j\right)$$

$$= \sum_{j=0}^{\infty}\sum_{s=0}^{\infty} \frac{(-1)^j a_j}{(s+1)!} \sum_{i=0}^{s} x^i\,(\mathrm{ad}\,x)^j(y)\,x^{s-i} \quad \left(\text{sums can be interchanged!}\right)$$

$$= \sum_{s=0}^{\infty} \frac{1}{(s+1)!} \sum_{i=0}^{s} x^i\, z\, x^{s-i}. \tag{1.37}$$

---

[19]John Edward Campbell; Lisburn (Ireland), 1862 – Oxford (England), 1924.

[20]Here and in what follows, we use different notations with respect to Cambell's.

The above computation (to be considered as holding true in the algebra of formal power series in x, y) easily gives

$$\begin{cases} (1 + \mu\, y)\, e^x = e^{x + \mu\, z} + \mathcal{O}(\mu^2), \\ e^{\mu\, y}\, e^x = e^{x + \mu\, z} + \mathcal{O}(\mu^2), \end{cases} \tag{1.38}$$

where $z = z(x, y)$ is as in (1.36), and the coefficients a_j are explicitly given by (1.35).[21]

Campbell's main aim is to give a direct proof of Lie's Second Theorem. He considers a set $X_1, \ldots, X_r$ of operators,[22] such that $[X_i, X_j] = \sum_{k=1}^{r} c_{ijk}\, X_k$. The goal is to show that, if X, Y belong to $V := \mathrm{span}\{X_1, \ldots, X_r\}$, then there exists $Z \in V$ such that $e^Z = e^Y \circ e^X$ (the notation in (1.7) is followed).

A crucial tool is provided by the following interesting fact. In the change of coordinates defined by $x' = e^{tX}(x)$, we have[23]

$$e^{t\,\mathrm{ad}\,X}(Y) = Y', \qquad \text{where} \quad \begin{cases} Y = \sum_{j=1}^{n} \eta_j(x)\, \frac{\partial}{\partial x_j} \quad \text{and} \\ Y' = \sum_{j=1}^{n} \eta_j(x'(x))\, \frac{\partial}{\partial x'_j}. \end{cases} \tag{1.39}$$

From this point onwards, Campbell's arguments in proving the Second Theorem become quite unclear, and they take on rather the form of a sketch than of a rigorous proof. Nonetheless they contain "in a nutshell" the forthcoming ideas by Baker and Hausdorff on iteration of the operator $x \mapsto z(x, y)$:

1. In determining $e^Y \circ e^X$, e^Y can be replaced by iterated applications of $1 + \mu\, Y$, where μ is so small that $\mathcal{O}(\mu^2)$ can be ignored, indeed

$$e^Y = \lim_{n \to \infty} (1 + Y/n)^n. \tag{1.40}$$

2. Now, by taking care of the proper substitutions, we have

$$(1 + \mu Y) \circ e^X(x) = (1 + \mu\, Y')(x'),$$

---

[21] As we have already showed, Poincaré will use similar formulas as a starting point for his ODEs techniques in attacking the exponential formula.

[22] That is, linear first order differential operators of class C^ω on some domain of $\mathbb{R}^n$.

[23] Here $\partial/\partial x'_j$ must be properly interpreted as

$$\frac{\partial}{\partial x'_j} = \sum_{i=1}^{n} \frac{\partial \mathbf{x}_i}{\partial x'_j}(\mathbf{x}'(x))\, \frac{\partial}{\partial x_i},$$

where $\mathbf{x}(x') = e^{-tX}(x')$ and $\mathbf{x}'(x) = e^{tX}(x)$.

where Y' has the same meaning as in (1.39). By the identity in the left-hand side of (1.39), it clearly appears that Y' is a Lie series in X, Y. Hence, Y' belongs to $V = \mathrm{span}\{X_1, \ldots, X_r\}$, since $X, Y \in V$ and V is a Lie algebra. Thus the Second Theorem follows if we can write $(1 + \mu Y')(x')$ as $e^Z(x)$.

3. Given any X, Y in V, the first identity in (1.38) ensures that

$$(1 + \mu Y)\, e^X = e^{X + \mu Z_1} + \mathcal{O}(\mu^2),$$

where $Z_1 = z(X, Y)$ is a series of brackets in X, Y, *in view of* (1.36). From the same reasoning as above, we deduce that $Z_1 \in V$. Set $X_1 := X + \mu Z_1$, we have $X_1 \in V$ too, and the above identity is rewritten as

$$(1 + \mu Y)\, e^X = e^{X_1} + \mathcal{O}(\mu^2).$$

4. By the same argument as in the previous step, we have

$$(1 + \mu Y)^2\, e^X = (1 + \mu Y)\, e^{X_1} + \mathcal{O}(\mu^2) \stackrel{(1.36)}{=} e^{X_2} + 2\,\mathcal{O}(\mu^2),$$

where $X_2 := X_1 + \mu Z_2$ and $Z_2 = z(X_1, Y)$.

The above argument continues in [30], where the problems of the "convergence" of X_n to some X_∞ and the vanishing of $n \cdot \mathcal{O}(\mu^2)$ as $n \to \infty$ are studied. An overview of [30] is given in [3]. Here we confine ourselves to saying that (after several elaborate computations) Campbell derives a formula for x_∞ of the following form

$$e^y e^x = e^{x_\infty} \quad \text{where}$$

$$x_\infty = x + y + \frac{1}{2}\, [y, x] + \sum_{j=1}^{\infty} a_{2j}\, (\mathrm{ad}\, x)^{2j}(y)$$

$$+ \frac{1}{2!} \sum_{p,q=0}^{\infty} b_{p,q} \left[(\mathrm{ad}\, x)^p(y), (\mathrm{ad}\, x)^q(y) \right] \tag{1.41}$$

$$+ \frac{1}{3!} \sum_{p,q,r=0}^{\infty} b_{p,q,r} \left[\left[(\mathrm{ad}\, x)^p(y), (\mathrm{ad}\, x)^q(y) \right], (\mathrm{ad}\, x)^r(y) \right] + \cdots$$

where the constants $b_{p,q}, b_{p,q,r}, \ldots$ can be deduced by a universal recursion formula, based on the a_j in (1.35). For example,

$$b_{p,q} = (-1)^{p+q} \sum_{r=p+1}^{p+q+1} a_r\, a_{p+q+1-r} \binom{r}{p}.$$

This can be considered by rights (even despite Campbell's not completely cogent derivation) *the first form of the CBHD Theorem in the literature*.

Here we make the point that, whereas in the first paper [28] Campbell focussed on an exponential theorem within the setting of Lie groups of transformations, the second paper [30] does not mention any underlying group structure: a remarkable break with respect to [28]. It is in this abstract direction that Baker and Hausdorff will concentrate their attention, whereas Poincaré and Pascal never really separated the abstract and the group contexts.

1.1.3.2   Baker

Preceded by a series of papers [5–7] on exponential-type theorems in the context of matrix groups, Baker[24] devoted a single important paper [8] to the *"exponential theorem"* (incidentally, he is the first to use this expression) $e^{A'}e^{A} = e^{A''}$ for arbitrary noncommutative indeterminates.

The first section of [8] is devoted to describing (rather than introducing rigorously) a certain *formalism* concerned with bracketing. At a first reading this formalism looks decidedly mystifying. For Baker's purposes however, it "furnishes a compendious way of expressing the relations among alternants" [i.e., brackets]. Let us take a look at this formalism.

To each capital $A, B, C, \ldots$ (basis of some associative but noncommutative algebra) a lower-case $a, b, c, \ldots$ is associated in such a way that:

- a is called the *base* of A, and A is called the *derivative* of a.
- The map $a \mapsto \epsilon a := A$ is linear and injective.
- It is possible to extend this map to *the whole Lie algebra generated by capitals*: the base of $[A, B]$ is denoted by Ab and, more generally

$$A_1 A_2 \ldots A_n b \quad \text{denotes the base of} \quad [A_1, [A_2 \cdots [A_n, B] \cdots]];$$

- The pairing base–derivative can be extended to formal power series.

Baker provides some delicate identities concerning bases and derivatives, proving that his symbols obey some natural associative and distributive laws. This is done with the aim to get an analogous formalism for non-nested commutators: roughly speaking, skew-symmetry and Jacobi identities must be "encoded" in this formalism, e.g.,

$$Aa = 0, \quad Ab + Ba = 0, \quad ABc + BCa + CAb = 0. \tag{1.42}$$

---

[24]Henry Frederick Baker; Cambridge (England), 1866–1956.

This allows Baker to make computations[25] with his "disguised brackets" in a very fluent fashion.

A more delicate topic is the so-called "substitutional operation" treated in §2. This same operation will return in Hausdorff's paper (but with no mention of Baker's precursory study). Let us summarize it.

Given bases A, B with derivatives a, b, the symbol $b \frac{\partial}{\partial a}$ defines the operation replacing a and A by b and B (respectively) *one at a time*. For example

$$\left(b \frac{\partial}{\partial a} \right) A^2 Ca = BACa + ABCa + A^2 Cb.$$

In practice, Baker is defining a sort of mixed *algebra-derivation*, operating both on bases and derivatives. As a first main result on substitutions, Baker gives the following remarkable fact (the μ_j are arbitrary scalars)

$$\sum_{i=0}^{\infty} \frac{t^i}{i!} \left(b \frac{\partial}{\partial a} \right)^i \left(\sum_{j=0}^{\infty} \mu_j A^j \right) = \sum_{j=0}^{\infty} \mu_j \left\{ \sum_{i=0}^{\infty} \frac{t^i}{i!} \left(b \frac{\partial}{\partial a} \right)^i A \right\}^j, \tag{1.43}$$

where b is any base of the form $b = \left(\sum_{i=0}^{\infty} \lambda_i A^i \right) c$ where the λ_i are any scalars. Identity (1.43) can be interpreted as follows: since $\delta = b \frac{\partial}{\partial a}$ is a derivation, $\exp(t\delta) := \sum_{i \geq 0} t^i \delta^i / i!$ is an algebra morphism, and (1.43) follows by "continuity". Another momentous identity is the following one:[26]

$$\left(b \frac{\partial}{\partial a} \right) e^A = f(\mathrm{ad}\, A)(B)\, e^A, \tag{1.44}$$

where f denotes the following formal power series

$$f(z) := \sum_{j=1}^{\infty} z^{j-1} / j! = (e^z - 1)/z.$$

With this same f, Baker now makes the choice

$$b := \left(1 - \frac{A}{2} + \sum_{j=1}^{\infty} \frac{\varpi_j}{(2j)!} A^{2j} \right) a' = \left(\frac{1}{f}(A) \right) a', \tag{1.45}$$

---

[25]For example from the first identity in (1.42) with $A = [B, C]$ one gets $[B, C]Bc = 0$, the associative and distributive laws then give $BCBc - CB^2c = 0$, and by applying the map ϵ we derive a (not obvious) identity between nested commutators

$$[B, [C, [B, C]]] - [C, [B, [B, C]]] = 0.$$

[26]We explicitly remark that Baker proves (1.44) when $b = Ac$, and then he asserts – without proof – that the same holds for an arbitrary b.

where a' is any base and the ϖ_j are the Bernoulli numbers, according to Baker's notation.[27] This implies $a' = f(A)b$ so that, by passing to the associated derivatives one obtains

$$A' = f(\operatorname{ad} A)(B). \tag{1.46}$$

Gathering (1.44) and (1.46), we infer $\left(b \frac{\partial}{\partial a}\right) e^A = A' e^A$, and inductively

$$\left(b \frac{\partial}{\partial a}\right)^i e^A = (A')^i\, e^A, \quad \forall\, i \geq 0. \tag{1.47}$$

Next he puts (be careful: *this will turn out to be the CBHD series!*)

$$A'' := \sum_{i=0}^{\infty} \frac{1}{i!} \left(b \frac{\partial}{\partial a}\right)^i A. \tag{1.48}$$

If we now apply (1.43) with $t = 1$ and $\mu_j = 1/j!$, we immediately get (by exploiting the latter representation of A'')

$$e^{A''} = \sum_{i=0}^{\infty} \frac{1}{i!} \left(b \frac{\partial}{\partial a}\right)^i e^A \overset{(1.47)}{=} e^{A'} e^A, \quad \text{whence } e^{A''} = e^{A'} e^A, \tag{1.49}$$

and the exponential theorem is proved: it suffices to remark that A'' in (1.48) is a Lie series in A, A'. But this is true since A'' is the derivative of an infinite sum of bases,[28] and the relation "base$\leftrightarrow$derivative" has been defined only between Lie elements. Gathering together (1.45), (1.48) and (1.49), Baker has proved the following remarkable formula

$$e^{A'} e^A = e^{A''}, \quad \textit{where } A'' = \sum_{i=0}^{\infty} \frac{1}{i!} \left(B \frac{\partial}{\partial a}\right)^i A,$$

$$\textit{with} \quad B = A' - \frac{1}{2}[A, A'] + \sum_{j=1}^{\infty} \frac{\varpi_j}{(2j)!} (\operatorname{ad} A)^{2j}(A'), \tag{1.50}$$

$$\textit{where the } \varpi_j \textit{ are defined by} \quad \frac{z}{e^z - 1} = 1 - \frac{z}{2} + \sum_{j=1}^{\infty} \frac{\varpi_j}{(2j)!} z^{2j}.$$

---

[27] Baker's definition of the Bernoulli numbers ϖ_j is the following one

$$\frac{z}{e^z - 1} = 1 - \frac{z}{2} + \sum_{j=1}^{\infty} \frac{\varpi_j}{(2j)!} z^{2j}.$$

[28] Indeed A'' is the derivative of $a'' := \sum_{i=0}^{\infty} \frac{1}{i!} \left(b \frac{\partial}{\partial a}\right)^i a$.

Finally, the last two sections of the paper are devoted to an application of the exponential theorem to transformation groups, and in particular to a proof of the Second Theorem. Actually, Baker does not add too much to what Pascal (and Campbell) had already proved, nor does he consider in any case the problem of the convergence of the series he obtained.

1.1.3.3 Hausdorff

Hausdorff[29] devoted one single paper [78] to what he called "the symbolic exponential formula in the group theory", i.e., the study of the function $z = z(x, y)$ defined by the identity $e^x e^y = e^z$, a problem of "symbolic analysis", as he defines it.

In the foreword of his paper, Hausdorff provides a brief review of the work of his predecessors. Besides Hausdorff's comments on Schur, Poincaré and Pascal (which we already mentioned), he turns to Campbell and Baker.

Campbell is acknowledged as the first who "attempted" to give a proof of the Second Theorem of Lie with the aid of a symbolic exponential formula. Hausdorff's opinion is that Campbell's prolongation of the expansion of z "is based on a passage to the limit, neither completely clear nor simple".

Now, surprisingly, Hausdorff does not mention at all Baker's 1905 paper [8] (the only citation is to Baker's 1901 paper [5]), even though more than one year had elapsed between Baker's [8] and Hausdorff's [78] publications. It seems beyond doubt (which we cannot know, however) that if Hausdorff had known [8] he would have considered his own (independent) proof of the symbolic exponential formula as overlapping to a great extent with Baker's.[30]

In the first sections, the necessary algebraic objects are described. In modern words, Hausdorff introduces the following structures:

L_0: This is the associative algebra (over $\mathbb{R}$ or $\mathbb{C}$) of the polynomials P in a finite set of non-commuting symbols $x, y, z, u, \ldots$; the "dimension" of P is the *smallest* of the degrees of its monomials.

L: This is the associative algebra of the formal power series related to L_0; any infinite sum is allowed, provided it involves summands with *different* (hence increasing) dimensions.

K_0: This is the Lie subalgebra of L_0 consisting of the Lie polynomials in the basis symbols $x, y, z, u, \ldots$

K: This is the Lie subalgebra of L consisting of the Lie series associated to K_0.

---

[29]Felix Hausdorff; Breslau, 1868 (at that time, Silesia – Prussia; now Wrocław – Poland) – Bonn (Germany), 1942.

[30]A detailed analysis of the similarities between the papers [8] and [78] can be found in [3].

Then Hausdorff considers the same substitutional operation as Baker had introduced earlier, but in a more "differential sense": If $F \in L$ is momentarily thought of as a function of the basis symbol x, and u is a new symbol, we have the "Taylor expansion"

$$F(x+u) = F(x) + \left(u\frac{\partial}{\partial x}\right)F(x) + \frac{1}{2!}\left(u\frac{\partial}{\partial x}\right)^2 F(x) + \frac{1}{3!}\left(u\frac{\partial}{\partial x}\right)^3 F(x) + \cdots,$$

(1.51)

where $u\frac{\partial}{\partial x}$ is the derivation of L mapping x to u and leaving unchanged all the other basis symbols.[31] For example, if $F = F_0^x + F_1^x + F_2^x + \cdots$, where F_n^x contains x precisely n times, one has

$$\left(x\frac{\partial}{\partial x}\right)F(x) = F_1^x + 2\,F_2^x + \cdots + n\,F_n^x + \cdots$$

(1.52)

From (1.51) and the commutativity of $u\frac{\partial}{\partial x}$ and $v\frac{\partial}{\partial y}$, one also obtains

$$F(x+u, y+v) = F(x, y) + \sum_{n=1}^{\infty} \frac{1}{n!}\left(u\frac{\partial}{\partial x} + v\frac{\partial}{\partial y}\right)^n F(x, y).$$

(1.53)

In order to preserve many of Hausdorff's elegant formulas, we shall use the notation $[P]$ to denote *left-nested* iterated brackets: for example

$$[xy] = [x, y], \quad [xyz] = [[x, y], z], \quad [xyzu] = [[[x, y], z], u], \quad \cdots$$

(1.54)

With only these few prerequisites, §3 is devoted to the proof of Hausdorff's main result, Proposition B [78, page 29], which is the announced symbolical exponential formula: *The function z of x, y defined by $e^x e^y = e^z$ can be represented as an infinite series whose summands are obtained from x, y by bracketing operations, times a numerical factor.* Let us analyze Hausdorff's argument and compare it to the proofs of his predecessors.

We are in a position to say that $z = z(x, y)$ is a true function of x, y for in L the logarithm makes sense, so that

$$z = (e^x e^y - 1) - \frac{1}{2}(e^x e^y - 1)^2 + \frac{1}{3}(e^x e^y - 1)^3 + \cdots$$

We aim to prove that the above z actually belongs to K, not only to L.

Since $\left(u\frac{\partial}{\partial x}\right)e^x = \sum_{n=1}^{\infty}\frac{1}{n!}\sum_{i=0}^{n-1}x^i u x^{n-1-i}$, the substitution $u = [w, x]$ generates a telescopic sum, so that

$$\left([w, x]\frac{\partial}{\partial x}\right)e^x = we^x - e^x w.$$

(1.55)

---

[31]Equivalently, $\left(u\frac{\partial}{\partial x}\right)^n F(x)$ is the sum of all the summands of $F(x+u)$ containing u precisely n times.

Furthermore Hausdorff provides two other, now well known, formulas of noncommutative algebra[32] (see also the notation in (1.54)):

$$[wx^n] = \sum_{i=0}^{n}(-1)^i \binom{n}{i}x^i wx^{n-i}, \tag{1.56}$$

$$e^{-x}we^x = \sum_{n=0}^{\infty} \frac{1}{n!}[wx^n], \tag{1.57}$$

where (1.57) follows easily from (1.56) by a computation similar to Campbell's (1.37). As a consequence we obtain

$$\left([w,x]\frac{\partial}{\partial x}\right)e^x \overset{(1.55)}{=} e^x(e^{-x}we^x - w) \overset{(1.57)}{=} e^x\sum_{n=1}^{\infty}\frac{1}{n!}[wx^n].$$

An analogous formula with e^x as a right factor holds. This gives the following results: *If u is of the form $[w,x]$ for some $w \in L$, we have*[33]

$$\left(u\frac{\partial}{\partial x}\right)e^x = e^x\,\varphi(u,x) \quad where \quad \varphi(u,x) = \sum_{n=1}^{\infty}\frac{1}{n!}[ux^{n-1}], \tag{1.58}$$

$$\left(u\frac{\partial}{\partial x}\right)e^x = \psi(u,x)\,e^x \quad where \quad \psi(u,x) = \sum_{n=1}^{\infty}\frac{(-1)^{n-1}}{n!}[ux^{n-1}]. \tag{1.59}$$

We remark that identity (1.59) was already discovered by Baker, see (1.44). If we introduce the functions

$$h(z) = \frac{1 - e^{-z}}{z}, \qquad g(z) = \frac{1}{h(z)},$$

then φ and ψ can be rewritten as

$$\varphi(u,x) = h(\mathrm{ad}\,x)(u), \qquad \psi(u,x) = h(-\mathrm{ad}\,x)(u), \tag{1.60}$$

Furthermore, from (1.60) we get the inversion formulas

$$\begin{aligned} p &= \varphi(u,x) \;\Leftrightarrow\; u = \chi(p,x), \\ q &= \psi(u,x) \;\Leftrightarrow\; u = \omega(q,x), \end{aligned} \tag{1.61}$$

---

[32]Formula (1.56) also appears in Campbell [28, page 387] and in Baker [8, page 34], whereas formula (1.57) also appears in Campbell [28, page 386] and in Baker [8, page 38].
[33]Analogous identities hold when $u = x$ or when u is a series of summands of the form $[\cdot,x]$, and the restriction $u = [w,x]$ will be systematically omitted by Hausdorff.

where

$$\chi(p,x) := p - \frac{1}{2}\,[p\,x] + \sum_{n=1}^{\infty} \frac{(-1)^{n-1}B_n}{(2\,n)!}\,[p\,x^{2n}], \qquad \omega(q,x) := \chi(q,-x),$$

(1.62)

and we see how the Bernoulli numbers step in.[34] We thus get the following important formulas (note that $\varphi(\cdot,x)$ and $\psi(\cdot,x)$ are linear)

$$e^{x+\alpha u} = e^x\big(1 + \alpha\,\varphi(u,x) + \mathcal{O}(\alpha^2)\big), \qquad (1.63)$$

$$e^{x+\alpha u} = \big(1 + \alpha\,\psi(u,x) + \mathcal{O}(\alpha^2)\big)e^x, \qquad (1.64)$$

valid for every scalar α. We remark that a proof of (1.63) is contained in Poincaré [142, page 244, 245] (see indeed the computation in (1.26)), but Hausdorff does not mention it.

At this point, Hausdorff's argument becomes somewhat opaque: he states that it is possible to leave z unchanged in $e^x e^y = e^z$ by adding αu to x and by accordingly adding to y a certain quantity $-\alpha v + \mathcal{O}(\alpha^2)$, so that the identity $e^{x+\alpha u}e^{y-\alpha v+\mathcal{O}(\alpha^2)} = e^z$ also holds. We prefer to modify Hausdorff's argument in the following way (similarly to Yosida, [183]): Let u,v be any pair of elements of L satisfying

$$\varphi(u,x) = \psi(v,y), \qquad (1.65)$$

and let $z(\alpha)$ be defined by

$$e^{z(\alpha)} := e^{x+\alpha u}e^{y-\alpha v}, \qquad \alpha \in \mathbb{R}.$$

For example, thanks to the inversion formulas (1.61), the choices

$$\{u = x, \ v = \omega(x,y)\} \quad \text{or} \quad \{v = y, \ u = \chi(y,x)\} \qquad (1.66)$$

do satisfy (1.65). We thus have the following computation:

$$e^{z(\alpha)} = e^x\big(1 + \alpha\,\varphi(u,x) + \mathcal{O}(\alpha^2)\big)\big(1 - \alpha\,\psi(v,y) + \mathcal{O}(\alpha^2)\big)e^y$$

$$= e^x\big(1 + \alpha(\varphi(u,x) - \psi(v,y)) + \mathcal{O}(\alpha^2)\big)e^y$$

$$\overset{(1.65)}{=} e^x(1 + \mathcal{O}(\alpha^2))e^y = e^x e^y(1 + \mathcal{O}(\alpha^2)).$$

---

[34]Indeed we have

$$g(z) = \frac{z}{1-e^{-z}} = 1 + \frac{z}{2} + \sum_{n=1}^{\infty} \frac{(-1)^{n-1}B_n}{(2\,n)!}\,z^{2n},$$

where, according to Hausdorff's notation, the following definition of Bernoulli numbers B_n holds: $\frac{z}{e^z-1} = 1 - \frac{z}{2} + \sum_{n=1}^{\infty} \frac{(-1)^{n-1}B_n}{(2\,n)!}\,z^{2n}$.

From the above expansion it is easily derived that $\dot{z}(0) = 0$. On the other hand, by applying the expansion (1.53) to $F(x,y) := \log(e^x e^y)$, we get

$$z(\alpha) = F(x + \alpha u, y - \alpha v) = z(x,y) + \alpha \left(u \frac{\partial}{\partial x} - v \frac{\partial}{\partial y} \right) z(x,y) + \mathcal{O}(\alpha^2).$$

Hence, $\dot{z}(0) = 0$ ensures that $z(x,y) = e^x e^y$ *satisfies the following PDE*

$$\left(u \frac{\partial}{\partial x} \right) z = \left(v \frac{\partial}{\partial x} \right) z, \tag{1.67}$$

for any u, v satisfying (1.65). We are thus allowed to make, e.g., the choices in (1.66), which respectively give

$$\left(x \frac{\partial}{\partial x} \right) z = \left(\omega(x,y) \frac{\partial}{\partial y} \right) z, \tag{1.68}$$

$$\left(y \frac{\partial}{\partial y} \right) z = \left(\chi(y,x) \frac{\partial}{\partial x} \right) z, \tag{1.69}$$

and each of these suffices to determine z. Indeed, by writing $z = z_0^x + z_1^x + \cdots$ (where z has been ordered with respect to increasing powers of x) and by using (1.52), Hausdorff derives from (1.68) the following remarkable formula

$$z_n^x = \frac{1}{n!} \left(\omega(x,y) \frac{\partial}{\partial y} \right)^n y, \quad n \geq 0. \tag{1.70}$$

Here we have used the fact that an application of the operator $\omega(x,y) \frac{\partial}{\partial y}$ increases the degree in x by one unit. Analogously, from (1.69) one gets

$$z_n^y = \frac{1}{n!} \left(\chi(y,x) \frac{\partial}{\partial x} \right)^n x, \quad n \geq 0. \tag{1.71}$$

Since x, y and $\omega(x,y), \chi(y,x)$ are all Lie series (see (1.62)), this proves that any z_n^x, z_n^y is a Lie polynomial and the exponential formula is proved.

From (1.70) and the definition of $\omega(x,y)$, it follows that Hausdorff has proved the following result (note that $[x\, y^{2n}] = (\operatorname{ad} y)^{2n}(x)$)

$$e^x e^y = e^z, \quad \text{where } z = \sum_{n=0}^{\infty} \frac{1}{n!} \left(\omega(x,y) \frac{\partial}{\partial y} \right)^n y,$$

$$\text{with} \quad \omega(x,y) = x + \frac{1}{2}[x,y] + \sum_{n=1}^{\infty} \frac{(-1)^{n-1} B_n}{(2n)!} (\operatorname{ad} y)^{2n}(x), \tag{1.72}$$

$$\text{where the } B_n \text{ are defined by} \quad \frac{z}{e^z - 1} = 1 - \frac{z}{2} + \sum_{n=1}^{\infty} \frac{(-1)^{n-1} B_n}{(2n)!} z^{2n}.$$

We observe that *this is exactly Baker's formula* (1.50), proved one year earlier.[35]

Furthermore, Hausdorff provides a new recursion formula, allowing us to obtain the *homogeneous* summands of z, ordered with respect to the *joint* degree in x, y. His argument is based on his previous techniques, more specifically, by deriving an ODE for $z(\alpha)$ defined by $e^{z(\alpha)} = e^{x+\alpha x}e^y$. Hausdorff thus obtains the remarkable formula (see [78, eq. (29), page 31])

$$\left(x\frac{\partial}{\partial x}\right)z + \left(y\frac{\partial}{\partial y}\right)z = [x, z] + \chi(x + y, z). \tag{1.73}$$

Inserting in (1.73) the expansion $z = z_1^{x,y} + z_2^{x,y} + \cdots$, where $z_n^{x,y}$ has joint degree n in x, y we obtain a recursion formula for the summands $z_n^{x,y}$ which exhibits in a very limpid form their Lie-polynomial nature. This formula will return in modern proofs of the Campbell-Baker-Hausdorff Theorem, see Djoković [48] and Varadarajan [171] (see also Chap. 4 of this book) and – as shown in [171] – it can be profitably used to settle convergence questions.

From Sect. 4 onwards, Hausdorff turns his attention to the applications to groups of transformations. After having criticized his predecessors for this omission, Hausdorff's main concern is to solve the *convergence* problem. Of all his series expansions of z, he studies the one obtained by ordering according to increasing powers of y in (1.71).

To this end, let $t_1, \ldots, t_r$ be a basis of a set of infinitesimal transformations, with structure constants given by $[t_\rho, t_\sigma] = \sum_{\lambda=1}^r c_{\rho\sigma\lambda}t_\lambda$. Let $x = \sum_\rho \xi_\rho t_\rho$, $y = \sum_\rho \eta_\rho t_\rho$ and suppose that the series $z(x, y) = \sum_{n=0}^\infty z_n^y$ defined by (1.71) converges to $z = \sum_\rho \zeta_\rho t_\rho$. Set $z_1^y = \chi(y, x)$, and suppose that this converges to $u = \sum_\rho \vartheta_\rho t_\rho$. From the definition of $u = \chi(y, x)$, we see that u depends linearly on y and vice versa. Passing to coordinates with respect to $\{t_1, \ldots, t_r\}$, we infer the existence of a matrix $A(\xi) = \big(\alpha_{\rho\sigma}(\xi)\big)$, such that $\vartheta_\sigma = \sum_\rho \alpha_{\rho\sigma}(\xi)\eta_\rho$. Thus, the identity $u = \chi(y, x)$ is rewritten compactly as

$$\sum_\sigma \alpha_{\rho\sigma}(\xi)t_\sigma = \chi\big(t_\rho, \sum_\lambda \xi_\lambda t_\lambda\big), \quad \rho = 1, \ldots, r. \tag{1.74}$$

---

[35]This might lead us to suppose that Hausdorff did not ignore Baker's results in [8]. Nonetheless, it is beyond doubt that Hausdorff's argument, devoid of the intricate formalism of Baker, is the first totally perspicuous proof of the exponential formula, with the merit to join together – in the most effective way – *the right contributions* from his predecessors: some algebraic computations from Pascal and Campbell; Poincaré's technique of deriving ODEs for $z(\alpha)$; the use of Baker's substitutional operator.

This will allow Hausdorff to obtain an explicit series for each of the functions $\alpha_{\rho\sigma}(\xi)$ and, consequently, for each of the functions ϑ_ρ. To this aim, let us introduce the structure matrix[36]

$$\Xi(\xi) := \left(\xi_{\rho\sigma}\right)_{\rho\sigma}, \quad \text{where } \xi_{\rho\sigma} := \sum_{\lambda=1}^{r} c_{\rho\lambda\sigma}\xi_\lambda.$$

With the aid of the matrix Ξ, Hausdorff recognizes that (1.74) can be elegantly rewritten as (Hausdorff cites Schur for a less compact version of this identity)

$$A = 1 - \frac{1}{2}\Xi + \frac{B_1}{2!}\Xi^2 - \frac{B_2}{4!}\Xi^4 + \cdots,$$

$$\text{that is,} \quad A(\xi) = f(\Xi(\xi)), \quad \text{where } f(z) = \frac{z}{e^z - 1}. \tag{1.75}$$

It is now very simple to deduce from (1.75) a domain of convergence for the series expressing the functions $\alpha_{\rho\sigma}(\xi)$: If M is an upper bound for all of the quantities $|\xi_{\rho\sigma}|$, it suffices to have $rM < 2\pi$, since the complex function $f(z)$ is holomorphic in the disc about 0 of radius 2π. This produces a domain of convergence for each $\vartheta_\sigma = \sum_\rho \alpha_{\rho\sigma}(\xi)\eta_\rho$ and hence for $u = \sum_\rho \vartheta_\rho t_\rho$, which is the first summand $\chi(y, x)$ in the expansion for $z(x, y)$.

As for the other summands in the expansion

$$z = x + \chi(y, x) + \frac{1}{2!}\left(\chi(y, x)\frac{\partial}{\partial x}\right)\chi(y, x) + \frac{1}{3!}\left(\chi(y, x)\frac{\partial}{\partial x}\right)^2\chi(y, x) + \cdots, \tag{1.76}$$

it suffices to discover what the operator $u\frac{\partial}{\partial x}$ looks like in coordinates. Setting $F(x) = \sum_\rho f_\rho(\xi) t_\rho$, we have the usual Taylor expansion

$$F(x + u) = \sum_\rho f_\rho(\xi + \vartheta) t_\rho = \sum_\rho \left(f_\rho(\xi) + \sum_\sigma \vartheta_\sigma \frac{\partial f_\rho}{\partial \xi_\sigma}(\xi) + \mathcal{O}(|\vartheta|^2)\right)t_\rho,$$

which, compared to the expansion $F(x + u) = F(x) + u\frac{\partial}{\partial x}F(x) + \mathcal{O}(u^2)$ in (1.51), shows that the operator $u\frac{\partial}{\partial x}$ has the same meaning as the infinitesimal transformation $\Lambda = \sum_{\sigma=1}^{r} \vartheta_\sigma(\xi, \eta)\frac{\partial}{\partial \xi_\sigma}$, that is,

$$\Lambda = \sum_{\rho,\sigma=1}^{r} \eta_\rho \alpha_{\rho\sigma}(\xi)\frac{\partial}{\partial \xi_\sigma}. \tag{1.77}$$

This proves that the expansion (1.76) becomes, in coordinates $z = \sum_\rho \zeta_\rho t_\rho$,

$$\zeta_\rho = \xi_\rho + \Lambda\xi_\rho + \frac{1}{2!}\Lambda^2\xi_\rho + \frac{1}{3!}\Lambda^3\xi_\rho + \cdots =: e^\Lambda(\xi_\rho). \tag{1.78}$$

---

[36]This is nothing but the transpose of the matrix representing the right-adjoint map $Y \mapsto [Y, \sum_\lambda \xi_\lambda t_\lambda]$.

Now, the crucial device is to observe that *this is precisely the expansion of the solution $t \mapsto \zeta(t)$ to the following ODE system*

$$\begin{cases} \dot{\zeta}_\sigma(t) = \sum_\rho \eta_\rho \, \alpha_{\rho\sigma}(\zeta(t)) \\ \zeta_\sigma(0) = \xi_\sigma \end{cases} \qquad \sigma = 1, \ldots, r.$$

Hence, since it has already been proved that the functions $\alpha_{\rho\sigma}$ are analytic in a neighborhood of the origin, *the convergence of ζ_ρ in* (1.78) *is a consequence of the general theory of ODEs*. Note the similarity with Poincaré's convergence argument, see (1.25).

This all leads to the proof of the *group exponential theorem*, Proposition C [78, page 39]: *Given the structure equations* $[t_\rho, t_\sigma] = \sum_\lambda c_{\rho\sigma\lambda} t_\lambda$, *by means of the exponential formula*

$$e^{\sum \xi_\rho \, t_\rho} \, e^{\sum \eta_\rho \, t_\rho} = e^{\sum \zeta_\rho \, t_\rho},$$

the functions $\zeta_\rho = \zeta_\rho(\xi, \eta)$ are well defined and analytic in a suitable neighborhood of $\xi = \eta = 0$.

Section 7 is devoted to the derivation of the Second and Third Theorem of Lie by means of the above Proposition C. As a fact, Hausdorff's argument does not add too much to what his predecessors, from Poincaré to Baker, had said about the same topic. There is no better way to end this brief review of Hausdorff's proof of the *Exponentialformel* than quoting his words (see [78, page 44]): the symbolic exponential theorem "Proposition B is the *nervus probandi* of the fundamental theorems of group theory".[37]

1.1.3.4 Dynkin

After Hausdorff's 1906 paper [78], forty years elapsed before the problem of providing a truly explicit representation of the series $\log(e^x e^y)$ was solved. This question was first answered by Dynkin[38] in his 1947 paper [54].[39]

Starting from what Dynkin calls "the theorem of Campbell and Hausdorff", i.e., the result stating that $\log(e^x e^y)$ is a series of Lie polynomials

---

[37]The – not so current – Latin expression "nervus probandi" (occurring frequently e.g., in Immanuel Kant's philosophic treatises) means, literally, "the sinews of what has to be proved", that is, *the crucial argument of the proof*. This Latin expression describes very well the paramount rôle played by the exponential theorem in Hausdorff's arguments, as the real cornerstone in the proof of many results of group theory.

[38]Eugene Borisovich Dynkin; Leningrad (Russia), 1924.

[39]In what follows, we will quote the 2000 English translation of the Russian paper [54], contained in [57, pages 31–34].

in x, y, formula (12) of [57] provides the following explicit representation, later known as *Dynkin's Formula* (for the Campbell-Baker-Hausdorff series)

$$\log(e^x e^y) = \sum \frac{(-1)^{k-1}}{k} \frac{1}{p_1! q_1! p_2! q_2! \cdots p_k! q_k!} \left(x^{p_1} y^{q_1} x^{p_2} y^{q_2} \cdots x^{p_k} y^{q_k} \right)^0,$$
(1.79)

where the sum runs over $k \in \mathbb{N}$ and all non-vanishing couples (p_i, q_i), with $i = 1, \ldots, k$. Most importantly, the map $P \mapsto P^0$ – which we now describe – is introduced, where P is any polynomial in a finite set of non-commuting indeterminates.

Let $\mathcal{R}$ denote the algebra of the polynomials in the non-commuting indeterminates $x_1, \ldots, x_n$ over a field of characteristic zero. For $P, Q \in \mathcal{R}$, let $[P, Q] = PQ - QP$ denote the usual commutator,[40] and let $\mathcal{R}^0$ be the smallest Lie subalgebra of $\mathcal{R}$ containing $x_1, \ldots, x_n$. Finally, consider the unique linear map from $\mathcal{R}$ to $\mathcal{R}^0$ mapping $P = x_{i_1} x_{i_2} \cdots x_{i_k}$ to P^0, where

$$P^0 = \frac{1}{k} \left[\cdots [[x_{i_1}, x_{i_2}], x_{i_3}] \cdots x_{i_k} \right].$$

Then (see [57, Theorem at page 32]) Dynkin proves that

$$P \in \mathcal{R}^0 \text{ if and only if } P = P^0. \tag{1.80}$$

This theorem, later referred to as the Dynkin-Specht-Wever Theorem (see also Specht, 1948 [161], and Wever, 1949 [179]), is one of the main characterizations of Lie polynomials.

With this result at hands, the derivation of the representation (1.79) is almost trivial. Indeed, the very definitions of $\log, \exp$ give

$$\log(e^x e^y) = \sum \frac{(-1)^{k-1}}{k} \frac{1}{p_1! q_1! p_2! q_2! \cdots p_k! q_k!} x^{p_1} y^{q_1} x^{p_2} y^{q_2} \cdots x^{p_k} y^{q_k},$$

where the sum is as in (1.79). *If we assume that the exponential-theorem holds* (that is, that the above series is indeed a Lie series in x, y), an application of the map $P \mapsto P^0$ (naturally extended to series) leaves unchanged $\log(e^x e^y)$ so that (1.79) holds.

Three other fundamental results are contained in Dynkin's paper [54]:

1. If $\mathbb{K}$ is $\mathbb{R}$ or $\mathbb{C}$ and R is any finite dimensional Lie algebra over $\mathbb{K}$, then the series in the right-hand side of (1.79), say $\Phi^0(x, y)$, converges for every x, y

---

[40]Dynkin used the notation $P \circ Q := PQ - QP$; also $x_{i_1} \circ x_{i_2} \circ \cdots \circ x_{i_k}$ denotes the left-nested commutator $(\cdots((x_{i_1} \circ x_{i_2}) \circ x_{i_3}) \circ \cdots \circ x_{i_k})$. We allowed ourselves to use the bracketing notation, as in the rest of the book.

in a neighborhood of $0 \in R$. Indeed, *thanks to the very explicit expression of Dynkin's series*, the following direct computation holds: if $\|\cdot\|$ is any norm on R compatible with the Lie bracket,[41] then

$$\left\| \left(x^{p_1} y^{q_1} x^{p_2} y^{q_2} \cdots x^{p_k} y^{q_k} \right)^0 \right\| \leq \|x\|^{p_1 + \cdots + p_k} \cdot \|y\|^{q_1 + \cdots + q_k}.$$

Consequently, as for the study of the total convergence of the series $\Phi^0(x, y)$, an upper bound is given by

$$\sum_k \frac{(-1)^{k-1}}{k} \left(e^{\|x\|} e^{\|y\|} - 1 \right)^k = \log(e^{\|x\|} e^{\|y\|}) = \|x\| + \|y\| < \infty,$$

provided that $\|x\| + \|y\| < \log 2$. As a matter of fact, *this is the first argument – in the history of the CBHD Theorem – for a direct proof of the convergence problem*, an argument much more natural and direct than those given by Poincaré, by Pascal, and even by Hausdorff. Also, Dynkin's proof works *for any finite dimensional Lie algebra*, hence in particular for the algebra of the infinitesimal transformations of a Lie group, thus comprising all the related results by his predecessors.

2. The same arguments as above can be straightforwardly generalized to the so-called *Banach-Lie algebras*,[42] thus anticipating a new research field. Dynkin will extensively return to this generalization in [56].

3. Dynkin's series (1.79), together with the obvious (local) associativity of the operation $x * y := \Phi^0(x, y)$, allows us to prove Lie's Third Theorem in its local version (a concern for Schur, Poincaré, Pascal and Hausdorff) in a very simple way: indeed $*$ defines a *local group* on a neighborhood of the origin of every finite dimensional (real or complex) Lie algebra, with prescribed structure constants.

As observed above, Dynkin provides a commutator-formula for $\log(e^x e^y)$, yet his proof assumes its commutator-nature in advance. Two years later in [55], Dynkin will give another proof of the fact that $\log(e^x e^y)$ is a Lie series, completely independent of the arguments of his predecessors, and mainly based on his theorem (1.80) and on *combinatorial algebra*. Following

---

[41]This means that $\|[x, y]\| \leq \|x\| \cdot \|y\|$ for every $x, y \in R$. Note that such a norm always exists, thanks to the continuity and bilinearity of the map $(x, y) \mapsto [x, y]$ and the finite dimensionality of R.

[42]A Banach-Lie algebra is a Banach space (over $\mathbb{R}$ or $\mathbb{C}$) endowed with a Lie algebra structure such that $A \times A \ni (x, y) \mapsto [x, y] \in A$ is continuous. In this case, if $\|\cdot\|$ is the norm of A, there exists a positive constant M such that $\|[x, y]\| \leq M \|x\| \cdot \|y\|$ for every $x, y \in A$, so that the norm $M\|\cdot\|$ is compatible with the Lie bracket of A and Dynkin's arguments – this time also appealing to the completeness of A – generalize directly.

Bose [23, page 2035], with respect to the recursive formulas for $\log(e^x e^y)$ proved by Baker and Hausdorff, "Dynkin radically simplified the problem", by deriving "an effective procedure" for determining the BCH series.

Dynkin's original proof provides a major contribution to the understanding of the combinatorial aspects hidden behind the composition of exponentials. Crucial ideas contained in [55] will return over the subsequent 60 years of life of the CBHD Theorem, e.g. in the study of (numerical) algorithms for obtaining efficiently simplified expansions of the series representation of $\log(e^x e^y)$. For this reason (and for the fact that Dynkin provided a completely new and self contained proof of the theorem of Campbell, Baker, Hausdorff), we considered the acronym CBHD as the appropriate title for our book.

An overview of Dynkin's proof in [55] can be found in [3]. We here confine ourselves to writing the new representation found in [55]. First some notation: in the $(n+1)$-tuple $I = (i_0, \ldots, i_n)$ of pairwise distinct integers, a couple of consecutive indices $i_\beta, i_{\beta+1}$ is called *regular* (nowadays also known as a rise) if $i_\beta < i_{\beta+1}$ and *irregular* (a fall) otherwise; denote by s_I the number of regular couples of I and by t_I the number of irregular ones. Then it holds

$$\log(e^x e^y) = \sum_{p,q=0}^{\infty} \frac{1}{p!\, q!}\, P(\underbrace{x, \ldots, x}_{p \text{ times}}; \underbrace{y, \ldots, y}_{q \text{ times}}), \quad \text{where} \qquad (1.81)$$

$$P(x_0, \ldots, x_n) = \frac{1}{(n+1)!} \sum_J (-1)^{t_{0J}}\, s_{0J}!\, t_{0J}!\, [[x_0, x_{j_1}] \cdots x_{j_n}],$$

where $0J$ means $(0, j_1, \ldots, j_n)$ and $J = (j_1, \ldots, i_n)$ runs over the permutations of $\{1, \ldots, n\}$ (the meaning of t_{0J}, s_{0J} is explained above).

Finally, in [56] Dynkin studies in great detail the applications of his representation formula for $\log(e^x e^y)$ to normed Lie algebras and to analytic groups. The starting points are not transformation groups, but Lie algebras. The main result is thus the *construction of a local topological group attached to every Banach-Lie algebra by means of the explicit series* of $\log(e^x e^y)$. The theory of groups and algebras was meanwhile advanced sufficiently to make it possible to use more general notions and provide broader generalizations (non-Archimedean fields are considered, normed spaces and normed algebras are involved, together with local topological or analytic groups).

As for the history of the CBHD Theorem, this paper paved the way, by happenstance, for the study of other possible representations of $\log(e^x e^y)$ and therefore for the problem of *improved domains of convergence* for the representations of $\log(e^x e^y)$. It is therefore to be considered as opening the "modern era" of the CBHD Theorem, the subject of the next section.

1.2   The "Modern Era" of the CBHD Theorem

The second span of life of the CBHD Theorem (1950-today), which we decided to name its "modern era", can be thought of as starting with the re-formalization of Algebra operated by the Bourbakist school. Indeed, in Bourbaki [27] (see in particular II, §6–§8) the well-behaved properties of the "Hausdorff series" and of the "Hausdorff group" are derived as byproducts of general results of Lie algebra theory. In particular, the main tool to prove that $\log(e^x e^y)$ is a Lie series in x, y is the following characterization of Lie elements $L(X)$ within the free associative algebra $A(X)$ of the polynomials in the letters of a set X:

$$L(X) = \big\{ t \in A(X) \ : \ \delta(t) = t \otimes 1 + 1 \otimes t \big\}.$$

Here $\delta \ : \ A(X) \to A(X) \otimes A(X)$ is the unique algebra-morphism such that $\delta(x) = x \otimes 1 + 1 \otimes x$, for all $x \in X$. This result (frequently named after Friedrichs[43] [64]) or equivalent versions of it are employed also by other coeval books [85, 99, 159], framing the CBHD Theorem within a vast algebraic theory. The *ad hoc* techniques of the early proofs from the previous period 1890–1950 do not play any rôle in this new approach and any reference to the theory of Sophus Lie's groups of transformations becomes immaterial.

A compelling justification of the interest in the study of product of exponentials (possibly, of more general objects too, such as operators on Hilbert spaces or on Banach algebras) comes from modern (actually 1960–1970s) Physics, especially from Quantum and Statistical Mechanics. The applications of this algebraic result – mainly named after Baker, Campbell, Hausdorff – cover many Physical disciplines: mathematical physics, theoretical physics (perturbation theory, transformation theory), quantum and statistical Mechanics (the study of quantum mechanical systems with time-dependent Hamiltonians, linear stochastic motions) and many references can be provided (see [51, 64, 65, 104, 119, 128, 178, 180, 181] and the references therein). It is therefore not surprising that a great part of the results available on the CBHD formula have been published in journals of Physics.

Along with applications in Physics, the CBHD Theorem produces, unquestionably, the most remarkable results in the structure theory of *Lie groups*, which is the reason why it is often popularized – in undergraduate and graduate courses – as a result of Lie group theory. Indeed, as is well known, in this context it allows us to prove a great variety of results: the universal expressibility of the composition law in logarithmic coordinates

---

[43]See Reutenauer [144, Notes 1.7 on Theorem 1.4] for a comprehensive list of references for this theorem.

around the unit element; the effective analytic regularity of all smooth Lie groups (an old result of Schur!) and of all continuous homomorphisms; the local "reconstruction" of the group law via the bracket in the Lie algebra, the existence of a local group homomorphism with prescribed differential, the local isomorphism of two Lie groups with isomorphic Lie algebras; the possibility to fully classify the simply connected Lie groups by their Lie algebras, together with the possibility of "explicitly" writing the group law in nilpotent Lie groups; the local version of Lie's Third Theorem (the existence of a local Lie group attached to every finite dimensional Lie algebra), and many others.

For this reason, all[44] major books in Lie group theory starting from the sixties contain the CBHD Theorem (mainly named after Campbell, Hausdorff or Baker, Campbell, Hausdorff), see Table 1.2.

Table 1.2 Some books on Lie group/Lie algebra theories comprising CBHD

Year	Author	book
1962	Jacobson	[99]
1965	Hochschild	[85]
1965	Serre	[159]
1968	Hausner, Schwartz	[79]
1972	Bourbaki	[27]
1973	Sagle, Walde	[151]
1974	Varadarajan	[171]
1982	Godement	[70]
1991	Hilgert, Neeb	[84]
1993	Reutenauer	[144]
1997	Gorbatsevich, Onishchik, Vinberg	[72]
1998	Hofmann, Morris	[91]
2000	Duistermaat, Kolk	[52]
2002	Rossmann	[149]
2003	Hall	[77]
2007	Abbaspour, Moskowitz	[1]
2007	Sepanski	[158]

It is interesting to observe that the proofs of the CBHD Theorem in these books are often quite different from one other and ideas from the early period often reappear. For example, the ODE technique (going back to Poincaré and Hausdorff) is exploited – and carried forward – in [1, 52, 70, 79, 171]; the old ideas of Baker, Hausdorff on polar differentiation are formalized in [144]; Eichler's [59] algebraic approach is followed in [151].

---

[44]Exceptions are the very influential book by Chevalley [38], which is actually older than the other books cited (1946), and Helgason [81], where only expansions up to the second order are used.

Indeed, even if in the setting of a Lie group $(G, \cdot)$ the (local) operation

$$X * Y := \mathrm{Log}_G((\mathrm{Exp}_G X) \cdot (\mathrm{Exp}_G Y)), \quad X, Y \in \mathrm{Lie}(G)$$

and its universal Lie series representation may be studied from a purely algebraic point of view, it turns out that, by fully exploiting differential (and integral) calculus on G and on $\mathrm{Lie}(G)$, more fitting arguments can be given. For instance, arguments exploiting ODE technique have a genuine meaning in this context and many formal series identities (especially those involving adjoint maps) acquire a full consistency and new significance (see e.g., [77, 158] where matrix groups are involved).

As for journal papers containing new algebraic proofs of the exponential theorem, the new era of the CBHD Theorem enumerates plenty of them, see e.g., Cartier [33, 1956], Eichler [59, 1968], Djoković [48, 1975], Veldkamp [174, 1980], Tu [169, 2004]. Furthermore, in the group setting (possibly, infinite-dimensional), other interesting results of extendibility and non-extendibility of the above local operation $*$ have been considered (see [47,49,50,58,69,88, 89,93,170,182]).

A remarkable turning point for the history of the CBHD Theorem is provided by Magnus's 1954 paper [112], whose applications to the applied sciences were soon revealed to be paramount. For a comprehensive treatise on the Magnus expansion, the Reader is referred to Blanes, Casas, Oteo, Ros, 2009 [16] (and to the detailed list of references therein). We here confine ourselves to a short description of its interlacement with the CBHD Formula. In studying the exponential form $\exp(\Omega(t))$ under which the solution $Y(t)$ to the nonautonomous linear ODE system $Y'(t) = A(t)Y(t)$ can be represented, Magnus introduced a formula for expanding $\Omega(t)$, later also referred to as the *continuous Campbell-Baker-Hausdorff Formula*. (See also Chen [37], 1957.) Indeed, the *Ansatz* $Y(t) = \exp(\Omega(t))$ turns the linear equation $Y'(t) = A(t)Y(t)$ into the nonlinear equation for $\Omega(t)$

$$\Omega'(t) = f(\mathrm{ad}\,\Omega(t))(A(t)), \quad \text{where} \quad f(z) = \frac{z}{e^z - 1}.$$

The above right-hand side is to be interpreted as the usual Lie series

$$\sum_{k=0}^{\infty} \frac{B_k}{k!}\,(\mathrm{ad}\,\Omega(t))^k(A(t)),$$

where the B_k are the Bernoulli numbers. This procedure makes sense in various contexts, from the simplest of matrix algebras to the more general, *mutatis mutandis*, of Hilbert spaces or Banach algebras. If we add the initial value condition $Y(0) = I$ (I is the identity, depending on the context), then $\Omega(0) = 0$ and the well-know Picard's Iteration Theorem for ODE's gives $\Omega(t) = \sum_{n=1}^{\infty} \Omega_n(t)$, where $\Omega_1(t) = \int_0^t A(s)\,\mathrm{d}s$, and inductively for $n \geq 1$

$$\Omega_{n+1}(t) = \sum_{j=1}^{n} \frac{B_j}{j!} \sum_{k_1+\cdots+k_j=n} \int_0^t [\Omega_{k_1}(s), [\Omega_{k_2}(s) \cdots [\Omega_{k_j}(s), A(s)] \cdots]] \, ds.$$

As a matter of fact, in suitable settings $t \mapsto A(t)$ is also allowed to be discontinuous (in which case the above differential equations must be replaced by the corresponding Volterra integral forms). Namely, one may consider

$$[0,2] \ni t \mapsto A(t) = \begin{cases} x, & \text{if } t \in [0,1], \\ y, & \text{if } t \in (1,2]. \end{cases}$$

In this case the unique continuous solution Y to $Y(t) = I + \int_0^t A(s)Y(s)\,ds$ is

$$Y(t) = \begin{cases} e^{tx}, & \text{if } t \in [0,1], \\ e^{(t-1)y}e^x, & \text{if } t \in (1,2], \end{cases}$$

so that $Y(2) = e^y e^x$. If, on the other hand, the exponential representation $Y(t) = \exp(\Omega(t))$ holds and the above expansion (the Magnus Expansion) is convergent, then $z := \Omega(2) = \sum_{n=1}^{\infty} \Omega_n(2)$ satisfies $e^z = Y(2) = e^y e^x$. Note that, when A is the above step function, the inductive representation of Ω_n ensures that $\Omega_n(2)$ is a Lie polynomial in x, y of length n.

The above sketch shows the intertwinement between the Magnus expansion and the CBHD Theorem, whence *convergence results* for the former provide analogous results for the latter. *De facto*, in the study of convergence domains for the CBHD series (to which more recent literature – mainly from the 1980s – has paid great attention), the use of the Magnus expansion has proved to be momentous. Despite this fact, the problem of the best domain of convergence for the CBHD series in the setting of arbitrary *Banach algebras* and – more generally – in *Banach-Lie algebras* is still open, though many optimal results exist for matrix algebras and in the setting of Hilbert spaces, see the references in Sect. 5.7 on page 359.

The problem is obviously enriched by the fact that many presentations of the series for $\log(e^x e^y)$ (in commutator or non-commutator forms) exist, further complicated by the fact that absolute and conditional convergence provide very different results. Also, the problem of finding efficient algorithms for computing the terms of this series (in suitable bases for free Lie algebras and/or under minimal numbers of commutators) has played a major rôle during the modern age of the CBHD Formula. For related topics, see e.g., [14–16, 23, 35, 36, 46, 61, 103, 109, 118, 122, 123, 131, 134, 145, 146, 166, 166, 167, 177].

In parallel with the applications in Physics and Geometry, starting from the mid seventies, the CBHD Theorem has been crucially employed also in Analysis within the study of wide classes of PDE's, mostly of *subelliptic* type, especially those involving so-called *Hörmander systems of vector fields*. See, for example, the use of the CBHD Theorem in the following papers: Christ, Nagel, Stein, Wainger [39], Folland [62], Folland, Stein [63], Hörmander [94], Rothschild, Stein [150], Nagel, Stein, Wainger [129], Varopoulos, Saloff-Coste, Coulhon [172].

Due to our own interest in these kinds of applications,[45] we would like to make explicit, for the convenience of the interested Reader, the kind of statements used in the analysis of the mentioned PDE's. First we recall some notation. Let X be a smooth vector field on an open set $\Omega \subseteq \mathbb{R}^N$, that is, X is a linear first order partial differential operator of the form $X = \sum_{j=1}^{N} a_j \, \partial_j$, where a_j is a smooth real valued function on Ω. Fixing $x \in \Omega$ and a smooth vector field X on Ω, we denote by $t \mapsto \exp(tX)(x)$ the integral curve of X starting at x, that is, the (unique maximal) solution $t \mapsto \gamma(t)$ to the ODE system

$$\dot{\gamma}(t) = \big(a_1(\gamma(t)), \ldots, a_N(\gamma(t))\big), \quad \gamma(0) = x.$$

The very well-known equivalent notations $e^{tX}(x)$, $\exp(tX)(x)$ are motivated by the fact that, given any smooth real or vector valued function f on Ω, the Taylor expansion of $f(\gamma(t))$ at $t = 0$ is obviously given by

$$\sum_{k=0}^{\infty} \frac{t^k}{k!} \, (X^k f)(x).$$

This kind of exponential-type maps play a central rôle in sub-Riemannian geometries. For example, when $\{X_1, \ldots, X_m\}$ is a system of smooth vector fields on $\mathbb{R}^N$ satisfying the so-called *Hörmander's rank condition*[46] the so-called *Carathéodory-Chow-Rashevsky Connectivity Theorem* (see Gromov [75]; see also [21, Chapter 19]) ensures that any two points of $\mathbb{R}^N$ can be joined by a finite number of pieces of paths, each of the form $x \mapsto e^{\pm X_j}(x)$. The relevance of this kind of map is also motivated by the fact that, for two smooth vector fields X, Y on Ω, it holds that

$$\lim_{t \to 0} \frac{e^{-tY} \circ e^{-tX} \circ e^{tY} \circ e^{tX}(x) - x}{t^2} = [X, Y](x), \quad x \in \Omega,$$

a remarkable geometric interpretation of the commutator.

We are ready for the statement of the "ODE version" of the CBHD Theorem used in [129, Proposition 4.3, page 146]: Let $\Omega \subseteq \mathbb{R}^N$ be an open

---

[45]See [21, Chapters 15, 19], and [17, 20, 22].

[46]That is, the dimension of the vector space

$$\mathrm{span}\{X(x) \, : \, X \in \mathrm{Lie}(\{X_1, \ldots, X_m\})\}$$

equals N, for every $x \in \mathbb{R}^N$.

set. Let $Y = \{Y_1, \dots, Y_m\}$ be a family of smooth vector fields on Ω. Then, for every compact subset K of Ω and every $M \in \mathbb{N}$, there exist positive constants C, ε depending on M, K, Y, Ω such that

$$\left| \exp \Big(\sum_{j=1}^m s_j Y_j \Big) \circ \exp \Big(\sum_{j=1}^m t_j Y_j \Big)(x) - \exp \big(Z_M(s,t) \big)(x) \right| \leq C \big(|s|^M + |t|^M \big),$$

for every $x \in K$ and for every $s, t \in \mathbb{R}^m$ such that $|s|, |t| \leq \varepsilon$. Here $|\cdot|$ is the Euclidean norm on $\mathbb{R}^N$ and $Z_M(s,t) = \eta_M \big(\sum_{j=1}^m s_j Y_j, \sum_{j=1}^m t_j Y_j \big)$, where

$$\eta_M(A, B) = \sum_{n=1}^M \frac{(-1)^{n+1}}{n} \sum_{\substack{(h_1, k_1), \dots, (h_n, k_n) \neq (0,0) \\ h_1 + k_1 + \cdots + h_n + k_n \leq M}} \frac{1}{h!\, k!\, \big(\sum_{i=1}^n (h_i + k_i) \big)}$$

$$\times \underbrace{[A \cdots [A}_{h_1 \text{ times}} [\underbrace{B \cdots [B}_{k_1 \text{ times}} \cdots \underbrace{[A \cdots [A}_{h_n \text{ times}} [\underbrace{B \cdots [B}_{k_n \text{ times}}, B]]]]]]]].$$

We recognize that $\eta_M(A, B)$ is the M-th partial sum of the usual CBHD formal series for $\log(e^A e^B)$. Other delicate estimates involving smooth vector fields and their flows are obtained by Hörmander by making use of the CHBD formula (in fact, the degree-two expansion $x + y + \frac{1}{2}[x, y] + \cdots$ suffices), used to derived Zassenhaus-type decompositions (see e.g., [162]). Indeed, in [94, pages 160-161], it is proved that, for every $k \geq 2$, the following decomposition holds (in the $\mathbb{Q}$-algebra of the formal power series in x, y)

$$e^{x+y} = e^x e^y e^{z_2} e^{z_3} \cdots e^{z_k} e^{r_{k+1}},$$

where (for every $n = 2, \dots, k$) z_n is a Lie polynomial in x, y of length n, whilst r_{k+1} is a formal power series of Lie polynomials in x, y of lengths $\geq k + 1$. Analogously, starting from the cited degree-two expansion, Hörmander derives a remarkable result – corollary of the CBHD Formula – which is the key tool for the cited Carathéodory-Chow-Rashevsky Connectivity Theorem: this is based on an iteration of the important identity for the "group-like commutator" $e^{-x} e^{-y} e^x e^y$ as in

$$e^{-x} e^{-y} e^x e^y = \exp \big([x, y] + \{\text{brackets of heights} \geq 3\} \big),$$

the iteration being aimed to obtain a decomposition of

$$\exp \big([[[x_1, x_2], x_3] \cdots x_n] \big)$$

as a suitable universally expressible product of elementary exponentials $e^{\pm x_j}$, $j = 1, \dots, n$ (plus a remainder). We refer the Reader directly to [94, page 162] for details (or to Lemma 5.45 on page 326 of this book).

Analogous decompositions are frequently used in PDEs, see e.g., Folland [62, §5, page 193] and Varopoulos, Saloff-Coste, Coulhon [172, §III.3, pages 34-39] (see also the recent papers [22, 40, 110, 127]).

Another application of the CBHD Formula occurs in the seminal paper by Rothschild and Stein [150] on the so-called Lifting Theorem. For example, the following formula is used, see [150, §10, page 279]: Given smooth vector fields $W_1, \ldots, W_m$ on an open subset Ω of $\mathbb{R}^N$ and set $u \cdot W := \sum_{j=1}^m u_j W_j$, it holds that (for any integer $l \geq 2$)

$$\exp\left(u \cdot W\right) \circ \exp(\tau\, W_1)(\xi)$$
$$= \exp\left(u \cdot W + \tau\, W_1 + \tau \sum_{1 \leq p < l} c_p \left(\operatorname{ad}\left(u \cdot W\right)\right)^p (W_1)\right)(\xi) + \mathcal{O}(|u|^l, \tau^2),$$

for $\xi \in \Omega$ and small τ, u. (From the early papers on the exponential formula, we also know the actual value of the constant c_p, viz $c_p = \frac{B_p}{p!}$, where the B_p are the Bernoulli numbers: $\frac{z}{e^z - 1} = \sum_{p=0}^\infty \frac{B_p}{p!} z^p$.)

The CBHD Formula has not ceased to provide a useful tool in Analysis. For example, we cite the recent paper by Christ, Nagel, Stein, Wainger [39], where the following version for smooth vector fields is used: Let $X_1, \ldots, X_p$, $Y_1, \ldots, Y_p$ be smooth vector fields on an open subset of $\mathbb{R}^n$; for $u, v \in \mathbb{R}^p$ and $m \in \mathbb{N}$ define

$$P(u, X) = \sum_{0 < |\alpha| \leq m} u_1^{\alpha_1} \cdots u_p^{\alpha_p} [X_{\alpha_1} \cdots [X_{\alpha_{p-1}}, X_{\alpha_p}]],$$

and analogously for $Q(v, Y)$ (the v, Ys replacing the u, Xs in $P(u, X)$); then, for each $N \geq 1$ the following equality of local diffeomorphisms holds

$$\exp(Q(v, Y)) \circ \exp(P(u, X))$$
$$= \exp\left(\sum_{k=1}^N c_k(P(u, X), Q(v, Y))\right) + \mathcal{O}((|u| + |v|)^{N+1}), \quad \text{as } |u| + |v| \to 0,$$

where the $c_k = c_k(a, b)$ are the Lie polynomials (homogeneous of bi-degree k) defined by the usual CBHD series $e^{ta} e^{tb} = \exp(\sum_{k=1}^\infty c_k(a, b) t^k)$.

All the above results can be proved without difficulty starting from the general results on the CBHD Theorem contained in this book, as it is shown in detail e.g., in [21, Section 15.4].

The rôle of the CBHD Theorem is not only prominent for usual Lie groups, but also for *infinite dimensional* Lie groups. For a detailed survey and references (a comprehensive bibliography is unfortunately out of our scope here), see Neeb [130]. As for the topics of this book, the notion of

BCH-group (Baker-Campbell-Hausdorff group) is particularly significant. For some related topics, see, e.g., [12, 13, 24, 42, 43, 56, 66–68, 73, 83, 86, 87, 92, 130, 133, 147, 148, 152, 173].

Finally, to put to an end our excursus on the modern applications of the CBHD Theorem, we point out that the years 2000 plus have seen a renewed interest in CBHD-type theorems (both continuous and discrete) within yet another field of application: that of so-called *geometric integration*, a recent branch of Numerical Analysis (see e.g. [76, 97, 98, 101, 114]).

1.3   The "Name of the Game"

We reserve a few lines to discuss our choice of the acronym "CBHD" and to recall the other choices from the existing literature.

As it appears from Table 1.3 (which collects the different titles used for the "exponential theorem" in the about 180 or so related references quoted in this book), there is definitely no agreement on the provenance of the theorem to which this book is entirely devoted. Certainly, custom and tradition have consolidated the usage of some set-expressions (such as "Campbell-Hausdorff Formula"), which cannot now be uprooted, even if they do not seem adequate after a brief historical investigation.

For example, Baker's contribution matches with Hausdorff's to such an extent that, if we had to choose between "Baker-Hausdorff" or "Campbell-Hausdorff", we would choose the former expression. Furthermore, what is this "Formula" after all? If the term "formula" refers – as seems plausible – to identities like

$$e^x e^y = \exp\left(x + y + \tfrac{1}{2}[x, y] + \tfrac{1}{12}[x[x, y]] + \tfrac{1}{12}[y[y, x]] + \cdots\right),$$

then it would be more appropriate to speak of "Dynkin's Formula".

As it emerges from the historical overview of the present chapter, the possible phraseologies may be even richer; we propose some of them:

1. *"Campbell's problem"* has been solved (almost completely) by the *"Poincaré-Pascal Theorem"* and (completely) by the *"Baker-Hausdorff Theorem"*.
2. The same *recursion formula* for a series expressing z in the identity $e^x e^y = e^z$ has been given by Baker and by Hausdorff (see (1.50) and (1.72): the *"Baker-Hausdorff series"* of *"Baker-Hausdorff Formula"*); Hausdorff gave another recursion formula for z (see (1.71)) and – implicitly – yet another one (contained in (1.73) and destined for a great success). Thus, the naming of *"the Hausdorff series"* (also widely used) may be misleading.
3. A result providing an *explicit series expansion* for z is first given by *"Dynkin's series"* (also called *"Dynkin's Formula"*), see (1.79).

If one further considers that – among Dynkin's results – we can also enumerate a new solution to Campbell's problem (see [55]), and two other explicit series expansions (see (1.81) and [55, eq. (19)', page 162]), then our choice "CBHD Theorem" seems justified.

[As a matter of fact, as can be seen from Table 1.3, the four-name choice is not commonly accepted and the reference to Dynkin is often limited to the cases when the actual Dynkin series (1.79) is involved. This fact, though neglecting the original contributions to the Campbell-Baker-Hausdorff Theorem given by Dynkin in [54–56], is so deeply entrenched that we cannot propose the acronym "CBHD" as final, but as our personal point of view, instead. The spotlight on Dynkin's series in the present book is so evident that we found it more appropriate for the title of our book to commemorate the contributions of all four Mathematicians.]

Finally, a couple of remarks on the history of the name. The expression "Campbell-Hausdorff Formula" is the one commonly employed by analysts, whereas geometers and physicists widely use the three names (differently combined) of Baker, Campbell, Hausdorff. Apparently, the first book to use the two-name expression is Jacobson's [99], whereas the first book using the three-name one seems to be Hausner and Schwartz's [79].

Table 1.3 A cross-section of the naming used for the Theorem of Campbell (C), Baker (B), Hausdorff (H) and Dynkin (D), according to the literature cited in the List of References of this book

CH
BCH
CBH
H
BH
CBHD
BDCH
BCDH
BCHD
others

Part I
Algebraic Proofs of the Theorem of Campbell, Baker, Hausdorff and Dynkin

Chapter 2
Background Algebra

THE aim of this chapter is to recall the main algebraic prerequisites and all the notation and definitions used throughout the Book. All main proofs are deferred to Chap. 7. This chapter (and its counterpart Chap. 7) is intended for a Reader having only a basic undergraduate knowledge in Algebra; a Reader acquainted with a more advanced knowledge of Algebra may pass directly to Chap. 3.

Our main objects of interest for this chapter are:

- Free vector spaces, unital associative algebras, tensor products
- Free objects over a set X: the free magma, the free monoid, the free (associative and non-associative) algebra over X
- Free Lie algebras
- Completions of metric spaces and of graded algebras; formal power series
- The universal enveloping algebra of a Lie algebra

2.1 Free Vector Spaces, Algebras and Tensor Products

2.1.1 Vector Spaces and Free Vector Spaces

Throughout this section, $\mathbb{K}$ will denote a field, while V will denote a vector space over $\mathbb{K}$. Moreover, when referring to linear maps, spans, basis, generators, linear independence, etc., we shall tacitly mean[1] "with respect to $\mathbb{K}$".

---

[1]For instance, "let U, V be vector spaces" means that both U and V are vector spaces over the *same* field $\mathbb{K}$.

A. Bonfiglioli and R. Fulci, *Topics in Noncommutative Algebra*, Lecture Notes in Mathematics 2034, DOI 10.1007/978-3-642-22597-0_2,
© Springer-Verlag Berlin Heidelberg 2012

We recall the well known fact that any vector space possesses a basis. More generally, we shall have occasion to apply the following result, which can be easily proved by means of Zorn's Lemma (as in [108, Theorem 5.1]):

Let $V \neq \{0\}$ be a vector space. Let I, G be subsets of V such that $I \subseteq G$, I is linearly independent and G generates V. Then there exists a basis $\mathcal{B}$ of V with $I \subseteq \mathcal{B} \subseteq G$.

Bases of vector spaces will always assumed to be *indexed*. Let $\mathcal{B} = \{v_i\}_{i \in \mathcal{J}}$ be a basis of V (indexed over the nonempty set $\mathcal{J}$). Then for every $v \in V$ there exists a *unique* family $\{c_i(v)\}_{i \in \mathcal{J}} \subset \mathbb{K}$ such that $c_i(v) \neq 0$ for all but finitely many indices i in $\mathcal{J}$ and such that $v = \sum_{i \in \mathcal{J}} c_i(v)\, v_i$ (the sum being well posed since it runs over a finite set). Occasionally, the subset $\mathcal{J}' \subseteq \mathcal{J}$ such that $c_i(v) \neq 0$ for every $i \in \mathcal{J}'$ will be denoted by $\mathcal{J}(v)$. When $v = 0$, or equivalently $\mathcal{J}(v) = \emptyset$, the notation $\sum_{i \in \emptyset} c_i\, v_i := 0$ applies. Note that, for every fixed $v \in V$, the following formula

$$c : \mathcal{J} \to \mathbb{K}, \quad i \mapsto c_i(v)$$

defines a well posed *function*, uniquely depending on v.

We obviously have the following result.

Proposition 2.1. *Let V be a vector space and let $\mathcal{B}$ be a basis of V. Then for every vector space X and every function $L : \mathcal{B} \to X$, there exists a unique linear map $\overline{L} : V \to X$ prolonging L.*

If $\mathcal{B} = \{v_i\}_{i \in \mathcal{J}}$, it suffices to set

$$\overline{L}(v) := \sum_{i \in \mathcal{J}(v)} c_i(v)\, L(v_i).$$

The above proposition asserts that there always exists a unique linear map $\overline{L}$ making the following diagram commute:

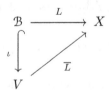

Here and in the sequel, when the context is understood, ι will always denote the *inclusion* map of a set $A \subseteq B$ into a set B.

The following are well known standard facts from Linear Algebra and are stated without proofs for the sake of future reference.

Proposition 2.2. (i). *Let V, X be vector spaces and let W be a vector subspace of V. Suppose also that $L : V \to X$ is a linear map such that $W \subseteq \ker(L)$ and let $\pi : V \to V/W$ denote the natural projection map.*

Then there exists a unique linear map $\widetilde{L} : V/W \to X$ such that

$$\widetilde{L}(\pi(v)) = L(v) \quad \text{for every } v \in V, \tag{2.1}$$

thus making the following a commutative diagram:

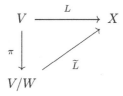

(ii). *Let V, X be vector spaces and let $L : V \to X$ be a linear map. Then the map*

$$\widetilde{L} : V/\ker(L) \to L(V), \quad [v]_{\ker(L)} \mapsto L(v)$$

is an isomorphism of vector spaces.

Actually, (2.1) also defines $\widetilde{L}$ uniquely, the definition being well posed thanks to the hypothesis $W \subseteq \ker(L)$ (indeed, $\pi(v) = \pi(v')$ iff $v - v' \in W$, so that $\pi(v) = \pi(v - v') + \pi(v') = \pi(v')$).

Definition 2.3 (Free Vector Space). Let S be any nonempty set. We denote by $\mathbb{K}\langle S \rangle$ the vector space of the $\mathbb{K}$-valued functions on S non-vanishing only on a finite (possibly empty) subset of S. The set $\mathbb{K}\langle S \rangle$ is called *the free vector space over S*.

Occasionally, a function $f : S \to \mathbb{K}$ non-vanishing only on a finite subset of S will be said to have "compact support".

Remark 2.4. Let $v \in S$ be fixed. We denote by

$$\chi(v) : S \to \mathbb{K}, \quad \chi(v)(s) := \begin{cases} 1, & \text{if } s = v \\ 0, & \text{if } s \neq v \end{cases} \tag{2.2}$$

the characteristic function of $\{v\}$ on S. With this notation at hand, it is easily seen that one has

$$\mathbb{K}\langle S \rangle = \text{span}\{\chi(v) \,|\, v \in S\}, \tag{2.3}$$

so that the generic element of $\mathbb{K}\langle S \rangle$ is of the form

$$\sum_{j=1}^{n} \lambda_j \, \chi(v_j), \quad \text{where } n \in \mathbb{N}, \lambda_1, \dots, \lambda_n \in \mathbb{K}, v_1, \dots, v_n \in S.$$

In the sequel, when there is no possibility of confusion, we shall identify $v \in S$ with $\chi(v) \in \mathbb{K}\langle S \rangle$, so that the generic element of $\mathbb{K}\langle S \rangle$ is of the form $\sum_{j=1}^{n} \lambda_j v_j$ (with n, λ_j and v_j as above), that is, $\mathbb{K}\langle S \rangle$ can be thought of as the set of the "formal linear combinations" of elements of S. Thus S can be viewed as a subset (actually, a basis) of $\mathbb{K}\langle S \rangle$. Occasionally, we shall also write an element f of $\mathbb{K}\langle S \rangle$ as

$$f = \sum_{s \in S} f(s)\, \chi(s) \qquad \left(\text{or } f = \sum_{s \in S} f_s\, \chi(s) \right), \tag{2.4}$$

the sum being finite, for $f : S \to \mathbb{K}$ has compact support.

Remark 2.5. With the above notation, *the set* $\chi(S) := \{\chi(v)\,|\,v \in S\}$ *is a linear basis of* $\mathbb{K}\langle S \rangle$. Indeed, let $\lambda_1, \ldots, \lambda_n \in \mathbb{K}$ and let $v_1, \ldots, v_n$ be pairwise distinct elements of S and suppose $\sum_{j=1}^{n} \lambda_j\, \chi(v_j) = 0$ in $\mathbb{K}\langle S \rangle$. For any fixed $i \in \{1, \ldots, n\}$ we then have[2]

$$0 = \left(\sum_{j=1}^{n} \lambda_j\, \chi(v_j) \right)(v_i) = \sum_{j=1}^{n} \lambda_j\, \chi(v_j)\, \delta_{i,j} = \lambda_i\, 1,$$

whence $\chi(v_1), \ldots, \chi(v_n)$ are linearly independent. Moreover (2.3) proves that $\chi(S)$ generates $\mathbb{K}\langle S \rangle$.

We remark that the linear independence of the set $\chi(S)$ implies in particular that $\chi : S \to \mathbb{K}\langle S \rangle$ *is an injective map.*

As a consequence, $\mathbb{K}\langle S \rangle$ is finite dimensional iff S is finite. In this case, if $S = \{v_1, \ldots, v_N\}$, we also use the brief notation $\mathbb{K}\langle v_1, \ldots, v_N \rangle := \mathbb{K}\langle S \rangle$.

In the rest of this Book, the following result will be used many times. This is the first of a series of *universal properties* of algebraic objects, which we shall encounter frequently.

Theorem 2.6 (Universal Property of the Free Vector Space).

(i) *Let S be any set. Then for every vector space X and every map $F : S \to X$ there exists a unique linear map $F^\chi : \mathbb{K}\langle S \rangle \to X$ such that*

$$F^\chi(\chi(v)) = F(v) \quad \text{for every } v \in S, \tag{2.5}$$

thus making the following a commutative diagram:

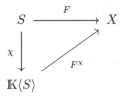

(ii) *Vice versa, suppose V, φ are respectively a vector space and a map $\varphi : S \to V$ with the following property: For every vector space X and every map $F{:}S{\to}X$ there exists a unique linear map $F^{\varphi} : V \to X$ such that*

$$F^{\varphi}(\varphi(v)) = F(v) \quad \text{for every } v \in S, \tag{2.6}$$

thus making the following a commutative diagram:

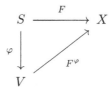

Then V is canonically isomorphic to $\mathbb{K}\langle S \rangle$, the isomorphism being φ^{χ} : $\mathbb{K}\langle S \rangle \to V$ and its inverse being $\chi^{\varphi} : V \to \mathbb{K}\langle S \rangle$. Furthermore φ is injective and the set $\varphi(S)$ is a basis of V. Actually, it holds that $\varphi = \varphi^{\chi} \circ \chi$.

When the identification $S \ni v \equiv \chi(v) \in \mathbb{K}\langle S \rangle$ applies, the above map χ is the associated inclusion $\iota : S \hookrightarrow \mathbb{K}\langle S \rangle$, so that we may think of F^{χ} as a "prolongation" of F.

Proof. See page 393 in Chap. 7. □

We recall the definitions of (external) direct sum and of product of a family of vector spaces. Let $\{V_i\}_{i \in \mathfrak{I}}$ be a family of vector spaces (indexed over a set $\mathfrak{I}$, finite, denumerable or not). We set

$$\prod_{i \in \mathfrak{I}} V_i := \left\{ (v_i)_{i \in \mathfrak{I}} \,\middle|\, v_i \in V_i \text{ for every } i \in \mathfrak{I} \right\},$$

$$\bigoplus_{i \in \mathfrak{I}} V_i := \left\{ (v_i)_{i \in \mathfrak{I}} \,\middle|\, v_i \in V_i \text{ for every } i \in \mathfrak{I} \text{ and } v_i \neq 0 \text{ for finitely many } i \right\}.$$

The former is called the *product space* of the vector spaces V_i, the latter is called the *(external) direct sum* of the spaces V_i. More precisely, we use a "sequence-style" notation $(v_i)_{i \in \mathfrak{I}}$ to mean a *function* $v : \mathfrak{I} \to \bigcup_{i \in \mathfrak{I}} V_i$, $v(i) =: v_i$ with $v_i \in V_i$ for every $i \in \mathfrak{I}$. In other words

$$(v_i)_{i \in \mathfrak{I}} = (v_i')_{i \in \mathfrak{I}} \quad \Longleftrightarrow \quad \left(\text{for all } i \in \mathfrak{I}, \, v_i, v_i' \in V_i \text{ and } v_i = v_i' \right). \tag{2.7}$$

Occasionally, *when $\mathfrak{I}$ is at most denumerable* we may also use the notation $\sum_{i \in \mathfrak{I}} v_i$ instead of $(v_i)_{i \in \mathfrak{I}}$. For example, according to this notation when

$\mathfrak{I} = \mathbb{N}$, the generic element of $\bigoplus_{n \in \mathbb{N}} V_n$ is of the form $v_1 + \ldots + v_p$ where $p \in \mathbb{N}$ and $v_n \in V_n$ for every $n = 1, \ldots, p$.

This notation is justified by the fact that the product space and the external direct sum of the spaces V_i are naturally endowed with a vector space structure (simply by defining the vector space operations *componentwise*). Obviously $\bigoplus_{i \in \mathfrak{I}} V_i$ is a subspace of $\prod_{i \in \mathfrak{I}} V_i$.

Remark 2.7. With the above notation, for any fixed $j \in \mathfrak{I}$ let

$$\widetilde{V}_j := \prod_{i \in \mathfrak{I}} V_i' \quad \text{where} \quad \begin{cases} V_i' := V_j, & \text{for } i = j, \\ V_i' := \{0\}, & \text{for } i \neq j. \end{cases}$$

Note that, for every $j \in \mathfrak{I}$, $\widetilde{V}_j$ is a vector *subspace* of $\bigoplus_{i \in \mathfrak{I}} V_i$ (hence of $\prod_{i \in \mathfrak{I}} V_i$). We now leave to the Reader the simple verification that the spaces $\widetilde{V}_j$ have the following property: Any $v \in \bigoplus_{i \in \mathfrak{I}} V_i$ can be written in a *unique* way as a (finite) sum $\sum_{i \in \mathfrak{I}} v_i$ with $v_i \in \widetilde{V}_i$ for every $i \in \mathfrak{I}$. Consequently, $\bigoplus_{i \in \mathfrak{I}} V_i$ is the (usual) *direct sum of its subspaces* $\{\widetilde{V}_i\}_{i \in \mathfrak{I}}$ (and the name "external direct sum" is thus well justified).

If, for every fixed $j \in \mathfrak{I}$, we consider the linear map

$$\iota_j : V_j \to \bigoplus_{i \in \mathfrak{I}} V_i, \quad V_j \ni v \overset{\iota_j}{\mapsto} (v_i')_{i \in \mathfrak{I}} \quad \text{where} \quad \begin{cases} v_i' := v, & \text{for } i = j, \\ v_i' := 0, & \text{for } i \neq j, \end{cases}$$

it is easily seen that $\iota_j(V_j) = \widetilde{V}_j$. Moreover, ι_j is an isomorphism of V_j onto its image $\widetilde{V}_j$, so that $\widetilde{V}_j \simeq V_j$ for every $j \in \mathfrak{I}$. As claimed above, using also (2.7), for any $v = (v_i)_{i \in \mathfrak{I}} \in \bigoplus_{i \in \mathfrak{I}} V_i$ we have the decomposition

$$v = \sum_{i \in \mathfrak{I}} \iota_i(v_i) \quad \left(\text{with } \iota_i(v_i) \in \widetilde{V}_i \text{ for all } i \in \mathfrak{I} \right). \tag{2.8}$$

Hence, throughout the sequel *we shall always identify any V_j as a subspace of $\bigoplus_{i \in \mathfrak{I}} V_i$ (or of $\prod_{i \in \mathfrak{I}} V_i$) by the canonical identification $V_j \simeq \widetilde{V}_j$ via ι_j.*

The following simple fact holds:

Theorem 2.8 (Universal Property of the External Direct Sum).

(i) *Let $\{V_i\}_{i \in \mathfrak{I}}$ be an indexed family of vector spaces. Then, for every vector space X, and every family of linear maps $\{F_i\}_{i \in \mathfrak{I}}$ (also indexed over $\mathfrak{I}$) with $F_i : V_i \to X$ (for every $i \in \mathfrak{I}$) there exists a unique linear map $F_\Sigma : \bigoplus_{i \in \mathfrak{I}} V_i \to X$ prolonging F_i, for every $i \in \mathfrak{I}$. More precisely it holds that*

$$F_\Sigma(\iota_i(v)) = F_i(v) \quad \text{for every } i \in \mathfrak{I} \text{ and every } v \in V_i, \tag{2.9}$$

thus making the following a family (over $i \in \mathfrak{I}$) of commutative diagrams:

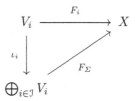

(The notation $\oplus_{i \in \mathfrak{I}} F_i$ for F_Σ will also be allowed.)

(ii) *Conversely, suppose $V, \{\varphi_i\}_{i \in \mathfrak{I}}$ are respectively a vector space and a family of linear maps $\varphi_i : V_i \to V$ with the following property: For every vector space X and every family of linear maps $\{F_i\}_{i \in \mathfrak{I}}$ with $F_i : V_i \to X$ (for every $i \in \mathfrak{I}$) there exists a unique linear map $F_\varphi : V \to X$ such that*

$$F_\varphi(\varphi_i(v)) = F_i(v) \quad \text{for every } i \in \mathfrak{I} \text{ and every } v \in V_i, \tag{2.10}$$

thus making the following a family (over $i \in \mathfrak{I}$) of commutative diagrams:

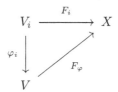

(The notation $\Phi_{i \in \mathfrak{I}} F_i$ for F_φ will also be allowed.) Then V is canonically isomorphic to $\bigoplus_{i \in \mathfrak{I}} V_i$, the isomorphism being $\oplus_{i \in \mathfrak{I}} \varphi_i : \bigoplus_{i \in \mathfrak{I}} V_i \to V$ and its inverse being $\Phi_{i \in \mathfrak{I}} \iota_i : V \to \bigoplus_{i \in \mathfrak{I}} V_i$. Furthermore any φ_i is injective and $V = \bigoplus_{i \in \mathfrak{I}} \varphi_i(V_i)$ (direct sum of subspaces of V). Actually, it holds that $\varphi_i \equiv (\oplus_{i \in \mathfrak{I}} \varphi_i) \circ \iota_i$.

Proof. (i) follows from (2.8), by setting (here $v_i \in V_i$ for all i)

$$F_\Sigma : \bigoplus_{i \in \mathfrak{I}} V_i \to X, \quad F_\Sigma\Big(\sum_{i \in \mathfrak{I}} \iota_i(v_i)\Big) := \sum_{i \in \mathfrak{I}} F_i(v_i).$$

A simple verification shows that this map is linear and obviously it is the unique linear map satisfying (2.9).

(ii) follows by arguing as in the proof of Theorem 2.6 (see page 393). The fact that $V = \bigoplus_{i \in \mathfrak{I}} \varphi_i(V_i)$ derives from the following ingredients:

- The decomposition of $\bigoplus_{i \in \mathfrak{I}} V_i$ into the direct sum of its subspaces $\tilde{V}_i = \iota_i(V_i)$.
- The isomorphism $\oplus_{i \in \mathfrak{I}} \varphi_i : \bigoplus_{i \in \mathfrak{I}} V_i \to V$.
- The set equality $\big(\oplus_{i \in \mathfrak{I}} \varphi_i\big)(\iota_i(\tilde{V}_i)) = \varphi_i(V_i)$. $\square$

The following is easily seen to hold.

Proposition 2.9. *Let* $\{V_i\}_{i\in\mathcal{I}}$ *be a family of vector spaces. For every* $i \in \mathcal{I}$, *let* $\mathcal{B}_i$ *be a basis of* V_i. *Then the following is a basis for the external direct sum* $\bigoplus_{i\in\mathcal{I}} V_i$:

$$\left\{ (w_i)_{i\in\mathcal{I}} \,\middle|\, w_i \in \mathcal{B}_i \text{ for every } i \in \mathcal{I} \text{ and } \exists! \ i_0 \in \mathcal{I} \text{ such that } w_{i_0} \neq 0 \right\}.$$

2.1.2　Magmas, Algebras and (Unital) Associative Algebras

2.1.2.1　Some Structures and Their Morphisms

Since there are no universal agreements for names, we make explicit our convention to say that a set A is:

1. *A magma,* if on A there is given a binary operation $A \times A \to A$, $(a, a') \mapsto a * a'$.
2. *A monoid,* if $(A, *)$ is a magma, $*$ is associative and endowed with a unit element.
3. *An algebra,* if $(A, *)$ is a magma, A is a vector space and $*$ is bilinear.
4. *An associative algebra,* if $(A, *)$ is an algebra and $*$ is associative.
5. *A unital associative algebra* (UA algebra, for brevity), if $(A, *)$ is an associative algebra and $*$ is endowed with a unit element.
6. *A Lie algebra,* if $(A, *)$ is an algebra, $*$ is skew-symmetric and the following *Jacobi identity* holds

$$a * (b * c) + b * (c * a) + c * (a * b) = 0, \quad \text{for all } a, b, c \in A.$$

As usual, in the context of Lie algebras, the associated operation will be denoted by $(a, a') \mapsto [a, a']$ (occasionally, $[a, a']_A$) and it will be called *the Lie bracket* (or simply, *bracket* or, sometimes, *commutator*[3]) of A.

Other structures (which we shall use less frequently) are recalled in the following (self-explanatory) table:

$(A, *)$	$*$ Binary	$*$ Associative	$*$ Has a unit	$*$ Bilinear (A vector space)
Magma	✓			
Unital magma	✓		✓	
Semigroup	✓	✓		
Monoid	✓	✓	✓	
Algebra	✓			✓
Associative algebra	✓	✓		✓
UA algebra	✓	✓	✓	✓

---

[3]In the literature, the term "commutator" is commonly used as a synonym of "bracket". In this Book we shall use the term commutator only for a special kind of bracket: that obtained from an underlying associative algebra structure.

If $(A, \circledast)$, $(B, \odot)$ are two magmas (respectively, two monoids, two algebras, two unital associative algebras, two Lie algebras), we say that a given map $\varphi : A \to B$ is:

1. *A magma morphism*, if $\varphi(a \circledast a') = \varphi(a) \odot \varphi(a')$, for every $a, a' \in A$.
2. *A monoid morphism*, if φ is a magma morphism mapping the unit of A into the unit of B.
3. *An algebra morphism*, if φ is a linear magma morphism.
4. *A morphism of unital associative algebras* (UAA morphism, in short), if φ is a linear monoid morphism, or equivalently, if φ is an algebra morphism mapping the unit of A into the unit of B.
5. *A Lie algebra morphism* (LA morphism, in short), if φ is an algebra morphism, i.e. (with the alternative notation for the algebra operation)

$$\varphi([a, a']_A) = [\varphi(a), \varphi(a')]_B, \quad \text{for every } a, a' \in A.$$

The prefix "iso" applies to any of the above notions of morphism φ, when φ is also a bijection. Plenty of examples of the above algebraic structures will be given in the next sections. The following definitions will also be used in the sequel:

1. Let $(M, *)$ be a magma (possibly, a monoid) and let $U \subseteq M$; we say that U is a *set of magma-generators* for M (or that U *generates M as a magma*) if every element of M can be written as an iterated $*$-product (with any coherent insertion of parentheses) of finitely many elements of U. In the presence of associativity, this amounts to saying that every element of M can be written in the form $u_1 * \cdots * u_k$, for some $k \in \mathbb{N}$ and $u_1, \ldots, u_k \in U$. When M is a monoid, the locution U *generates M as a monoid* will also apply.
2. Let $(A, *)$ be an algebra (associative or not, unital or not) and let $U \subseteq A$; we say that U is a *set of algebra-generators* for A (or that U *generates A as an algebra*) if every element of A can be written as a *finite linear combination* of iterated $*$-products (with coherent insertions of parentheses) of finitely many elements of U.
3. When $(A, [\cdot, \cdot])$ is a Lie algebra, in case (2) we say that U is a *set of Lie-generators* for A (or that U *Lie-generates A*). In this case (see Theorem 2.15 at the end of the section), this is equivalent to saying that every element of A can be written as a *finite linear combination* of nested elements of the form $[u_1 \cdots [u_{k-1}, u_k] \cdots]$, for $k \in \mathbb{N}$ and $u_1, \ldots, u_k \in U$.

Definition 2.10 (Derivation of an Algebra). If $(A, *)$ is an algebra, we say that a map $D : A \to A$ is a *derivation* of A if D is linear and it holds that

$$D(a * b) = (Da) * b + a * (Db), \quad \text{for every } a, b \in A.$$

When A is a Lie algebra, this can be rewritten

$$D[a, b] = [Da, b] + [a, Db], \quad \text{for every } a, b \in A.$$

Here is another definition that will play a central rôle.

Definition 2.11 (Graded and Filtered Algebras).

Graded Algebra: We say that an algebra $(A, *)$ is a *graded algebra* if it admits a decomposition of the form $A = \bigoplus_{j=1}^{\infty} A_j$, where the A_j are vector subspaces of A such that $A_i * A_j \subseteq A_{i+j}$ for every $i, j \geq 1$. In this case, the family $\{A_j\}_{j \geq 1}$ will be called a *grading* of A.

Filtered Algebra: We say that an algebra $(A, *)$ is a *filtered algebra* if $A = \bigcup_{j=1}^{\infty} F_j$, where the sets F_j are vector subspaces of A such that $F_i * F_j \subseteq F_{i+j}$ for every $i, j \geq 1$ and

$$F_j \subseteq F_{j+1}, \quad \text{for every } j \in \mathbb{N}.$$

In this case, the family $\{F_j\}_{j \geq 1}$ will be called a *filtration* of A.

For example, in the case of Lie algebras, a graded Lie algebra $A = \bigoplus_{j=1}^{\infty} A_j$ fulfils $[A_i, A_j] \subseteq A_{i+j}$, for every $i, j \geq 1$. Note that if $\{A_j\}_{j \geq 1}$ is a grading of A then A admits the filtration $\{F_j\}_{j \geq 1}$, where $F_j := \bigoplus_{i=1}^{j} A_j$.

The following simple result will be applied frequently in this Book.

Proposition 2.12 (Quotient Algebra).   *Let $(A, *)$ be an algebra and let $I \subseteq A$ be a two-sided ideal[4] of A. Then the quotient vector space A/I is an algebra (called quotient algebra of A modulo I), when equipped with the operation*

$$\circledast : A/I \times A/I \to A/I, \qquad [a]_I \circledast [b]_I := [a * b]_I, \quad \forall\, a, b \in A.$$

Moreover, the associated projection $\pi : A \to A/I$ (i.e., $\pi(a) := [a]_I$ for every $a \in A$) is an algebra morphism. Finally, if $(A, *)$ is associative (respectively, unital), then the same is true of $(A/I, \circledast)$ (and respectively, its unit is $[1_A]_I$).

The proof is simple and we only remark that the well-posedness of $\circledast$ follows by this argument: if $[a]_I = [a']_I$ and $[b]_I = [b']_I$ then $a' = a + x$ and $b' = b + y$ with $x, y \in I$ so that

$$a' * b' = a * b + \underbrace{a * y + x * b + x * y}_{\in I}, \quad \text{whence } [a' * b']_I = [a * b]_I.$$

---

[4]We recall that this means that I is a vector subspace of A and that $a * i, i * a \in I$ for every $i \in I$ and every $a \in A$.

2.1.2.2 Some Notation on Lie Algebras

In this section, $(A, [\cdot, \cdot])$ denotes a Lie algebra. If $U, V \subseteq A$ we set

$$[U, V] := \mathrm{span}\{[u, v] \mid u \in U, \ v \in V\}.$$

Note that (unlike some customary notation) $[U, V]$ is not the set of brackets $[u, v]$ with $u \in U$, $v \in V$, but the *span* of these.

Let $U \subseteq A$. We say that the elements of U are brackets of length 1 of U. Inductively, once brackets of length $1, \ldots, k - 1$ have been defined, we say that $[u, v]$ is a bracket of length k of U, if u, v are, respectively, brackets of lengths i, j of U and $i + j = k$. As synonyms for "length", we shall also use *height* or *order*. For example, if $u_1, \ldots, u_7 \in U$, then

$$[[u_1, u_2], [[[u_3, [u_4, u_5]], u_6], u_7]], \qquad [[[u_1, [[u_2, u_3], u_4]], u_5], [u_6, u_7]]$$

are brackets of length 7 of U. Note that an element of a Lie algebra may have more than one length (or even infinitely many!). For example, if A is the Lie algebra of the smooth vector fields on $\mathbb{R}^1$ and $X = \partial_x$, $Y = x\,\partial_x$, then

$$X = [\cdots [X, \underbrace{Y] \cdots Y}_{k \text{ times}}], \quad \forall\, k \in \mathbb{N},$$

so that X is a bracket of length k of $U = \{X, Y\}$, for every $k \in \mathbb{N}$.

When $u_1, \ldots, u_k \in U$, brackets of the form

$$[u_1, [u_2 \cdots [u_{k-1}, u_k] \cdots]], \qquad [[\cdots [u_1, u_2] \cdots u_{k-1}], u_k]$$

are called *nested* (respectively, *right-nested* and *left-nested*). The following result shows that the right-nested brackets span the brackets of any order. First we give a definition.

Definition 2.13 (Lie Subalgebra Generated by a Set). Let A be a Lie algebra and let $U \subseteq A$. We denote by $\mathrm{Lie}\{U\}$ the smallest Lie subalgebra of A containing U and we call it *the Lie algebra generated by U in A*. More precisely, $\mathrm{Lie}\{U\} = \bigcap \mathfrak{h}$, where the spaces $\mathfrak{h}$ run over the set of subalgebras of A containing U.

Remark 2.14. With the above notation, it is easily seen that $\mathrm{Lie}\{U\}$ coincides with the span of the brackets of U of any order. More precisely, if W_k denotes the span of the brackets of U of order k, it holds that $\mathrm{Lie}\{U\} = \biguplus_{k \in \mathbb{N}} W_k$, where $\biguplus$ denotes the sum of vector subspaces of A. Equivalently,

$$\mathrm{Lie}\{U\} = \mathrm{span}\{W_k \mid k \in \mathbb{N}\}$$

$$= \mathrm{span}\{w \mid w \text{ is a bracket of order } k \text{ of } U, \text{ with } k \in \mathbb{N}\}.$$

Theorem 2.15 (Nested Brackets). *Let A be a Lie algebra and $U \subseteq A$. Set*

$$U_1 := \text{span}\{U\}, \qquad U_n := [U, U_{n-1}], \quad n \geq 2.$$

Then we have $\text{Lie}\{U\} = \text{span}\{U_n \mid n \in \mathbb{N}\}$. *Moreover, it holds that*

$$[U_i, U_j] \subseteq U_{i+j}, \quad \text{for every } i, j \in \mathbb{N}. \tag{2.11}$$

We remark that, from the definition of U_n, the elements of U_n are linear combination of right-nested brackets of length n of U. The above theorem states that *every element of* $\text{Lie}\{U\}$ *is in fact a linear combination of right-nested brackets* (an analogous statement holding for the left case).

To show the idea behind the proof (which is a consequence of the Jacobi identity and the skew-symmetry of the bracket), let us take $u_1, u_2,$ $v_1, v_2 \in U$ and prove that $[[u_1, u_2], [v_1, v_2]]$ is a linear combination of right-nested brackets of length 4. By the Jacobi identity $[X, [Y, Z]] = -[Y, [Z, X]] - [Z, [X, Y]]$ one has

$$[\underbrace{[u_1, u_2]}_{X}, [\underbrace{v_1}_{Y}, \underbrace{v_2}_{Z}]] = -[v_1, [v_2, [u_1, u_2]]] - [v_2, [[u_1, u_2], v_1]]$$

$$= -[v_1, [v_2, [u_1, u_2]]] + [v_2, [v_1, [u_1, u_2]]] \in U_4.$$

Proof (of Theorem 2.15). We set $U^* := \text{span}\{U_n \mid n \in \mathbb{N}\}$. Obviously, U^* contains U and is contained in any Lie subalgebra of A which contains U. Hence, we are left to prove that U^* is closed under the bracket operation. Obviously, it is enough to show that, for any $i, j \in \mathbb{N}$ and for any $u_1, \ldots, u_i,$ $v_1, \ldots, v_j \in U$ we have

$$\Big[[u_1[u_2[\cdots[u_{i-1}, u_i]\cdots]]]; [v_1[v_2[\cdots[v_{j-1}, v_j]\cdots]]]\Big] \in U_{i+j}.$$

We argue by induction on $k := i + j \geq 2$. For $k = 2$ and 3 the assertion is obvious whilst for $k = 4$ we proved it after the statement of this theorem. Let us now suppose that the result holds for every $i + j \leq k$, with $k \geq 4$, and prove it then holds when $i + j = k + 1$. We can assume, by skew-symmetry, that $j \geq 3$. Exploiting repeatedly the induction hypothesis, the Jacobi identity and skew-symmetry, we have

$$\Big[u; [v_1[v_2[\cdots[v_{j-1}, v_j]\cdots]]]\Big]$$

$$= -[v_1, \underbrace{[[v_2, [v_3, \cdots]], u]}_{\text{length } k}] - [[v_2, [v_3, \cdots]], [u, v_1]]$$

$$= \{\text{element of } U_{k+1}\} - [[v_1, u], [v_2, [v_3, \cdots]]]$$

$$= \{\text{element of } U_{k+1}\} + [v_2, \underbrace{[[v_3, \cdots], [v_1, u]]]}_{\text{length } k} + [[v_3, \cdots], [[v_1, u]v_2]]$$

$$= \{\text{element of } U_{k+1}\} + [[v_2, [v_1, u]], [v_3, \cdots]]$$

(after finitely many steps)

$$= \{\text{element of } U_{k+1}\} + (-1)^{j-1}[[v_{j-i}, [v_{j-2}, \cdots [v_1, u]]], v_j]$$

$$= \{\text{element of } U_{k+1}\} + (-1)^j [v_j, [v_{j-i}, [v_{j-2}, \cdots [v_1, u]]]]$$

$$\in U_{k+1}.$$

This ends the proof. □

The previous proof shows something more: An arbitrary bracket u of length k of $\{u_1, \ldots, u_k\}$ (the minimal set of elements appearing in u) is a linear combination (with coefficients in $\{-1, 1\}$) of right-nested brackets of length k of the same set $\{u_1, \ldots, u_k\}$ and in any such summand there appear all the u_i for $i = 1, \ldots, k$. (An analogous result also holds for left-nested brackets.)

Definition 2.16. Let $(A, *)$ be an associative algebra. Let us set

$$[a, b]_* := a * b - b * a, \quad \text{for every } a, b \in A. \tag{2.12}$$

Then $(A, [\cdot, \cdot]_*)$ is a Lie algebra, called *the Lie algebra related to A*.

The Lie bracket defined in (2.12) will be referred to as the *commutator related to A (or the $*$-commutator)* and the Lie algebra $(A, [\cdot, \cdot]_*)$ will also be called *the commutator-algebra related to A*. The notation $[\cdot, \cdot]_A$ will occasionally apply instead of $[\cdot, \cdot]_*$ when confusion may not arise.

Even if authors often use the term "commutator" as a synonym for "bracket", we shall reserve it for brackets obtained from an associative multiplication as in (2.12).

Due to the massive use of commutators throughout the Book, we exhibit here the proof of the Jacobi identity (anti-symmetry and bilinearity being trivial):

$$[a, [b, c]_*]_* + [b, [c, a]_*]_* + [c, [a, b]_*]_*$$

$$= \underline{a * b * c} - \underline{a * c * b} - \underline{b * c * a} + \overline{c * b * a} + \underline{\underline{b * c * a}} - \overline{\overline{b * a * c}} +$$

$$- \overline{\overline{c * a * b}} + \underline{a * c * b} + \overline{c * a * b} - \underline{\underline{c * b * a}} - \underline{a * b * c} + \overline{b * a * c}$$

$$= 0 \quad \text{(summands canceling as over-/under-lined.)}$$

It will be via the Poincaré-Birkhoff-Witt Theorem (a highly nontrivial result) that we shall be able to prove that (roughly speaking) *every* Lie bracket can be realized as a suitable commutator (see Sect. 2.4).

Convention. *Let $(A, *)$ be an associative algebra. When a Lie algebra structure on A is invoked, unless otherwise stated, we refer to the Lie algebra on A which is induced by the associated $*$-commutator.* So, for example, if $(\mathfrak{g}, [\cdot, \cdot]_\mathfrak{g})$ is a Lie algebra, $(A, *)$ is an associative algebra and $\varphi : \mathfrak{g} \to A$ is a map, when we say that "φ is a Lie algebra morphism", we mean that φ is linear and that it satisfies $\varphi([a, b]_\mathfrak{g}) = \varphi(a) * \varphi(b) - \varphi(b) * \varphi(a)$, for every $a, b \in \mathfrak{g}$.

Remark 2.17. Let $(A, \circledast)$, $(B, \odot)$ be associative algebras and let $\varphi : A \to B$ be an algebra morphism. Then φ is also a Lie algebra morphism of the associated commutator-algebras. Indeed, for every $a, a' \in A$ one has

$$\varphi([a, a']_\circledast) = \varphi(a \circledast a' - a' \circledast a) = \varphi(a) \odot \varphi(a') - \varphi(a') \odot \varphi(a)$$
$$= [\varphi(a), \varphi(a')]_\odot.$$

Remark 2.18. Let $(A, *)$ be an associative algebra and let $D : A \to A$ be a derivation of A. Then D is also a derivation of the commutator-algebra related to A. Indeed, for every $a, a' \in A$ one has

$$D([a, a']_*) = D(a * a' - a' * a)$$
$$= D(a) * a' + a * D(a') - D(a') * a - a' * D(a)$$
$$= \big(D(a) * a' - a' * D(a)\big) + \big(a * D(a') - D(a') * a\big)$$
$$= [D(a), a']_* + [a, D(a')]_*.$$

2.1.2.3 Free Magma and Free Monoid

The remainder of this section is devoted to the construction of the free magma, the free monoid and the free algebra (associative or not) generated by a set. These structures will turn out to be of fundamental importance when we shall be dealing with the construction of free Lie algebras, without the use of the Poincaré-Birkhoff-Witt Theorem (see Sect. 2.2).

We begin with the construction of a free magma generated by a set. We follow the construction in [26, I, §7, n.1]. Henceforth, X will denote a fixed set.

To begin with, we inductively set $M_1(X) := X$, and (if $\coprod$ denotes disjoint union[5] of sets)

---

[5]We recall the relevant definition: let $\{A_i\}_{i \in \mathfrak{I}}$ be an indexed family of sets ($\mathfrak{I}$ may be finite, denumerable or not). By $\coprod_{i \in \mathfrak{I}} A_i$ we mean the set of the *ordered couples* (i, a) where $i \in \mathfrak{I}$ and $a \in A_i$, and we call it *the disjoint union of (the indexed family of) sets* $\{A_i\}_{i \in \mathfrak{I}}$. As a common habit, the first entry of the couple is dropped, but care must be paid since the same element a possibly belonging to A_i and A_j with $i \neq j$ gives rise to *distinct* elements in $\coprod_i A_i$.

$$M_2(X) := X \times X, \quad M_3(X) := \big(M_2(X) \times M_1(X) \big) \coprod \big(M_1(X) \times M_2(X) \big),$$

$$M_n(X) := \coprod_{p \in \{1,\dots,n-1\}} M_{n-p}(X) \times M_p(X), \quad \text{for every } n \geq 2; \tag{2.13}$$

$$M(X) := \coprod_{n \in \mathbb{N}} M_n(X). \tag{2.14}$$

Equivalently, we can drop the sign of disjoint union and replace it with standard set-union, *provided we consider as distinct the Cartesian products*

$$\underbrace{(X \times \cdots \times X)}_{n \text{ times}} \times \underbrace{(X \times \cdots \times X)}_{m \text{ times}} \neq \underbrace{X \times \cdots \times X}_{n + m \text{ times}}.$$

Hence, we have

$$M_1(X) = X, \quad M_2(X) = X \times X,$$

$$M_3(X) = ((X \times X) \times X) \cup (X \times (X \times X)),$$

$$M_4(X) = (((X \times X) \times X) \times X) \cup ((X \times (X \times X)) \times X) \cup$$

$$\cup ((X \times X) \times (X \times X)) \cup (X \times ((X \times X) \times X)) \cup (X \times (X \times (X \times X))),$$

$$\vdots$$

$$M_n(X) := \bigcup_{p \in \{1,\dots,n-1\}} M_{n-p}(X) \times M_p(X), \quad \text{for every } n \geq 2,$$

and $M(X) := \bigcup_{n \in \mathbb{N}} M_n(X)$.

Roughly, $M(X)$ is the set of *non-commutative and non-associative* words on the letters of X, where parentheses are inserted in any coherent way (different parentheses defining different words). For brevity, we set $M_n := M_n(X)$. For example, if $x \in X$, the following are distinct elements of M_7:

$$\Big((x,x), \Big(\big((x,(x,x)), x \big), x \Big) \Big), \qquad \Big(\big((x,((x,x),x)), x \big), (x,x) \Big)$$

Via the natural injection $X \equiv M_1 \subset M(X)$, we consider X as a subset of $M(X)$ (and the same is done for every M_n). For every $w \in M(X)$ there exists a unique $n \in \mathbb{N}$ such that $w \in M_n$, which is denoted by $n = \ell(w)$ and called the *length* of w. Note that any $w \in M(X)$ with $\ell(w) \geq 2$ is of the form $w = (w', w'')$ for unique $w', w'' \in M(X)$ satisfying $\ell(w') + \ell(w'') = \ell(w)$. For any $w, w' \in M(X)$ with $w \in M_n$ and $w' \in M_{n'}$, we denote by $w.w'$ the (unique) element of $M_{n+n'}$ corresponding to (w, w') in the canonical injections $M_n \times M_{n'} \subset M_{n+n'} \subset M(X)$. The binary operation $(w, w') \mapsto w.w'$ endows $M(X)$ with the structure of a magma, called the *free magma over X*.

Remark 2.19. Obviously, X is a set of magma-generators for $M(X)$. More-over, we have a sort of "grading" on $M(X)$ ($M(X)$ has no vector space struc-ture though), for it holds that $M(X) = \bigcup_{n \in \mathbb{N}} M_n(X)$ and $M_i(X).M_j(X) \subseteq M_{i+j}(X)$, for every $i, j \geq 1$.

Lemma 2.20 (Universal Property of the Free Magma). *Let X be any set.*

(i) *For every magma M and every function $f : X \to M$, there exists a unique magma morphism $\overline{f} : M(X) \to M$ prolonging f, thus making the following a commutative diagram:*

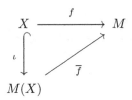

(ii) *Vice versa, suppose N, φ are respectively a magma and a function $\varphi : X \to N$ with the following property: For every magma M and every function $f : X \to M$, there exists a unique magma morphism $f^{\varphi} : N \to M$ such that*

$$f^{\varphi}(\varphi(x)) = f(x), \quad \text{for every } x \in X,$$

thus making the following a commutative diagram:

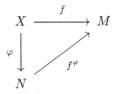

Then N is canonically magma-isomorphic to $M(X)$, the magma isomorphism being (see the notation in part (i) above) $\overline{\varphi} : M(X) \to N$ and its inverse being $\iota^{\varphi} : N \to M(X)$. Furthermore φ is injective and N is generated, as a magma, by $\varphi(X)$. Actually, it holds that $\varphi = \overline{\varphi} \circ \iota$. Finally, we have $N \simeq M(\varphi(X))$.

Proof. (i) The map $\overline{f}$ is defined as follows: Let $*$ be the operation on M and let us consider the maps f_n defined by

$f_1 : M_1 \to M, \; f_1(x) := f(x), \; \forall \, x \in X,$

$f_2 : M_2 \to M, \; f_2(x_1.x_2) := f(x_1) * f(x_2), \; \forall \, x_1, x_2 \in X,$

$f_3 : M_3 \to M, \; \begin{cases} f_3((x_1.x_2).x_3) := (f(x_1) * f(x_2)) * f(x_3) \\ f_3(x_1.(x_2.x_3)) := f(x_1) * (f(x_2) * f(x_3)) \end{cases} \; \forall \, x_1, x_2, x_3 \in X,$

and, inductively, $f_n : M_n \to M$ is defined by setting $f_n(w.w') := f_{n-p}(w) * f_p(w')$, for every $p \in \{1, \ldots, n-1\}$ and every $(w, w') \in M_{n-p} \times M_p$. Finally, let $\overline{f} : M(X) \to M$ be the unique map such that $\overline{f}|_{M_n}$ coincides with f_n. It is easily seen that $\overline{f}$ is a magma morphism and that it is the only morphism fulfilling (i).

(ii) follows by arguing as in the proof of Theorem 2.6 (see page 393). We recall the scheme of the proof. We have the commutative diagrams

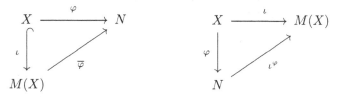

Obviously, the following are commutative diagrams too

The maps $\iota^\varphi \circ \overline{\varphi} : M(X) \to M(X)$, $\overline{\varphi} \circ \iota^\varphi : N \to N$ are magma morphisms such that

$$(\iota^\varphi \circ \overline{\varphi})(\iota(x)) = \iota(x) \quad \forall x \in X, \qquad (\overline{\varphi} \circ \iota^\varphi)(\varphi(x)) = \varphi(x) \quad \forall x \in X.$$

Hence, by the uniqueness of the morphisms represented by the "diagonal" arrows in the last couples of commutative diagrams above, we have

$$\iota^\varphi \circ \overline{\varphi} \equiv \mathrm{id}_{M(X)}, \qquad \overline{\varphi} \circ \iota^\varphi \equiv \mathrm{id}_N.$$

The rest of the proof is straightforward. $\qquad\qquad\qquad\qquad\qquad\qquad\qquad\quad$ □

We next construct the free monoid over X. We could realize it as a quotient of the free magma $M(X)$ by identifying any two elements in M_n which are obtained by inserting parentheses to the same ordered n-tuple of elements of X. Alternatively, we proceed as follows (which allows us to introduce in a rigorous way the important notion of a *word over a set*).

Let X be any fixed set. Any ordered n-tuple $w = (x_1, \ldots, x_n)$ of elements of X is called a *word on X* and $n =: \ell(w)$ is called its *length*. By convention, the empty set is called the *empty word*, it is denoted by e and its length is taken to be 0. The set of all words of length n is denoted by W_n and we set

$$\mathrm{Mo}(X) := \bigcup_{n \geq 0} W_n.$$

Obviously, X is identified with the set of words in $\mathrm{Mo}(X)$ whose length is
1. If $w = (x_1, \ldots, x_n)$ and $w' = (x'_1, \ldots, x'_{n'})$ are two words on X, we define
a new word $w'' = (x''_1, \ldots, x''_{n+n'})$ (by juxtaposition of w and w') by setting

$$
x''_j := \begin{cases} x_j, & \text{for } j = 1, \ldots, n, \\ x'_{j-n}, & \text{for } j = n+1, \ldots, n+n'. \end{cases}
$$

With the above definition, we set $w.w' := w''$. It then holds $\ell(w.w') = \ell(w) + \ell(w')$ so that $W_n.W_{n'} = W_{n+n'}$ for every $n, n' \geq 0$. Any word
$w = (x_1, \ldots, x_n)$ (with $x_1, \ldots, x_n \in X$) is written in a unique way as
$w = x_1.x_2. \cdots .x_n$, so that

$$
W_0 = \{e\}, \qquad W_n = \{x_1.x_2. \cdots .x_n \mid x_1, \ldots, x_n \in X\}, \quad n \in \mathbb{N}. \qquad (2.15)
$$

Obviously, one has $e.w = w.e = w$ for every $w \in \mathrm{Mo}(X)$.

If $w, w', w'' \in \mathrm{Mo}(X)$, then $(w.w').w''$ and $w.(w'.w'')$ are both equal to the
word $w''' = (x'''_1, \ldots, x'''_h)$ where $h = \ell(w) + \ell(w') + \ell(w'')$ and

$$
x'''_j := \begin{cases} x_j, & j = 1, \ldots, \ell(w), \\ x'_{j-\ell(w)}, & j = \ell(w) + 1, \ldots, \ell(w) + \ell(w'), \\ x''_{j-\ell(w)-\ell(w')}, & j = \ell(w) + \ell(w') + 1, \ldots, \ell(w) + \ell(w') + \ell(w''). \end{cases}
$$

As a result, $(\mathrm{Mo}(X), .)$ is a monoid, called *the free monoid over X*.

Remark 2.21. Obviously, $\{e\} \cup X$ is a set of generators for $\mathrm{Mo}(X)$ as a
monoid. Note that $\mathrm{Mo}(X) \setminus \{e\}$ *is a semigroup, i.e., an associative magma*
(which is not unital, though) and that X is a set of magma-generators for
$\mathrm{Mo}(X) \setminus \{e\}$ (i.e., every element of $\mathrm{Mo}(X) \setminus \{e\}$ can be written as a finite –
nonempty – product of elements of X).

Moreover, we have a sort of "grading" on $\mathrm{Mo}(X)$ (though $\mathrm{Mo}(X)$ is not
a vector space), for it holds that $\mathrm{Mo}(X) = \bigcup_{n \geq 0} W_n$ and $W_i.W_j \subseteq W_{i+j}$, for
every $i, j \geq 0$.

The adjective "free" is justified by the following universal property, whose
proof is completely analogous to that of Lemma 2.20.

Lemma 2.22 (Universal Property of the Free Monoid). *Let X be any set.*

(i) *For every monoid M and every function $f : X \to M$, there exists a unique
    monoid morphism $\overline{f} : \mathrm{Mo}(X) \to M$ prolonging f, thus making the following
    a commutative diagram:*

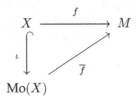

(ii) *Conversely, suppose N, φ are respectively a monoid and a function $\varphi : X \to N$ with the following property: For every monoid M and every function $f : X \to M$, there exists a unique monoid morphism $f^{\varphi} : N \to M$ such that*

$$f^{\varphi}(\varphi(x)) = f(x), \quad \text{for every } x \in X,$$

thus making the following a commutative diagram:

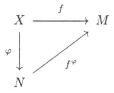

Then N is canonically monoid-isomorphic to $\mathrm{Mo}(X)$, the monoid isomorphism being (see the notation in part (i) above) $\overline{\varphi} : \mathrm{Mo}(X) \to N$ and its inverse being $\iota^{\varphi} : N \to \mathrm{Mo}(X)$. Furthermore φ is injective and N is generated, as a monoid, by $\varphi(X)$. Actually, it holds that $\varphi = \overline{\varphi} \circ \iota$. Finally, we have $N \simeq \mathrm{Mo}(\varphi(X))$.

2.1.2.4 Free Associative and Non-associative Algebras

We now associate to each of $M(X), \mathrm{Mo}(X)$ of the previous section an algebra (over $\mathbb{K}$). Let, in general, $(M, .)$ be a magma. Let M_{alg} be the free vector space over M (see Definition 2.3), i.e.,

$$M_{\mathrm{alg}} := \mathbb{K}\langle M \rangle.$$

With reference to the map χ in Remark 2.4, we know from Remark 2.5 that $\{\chi(m) \mid m \in M\}$ is a basis for M_{alg}. We now define on M_{alg} an algebra structure, compatible with the underlying structure $(M, .)$. With this aim we set

$$\left(\sum_{i=1}^{p} \lambda_i \, \chi(m_i) \right) * \left(\sum_{i'=1}^{p'} \lambda'_{i'} \, \chi(m'_{i'}) \right) := \sum_{1 \le i \le p, \, 1 \le i' \le p'} \lambda_i \lambda'_{i'} \, \chi(m_i . m'_{i'}),$$

for any arbitrary $p, p' \in \mathbb{N}$, $\lambda_1, \ldots, \lambda_p \in \mathbb{K}$, $\lambda'_1, \ldots, \lambda'_{p'} \in \mathbb{K}$, $m_1, \ldots, m_p \in M$, $m'_1, \ldots, m'_{p'} \in M$. Following the notation in (2.4), the $*$ operation can be rewritten (w.r.t. the basis $\chi(M)$) as

$$f * f' = \sum_{m \in M} \left(\sum_{a, a' \in M: \, a.a'=m} f(a) f'(a') \right) \chi(m), \qquad \forall \, f, f' \in M_{\mathrm{alg}}$$

(having set $f = \sum_{a \in M} f(a) \, \chi(a)$, $f' = \sum_{a' \in M} f'(a') \, \chi(a')$).

It is easy to prove that $(M_{\mathrm{alg}}, *)$ is an algebra (when M is a magma), an associative algebra (when M is a semigroup) and a UA algebra (when M is a monoid) with unit $\chi(e)$ (e being the unit of M), called *the algebra of M*. Clearly $m * m' = m.m'$ for every $m, m' \in M$ (by identifying $m \equiv \chi(m)$, $m' \equiv \chi(m')$) so that $*$ can be viewed as a prolongation of the former . operation.

Remark 2.23. If $(M,.)$ is a magma (resp., a monoid), then the *injective* map

$$\chi : (M,.) \to (M_{\mathrm{alg}}, *)$$

is a magma morphism (resp., a monoid morphism). Indeed, one has $\chi(m) * \chi(m') = \chi(m.m')$, for every $m, m' \in M$ by the definition of $*$ (together with the fact that $\chi(e)$ is the unit of M_{alg} when e is the unit of the monoid M).

The passage from M to the corresponding M_{alg} has a universal property:

Lemma 2.24 (Universal Property of the Algebra of a Magma, of a Monoid). *Let M be a magma.*

(i) *For every algebra A and every magma morphism $f : M \to A$ (here A is equipped only with its magma structure), there exists a unique algebra morphism $f^\chi : M_{\mathrm{alg}} \to A$ with the following property*

$$f^\chi(\chi(m)) = f(m), \quad \text{for every } m \in M, \tag{2.16}$$

thus making the following a commutative diagram:

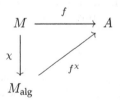

(ii) *Vice versa, suppose N, φ are respectively an algebra and a magma morphism $\varphi : M \to N$ with the following property: For every algebra A and every magma morphism $f : M \to A$, there exists a unique algebra morphism $f^\varphi : N \to A$ such that*

$$f^\varphi(\varphi(m)) = f(m), \quad \text{for every } m \in M,$$

thus making the following a commutative diagram:

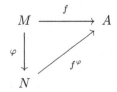

Then N is canonically algebra-isomorphic to M_{alg}, the algebra isomorphism being (see the notation in part (i) above) $\varphi^\chi : M_{\mathrm{alg}} \to N$ and its inverse being $\chi^\varphi : N \to M_{\mathrm{alg}}$. Furthermore φ is injective and $\varphi(M)$ is a linear basis for N. Actually, it holds that $\varphi = \varphi^\chi \circ \chi$. Finally, it also holds $N \simeq (\varphi(M))_{\mathrm{alg}} = \mathbb{K}\langle\varphi(M)\rangle$, the algebra of the magma $\varphi(M)$ or, equivalently, the free vector space over the set $\varphi(M)$.

(iii) *Statements analogous to (i) and (ii) hold when M is a monoid, by replacing, respectively, the above algebras A, N, the magma morphisms f, φ and the algebra morphisms f^χ, f^φ by, respectively, UA algebras A, N, monoid morphisms f, φ and UA algebra morphisms f^χ, f^φ.*

Proof. See page 396 in Chap. 7. □

In the particular case when $M = M(X)$ is the free magma over the set X, we set $\mathrm{Lib}(X) := (M(X))_{\mathrm{alg}}$ and we call it *the free (non-associative) algebra over X.* Moreover, when $M = \mathrm{Mo}(X)$ is the free monoid over X, we set $\mathrm{Libas}(X) := (\mathrm{Mo}(X))_{\mathrm{alg}}$ and we call it *the free UA algebra over X.*

More explicitly, we have

$$\mathrm{Lib}(X) := \mathbb{K}\langle M(X)\rangle, \qquad \mathrm{Libas}(X) := \mathbb{K}\langle\mathrm{Mo}(X)\rangle, \qquad (2.17)$$

i.e., *the free (non-associative) algebra over X is the free vector space related to the free magma over X and the free UA algebra over X is the free vector space related to the free monoid over X,* both endowed with the associated algebra structure introduced at the beginning of this section.

It is customary to identify $M(X)$ (resp., $\mathrm{Mo}(X)$) with a subset of $\mathrm{Lib}(X)$ (resp., of $\mathrm{Libas}(X)$) via the associated map χ, and we shall do this when confusion does not arise. Hence, it is customary to consider X as a subset of $\mathrm{Lib}(X)$ and of $\mathrm{Libas}(X)$. (But within special commutative diagrams we shall often preserve the map χ.)

Remark 2.25. By an abuse of notation, we shall use the same symbol $\chi|_X$ in the following statements, whose proof is straightforward:

1. The map $\chi|_X : X \to \mathrm{Lib}(X)$ obtained by composing the maps $X \hookrightarrow M(X) \xrightarrow{\chi} \mathbb{K}\langle M(X)\rangle = \mathrm{Lib}(X)$ is injective and $\chi(X)$ generates $\mathrm{Lib}(X)$ as an algebra (in the non-associative case).
2. The map $\chi|_X : X \to \mathrm{Libas}(X)$ obtained by composing the maps $X \hookrightarrow \mathrm{Mo}(X) \xrightarrow{\chi} \mathbb{K}\langle\mathrm{Mo}(X)\rangle = \mathrm{Libas}(X)$ is injective and $\{\chi(e)\} \cup \chi(X)$ generates $\mathrm{Libas}(X)$ as an algebra (in the associative case).

Remark 2.26. 1. *The set $\chi(X)$ is a set of generators for $\mathrm{Lib}(X)$, as an algebra* (this follows from Remark 2.19). Identifying $M(X)$ with $\chi(M(X))$, we shall also say that X is a set of generators for $\mathrm{Lib}(X)$, as an algebra. If we set (M_n being defined in (2.13))

$$\mathrm{Lib}_n(X) := \mathrm{span}\{\chi(M_n(X))\}, \quad n \in \mathbb{N}, \qquad (2.18)$$

then $\text{Lib}(X)$ is a graded algebra, for it holds that

$$\text{Lib}(X) = \bigoplus_{n \geq 1} \text{Lib}_n(X), \qquad \text{Lib}_i(X) * \text{Lib}_j(X) \subseteq \text{Lib}_{i+j}(X), \quad i, j \geq 1,$$

(2.19)

where $*$ here denotes the algebra structure on $\text{Lib}(X)$ induced by the magma $(M(X), .)$.

2. Let e denote the empty word, i.e., the unit of $\text{Mo}(X)$. Then *the set $\{\chi(e)\} \cup \chi(X)$ is a set of generators for* $\text{Libas}(X)$, *as an algebra* (this follows from Remark 2.21). With the identification of $\text{Mo}(X)$ with $\chi(\text{Mo}(X))$, we shall also say that $\{e\} \cup X$ is a set of generators for $\text{Libas}(X)$, as an algebra. If we set (W_n being defined in (2.15))

$$\text{Libas}_n(X) := \text{span}\{\chi(W_n)\}, \quad n \geq 0,$$

(2.20)

then $\text{Libas}(X)$ is a graded algebra, for it holds that

$$\text{Libas}(X) = \bigoplus_{n \geq 0} \text{Libas}_n(X),$$
$$\text{Libas}_i(X) * \text{Libas}_j(X) \subseteq \text{Libas}_{i+j}(X), \quad i, j \geq 0,$$

(2.21)

where $*$ here denotes the algebra structure on $\text{Libas}(X)$ induced by the monoid $(\text{Mo}(X), .)$.

The above Lemma 2.24 produces the following results, which we explicitly state for the sake of future reference.

Theorem 2.27 (Universal Property of the Free Algebra). *Let X be a set.*

(i) *For every algebra A and every function $f : X \to A$, there exists a unique algebra morphism $f^\chi : \text{Lib}(X) \to A$ with the following property*

$$f^\chi(\chi(x)) = f(x), \quad \text{for every } x \in X,$$

(2.22)

thus making the following a commutative diagram:

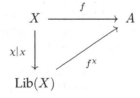

(Here $\chi|_X : X \to \text{Lib}(X)$ is the composition of maps $X \hookrightarrow M(X) \xrightarrow{\chi} \mathbb{K}\langle M(X)\rangle = \text{Lib}(X)$.)

(ii) *Conversely, suppose N, φ are respectively an algebra and a map $\varphi : X \to N$ with the following property: For every algebra A and every function $f : X \to A$, there exists a unique algebra morphism $f^\varphi : N \to A$ such that*

$$f^\varphi(\varphi(x)) = f(x), \quad \text{for every } x \in M,$$

thus making the following a commutative diagram:

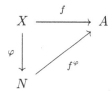

Then N is canonically algebra-isomorphic to $\text{Lib}(X)$, the algebra isomorphism being (see the notation in part (i) above) $\varphi^\chi : \text{Lib}(X) \to N$ and its inverse being $(\chi|_X)^\varphi : N \to \text{Lib}(X)$. Furthermore φ is injective and $\varphi(X)$ generates N as an algebra. Actually, it holds that $\varphi = \varphi^\chi \circ (\chi|_X)$. Finally, it also holds $N \simeq \text{Lib}(\varphi(X))$, the free non-associative algebra over the set $\varphi(X)$.

Proof. (i): From Lemma 2.20-(i), there exists a magma morphism $\overline{f}$: $M(X) \to A$ prolonging f. From Lemma 2.24-(i), there exists an algebra morphism

$$\overline{f}^\chi : (M(X))_{\text{alg}} = \text{Lib}(X) \to A$$

such that $\overline{f}^\chi(\chi(m)) = \overline{f}(m)$ for every $m \in M(X)$. The choice $f^\chi := \overline{f}^\chi$ does the job. The uniqueness part of the thesis derives from the fact that $\chi(X)$ generates $\text{Lib}(X)$ as an algebra.

Part (ii) is standard (it makes use of Remark 2.25-1). $\qquad\square$

Theorem 2.28 (Universal Property of the Free UA Algebra). *Let X be a set.*

(i) *For every UA algebra A and every function $f : X \to A$, there exists a unique UAA morphism $f^\chi : \text{Libas}(X) \to A$ with the following property*

$$f^\chi(\chi(x)) = f(x), \quad \text{for every } x \in X, \tag{2.23}$$

thus making the following a commutative diagram:

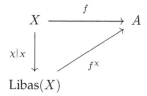

(Here $\chi|_X : X \to \mathrm{Libas}(X)$ is the composition of maps $X \hookrightarrow \mathrm{Mo}(X) \xrightarrow{\chi}$ $\mathbb{K}\langle \mathrm{Mo}(X)\rangle = \mathrm{Libas}(X).)$

(ii) *Vice versa, suppose N, φ are respectively a UA algebra and a map $\varphi : X \to N$ with the following property: For every UA algebra A and every function $f : X \to A$, there exists a unique UAA morphism $f^\varphi : N \to A$ such that*

$$f^\varphi(\varphi(x)) = f(x), \quad \text{for every } x \in X,$$

thus making the following a commutative diagram:

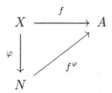

Then N is canonically isomorphic to $\mathrm{Libas}(X)$ as UA algebras, the UAA isomorphism being (see the notation in part (i) above) $\varphi^\chi : \mathrm{Libas}(X) \to N$ and its inverse being $(\chi|_X)^\varphi : N \to \mathrm{Libas}(X)$. Furthermore φ is injective and $\{e_N\} \cup \varphi(X)$ generates N as an algebra (e_N denoting the unit of N). Actually, it holds that $\varphi = \varphi^\chi \circ (\chi|_X)$. Finally, it holds that $N \simeq \mathrm{Libas}(\varphi(X))$, the free UA algebra over $\varphi(X)$.

Proof. The proof is analogous to that of Theorem 2.27, making use of Lemma 2.22-(i), Lemma 2.24-(iii) and Remark 2.25-2. $\qquad\qquad\square$

2.1.3 Tensor Product and Tensor Algebra

Let $n \in \mathbb{N}$, $n \geq 2$ and $V_1, \ldots, V_n$ be vector spaces. Let us consider the Cartesian product $V_1 \times \cdots \times V_n$ (which we *do not* endow with a vector space structure!) and the corresponding free vector space $\mathbb{K}\langle V_1 \times \cdots \times V_n \rangle$ (see Definition 2.3). The notation $\chi(v_1, \ldots, v_n)$ agrees with the one given in Remark 2.4.

Let us consider the subspace of $\mathbb{K}\langle V_1 \times \cdots \times V_n \rangle$, say W, spanned by the elements of the following form

$$\chi(v_1, \ldots, a\, v_i, \ldots, v_n) - a\, \chi(v_1, \ldots, v_i, \ldots, v_n),$$
$$\chi(v_1, \ldots, v_i + v_i', \ldots, v_n) - \chi(v_1, \ldots, v_i, \ldots, v_n) - \chi(v_1, \ldots, v_i', \ldots, v_n),$$

$$(2.24)$$

where $a \in \mathbb{K}$, $i \in \{1, \ldots, n\}$, $v_j, v_j' \in V_j$ for every $j \in \{1, \ldots, n\}$. The main definition of this section is the following one:

$$V_1 \otimes \cdots \otimes V_n := \mathbb{K}\langle V_1 \times \cdots \times V_n \rangle / W.$$

We say that $V_1 \otimes \cdots \otimes V_n$ is the *tensor product* of the (ordered) vector spaces $V_1, \ldots, V_n$ (orderly). Moreover, if $\pi : \mathbb{K}\langle V_1 \times \cdots \times V_n \rangle \to V_1 \otimes \cdots \otimes V_n$ is the associated projection, we also set

$$v_1 \otimes \cdots \otimes v_n := \pi(\chi(v_1, \ldots, v_n)), \quad \forall\, v_1 \in V_1, \ldots, \forall\, v_n \in V_n.$$

The element $v_1 \otimes \cdots \otimes v_n$ of the tensor product $V_1 \otimes \cdots \otimes V_n$ is called an *elementary tensor* of $V_1 \otimes \cdots \otimes V_n$. Not every element of $V_1 \otimes \cdots \otimes V_n$ is elementary, but every element of $V_1 \otimes \cdots \otimes V_n$ is a linear combination of elementary tensors. Finally we introduce the notation

$$\psi : V_1 \times \cdots \times V_n \to V_1 \otimes \cdots \otimes V_n, \quad \psi(v_1, \ldots, v_n) := v_1 \otimes \cdots \otimes v_n.$$

In other words $\psi = \pi \circ \chi$.

Remark 2.29. With the above notation, ψ is *n-linear* and $\psi(V_1 \times \cdots \times V_n)$ *generates* $V_1 \otimes \cdots \otimes V_n$. The last statement is obvious, whilst the former follows from the computation:

$$\psi(v_1, \ldots, a\, v_i + a'\, v_i', \ldots, v_n) = \left[\chi(v_1, \ldots, a\, v_i + a'\, v_i', \ldots, v_n)\right]_W$$

$$= \Big[\chi(v_1, \ldots, a\, v_i + a'\, v_i', \ldots, v_n)$$

$$- \chi(v_1, \ldots, a\, v_i, \ldots, v_n) - \chi(v_1, \ldots, a'\, v_i', \ldots, v_n)\Big]_W$$

$$+ \left[\chi(v_1, \ldots, a\, v_i, \ldots, v_n) + \chi(v_1, \ldots, a'\, v_i', \ldots, v_n)\right]_W$$

$$= 0 + \left[\chi(v_1, \ldots, a\, v_i, \ldots, v_n) - a\,\chi(v_1, \ldots, v_i, \ldots, v_n)\right]_W$$

$$+ \left[\chi(v_1, \ldots, a'\, v_i', \ldots, v_n) - a'\,\chi(v_1, \ldots, v_i', \ldots, v_n)\right]_W$$

$$+ a\,[\chi(v_1, \ldots, v_i, \ldots, v_n)]_W + a'\,[\chi(v_1, \ldots, v_i', \ldots, v_n)]_W$$

$$= 0 + 0 + 0 + a\,\psi(v_1, \ldots, v_i, \ldots, v_n) + a'\,\psi(v_1, \ldots, v_i', \ldots, v_n).$$

Using the "$\otimes$" notation instead of ψ, the previous remark takes the form

$$v_1 \otimes \cdots \otimes (a\, v_i + a'\, v_i') \otimes \cdots \otimes v_n = a\left(v_1 \otimes \cdots \otimes v_i \otimes \cdots \otimes v_n\right)$$

$$+ a'\left(v_1 \otimes \cdots \otimes v_i' \otimes \cdots \otimes v_n\right),$$

for every $a, a' \in \mathbb{K}$, every $i \in \{1, \ldots, n\}$ and every $v_j, v_j' \in V_j$ for $j = 1, \ldots, n$.

We are ready for another universal-property theorem of major importance.

Theorem 2.30 (Universal Property of the Tensor Product).

(i) *Let $n \in \mathbb{N}$, $n \geq 2$ and let $V_1, \ldots, V_n$ be vector spaces. Then, for every vector space X and every n-linear map $F : V_1 \times \cdots \times V_n \to X$, there exists a unique linear map $F^\psi : V_1 \otimes \cdots \otimes V_n \to X$ such that*

$$F^\psi(\psi(v)) = F(v) \quad \text{for every } v \in V_1 \times \cdots \times V_n, \qquad (2.25)$$

thus making the following a commutative diagram:

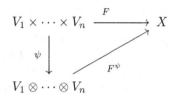

(ii) *Conversely, suppose that V, φ are respectively a vector space and an n-linear map $\varphi : V_1 \times \cdots \times V_n \to V$ with the following property: for every vector space X and every n-linear map $F : V_1 \times \cdots \times V_n \to X$, there exists a unique linear map $F^\varphi : V \to X$ such that*

$$F^\varphi(\varphi(v)) = F(v) \quad \text{for every } v \in V_1 \times \cdots \times V_n, \qquad (2.26)$$

thus making the following a commutative diagram:

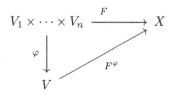

Then V is canonically isomorphic to $V_1 \otimes \cdots \otimes V_n$, the isomorphism in one direction being $\varphi^\psi : V_1 \otimes \cdots \otimes V_n \to V$ with its inverse being $\psi^\varphi : V \to V_1 \otimes \cdots \otimes V_n$. Furthermore the set $\varphi(S)$ is a set of generators for V.

Proof. See page 396 in Chap. 7.   □

Some natural properties of tensor products are now in order.

Theorem 2.31 (Basis of the Tensor Product). *Let V, W be vector spaces with bases $\{v_i\}_{i \in \mathfrak{I}}$ and $\{w_k\}_{k \in \mathcal{K}}$ respectively. Then*

$$\left\{ v_i \otimes w_k \right\}_{(i,k) \in \mathfrak{I} \times \mathcal{K}}$$

is a basis of $V \otimes W$.

Proof. The proof of this expected result is unexpectedly delicate: See page 399 in Chap. 7. ☐

Proposition 2.32 ("Associativity" of $\otimes$). *Let $n, m \in \mathbb{N}$, $n, m \geq 2$ and let $V_1, \ldots, V_n, W_1, \ldots, W_m$ be vector spaces. Then we have the isomorphism (of vector spaces)*

$$(V_1 \otimes \cdots \otimes V_n) \otimes (W_1 \otimes \cdots \otimes W_m) \simeq V_1 \otimes \cdots \otimes V_n \otimes W_1 \otimes \cdots \otimes W_m.$$

To this end, we can consider the canonical *isomorphism mapping*

$$(v_1 \otimes \cdots \otimes v_n) \otimes (w_1 \otimes \cdots \otimes w_m)$$

into $v_1 \otimes \cdots \otimes v_n \otimes w_1 \otimes \cdots \otimes w_m$.

Proof. See page 403 in Chap. 7. ☐

If V is a vector space and $k \in \mathbb{N}$, we set

$$\mathscr{T}_k(V) := \underbrace{V \otimes \cdots \otimes V}_{k \text{ times}}.$$

Thus, the generic element of $\mathscr{T}_k(V)$ is a finite linear combination of tensors of the form $v_1 \otimes \cdots \otimes v_k$, with $v_1, \ldots, v_k \in V$. We also set $\mathscr{T}_0(V) := \mathbb{K}$. The elements of $\mathscr{T}_k(V)$ are referred to as being tensors of *degree* (or *order*, or *length*) k on V. We are in a position to introduce a fundamental definition.

Definition 2.33 (Tensor Algebra of a Vector Space). Let V be a vector space. We set $\mathscr{T}(V) := \bigoplus_{k \in \mathbb{N} \cup \{0\}} \mathscr{T}_k(V)$ (in the sense of external direct sums).
On $\mathscr{T}(V)$ we consider the operation defined by

$$(v_i)_{i \geq 0} \cdot (w_j)_{j \geq 0} := \left(\sum_{j=0}^{k} v_{k-j} \otimes w_j \right)_{k \geq 0}, \tag{2.27}$$

where $v_i, w_i \in \mathscr{T}_i(V)$ for every $i \geq 0$. Here, we identify any tensor product

$$\mathscr{T}_{k-j}(V) \otimes \mathscr{T}_j(V)$$

with $\mathscr{T}_k(V)$, for every $k \in \mathbb{N} \cup \{0\}$ and every $j = 0, \ldots, k$ (thanks to Proposition 2.32). We call $\mathscr{T}(V)$ (equipped with this operation) *the tensor algebra of V.*

Throughout the Book, we consider any $\mathscr{T}_k(V)$ as a subset of $\mathscr{T}(V)$ as described in Remark 2.7. Moreover, we make the identification $V \equiv \mathscr{T}_1(V)$ so that V is considered as a subset of its tensor algebra. When there is no possibility of confusion, we denote $\mathscr{T}_k(V)$ and $\mathscr{T}(V)$ simply by $\mathscr{T}_k$ and $\mathscr{T}$.

If $v = (v_i)_{i \geq 0} \in \mathscr{T}(V)$ (being $v_i \in \mathscr{T}_i(V)$ for every $i \geq 0$), we say that v_i is the *homogeneous component* of v of degree (or order, or length) i. Moreover, in writing $v = (v_i)_{i \geq 0}$ for an element $v \in \mathscr{T}(V)$ we tacitly mean that $v_i \in \mathscr{T}_i(V)$ for every $i \in \mathbb{N} \cup \{0\}$. The notation $\sum_{i \geq 0} v_i$ for $(v_i)_{i \geq 0}$ will sometimes apply.

Remark 2.34. We have the following remarks.

1. The operation $\cdot$ on $\mathscr{T}(V)$ is the only bilinear operation on $\mathscr{T}(V)$ whose restriction to $\mathscr{T}_{k-j} \times \mathscr{T}_j$ coincides with the map

$$\mathscr{T}_{k-j}(V) \times \mathscr{T}_j(V) \ni (v_{k-j}, w_j) \mapsto v_{k-j} \otimes w_j \in \mathscr{T}_k(V),$$

for every $k \in \mathbb{N} \cup \{0\}$ and every $j = 0, \ldots, k$. Equivalently, it holds that $v \cdot w = v \otimes w$, whenever $v \in \mathscr{T}_i(V)$ and $\mathscr{T}_j(V)$ for some $i, j \geq 0$. Note that $\mathscr{T}(V)$ is generated, as an algebra, by the elements of V (or of a basis of V) through iterated $\otimes$ operations (or equivalently, iterated $\cdot$ operations).

2. The name "tensor algebra" is motivated by the fact that $(\mathscr{T}(V), \cdot)$ *is a unital associative algebra*. The unit is $1_{\mathbb{K}} \in \mathscr{T}_0(V)$. As for the other axioms of UA algebra, we leave them all to the Reader, apart from the associativity of $\cdot$, which we prove explicitly as follows:

$$(u_i)_{i \geq 0} \cdot \Big((v_i)_{i \geq 0} \cdot (w_i)_{i \geq 0} \Big) = (u_i)_{i \geq 0} \cdot \Big(\sum_{j=0}^{i} v_{i-j} \otimes w_j \Big)_{i \geq 0}$$

$$= \Big(\sum_{h=0}^{i} u_{i-h} \otimes \big(\sum_{j=0}^{h} v_{h-j} \otimes w_j \big) \Big)_{i \geq 0} = \Big(\sum_{h=0}^{i} \sum_{j=0}^{h} u_{i-h} \otimes v_{h-j} \otimes w_j \Big)_{i \geq 0}$$

(we interchange the sums and then rename the dummy index $h - j =: k$)

$$= \Big(\sum_{j=0}^{i} \sum_{h=j}^{i} \cdots \Big)_{i \geq 0} = \Big(\sum_{j=0}^{i} \sum_{k=0}^{i-j} u_{i-j-k} \otimes v_k \otimes w_j \Big)_{i \geq 0}$$

$$= \Big(\sum_{j=0}^{i} \big((u_i)_{i \geq 0} \cdot (v_i)_{i \geq 0} \big)_{i-j} \otimes w_j \Big)_{i \geq 0} = \Big((u_i)_{i \geq 0} \cdot (v_i)_{i \geq 0} \Big) \cdot (w_i)_{i \geq 0}.$$

3. By the very definition of $\mathscr{T}(V)$, we have

$$\mathscr{T}(V) = \bigoplus_{i \geq 0} \mathscr{T}_i(V), \quad \text{and} \quad \mathscr{T}_i(V) \cdot \mathscr{T}_j(V) \subseteq \mathscr{T}_{i+j}(V) \text{ for every } i, j \geq 0.$$

$$(2.28)$$

In particular, $\mathscr{T}(V)$ *is a graded algebra*. We next introduce a notation which will be used repeatedly in the sequel: for $k \in \mathbb{N} \cup \{0\}$ we set

$$U_k(V) := \bigoplus_{i \geq k} \mathscr{T}_i(V), \quad \mathscr{T}_+(V) := U_1(V) = \bigoplus_{i \geq 1} \mathscr{T}_i(V). \qquad (2.29)$$

The notation U_k, $\mathscr{T}_+$ will also apply. We have the following properties:

a. Every U_k is an ideal in $\mathscr{T}(V)$ containing $\mathscr{T}_k(V)$.
b. $\mathscr{T}(V) = U_0(V) \supset U_1(V) \supset \cdots U_k(V) \supset U_{k+1}(V) \supset \cdots$.
c. $U_i(V) \cdot U_j(V) \subseteq U_{i+j}(V)$, for every $i, j \geq 0$.
d. $\bigcap_{i \geq 0} U_i(V) = \{0\}$ and, more generally, $\bigcap_{i \geq k} U_i(V) = \{0\}$ for every $k \in \mathbb{N} \cup \{0\}$.

Note that $\mathscr{T}_+(V)$ (also, any $U_k(V)$ with $k \geq 1$) is an associative algebra with the operation $\cdot$, but *it is not a unital associative algebra*.

Proposition 2.35 (Basis of the Tensor Algebra). *Let V be a vector space and let $\mathcal{B} = \{e_i\}_{i \in \mathcal{I}}$ be a basis of V. Then the following facts hold:*

1. *For every fixed $k \in \mathbb{N}$, the system $\mathcal{B}_k := \left\{ e_{i_1} \otimes \cdots \otimes e_{i_k} \mid i_1, \ldots, i_k \in \mathcal{I} \right\}$ is a basis of $\mathscr{T}_k(V)$ (which we call* induced by $\mathcal{B}$*).*
2. *The system*

$$\{1_{\mathbb{K}}\} \cup \bigcup_{k \in \mathbb{N}} \mathcal{B}_k = \left\{ 1_K, \ e_{i_1} \otimes \cdots \otimes e_{i_k} \mid k \in \mathbb{N}, \ i_1, \ldots, i_k \in \mathcal{I} \right\}$$

is a basis of $\mathscr{T}(V)$ (which we call induced by $\mathcal{B}$*).*

Proof. (1) follows from Theorem 2.31, whilst (2) follows from (1) together with Proposition 2.9. $\qquad\square$

Remark 2.36. Also, the following are systems of generators for $\mathscr{T}(V)$:

$$\{1_K\} \cup \left\{ v_1^n \otimes \cdots \otimes v_n^n \text{ where } n \in \mathbb{N}, \ v_i^n \in V \text{ for } i \leq n \right\};$$

$$\text{and} \quad \left\{ \left(1_{\mathbb{K}}, v_1^1, v_1^2 \otimes v_2^2, v_1^3 \otimes v_2^3 \otimes v_3^3, \ldots, v_1^n \otimes \cdots \otimes v_n^n, 0, \ldots \right), \right.$$

$$\left. \text{where} \quad n \in \mathbb{N}, \ v_i^j \in V \text{ for every } j \leq n \text{ and } i \leq j \right\}.$$

Remark 2.37. The previous remark shows that V *generates* $\mathscr{T}_+(V)$ *as an algebra* and that *the set $\{1_{\mathbb{K}}\} \cup V$ generates $\mathscr{T}(V)$ as an algebra.* (Indeed, if $v_1, \ldots, v_n \in V$ we have $v_1 \cdot \ldots \cdot v_n = v_1 \otimes \cdots \otimes v_n$.)

Together with the fact that $\chi(X)$ generates $\mathbb{K}\langle X \rangle$ as a vector space, we get that $\{1_{\mathbb{K}}\} \cup \chi(X)$ generates $\mathscr{T}(\mathbb{K}\langle X \rangle)$ as an algebra (and $\chi(X)$ generates $\mathscr{T}_+(\mathbb{K}\langle X \rangle)$). By identifying X and $\chi(X)$, this last fact amounts simply to saying that the letters of X and the unit $1_{\mathbb{K}}$ generate $\mathscr{T}(\mathbb{K}\langle X \rangle)$, the *free UA algebra of the words on X.*

The following result will be used again and again in this Book.

Theorem 2.38 (Universal Property of the Tensor Algebra).
Let V be a vector space.

(i) *For every associative algebra A and every linear map $f : V \to A$, there exists a unique algebra morphism $\overline{f} : \mathscr{T}_+(V) \to A$ prolonging f, thus making the following a commutative diagram:*

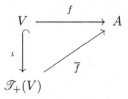

(ii) *For every unital associative algebra A and every linear map $f : V \to A$, there exists a unique UAA morphism $\overline{f} : \mathscr{T}(V) \to A$ prolonging f, thus making the following a commutative diagram:*

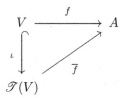

(iii) *Vice versa, suppose W, φ are respectively a UA algebra and a linear map $\varphi : V \to W$ with the following property: For every UA algebra A and every linear map $f : V \to A$, there exists a unique UAA morphism $f^\varphi : W \to A$ such that*

$$f^\varphi(\varphi(v)) = f(v) \quad \text{for every } v \in V, \tag{2.30}$$

thus making the following a commutative diagram:

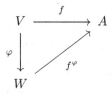

Then W is canonically isomorphic, as UA algebra, to $\mathscr{T}(V)$, the isomorphism being (see the notation in (ii) above) $\overline{\varphi} : \mathscr{T}(V) \to W$ and its inverse being $\iota^\varphi : W \to \mathscr{T}(V)$. Furthermore, φ is injective and W is generated, as an algebra, by the set $\{1_W\} \cup \varphi(V)$. Actually it holds that $\varphi = \overline{\varphi} \circ \iota$. Finally we have $W \simeq \mathscr{T}(\varphi(V))$, canonically.

Proof. Explicitly, if $\star$ is the operation on A, $\overline{f}$ in (i) is the unique linear map such that

$$\overline{f}(v_1 \otimes \cdots \otimes v_k) = f(v_1) \star \cdots \star f(v_k), \tag{2.31}$$

for every $k \in \mathbb{N}$ and every $v_1, \ldots, v_k \in V$. Also, if e_A is the unit of the UA algebra A, the map $\overline{f}$ in (ii) is the unique linear map such that

$$\overline{f}(1_K) = e_A, \quad \overline{f}(v_1 \otimes \cdots \otimes v_k) = f(v_1) \star \cdots \star f(v_k), \tag{2.32}$$

for every $k \in \mathbb{N}$ and every $v_1, \ldots, v_k \in V$. For the proof of this theorem, see page 404 in Chap. 7. $\square$

Remark 2.39. Let V, W be two isomorphic vector spaces with isomorphism $\Psi : V \to W$. Then $\mathscr{T}(V)$ and $\mathscr{T}(W)$ are isomorphic as UA algebras, via the UAA isomorphism $\widetilde{\Psi} : \mathscr{T}(V) \to \mathscr{T}(W)$ such that $\widetilde{\Psi}(v) = \Psi(v)$ for every $v \in V$ and

$$\widetilde{\Psi}(v_1 \otimes \cdots \otimes v_k) = \Psi(v_1) \otimes \cdots \otimes \Psi(v_k), \tag{2.33}$$

for every $k \in \mathbb{N}$ and every $v_1, \ldots, v_k \in V$. [Indeed, the above map $\widetilde{\Psi}$ is the unique UAA morphism prolonging the linear map $V \xrightarrow{\Psi} W \hookrightarrow \mathscr{T}(W)$; $\widetilde{\Psi}$ is an isomorphism, for its inverse is the unique UAA morphism from $\mathscr{T}(W)$ to $\mathscr{T}(V)$ prolonging the linear map $W \xrightarrow{\Psi^{-1}} V \hookrightarrow \mathscr{T}(V)$.]

The following theorem describes one of the distinguished properties of $\mathscr{T}(V)$ as the "container" of several of our universal objects (we shall see later that it contains the free Lie algebra of V and the symmetric algebra of V, too).

Theorem 2.40 ($\mathscr{T}(\mathbb{K}\langle X \rangle)$ is isomorphic to Libas(X)). *Let X be any set and $\mathbb{K}$ a field.*

(1). The tensor algebra $\mathscr{T}(\mathbb{K}\langle X \rangle)$ of the free vector space $\mathbb{K}\langle X \rangle$ is isomorphic, as UA algebra, to Libas(X), *the free unital associative algebra over X. As a (canonical) UAA isomorphism, we can consider the linear map $\Psi : \mathscr{T}(\mathbb{K}\langle X \rangle) \to$* Libas($X$) *such that[6]*

$$\Psi(x_1 \otimes \cdots \otimes x_k) = x_1. \cdots . x_k, \quad \text{for every } k \in \mathbb{N} \text{ and every } x_1, \ldots, x_k \in X,$$

and such that $\Psi(1_{\mathbb{K}}) = e$, e being the unit of Libas(X).

(2). More precisely, the couple $(\mathscr{T}(\mathbb{K}\langle X \rangle), \varphi)$ satisfies the universal property of the free UA algebra over X, where $\varphi : X \to \mathscr{T}(\mathbb{K}\langle X \rangle)$ denotes the canonical injection

$$X \xrightarrow{\chi} \mathbb{K}\langle X \rangle \xhookrightarrow{\iota} \mathscr{T}(\mathbb{K}\langle X \rangle).$$

---

[6]Here we are thinking of X (respectively, Mo(X)) as a subset of $\mathbb{K}\langle X \rangle \hookrightarrow \mathscr{T}(\mathbb{K}\langle X \rangle)$ (of $\mathbb{K}\langle$Mo(X)$\rangle =$ Libas(X), respectively).

This means that, for every UA algebra A and every function $f : X \to A$, there exists a unique UAA morphism $f^{\varphi} : \mathscr{T}(\mathbb{K}\langle X\rangle) \to A$ with the following property

$$f^{\varphi}(\varphi(x)) = f(x), \quad \text{for every } x \in X, \tag{2.34}$$

thus making the following a commutative diagram:

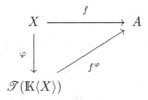

Proof. In view of Theorem 2.28-(ii), it is enough to show that the above couple $(\mathscr{T}(\mathbb{K}\langle X\rangle), \varphi)$ satisfies the universal property of the free UA algebra over X. With this aim, let A be a UA algebra and let $f : X \to A$ be any map. By Theorem 2.6-(i), there exists a linear map $f^{\chi} : \mathbb{K}\langle X\rangle \to A$ such that $f^{\chi}(\chi(x)) = f(x)$ for every $x \in X$. Then, by Theorem 2.38-(ii), there exists a UAA morphism $\overline{f^{\chi}} : \mathscr{T}(\mathbb{K}\langle X\rangle) \to A$ such that $\overline{f^{\chi}}(\iota(v)) = f^{\chi}(v)$, for every $v \in \mathbb{K}\langle X\rangle$. Setting $f^{\varphi} := \overline{f^{\chi}}$, one obviously has

$$f^{\varphi}(\varphi(x)) = \overline{f^{\chi}}((\iota \circ \chi)(x)) = f^{\chi}(\chi(x)) = f(x), \quad \forall\, x \in X.$$

Moreover f^{φ} is the unique UAA morphism such that $f^{\varphi}(\varphi(x)) = f(x)$, for every $x \in X$, since $\varphi(X) = \chi(X)$ and $\{1\} \cup \chi(X)$ generates $\mathscr{T}(\mathbb{K}\langle X\rangle)$, as an algebra.

By Theorem 2.28-(ii), we thus have $\mathscr{T}(\mathbb{K}\langle X\rangle) \simeq \text{Libas}(X)$ via the (unique) UAA isomorphism $\Psi : \mathscr{T}(\mathbb{K}\langle X\rangle) \to \text{Libas}(X)$ mapping $x \equiv \chi(x) \in \mathbb{K}\langle X\rangle$ into $x \equiv \chi(x) \in \mathbb{K}\langle \text{Mo}(X)\rangle$. The theorem is proved. $\qquad\square$

2.1.3.1 Tensor Product of Algebras

Let $(A, \circledast)$ and $(B, \odot)$ be two UA algebras (over $\mathbb{K}$). We describe a natural way to equip $A \otimes B$ with a UA algebra structure. Consider the Cartesian product $A \times B \times A \times B$ and the map

$$F : A \times B \times A \times B \to A \otimes B, \quad (a_1, b_1, a_2, b_2) \mapsto (a_1 \circledast a_2) \otimes (b_1 \odot b_2).$$

We fix $(a_2, b_2) \in A \times B$ and we consider the restriction of F defined by $A \times B \ni (a_1, b_1) \mapsto F(a_1, b_1, a_2, b_2)$. This map is clearly bilinear. Hence, by the universal property of the tensor product in Theorem 2.30, there exists a unique *linear* map $G_{a_2,b_2} : A \otimes B \to A \otimes B$ such that

$$G_{a_2,b_2}(a \otimes b) = F(a, b, a_2, b_2) = (a \circledast a_2) \otimes (b \odot b_2), \quad \forall\, (a, b) \in A \times B. \tag{2.35}$$

Then we fix $c_1 \in A \otimes B$ and we consider the map

$$\alpha_{c_1} : A \times B \to A \otimes B, \quad (a_2, b_2) \mapsto G_{a_2, b_2}(c_1).$$

It is not difficult to prove that this map is bilinear. Hence, again by Theorem 2.30, there exists a unique *linear* map $\beta_{c_1} : A \otimes B \to A \otimes B$ such that

$$\beta_{c_1}(a \otimes b) = \alpha_{c_1}(a, b) = G_{a,b}(c_1), \qquad \forall\, (a, b) \in A \times B. \tag{2.36}$$

Furthermore, we set

$$H : (A \otimes B) \times (A \otimes B) \to A \otimes B, \quad H(c_1, c_2) := \beta_{c_1}(c_2).$$

By (2.35) and (2.36), we have $H(a_1 \otimes b_1, a_2 \otimes b_2) = (a_1 \circledast a_2) \otimes (b_1 \odot b_2)$. Finally, we define a composition $\bullet$ on $A \otimes B$ as follows:

$$c_1 \bullet c_2 := H(c_1, c_2), \quad \forall\, c_1, c_2 \in A \otimes B.$$

With the above definitions, we have the following fact:

$$(A \otimes B, \bullet) \text{ is a unital associative algebra.}$$

The (tedious) proof of this fact is omitted: the Reader will certainly have no problem in deriving it. Hence, the following result follows:

Proposition 2.41. *Let $(A, \circledast)$ and $(B, \odot)$ be two UA algebras (over $\mathbb{K}$). Then $A \otimes B$ can be equipped with a UA algebra structure by an operation $\bullet$ which is characterized (in a unique way) by its action on elementary tensors as follows:*

$$(a_1 \otimes b_1) \bullet (a_2 \otimes b_2) = (a_1 \circledast a_2) \otimes (b_1 \odot b_2), \quad \forall\, (a_1 \otimes b_1), (a_2 \otimes b_2) \in A \otimes B. \tag{2.37}$$

2.1.3.2 The Algebra $\mathscr{T}(V) \otimes \mathscr{T}(V)$

Let V be a vector space. Following the above section, the tensor product $\mathscr{T}(V) \otimes \mathscr{T}(V)$ can be equipped with a UA algebra structure by means of the operation $\bullet$ such that

$$(a \otimes b) \bullet (a' \otimes b') = (a \cdot a') \otimes (b \cdot b'), \qquad (a, b), (a', b') \in \mathscr{T}(V) \otimes \mathscr{T}(V), \tag{2.38}$$

where $\cdot$ is as in (2.27). Obviously, extended by bilinearity to $\mathscr{T}(V) \otimes \mathscr{T}(V)$, (2.38) characterizes $\bullet$. For any $i, j \in \mathbb{N} \cup \{0\}$, we set[7]

$$\mathscr{T}_{i,j}(V) := \mathscr{T}_i(V) \otimes \mathscr{T}_j(V) \qquad \text{(as a subset of } \mathscr{T}(V) \otimes \mathscr{T}(V)). \tag{2.39}$$

---

[7]The Reader will have care, this time, not to identify

$$\mathscr{T}_i(V) \otimes \mathscr{T}_j(V) = \underbrace{V \otimes \cdots \otimes V}_{i \text{ times}} \otimes \underbrace{V \otimes \cdots \otimes V}_{j \text{ times}} \quad \text{with } \underbrace{V \otimes \cdots \otimes V}_{i + j \text{ times}},$$

as we had to do in Definition 2.33.

Given $\mathscr{T} = \bigoplus_{i\geq0} \mathscr{T}_i$, one obviously has

$$\mathscr{T}(V) \otimes \mathscr{T}(V) = \bigoplus_{i,j\geq0} \mathscr{T}_{i,j}(V). \qquad (2.40)$$

Thanks to the definition of $\bullet$ in (2.38), it holds that

$$\mathscr{T}_{i,j}(V) \bullet \mathscr{T}_{i',j'}(V) \subseteq \mathscr{T}_{i+i',j+j'}(V), \qquad \forall\ i,j,i',j' \geq 0. \qquad (2.41)$$

Occasionally, we will also invoke the following direct-sum decomposition:

$$\mathscr{T}(V) \otimes \mathscr{T}(V) = \bigoplus_{k\geq0} K_k(V), \quad \text{where } K_k(V) := \bigoplus_{i+j=k} \mathscr{T}_{i,j}(V). \qquad (2.42)$$

More explicitly,

$$\mathscr{T}(V) \otimes \mathscr{T}(V) =$$
$$= \underbrace{\mathscr{T}_0 \otimes \mathscr{T}_0}_{\mathscr{T}_{0,0}} \oplus \underbrace{\mathscr{T}_1 \otimes \mathscr{T}_0}_{\mathscr{T}_{1,0}} \oplus \underbrace{\mathscr{T}_0 \otimes \mathscr{T}_1}_{\mathscr{T}_{0,1}} \oplus \underbrace{\mathscr{T}_2 \otimes \mathscr{T}_0}_{\mathscr{T}_{2,0}} \oplus \underbrace{\mathscr{T}_1 \otimes \mathscr{T}_1}_{\mathscr{T}_{1,1}} \oplus \underbrace{\mathscr{T}_0 \otimes \mathscr{T}_2}_{\mathscr{T}_{0,2}} \oplus \cdots$$
$$= \underbrace{\mathscr{T}_0 \otimes \mathscr{T}_0}_{K_0} \oplus \underbrace{(\mathscr{T}_1 \otimes \mathscr{T}_0 \oplus \mathscr{T}_0 \otimes \mathscr{T}_1)}_{K_1} \oplus \underbrace{(\mathscr{T}_2 \otimes \mathscr{T}_0 \oplus \mathscr{T}_1 \otimes \mathscr{T}_1 \oplus \mathscr{T}_0 \otimes \mathscr{T}_2)}_{K_2} \oplus \cdots$$

In particular, with the decomposition (2.42), $\mathscr{T}(V) \otimes \mathscr{T}(V)$ *is a graded algebra*: Indeed, (2.41) proves that

$$K_k(V) \bullet K_{k'}(V) \subseteq K_{k+k'}(V) \qquad \text{for every } k, k' \geq 0. \qquad (2.43)$$

We next introduce a notation analogous to (2.29), which will be used repeatedly in the sequel: for $k \in \mathbb{N} \cup \{0\}$ we set

$$W_k(V) := \bigoplus_{i+j\geq k} \mathscr{T}_{i,j}(V), \quad (\mathscr{T} \otimes \mathscr{T})_+(V) := W_1(V) = \bigoplus_{i+j\geq1} \mathscr{T}_{i,j}(V). \quad (2.44)$$

Note that, with reference to K_k in (2.42), we have $W_k(V) = \bigoplus_{i\geq k} K_i(V)$, for every $k \geq 0$. The notation $W_k, (\mathscr{T} \otimes \mathscr{T})_+$ will also apply.

Remark 2.42. The following facts are easily seen to hold true:

1. Every W_k is an ideal in $\mathscr{T}(V) \otimes \mathscr{T}(V)$ containing $\mathscr{T}_{i,j}(V)$ for $i + j = k$.
2. $\mathscr{T} \otimes \mathscr{T} = W_0(V) \supset W_1(V) \supset \cdots W_k(V) \supset W_{k+1}(V) \supset \cdots$.
3. $W_i(V) \bullet W_j(V) \subseteq W_{i+j}(V)$, for every $i, j \geq 0$.
4. $\bigcap_{i\geq0} W_i(V) = \{0\}$ and, more generally, $\bigcap_{i\geq k} W_i(V) = \{0\}$ for every $k \in \mathbb{N} \cup \{0\}$.

To avoid confusion in the notation (as it appears from the note at page 81), we decided to apply the following conventional notation:

Convention. *When the tensor products in the sets $\mathscr{T}_i(V) = V \otimes \cdots \otimes V$ (i times) and the tensor product of $\mathscr{T}(V) \otimes \mathscr{T}(V)$ simultaneously arise, with the consequent risk of confusion, we use the larger symbol "$\bigotimes$" for the latter.*

For example, if $u, v, w \in V$ then

$$(u \otimes v) \bigotimes w \in \mathscr{T}_{2,1}(V), \qquad \text{whereas} \quad u \bigotimes (v \otimes w) \in \mathscr{T}_{1,2}(V),$$

and the above tensors are distinct in $\mathscr{T}(V) \otimes \mathscr{T}(V)$. Instead, $(u \otimes v) \otimes w$ and $u \otimes (v \otimes w)$ denote the same element $u \otimes v \otimes w \in \mathscr{T}_3(V)$ in $\mathscr{T}(V)$.

Proposition 2.43 (Basis of $\mathscr{T}(V) \otimes \mathscr{T}(V)$). *Let V be a vector space and let $\mathcal{B} = \{e_h\}_{h \in \mathfrak{I}}$ be a basis of V. Then the following facts hold[8]:*

1. *For every fixed $i, j \geq 0$, the system*

$$\mathcal{B}_{i,j} := \left\{ (e_{h_1} \otimes \cdots \otimes e_{h_i}) \bigotimes (e_{k_1} \otimes \cdots \otimes e_{k_j}) \,\middle|\, h_1, \ldots, h_i, \ k_1, \ldots, k_j \in \mathfrak{I} \right\}$$

   is a basis of $\mathscr{T}_{i,j}(V)$ (which we call induced by $\mathcal{B}$*).*

2. *The system $\bigcup_{i,j \geq 0} \mathcal{B}_{i,j}$, i.e.,*

$$\left\{ (e_{h_1} \otimes \cdots \otimes e_{h_i}) \bigotimes (e_{k_1} \otimes \cdots \otimes e_{k_j}) \,\middle|\, i, j \geq 0, \ h_1, \ldots, h_i, \ k_1, \ldots, k_j \in \mathfrak{I} \right\}$$

   is a basis of $\mathscr{T}(V) \otimes \mathscr{T}(V)$ (which we call induced by $\mathcal{B}$*).*

Proof. It follows from Theorem 2.31, and Propositions 2.9 and 2.35. □

Remark 2.44. Thanks to Remark 2.36, the following is a system of generators for $\mathscr{T}(V) \otimes \mathscr{T}(V)$:

$$\left\{ (u_1 \otimes \cdots \otimes u_i) \bigotimes (v_1 \otimes \cdots \otimes v_j) \,\middle|\, i, j \geq 0, \ u_1, \ldots, u_i, \ v_1, \ldots, v_j \in V \right\},$$

where the convention $u_1 \otimes \cdots \otimes u_i = 1_{\mathbb{K}} = v_1 \otimes \cdots \otimes v_j$ applies when $i, j = 0$.

Following the decomposition in (2.40) (and the notation we used in direct sums), an element of $\mathscr{T} \otimes \mathscr{T}$ will be also denoted with a double-sequence styled notation:

$$(t_{i,j})_{i,j \geq 0}, \quad \text{where } t_{i,j} \in \mathscr{T}_{i,j}(V) \text{ for every } i, j \geq 0.$$

---

[8]When $i = 0$, the term $e_{h_1} \otimes \cdots \otimes e_{h_i}$ has to be read as $1_{\mathbb{K}}$.

The notation $(t_{i,j})_{i,j}$ will equally apply (and there will be no need to specify that $t_{i,j} \in \mathscr{T}_{i,j}(V)$). Then the $\bullet$ operation in (2.38) is recast in Cauchy form as follows (thanks to (2.41)):

$$(t_{i,j})_{i,j} \bullet (\widetilde{t}_{i,j})_{i,j} = \left(\sum_{r+\widetilde{r}=i,\ s+\widetilde{s}=j} t_{r,s} \bullet \widetilde{t}_{\widetilde{r},\widetilde{s}} \right)_{i,j \geq 0}. \tag{2.45}$$

We now introduce a selected subspace of $\mathscr{T} \otimes \mathscr{T}$, which will play a central rôle in Chap. 3: we set

$$K := \left\{ v \otimes 1 + 1 \otimes w \mid v, w \in V \right\} \subset \mathscr{T}(V) \otimes \mathscr{T}(V). \tag{2.46}$$

Here and henceforth, 1 will denote the unit in $\mathbb{K}$, which is also the identity element of the algebra $\mathscr{T}(V)$. By the bilinearity of $\otimes$, we have

$$K = \mathscr{T}_{1,0}(V) \oplus \mathscr{T}_{0,1}(V) = K_1(V). \tag{2.47}$$

The following computations are simple consequences of (2.38):

$$(u_1 \otimes 1) \bullet \cdots \bullet (u_i \otimes 1) = (u_1 \otimes \cdots \otimes u_i) \bigotimes 1, \tag{2.48a}$$
$$(1 \otimes v_1) \bullet \cdots \bullet (1 \otimes v_j) = 1 \bigotimes (v_1 \otimes \cdots \otimes v_j),$$

$$(u_1 \otimes 1) \bullet \cdots \bullet (u_i \otimes 1) \bullet (1 \otimes v_1) \bullet \cdots \bullet (1 \otimes v_j)$$
$$= (u_1 \otimes \cdots \otimes u_i) \bigotimes (v_1 \otimes \cdots \otimes v_j), \tag{2.48b}$$

for every $i, j \in \mathbb{N}$, and every $u_1, \ldots, u_i, v_1, \ldots, v_j \in V$. From (2.48b) and Remark 2.44, we derive the next proposition:

Proposition 2.45. *The following is a system of generators for* $\mathscr{T}(V) \otimes \mathscr{T}(V)$:

$$\left\{ (u_1 \otimes 1) \bullet \cdots \bullet (u_i \otimes 1) \bullet (1 \otimes v_1) \bullet \cdots \bullet (1 \otimes v_j) \,\middle|\, i, j \geq 0, u_1, \ldots, u_i, v_1, \ldots, v_j \in V \right\},$$

where the convention $u_1 \otimes \cdots \otimes u_i = 1_{\mathbb{K}} = v_1 \otimes \cdots \otimes v_j$ *applies, when* $i, j = 0$. *Moreover, if* $\mathcal{B} = \{e_h\}_{h \in \mathcal{I}}$ *is a basis of* V, *the following is a basis of* $\mathscr{T}(V) \otimes \mathscr{T}(V)$:

$$\Big\{ 1 \otimes 1, \quad (e_{h_1} \otimes 1) \bullet \cdots \bullet (e_{h_i} \otimes 1), \quad (1 \otimes e_{k_1}) \bullet \cdots \bullet (1 \otimes e_{k_j}),$$

$$(e_{\alpha_1} \otimes 1) \bullet \cdots \bullet (e_{\alpha_a} \otimes 1) \bullet (1 \otimes e_{\beta_1}) \bullet \cdots \bullet (1 \otimes e_{\beta_b}),$$

where $i, j, a, b \in \mathbb{N}$ *and* $h_1, \ldots, h_i, k_1, \ldots, k_j, \alpha_1, \ldots, \alpha_a, \beta_1, \ldots, \beta_b \in \mathcal{I} \Big\}$.

2.1.3.3 The Lie Algebra $\mathcal{L}(V)$

The aim of this section is to describe another distinguished subset of $\mathscr{T}(V)$ having important features in Lie Algebra Theory.

First the relevant definition.

Definition 2.46 (Free Lie Algebra Generated by a Vector Space). Let V be a vector space and consider its tensor algebra $(\mathscr{T}(V), \cdot)$. We equip $\mathscr{T}(V)$ with the Lie algebra structure related to the corresponding commutator (see Definition 2.16).

We denote by $\mathcal{L}(V)$ the Lie algebra generated by the set V in $\mathscr{T}(V)$ (according to Definition 2.13) and we call it *the free Lie algebra generated by V*. Namely, $\mathcal{L}(V)$ is the smallest Lie subalgebra of the (commutator-) Lie algebra $\mathscr{T}(V)$ containing V.

The above adjective "free" will be soon justified in Theorem 2.49 below (though its proof requires a lot of work and will be deferred to Sect. 2.2). We straightaway remark that we are not using the phrasing "free Lie algebra over V" (which, according to previous similar expressions in this Book, would – and will – mean a free object *over the set V*). All will be clarified in Sect. 2.2.

Convention. To avoid the (proper) odd notation $[u, v].$ for the commutator related to $(\mathscr{T}(V), \cdot)$, we shall occasionally make use of the abuse of notation $[u, v]_{\otimes}$ for $u \cdot v - v \cdot u$ (when $u, v \in \mathscr{T}(V)$). This notation becomes particularly suggestive when applied to elementary tensors u, v of the form $w_1 \otimes \cdots \otimes w_n$, for in this case the $\cdot$ product coincides with $\otimes$.

Proposition 2.47. *Let V be a vector space and let the notation in Definition 2.46 apply. We set $\mathcal{L}_1(V) := V$ and, for every $n \in \mathbb{N}$, $n \geq 2$,*

$$\mathcal{L}_n(V) := \underbrace{[V \cdots [V, V] \cdots]}_{n \; times} = \mathrm{span}\Big\{ [v_1 \cdots [v_{n-1}, v_n] \cdots] \,\Big|\, v_1, \ldots, v_n \in V \Big\}.$$
$$(2.49)$$

Then $\mathcal{L}_n(V) \subseteq \mathscr{T}_n(V)$ for every $n \in \mathbb{N}$, and we have the direct sum decomposition

$$\mathcal{L}(V) = \bigoplus_{n \geq 1} \mathcal{L}_n(V). \tag{2.50}$$

In particular, the set V Lie generates $\mathcal{L}(V)$. Moreover, $\mathcal{L}(V)$ is a graded Lie algebra, for it holds that

$$[\mathcal{L}_i(V), \mathcal{L}_j(V)] \subseteq \mathcal{L}_{i+j}(V), \quad for \; every \; i, j \geq 1. \tag{2.51}$$

Proof. From Theorem 2.15, we deduce that $\bigcup_n \mathcal{L}_n(V)$ spans $\mathcal{L}(V)$ and that (2.51) holds (see (2.11)). Finally, (2.50) follows from $\mathcal{L}_n(V) \subseteq \mathscr{T}_n(V)$, which can be proved by an inductive argument, starting from:

$$[v_1, v_2] = v_1 \cdot v_2 - v_2 \cdot v_1 = v_1 \otimes v_2 - v_2 \otimes v_1 \in \mathscr{T}_2(V),$$

holding for every $v_1, v_2 \in V$, and using (2.28). This ends the proof.   □

Remark 2.48. Let V, W be isomorphic vector spaces and let $\Psi : V \to W$ be an isomorphism. Let $\widetilde{\Psi} : \mathscr{T}(V) \to \mathscr{T}(W)$ be the UAA isomorphism constructed in Remark 2.39. We claim that

$$\widetilde{\Psi}_{\mathcal{L}} := \widetilde{\Psi}|_{\mathcal{L}(V)} : \mathcal{L}(V) \to \mathcal{L}(W) \text{ is a Lie algebra isomorphism.}$$

Indeed, since $\widetilde{\Psi}$ is a UAA isomorphism, it is also a Lie algebra isomorphism, when $\mathscr{T}(V)$ and $\mathscr{T}(W)$ are equipped with the associated commutator-algebra structures (see Remark 2.17). As a consequence, the restriction of $\widetilde{\Psi}$ to $\mathcal{L}(V)$ is a Lie algebra isomorphism onto $\widetilde{\Psi}(\mathcal{L}(V))$ (recall that $\mathcal{L}(V)$ is a Lie subalgebra of the commutator-algebra of $\mathscr{T}(V)$). To complete the claim, we have to show that $\widetilde{\Psi}(\mathcal{L}(V)) = \mathcal{L}(W)$. To prove this, we begin by noticing that (in view of (2.33) in Remark 2.39) $\widetilde{\Psi}_{\mathcal{L}}(v) = \Psi(v)$ for every $v \in V$ and

$$\widetilde{\Psi}_{\mathcal{L}}\left([v_1 \cdots [v_{k-1}, v_k] \cdots]_{\mathscr{T}(V)}\right) = [\Psi(v_1) \cdots [\Psi(v_{k-1}), \Psi(v_k)] \cdots]_{\mathscr{T}(W)}, \tag{2.52}$$

for every $k \in \mathbb{K}$ and every $v_1, \ldots, v_k \in V$. Here we have denoted by $[\cdot, \cdot]_{\mathscr{T}(V)}$ the commutator related to the associative algebra $\mathscr{T}(V)$ (and analogously for $[\cdot, \cdot]_{\mathscr{T}(W)}$). Now, (2.52) shows that $\widetilde{\Psi}(\mathcal{L}(V)) \subseteq \mathcal{L}(W)$ (recall Proposition 2.47). To prove that "=" holds instead of "⊆", it suffices to recognize that the arbitrary element $[w_1 \cdots [w_{k-1}, w_k] \cdots]_{\mathscr{T}(W)}$ of $\mathcal{L}(W)$ (where $k \in \mathbb{K}$ and $w_1, \ldots, w_k \in V$) is the image via $\widetilde{\Psi}$ of

$$[\Psi^{-1}(w_1) \cdots [\Psi^{-1}(w_{k-1}), \Psi^{-1}(w_k)] \cdots]_{\mathscr{T}(V)}.$$

Theorem 2.49 (Universal Property of $\mathcal{L}(V)$). *Let V be a vector space.*

(i) *For every Lie algebra $\mathfrak{g}$ and every linear map $f : V \to \mathfrak{g}$, there exists a unique Lie algebra morphism $\overline{f} : \mathcal{L}(V) \to \mathfrak{g}$ prolonging f, thus making the following a commutative diagram:*

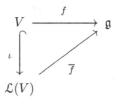

(ii) *Conversely, suppose L, φ are respectively a Lie algebra and a linear map $\varphi : V \to L$ with the following property: For every Lie algebra $\mathfrak{g}$ and every linear* *map $f : V \to \mathfrak{g}$, there exists a unique Lie algebra morphism $f^\varphi : L \to \mathfrak{g}$ such that*

$$f^\varphi(\varphi(v)) = f(v) \quad \text{for every } v \in V, \tag{2.53}$$

thus making the following a commutative diagram:

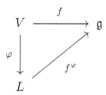

Then L is canonically isomorphic, as a Lie algebra, to $\mathcal{L}(V)$, the isomorphism being (see the notation in (i) above) $\overline{\varphi} : \mathcal{L}(V) \to L$ and its inverse being $\iota^\varphi : L \to \mathcal{L}(V)$. Furthermore, φ is injective and L is Lie-generated by the set $\varphi(V)$. Actually it holds that $\varphi = \overline{\varphi} \circ \iota$. Finally we have $L \simeq \mathcal{L}(\varphi(V))$, canonically.

Proof. Explicitly, if $[\cdot,\cdot]_\mathfrak{g}$ is the Lie bracket of $\mathfrak{g}$, $\overline{f}$ in (i) is the unique linear map such that

$$\overline{f}\big([v_1 \cdots [v_{k-1}, v_k]_\otimes \cdots]_\otimes\big) = [f(v_1) \cdots [f(v_{k-1}), f(v_k)]_\mathfrak{g} \cdots]_\mathfrak{g},$$

for every $k \in \mathbb{N}$ and every $v_1, \dots, v_k \in V$.

Unfortunately, the proof of this theorem requires the results of Sect. 2.2, on the existence of $\mathrm{Lie}(X)$, the free Lie algebra related to a set X (together with the characterization $\mathrm{Lie}(X) \simeq \mathcal{L}(\mathbb{K}\langle X\rangle)$). Alternatively, it can be proved by means of the fact that every Lie algebra $\mathfrak{g}$ can be embedded in its universal enveloping algebra $\mathscr{U}(\mathfrak{g})$ (a corollary of the Poincaré-Birkhoff-Witt Theorem, see Sect. 2.4). Hence, we shall furnish two proofs of Theorem 2.49, see pages 92 and 112. □

2.2 Free Lie Algebras

The aim of this section is to prove the existence of the so-called free Lie algebra $\mathrm{Lie}(X)$ related to a set X. Classically, the existence of $\mathrm{Lie}(X)$ follows as a trivial corollary of a highly nontrivial theorem, the Poincaré-Birkhoff-Witt Theorem. For a reason that will become apparent in later chapters concerning with the CBHD Theorem, our aim here is *to prove the existence of* $\mathrm{Lie}(X)$ *without the aid of the Poincaré-Birkhoff-Witt Theorem.*

Moreover, for the aims of this Book, it is also a central fact to obtain the isomorphism of $\mathrm{Lie}(X)$ with $\mathcal{L}(\mathbb{K}\langle X\rangle)$, the smallest Lie subalgebra of the tensor algebra over the free vector space $\mathbb{K}\langle X\rangle$.

The main reference for the topics of this section is Bourbaki [27, Chapitre II, §2 n.2 and §3 n.1]. Unfortunately, there is a feature in [27] which does not allow us to simply rerun Bourbaki's arguments: Indeed, the isomorphism $\mathrm{Lie}(X) \simeq \mathcal{L}(\mathbb{K}\langle X \rangle)$ is proved in [27, Chapitre II, §3 n.1] as a consequence[9] of the Poincaré-Birkhoff-Witt Theorem. So we are forced to present a new argument, which bypasses this inconvenience.

To avoid confusion between the notion of free Lie algebra *generated by a vector space* (see Definition 2.46) and the new notion – we are giving here – of free Lie algebra *related to a set*, we introduce dedicated notations.

Definition 2.50 (Free Lie Algebra Related to a Set). Let X be any set. We say that the couple (L, φ) is a *free Lie algebra related to X*, if the following facts hold: L is a Lie algebra and $\varphi : X \to L$ is a map such that, for every Lie algebra $\mathfrak{g}$ and every map $f : X \to \mathfrak{g}$, there exists a unique Lie algebra morphism $f^\varphi : L \to \mathfrak{g}$, such that the following fact holds

$$f^\varphi(\varphi(x)) = f(x) \quad \text{for every } x \in X, \tag{2.54}$$

thus making the following a commutative diagram:

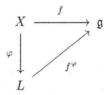

If, in the above definition, $X \subset L$ (set-theoretically) and $\varphi = \iota$ is the set inclusion, we say that (L, ι) is a *free Lie algebra over X*.

By abuse, if (L, φ) (respectively, (L, ι)) is as above, we shall also say that L itself is a free Lie algebra related to X (respectively, a free Lie algebra over X). It is easily seen that any two free Lie algebras related to X are canonically isomorphic. More precisely, the following facts hold.

Proposition 2.51. *Let X be a nonempty set.*

1. *If (L_1, φ_1), (L_2, φ_2) are two free Lie algebras related to the same set X, then L_1, L_2 are isomorphic Lie algebras via the isomorphisms (inverse to each other) $\varphi_2^{\varphi_1} : L_1 \to L_2$, $\varphi_1^{\varphi_2} : L_2 \to L_1$ and $\varphi_2 \equiv \varphi_2^{\varphi_1} \circ \varphi_1$ (analogously, $\varphi_1 \equiv \varphi_1^{\varphi_2} \circ \varphi_2$).*

---

[9]See [27, Chapitre II, §3, n.1, Théorème 1] where it is employed [25, Chapitre I, §2, n.7, Corollaire 3 du Théorème 1] which is the Poincaré-Birkhoff-Witt Theorem.

2. *If (L_1, φ_1) is a free Lie algebra related to X, if L_2 is a Lie algebra isomorphic to L_1 and $\psi : L_1 \to L_2$ is a Lie algebra isomorphism, then (L_2, φ_2) is another free Lie algebra related to X, where $\varphi_2 := \psi \circ \varphi_1$.*

Proof. (1). As usual, it suffices to consider the commutative diagrams

and to show that the diagonal arrows in the following commutative diagrams are respectively "closed" by the maps $\varphi_1^{\varphi_2} \circ \varphi_2^{\varphi_1}$ and $\varphi_2^{\varphi_1} \circ \varphi_1^{\varphi_2}$:

We conclude by the uniqueness of the "closing" morphism, as stated in the definition of free Lie algebra related to X. Part (2) of the proposition is a simple verification.   □

We next turn to the actual *construction* of a free Lie algebra related to the set X. First we need some preliminary results.

  Whereas ideals are usually defined in an associative setting, we need the following (non-standard) definition.

Definition 2.52 (Magma Ideal). Let $(M, *)$ be an algebra (not necessarily associative).

1. We say that $S \subseteq M$ is a *magma ideal in M*, if S is a subspace of the vector space M such that $s * m$ and $m * s$ belong to S, for every $s \in S$ and $m \in M$.
2. Let A be any subset of M. The smallest magma ideal in M containing A is called *the magma ideal generated by A*.

With the above definition, it is evident that the magma ideal generated by A coincides with $\bigcap S$, where the intersection runs over the magma ideals S in M containing A.

  Up to the end of this section, X will denote a fixed set. Let us now consider $\mathrm{Lib}(X)$, i.e., the free non-associative algebra over X, introduced in Sect. 2.1.2 (see (2.17)). We shall denote its operation by $*$, recalling that this is the bilinear map extending the operation of the free magma $(M(X), .)$

(and that $\mathrm{Lib}(X)$ is the free vector space of the formal linear combinations of elements of $M(X)$). Let us introduce the subset of $\mathrm{Lib}(X)$ defined as

$$A := \big\{ Q(a),\ J(a,b,c) \,\big|\, a,b,c \in \mathrm{Lib}(X) \big\}, \quad \text{where}$$

$$Q(a) := a * a, \ \ J(a,b,c) := a * (b * c) + b * (c * a) + c * (a * b). \tag{2.55}$$

We henceforth denote by $\mathfrak{a}$ the magma ideal in $\mathrm{Lib}(X)$ generated by A, according to Definition 2.52. We next consider the quotient vector space

$$\mathrm{Lie}(X) := \mathrm{Lib}(X)/\mathfrak{a}, \tag{2.56}$$

and the associated natural projection

$$\pi : \mathrm{Lib}(X) \to \mathrm{Lie}(X), \quad \pi(t) := [t]_{\mathfrak{a}}. \tag{2.57}$$

Then the following fact holds:

Proposition 2.53. *With all the above notation, the map* $[\cdot,\cdot] : \mathrm{Lie}(X) \to \mathrm{Lie}(X)$ *defined by*

$$[\pi(a), \pi(b)] := \pi(a * b) \quad \text{for every } a, b \in \mathrm{Lib}(X), \tag{2.58}$$

is well posed and it endows $\mathrm{Lie}(X)$ *with a* **Lie** *algebra structure. Moreover, the map* π *in (2.57) is an algebra morphism (when we consider* $\mathrm{Lie}(X)$ *as an algebra with the binary bilinear operation* $\mathrm{Lie}(X) \times \mathrm{Lie}(X) \ni (\ell, \ell') \mapsto [\ell, \ell'] \in \mathrm{Lie}(X)$).

Proof. The well posedness of $[\cdot, \cdot]$ follows from $\mathfrak{a}$ being a magma ideal[10], while the fact that it endows $\mathrm{Lie}(X)$ with a Lie algebra structure is a simple consequence[11] of the definition of A. Finally, π is an algebra morphism because it is obviously linear ($\mathrm{Lie}(X)$ is a quotient vector space and π is the associated projection!) and it satisfies (2.58). □

The Reader will take care not to confuse $\mathrm{Lie}(X)$ with $\mathrm{Lie}\{X\}$ (the latter being the smallest Lie subalgebra – of some Lie algebra $\mathfrak{g}$ – containing X, in case X is a subset of a pre-existing Lie algebra $\mathfrak{g}$). Obviously, there is an expected meaning for the similarity of the notation, which will soon be clarified (see Remark 2.55 below). We are ready to state the important fact that $\mathrm{Lie}(X)$ is a free Lie algebra related to X.

---

[10]Indeed, if $\pi(a) = \pi(a')$ and $\pi(b) = \pi(b')$ there exist $\alpha, \beta \in \mathfrak{a}$ such that $a' = a + \alpha$, $b' = b + \beta$. Hence $a' * b' = a * b + a * \beta + \alpha * b + \alpha * \beta \in a * b + a * \mathfrak{a} + \mathfrak{a} * b + \mathfrak{a} * \mathfrak{a} \subseteq a * b + \mathfrak{a}$, so that $\pi(a' * b') = \pi(a * b)$.

[11]For example, the Jacobi identity follows from $[\pi(a), [\pi(b), \pi(c)]] = \pi(a * (b * c))$ so that $[\pi(a), [\pi(b), \pi(c)]] + [\pi(b), [\pi(c), \pi(a)]] + [\pi(c), [\pi(a), \pi(b)]] = \pi(J(a,b,c)) = 0$.

Theorem 2.54 ($\mathrm{Lie}(X)$ **is a Free Lie Algebra Related to** X). *Let X be any set and, with the notation in (2.56) and (2.57), let us consider the map*

$$\varphi : X \to \mathrm{Lie}(X), \quad x \mapsto \pi(x), \tag{2.59}$$

i.e.,[12] *$\varphi \equiv \pi|_X$. Then the following facts hold:*

1. *The couple $(\mathrm{Lie}(X), \varphi)$ is a free Lie algebra related to X (see Definition 2.50).*
2. *The set $\{\varphi(x)\}_{x \in X}$ is independent in $\mathrm{Lie}(X)$, whence φ is injective.*
3. *The set $\varphi(X)$ Lie-generates $\mathrm{Lie}(X)$, that is, the smallest Lie subalgebra of $\mathrm{Lie}(X)$ containing $\varphi(X)$ coincides with $\mathrm{Lie}(X)$.*

Proof. See Scct. 8.1 (page 459) in Chap. 8. □

Remark 2.55. Part 3 of the statement of Theorem 2.54 says that $\mathrm{Lie}\{\varphi(X)\} = \mathrm{Lie}(X)$, the former being meant as the smallest subalgebra – of the latter – containing X (see Definition 2.13). This fact, together with the identification $X \equiv \varphi(X)$ (this is possible due to part 2 of Theorem 2.54) says that

$$\mathrm{Lie}\{X\} \equiv \mathrm{Lie}(X)$$

(which is extremely convenient given the abundance of notation for free Lie algebras generated by a set!).

Here is another (very!) desirable result concerning free Lie algebras.

Theorem 2.56 (The Isomorphism $\mathcal{L}(\mathbb{K}\langle X \rangle) \simeq \mathrm{Lie}(X)$). *Let X be any set and consider the free vector space $\mathbb{K}\langle X \rangle$ over X. Consider also $\mathcal{L}(\mathbb{K}\langle X \rangle)$, the smallest Lie subalgebra of $\mathscr{T}(\mathbb{K}\langle X \rangle)$ containing X.*

Then $\mathcal{L}(\mathbb{K}\langle X \rangle)$ and $\mathrm{Lie}(X)$ are isomorphic, as Lie algebras. More precisely, the pair $(\mathcal{L}(\mathbb{K}\langle X \rangle), \chi)$ is a free Lie algebra related to X.

When, occasionally, we shall allow ourselves to identify X with the subset $\chi(X)$ of $\mathbb{K}\langle X \rangle$ (via the injective map χ), the map $\chi : X \to \mathcal{L}(\mathbb{K}\langle X \rangle)$ becomes the map of set inclusion, whence Theorem 2.56 will permit us to say that $\mathcal{L}(\mathbb{K}\langle X \rangle)$ is a free Lie algebra *over* X.

Proof. If φ is as in (2.59), we know from Theorem 2.54 that $(\mathrm{Lie}(X), \varphi)$ is a free Lie algebra related to X. Hence, considering the map $X \ni x \mapsto \chi(x) \in \mathcal{L}(\mathbb{K}\langle X \rangle)$, there exists a unique Lie algebra morphism (see the notation in Definition 2.50) χ^{φ}, say f for short, such that

$$f : \mathrm{Lie}(X) \to \mathcal{L}(\mathbb{K}\langle X \rangle) \quad \text{and } f(\varphi(x)) = \chi(x), \text{ for every } x \in X. \tag{2.60}$$

We claim that *f is a Lie algebra isomorphism*. This claim is proved in Sect. 8.1 in Chap. 8 (precisely in Corollary 8.6, page 469). Hence, by Proposition 2.51-2,

---

[12]More precisely, the map φ is the composition

$$X \xrightarrow{\iota} M(X) \xrightarrow{\chi} \mathrm{Lib}(X) \xrightarrow{\pi} \mathrm{Lie}(X).$$

Via the identification $X \equiv \chi(X) \xrightarrow{\iota} \mathrm{Lib}(X)$ we can write $\varphi \equiv \pi|_X$.

$(\mathcal{L}(\mathbb{K}\langle X\rangle), \varphi_2)$ is a free Lie algebra related to X, with $\varphi_2 = f \circ \varphi \equiv \chi$ on X (where we also used (2.60)).   □

Collecting together Theorems 2.54 and 2.56 (and Definition 2.50), we can deduce that, if X is any set, $\mathfrak{g}$ any Lie algebra and $f : X \to \mathfrak{g}$ any map, there exist Lie algebra morphisms

$$f^\varphi : \mathrm{Lie}(X) \to \mathfrak{g}, \qquad f^\chi : \mathcal{L}(\mathbb{K}\langle X\rangle) \to \mathfrak{g}$$

such that

$$f^\varphi(\varphi(x)) = f(x) = f^\chi(\chi(x)), \qquad \forall\, x \in X$$

and, more explicitly, these morphisms act – on typical elements of their respective domains – as follows:

$$f^\varphi\Big([\varphi(x_1) \cdots [\varphi(x_{k-1}), \varphi(x_k)]_{\mathrm{Lie}(X)} \cdots]_{\mathrm{Lie}(X)} \Big)$$
$$= f^\chi\big([\chi(x_1) \cdots [\chi(x_{k-1}), \chi(x_k)]_\otimes \cdots]_\otimes \big)$$
$$= [f(x_1) \cdots [f(x_{k-1}), f(x_k)]_\mathfrak{g} \cdots]_\mathfrak{g},$$

for every $x_1, \ldots, x_k \in X$ and every $k \in \mathbb{N}$. Here

$$[\cdot, \cdot]_{\mathrm{Lie}(X)}, \quad [\cdot, \cdot]_\otimes, \quad [\cdot, \cdot]_\mathfrak{g}$$

are, respectively, the Lie brackets of $\mathrm{Lie}(X)$, of $\mathcal{L}(\mathbb{K}\langle X\rangle)$ (with Lie bracket inherited from the commutator on $\mathscr{T}(\mathbb{K}\langle X\rangle)$) and of $\mathfrak{g}$.

With Theorem 2.56 at hand, we are ready to provide the following:

Proof (of Theorem 2.49, page 86). Since (ii) is standard, we restrict our attention to the proof of (i). Let $\mathfrak{g}$ be a Lie algebra and let $f : V \to \mathfrak{g}$ be any linear map. We have to prove that there exists a unique Lie algebra morphism $\overline{f} : \mathcal{L}(V) \to \mathfrak{g}$ prolonging f. Since $\mathcal{L}(V)$ is Lie-generated by V (see e.g., Proposition 2.47) the uniqueness of $\overline{f}$ will follow from its existence. To prove this latter fact, we make use of a basis of V (the "non-canonical" nature of this argument being completely immaterial). See also the diagram below:

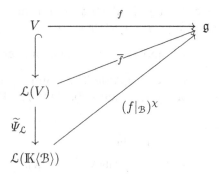

With this aim, let $\mathcal{B} = \{b_i\}_{i \in \mathcal{I}}$ be a basis of V. Then V is isomorphic (as a vector space) to the free vector space $\mathbb{K}\langle\mathcal{B}\rangle$, via the (unique) linear map $\Psi : V \to \mathbb{K}\langle\mathcal{B}\rangle$ mapping $b_i \in V$ into $\chi(b_i) \in \mathbb{K}\langle\mathcal{B}\rangle$ for every $i \in \mathcal{I}$ (recall the notation in (2.2)): more explicitly

$$\Psi\Big(\sum_{i \in \mathcal{I}'} \lambda_i \, b_i\Big) = \sum_{i \in \mathcal{I}'} \lambda_i \, \chi(b_i), \tag{2.61}$$

where $\mathcal{I}'$ is any finite subset of $\mathcal{I}$ and the coefficients λ_i are arbitrary scalars. Since $\Psi : V \to \mathbb{K}\langle\mathcal{B}\rangle$ is an isomorphism, by Remark 2.48 we can deduce that $\mathcal{L}(V)$ and $\mathcal{L}(\mathbb{K}\langle\mathcal{B}\rangle)$ are isomorphic via the unique LA isomorphism $\widetilde{\Psi}_{\mathcal{L}} : \mathcal{L}(V) \to \mathcal{L}(\mathbb{K}\langle\mathcal{B}\rangle)$ such that

$$\widetilde{\Psi}_{\mathcal{L}}(v) = \Psi(v), \quad \text{for every } v \in V. \tag{2.62}$$

Since the pair $(\mathcal{L}(\mathbb{K}\langle\mathcal{B}\rangle), \chi)$ is a free Lie algebra related to $\mathcal{B}$ (see Theorem 2.56), considering the map $f|_{\mathcal{B}} : \mathcal{B} \to \mathfrak{g}$, there exists an LA morphism $(f|_{\mathcal{B}})^{\chi} : \mathcal{L}(\mathbb{K}\langle\mathcal{B}\rangle) \to \mathfrak{g}$ such that

$$(f|_{\mathcal{B}})^{\chi}(\chi(b_i)) = f(b_i), \quad \forall \, i \in \mathcal{I}. \tag{2.63}$$

We claim that $\overline{f} := (f|_{\mathcal{B}})^{\chi} \circ \widetilde{\Psi}_{\mathcal{L}} : \mathcal{L}(V) \to \mathfrak{g}$ prolongs f (see the diagram above). Indeed, if $v \in V$, say $v = \sum_{i \in \mathcal{I}'} \lambda_i \, b_i$, we have

$$\overline{f}(v) = (f|_{\mathcal{B}})^{\chi}(\widetilde{\Psi}_{\mathcal{L}}(v)) \overset{(2.62)}{=} (f|_{\mathcal{B}})^{\chi}(\Psi(v)) \overset{(2.61)}{=} (f|_{\mathcal{B}})^{\chi}\Big(\sum_{i \in \mathcal{I}'} \lambda_i \, \chi(b_i)\Big)$$

$$\overset{(2.63)}{=} \sum_{i \in \mathcal{I}'} \lambda_i \, f(b_i) = f\Big(\sum_{i \in \mathcal{I}'} \lambda_i \, b_i\Big) = f(v).$$

This ends the proof. □

2.3 Completions of Graded Topological Algebras

The aim of this section is to equip a certain class of algebras A with a topology endowing A with the structure of a topological algebra. It will turn out that a structure of metric space will also be available in this setting. Then we shall describe the general process of completion of a metric space. Finally, we shall focus on graded algebras and the concept of formal power series will be closely investigated. All these topics will be of relevance when we shall deal with the CBHD Formula (and convergence aspects concerned with it).

2.3.1 Topology on Some Classes of Algebras

Definition 2.57. Let $(A, *)$ be an associative algebra. We say that $\{\Omega_k\}_{k \in \mathbb{N}}$ is *a topologically admissible family in* A if the sets Ω_k are subsets of A satisfying the properties:

(H1.) Ω_k is an ideal of A, for every $k \in \mathbb{N}$.
(H2.) $\Omega_1 = A$ and $\Omega_k \supseteq \Omega_{k+1}$, for every $k \in \mathbb{N}$.
(H3.) $\Omega_h * \Omega_k \subseteq \Omega_{h+k}$, for every $h, k \in \mathbb{N}$.
(H4.) $\bigcap_{k \in \mathbb{N}} \Omega_k = \{0\}$.

The main aim of this section is to prove the following theorem.

Theorem 2.58. *Let* $(A, *)$ *be an associative algebra and suppose that* $\{\Omega_k\}_{k \in \mathbb{N}}$ *is a topologically admissible family of subsets of* A. *Then the family*

$$\emptyset \cup \left\{ a + \Omega_k \right\}_{a \in A, \, k \in \mathbb{N}} \tag{2.64}$$

is a basis for a topology Ω *on* A *endowing* A *with the structure of a topological algebra.*[13] *Even more, the topology* Ω *is induced by the metric* $d : A \times A \to [0, \infty)$ *defined as follows* $(\exp(-\infty) := 0$ *applies)*

$$d(x, y) := \exp(-\nu(x - y)), \quad \text{for all } x, y \in A, \tag{2.65}$$

where $\nu : A \to \mathbb{N} \cup \{0, \infty\}$ *is defined by* $\nu(z) := \sup \left\{ n \geq 1 \mid z \in \Omega_n \right\}$, *or more precisely*

$$\nu(z) := \begin{cases} \text{if } z \neq 0, & \max \left\{ n \geq 1 \mid z \in \Omega_n \right\} \\ \text{if } z = 0, & \infty. \end{cases} \tag{2.66}$$

The triangle inequality for d *holds in the stronger form*[14]:

$$d(x, y) \leq \max\{d(x, z), d(z, y)\}, \quad \text{for every } x, y, z \in A. \tag{2.67}$$

Proof. See page 407 in Chap. 7.   □

---

[13]We recall that a topological algebra is a pair (A, Ω) where $(A, +, *)$ is an algebra and Ω is a topology on A such that the maps

$$A \times A \ni (a, b) \mapsto a + b, a * b \in A, \quad \mathbb{K} \times A \ni (k, a) \mapsto k\,a \in A$$

are continuous (with the associated product topologies, $\mathbb{K}$ being equipped with the discrete topology) and such that (A, Ω) is a Hausdorff topological space.

[14]A metric space (A, d) whose distance satisfies (2.67) (called the *strong triangle inequality* or *ultrametric inequality*) is usually referred to as an *ultrametric space*. Hence, a topologically admissible family of subsets of an algebra A endows A with the structure of an ultrametric space.

Remark 2.59. In the notation of the previous theorem, (2.67) easily implies the following peculiar fact: *A sequence $\{a_n\}_n$ in A is a Cauchy sequence in (A, d) if and only if $\lim_{n\to\infty} d(a_n, a_{n+1}) = 0$.*

Indeed, given a sequence $\{a_n\}_n$ in A, as a consequence of (2.67) the following telescopic estimate applies, for every $n, p \in \mathbb{N}$:

$$d(a_n, a_{n+p}) \leq \max\{d(a_n, a_{n+1}), d(a_{n+1}, a_{n+p})\}$$

$$\leq \max\{d(a_n, a_{n+1}), \max\{d(a_{n+1}, a_{n+2}), d(a_{n+2}, a_{n+p})\}\}$$

$$= \max\{d(a_n, a_{n+1}), d(a_{n+1}, a_{n+2}), d(a_{n+2}, a_{n+p})\}$$

(after finitely many steps)

$$\leq \max\{d(a_n, a_{n+1}), d(a_{n+1}, a_{n+2}), \ldots, d(a_{n+p-1}, a_{n+p})\}.$$

This shows that $\{a_n\}_n$ is a Cauchy sequence in (A, d) if and only if

$$\lim_{n\to\infty} d(a_n, a_{n+1}) = 0.$$

Definition 2.60. If A is an associative algebra and if $\{\Omega_k\}_{k\in\mathbb{N}}$ is a topologically admissible family of subsets of A, the topology Ω (respectively, the metric d) in Theorem 2.58 will be called *the topology on A induced by $\{\Omega_k\}_{k\in\mathbb{N}}$* (respectively, *the metric on A induced by $\{\Omega_k\}_{k\in\mathbb{N}}$*).

Remark 2.61. When A is an associative algebra and d is the metric on A induced by a topologically admissible family $\{\Omega_k\}_{k\in\mathbb{N}}$, we have the following *algebraic properties* of the metric d in (2.65) (proved in due course within Chap. 7, see page 415):

1. $d(x, y) = d(x + z, y + z)$, for every $x, y, z \in A$.
2. $d(k\, x, k\, y) = d(x, y)$, for every $k \in \mathbb{K} \setminus \{0\}$ and every $x, y \in A$.
3. $d(x * y, \xi * \eta) \leq \max\{d(x, \xi), d(y, \eta)\}$, for every $x, y, \xi, \eta \in A$.

Remark 2.62. In the notation of Theorem 2.58, we have the following fact: *A sequence $\{a_n\}_n$ in A is a Cauchy sequence in the metric space (A, d) sequence if and only if $\lim_{n\to\infty}(a_{n+1} - a_n) = 0$.*

Consequently, *a series $\sum_{n=1}^{\infty} a_n$ consisting of elements in A is a Cauchy sequence in (A, d) if and only if $\lim_{n\to\infty} a_n = 0$ in (A, d).*

Indeed, by Remark 2.59 $\{a_n\}_n$ is Cauchy in (A, d) iff $\lim_{n\to\infty} d(a_n, a_{n+1}) = 0$. In its turn, by (1) in Remark 2.61, we see that this latter fact coincides with $\lim_{n\to\infty} d(0, a_{n+1} - a_n) = 0$. Finally, this is the definition of $\lim_{n\to\infty}(a_{n+1} - a_n) = 0$ in (A, d).

The above remark shows how different are the metrics in Theorem 2.58 (indeed, all ultrametrics), if they are compared to the usual Euclidean metric in $\mathbb{R}^n$, where the above facts are false (as shown by the trivial example $\sum_{n=1}^{\infty} 1/n = \infty$ in the usual Euclidean space $\mathbb{R}$).

Remark 2.63. With the notation of Theorem 2.58, by unraveling the definition of d we have that, for two points $x, y \in A$ and a positive real number ε, the condition $d(x, y) < \varepsilon$ is equivalent to $\sup \left\{ n \geq 1 \mid z \in \Omega_n \right\} > \ln(1/\varepsilon)$, that is,

$$\left(x, y \in A \quad d(x, y) < \varepsilon \right) \quad \Longleftrightarrow \quad \left(\begin{array}{c} \text{there exists } n \in \mathbb{N}, \text{ with } n > \ln(1/\varepsilon) \\ \text{such that } x - y \in \Omega_n \end{array} \right).$$
(2.68)

Example 2.64. Before proceeding, we make explicit some examples of topologically admissible families, useful for the sequel.

1. Let $(A, *)$ be an associative algebra and let $I \subseteq A$ be an ideal. Let us set $\Omega_0 := A$ and, for $k \in \mathbb{N}$, let

$$\Omega_k := \text{ideal generated by } \{ I * \cdots * I \ (k \text{ times}) \}$$

$$= \left\{ \begin{array}{c} \text{set of the finite sums of elements of the form } r * i_1 * \cdots * i_k * \rho \\ \text{where } r, \rho \in A \text{ and } i_1, \ldots, i_k \in I \end{array} \right\}.$$

Then it is easily seen that the family $\{\Omega_k\}_{k \geq 0}$ fulfils hypotheses (H1), (H2) and (H3) in Definition 2.57. Hence, whenever $\{\Omega_k\}_{k \geq 0}$ fulfils also hypothesis (H4), it is a topologically admissible family in A.

2. Suppose $(A, *)$ is an associative algebra which is also *graded* (see Definition 2.11). We set $A_+ := \bigoplus_{j=1}^{\infty} A_j$. Also, let $\Omega_0 := A$ and, for $k \in \mathbb{N}$,

$$\Omega_k := \text{span}\left\{ a_1 * \cdots * a_k \mid a_1, \ldots, a_k \in A_+ \right\}.$$

It is not difficult to show that $\{\Omega_k\}_{k \geq 0}$ is a topologically admissible family in A. For example, we prove (H4): First note that $\Omega_k \subseteq \bigoplus_{j=k}^{\infty} A_j$ for any $k \in \mathbb{N} \cup \{0\}$ (indeed equality holds); hence we have

$$\{0\} \subseteq \bigcap_{k \geq 0} \Omega_k \subseteq \bigcap_{k \geq 0} \bigoplus_{j=k}^{\infty} A_j = \{0\}.$$

The last equality is proved as follows: if $a \in A$ and $a \neq 0$, then we can write $a = \sum_{i=1}^{N} a_i$ with $a_i \in A_i$ and $a_N \neq 0$; in this case $a \notin \bigoplus_{j=N+1}^{\infty} A_j$ and the assertion follows. Note that also

$$\Omega_k = \bigoplus_{j \geq k} A_j, \qquad \forall \, k \in \mathbb{N} \cup \{0\}.$$
(2.69)

As for hypotheses (H1)-(H2)-(H3), they follow by the previous Example 1, since it can be easily seen that, for all $k \in \mathbb{N}$,

$$\Omega_k = \text{ideal generated by } \{ A_+ * \cdots * A_+ \ (k \text{ times}) \}.$$
(2.70)

3. Let V be a vector space and let $A = \mathscr{T}(V)$. We can construct the family $\{\Omega_k\}_{k \in \mathbb{N}}$ according to the previous example, with respect to the usual

grading $\mathcal{T} = \bigoplus_{j=0}^{\infty} \mathcal{T}_j$. By (2.69), we have $\Omega_k = U_k$ for every $k \in \mathbb{N}$, where U_k has been defined in (2.29). Hence $\{U_k\}_{k \in \mathbb{N}}$ *is a topologically admissible family in* $\mathcal{T}(V)$, *thus equipping* $\mathcal{T}(V)$ *with both a metric space and a topological algebra structure.* (This same fact also follows from the results in Remark 2.34-(3).) As stated in Example 2, we can view this example as a particular case of Example 1 above, since U_k coincides with the ideal generated by the k-products of $I = \mathcal{T}_+(V)$ (or equivalently, of $I = V$).

4. Let V be a vector space, and on the tensor product $\mathcal{T}(V) \otimes \mathcal{T}(V)$ let us consider the family of subsets $\{W_k\}_{k \in \mathbb{N}}$ introduced in (2.44). Then, by the results in Remark 2.42, $\{W_k\}_{k \in \mathbb{N}}$ *is a topologically admissible family in* $\mathcal{T}(V) \otimes \mathcal{T}(V)$, *thus equipping* $\mathcal{T}(V) \otimes \mathcal{T}(V)$ *with both a metric space and a topological algebra structure.* Analogously, the fact that $\{W_k\}_{k \in \mathbb{N}}$ is a topologically admissible family can be proved by Example 2 above, by considering the grading $\mathcal{T} \otimes \mathcal{T} = \bigoplus_{k \geq 0} K_k$ as in (2.42). Indeed, $W_k = \mathrm{span}\{a_1 \bullet \cdots \bullet a_k \mid a_1, \ldots, a_k \in \bigoplus_{j \geq 1} \bar{K}_j = (\mathcal{T} \otimes \mathcal{T})_+\}$. As stated in Example 2, we can also view this example as a particular case of Example 1 above, since W_k coincides with the ideal generated by the k-products of $I = (\mathcal{T} \otimes \mathcal{T})_+$ (or equivalently, of $I := K$ introduced in (2.46)).

Since we are mainly interested in graded algebras, for the sake of future reference, we collect many of the aforementioned results in the following proposition, and we take the opportunity to prove some further facts.

Proposition 2.65 (Metric Related to a Graded Algebra). *Let* $(A, *)$ *be an associative graded algebra with grading* $\{A_j\}_{j \geq 0}$. *For every* $k \in \mathbb{N} \cup \{0\}$, *set* $\Omega_k := \bigoplus_{j \geq k} A_j$.

(a) *Then* $\{\Omega_k\}_{k \geq 0}$ *is a topologically admissible family of A, thus endowing A with both a metric space and a topological algebra structure (both structures are referred to as "related to the grading* $\{A_j\}_{j \geq 0}$"*).*

(b) *The induced metric d has the algebraic (and ultrametric) properties*

$$d(x, y) = d(x + z, y + z), \qquad d(k\,x, k\,y) = d(x, y)$$
$$d(x * y, \xi * \eta) \leq \max\{d(x, \xi), d(y, \eta)\}, \tag{2.71}$$

for every $x, y, z, \xi, \eta \in A$ and every $k \in \mathbb{K} \setminus \{0\}$.

(c) *A sequence $\{a_n\}_n$ of elements of A is a Cauchy sequence in (A, d) if and only if $\lim_{n \to \infty}(a_{n+1} - a_n) = 0$ in (A, d). Moreover a series $\sum_{n=1}^{\infty} a_n$ of elements of A is a Cauchy sequence in (A, d) if and only if $\lim_{n \to \infty} a_n = 0$ in (A, d).*

(d) *For every $z = (z_j)_{j \geq 0} \in A$, we have*

$$d(z) = \begin{cases} \exp\left(-\min\{j \geq 0 : z_j \neq 0\}\right), & \text{if } z \neq 0, \\ 0, & \text{if } z = 0 \end{cases}$$
$$= \begin{cases} \max\{e^{-j} : z_j \neq 0\}, & \text{if } z \neq 0, \\ 0, & \text{if } z = 0. \end{cases} \tag{2.72}$$

(e) Let $\{b_n\}_{n\in\mathbb{N}}$ be a sequence of elements in A and let $\beta \in A$; let us write

$$b_n = (a_j^{(n)})_{j\geq 0} \quad and \quad \beta = (a_j)_{j\geq 0},$$

with $a_j^{(n)}, a_j \in A_j$ for every $j \geq 0$ and every $n \in \mathbb{N}$.
Then we have $\lim_{n\to\infty} b_n = \beta$ in (A, d) if and only if

$$\forall J \geq 0 \ \exists N_J \in \mathbb{N} : \quad n \geq N_J \text{ implies } a_j^{(n)} = a_j \quad for \ 0 \leq j \leq J. \quad (2.73)$$

Proof. See page 416 in Chap. 7. □

2.3.2 Completions of Graded Topological Algebras

We begin by recalling some classical results of Analysis concerning metric spaces. As usual, the associated proofs are postponed to Chap. 7.

Definition 2.66. Let (X, d) be a metric space. We say that (Y, δ) is an *isometric completion* of (X, d), if the following facts hold:

1. (Y, δ) is a complete metric space.
2. There exists a metric subspace X_0 of Y which is dense in Y and such that (X_0, δ) is isometric (in the sense of metric spaces[15]) to (X, d).

The following simple fact holds, highlighting the fact that the notion of isometric completion is unique, up to isomorphism.

Proposition 2.67. Let (X, d) be a metric space. If (Y_1, δ_1) and (Y_2, δ_2) are isometric completions of (X, d), then they are (canonically) isomorphic.

Proof. See page 417 in Chap. 7. □

The following remarkable result states that every metric space always admits an isometric completion.

Theorem 2.68 (Completion of a Metric Space). Let (X, d) be a metric space. Then there exists an isometric completion $(\widetilde{X}, \widetilde{d})$ of (X, d), which can be constructed as follows. We first consider the set $\mathcal{C}$ of all the Cauchy sequences $\widetilde{x} = (x_n)_n$ in (X, d). We introduce in $\mathcal{C}$ an equivalence relation by setting

$$(x_n)_n \sim (x_n')_n \quad iff \quad \lim_{n\to\infty} d(x_n, x_n') = 0. \quad (2.74)$$

---

[15]We recall that, given two metric spaces $(Y_1, d_1), (Y_2, d_2)$, a map $\Phi : Y_1 \to Y_2$ is called an isomorphism of metric spaces if Φ is bijective and such that $d_2(\Phi(y), \Phi(y')) = d_1(y, y')$ for every $y, y' \in Y_1$ (note that this last condition implicitly contains the injectivity of Φ together with the fact that Φ is a homeomorphism of the associated topological spaces).

We take as $\widetilde{X}$ the quotient set $\mathcal{C}/\sim$, with the metric defined by

$$\widetilde{d}\Big(\big[(x_n)_n\big]_\sim, \big[(y_n)_n\big]_\sim\Big) := \lim_{n\to\infty} d(x_n, y_n). \tag{2.75}$$

Furthermore (according to the notation in Definition 2.66), we take as X_0 (say, the isometric copy of X inside $\widetilde{X}$) the quotient set of the constant sequences $(x)_n$ with $x \in X$ and the associated isometry is the map

$$\alpha : X \to X_0, \quad x \mapsto [(x_n)_n]_\sim \quad \text{with } x_n = x \text{ for every } n \in \mathbb{N}. \tag{2.76}$$

Proof. See page 418 in Chap. 7.   □

In the sequel, when dealing with isometric completions of a given metric space X, we shall reserve the notation $\widetilde{X}$ for the metric space introduced in Theorem 2.68. The following result states that the passage to the isometric completion preserves many of the underlying algebraic structures, in a very natural way.

Theorem 2.69 (Algebraic Structure on the Isometric Completion of a UAA). *Let $(A, +, *)$ be a UA algebra. Suppose $\{\Omega_k\}_{k\in\mathbb{N}}$ is a topologically admissible family in A and let d be the metric on A induced by $\{\Omega_k\}_{k\in\mathbb{N}}$. Finally, consider the isometric completion $\widetilde{A}$ of (A, d) as in Theorem 2.68 and let $A_0 \subseteq \widetilde{A}$ be the set containing the equivalence classes of the constant sequences.*

Then $\widetilde{A}$ can be equipped with a structure of a UA algebra $(\widetilde{A}, \widetilde{+}, \widetilde{*})$, which is also a topological algebra containing A_0 as a (dense) subalgebra isomorphic to A. More precisely, the map α in (2.76) is an isomorphism of metric spaces and of UA algebras. The relevant operations on $\widetilde{A}$ are defined as follows:

$$\begin{aligned}
\big[(x_n)_n\big]_\sim \widetilde{+} \big[(y_n)_n\big]_\sim &:= \big[(x_n + y_n)_n\big]_\sim, \\
\big[(x_n)_n\big]_\sim \widetilde{*} \big[(y_n)_n\big]_\sim &:= \big[(x_n * y_n)_n\big]_\sim, \\
k \big[(x_n)_n\big]_\sim &:= \big[(k\, x_n)_n\big]_\sim, \quad k \in \mathbb{K}, \\
1_{\widetilde{A}} &:= [(1_A)_n]_\sim.
\end{aligned} \tag{2.77}$$

Proof. See page 422 in Chap. 7.   □

Remark 2.70. Let $(A, *), \{\Omega_k\}_k, d, \widetilde{A}$ be as in Theorem 2.69. Suppose B is equipped with a UAA structure by the operation $\star$, that it is equipped with a metric space structure by the metric δ and that the following properties hold:

1. A is a *subset* of B.
2. $\star$ coincides with $*$ on $A \times A$.
3. δ coincides with d on $A \times A$.
4. A is dense in B.
5. B is a complete metric space.

Then B and $\widetilde{A}$ are not only isomorphic as metric spaces (according to Proposition 2.67) but also as UA algebras (via the same isomorphism). (See the proof in Chap. 7, page 425.)

By collecting together some results obtained so far (and derived within the proofs of some of the previous results), we obtain the following further characterization of the isometric completion $\widetilde{A}$ of A.

Theorem 2.71 (Characterizations of the Isometric Completion of a UAA).
Let $(A, +, *)$ be a UA algebra. Suppose $\{\Omega_k\}_{k\in\mathbb{N}}$ is a topologically admissible family in A and let d be the metric on A induced by $\{\Omega_k\}_{k\in\mathbb{N}}$. Finally, consider the isometric completion $\widetilde{A}$ of (A, d) as in Theorem 2.68.

If $\alpha \in \widetilde{A}$ is represented by the (Cauchy) sequence $(a_n)_n$ in A (that is, $\alpha = [(a_n)_n]_\sim$), we have

$$\alpha = \lim_{n\to\infty} a_n \quad in \; (\widetilde{A}, \widetilde{d}),$$

where each $a_n \in A$ is identified with an element of $\widetilde{A}$ via the map α in (2.76). Hence, roughly, $\widetilde{A}$ can be thought of as the set of the "limits" of the Cauchy sequences in A, more precisely

$$\widetilde{A} = \left\{ \lim_{j\to\infty} \left[(a_j, a_j, \cdots) \right]_\sim \middle| (a_n)_n \text{ is a Cauchy sequence in } A \right\}. \qquad (2.78a)$$

Equivalently (see also Proposition 2.65-(c)), $\widetilde{A}$ can also be thought of as the set of the A-valued series associated to a vanishing sequence, more precisely

$$\widetilde{A} = \left\{ \sum_{j=1}^\infty \left[(b_j, b_j, \cdots) \right]_\sim \middle| (b_n)_n \text{ is a sequence converging to zero in } (A, d) \right\}. \tag{2.78b}$$

Here is a very natural result on the relation $A \mapsto \widetilde{A}$.

Lemma 2.72. *Let A, B be two isomorphic UA algebras. Suppose $\varphi : A \to B$ is a UAA isomorphism and suppose that $\{\Omega_k\}_{k\in\mathbb{N}}$ is a topologically admissible family in A. Set $\widetilde{\Omega}_k := \varphi(\Omega_k)$, for every $k \in \mathbb{N}$.*

Then the family $\{\widetilde{\Omega}_k\}_{k\in\mathbb{N}}$ is a topologically admissible family in B. Moreover the metric spaces induced on A and on B respectively by the families $\{\Omega_k\}_{k\in\mathbb{N}}$ and $\{\widetilde{\Omega}_k\}_{k\in\mathbb{N}}$ are isomorphic metric spaces and φ can be uniquely prolonged to a continuous map $\widetilde{\varphi} : \widetilde{A} \to \widetilde{B}$ which is both a metric isomorphism and a UAA isomorphism.

Proof. As claimed, as an isomorphism of metric spaces we can take the map

$$\widetilde{\varphi} : \widetilde{A} \to \widetilde{B} \text{ such that } \quad \widetilde{\varphi}([(a_n)_n]_\sim) := [(\varphi(a_n))_n]_\sim. \tag{2.79}$$

Here, we used the following notation: Let (A, d) denote the metric induced on A by the family $\{\Omega_k\}_{k\in\mathbb{N}}$, and let (B, δ) denote the metric induced on B by the family $\{\widetilde{\Omega}_k\}_{k\in\mathbb{N}}$; in (2.79), $(a_n)_n$ is a Cauchy sequence in (A, d) while the two classes $[\cdot]_\sim$ from left to right in (2.79) are the equivalence classes as in (2.74) related respectively to the equivalence relations induced by the metrics d and δ. See page 426 in Chap. 7 for the complete proof. $\qquad\qquad\square$

2.3.3 Formal Power Series

Throughout this section, $(A, *)$ is a UA *graded* algebra with a fixed grading $\{A_j\}_{j\geq 0}$. Following the notation in the previous section, for $k \geq 0$ we set

$$\Omega_k := \bigoplus_{j\geq k} A_j. \tag{2.80}$$

We know from Proposition 2.65 that $\{\Omega_k\}_{k\in\mathbb{N}}$ is a topologically admissible family of A, thus endowing A with both a metric space and a topological algebra structure. We aim to give a very explicit realization of an isometric completion of (A, d), as the set of the so-called formal power series on A (w.r.t. the grading $\{A_j\}_j$). We begin with the relevant definitions.

Definition 2.73 (Formal Power Series on A). Let $A = \bigoplus_{j\geq 0} A_j$ be a UA graded algebra. We set

$$\widehat{A} := \prod_{j\geq 0} A_j, \tag{2.81}$$

and we call $\widehat{A}$ the space of *formal power series on A (w.r.t. the grading $\{A_j\}_j$)*. On $\widehat{A}$ we consider the operation $\widehat{*}$ defined by

$$(a_j)_j \mathbin{\widehat{*}} (b_j)_j := \left(\sum_{k=0}^{j} a_{j-k} * b_k \right)_{j\geq 0} \qquad (a_j, b_j \in A_j, \ \forall\, j \geq 0). \tag{2.82}$$

Then $(\widehat{A}, \widehat{*})$ is a UA algebra, called *the algebra of the formal power series on A*.

Remark 2.74. Note that (2.82) is well posed thanks to the fact that $a_{j-k} * b_k \in A_{j-k} * A_k \subseteq A_j$ for every $j \geq 0$ and every $k = 0, \ldots, j$. Obviously, A is a subset of $\widehat{A}$ and it is trivially seen that

$$a, b \in A \quad\Longrightarrow\quad a \mathbin{\widehat{*}} b = a * b. \tag{2.83}$$

We now introduce on $\widehat{A}$ a distinguished topology, by introducing a suitable topologically admissible family. To this aim, we set

$$\widehat{\Omega}_k := \prod_{j\geq k} A_j, \quad k \in \mathbb{N} \cup \{0\}, \tag{2.84}$$

naturally considered as subspaces of $\widehat{A}$. The following facts hold:

1. Every $\widehat{\Omega}_k$ is an ideal in $\widehat{A}$.
2. $\widehat{A} = \widehat{\Omega}_0 \supseteq \widehat{\Omega}_1 \supseteq \cdots \widehat{\Omega}_k \supseteq \widehat{\Omega}_{k+1} \supseteq \cdots$.
3. $\widehat{\Omega}_i \, \widehat{*} \, \widehat{\Omega}_j \subseteq \widehat{\Omega}_{i+j}$, for every $i, j \geq 0$.
4. $\bigcap_{i \geq 0} \widehat{\Omega}_i = \{0\}$.

As a consequence, $\{\widehat{\Omega}_k\}_{k \geq 0}$ is a topologically admissible family of $\widehat{A}$. By means of Theorem 2.58 we can deduce that $\{\widehat{\Omega}_k\}_{k \geq 0}$ endows $\widehat{A}$ with a topology $\widehat{\Omega}$ (more, with the structures of a topological algebra and of a metric space) and we call $(\widehat{A}, \widehat{\Omega})$ the *topological space of the formal power series* (related to the given grading). Note that

$$\Omega_k = A \cap \widehat{\Omega}_k, \quad \forall \, k \in \mathbb{N} \cup \{0\}, \tag{2.85}$$

whence the inclusion $A \hookrightarrow \widehat{A}$ is continuous (here A has the topology induced by $\{\Omega_k\}_k$ and $\widehat{A}$ has the topology induced by $\{\widehat{\Omega}_k\}_k$). We have the following important result.

Theorem 2.75 (The Isometry $\widehat{A} \simeq \widetilde{A}$). *Let $(A, *)$ be a UA graded algebra with grading $\{A_j\}_{j \geq 0}$. Let Ω_k and $\widehat{\Omega}_k$ be defined, respectively, as in (2.80) and (2.84).*

Then the space $\widehat{A}$ (with the metric induced by $\{\widehat{\Omega}_k\}_{k \geq 0}$) is a complete metric space and it is an isometric completion of A (with the metric induced by $\{\Omega_k\}_{k \geq 0}$). The natural inclusion $A \hookrightarrow \widehat{A}$ is both an isometry and a UAA isomorphism, and A is dense in $\widehat{A}$.

In particular, denoting by d (resp. by $\widehat{d}$) the metric on A (resp. on $\widehat{A}$) induced by the family $\{\Omega_k\}_{k \geq 0}$ (resp. by the family $\{\widehat{\Omega}_k\}_{k \geq 0}$) we have that the restriction of $\widehat{d}$ to $A \times A$ coincides with d.

Proof. See page 428 in Chap. 7. $\square$

Remark 2.76. We have the following results.

1. By Proposition 2.65-(d), we get:
   A sequence $w_k = (u_0^k, u_1^k, \ldots)$ in $\widehat{A}$ converges to $w = (u_0, u_1, \ldots)$ in $\widehat{A}$ if and only if for every $N \in \mathbb{N}$ there exists $k(N) \in \mathbb{N}$ such that, for all $k \geq k(N)$, it holds that

$$w_k = \left(u_0, u_1, u_2, u_3, \ldots, u_N, \, u_{N+1}^k, u_{N+2}^k, u_{N+3}^k, \ldots \right).$$

2. With all the above notation, if $a = (a_j)_j \in \widehat{A} = \prod_{j \geq 0} A_j$ (with $a_j \in A_j$ for every $j \geq 0$) then we have the limit

$$A \ni \sum_{j=0}^{N} a_j \equiv (a_0, a_1, \ldots, a_N, 0, 0, \ldots) \xrightarrow[N \to \infty]{} a, \tag{2.86}$$

the limit being taken in the metric space $\widehat{A}$. We can thus represent the elements of $\widehat{A}$ as series $\sum_{j=0}^{\infty} a_j$ (with $a_j \in A_j$ for every $j \geq 0$).

3. Furthermore, any series $\sum_{n=1}^{\infty} b_n$ of elements of A converges in $\widehat{A}$ if and only if it is Cauchy (completeness of $\widehat{A}$), which is equivalent to $\lim_{n \to \infty} b_n = 0$ in $\widehat{A}$ (see Remark 2.62), which, in its turn, is equivalent to $\lim_{n \to \infty} b_n = 0$ in A (see (2.85)). For example, if $a_n \in A_n$ for every $n \geq 0$, the series $\sum_{n=1}^{\infty} a_n$ is convergent in $\widehat{A}$.

Analogously, *any series $\sum_{n=1}^{\infty} b_n$ of elements of $\widehat{A}$ converges in $\widehat{A}$ if and only if $\lim_{n \to \infty} b_n = 0$ in $\widehat{A}$* (again by an application of Remark 2.62).

4. Any set $\widehat{\Omega}_k$ ($k \in \mathbb{N} \cup \{0\}$) is both open and closed in $\widehat{A}$. Thus, by (2.85), the same is true of any Ω_k in A. More generally, see Proposition 2.77 below.

Proposition 2.77. *Let J be any fixed subset of $\mathbb{N} \cup \{0\}$. Then the set*

$$H := \left\{ (u_j)_j \in \widehat{A} \mid u_j \in A_j \text{ for every } j \geq 0 \text{ and } u_j = 0 \text{ for every } j \in J \right\}$$

is closed in the topological space $\widehat{A}$.

Proof. Suppose $\{w_k\}_k$ is a sequence in H converging to w in $\widehat{A}$. We use for w_k and w the notation in Remark 2.76-(1). Let $j_0 \in J$ be fixed. By the cited remark, there exists $k(j_0) \in \mathbb{N}$ such that, for all $k \geq k(j_0)$,

$$w_k = \left(u_0, u_1, u_2, \ldots, u_{j_0}, u_{j_0+1}^k, u_{j_0+2}^k, \ldots \right). \tag{2.87}$$

Since $w_k \in H$ for every k, its j_0-component is null. By (2.87), this j_0-component equals the j_0-component of w. Since j_0 is arbitrary in H, this proves that $w \in H$. $\qquad\qquad\square$

Remark 2.78. For example, we can apply Proposition 2.77 in the cases when $J = \{0, 1, \ldots, k-1\}$, or $J = \{0\}$ or $J = (\mathbb{N} \cup \{0\}) \setminus \{k\}$, in which cases we obtain respectively the closed sets $\widehat{\Omega}_k$, $\widehat{A}_+ := \prod_{j \geq 1} A_j$, and A_k.

The following lemma will be used frequently in the sequel.

Lemma 2.79 (Prolongation Lemma). *Suppose $A = \bigoplus_{j \geq 0} A_j$ and $B = \bigoplus_{j \geq 0} B_j$ are graded UA algebras and let $\widehat{A}, \widehat{B}$ be the corresponding topological spaces of their formal power series.*

Following (2.80), we use the notation $\Omega_k^A := \bigoplus_{j \geq k} A_j$ and $\Omega_k^B := \bigoplus_{j \geq k} B_j$. Suppose $\varphi : A \to B$ is a linear map with the following property:

There exists a sequence $\{k_n\}_n$ in $\mathbb{N}$ such that $\lim_{n \to \infty} k_n = \infty$ and

$$\varphi\left(\Omega_n^A\right) \subseteq \Omega_{k_n}^B \quad \text{for every } n \in \mathbb{N}. \tag{2.88}$$

Then φ is uniformly continuous (considering A, B as subspaces of the metric spaces $\widehat{A}, \widehat{B}$, respectively[16]). Hence, φ can be extended in a unique way to a continuous linear map $\widehat{\varphi} : \widehat{A} \to \widehat{B}$. Moreover, if φ is a UAA morphism, the same is true of $\widehat{\varphi}$.

Proof. See page 430 in Chap. 7. $\square$

Remark 2.80. Theorem 2.75 can be applied to the graded algebras

$$\mathscr{T}(V) = \bigoplus_{j \geq 0} \mathscr{T}_j(V) \quad \text{and} \quad \mathscr{T}(V) \otimes \mathscr{T}(V) = \bigoplus_{j \geq 0} K_j(V)$$

(see (2.28) and (2.42), respectively). Thus, on the algebras $\mathscr{T}$ and $\mathscr{T} \otimes \mathscr{T}$ we are given metric space structures induced respectively by the topologically admissible families $\{U_k\}_k$ and $\{W_k\}_k$, where

$$U_k = \bigoplus_{j \geq k} \mathscr{T}_j(V), \qquad W_k = \bigoplus_{i+j \geq k} \mathscr{T}_{i,j}(V).$$

The formal power series related to the graded algebras $\mathscr{T}$ and $\mathscr{T} \otimes \mathscr{T}$, denoted henceforth by $\widehat{\mathscr{T}}(V)$ and $\widehat{\mathscr{T} \otimes \mathscr{T}}(V)$ (or, shortly, by $\widehat{\mathscr{T}}$ and $\widehat{\mathscr{T} \otimes \mathscr{T}}$), are the algebras

$$\widehat{\mathscr{T}}(V) = \prod_{j \geq 0} \mathscr{T}_j(V), \qquad \widehat{\mathscr{T} \otimes \mathscr{T}}(V) = \prod_{i+j \geq 0} \mathscr{T}_{i,j}(V), \qquad (2.89)$$

with operations as in (2.82) (respectively inherited from the operations on $(\mathscr{T}, \cdot)$ and on $(\mathscr{T} \otimes \mathscr{T}, \bullet)$), respectively equipped with the metric space structures induced by the topologically admissible families $\{\widehat{U}_k\}_k$ and $\{\widehat{W}_k\}_k$, where

$$\widehat{U}_k := \prod_{j \geq k} \mathscr{T}_j(V), \qquad \widehat{W}_k := \prod_{i+j \geq k} \mathscr{T}_{i,j}(V). \qquad (2.90)$$

Convention. In order to avoid heavy notation, the operations $\widehat{\cdot}$ and $\widehat{\bullet}$ (see the notation in (2.82)) will usually appear without the "$\widehat{}$" sign. This slight abuse of notation is in accordance with (2.83).

---

[16]Which is the same as considering A, B as metric spaces with metrics induced by the families $\{\Omega_k^A\}_k$ and $\{\Omega_k^B\}_k$, respectively (see (2.85)).

As a consequence, we have

$$\cdot : \widehat{\mathcal{T}} \times \widehat{\mathcal{T}} \to \widehat{\mathcal{T}}, \quad (a_j)_j \cdot (b_j)_j = \left(\sum_{k=0}^{j} a_{j-k} \cdot b_k \right)_{j \geq 0} \tag{2.91}$$

where $a_j, b_j \in \mathcal{T}_j(V)$ for all $j \geq 0$;

$$\bullet : \widehat{\mathcal{T} \otimes \mathcal{T}} \times \widehat{\mathcal{T} \otimes \mathcal{T}} \to \widehat{\mathcal{T} \otimes \mathcal{T}}, \quad (a_j)_j \bullet (b_j)_j = \left(\sum_{k=0}^{j} a_{j-k} \bullet b_k \right)_{j \geq 0} \tag{2.92}$$

where $a_j, b_j \in \bigoplus_{h+k=j} \mathcal{T}_{h,k}(V)$ for all $j \geq 0$.

The operation $\bullet$ on $\widehat{\mathcal{T} \otimes \mathcal{T}}$ can also be rewritten by using the double-sequenced notation $(u_{i,j})_{i,j}$ for the elements of $\widehat{\mathcal{T} \otimes \mathcal{T}}$ (this means that $u_{i,j} \in \mathcal{T}_{i,j}(V) = \mathcal{T}_i(V) \otimes \mathcal{T}_j(V)$ for every $i, j \geq 0$): indeed, following (2.45), we have

$$\bullet : \widehat{\mathcal{T} \otimes \mathcal{T}} \times \widehat{\mathcal{T} \otimes \mathcal{T}} \to \widehat{\mathcal{T} \otimes \mathcal{T}}$$

$$(t_{i,j})_{i,j} \bullet (\widetilde{t}_{i,j})_{i,j} = \left(\sum_{r+\widetilde{r}=i, \ s+\widetilde{s}=j} t_{r,s} \bullet \widetilde{t}_{\widetilde{r},\widetilde{s}} \right)_{i,j \geq 0}, \tag{2.93}$$

where $t_{i,j}, \widetilde{t}_{i,j} \in \mathcal{T}_i(V) \otimes \mathcal{T}_j(V)$ for all $i, j \geq 0$.

Remark 2.81. When expressed in coordinate form on the product space $\widehat{\mathcal{T}} = \prod_{j \geq 0} \mathcal{T}_j$, the Lie bracket operation takes a particularly easy form: Indeed, if $u, v \in \widehat{\mathcal{T}}$ and $u = (u_j)_j$ and $v = (v_j)_j$, with $u_j, v_j \in \mathcal{T}_j(V)$ (for every $j \in \mathbb{N}$), we have

$$[u, v] = [(u_j)_j, (v_j)_j] = \left(\sum_{h+k=j} [u_h, v_k] \right)_{j \geq 0}. \tag{2.94}$$

Indeed, the following computation holds

$$[(u_j)_j, (v_j)_j] = (u_j)_j \cdot (v_j)_j - (v_j)_j \cdot (u_j)_j$$

$$\overset{(2.91)}{=} \left(\sum_{k=0}^{j} u_{j-k} \cdot v_k \right)_{j \geq 0} - \left(\sum_{k=0}^{j} v_{j-k} \cdot u_k \right)_{j \geq 0}$$

(change the dummy index in the second sum)

$$= \left(\sum_{k=0}^{j} (u_{j-k} \cdot v_k - v_k \cdot u_{j-k}) \right)_{j \geq 0} = \left(\sum_{k=0}^{j} [u_{j-k}, v_k] \right)_{j \geq 0}.$$

Now note that the last term in the above chain of equalities is indeed the coordinate expression of $[u, v]$, since (as $u_{j-k} \in \mathcal{T}_{j-k}$, $v_k \in \mathcal{T}_k$) one has $[u_{j-k}, v_k] \in [\mathcal{T}_{j-k}, \mathcal{T}_k] \subseteq \mathcal{T}_{j-k} \otimes \mathcal{T}_k = \mathcal{T}_j$. $\qquad \square$

In Chap. 3, we will have occasion to apply the following result.

Proposition 2.82 ($\widehat{\mathscr{T}} \otimes \widehat{\mathscr{T}}$ as a Subalgebra of $\widehat{\mathscr{T} \otimes \mathscr{T}}$). *Let V be a vector space. With the notation of this section, the tensor product $\widehat{\mathscr{T}}(V) \otimes \widehat{\mathscr{T}}(V)$ can be identified with a subalgebra of $\widehat{\mathscr{T} \otimes \mathscr{T}}(V)$.*

Indeed, we can identify the element $(u_i)_i \otimes (v_j)_j$ of $\widehat{\mathscr{T}} \otimes \widehat{\mathscr{T}}$ (where $u_i, v_i \in \mathscr{T}_i(V)$ for every $i \geq 0$) with the element $(u_i \otimes v_j)_{i,j}$ of $\widehat{\mathscr{T} \otimes \mathscr{T}}$, this identification being a UAA morphism. Here, $\widehat{\mathscr{T}}(V) \otimes \widehat{\mathscr{T}}(V)$ is equipped with the UA algebra structure obtained, as in Proposition 2.41, from the UAA structure of $(\widehat{\mathscr{T}}(V), \cdot)$. Hereafter, when writing $\widehat{\mathscr{T}}(V) \otimes \widehat{\mathscr{T}}(V) \hookrightarrow \widehat{\mathscr{T} \otimes \mathscr{T}}(V)$, we shall understand the previously mentioned immersion:

$$\widehat{\mathscr{T}}(V) \otimes \widehat{\mathscr{T}}(V) \ni (u_i)_i \otimes (v_j)_j \mapsto (u_i \otimes v_j)_{i,j} \in \widehat{\mathscr{T} \otimes \mathscr{T}}(V). \tag{2.95}$$

Proof. See page 433 in Chap. 7.   □

Remark 2.83. Let $\{\alpha_k\}_k$ and $\{\beta_k\}_k$ be two sequences of elements in $\widehat{\mathscr{T}}(V)$ such that $\lim_{k \to \infty} \alpha_k = \alpha$ and $\lim_{k \to \infty} \beta_k = \beta$ in $\widehat{\mathscr{T}}(V)$. Then

$$\lim_{k \to \infty} \alpha_k \otimes \beta_k = \alpha \otimes \beta \quad in \ \widehat{\mathscr{T} \otimes \mathscr{T}}(V),$$

where we consider $\alpha \otimes \beta$ and any $\alpha_k \otimes \beta_k$ as elements of $\widehat{\mathscr{T} \otimes \mathscr{T}}(V)$ (according to (2.95) in Proposition 2.82). For the proof, see page 434 in Chap. 7.

Remark 2.84. Following the notation in (2.91) and (2.92) and by using the immersion $\widehat{\mathscr{T}}(V) \otimes \widehat{\mathscr{T}}(V) \hookrightarrow \widehat{\mathscr{T} \otimes \mathscr{T}}(V)$ in (2.95), it is not difficult to prove that

$$(a \otimes b) \bullet (\alpha \otimes \beta) = (a \cdot b) \otimes (b \cdot \beta), \qquad for \ every \ a, b, \alpha, \beta \in \widehat{\mathscr{T}}(V), \tag{2.96}$$

where this is meant as an equality of elements of $\widehat{\mathscr{T} \otimes \mathscr{T}}(V)$.

2.3.4 Some More Notation on Formal Power Series

Let $n \in \mathbb{N}$ and let $S = \{x_1, \ldots, x_n\}$ be a set of cardinality n. The free vector space $\mathbb{K}\langle S \rangle$ will be denoted by

$$\mathbb{K}\langle x_1, \ldots, x_n \rangle.$$

The algebras $\mathscr{T}(\mathbb{K}\langle x_1, \ldots, x_n \rangle)$ and $\widehat{\mathscr{T}}(\mathbb{K}\langle x_1, \ldots, x_n \rangle)$ can be thought of as, respectively, the *algebra of polynomials in the n non-commuting indeterminates* $x_1, \ldots, x_n$ and the *algebra of formal power series in the n non-commuting indeterminates* $x_1, \ldots, x_n$.

Recall that $\mathscr{T}(\mathbb{K}\langle x_1,\ldots,x_n\rangle)$ is isomorphic to $\mathrm{Libas}(\{x_1,\ldots,x_n\})$, the free UAA over $\{x_1,\ldots,x_n\}$ (see Theorem 2.40). Analogously, the Lie algebra $\mathcal{L}(\mathbb{K}\langle x_1,\ldots,x_n\rangle)$ can be thought of as the *Lie algebra of the Lie-polynomials in the n non-commuting indeterminates $x_1,\ldots,x_n$.* Recall that $\mathcal{L}(\mathbb{K}\langle x_1,\ldots,x_n\rangle)$ is a free Lie algebra related to the set $\{x_1,\ldots,x_n\}$, being isomorphic to $\mathrm{Lie}(\{x_1,\ldots,x_n\})$ (see Theorem 2.56).

When $n=1$, it is customary to write

$$\mathbb{K}[x] := \mathscr{T}(\mathbb{K}\langle x\rangle) \quad \text{and} \quad \mathbb{K}[[x]] := \widehat{\mathscr{T}}(\mathbb{K}\langle x\rangle).$$

(Note that $\mathscr{T}(\mathbb{K}\langle x\rangle)$ and $\widehat{\mathscr{T}}(\mathbb{K}\langle x\rangle)$ are commutative algebras!) Some very important features of $\mathbb{K}[[x]]$ will be stated in Sect. 4.3 of Chap. 4 (and proved in Chap. 9).

Finally, when writing expressions like

$$\mathscr{T}(\mathbb{K}\langle x,y\rangle), \quad \widehat{\mathscr{T}}(\mathbb{K}\langle x,y\rangle), \quad \mathscr{T}(\mathbb{K}\langle x,y,z\rangle), \quad \widehat{\mathscr{T}}(\mathbb{K}\langle x,y,z\rangle),$$

we shall always mean (possibly without the need to say it explicitly) that the sets $\{x,y\}$ and $\{x,y,z\}$ have cardinality, respectively, two and three.

For the sake of future reference, we explicitly state the contents of Theorem 2.40 and 2.56 in the cases of two $\{x,y\}$ and three $\{x,y,z\}$ non-commuting indeterminates. We also seize the opportunity to introduce a new notation $\Phi_{a,b}$. In what follows, by an abuse of notation, we identify the canonical injection $\varphi : X \to \mathscr{T}(\mathbb{K}\langle X\rangle)$ defined by

$$X \xrightarrow{\;x\;} \mathbb{K}\langle X\rangle \overset{\iota}{\hookrightarrow} \mathscr{T}(\mathbb{K}\langle X\rangle),$$

with the set inclusion $X \hookrightarrow \mathscr{T}(\mathbb{K}\langle X\rangle)$.

Theorem 2.85. *The following universal properties are satisfied.*

(1a). *For every UA algebra A and every pair of elements $a,b \in A$, there exists a unique UAA morphism $\Phi_{a,b} : \mathscr{T}(\mathbb{K}\langle x,y\rangle) \to A$ such that*

$$\Phi_{a,b}(x) = a \quad \text{and} \quad \Phi_{a,b}(y) = b. \tag{2.97}$$

(1b). *For every Lie algebra $\mathfrak{g}$ and every pair of elements $a,b \in \mathfrak{g}$, there exists a unique LA morphism $\Phi_{a,b} : \mathcal{L}(\mathbb{K}\langle x,y\rangle) \to \mathfrak{g}$ such that (2.97) holds.*

(2a). *For every UA algebra A and every triple of elements $a,b,c \in A$, there exists a unique UAA morphism $\Phi_{a,b,c} : \mathscr{T}(\mathbb{K}\langle x,y,z\rangle) \to A$ such that*

$$\Phi_{a,b,c}(x) = a, \quad \Phi_{a,b,c}(y) = b, \quad \text{and} \quad \Phi_{a,b,c}(z) = c. \tag{2.98}$$

(2b). *For every Lie algebra $\mathfrak{g}$ and every triple of elements $a,b,c \in \mathfrak{g}$, there exists a unique LA morphism $\Phi_{a,b,c} : \mathcal{L}(\mathbb{K}\langle x,y,z\rangle) \to \mathfrak{g}$ such that (2.98) holds.*

2.4   The Universal Enveloping Algebra

The aim of this section is to introduce the so-called universal enveloping algebra $\mathscr{U}(\mathfrak{g})$ of a Lie algebra $\mathfrak{g}$ and to collect some useful related results. In particular, we will present the remarkable Poincaré-Birkhoff-Witt Theorem.

Throughout this section, $\mathfrak{g}$ will denote a fixed Lie algebra and its Lie bracket is denoted by $[\cdot,\cdot]_{\mathfrak{g}}$ (or simply by $[\cdot,\cdot]$). As usual, $(\mathscr{T}(\mathfrak{g}),\cdot)$ is the tensor algebra of (the vector space of) $\mathfrak{g}$. We denote by $\mathscr{J}(\mathfrak{g})$ (sometimes $\mathscr{J}$ for short) the two-sided ideal in $\mathscr{T}(\mathfrak{g})$ generated by the set

$$\{x \otimes y - y \otimes x - [x,y]_{\mathfrak{g}} \ : \ x, y \in \mathfrak{g}\}.$$

More explicitly, we have

$$\mathscr{J}(\mathfrak{g}) = \mathrm{span}\Big\{t \cdot \big(x \otimes y - y \otimes x - [x,y]_{\mathfrak{g}}\big) \cdot t' \ \Big| \ x, y \in \mathfrak{g}, \ t, t' \in \mathscr{T}(\mathfrak{g})\Big\}. \tag{2.99}$$

Remark 2.86. We remark that *the ideal $\mathscr{J}(\mathfrak{g})$ is not homogeneous (in the natural grading of $\mathscr{T}(\mathfrak{g})$).* Indeed, in the sequence-style notation $(t_k)_{k\geq 0}$ for the elements of $\mathscr{T}(\mathfrak{g}) = \bigoplus_{k\geq 0}\mathscr{T}_k(\mathfrak{g})$, the element $x \otimes y - y \otimes x - [x,y]_{\mathfrak{g}}$ is rewritten as

$$\big(0, -[x,y]_{\mathfrak{g}}, x \otimes y - y \otimes x, 0, 0\cdots\big) \in \mathscr{T}_1 \oplus \mathscr{T}_2. \tag{2.100}$$

Definition 2.87 (Universal Enveloping Algebra). With all the above notation, we consider the quotient space

$$\mathscr{U}(\mathfrak{g}) := \mathscr{T}(\mathfrak{g})/\mathscr{J}(\mathfrak{g})$$

and we call it *the universal enveloping algebra of* $\mathfrak{g}$. We denote by

$$\pi : \mathscr{T}(\mathfrak{g}) \to \mathscr{U}(\mathfrak{g}), \qquad \pi(t) := [t]_{\mathscr{J}(\mathfrak{g})}, \ t \in \mathscr{T}(\mathfrak{g}) \tag{2.101}$$

the associated projection. The natural operation[17] on $\mathscr{U}(\mathfrak{g})$

$$\mathscr{U}(\mathfrak{g}) \times \mathscr{U}(\mathfrak{g}) \ni (\pi(t), \pi(t')) \mapsto \pi(t \cdot t'), \qquad (t, t' \in \mathscr{T}(\mathfrak{g})),$$

which equips $\mathscr{U}(\mathfrak{g})$ with the structure of a UA algebra (see Proposition 2.12 on page 58), will be simply denoted by juxtaposition.

The natural injection $\mathfrak{g} \hookrightarrow \mathscr{T}(\mathfrak{g})$ induces a linear map

$$\mu : \mathfrak{g} \to \mathscr{U}(\mathfrak{g}), \qquad \mu(x) := [x]_{\mathfrak{g}} \quad (x \in \mathfrak{g}), \tag{2.102}$$

---

[17]This operation is well-posed because $\mathscr{J}(\mathfrak{g})$ is an ideal of $\mathscr{T}(\mathfrak{g})$.

that is, $\mu = \pi|_{\mathfrak{g}}$. The following important proposition proves that the Lie bracket of $\mathfrak{g}$ is turned by μ into the commutator of $\mathscr{U}(\mathfrak{g})$. As soon as we will know that μ is injective (a corollary of the Poincaré-Birkhoff-Witt Theorem), this will prove that (up to an identification) *every Lie bracket is a commutator* (in the very meaning used in this Book).

As usual, if $\mathscr{U}(\mathfrak{g})$ is involved as a Lie algebra, it is understood to be equipped with the associated commutator, which we denote by $[\cdot,\cdot]_{\mathscr{U}}$.

Remark 2.88. By its very definition, *the map* $\pi : \mathscr{T}(\mathfrak{g}) \to \mathscr{U}(\mathfrak{g})$ *is a UAA morphism, whence it is a Lie algebra morphism,* when $\mathscr{T}(\mathfrak{g})$ and $\mathscr{U}(\mathfrak{g})$ are equipped with their appropriate commutators (see Remark 2.17). Note that this does *not* prove (yet) that μ is a Lie algebra morphism, since $\mathfrak{g}$ (equipped with its *intrinsic* Lie bracket) is not a Lie subalgebra of $\mathscr{T}(\mathfrak{g})$ (equipped with its commutator).

Remark 2.89. The set $\{\pi(1)\} \cup \mu(\mathfrak{g})$ generates $\mathscr{U}(\mathfrak{g})$ as an algebra. (This follows from the fact that $\{1\} \cup \mathfrak{g}$ generates $\mathscr{T}(\mathfrak{g})$ as an algebra, together with the fact that π is a UAA morphism.)

Proposition 2.90. *With the above notation, the map* μ *in (2.102) is a Lie algebra morphism, i.e.,*

$$\mu([x,y]_{\mathfrak{g}}) = \mu(x)\mu(y) - \mu(y)\mu(x), \qquad \text{for every } x, y \in \mathfrak{g}. \tag{2.103}$$

In particular, $\mu(\mathfrak{g})$ *is a Lie subalgebra of* $\mathscr{U}(\mathfrak{g})$*, equipped with the associated commutator-algebra structure.*

Note that (2.103) can be rewritten as

$$\mu([x,y]_{\mathfrak{g}}) = [\mu(x), \mu(y)]_{\mathscr{U}}, \quad \text{for every } x, y \in \mathfrak{g}. \tag{2.104}$$

Proof. First we remark that (2.103) is equivalent to $\pi([x,y]_{\mathfrak{g}}) = \pi(x \otimes y - y \otimes x)$, which in its turn is equivalent to $x \otimes y - y \otimes x - [x,y]_{\mathfrak{g}} \in \mathscr{J}(\mathfrak{g})$. This is true (for any $x, y \in \mathfrak{g}$) by the definition of $\mathscr{J}(\mathfrak{g})$. $\qquad\square$

Remark 2.91. Via the map π, the grading $\mathscr{T}(\mathfrak{g}) = \bigoplus_{k \geq 0} \mathscr{T}_k(\mathfrak{g})$ turns into $\mathscr{U}(\mathfrak{g}) = \biguplus_{k \geq 0} \pi(\mathscr{T}_k(\mathfrak{g}))$ (in the sense of sum of vector subspaces) but the family of vector spaces $\{\pi(\mathscr{T}_k(\mathfrak{g}))\}_{k \geq 0}$ does not furnish a *direct sum* decomposition of $\mathscr{U}(\mathfrak{g})$. Indeed, if $x, y \in \mathfrak{g}$ we have

$$\underbrace{\pi([x,y]_{\mathfrak{g}})}_{\in \pi(\mathscr{T}_1(\mathfrak{g}))} = \underbrace{\pi(x \otimes y - y \otimes x)}_{\in \pi(\mathscr{T}_2(\mathfrak{g}))}$$

(and we shall see explicit examples where this does not vanish). This is obviously due to the non-homogeneity of $\mathscr{J}(\mathfrak{g})$.

As expected, $\mathcal{U}(\mathfrak{g})$ has a universal property:

Theorem 2.92 (Universal Property of the Universal Enveloping Algebra).
Let $\mathfrak{g}$ be a Lie algebra and let $\mathcal{U}(\mathfrak{g})$ be its universal enveloping algebra.

(i) *For every UA algebra $(A, *)$ and for every Lie algebra morphism $f : \mathfrak{g} \to A$, there exists a unique UAA morphism $f^{\mu} : \mathcal{U}(\mathfrak{g}) \to A$ such that*

$$f^{\mu}(\mu(x)) = f(x) \quad \textit{for every } \ddot{x} \in \mathfrak{g}, \tag{2.105}$$

thus making the following a commutative diagram:

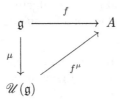

(ii) *Vice versa, suppose U, φ are respectively a UA algebra and a Lie algebra morphism $\varphi : \mathfrak{g} \to U$ with the following property: For every UA algebra $(A, *)$ and for every Lie algebra morphism $f : \mathfrak{g} \to A$, there exists a unique UAA morphism $f^{\varphi} : U \to A$ such that*

$$f^{\varphi}(\varphi(x)) = f(x) \quad \textit{for every } x \in \mathfrak{g}, \tag{2.106}$$

thus making the following a commutative diagram:

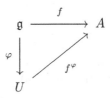

Then U is canonically isomorphic to $\mathcal{U}(\mathfrak{g})$, the isomorphism being $\varphi^{\mu} : \mathcal{U}(\mathfrak{g}) \to U$ and its inverse being $\mu^{\varphi} : U \to \mathcal{U}(\mathfrak{g})$. Moreover, $\varphi = \varphi^{\mu} \circ \mu$. Furthermore (if 1_U denotes the unit of U) the set $\{1_U\} \cup \varphi(\mathfrak{g})$ is a set of algebra generators for U and $U \simeq \mathcal{U}(\varphi(\mathfrak{g}))$, canonically as UA algebras.

Proof. Explicitly, the map f^{μ} is defined by

$$f^{\mu} : \mathcal{U}(\mathfrak{g}) \to A, \qquad \pi(t) \mapsto \overline{f}(t) \quad (t \in \mathscr{T}(\mathfrak{g})), \tag{2.107}$$

where $\overline{f} : \mathscr{T}(\mathfrak{g}) \to A$ is the unique UAA morphism extending $f : \mathfrak{g} \to A$.
For the rest of the proof, see page 435 in Chap. 7.   □

We are in a position to prove a useful result on the enveloping algebra of the *free* Lie algebra generated by a vector space.

Proposition 2.93. *Let X be any set. Let $V := \mathbb{K}\langle X \rangle$ denote the free vector space over X. Let $\mathcal{L}(V)$ be the free Lie algebra generated by the vector space V (i.e., $\mathcal{L}(V)$ is the smallest Lie subalgebra of $\mathcal{T}(V)$ containing V).*

Then $\mathcal{U}(\mathcal{L}(V))$ and $\mathcal{T}(V)$ are isomorphic (as unital associative algebras).

Proof. More explicitly, we can take as isomorphism $j : \mathcal{U}(\mathcal{L}(V)) \to \mathcal{T}(V)$ the only UAA morphism such that

$$j(\pi(t)) = \iota(t), \quad \text{for every } t \in \mathcal{L}(V). \tag{2.108}$$

See page 437 in Chap. 7 for the proof. We remark that in that proof we will not use explicitly the fact that $\mathcal{L}(\mathbb{K}\langle X \rangle)$ is a free Lie algebra related to X (proved in Theorem 2.56). $\qquad\square$

Here we have the fundamental result on the universal enveloping algebra.

Theorem 2.94 (Poincaré-Birkhoff-Witt). *Let $\mathfrak{g}$ be a Lie algebra and let $\mathcal{U}(\mathfrak{g})$ be its universal enveloping algebra. Let 1 denote the unit of $\mathcal{U}(\mathfrak{g})$ and let μ be the map in (2.102). Suppose $\mathfrak{g}$ is endowed with an indexed (linear) basis $\{x_i\}_{i \in \mathcal{I}}$, where $\mathcal{I}$ is totally ordered by the relation $\preccurlyeq$. Set $X_i := \mu(x_i)$, for $i \in \mathcal{I}$.*

Then the following elements form a linear basis of $\mathcal{U}(\mathfrak{g})$:

$$1, \quad X_{i_1} \cdots X_{i_n}, \quad \text{where} \quad n \in \mathbb{N}, \ i_1, \dots, i_n \in \mathcal{I}, \ i_1 \preccurlyeq \dots \preccurlyeq i_n. \tag{2.109}$$

Proof. The (laborious) proof of this key result is given in Chap. 7 (starting from page 438). For other proofs, the Reader is referred for example to [25, 85, 95, 99, 159, 171]. $\qquad\square$

In the sequel, the Poincaré-Birkhoff-Witt Theorem will be referred to as PBW for short. Apparently until 1956, the theorem was only referred to as the "Birkhoff-Witt Theorem": see Schmid [153], Grivel [74], Ton-That, Tran [168] for a historical overview on this topic and for a description of (the long forgotten) contribution of Poincaré to this theorem, dated back to 1900.

Corollary 2.95. *Let $\mathfrak{g}$ be a Lie algebra and let $\mathcal{U}(\mathfrak{g})$ be its universal enveloping algebra. Then the map μ in (2.102) is injective, so that $\mu : \mathfrak{g} \to \mu(\mathfrak{g})$ is a Lie algebra isomorphism.*

As a consequence, every Lie algebra can be identified with a Lie subalgebra of a UA algebra (endowed with the commutator), in the following way:

$$(\mathfrak{g}, [\cdot, \cdot]_{\mathfrak{g}}) \equiv (\mu(\mathfrak{g}), [\cdot, \cdot]_{\mathcal{U}}) \hookrightarrow \mathcal{U}(\mathfrak{g}) \quad \begin{pmatrix} \text{both a UA algebra} \\ \text{and a commutator-algebra} \end{pmatrix}.$$

Proof. Let $x \in \mathfrak{g}$ be such that $\mu(x) = 0$. With the notation of Theorem 2.94, we have $x = \sum_{i \in \mathcal{J}'} \lambda_i x_i$, where $\mathcal{J}' \subseteq \mathcal{J}$ is finite and the λ_i are scalars. Thus $0 = \mu(\sum_{i \in \mathcal{J}'} \lambda_i x_i) = \sum_{i \in \mathcal{J}'} \lambda_i X_i$, which is possible iff $\lambda_i = 0$ for every $i \in \mathcal{J}'$, since the vectors X_i appear in the basis (2.109) of $\mathscr{U}(\mathfrak{g})$, i.e., $x = 0$.

Hence, the map $\mu : \mathfrak{g} \to \mu(\mathfrak{g})$ is a bijection and it is also a Lie algebra morphism, in view of Proposition 2.90, when $\mu(\mathfrak{g})$ is equipped with the commutator from the UA algebra $\mathscr{U}(\mathfrak{g})$.   □

By means of the PBW Theorem, we are able to give a short proof of the existence of free Lie algebras generated by a vector space.

Proof (of Theorem 2.49, page 86). Let V be a vector space and let $f : V \to \mathfrak{g}$ be a linear map, $\mathfrak{g}$ being a Lie algebra. We need to prove that there exists a unique LA morphism $\bar{f} : \mathcal{L}(V) \to \mathfrak{g}$ prolonging f. The uniqueness is trivial, once existence is proved. To this end, let us consider the LA morphism $\mu : \mathfrak{g} \to \mathscr{U}(\mathfrak{g})$ in (2.102). Since the map $\mu \circ f : V \to \mathscr{U}(\mathfrak{g})$ is linear and $\mathscr{U}(\mathfrak{g})$ is a UA algebra, by Theorem 2.38 there exists a UAA morphism $\overline{\mu \circ f} : \mathscr{T}(V) \to \mathscr{U}(\mathfrak{g})$ prolonging $\mu \circ f$. Now we restrict $\overline{\mu \circ f}$ both in domain and codomain, by considering the map

$$\widehat{f} : \mathcal{L}(V) \to \mu(\mathfrak{g}), \qquad \widehat{f}(t) := \overline{\mu \circ f}(t) \quad (t \in \mathcal{L}(V)).$$

To prove that $\widehat{f}$ is well posed, we need to show that

$$\overline{\mu \circ f}(t) \in \mu(\mathfrak{g}) \quad \text{for every } t \in \mathcal{L}(V). \tag{2.110}$$

Since $\mathcal{L}(V)$ is Lie-generated by V (see Proposition 2.47) it suffices to prove (2.110) when $t = [v_1 \cdots [v_{n-1}, v_n] \cdots]$, for any $n \in \mathbb{N}$ and $v_1, \ldots, v_n \in V$. To this end (denoting by $[\cdot, \cdot]_\mathscr{U}$ the commutator of $\mathscr{U}(\mathfrak{g})$), we argue as follows:

$$\overline{\mu \circ f}(t) = [\overline{\mu \circ f}(v_1) \cdots [\overline{\mu \circ f}(v_{n-1}), \overline{\mu \circ f}(v_n)]_\mathscr{U} \cdots]_\mathscr{U}$$
$$= [\mu(f(v_1)) \cdots [\mu(f(v_{n-1})), \mu(f(v_n))]_\mathscr{U} \cdots]_\mathscr{U}$$

In the first equality we applied the fact that $\overline{\mu \circ f}$ is a UAA morphism and in the second equality the fact that $\overline{\mu \circ f}$ coincides with $\mu \circ f$ on V. Now the above right-hand side is an element of $\mu(\mathfrak{g})$ since $f(v_i) \in \mathfrak{g}$ for every $i = 1, \ldots, n$ and $\mu(\mathfrak{g})$ is a Lie subalgebra of $\mathscr{U}(\mathfrak{g})$ (see Proposition 2.90). This proves (2.110). We now remark that $\widehat{f}$ is an LA morphism (of the associated commutator-algebras) since it is the restriction of $\overline{\mu \circ f}$, which is an LA morphism (being a UAA morphism).

Since $\mu : \mathfrak{g} \to \mu(\mathfrak{g})$ is a Lie algebra isomorphism (thanks to Corollary 2.95), the map

$$\mu^{-1} \circ \widehat{f} : \mathcal{L}(V) \xrightarrow{\widehat{f}} \mu(\mathfrak{g}) \xrightarrow{\mu^{-1}} \mathfrak{g}$$

is an LA morphism, since both μ^{-1} and $\widehat{f}$ are. We set $\overline{f} := \mu^{-1} \circ \widehat{f}$. It remains to show that $\overline{f}$ prolongs f. This follows immediately from

$$\overline{f}(v) = \mu^{-1}(\overline{\mu \circ f}(v)) = (\mu^{-1} \circ \mu \circ f)(v) = f(v), \quad \forall\, v \in V.$$

This ends the proof.   □

The following diagram describes the maps in the above argument:

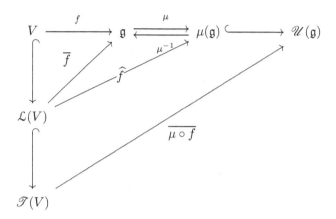

We end the section with an example of how the injective map $\mu : \mathfrak{g} \to \mathscr{U}(\mathfrak{g})$ can be used to perform computations involving the Lie bracket of a Lie algebra (without "explicit knowledge" of the Lie bracket on $\mathfrak{g}$).

Example 2.96. We prove that, for every Lie algebra $\mathfrak{g}$, one has

$$[a, [b, [a, b]]]_{\mathfrak{g}} = -[b, [a, [b, a]]]_{\mathfrak{g}}, \quad \text{for every } a, b \in \mathfrak{g}. \qquad (2.111)$$

Obviously, this computation can be a consequence only of the skew-symmetry and the Jacobi identity, but it may not at first be obvious how to perform the computation.[18]

Let us use, instead, the injection $\mu : \mathfrak{g} \to \mathscr{U}(\mathfrak{g})$. Given arbitrary $a, b \in \mathfrak{g}$, we set $A := \mu(a)$ and $B := \mu(b)$. We begin by showing that

$$[A, [B, [A, B]]]_{\mathscr{U}} = -[B, [A, [B, A]]]_{\mathscr{U}}, \quad \text{in } \mathscr{U}(\mathfrak{g}). \qquad (2.112)$$

---

[18]Indeed, (2.111) follows from the following argument: Set $x := [a, b]$, $y := a$, $z := b$ and write the Jacobi identity $[x, [y, z]] + [y, [z, x]] + [z, [x, y]] = 0$; the first summand is $[[a, b], [a, b]]$ which is null by skew-symmetry. Hence we get $[a, [b, [a, b]]] + [b, [[a, b], a]] = 0$ which leads directly to (2.111), again by skew-symmetry.

Indeed, unraveling the commutators (and dropping the subscript $\mathscr{U}$)

$$[A, [B, [A, B]]] = [A, [B, AB - BA]] = [A, BAB - B^2 A - AB^2 + BAB]$$
$$= ABAB - AB^2 A - A^2 B^2 + ABAB+$$
$$- BABA + B^2 A^2 + AB^2 A - BABA$$
$$= 2 ABAB - 2 BABA + B^2 A^2 - A^2 B^2.$$

Hence, by interchanging A and B we get

$$[B, [A, [B, A]]] = 2 BABA - 2 ABAB + A^2 B^2 - B^2 A^2,$$

which proves (2.112). By exploiting (2.104), we thus get

$$\mu([a, [b, [a, b]]]_{\mathfrak{g}}) \stackrel{(2.104)}{=} [\mu(a), [\mu(b), [\mu(a), \mu(b)]]]_{\mathscr{U}} = [A, [B, [A, B]]]_{\mathscr{U}}$$
$$\stackrel{(2.112)}{=} -[B, [A, [B, A]]]_{\mathscr{U}} = -[\mu(b), [\mu(a), [\mu(b), \mu(a)]]]_{\mathscr{U}}$$
$$\stackrel{(2.104)}{=} -\mu([b, [a, [b, a]]]_{\mathfrak{g}}).$$

This yields the identity

$$\mu([a, [b, [a, b]]]_{\mathfrak{g}}) = \mu(-[b, [a, [b, a]]]_{\mathfrak{g}}).$$

The injectivity of μ now gives the claimed formula in (2.111).   □

Chapter 3
The Main Proof of the CBHD Theorem

THE aim of this chapter is to present the main proof of the Campbell-Baker-Hausdorff-Dynkin Theorem (CBHD for short), the topic of this Book. The proof is split into two very separate parts. On the one hand, we have the Theorem of Campbell-Baker-Hausdorff, stating that (in a context which will be clarified in due course)

$$x \blacklozenge y := \mathrm{Log}(\mathrm{Exp}(x) \cdot \mathrm{Exp}(y))$$

belongs to the closure of the Lie algebra generated by $\{x, y\}$, that is, it is a series of Lie polynomials in x, y. Roughly speaking, this is a "qualitative" result, disclosing an unexpected property of $x \blacklozenge y$. On the other hand, we have the contribution of Dynkin, which exhibits an explicit "quantitative" formula for $\mathrm{Log}(\mathrm{Exp}(x) \cdot \mathrm{Exp}(y))$, *once it is known that this is a series of Lie polynomials.*

Whereas in proving the Dynkin representation of the aforementioned series, we follow quite closely[1] the original ideas by Dynkin in his 1947 paper [54], in proving the Theorem of Campbell-Baker-Hausdorff we do *not* follow the original proofs by any of the mathematicians whose names come with the Theorem.[2]

Instead, the arguments presented in this chapter (precisely, in Sect. 3.2) are inspired by those in Hochschild [85, Chapter X] and in Bourbaki [27, Chapitre II, §6] (both dating back to the late sixties, early seventies). This

---

[1] As a matter of fact, the proof of the so-called Dynkin-Specht-Wever Lemma that we present in this chapter is not the original one given by Dynkin in [54]. Instead, in the derivation of the explicit Dynkin series starting from the Theorem of Campbell-Baker-Hausdorff, we exploit the technique in [54].

[2] A critical exposition of the original proofs of Campbell, Baker, Hausdorff – along with those of other mathematicians mostly coeval with these three authors – can be found in [3]; for a résumé of these topics, see Chap. 1 of this Book.

A. Bonfiglioli and R. Fulci, *Topics in Noncommutative Algebra*, Lecture Notes in Mathematics 2034, DOI 10.1007/978-3-642-22597-0_3,
© Springer-Verlag Berlin Heidelberg 2012

approach can undoubtedly be considered as the most modern one and it has the further advantage of inscribing the CBHD Theorem within wider contexts. For instance, within this proof we shall have occasion to see in action all the algebraic background from Chap. 2 (tensor algebras, formal power series over graded algebras, free Lie algebras, the Poincaré-Birkhoff-Witt Theorem) and to introduce some useful results (well established in literature):

– Friedrichs's characterization of Lie elements (see Sect. 3.2.1).
– The Lemma of Dynkin, Specht, Wever (see Sect. 3.3.1).

A rough summary of the contents of this chapter is finally in order:

1. We introduce the Exp and Log maps related to the completion of any graded algebra. In particular we are mainly concerned with the cases of the formal power series over $\mathscr{T}(V)$ and over $\mathscr{T}(V) \otimes \mathscr{T}(V)$.
2. We introduce the operation $u \blacklozenge v := \mathrm{Log}(\mathrm{Exp}(u) \cdot \mathrm{Exp}(v))$ on $\widehat{\mathscr{T}_+}(V)$.
3. By means of Friedrichs's characterization of Lie elements, we characterize $\mathcal{L}(V)$ as the set of primitive elements of $\mathscr{T}(V)$:

$$\mathcal{L}(V) = \{t \in \mathscr{T}(V) \,\big|\, \delta(t) = t \otimes 1 + 1 \otimes t\},$$

where $\delta : \mathscr{T}(V) \to \mathscr{T}(V) \otimes \mathscr{T}(V)$ is the UAA morphism such that $\delta(v) = v \otimes 1 + 1 \otimes v$, for all $v \in V$. To do this, we use the Poincaré-Birkhoff-Witt Theorem.
4. With this crucial result at hand (plus some machinery established in due course) we are able to prove that $u \blacklozenge v$ belongs to the closure of $\mathcal{L}(V)$ in $\widehat{\mathscr{T}}(V)$. This is the Theorem of Campbell, Baker, Hausdorff.
5. We next consider the following series of Lie polynomials ($u, v \in \widehat{\mathscr{T}_+}(V)$):

$$u \diamond v := \sum_{j=1}^{\infty} \left(\sum_{n=1}^{j} \frac{(-1)^{n+1}}{n} \sum_{\substack{(h_1,k_1),\ldots,(h_n,k_n)\neq(0,0) \\ h_1+k_1+\cdots+h_n+k_n=j}} \right.$$
$$\left. \times \frac{(\mathrm{ad}\,u)^{h_1}(\mathrm{ad}\,v)^{k_1} \cdots (\mathrm{ad}\,u)^{h_n}(\mathrm{ad}\,v)^{k_n-1}(v)}{h_1!\cdots h_n!\,k_1!\cdots k_n!\,(\sum_{i=1}^{n}(h_i+k_i))} \right).$$

6. By means of the Lemma of Dynkin, Specht and Wever, we can construct the projection P of $\mathscr{T}(V)$ onto $\mathcal{L}(V)$ such that

$$P(v_1 \otimes \cdots \otimes v_k) = k^{-1}\,[v_1, \ldots [v_{k-1}, v_k] \ldots],$$

for any $v, v_1, \ldots, v_k \in V$ and any $k \in \mathbb{N}$. This this gives another characterization of the Lie elements in $\mathscr{T}(V)$:

$$\mathcal{L}(V) = \{t \in \mathscr{T}(V) \,\big|\, P(t) = t\}.$$

7. By means of this projection, we are finally able to prove that

$$x \blacklozenge y = x \diamond y,$$

an identity in the topological space of the formal power series related to the algebra $\mathscr{T}(\mathbb{Q}\langle x, y \rangle)$ of the polynomials in two non-commuting indeterminates x, y and coefficients in $\mathbb{Q}$.

8. By the universal property of $\mathscr{T}(\mathbb{Q}\langle x, y \rangle)$, this easily gives the CBHD Theorem, namely

$$u \blacklozenge v = u \diamond v, \quad \text{for every } u, v \in \widehat{\mathscr{T}_+}(V),$$

where V is any vector space over a field of characteristic zero. This identity has the well known explicit and suggestive form:

$$\mathrm{Exp}(u) \cdot \mathrm{Exp}(v) = \mathrm{Exp}\left(\sum_{j=1}^{\infty} \left(\sum_{n=1}^{j} \frac{(-1)^{n+1}}{n} \sum_{\substack{(h_1,k_1),\ldots,(h_n,k_n) \neq (0,0) \\ h_1+k_1+\cdots+h_n+k_n=j}} \right.\right.$$
$$\left.\left. \times \frac{(\mathrm{ad}\, u)^{h_1}(\mathrm{ad}\, v)^{k_1} \cdots (\mathrm{ad}\, u)^{h_n}(\mathrm{ad}\, v)^{k_n-1}(v)}{h_1! \cdots h_n!\, k_1! \cdots k_n!\, (\sum_{i=1}^{n}(h_i+k_i))} \right)\right),$$

valid for all $u, v \in \widehat{\mathscr{T}_+}(V)$ and every vector space V over a field of characteristic zero.

3.1 Exponential and Logarithm

From now on throughout this Book, $\mathbb{K}$ will be a field subject to the following convention.

Convention. $\mathbb{K}$ *will always denote a field of characteristic zero.*

This is justified by the need for a well-posed definition of the soon-to-come exponential and logarithmic series. To define these series, we consider a graded UA algebra A, with grading $\{A_j\}_{j \geq 0}$. We also assume that $A_0 = \mathbb{K}$. [This will not be a restrictive assumption since we shall be soon interested only in the cases when A is $\mathscr{T}(V)$ or $\mathscr{T}(V) \otimes \mathscr{T}(V)$.] As usual, $\widehat{A}$ denotes the topological algebra of the formal power series on A. We also set

$$\Omega_k := \bigoplus_{j \geq k} A_j, \qquad \widehat{\Omega}_k := \prod_{j \geq k} A_j \qquad \widehat{A}_+ := \widehat{\Omega}_1 = \prod_{j \geq 1} A_j. \qquad (3.1)$$

Moreover, $1_A + \widehat{A}_+$ denotes the subset of $\widehat{A}$ consisting of formal power series whose zero-degree term is 1_A, the unit of A. More explicitly:

$$1_A + \widehat{A}_+ := \left\{ (a_n)_n \in \widehat{A} \mid a_0 = 1_A, \ a_n \in A_n \text{ for all } n \in \mathbb{N} \right\}. \qquad (3.2)$$

As usual, if $*$ denotes the multiplication on A, then $\widehat{*}$ denotes the associated operation on $\widehat{A}$ (see (2.82), page 101).

Lemma 3.1. *With the above notation, $1_A + \widehat{A}_+$ is a subgroup (called the* Magnus group*) of the multiplicative group of $\widehat{A}$. For instance, given $a \in 1_A + \widehat{A}_+$, we have*

$$a^{-1} = \sum_{n=0}^{\infty} (1_A - a)^{\widehat{*} \, n}, \qquad (3.3)$$

the series on the right-hand side converging in $\widehat{A}$ to an element of $1_A + \widehat{A}_+$.

Proof. We set for brevity $\Gamma := 1_A + \widehat{A}_+$. We divided the proof into four steps.

  I. Given $a = (a_n)_n$ and $b = (b_n)_n$ in Γ we have

$$a \,\widehat{*}\, b = \big(a_0 * b_0, a_1 * b_0 + a_0 * b_1, \cdots \big) = \big(1_A, a_1 * b_0 + a_0 * b_1, \cdots \big) \in \Gamma.$$

 II. Let $a \in \Gamma$. Then $1_A - a \in \widehat{A}_+ = \widehat{\Omega}_1$ so that[3] $(1_A - a)^{\widehat{*} \, n} \in \widehat{\Omega}_n$ for every $n \in \mathbb{N}$. This gives $\lim_{n \to \infty} (1_A - a)^{\widehat{*} \, n} = 0$ in the usual topology of $\widehat{A}$ and, by well-known properties of this topology,[4] the series $\tilde{a} := \sum_{n=0}^{\infty} (1_A - a)^{\widehat{*} \, n}$ converges in $\widehat{A}$.

III. With the above notation, we have $\tilde{a} \in \Gamma$ for every $a \in \Gamma$. Indeed,

$$\tilde{a} = 1_A + \underbrace{\sum_{n=1}^{\infty} (1_A - a)^{\widehat{*} \, n}}_{\in \widehat{\Omega}_n} \in 1_A + \widehat{\Omega}_1 = \Gamma.$$

IV. We are left to prove that $a \,\widehat{*}\, \tilde{a} = \tilde{a} \,\widehat{*}\, a = 1_A$, that is, $\tilde{a} = a^{-1}$ (the inversion is intended to be seen as applying within the multiplicative subgroup of $\widehat{A}$). Recalling that $(\widehat{A}, \widehat{*})$ is a topological algebra, we have:

---

[3] Recall that $\widehat{\Omega}_i \,\widehat{*}\, \widehat{\Omega}_j \subseteq \widehat{\Omega}_{i+j}$, for every $i, j \geq 0$.
[4] See Remark 2.76-3 on page 102.

$$a \mathbin{\widehat{*}} \tilde{a} = -\big((-a) \mathbin{\widehat{*}} \tilde{a}\big) = -\big((-1_A + 1_A - a) \mathbin{\widehat{*}} \tilde{a}\big) = \tilde{a} - \big((1_A - a) \mathbin{\widehat{*}} \tilde{a}\big)$$

$$= \tilde{a} - \sum_{n=0}^{\infty} (1_A - a) \mathbin{\widehat{*}} (1_A - a)^{\widehat{*}\,n} = \tilde{a} - \sum_{n=0}^{\infty} (1_A - a)^{\widehat{*}\,n+1}$$

$$= \tilde{a} - \Big(\sum_{m=0}^{\infty} (1_A - a)^{\widehat{*}\,m} - 1_A \Big) = \tilde{a} - \tilde{a} + 1_A = 1_A.$$

The computation leading to $\tilde{a} \mathbin{\widehat{*}} a = 1_A$ is completely analogous and the proof is complete. □

Occasionally, when there is no possibility of confusion, if $(A, \circledast)$ is a UA algebra and $n \geq 0$, we shall denote the n-th power $a^{\circledast\,n}$ of $a \in A$ simply by a^n. Moreover, 1 will denote the unit 1_A in any UA algebra A. So, for example, (3.3) can be rewritten in the more concise (von Neumann) form $a^{-1} = \sum_{n=0}^{\infty} (1 - a)^n$.

3.1.1 Exponentials and Logarithms

Definition 3.2 (Exponential and Logarithm). If A is a graded UA algebra and $\widehat{A}_+$ is as in (3.2), we set

$$\mathrm{Exp} : \widehat{A}_+ \longrightarrow 1_A + \widehat{A}_+, \quad \mathrm{Exp}(u) := \sum_{k=0}^{\infty} \frac{1}{k!} u^{\widehat{*}\,k},$$

$$\mathrm{Log} : 1_A + \widehat{A}_+ \longrightarrow \widehat{A}_+, \quad \mathrm{Log}(1_A + u) = \sum_{k=1}^{\infty} \frac{(-1)^{k+1}}{k} u^{\widehat{*}\,k}.$$

The fact that the above defined Exp and Log are well posed maps follows from the following simple facts (which we shall frequently use without mention) together with an application of Remark 2.76-(2,3) on page 102:

$$\left.\begin{array}{l} \big(u \in \widehat{A}_+, \; n \in \mathbb{N}\big) \;\; \Longrightarrow \;\; u^{\widehat{*}\,n} \in \prod_{j \geq n} A_j = \widehat{\Omega}_n, \\[2mm] u \in \widehat{A}_+ \;\; \Longrightarrow \;\; \lim_{n \to \infty} u^{\widehat{*}\,n} = 0, \\[2mm] \big(u \in \widehat{A}_+, \; c_n \in \mathbb{K} \text{ for all } n \in \mathbb{N}\big) \Longrightarrow \sum_{n=1}^{\infty} c_n\, u^{\widehat{*}\,n} \text{ converges in } \widehat{A}_+. \end{array}\right\} \quad (3.4)$$

The notation Exp_* and Log_* will also occur (here $*$ denotes the algebra operation on A), when there is some possibility of misunderstanding. In forthcoming sections, when A is given by some very special graded UA algebra (for example the tensor algebra related to a free vector space over two or three non-commuting indeterminates), we shall also admit other more common notation:

$$\mathbf{e}^u, \quad \mathbf{log}(1+u), \quad \exp(u), \quad \log(1+u), \quad \ldots$$

We now collect some remarkable (and expected) results on the Exp and Log maps. Throughout, $A = \bigoplus_{j \geq 0} A_j$ is a graded UA algebra.

Lemma 3.3. *With the hypothesis and the notation in Definition 3.2, the functions* Exp, Log *are continuous on their corresponding domains.*

Proof. This follows from the fact that the series defining these functions are uniformly convergent series (on the corresponding domains, subsets of the *metric* space $\widehat{A}$) of continuous functions, hence they are continuous (from well-known general results on metric spaces). Indeed:

1. Any polynomial function $\widehat{A} \ni u \mapsto c_0 + c_1 u + \cdots + c_N u^{\widehat{*} N}$ is continuous on $\widehat{A}$ for $(\widehat{A}, \widehat{*})$ is a topological algebra.
2. The series for Exp and Log converge uniformly on $\widehat{A}_+$ and $1 + \widehat{A}_+$ respectively. We prove the former fact, the latter being analogous. We denote by $\widehat{d}$ the metric on $\widehat{A}$ induced by the family $\{\widehat{\Omega}_k\}_k$. For every $N, P \in \mathbb{N}$ we have

$$\sup_{u \in \widehat{A}_+} \widehat{d}\left(\sum_{k=N}^{N+P} \tfrac{1}{k!} u^{\widehat{*} k}, 0 \right)$$

$$= \sup_{u \in \widehat{A}_+} \exp\left(- \max\left\{ n \geq 1 \,\Big|\, \sum_{k=N}^{N+P} \tfrac{1}{k!} u^{\widehat{*} k} \in \widehat{\Omega}_k \right\} \right)$$

(use (3.4) and recall that $\widehat{\Omega}_N \supseteq \cdots \supseteq \widehat{\Omega}_{N+P}$ to estimate the inner max)

$$\leq \sup_{u \in \widehat{A}_+} \exp(-N) = \exp(-N) \xrightarrow[N \to \infty]{} 0, \quad \text{uniformly in } P \geq 1.$$

This proves that the sequence $\left\{ \sum_{k=0}^{N} \tfrac{1}{k!} u^{\widehat{*} k} \right\}_N$ converges uniformly (on $\widehat{A}_+$) to its limit, namely $\mathrm{Exp}(u)$. □

Proposition 3.4. *The functions* Exp *and* Log *introduced in Definition 3.2 are inverse to each other, so that*

$$\mathrm{Exp}(\mathrm{Log}(1+u)) = 1+u, \quad \mathrm{Log}(\mathrm{Exp}(u)) = u, \qquad \text{for every } u \in \widehat{A}_+. \quad (3.5)$$

Proof. Let us set, for brevity,

$$b_0 := 1, \quad b_n := \frac{1}{n!}, \qquad c_n := \frac{(-1)^{n+1}}{n} \quad \forall\, n \in \mathbb{N},$$

so that $\mathrm{Exp}(w) = \sum_{n=0}^{\infty} b_n w^{\widehat{*} n}$ and $\mathrm{Log}(1+w) = \sum_{n=1}^{\infty} c_n w^{\widehat{*} n}$ for every $w \in \widehat{A}_+$. Let $u \in \widehat{A}_+$ be fixed. Then we have:

$$\mathrm{Exp}(\mathrm{Log}(1+u)) = \sum_{n=0}^{\infty} b_n \left(\sum_{k=1}^{\infty} c_k \, w^{\widehat{*}\,k} \right)^{\widehat{*}\,n}$$

(recall that $(\widehat{A}, \widehat{*})$ is a topological algebra)

$$= 1 + \sum_{n=1}^{\infty} b_n \sum_{k_1,\ldots,k_n \geq 1} c_{k_1} \cdots c_{k_n} \, w^{\widehat{*}\,k_1 + \cdots + k_n}$$

(a simple reordering argument)

$$= 1 + \sum_{j=1}^{\infty} \left(\sum_{n=1}^{j} b_n \sum_{\substack{k_1,\ldots,k_n \geq 1 \\ k_1 + \cdots + k_n = j}} c_{k_1} \cdots c_{k_n} \right) w^{\widehat{*}\,j} = 1 + w.$$

Indeed, in the last equality we used the identities

$$b_1 c_1 = 1, \qquad \sum_{n=1}^{j} b_n \sum_{\substack{k_1,\ldots,k_n \geq 1 \\ k_1 + \cdots + k_n = j}} c_{k_1} \cdots c_{k_n} = 0 \quad j \geq 2,$$

proved in (9.26), page 491 of Chap. 9, devoted to the formal power series in one indeterminate. The second equality in (3.5) follows analogously, by means of the dual identities

$$c_1 b_1 = 1, \qquad \sum_{n=1}^{j} c_n \sum_{\substack{k_1,\ldots,k_n \geq 1 \\ k_1 + \cdots + k_n = j}} b_{k_1} \cdots b_{k_n} = 0, \quad j \geq 2,$$

also proved in (9.26). This ends the proof.   □

The following is a trivial version of our CBHD Theorem: *the CBHD Theorem in a commutative setting.*

Proposition 3.5 (Commutative CBHD Theorem). *If $u, v \in \widehat{A}_+$ and $u \,\widehat{*}\, v = v \,\widehat{*}\, u$, then we have the identity*

$$\mathrm{Exp}(u) \,\widehat{*}\, \mathrm{Exp}(v) = \mathrm{Exp}(u + v). \tag{3.6}$$

Proof. The proof is completely analogous to that of Lemma 4.8, page 190, to which the Reader is directly referred.   □

Theorem 3.6. *Let $(A, \circledast)$ and $(B, \odot)$ be two graded UA algebras and let $\varphi : \widehat{A} \to \widehat{B}$ be a continuous UAA morphism, of the associated algebras of formal power series $(\widehat{A}, \widehat{\circledast})$, $(\widehat{B}, \widehat{\odot})$, with the additional property*

$$\varphi(\widehat{A}_+) \subseteq \widehat{B}_+. \tag{3.7}$$

Then we have

$$\varphi \circ \mathrm{Exp}_\circledast = \mathrm{Exp}_\circledcirc \circ \varphi \qquad \text{on } \widehat{A}_+,$$

$$\varphi \circ \mathrm{Log}_\circledast = \mathrm{Log}_\circledcirc \circ \varphi \qquad \text{on } 1_A + \widehat{A}_+. \tag{3.8}$$

Proof. The hypothesis $\varphi(\widehat{A}_+) \subseteq \widehat{B}_+$ ensures that $\varphi(1_A + \widehat{A}_+) \subseteq 1_B + \widehat{B}_+$, so that the identities in (3.8) are well posed. We will only prove the first identity in (3.8), as the second may be done in an analogous fashion. The following argument (explained below) completes the proof: For any $u \in \widehat{A}_+$, we have

$$(\varphi \circ \mathrm{Exp}_\circledast)(u) \stackrel{(1)}{=} \sum_{n=0}^{\infty} \varphi\Big(\frac{1}{n!}\, u^{\circledast\, n}\Big) \stackrel{(2)}{=} \sum_{n=0}^{\infty} \frac{1}{n!}\, (\varphi(u))^{\circledcirc\, n} \stackrel{(3)}{=} (\mathrm{Exp}_\circledcirc \circ \varphi)(u).$$

Here we used the following:

(1): Definition of $\mathrm{Exp}_\circledast$ and continuity of φ.
(2): φ is a UAA morphism (in particular, linear).
(3): Definition of $\mathrm{Exp}_\circledcirc$.   □

Particularly important for our purposes are the cases when the graded algebra A is, respectively, $(\mathscr{T}(V), \cdot)$ and $(\mathscr{T}(V) \otimes \mathscr{T}(V), \bullet)$. In these cases, instead of the somewhat awe-inspiring notation

$$\mathrm{Exp}_{\widehat{\cdot}} \qquad \mathrm{Log}_{\widehat{\cdot}}$$

we shall use (admittedly with some abuse) the notation

$$\mathrm{Exp}_\otimes \qquad \mathrm{Log}_\otimes$$

for the exponential/logarithmic maps related to $\mathscr{T}(V)$. Analogously,

$$\mathrm{Exp}_{\widehat{\bullet}} \qquad \mathrm{Log}_{\widehat{\bullet}}$$

will be replaced by

$$\mathrm{Exp}_\bullet \qquad \mathrm{Log}_\bullet,$$

denoting the exponential/logarithmic maps related to $\mathscr{T}(V) \otimes \mathscr{T}(V)$. More explicitly, we have

$$\mathrm{Exp}_\otimes : \widehat{\mathscr{T}_+} \longrightarrow 1 + \widehat{\mathscr{T}_+} \qquad\qquad \mathrm{Log}_\otimes : 1 + \widehat{\mathscr{T}_+} \longrightarrow \widehat{\mathscr{T}_+}$$

$$u \mapsto \sum_{k=0}^{\infty} \frac{1}{k!}\, u^{\widehat{\cdot}\, k} \qquad\qquad 1 + w \mapsto \sum_{k=1}^{\infty} \frac{(-1)^{k+1}}{k}\, w^{\widehat{\cdot}\, k}.$$

Analogously,

$$\mathrm{Exp}_\bullet : \widehat{\mathscr{T}{\otimes}\mathscr{T}}_+ \longrightarrow 1 + \widehat{\mathscr{T}{\otimes}\mathscr{T}}_+ \qquad \mathrm{Log}_\bullet : 1 + \widehat{\mathscr{T}{\otimes}\mathscr{T}}_+ \longrightarrow \widehat{\mathscr{T}{\otimes}\mathscr{T}}_+$$

$$u \mapsto \sum_{k=0}^\infty \frac{1}{k!} u^{\hat{\bullet} k} \qquad\qquad 1 + w \mapsto \sum_{k=1}^\infty \frac{(-1)^{k+1}}{k} w^{\hat{\bullet} k}.$$

Here we have a lemma concerning the relationship between these maps in the cases of $\mathscr{T}$ and $\mathscr{T} \otimes \mathscr{T}$. Henceforth, V is a fixed vector space and 1 will denote the identity of $\mathscr{T}(V)$. The Reader is also invited to review the identification of $\widehat{\mathscr{T}} \otimes \widehat{\mathscr{T}}$ as a subalgebra of $\widehat{\mathscr{T}{\otimes}\mathscr{T}}$ in Proposition 2.82 on page 106, performed by the natural map

$$\widehat{\mathscr{T}}(V) \otimes \widehat{\mathscr{T}}(V) \ni (u_i)_i \otimes (v_j)_j \mapsto (u_i \otimes v_j)_{i,j} \in \widehat{\mathscr{T}{\otimes}\mathscr{T}}(V). \qquad (3.9)$$

We shall always tacitly assume this identification to be made. Note that this gives[5]

$$(a \otimes b) \,\hat{\bullet}\, (\alpha \otimes \beta) = (a \,\hat{\cdot}\, b) \otimes (b \,\hat{\cdot}\, \beta), \qquad \text{for every } a, b, \alpha, \beta \in \widehat{\mathscr{T}}(V). \ (3.10)$$

Lemma 3.7. *With the above notation, we have*

$$\mathrm{Exp}_\bullet(z \otimes 1) = \mathrm{Exp}_\otimes z \otimes 1, \quad \mathrm{Exp}_\bullet(1 \otimes z) = 1 \otimes \mathrm{Exp}_\otimes z, \qquad (3.11\mathrm{a})$$

$$\mathrm{Exp}_\bullet(z \otimes 1 + 1 \otimes z) = \mathrm{Exp}_\otimes z \otimes \mathrm{Exp}_\otimes z, \qquad (3.11\mathrm{b})$$

for every $z \in \widehat{\mathscr{T}}_+(V)$. Dually, we have

$$\mathrm{Log}_\bullet(w \otimes 1) = \mathrm{Log}_\otimes w \otimes 1, \quad \mathrm{Log}_\bullet(1 \otimes w) = 1 \otimes \mathrm{Log}_\otimes w, \qquad (3.12\mathrm{a})$$

$$\mathrm{Log}_\bullet(w \otimes w) = \mathrm{Log}_\otimes w \otimes 1 + 1 \otimes \mathrm{Log}_\otimes w, \qquad (3.12\mathrm{b})$$

for every $w \in 1 + \widehat{\mathscr{T}}_+(V)$.

Proof. It suffices to prove (3.11a)–(3.11b), for (3.12a)–(3.12b) follow from these, together with Proposition 3.4.

(3.11a): We prove the first identity in (3.11a), the proof of the other one being analogous. First note that

$$z \otimes 1, \ 1 \otimes z \in \widehat{\mathscr{T}{\otimes}\mathscr{T}}_+, \qquad \text{for every } z \in \widehat{\mathscr{T}}_+.$$

---

[5] As we proved in (2.96), page 106, with a concise notation dropping the hat $\hat{\ }$.

Indeed, one has $z = (z_n)_n$ with $z_0 - 0$ and $z_n \in \mathscr{T}_n$ for every $n \in \mathbb{N}$, so that (setting $1 = (\delta_{0,n})_n$ with $\delta_{0,0} = 1$ and $\delta_{0,n} = 0$ for every $n \geq 1$)

$$z \otimes 1 \overset{(3.9)}{\equiv} (z_i \otimes \delta_{0,j})_{i,j}$$

$$= \Big(\underbrace{0}_{\text{entry } (0,0)} , \underbrace{z_1 \otimes 1}_{\text{entry } (1,0)} , \underbrace{0}_{\text{entry } (0,1)} , \underbrace{z_2 \otimes 1}_{\text{entry } (2,0)} , \underbrace{0}_{\text{entry } (1,1)} , \underbrace{0}_{\text{entry } (0,2)} , \cdots \Big).$$

This is clearly an element of $\widehat{\mathscr{T} \otimes \mathscr{T}}_+$. Let now $z \in \widehat{\mathscr{T}_+}(V)$. The following computation applies:

$$\mathrm{Exp}_{\bullet}(z \otimes 1) = \sum_{k=0}^{\infty} \frac{1}{k!} (z \otimes 1)^{\hat{\bullet} \, k} \overset{(3.10)}{=} \sum_{k=0}^{\infty} \frac{1}{k!} z^{\hat{\cdot} \, k} \otimes 1$$

$$= \Big(\sum_{k=0}^{\infty} \frac{1}{k!} z^{\hat{\cdot} \, k} \Big) \otimes 1 = \mathrm{Exp}_{\otimes} z \otimes 1.$$

(3.11b):   Let $x, y \in \widehat{\mathscr{T}_+}(V)$. By the above remarks we have $x \otimes 1, 1 \otimes y \in \widehat{\mathscr{T} \otimes \mathscr{T}}_+$ and the same is true of $x \otimes 1 + 1 \otimes y$. Note that $x \otimes 1$ and $1 \otimes y$ commute w.r.t. $\hat{\bullet}$, for one has, thanks to (3.10),

$$(x \otimes 1) \, \hat{\bullet} \, (1 \otimes y) = x \otimes y = (1 \otimes y) \, \hat{\bullet} \, (x \otimes 1).$$

Hence we are entitled to apply Proposition 3.5 when $A = (\mathscr{T} \otimes \mathscr{T}, \bullet)$ (which is a graded algebra!). We then have

$$\mathrm{Exp}_{\bullet}(x \otimes 1 + 1 \otimes y) = \mathrm{Exp}_{\bullet}(x \otimes 1) \, \hat{\bullet} \, \mathrm{Exp}_{\bullet}(1 \otimes y)$$

$$\overset{(3.11a)}{=} (\mathrm{Exp}_{\otimes} x \otimes 1) \, \hat{\bullet} \, (1 \otimes \mathrm{Exp}_{\otimes} y)$$

$$\overset{(3.10)}{=} (\mathrm{Exp}_{\otimes} x \, \hat{\cdot} \, 1) \otimes (1 \, \hat{\cdot} \, \mathrm{Exp}_{\otimes} y) = \mathrm{Exp}_{\otimes} x \otimes \mathrm{Exp}_{\otimes} y.$$

Finally (3.11b) follows by taking $x = y = z$.   □

Note that the derivation of (3.11b) needs the trivial version of the CBHD Theorem, proved in Proposition 3.5.

3.1.2   The Statement of Our Main CBHD Theorem

From now on, V will denote a fixed vector space over a field $\mathbb{K}$ (of characteristic zero). We denote by $\mathscr{T}(V)$ the tensor algebra of V and by $\widehat{\mathscr{T}}(V)$ the corresponding topological algebra of formal power series. Moreover, $\widehat{\mathscr{T}_+}(V)$

denotes the ideal of $\widehat{\mathscr{T}}(V)$ whose elements have vanishing component of degree 0 component. Finally, $\mathcal{L}(V)$ is the smallest Lie sub-algebra of $\mathscr{T}(V)$ containing V (i.e., $\mathcal{L}(V)$ is the free Lie algebra generated by V) and

$$\overline{\mathcal{L}(V)}$$

is the closure of $\mathcal{L}(V)$ in the topological (and metric) space $\widehat{\mathscr{T}}(V)$. Note that $\overline{\mathcal{L}(V)} \subset \widehat{\mathscr{T}_+}(V)$.

We are ready to state the central result of this Book, the *Campbell, Baker, Hausdorff, Dynkin Theorem*. To this end, a last bit of new notation is required. As usual, in any Lie algebra $\mathfrak{g}$, we set $\operatorname{ad} x(y) := [x, y]_{\mathfrak{g}}$. We introduce the following convenient (but not conventional) notation:

If $u, v \in \mathfrak{g}$, if $h_1, k_1, \ldots, h_n, k_n \in \mathbb{N} \cup \{0\}$ (with $(h_1, k_1, \ldots, h_n, k_n)$ non-identically null), we set

$$\left[u^{h_1} v^{k_1} \cdots u^{h_n} v^{k_n} \right]_{\mathfrak{g}} := (\operatorname{ad} u)^{h_1} \circ (\operatorname{ad} v)^{k_1} \circ \cdots \circ (\operatorname{ad} u)^{h_n} \circ (\operatorname{ad} v)^{k_n - 1}(v) \quad (3.13)$$

(when $k_n = 0$, this has the obvious meaning "$\cdots \circ (\operatorname{ad} u)^{h_n - 1}(u)$" and so on). The expression in (3.13) will be called a *right-nested* bracket of u and v. Indeed, we have

$$\left[u^{h_1} v^{k_1} \cdots u^{h_n} v^{k_n} \right]_{\mathfrak{g}} = [\underbrace{u \cdots [u}_{h_1 \text{ times}} [\underbrace{v \cdots [v}_{k_1 \text{ times}} \cdots [\underbrace{u \cdots [u}_{h_h \text{ times}} [\underbrace{v[\cdots v}_{k_n \text{ times}}]]]]]]]_{\mathfrak{g}}.$$

On the occasion, when it is understood, the subscript "$\mathfrak{g}$" may be omitted (as we did above in the majority of the "]" signs).

Convention. When equipped with their commutator-algebra structure, for both $\mathscr{T}(V)$ and $\widehat{\mathscr{T}}(V)$ we shall use the notation $[\cdot, \cdot]_{\otimes}$ for the associated commutator and also for the right-nested brackets $\left[u^{h_1} v^{k_1} \cdots u^{h_n} v^{k_n} \right]_{\otimes}$.

We are ready to state the main theorem of this Book.

Theorem 3.8 (Campbell, Baker, Hausdorff, Dynkin). *Let V be a vector space over the field $\mathbb{K}$ (of characteristic zero). Let $\widehat{\mathscr{T}}(V) = \prod_{k=0}^{\infty} \mathscr{T}_k(V)$ be the (usual) completion of the tensor algebra $\mathscr{T}(V)$ (i.e., $\widehat{\mathscr{T}}(V)$ is the algebra of the formal power series of the tensor algebra of V).*

For $u \in \widehat{\mathscr{T}_+}(V) = \prod_{k=1}^{\infty} \mathscr{T}_k(V)$, we set $\operatorname{Exp}_{\otimes}(u) = \sum_{k=0}^{\infty} \frac{1}{k!} u^k$.

Then we have the Campbell-Baker-Hausdorff-Dynkin Formula

$$\operatorname{Exp}_{\otimes}(u) \cdot \operatorname{Exp}_{\otimes}(v) = \operatorname{Exp}_{\otimes}(Z(u, v)), \qquad \forall\, u, v \in \widehat{\mathscr{T}_+}(V), \qquad (3.14)$$

where $Z(u, v) - \sum_{j=1}^{\infty} Z_j(u, v)$, and $Z_j(u, v)$ is an element of $\mathrm{Lie}\{u, v\}$, *i.e., the smallest Lie subalgebra of* $\mathscr{T}(V)$ *containing* u, v *(Campbell-Baker-Hausdorff Theorem). Moreover $Z_j(u, v)$ is homogeneous of degree j in u, v jointly, with the "universal" expression (Dynkin's Theorem)*

$$Z_j(u, v)$$

$$= \sum_{n=1}^{j} \frac{(-1)^{n+1}}{n} \sum_{\substack{(h_1,k_1),\dots,(h_n,k_n)\neq(0,0) \\ h_1+k_1+\cdots+h_n+k_n=j}} \frac{\left[u^{h_1} v^{k_1} \cdots u^{h_n} v^{k_n}\right]_{\otimes}}{h_1!\cdots h_n!\,k_1!\cdots k_n!\,(\sum_{i=1}^{n}(h_i+k_i))}.$$

$$(3.15)$$

Here $[u^{h_1} \cdots v^{k_n}]_{\otimes}$ is a right-nested bracket in the Lie algebra associated to $\widehat{\mathscr{T}}(V)$.

[We recall that, to unburden ourselves of heavy notation, the dot $\cdot$ replaces $\widehat{}$ for the operation on $\widehat{\mathscr{T}}(V)$.]

Throughout this Book, the above theorem is denoted the CBHD Theorem for short, or simply by CBHD. Also, in distinguishing the more "qualitative" part of the theorem (that is, the fact that $Z(u, v)$ belongs to the closure of $\mathrm{Lie}\{u, v\}$) from the "quantitative" actual series representation of $Z(u, v)$, the former will also be abbreviated as the CBH Theorem (and the latter will be referred to as the Dynkin series).

3.1.3 The Operation ♦ on $\widehat{\mathscr{T}}_+(V)$

Via the $\mathrm{Exp}/\mathrm{Log}$ maps, we can define an important composition law on $\widehat{\mathscr{T}}_+(V)$, namely:

$$u \blacklozenge v := \mathrm{Log}_{\otimes}\big(\mathrm{Exp}_{\otimes}(u) \cdot \mathrm{Exp}_{\otimes}(v)\big), \quad u, v \in \widehat{\mathscr{T}}_+(V). \qquad (3.16)$$

Note that ♦ is well posed since

$$\mathrm{Exp}_{\otimes}(\widehat{\mathscr{T}}_+) \cdot \mathrm{Exp}_{\otimes}(\widehat{\mathscr{T}}_+) = (1 + \widehat{\mathscr{T}}_+) \cdot (1 + \widehat{\mathscr{T}}_+) = 1 + \widehat{\mathscr{T}}_+.$$

Moreover $u \blacklozenge v \in \widehat{\mathscr{T}}_+$ for every $u, v \in \widehat{\mathscr{T}}_+$, since $\mathrm{Log}_{\otimes} w \in \widehat{\mathscr{T}}_+$ whenever $w \in 1 + \widehat{\mathscr{T}}_+$. As a consequence ♦ *defines a binary operation on* $\widehat{\mathscr{T}}_+(V)$.

Remark 3.9. It is immediately seen from its very definition that *the ♦ operation is associative* (since the operation $\cdot$ is).

The ♦ composition can be explicitly written as:

$$u \blacklozenge v = \sum_{n=1}^{\infty} \frac{(-1)^{n+1}}{n} \sum_{(h_1,k_1),\dots,(h_n,k_n)\neq(0,0)} \frac{u^{h_1} \cdot v^{k_1} \cdots u^{h_n} \cdot v^{k_n}}{h_1!\cdots h_n!\,k_1!\cdots k_n!}. \qquad (3.17)$$

Indeed, by taking into account the explicit definitions of Exp and Log, for every $u, v \in \widehat{\mathscr{T}_+}(V)$, we have

$$u \blacklozenge v = \mathrm{Log}_\otimes \left(\sum_{h=0}^\infty \frac{u^h}{h!} \cdot \sum_{k=0}^\infty \frac{v^k}{k!} \right) = \mathrm{Log}_\otimes \left(1 + \sum_{(h,k) \neq (0,0)} \frac{u^h \cdot v^k}{h! \, k!} \right)$$

$$= \sum_{n=1}^\infty \frac{(-1)^{n+1}}{n} \left(\sum_{(h,k) \neq (0,0)} \frac{u^h \cdot v^k}{h! \, k!} \right)^n \qquad (3.18)$$

$$= \sum_{n=1}^\infty \frac{(-1)^{n+1}}{n} \sum_{(h_1,k_1),\dots,(h_n,k_n) \neq (0,0)} \frac{u^{h_1} \cdot v^{k_1} \cdots u^{h_n} \cdot v^{k_n}}{h_1! \cdots h_n! \, k_1! \cdots k_n!}.$$

The following important facts hold:

Proposition 3.10. *If $\blacklozenge$ is as in (3.16), then $(\widehat{\mathscr{T}_+}(V), \blacklozenge)$ is a group. Moreover,*

$$\mathrm{Exp}_\otimes(u \blacklozenge v) = \mathrm{Exp}_\otimes(u) \cdot \mathrm{Exp}_\otimes(v), \qquad \text{for every } u, v \in \widehat{\mathscr{T}_+}(V). \qquad (3.19)$$

Proof. The fact that $(\widehat{\mathscr{T}_+}(V), \blacklozenge)$ is a group follows from the fact that $(1 + \widehat{\mathscr{T}_+}(V), \cdot)$ is a multiplicative subgroup of $\widehat{\mathscr{T}_+}(V)$ (see Lemma 3.1) together with the fact that

$$\mathrm{Exp}_\otimes : \widehat{\mathscr{T}_+} \to 1 + \widehat{\mathscr{T}_+}, \quad \mathrm{Log}_\otimes : 1 + \widehat{\mathscr{T}_+} \to \widehat{\mathscr{T}_+}$$

are inverse to each other. Finally, (3.19) follows from the definition of $\blacklozenge$. □

When dealing with the CBHD formula (3.15), it is convenient to fix some notation. We shall use the multi-index notation

$$|h| = \sum_{i=1}^n h_i, \qquad h! = h_1! \cdots h_n!$$

if $h = (h_1, \dots, h_n)$ with $h_1, \cdots, h_n \in \mathbb{N} \cup \{0\}$. Moreover, we make the following useful abbreviations: for any $n \in \mathbb{N}$ we set

$$\mathcal{N}_n := \left\{ (h,k) \, \big| \, h, k \in (\mathbb{N} \cup \{0\})^n, \ (h_1,k_1),\dots,(h_n,k_n) \neq (0,0) \right\},$$

$$c_n := \frac{(-1)^{n+1}}{n}, \quad \mathbf{c}(h,k) := \frac{1}{h! \, k! \, (|h| + |k|)}. \qquad (3.20)$$

With this notation and with the definition of $\blacklozenge$ as in (3.16), the CBHD formula in (3.15) takes the form

$$u \blacklozenge v \overset{\text{CBHD}}{=} \sum_{j=1}^{\infty} \left(\sum_{n=1}^{j} c_n \sum_{(h,k) \in \mathcal{N}_n : |h|+|k|=j} \mathbf{c}(h,k) \left[u^{h_1} v^{k_1} \cdots u^{h_n} v^{k_n} \right]_{\otimes} \right).$$

$$(3.21)$$

(For other different ways to write the series on the right-hand side of (3.21), see the end of Sect. 3.1.4.1.) Also, by using the same notation we can rewrite the $\blacklozenge$ composition as follows:

$$u \blacklozenge v = \sum_{n=1}^{\infty} c_n \sum_{(h,k) \in \mathcal{N}_n} \frac{u^{h_1} \cdot v^{k_1} \cdots u^{h_n} \cdot v^{k_n}}{h! \, k!}, \qquad (3.22)$$

or, by grouping together terms with "similar homogeneity", we can rewrite:

$$u \blacklozenge v = \sum_{j=1}^{\infty} \left(\sum_{n=1}^{j} c_n \sum_{(h,k) \in \mathcal{N}_n : |h|+|k|=j} \frac{u^{h_1} \cdot v^{k_1} \cdots u^{h_n} \cdot v^{k_n}}{h! \, k!} \right). \qquad (3.23)$$

One of the most important features of the $\blacklozenge$ operation will be proved in Corollary 3.21, by making use of the Campbell, Baker, Hausdorff Theorem 3.20. Namely, *restriction of the $\blacklozenge$ operation to $\overline{\mathcal{L}(V)} \times \overline{\mathcal{L}(V)}$ defines a binary operation on $\overline{\mathcal{L}(V)}$.*

3.1.4   The Operation $\diamond$ on $\widehat{\mathcal{T}_+}(V)$

Let us consider, for every $u, v \in \widehat{\mathcal{T}_+}(V)$, the series $\sum_{j=1}^{\infty} Z_j(u,v)$ appearing in our CBHD Theorem 3.8, namely the following series

$$u \diamond v := \sum_{j=1}^{\infty} \left(\sum_{n=1}^{j} \frac{(-1)^{n+1}}{n} \right.$$

$$\left. \times \sum_{\substack{(h_1,k_1),\dots,(h_n,k_n) \neq (0,0) \\ h_1+k_1+\cdots+h_n+k_n=j}} \frac{\left[u^{h_1} v^{k_1} \cdots u^{h_n} v^{k_n} \right]_{\otimes}}{h_1! \cdots h_n! \, k_1! \cdots k_n! \left(\sum_{i=1}^{n} (h_i + k_i) \right)} \right).$$

$$(3.24)$$

With the notation in the previous section, this takes the shorter form

$$u \diamond v = \sum_{j=1}^{\infty} \left(\sum_{n=1}^{j} c_n \sum_{(h,k) \in \mathcal{N}_n : |h|+|k|=j} \mathbf{c}(h,k) \left[u^{h_1} v^{k_1} \cdots u^{h_n} v^{k_n} \right]_{\otimes} \right),$$

We remark that *for every $u, v \in \widehat{\mathscr{T}}_+(V)$ this series is convergent to an element of $\widehat{\mathscr{T}}_+(V)$, whence $\diamond$ defines a binary operation on $\widehat{\mathscr{T}}_+(V)$.* Indeed, if $u, v \in \widehat{\mathscr{T}}_+(V) = \widehat{U}_1$, then,

$$\left[u^{h_1} v^{k_1} \cdots u^{h_n} v^{k_n} \right]_\otimes \in \widehat{U}_{|h|+|k|}.$$

As a consequence, the summand in parentheses in (3.24) – which is $Z_j(u, v)$ – belongs to $\widehat{U}_j$. Thus, the series converges in $\widehat{\mathscr{T}}(V)$, for $\widehat{U}_j \ni Z_j(u, v) \to 0$ as $j \to \infty$ (hence, we can apply Remark 2.76-3, page 102, on the convergence of series). Clearly $\diamond$ is binary on $\widehat{\mathscr{T}}_+$ since we have, for every $u, v \in \widehat{\mathscr{T}}_+$,

$$u \diamond v = \sum_{j=1}^\infty \underbrace{Z_j(u, v)}_{\in \widehat{U}_j \subseteq \widehat{U}_1} \in \widehat{U}_1 = \widehat{\mathscr{T}}_+(V)$$

(recall that the spaces $\widehat{U}_k$ are closed, by Remark 2.78).

Remark 3.11. The $\diamond$ operation has the following important feature: *The restriction of the $\diamond$ operation to $\overline{\mathcal{L}(V)} \times \overline{\mathcal{L}(V)}$ defines a binary operation on $\overline{\mathcal{L}(V)}$.* Indeed, since $\overline{\mathcal{L}(V)}$ is a Lie subalgebra of $\widehat{\mathscr{T}}_+(V)$ (see Remark 3.17), we have

$$\left[u^{h_1} v^{k_1} \cdots u^{h_n} v^{k_n} \right]_\otimes \in \overline{\mathcal{L}(V)}, \quad \text{for every } u, v \in \overline{\mathcal{L}(V)}.$$

Hence, if $u, v \in \overline{\mathcal{L}(V)}$, then $u \diamond v$ is expressed by a converging sequence of elements of $\overline{\mathcal{L}(V)}$ and it is therefore an element of $\overline{\mathcal{L}(V)}$ (which is obviously closed in $\widehat{\mathscr{T}}(V)$!).

The other fundamental property of $\diamond$, which will be proved only after Dynkin's Theorem 3.30, is that it satisfies

$$\mathrm{Exp}_\otimes(u \diamond v) = \mathrm{Exp}_\otimes(u) \cdot \mathrm{Exp}_\otimes(v), \quad \text{for every } u, v \in \widehat{\mathscr{T}}_+(V).$$

With the operations $\blacklozenge$ and $\diamond$ at hands, we can restate the CBHD Theorem as follows.

Theorem 3.12 (Campbell, Baker, Hausdorff, Dynkin). *Let V be a vector space over field $\mathbb{K}$ (of characteristic zero). Let $\blacklozenge$ and $\diamond$ be the operations on $\widehat{\mathscr{T}}_+(V)$ introduced in (3.16) and (3.24), respectively. Then these operations coincide on $\widehat{\mathscr{T}}_+(V)$, i.e., we have the* Campbell-Baker-Hausdorff-Dynkin Formula

$$u \blacklozenge v = u \diamond v, \quad \text{for every } u, v \in \widehat{\mathscr{T}}_+(V). \tag{3.25}$$

3.1.4.1 Other Ways to Write the Dynkin Series

There are other ways of rewriting the series expressing the operation $\diamond$ in (3.77), that is,

$$u \diamond v = \sum_{j=1}^{\infty} \left(\sum_{n=1}^{j} c_n \sum_{(h,k) \in \mathcal{N}_n : |h| + |k| = j} \mathbf{c}(h,k) \left[u^{h_1} v^{k_1} \cdots u^{h_n} v^{k_n} \right]_{\otimes} \right), \quad (3.26)$$

a series which is convergent, as we already know, for every $u, v \in \widehat{\mathcal{T}}_+(V)$. Interchanging the summations over n and j, we get at once:

$$u \diamond v = \sum_{n=1}^{\infty} \left(c_n \sum_{(h,k) \in \mathcal{N}_n : |h| + |k| \geq n} \mathbf{c}(h,k) \left[u^{h_1} v^{k_1} \cdots u^{h_n} v^{k_n} \right]_{\otimes} \right),$$

or, shorter (since it is always true that $|h| + |k| \geq n$ for every $(h,k) \in \mathcal{N}_n$),

$$u \diamond v = \sum_{n=1}^{\infty} \left(c_n \sum_{(h,k) \in \mathcal{N}_n} \mathbf{c}(h,k) \left[u^{h_1} v^{k_1} \cdots u^{h_n} v^{k_n} \right]_{\otimes} \right)$$

$$=: \sum_{n=1}^{\infty} H_n(u,v). \quad (3.27)$$

Note that, unlike for the expression in (3.26), each term in parentheses in (3.27) (denoted $H_n(u,v)$) is an infinite sum *for every fixed* $n \in \mathbb{N}$, and it has to be interpreted as the following limit (in the complete metric space $\widehat{\mathcal{T}}(V)$)

$$H_n(u,v) := \lim_{N \to \infty} c_n \sum_{(h,k) \in \mathcal{N}_n : |h| + |k| \leq N} \mathbf{c}(h,k) \left[u^{h_1} v^{k_1} \cdots u^{h_n} v^{k_n} \right]_{\otimes}. \quad (3.28)$$

Here as usual we have fixed $u, v \in \widehat{\mathcal{T}}_+(V)$. This limit exists, for $H_n(u,v)$ can obviously be rewritten as a convergent series

$$H_n(u,v) = \lim_{N \to \infty} \sum_{j=n}^{N} c_n \sum_{(h,k) \in \mathcal{N}_n : |h| + |k| = j} \mathbf{c}(h,k) \left[u^{h_1} v^{k_1} \cdots u^{h_n} v^{k_n} \right]_{\otimes}$$

$$= \sum_{j=n}^{\infty} c_n \sum_{(h,k) \in \mathcal{N}_n : |h| + |k| = j} \mathbf{c}(h,k) \left[u^{h_1} v^{k_1} \cdots u^{h_n} v^{k_n} \right]_{\otimes}. \quad (3.29)$$

The series on the far right-hand side of (3.29) is convergent in $\widehat{\mathcal{T}}(V)$ (in view of, e.g., Remark 2.62, page 95), since one has, for all $u, v \in \widehat{\mathcal{T}}_+(V)$:

$$c_n \sum_{(h,k)\in\mathcal{N}_n:\,|h|+|k|=j} \mathbf{c}(h,k)\left[u^{h_1}v^{k_1}\cdots u^{h_n}v^{k_n}\right]_{\otimes} \in \prod_{k\geq j}\mathcal{T}_k(V).$$

Another useful representation of $u \diamond v$ can be obtained as follows: For every $N \in \mathbb{N}$ and every fixed $u, v \in \widehat{\mathcal{T}}_+(V)$, let us set

$$\eta_N(u,v) := \sum_{n=1}^{N} c_n \sum_{(h,k)\in\mathcal{N}_n:\,|h|+|k|\leq N} \mathbf{c}(h,k)\left[u^{h_1}v^{k_1}\cdots u^{h_n}v^{k_n}\right]_{\otimes}. \quad (3.30)$$

Then we have

$$u \diamond v = \lim_{N\to\infty} \eta_N(u,v). \quad (3.31)$$

This is an easy consequence of the following reordering of the sum defining $\eta_N(u,v)$:

$$\eta_N(u,v) = \sum_{n=1}^{N} c_n \sum_{j=n}^{N} \sum_{\substack{(h,k)\in\mathcal{N}_n:\\ |h|+|k|=j}} \mathbf{c}(h,k)\left[u^{h_1}v^{k_1}\cdots u^{h_n}v^{k_n}\right]_{\otimes}$$

$$= \sum_{j=1}^{N} \sum_{n=1}^{j} c_n \sum_{(h,k)\in\mathcal{N}_n:\,|h|+|k|=j} \mathbf{c}(h,k)\left[u^{h_1}v^{k_1}\cdots u^{h_n}v^{k_n}\right]_{\otimes} \quad (3.32)$$

$$= \sum_{j=1}^{N} Z_j(u,v) \qquad (\text{see } (3.15)),$$

which gives (by definition of the sum of a series)

$$\lim_{N\to\infty} \eta_N(u,v) = \sum_{j=1}^{\infty} \sum_{n=1}^{j} c_n \sum_{(h,k)\in\mathcal{N}_n:\,|h|+|k|=j} \mathbf{c}(h,k)\left[u^{h_1}v^{k_1}\cdots u^{h_n}v^{k_n}\right]_{\otimes}$$

$$= \sum_{j=1}^{\infty} Z_j(u,v) = u \diamond v.$$

Another way to write $u \diamond v$ is:

$$u \diamond v = \sum_{\substack{r,s\geq 0\\ (r,s)\neq(0,0)}} Z_{r,s}(u,v) = \sum_{\substack{r,s\geq 0\\ (r,s)\neq(0,0)}} \left(Z'_{r,s}(u,v) + Z''_{r,s}(u,v)\right), \quad (3.33)$$

where, for every nonnegative integers r, s with $(r, s) \neq (0, 0)$ we have set
$Z_{r,s}(u, v) = Z'_{r,s}(u, v) + Z''_{r,s}(u, v)$ with

$$
Z'_{r,s}(u, v) = \frac{1}{r+s} \sum_{n=1}^{r+s} \frac{(-1)^{n+1}}{n} \sum_{\substack{h_1+\cdots+h_{n-1}+h_n=r \\ k_1+\cdots+k_{n-1}=s-1 \\ (h_1,k_1),\ldots,(h_{n-1},k_{n-1})\neq(0,0)}}
$$

$$
\times \frac{(\operatorname{ad} u)^{h_1}(\operatorname{ad} v)^{k_1}\cdots(\operatorname{ad} u)^{h_{n-1}}(\operatorname{ad} v)^{k_{n-1}}(\operatorname{ad} u)^{h_n}(v)}{h_1!k_1!\cdots h_{n-1}!k_{n-1}!h_n!}
$$

and

$$
Z''_{r,s}(u, v) = \frac{1}{r+s} \sum_{n=1}^{r+s} \frac{(-1)^{n+1}}{n} \sum_{\substack{h_1+\cdots+h_{n-1}=r-1 \\ k_1+\cdots+k_{n-1}=s \\ (h_1,k_1),\ldots,(h_{n-1},k_{n-1})\neq(0,0)}}
$$

$$
\times \frac{(\operatorname{ad} u)^{h_1}(\operatorname{ad} v)^{k_1}\cdots(\operatorname{ad} u)^{h_{n-1}}(\operatorname{ad} v)^{k_{n-1}}(u)}{h_1!k_1!\cdots h_{n-1}!k_{n-1}!}.
$$

Note that

$$
Z_j(u, v) = \sum_{\substack{r,s\geq 0 \\ r+s=j}} Z_{r,s}(u, v) = \sum_{\substack{r,s\geq 0 \\ r+s=j}} Z'_{r,s}(u, v) + \sum_{\substack{r,s\geq 0 \\ r+s=j}} Z''_{r,s}(u, v), \qquad (3.34)
$$

for every $j \in \mathbb{N}$. Roughly, $Z_{r,s}(u, v)$ collects the summands of $u \diamond v$ which are homogeneous of degree r in u and homogeneous of degree s in v. Also, $Z'_{r,s}(u, v)$ collects the summands of $Z_{r,s}(u, v)$ which "start" with a v (in the innermost position), whereas $Z''_{r,s}(u, v)$ collects those starting with a u. Another possible presentation is

$$
u \diamond v = Z'(u, v) + Z''(u, v), \qquad (3.35)
$$

where $Z'(u, v) = \sum_{j=1}^{\infty} Z'_j(u, v)$ and $Z''(u, v) = \sum_{j=1}^{\infty} Z''_j(u, v)$, with

$$
Z'_j(u, v) = \sum_{\substack{r,s\geq 0 \\ r+s=j}} Z'_{r,s}(u, v), \quad Z''_j(u, v) = \sum_{\substack{r,s\geq 0 \\ r+s=j}} Z''_{r,s}(u, v).
$$

Note that (3.34) implies that $Z_j(u, v) = Z'_j(u, v) + Z''_j(u, v)$.

3.2 The Campbell, Baker, Hausdorff Theorem

The goal of this section is the proof of Theorem 3.20. To this end we need a characterization (first due to Friedrichs) of $\overline{\mathcal{L}(V)}$.

3.2.1 Friedrichs's Characterization of Lie Elements

A key rôle is played by Theorem 3.13 below, first due to Friedrichs,[6] [64]. It states that $\mathcal{L}(V)$ coincides with the *primitive elements* of $\mathscr{T}(V)$. It is in the proof of this result that we invoke the PBW Theorem.

Theorem 3.13 (Friedrichs's Characterization of $\mathcal{L}(V)$). *Let V be any vector space. Let $\delta : \mathscr{T}(V) \to \mathscr{T}(V) \otimes \mathscr{T}(V)$ be the unique UAA morphism such that $\delta(v) = v \otimes 1 + 1 \otimes v$, for all $v \in V$. Then*

$$\mathcal{L}(V) = \Big\{ t \in \mathscr{T}(V) \,\Big|\, \delta(t) = t \otimes 1 + 1 \otimes t \Big\}. \tag{3.36}$$

Proof. Let us denote by L the set on the right-hand side of (3.36). The bilinearity of $\otimes$ proves at once that L is a vector subspace of $\mathscr{T}$. Moreover $V \subseteq L$ trivially. By using the definition of δ and of the $\bullet$ operation on $\mathscr{T}(V) \otimes \mathscr{T}(V)$, it is easily checked that L is a Lie subalgebra of $\mathscr{T}(V)$ containing V: indeed, given $t_1, t_2 \in L$ we have

$$\delta([t_1, t_2]_\otimes) = \delta(t_1 \cdot t_2 - t_2 \cdot t_1) = \delta(t_1) \bullet \delta(t_2) - \delta(t_2) \bullet \delta(t_1)$$

$$= (t_1 \otimes 1 + 1 \otimes t_1) \bullet (t_2 \otimes 1 + 1 \otimes t_2) +$$

$$- (t_2 \otimes 1 + 1 \otimes t_2) \bullet (t_1 \otimes 1 + 1 \otimes t_1)$$

$$\overset{(2.38)}{=} (t_1 \cdot t_2) \otimes 1 + t_1 \otimes t_2 + t_2 \otimes t_1 + 1 \otimes (t_1 \cdot t_2) +$$

$$- (t_2 \cdot t_1) \otimes 1 - t_2 \otimes t_1 - t_1 \otimes t_2 - 1 \otimes (t_2 \cdot t_1)$$

$$= [t_1, t_2]_\otimes \otimes 1 + 1 \otimes [t_1, t_2]_\otimes.$$

This proves that $[t_1, t_2]_\otimes \in L$ so that L is closed w.r.t. the bracket operation. By the definition of $\mathcal{L}(V)$, which is the smallest Lie algebra containing V, we derive $\mathcal{L}(V) \subseteq L$.

Vice versa, let us denote by $\{t_\alpha\}_{\alpha \in \mathcal{A}}$ a basis for $\mathcal{L}(V)$, and we assume $\mathcal{A}$ to be totally ordered by the relation $\preccurlyeq$ (we write $a \prec b$ if $a \preccurlyeq b$ and $a \neq b$). If π is as in (2.101) on page 108, we set $T_\alpha := \pi(t_\alpha)$, for every $\alpha \in \mathcal{A}$. *In view of the PBW Theorem 2.94*, the set

$$\mathcal{B} = \left\{ \pi(1), \; T_{\alpha_1}^{k_1} \cdots T_{\alpha_n}^{k_n} \;\middle|\; \begin{array}{l} n \in \mathbb{N}, \; \alpha_1, \ldots, \alpha_n \in \mathcal{A}, \\ \alpha_1 \prec \cdots \prec \alpha_n, \; k_1, \ldots, k_n \in \mathbb{N} \end{array} \right\}$$

is a linear basis for $\mathscr{U}(\mathcal{L}(V))$. As a consequence, by Proposition 2.93, which asserts that the map

---

[6]See Reutenauer [144, Notes 1.7 on Theorem 1.4] for a comprehensive list of references for this theorem.

$$j : \mathscr{U}(\mathcal{L}(V)) \to \mathscr{T}(V), \qquad j(\pi(t)) = t, \quad \text{for every } t \in \mathcal{L}(V)$$

is a UAA morphism, we deduce that

$$j(\mathcal{B}) = \left\{ 1_{\mathbb{K}}, \ t_{\alpha_1}^{k_1} \cdots t_{\alpha_n}^{k_n} \ \middle| \ \begin{array}{l} n \in \mathbb{N}, \ \alpha_1, \ldots, \alpha_n \in \mathcal{A}, \\ \alpha_1 \prec \cdots \prec \alpha_n, \ k_1, \ldots, k_n \in \mathbb{N} \end{array} \right\}$$

is a linear basis for $\mathscr{T}(V)$. Thanks to Theorem 2.31 on page 74, this proves that $j(\mathcal{B}) \otimes j(\mathcal{B})$ is a basis for $\mathscr{T}(V) \otimes \mathscr{T}(V)$.

Since $\delta(t_\alpha) = t_\alpha \otimes 1 + 1 \otimes t_\alpha$ (as $\mathcal{L}(V) \subseteq L$), we claim that

$$\begin{aligned}
\delta(t_{\alpha_1}^{k_1} \cdots t_{\alpha_n}^{k_n}) &= (t_{\alpha_1} \otimes 1 + 1 \otimes t_{\alpha_1})^{\bullet \, k_1} \bullet \cdots \bullet (t_{\alpha_n} \otimes 1 + 1 \otimes t_{\alpha_n})^{\bullet \, k_n} \\
&= t_{\alpha_1}^{k_1} \cdots t_{\alpha_n}^{k_n} \otimes 1 + 1 \otimes t_{\alpha_1}^{k_1} \cdots t_{\alpha_n}^{k_n} + \qquad\qquad (3.37) \\
&\quad + \sum_{h_i} \mathbf{c}_{h_1, \ldots, h_n}^{(k_1, \ldots, k_n)} \, t_{\alpha_1}^{h_1} \cdots t_{\alpha_n}^{h_n} \otimes t_{\alpha_1}^{k_1 - h_1} \cdots t_{\alpha_n}^{k_n - h_n},
\end{aligned}$$

where the sum runs over the integers h_i such that $0 \leq h_i \leq k_i$ for $i = 1, \ldots, n$ and such that $0 < h_1 + \cdots + h_n < k_1 + \cdots + k_n$, and the constants $\mathbf{c}$ are *positive* integers (resulting from sums of binomial coefficients: note that $t_{\alpha_1} \otimes 1$ and $1 \otimes t_{\alpha_1}$ commute w.r.t. $\bullet$). Note also that the sum on the far right-hand side of (3.37) is empty iff $k_1 + \cdots + k_n = 1$.

We prove the claimed (3.37):

$$\delta(t_{\alpha_1}^{k_1} \cdots t_{\alpha_n}^{k_n}) = (t_{\alpha_1} \otimes 1 + 1 \otimes t_{\alpha_1})^{\bullet \, k_1} \bullet \cdots \bullet (t_{\alpha_n} \otimes 1 + 1 \otimes t_{\alpha_n})^{\bullet \, k_n}$$

(by Newton's binomial formula, since $t_{\alpha_i} \otimes 1, 1 \otimes t_{\alpha_i}$ $\bullet$-commute)

$$\begin{aligned}
&= \sum_{0 \leq j_1 \leq k_1, \cdots, 0 \leq j_n \leq k_n} \binom{k_1}{j_1} \cdots \binom{k_n}{j_n} \\
&\quad \times (t_{\alpha_1} \otimes 1)^{\bullet \, j_1} \bullet (1 \otimes t_{\alpha_1})^{\bullet \, k_1 - j_1} \bullet \cdots \bullet (t_{\alpha_n} \otimes 1)^{\bullet \, j_n} \bullet (1 \otimes t_{\alpha_n})^{\bullet \, k_n - j_n}
\end{aligned}$$

$$\begin{aligned}
\overset{(2.38)}{=} \ &\sum_{0 \leq j_1 \leq k_1, \cdots, 0 \leq j_n \leq k_n} \binom{k_1}{j_1} \cdots \binom{k_n}{j_n} \\
&\quad \times (t_{\alpha_1}^{j_1} \otimes 1) \bullet (1 \otimes t_{\alpha_1}^{k_1 - j_1}) \bullet \cdots \bullet (t_{\alpha_n}^{j_n} \otimes 1) \bullet (1 \otimes t_{\alpha_n}^{k_n - j_n})
\end{aligned}$$

(any $t_{\alpha_i}^{j_i} \otimes 1$ commutes with any $1 \otimes t_{\alpha_h}^{k_h - j_h}$)

$$\begin{aligned}
&= \sum_{0 \leq j_1 \leq k_1, \cdots, 0 \leq j_n \leq k_n} \binom{k_1}{j_1} \cdots \binom{k_n}{j_n} \\
&\quad \times (t_{\alpha_1}^{j_1} \otimes 1) \bullet \cdots \bullet (t_{\alpha_n}^{j_n} \otimes 1) \bullet (1 \otimes t_{\alpha_1}^{k_1 - j_1}) \bullet \cdots \bullet (1 \otimes t_{\alpha_n}^{k_n - j_n})
\end{aligned}$$

$$\overset{(2.38)}{=} \sum_{0 \le j_1 \le k_1, \cdots, 0 \le j_n \le k_n} \binom{k_1}{j_1} \cdots \binom{k_n}{j_n}$$

$$\times \left((t_{\alpha_1}^{j_1} \cdots t_{\alpha_n}^{j_n}) \otimes 1 \right) \bullet \left(1 \otimes (t_{\alpha_1}^{k_1 - j_1} \cdots t_{\alpha_n}^{k_n - j_n}) \right)$$

$$\overset{(2.38)}{=} \sum_{0 \le j_1 \le k_1, \cdots, 0 \le j_n \le k_n} \binom{k_1}{j_1} \cdots \binom{k_n}{j_n} (t_{\alpha_1}^{j_1} \cdots t_{\alpha_n}^{j_n}) \otimes (t_{\alpha_1}^{k_1 - j_1} \cdots t_{\alpha_n}^{k_n - j_n})$$

(we isolate the summands with $(j_1, \ldots, j_n) = (0, \ldots, 0); (k_1, \ldots, k_n)$)

$$= t_{\alpha_1}^{k_1} \cdots t_{\alpha_n}^{k_n} \otimes 1 + 1 \otimes t_{\alpha_1}^{k_1} \cdots t_{\alpha_n}^{k_n}$$

$$+ \sum_{\substack{0 \le j_1 \le k_1, \cdots, 0 \le j_n \le k_n \\ (0, \ldots, 0) \neq (j_1, \ldots, j_n) \neq (k_1, \ldots, k_n)}} \binom{k_1}{j_1} \cdots \binom{k_n}{j_n} (t_{\alpha_1}^{j_1} \cdots t_{\alpha_n}^{j_n}) \otimes (t_{\alpha_1}^{k_1 - j_1} \cdots t_{\alpha_n}^{k_n - j_n}).$$

We now decompose an arbitrary $t \in L$ w.r.t. the above basis $j(\mathcal{B})$:

$$t = \sum_{n, \, \alpha_i, \, k_i} C_{k_1, \ldots, k_n}^{(\alpha_1, \ldots, \alpha_n)} t_{\alpha_1}^{k_1} \cdots t_{\alpha_n}^{k_n} \quad \text{(the sum is finite and the } C \text{ are all scalars).}$$

By applying δ to this identity and by (3.37), we get (as $t \in L$)

$$t \otimes 1 + 1 \otimes t = \delta(t) = \sum_{n, \, \alpha_i, \, k_i} C_{k_1, \ldots, k_n}^{(\alpha_1, \ldots, \alpha_n)} \left(t_{\alpha_1}^{k_1} \cdots t_{\alpha_n}^{k_n} \otimes 1 + 1 \otimes t_{\alpha_1}^{k_1} \cdots t_{\alpha_n}^{k_n} \right) +$$

$$+ \sum_{n, \, \alpha_i, \, k_i} \sum_{h_i} \left\{ C_{k_1, \ldots, k_n}^{(\alpha_1, \ldots, \alpha_n)} \, \mathbf{c}_{h_1, \ldots, h_n}^{(k_1, \ldots, k_n)} t_{\alpha_1}^{h_1} \cdots t_{\alpha_n}^{h_n} \otimes t_{\alpha_1}^{k_1 - h_1} \cdots t_{\alpha_n}^{k_n - h_n} \right\}$$

$$= t \otimes 1 + 1 \otimes t + \sum_{n, \, \alpha_i, \, k_i} \sum_{h_i} \{ \cdots \}.$$

After canceling out $t \otimes 1 + 1 \otimes t$, we infer that the double sum on the far right-hand side above is null. Using the linear independence of different elements in the basis $j(\mathcal{B}) \otimes j(\mathcal{B})$ of $\mathscr{T}(V) \otimes \mathscr{T}(V)$, we derive that every single product of type $C \, \mathbf{c}$ is actually zero. Hence, as all the $\mathbf{c}$ are non-vanishing, one gets that the constants $C_{k_1, \ldots, k_n}^{(\alpha_1, \ldots, \alpha_n)}$ are zero whenever $k_1 + \cdots + k_n > 1$. As a consequence

$$t = \sum_{k_1 + \cdots + k_n = 1} C_{k_1, \ldots, k_n}^{(\alpha_1, \ldots, \alpha_n)} t_{\alpha_1}^{k_1} \cdots t_{\alpha_n}^{k_n}$$

$$= C_{1, 0, \ldots, 0}^{(\alpha_1, 0, \ldots, 0)} t_{\alpha_1} + \cdots + C_{0, \ldots, 0, 1}^{(0, \ldots, 0, \alpha_n)} t_{\alpha_n} \in \mathcal{L}(V).$$

This demonstrates that $L \subseteq \mathcal{L}(V)$ and the proof is complete. $\qquad \square$

We explicitly observe that the map δ in Theorem 3.13 is a UAA morphism acting on $v \in V$ as $v \mapsto v \otimes 1 + 1 \otimes v$; thus it may also be characterized as being the unique linear map such that

$$\delta : \mathscr{T}(V) \to \mathscr{T}(V) \otimes \mathscr{T}(V)$$

$$1 \mapsto 1 \otimes 1 \quad \text{and, for every } k \in \mathbb{N}, \tag{3.38}$$

$$v_1 \otimes \cdots \otimes v_k \mapsto (v_1 \otimes 1 + 1 \otimes v_1) \bullet \cdots \bullet (v_k \otimes 1 + 1 \otimes v_k).$$

As a consequence the following inclusion holds (see also the notation in (2.42), page 82 and the grading condition (2.43))

$$\delta(\mathscr{T}_k(V)) \subseteq K_k(V) = \bigoplus_{i+j=k} \mathscr{T}_{i,j}(V). \tag{3.39}$$

In particular, if $u = (u_k)_{k \geq 0} \in \mathscr{T}(V)$ (here $u_k \in \mathscr{T}_k$ for every $k \geq 0$), then

$$\delta(u) = \sum_{k \geq 0} \delta(u_k) \tag{3.40}$$

gives the expression of $\delta(u)$ in the grading $\bigoplus_{k \geq 0} K_k(V)$ for $\mathscr{T} \otimes \mathscr{T}$ (the sum in (3.40) being finite, for $\mathscr{T} = \bigoplus_{k \geq 0} \mathscr{T}_k$).

As a consequence of (3.39), it follows that (considering $\mathscr{T}(V)$ and $\mathscr{T}(V) \otimes \mathscr{T}(V)$ as subspaces of the metric spaces $\widehat{\mathscr{T}}$ and $\widehat{\mathscr{T} \otimes \mathscr{T}}$, respectively) δ is *uniformly continuous*. Indeed, (3.39) implies

$$\delta\left(\bigoplus_{k \geq n} \mathscr{T}_k(V) \right) \subseteq \bigoplus_{k \geq n} K_k(V) = \bigoplus_{i+j \geq n} \mathscr{T}_{i,j}(V),$$

so that we are in a position to apply Lemma 2.79, page 103 (it suffices to take $k_n = n$ in (2.88)). Hence, δ extends uniquely to a continuous map

$$\widehat{\delta} : \widehat{\mathscr{T}}(V) \to \widehat{\mathscr{T} \otimes \mathscr{T}}(V), \quad \text{which is also a UAA morphism.}$$

The map $\widehat{\delta}$ has the property

$$\widehat{\delta}\left(\prod_{k \geq n} \mathscr{T}_k(V) \right) \subseteq \prod_{i+j \geq n} \mathscr{T}_{i,j}(V). \tag{3.41}$$

These facts give the following representation for $\widehat{\delta}(u)$, when $u \in \widehat{\mathscr{T}} = \prod_{k \geq 0} \mathscr{T}_k$ is expressed in its coordinate form $u = (u_k)_k$ (where $u_k \in \mathscr{T}_k$ for every $k \geq 0$):

$$\widehat{\delta}(u) = \sum_{k \geq 0} \delta(u_k). \tag{3.42}$$

(Recall that $\widehat{\delta}(u_k) = \delta(u_k)$, since $u_k \in \mathscr{T}_k$ and $\widehat{\delta}$ prolongs δ.) Note that the right-hand side of (3.42) is a convergent series in $\widehat{\mathscr{T} \otimes \mathscr{T}}$, since $\delta(u_k) \to 0$

as $k \to \infty$ thanks to (3.39) (and use Remark 2.76-(3)). Also, (3.42) can be viewed as the decomposition of $\hat{\delta}(u)$ in the grading $\bigoplus_{k\geq 0} K_k(V)$ of $\widehat{\mathscr{T}\otimes\mathscr{T}}$ (see (2.42)).

Moreover, the particular case of (3.41) when $n = 1$ gives

$$\hat{\delta}(\widehat{\mathscr{T}_+}(V)) \subseteq \widehat{\mathscr{T}\otimes\mathscr{T}}_+(V). \tag{3.43}$$

Note that this ensures that $\mathrm{Exp}_{\bullet}(\hat{\delta}(u))$ makes sense for every $u \in \widehat{\mathscr{T}_+}$. Also this ensures that $\hat{\delta}(1 + \widehat{\mathscr{T}_+}) \subseteq 1 + \widehat{\mathscr{T}\otimes\mathscr{T}}_+$, so that $\mathrm{Log}_{\bullet}(\hat{\delta}(w))$ makes sense for every $w \in 1 + \widehat{\mathscr{T}_+}$.

If V is a vector space, we consider $\mathcal{L}(V)$, the free Lie algebra generated by V, as a subspace of the metric space $\widehat{\mathscr{T}}(V)$. Then we denote by $\overline{\mathcal{L}(V)}$ the closure of $\mathcal{L}(V)$ in $\widehat{\mathscr{T}}(V)$. The following is the representation of $\overline{\mathcal{L}(V)}$ analogous to that for $\mathcal{L}(V)$ in the notable Theorem 3.13.

Theorem 3.14 (Friedrichs's Characterization of $\overline{\mathcal{L}(V)}$). *Let V be any vector space. Let $\hat{\delta} : \widehat{\mathscr{T}}(V) \to \widehat{\mathscr{T}\otimes\mathscr{T}}(V)$ be the unique continuous UAA morphism prolonging the map $\delta : \mathscr{T} \to \mathscr{T} \otimes \mathscr{T}$ in (3.38). Then we have*

$$\overline{\mathcal{L}(V)} = \left\{ t \in \widehat{\mathscr{T}}(V) \,\big|\, \hat{\delta}(t) = t \otimes 1 + 1 \otimes t \right\}. \tag{3.44}$$

We remark that the equality "$\hat{\delta}(t) = t \otimes 1 + 1 \otimes t$" has to be understood as follows: we note that its right-hand side is an element of $\widehat{\mathscr{T}} \otimes \widehat{\mathscr{T}}$, whereas the left-hand side is an element of $\widehat{\mathscr{T}\otimes\mathscr{T}}$; hence we are here identifying $\widehat{\mathscr{T}} \otimes \widehat{\mathscr{T}}$ as a subset of $\widehat{\mathscr{T}\otimes\mathscr{T}}$, as in Proposition 2.82 on page 106.

Proof. We denote by $\widehat{L}$ the set on the right-hand side of (3.44). We split the proof in two parts:

$\overline{\mathcal{L}(V)} \subseteq \widehat{L}$: By Theorem 3.13, any element t of $\overline{\mathcal{L}(V)}$ is the limit in $\widehat{\mathscr{T}}$ of a sequence $t_k \in \mathscr{T}(V)$ such that

$$\delta(t_k) = t_k \otimes 1 + 1 \otimes t_k, \quad \text{for every } k \in \mathbb{N}.$$

Passing to the limit $k \to \infty$ and invoking Remark 2.83 on page 106 and the continuity of $\hat{\delta} : \widehat{\mathscr{T}} \to \widehat{\mathscr{T}\otimes\mathscr{T}}$ (the prolongation of δ), we get

$$\hat{\delta}(t) = t \otimes 1 + 1 \otimes t, \quad \text{that is, } t \in \widehat{L}.$$

$\widehat{L} \subseteq \overline{\mathcal{L}(V)}$: Let $t \in \widehat{L}$. As an element of $\widehat{\mathscr{T}}$, we have $t = \sum_{k=0}^{\infty} t_k$, with $t_k \in \mathscr{T}_k(V)$ for every $k \geq 0$. The following identities then hold true:

$$(\bigstar) \qquad \sum_{k=0}^{\infty} \delta(t_k) \overset{(3.40)}{=} \widehat{\delta}(t) = t \otimes 1 + 1 \otimes t = \sum_{k=0}^{\infty} (t_k \otimes 1 + 1 \otimes t_k).$$

Since, for every $k \geq 0$ we have (see (3.39))

$$\delta(t_k), \quad t_k \otimes 1 + 1 \otimes t_k \in K_k(V) = \bigoplus_{i+j=k} \mathscr{T}_{i,j}(V),$$

and since $\widehat{\mathscr{T} \otimes \mathscr{T}} = \prod_{k \geq 0} K_k(V)$, we are able to derive from $(\bigstar)$ that $\delta(t_k) = t_k \otimes 1 + 1 \otimes t_k$ for every $k \geq 0$, whence any t_k belongs to $\mathcal{L}(V)$ (thanks to Theorem 3.13). This gives

$$t = \sum_{k=0}^{\infty} \underbrace{t_k}_{\in \mathcal{L}(V)} \in \overline{\mathcal{L}(V)},$$

and the proof is complete. $\qquad\qquad\qquad\qquad\qquad\qquad\qquad\qquad\qquad\qquad\qquad\square$

For the sake of completeness, we provide another characterization of $\overline{\mathcal{L}(V)}$.

Proposition 3.15. *Let V be a vector space. Consider $\mathcal{L}(V)$, the free Lie algebra generated by V, as a subspace of the metric space $\widehat{\mathscr{T}}(V) = \prod_{n=0}^{\infty} \mathscr{T}_n(V)$. Then, for the closure $\overline{\mathcal{L}(V)}$ of $\mathcal{L}(V)$ in $\widehat{\mathscr{T}}(V)$, we have the equality*

$$\overline{\mathcal{L}(V)} = \prod_{n=1}^{\infty} \mathcal{L}_n(V), \tag{3.45}$$

where $\mathcal{L}_n(V)$ is as in Proposition 2.47 on page 85.

Proof. The inclusion $\prod_{n=1}^{\infty} \mathcal{L}_n(V) \subseteq \overline{\mathcal{L}(V)}$ is an easy consequence of $\mathcal{L}(V) = \bigoplus_{n \geq 1} \mathcal{L}_n(V)$ (see (2.50), page 85 and (2.86), page 102): Indeed, if $(\ell_n)_n \in \prod_{n=1}^{\infty} \mathcal{L}_n(V)$ (with $\ell_n \in \mathcal{L}_n(V)$ for every $n \in \mathbb{N}$) we have

$$(\ell_n)_n \overset{(2.86)}{=} \lim_{n \to \infty} \underbrace{(0, \ell_1, \ell_2, \ldots, \ell_n, 0, 0, \cdots)}_{\in \mathcal{L}(V) \text{ by } (2.50)} \in \overline{\mathcal{L}(V)}.$$

Conversely, if $\ell \in \overline{\mathcal{L}(V)}$, there exists a sequence $\{\omega^{(k)}\}_{k \in \mathbb{N}}$ of elements in $\mathcal{L}(V)$ such that $\ell = \lim_{k \to \infty} \omega^{(k)}$ in $\widehat{\mathscr{T}}(V)$. For every fixed $k \in \mathbb{N}$, $\omega^{(k)}$ admits a decomposition (as an element of $\mathcal{L}(V) = \prod_{n=1}^{\infty} \mathcal{L}_n(V)$) of the form

$$\omega^{(k)} = \left(\omega_n^{(k)}\right)_n \qquad \text{with } \omega_n^{(k)} \in \mathcal{L}_n(V) \text{ for every } n \in \mathbb{N}.$$

Analogously, as an element of $\overline{\mathcal{L}(V)} \subset \widehat{\mathscr{T}}(V) = \prod_{n=0}^{\infty} \mathscr{T}_n(V)$, ℓ admits a decomposition of the form

$$\ell = \left(\ell_n\right)_n \qquad \text{with } \ell_n \in \mathscr{T}_n(V) \text{ for every } n \in \mathbb{N} \cup \{0\}.$$

By Remark 2.76-(1), the fact that $\ell = \lim_{k\to\infty} \omega^{(k)}$, means that for every $N \in \mathbb{N}$ there exists $k(N) \in \mathbb{N}$ such that, for all $k \geq k(N)$,

$$\omega^{(k)} = \left(\ell_0, \ell_1, \ell_2, \ldots, \ell_N, \ \omega_{N+1}^{(k)}, \omega_{N+2}^{(k)}, \ldots\right).$$

But since the N-th component of any $\omega^{(k)}$ belongs to $\mathcal{L}_N(V)$, this proves (by the arbitrariness of N) that $\ell_N \in \mathcal{L}_N(V)$ for every $N \in \mathbb{N}$. Thus $\ell = (\ell_n)_n \in \prod_{n=0}^{\infty} \mathcal{L}_n(V)$. The arbitrariness of $\ell \in \overline{\mathcal{L}(V)}$ thus gives $\overline{\mathcal{L}(V)} \subseteq \prod_{n=1}^{\infty} \mathcal{L}_n(V)$. This completes the proof. □

The same proof shows the following fact: *Suppose that for every $n \in \mathbb{N} \cup \{0\}$ there be assigned a subspace B_n of $\mathscr{T}_n(V)$. Then we have*

$$\overline{\bigoplus_{n=0}^{\infty} B_n} = \prod_{n=0}^{\infty} B_n,$$

the closure being taken in $\widehat{\mathscr{T}}(V)$.

The following corollary is a restatement of Proposition 3.15.

Corollary 3.16. *Suppose that $\gamma_j \in \mathscr{T}_j(V)$ for every $j \in \mathbb{N}$ and that $\sum_{j=1}^{\infty} \gamma_j \in \overline{\mathcal{L}(V)}$. Then $\gamma_j \in \mathcal{L}_j(V)$ for every $j \in \mathbb{N}$ (whence, in particular, $\sum_{j=1}^{N} \gamma_j \in \mathcal{L}(V)$ for every $N \in \mathbb{N}$).*

Proof. This follows straightforwardly from Proposition 3.15, by recalling that (see Remark 2.76-3) the series $\sum_{j=1}^{\infty} \gamma_j$ is convergent in $\widehat{\mathscr{T}}(V)$, since $\gamma_j \in \mathscr{T}_j(V)$ for every $j \in \mathbb{N}$. □

Remark 3.17. The set $\overline{\mathcal{L}(V)}$ *is a Lie subalgebra of the commutator-algebra of $\widehat{\mathscr{T}}(V)$. Indeed, we first note that $\overline{\mathcal{L}(V)}$ is a vector subspace of $\widehat{\mathscr{T}}$ as it is the closure (in a topological vector space) of a vector subspace. Hence we are left to show that $[u, v] \in \overline{\mathcal{L}(V)}$ whenever $u, v \in \overline{\mathcal{L}(V)}$. To this end, by (3.45) we have $u = (u_j)_j$ and $v = (v_j)_j$, with $u_j, v_j \in \mathcal{L}_j(V)$ for every $j \in \mathbb{N}$. Hence (see Remark 2.81)*

$$[u, v] = \left[(u_j)_j, (v_j)_j\right] = \left(\sum_{h+k=j} [u_h, v_k]\right)_{j \geq 2}.$$

The above far right-hand side is an element of $\overline{\mathcal{L}(V)}$, again by (3.45). Indeed, we have (see (2.51) at page 85) $[u_h, v_k] \in [\mathcal{L}_h(V), \mathcal{L}_k(V)] \subseteq \mathcal{L}_{h+k}(V)$. □

3.2.2 The Campbell, Baker, Hausdorff Theorem

In this section, as usual, V will denote a fixed vector space on a field of characteristic zero and the map $\widehat{\delta} : \widehat{\mathscr{T}}(V) \to \widehat{\mathscr{T} \otimes \mathscr{T}}(V)$ is the continuous

UAA morphism introduced in the previous section. Note that we have

$$\overline{\mathcal{L}(V)} \subset \widehat{\mathcal{T}_+}(V). \tag{3.46}$$

This follows from

$$\mathcal{L}(V) \subset \mathcal{T}_+(V) = \bigoplus_{n \geq 1} \mathcal{T}_n(V) \subset \prod_{n \geq 1} \mathcal{T}_n(V) = \widehat{\mathcal{T}_+}(V)$$

(for the first inclusion see (2.50), the other inclusions being trivial) and the fact that $\widehat{\mathcal{T}_+}(V) = \widehat{U}_1$ is closed in $\widehat{\mathcal{T}}$ (see (2.78)). From (3.46) it follows that the set $\mathrm{Exp}_\otimes(\overline{\mathcal{L}(V)})$ is well defined and that we have

$$\mathrm{Exp}_\otimes\left(\overline{\mathcal{L}(V)}\right) \subseteq 1 + \widehat{\mathcal{T}_+}(V). \tag{3.47}$$

As a corollary of Theorem 3.6, we have the following result, exhibiting the relationship between the map $\widehat{\delta}$ and the maps $\mathrm{Exp}_\otimes, \mathrm{Exp}_\bullet$ and $\mathrm{Log}_\otimes, \mathrm{Log}_\bullet$.

Corollary 3.18. *With the above notation, we have*

$$\begin{aligned}
\widehat{\delta} \circ \mathrm{Exp}_\otimes &= \mathrm{Exp}_\bullet \circ \widehat{\delta} \qquad \text{on } \widehat{\mathcal{T}_+}, \\
\widehat{\delta} \circ \mathrm{Log}_\otimes &= \mathrm{Log}_\bullet \circ \widehat{\delta} \qquad \text{on } 1 + \widehat{\mathcal{T}_+}.
\end{aligned} \tag{3.48}$$

Proof. It suffices to apply Theorem 3.6 when $A = \mathcal{T}(V)$, $B = \mathcal{T}(V) \otimes \mathcal{T}(V)$ and $\varphi = \widehat{\delta}$. Indeed, note that (3.7) is implied by (3.43). $\qquad \square$

Corollary 3.18 says that the following diagrams are commutative:

$$
\begin{array}{ccc}
\widehat{\mathcal{T}_+} & \xrightarrow{\ \mathrm{Exp}_\otimes\ } & 1 + \widehat{\mathcal{T}_+} \\[2pt]
{\scriptstyle \widehat{\delta}|_{\mathcal{T}_+}}\Big\downarrow & & \Big\downarrow{\scriptstyle \widehat{\delta}|_{1+\mathcal{T}_+}} \\[2pt]
\widehat{\mathcal{T}\otimes\mathcal{T}_+} & \xrightarrow[\ \mathrm{Exp}_\bullet\ ]{} & 1 \otimes 1 + \widehat{\mathcal{T}\otimes\mathcal{T}_+}
\end{array}
$$

$$
\begin{array}{ccc}
\widehat{\mathcal{T}_+} & \xleftarrow{\ \mathrm{Log}_\otimes\ } & 1 + \widehat{\mathcal{T}_+} \\[2pt]
{\scriptstyle \widehat{\delta}|_{\mathcal{T}_+}}\Big\downarrow & & \Big\downarrow{\scriptstyle \widehat{\delta}|_{1+\mathcal{T}_+}} \\[2pt]
\widehat{\mathcal{T}\otimes\mathcal{T}_+} & \xleftarrow[\ \mathrm{Log}_\bullet\ ]{} & 1 \otimes 1 + \widehat{\mathcal{T}\otimes\mathcal{T}_+}
\end{array}
$$

The following is another very important relationship between the maps $\widehat{\delta}$ and Exp. We emphasize that in the following proof we make use of *Friedrichs's Characterization of $\overline{\mathcal{L}(V)}$*, Theorem 3.14.

Theorem 3.19. *The following formulas hold (Exp$_\otimes$ is here shortened to Exp):*

$$\widehat{\delta}(\text{Exp}\, u) = \text{Exp}\, u \otimes \text{Exp}\, u, \tag{3.49}$$

$$\widehat{\delta}(\text{Exp}\, u \cdot \text{Exp}\, v) = (\text{Exp}\, u \cdot \text{Exp}\, v) \otimes (\text{Exp}\, u \cdot \text{Exp}\, v). \tag{3.50}$$

for every $u, v \in \overline{\mathcal{L}(V)}$.

Proof. Let $u \in \overline{\mathcal{L}(V)}$. The following computation applies:

$$\widehat{\delta}(\text{Exp}_\otimes u) = \widehat{\delta}\Big(\sum_{k=0}^\infty \frac{1}{k!} u^{\widehat{}k} \Big) \overset{(1)}{=} \sum_{k=0}^\infty \frac{1}{k!} \big(\widehat{\delta}(u) \big)^{\widehat{\bullet}\, k} \overset{(2)}{=} \text{Exp}_{\bullet}(\widehat{\delta}(u))$$

$$\overset{(3)}{=} \text{Exp}_{\bullet}(u \otimes 1 + 1 \otimes u) = \text{Exp}_\otimes u \otimes \text{Exp}_\otimes u.$$

Here we have used:

1. The map $\widehat{\delta} : \widehat{\mathscr{T}} \to \widehat{\mathscr{T} \otimes \mathscr{T}}$ is continuous and it is a UAA morphism;
2. from (3.43), $\text{Exp}_{\bullet} \circ \widehat{\delta}$ makes sense on $\widehat{\mathscr{T}_+}$, hence in particular on $\overline{\mathcal{L}(V)}$;
3. by (3.44), *Friedrichs's Characterization of $\overline{\mathcal{L}(V)}$*, as $u \in \overline{\mathcal{L}(V)}$, we have $\widehat{\delta}(u) = u \otimes 1 + 1 \otimes u$;
4. as $u \in \overline{\mathcal{L}(V)} \subset \widehat{\mathscr{T}_+}$, we can apply (3.11b), according to which

$$\text{Exp}_{\bullet}(u \otimes 1 + 1 \otimes u) = \text{Exp}_\otimes u \otimes \text{Exp}_\otimes u.$$

This proves (3.49). As for (3.50), given $u, v \in \overline{\mathcal{L}(V)}$ we have

$$\widehat{\delta}(\text{Exp}_\otimes u \cdot \text{Exp}_\otimes v) = \widehat{\delta}(\text{Exp}_\otimes u)\, \widehat{\bullet}\, \widehat{\delta}(\text{Exp}_\otimes v)$$

$$\overset{(3.49)}{=} (\text{Exp}_\otimes u \otimes \text{Exp}_\otimes u)\, \widehat{\bullet}\, (\text{Exp}_\otimes v \otimes \text{Exp}_\otimes v)$$

$$\overset{(3.10)}{=} (\text{Exp}_\otimes u \,\widehat{}\, \text{Exp}_\otimes v) \otimes (\text{Exp}_\otimes u \,\widehat{}\, \text{Exp}_\otimes v).$$

This completes the proof.   □

With the characterization (3.44) at hand (and making use also of Theorem 3.19), we are now able to show that the following key result holds.

Theorem 3.20 (Campbell, Baker, Hausdorff). *Let V be a vector space. Then*

$$\text{Log}(\text{Exp}\, u \cdot \text{Exp}\, v) \in \overline{\mathcal{L}(V)}, \quad \textit{for every } u, v \in \overline{\mathcal{L}(V)}. \tag{3.51}$$

Corollary 3.21. *The restriction of the $\blacklozenge$ operation defined in (3.16) to $\overline{\mathcal{L}(V)} \times \overline{\mathcal{L}(V)}$ defines a binary operation on $\overline{\mathcal{L}(V)}$.*

Proof. Using (3.50) and (3.11b), we claim that

$$\widehat{\delta}(u \blacklozenge v) = (u \blacklozenge v) \otimes 1 + 1 \otimes (u \blacklozenge v), \quad \text{for every } u, v \in \overline{\mathcal{L}(V)}. \tag{3.52}$$

This shows that $u \blacklozenge v \in \overline{\mathcal{L}(V)}$, thanks to Friedrichs's characterization of $\overline{\mathcal{L}(V)}$ in (3.44). We prove the claimed (3.52):

$$\widehat{\delta}(u \blacklozenge v) =$$

$$\text{(by definition of } \blacklozenge \text{)} = \widehat{\delta}\big(\mathrm{Log}_\otimes(\mathrm{Exp}_\otimes u \cdot \mathrm{Exp}_\otimes v)\big)$$

$$\text{(by (3.48))} = \mathrm{Log}_\bullet\big(\widehat{\delta}(\mathrm{Exp}\, u \cdot \mathrm{Exp}\, v)\big)$$

$$\text{(by (3.50))} = \mathrm{Log}_\bullet\big((\mathrm{Exp}\, u \cdot \mathrm{Exp}\, v) \otimes (\mathrm{Exp}\, u \cdot \mathrm{Exp}\, v)\big)$$

$$\text{(by definition of } \blacklozenge \text{)} = \mathrm{Log}_\bullet\big(\mathrm{Exp}(u \blacklozenge v) \otimes \mathrm{Exp}(u \blacklozenge v)\big)$$

$$\text{(by (3.12b))} = \mathrm{Log}_\otimes\big(\mathrm{Exp}_\otimes(u \blacklozenge v)\big) \otimes 1 + 1 \otimes \mathrm{Log}_\otimes\big(\mathrm{Exp}_\otimes(u \blacklozenge v)\big)$$

$$= (u \blacklozenge v) \otimes 1 + 1 \otimes (u \blacklozenge v).$$

This ends the proof. $\square$

Remark 3.22. As a consequence of the above theorem and thanks to the decompositions $\widehat{\mathcal{T}}(V) = \prod_{j=0}^{\infty} \mathcal{T}_j(V)$ and $\overline{\mathcal{L}(V)} = \prod_{j=1}^{\infty} \mathcal{L}_j(V)$, we deduce that, for every $u, v \in \overline{\mathcal{L}(V)}$, if we write

$$\mathrm{Log}(\mathrm{Exp}\, u \cdot \mathrm{Exp}\, v) = \sum_{j=0}^{\infty} z_j(u, v), \quad \text{with } z_j(u, v) \in \mathcal{T}_j(V) \text{ for every } j \geq 0,$$

we have $z_0(u, v) = 0$ and $z_j(u, v) \in \mathcal{L}_j(V)$ for every $j \in \mathbb{N}$.

3.2.3 The Hausdorff Group

The Reader has certainly realized that in the previous section we have implicitly handled the following set (and its well-behaved properties):

$$\Gamma(V) := \Big\{ x \in 1 + \widehat{\mathcal{T}_+}(V) \,\big|\, \widehat{\delta}(x) = x \otimes x \Big\}. \tag{3.53}$$

(As usual, $x \otimes x \in \widehat{\mathcal{T}} \otimes \widehat{\mathcal{T}}$ has to be viewed as an element of $\widehat{\mathcal{T} \otimes \mathcal{T}}$, see Proposition 2.82.) We shall refer to the above $\Gamma(V)$ as the *Hausdorff group* (related to V). The most significant property of $\Gamma(V)$ is that it is a subgroup of the multiplicative group of $\widehat{\mathscr{T}}(V)$:

Theorem 3.23 ($\Gamma(V)$ is a Group). *Let V be a vector space and consider its Hausdorff group $\Gamma(V)$ as in (3.53). Then $\Gamma(V)$ is a subgroup of the multiplicative group of $\widehat{\mathscr{T}}(V)$.*

Proof. We split the proof into two steps:

I. We first prove that the multiplication operation on $\widehat{\mathscr{T}}$ (which we simply denote by $\cdot$ instead of $\widehat{\cdot}$ as usual) restricts to a binary operation on $\Gamma(V)$. Let $x, y \in \Gamma(V)$. Then we have $x \cdot y \in 1 + \widehat{\mathscr{T}_+}$ (see Lemma 3.1) and

$$
\begin{aligned}
\widehat{\delta}(x \cdot y) &= \quad (\widehat{\delta} \text{ is a UAA morphism}) \quad \widehat{\delta}(x) \,\widehat{\bullet}\, \widehat{\delta}(y) \\
&= \quad (\text{by } x, y \in \Gamma(V)) \quad (x \otimes x) \,\widehat{\bullet}\, (y \otimes y) \\
&\overset{(3.10)}{=} (x \cdot y) \otimes (x \cdot y),
\end{aligned}
$$

which proves that $x \cdot y \in \Gamma(V)$.
II. From $\Gamma(V) \subseteq 1 + \widehat{\mathscr{T}_+}$, we know by Lemma 3.1 that every $x \in \Gamma(V)$ is endowed with an inverse element x^{-1}, belonging to $1 + \widehat{\mathscr{T}_+}$. All we have to prove is that $x^{-1} \in \Gamma(V)$. To this end, we are left to show that

$$
\widehat{\delta}(x^{-1}) = x^{-1} \otimes x^{-1}, \quad \text{for every } x \in \Gamma(V). \tag{3.54}
$$

First, we claim that

$$
(x \otimes x)^{\widehat{\bullet}\,(-1)} = x^{-1} \otimes x^{-1}, \quad \text{for every } x \in 1 + \widehat{\mathscr{T}_+}(V). \tag{3.55}
$$

This immediately follows from (3.10):

$$
(x \otimes x) \,\widehat{\bullet}\, (x^{-1} \otimes x^{-1}) \overset{(3.10)}{=} (x \cdot x^{-1}) \,\widehat{\bullet}\, (x \cdot x^{-1}) = 1 \bullet 1 = 1_{\widehat{\mathscr{T} \otimes \mathscr{T}}}.
$$

Finally, we are able to prove (3.54): given $x \in \Gamma(V)$, one has

$$
\widehat{\delta}(x^{-1}) = \quad (\widehat{\delta} \text{ is a UAA morphism}) \quad (\widehat{\delta}(x))^{\widehat{\bullet}\,(-1)}
$$

$$
(\text{recall that } x \in \Gamma(V), \text{ whence } \widehat{\delta}(x) = x \otimes x)
$$

$$
= (x \otimes x)^{\widehat{\bullet}\,(-1)} \overset{(3.55)}{=} x^{-1} \otimes x^{-1}.
$$

This demonstrates (3.54), thus completing the proof.   □

From (3.47) and (3.49) (and the very definition of $\Gamma(V)$) it follows that

$$\mathrm{Exp}_\otimes(\overline{\mathcal{L}(V)}) \subseteq \Gamma(V). \tag{3.56}$$

Note that, in invoking (3.49), we are implicitly applying Friedrichs's Characterization Theorem 3.14. Actually, the reverse inclusion holds too. To prove this, we first observe that

$$\widehat{\delta}(\mathrm{Log}_\otimes x) = \mathrm{Log}_\otimes x \otimes 1 + 1 \otimes \mathrm{Log}_\otimes x, \quad \text{for every } x \in \Gamma(V). \tag{3.57}$$

Indeed, if $x \in \Gamma(V)$, we have

$$\widehat{\delta}(\mathrm{Log}_\otimes x) \overset{(1)}{=} \mathrm{Log}_\bullet(\widehat{\delta}(x)) \overset{(2)}{=} \mathrm{Log}_\bullet(x \otimes x) \overset{(3)}{=} \mathrm{Log}_\otimes x \otimes 1 + 1 \otimes \mathrm{Log}_\otimes x.$$

Here we have applied the following results:

1. The second identity of (3.48) together with $\Gamma(V) \subseteq 1 + \widehat{\mathscr{T}_+}(V)$.
2. The definition of $\Gamma(V)$.
3. Identity (3.12b) and again $\Gamma(V) \subseteq 1 + \widehat{\mathscr{T}_+}(V)$.

Now, once again *thanks to Friedrichs's characterization of* $\overline{\mathcal{L}(V)}$ (see (3.44) of Theorem 3.14), (3.57) proves that $\mathrm{Log}_\otimes x \in \overline{\mathcal{L}(V)}$.

The arbitrariness of $x \in \Gamma(V)$ hence shows that

$$\mathrm{Log}_\otimes(\Gamma(V)) \subseteq \overline{\mathcal{L}(V)}. \tag{3.58}$$

As a consequence, gathering (3.56) and (3.58), from the fact that the maps

$$\widehat{\mathscr{T}_+}(V) \; \underset{\mathrm{Log}_\otimes}{\overset{\mathrm{Exp}_\otimes}{\rightleftarrows}} \; 1 + \widehat{\mathscr{T}_+}(V)$$

are inverse to each other, together with $\overline{\mathcal{L}(V)} \subseteq \widehat{\mathscr{T}_+}(V)$ and $\Gamma(V) \subseteq 1 + \widehat{\mathscr{T}_+}(V)$, we infer the following result.

Theorem 3.24. *Let V be a vector space and let $\Gamma(V)$ be its Hausdorff group, according to the definition in (3.53). Then we have*

$$\mathrm{Exp}(\overline{\mathcal{L}(V)}) = \Gamma(V) \quad \text{and} \quad \mathrm{Log}(\Gamma(V)) = \overline{\mathcal{L}(V)},$$

so that the maps

$$\overline{\mathcal{L}(V)} \; \underset{\mathrm{Log}|_{\Gamma(V)}}{\overset{\mathrm{Exp}|_{\overline{\mathcal{L}(V)}}}{\rightleftarrows}} \; \Gamma(V)$$

are inverse to each other. As a consequence, since $(\Gamma(V), \cdot)$ *is a group (see Theorem 3.23), the operation*

$$u \diamond v = \mathrm{Log}\big(\mathrm{Exp}\, u \cdot \mathrm{Exp}\, v\big), \quad u, v \in \overline{\mathcal{L}(V)}$$

defines on $\overline{\mathcal{L}(V)}$ *a group isomorphic to* $(\Gamma(V), \cdot)$ *via* $\mathrm{Exp}|_{\overline{\mathcal{L}(V)}}$.

In particular, this proves that the map $\diamond$ is a binary operation on $\overline{\mathcal{L}(V)}$, whence the Campbell, Baker, Hausdorff Formula (3.51) holds.

Remark 3.25. We warn the Reader that the above derivation of the CBHD Theorem is not simpler than the one in Sect. 3.2.2. Indeed, the main ingredients are the very same in both proofs, since the proof that $\mathrm{Exp}(\overline{\mathcal{L}(V)}) = \Gamma(V)$ requires Friedrichs's characterization of $\mathcal{L}(V)$.

3.3 Dynkin's Formula

Let us now turn to the derivation of an explicit formula for $u \diamond v$. We first need a crucial result, referred to as the Lemma of Dynkin, Specht and Wever, which is the topic of the next section.

3.3.1 The Lemma of Dynkin, Specht, Wever

As usual, $\mathbb{K}$ is a field of characteristic 0 and linearity properties are always meant to be understood with respect to $\mathbb{K}$. In the results below we agree to denote by $[\cdot, \cdot]$ the commutator related to the tensor algebra $\mathscr{T}(V)$ of a vector space V.

Lemma 3.26 (Dynkin, Specht, Wever). *Let V be a vector space. Consider the (unique) linear map $P : \mathscr{T}(V) \to \mathcal{L}(V)$ such that*

$$\begin{aligned} P(1) &= 0, \\ P(v) &= v, \\ P(v_1 \otimes \cdots \otimes v_k) &= k^{-1}\,[v_1, \ldots [v_{k-1}, v_k] \ldots], \quad \forall\, k \geq 2 \end{aligned} \tag{3.59}$$

for any $v, v_1, \ldots, v_k \in V$. Then P is surjective and it is the identity on $\mathcal{L}(V)$.

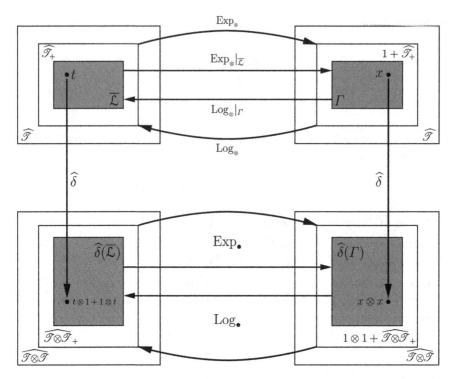

Fig. 3.1 Figure of the proof of the Campbell-Baker-Hausdorff Theorem 3.20

Hence $P^2 = P$ and P is a projection onto $\mathcal{L}(V)$ and this gives another characterization of the Lie elements in $\mathscr{T}(V)$, as follows:

Corollary 3.27. *Let V and P be as in Lemma 3.26 above. Then we have*

$$\mathcal{L}(V) = \left\{ t \in \mathscr{T}(V) \mid P(t) = t \right\}. \tag{3.60}$$

Proof. If $t \in \mathcal{L}(V)$ then $P(t) = t$, since P is the identity on $\mathcal{L}(V)$, by Lemma 3.26. Vice versa, if $t = P(t)$ then $t \in \mathcal{L}(V)$, since $P(\mathscr{T}(V)) = \mathcal{L}(V)$.   □

For the original proofs of Lemma 3.26, see [54, 161, 179].

This is a key result, and it is proved also in Hochschild [85, Proposition 2.2], Jacobson [99, Chapter V, §4, Theorem 8], Reutenauer [144, Theorem 1.4], Serre [159, Chapter IV, §8, LA 4.15].

Proof (of Lemma 3.26). The proof in split into several steps.

STEP I. *Well Posedness.* Let $k \geq 2$ be fixed. The map

$$\underbrace{V \times \cdots \times V}_{k \text{ times}} \longrightarrow \mathcal{L}(V), \quad (v_1, \ldots, v_k) \mapsto k^{-1} \left[v_1, \ldots \left[v_{k-1}, v_k \right] \ldots \right]$$

is well-posed (recall that $\mathbb{K}$ has characteristic zero) and is obviously k-linear. Hence, by Theorem 2.30-i (page 74), there exists a unique linear map

$$P_k : \mathscr{T}_k(V) \longrightarrow \mathcal{L}(V), \quad P_k(v_1 \otimes \cdots \otimes v_k) = k^{-1} [v_1, \ldots [v_{k-1}, v_k] \ldots].$$

Set also $P_0 : \mathscr{T}_0(V) \to \mathcal{L}(V)$, $P_0(k) := 0$ and $P_1 : \mathscr{T}_1(V) \to \mathcal{L}(V)$, $P_1(v) := v$, we can apply Theorem 2.8-i (page 54) to the family of linear maps $\{P_k\}_{k \geq 0}$ to obtain the unique linear map $P : \bigoplus_{k \geq 0} \mathscr{T}_k(V) = \mathscr{T}(V) \to \mathcal{L}(V)$ that prolongs all the maps P_k. This is exactly the map P in (3.59).

STEP II. *Surjectivity.* The surjectivity of P immediately follows from (2.49) and (2.50) in Proposition 2.47 on page 85.

STEP III. *P projects onto $\mathcal{L}(V)$.* To end the proof, we must demonstrate that $P|_{\mathcal{L}(V)}$ is the identity map of $\mathcal{L}(V)$, i.e.,

$$P(\ell) = \ell, \quad \text{for every } \ell \in \mathcal{L}(V). \tag{3.61}$$

This is the main task of the proof and it requires some work.

STEP III.i If we consider $\mathscr{T}(V)$ equipped with its commutator, it becomes a Lie algebra so that, for every fixed $t \in \mathscr{T}(V)$, the map

$$\mathrm{ad}\,(t) : \mathscr{T}(V) \to \mathscr{T}(V), \qquad \mathrm{ad}\,(t)(t') := [t, t'], \quad \text{for every } t' \in \mathscr{T}(V)$$

is an endomorphism of $\mathscr{T}(V)$. In the sequel, we denote by $\mathrm{End}(\mathscr{T}(V))$ the vector space of the endomorphisms of $\mathscr{T}(V)$; we recall that $\mathrm{End}(\mathscr{T}(V))$ is equipped with the structure of UA algebra with the operation $\circ$ of composition of maps. Obviously, $\mathscr{T}(V)$ and $\mathrm{End}(\mathscr{T}(V))$ are also equipped with Lie algebra structures by the associated commutators. With these structures at hand, we know from Lemma 3.28 below that

$$\mathrm{ad} : \mathscr{T}(V) \to \mathrm{End}(\mathscr{T}(V)) \quad \text{is a Lie algebra morphism.}$$

[Note that ad is not in general a UAA morphism though.] Let us now consider the map
$$V \to \mathrm{End}(\mathscr{T}(V)), \quad v \mapsto \mathrm{ad}\,(v).$$

This map is obviously linear. Then by Theorem 2.38-ii (page 78) there exists a unique UAA morphism

$$\varrho : \mathscr{T}(V) \to \mathrm{End}(\mathscr{T}(V))$$

prolonging the above map. Taking into account (2.32) on page 79, this map is the unique linear map such that

$$\varrho(1_{\mathbb{K}}) = \mathrm{Id}_{\mathscr{T}(V)}, \quad \varrho(v_1 \otimes \cdots \otimes v_k) = \mathrm{ad}\,(v_1) \circ \cdots \circ \mathrm{ad}\,(v_k), \tag{3.62}$$

for every $k \in \mathbb{N}$ and every $v_1, \ldots, v_k \in V$. Since ϱ is a UAA morphism, it is also a Lie algebra morphism of the associated commutator-algebra structures (see Remark 2.17 on page 62).

With the above notation, we claim that *the Lie algebra morphisms ϱ and* ad *do coincide on the Lie subalgebra $\mathcal{L}(V)$*, i.e.:

$$\varrho(\ell) = \mathrm{ad}(\ell), \quad \text{for every } \ell \in \mathcal{L}(V). \tag{3.63}$$

Since $\mathcal{L}(V)$ is the Lie subalgebra of $\mathscr{T}(V)$ Lie-generated by V, (3.63) follows from the fact that ϱ and ad actually coincide on V (and the fact that ϱ, ad are both LA morphisms).

STEP III.ii Arguing exactly as in STEP I of this proof, we can prove the existence of a unique linear map $P^* : \mathscr{T}(V) \to \mathscr{T}(V)$ such that

$$\begin{aligned}
&P^*(1) = 0, \\
&P^*(v) = v, \\
&P^*(v_1 \otimes \cdots \otimes v_k) = [v_1, \ldots [v_{k-1}, v_k] \ldots], \quad \forall\, k \geq 2
\end{aligned} \tag{3.64}$$

for any $v, v_1, \ldots, v_k \in V$. We obviously have

$$P^*(t) = k\, P(t) \quad \text{for every } t \in \mathscr{T}_k(V) \text{ and every } k \in \mathbb{N} \cup \{0\}. \tag{3.65}$$

The link between P^* and the map ϱ of STEP III.i is the following:

$$P^*(t \cdot t') = \varrho(t)(P^*(t')), \quad \text{for every } t \in \mathscr{T}(V) \text{ and } t' \in \mathscr{T}_+(V), \tag{3.66}$$

where as usual $\mathscr{T}_+(V) = \bigoplus_{k \geq 1} \mathscr{T}_k(V)$. If $t = k \in \mathscr{T}_0(V)$, (3.66) is trivially true: indeed we have $P^*(k \cdot t') = P^*(k\, t') = k\, P^*(t')$ (since P^* is linear) and

$$\varrho(k)(P^*(t')) \overset{(3.62)}{=} k\,\mathrm{Id}_{\mathscr{T}(V)}(P^*(t')) = k\, P^*(t').$$

Thus we are left to prove (3.66) when both t, t' belong to $\mathscr{T}_+$; moreover, by linearity, we can assume without loss of generality that $t = v_1 \otimes \cdots \otimes v_k$ and $t' = w_1 \otimes \cdots \otimes w_h$ with $h, k \geq 1$. We have

$$\begin{aligned}
P^*(t \cdot t') &= P^*\big(v_1 \otimes \cdots \otimes v_k \otimes w_1 \otimes \cdots \otimes w_h\big) \\
&\overset{(3.64)}{=} [v_1, \ldots [v_k, [w_1, \ldots [w_{h-1}, w_h] \ldots]] \ldots] \\
&= \mathrm{ad}(v_1) \circ \cdots \circ \mathrm{ad}(v_k)\big([w_1, \ldots [w_{h-1}, w_h] \ldots]\big) \\
&\quad \text{(by (3.62) and (3.64), the cases } h = 1 \text{ and } h > 1 \text{ being analogous)} \\
&= \varrho(t)(P^*(t')).
\end{aligned}$$

STEP III.iii With the notation of the previous step, note that the restriction $P^*|_{\mathcal{L}(V)}$ is an endomorphism of $\mathcal{L}(V)$ (since $P^*(\mathcal{T}(V)) \subseteq \mathcal{L}(V)$). We claim that *the restriction of P^* to $\mathcal{L}(V)$ is a derivation of the Lie algebra $\mathcal{L}(V)$*, i.e.:

$$P^*([\ell, \ell']) = [P^*(\ell), \ell'] + [\ell, P^*(\ell')], \quad \text{for every } \ell, \ell' \in \mathcal{L}(V). \tag{3.67}$$

Indeed, when $\ell, \ell' \in \mathcal{L}(V)$ we have

$$\begin{aligned}
P^*([\ell, \ell']) &= P^*(\ell \cdot \ell' - \ell' \cdot \ell) \\
&\quad (P^* \text{ is linear, (3.66) holds and } \mathcal{L}(V) \subset \mathcal{T}_+(V)) \\
&= \varrho(\ell)(P^*(\ell')) - \varrho(\ell')(P^*(\ell)) \\
&\overset{(3.63)}{=} \operatorname{ad}(\ell)(P^*(\ell')) - \operatorname{ad}(\ell')(P^*(\ell)) = [\ell, P^*(\ell')] - [\ell', P^*(\ell)] \\
&= [\ell, P^*(\ell')] + [P^*(\ell), \ell'],
\end{aligned}$$

and (3.67) follows.

STEP III.iv We claim that the following fact holds:

$$P^*(\ell) = k\,\ell, \quad \text{for every } \ell \in \mathcal{L}_k(V) \text{ and every } k \geq 1. \tag{3.68}$$

Since $\mathcal{L}_k(V)$ is spanned by right-nested brackets of length k of elements of V, we can restrict to prove (3.68) when

$$\ell = \begin{cases} v_1, & \text{when } k = 1, \\ [v_1, v_2], & \text{when } k = 2, \\ [v_1, [v_2 \ldots [v_{k-1}, v_k] \ldots]], & \text{when } k \geq 3, \end{cases}$$

where $v_1, \ldots, v_k \in V$. We argue by induction on $k \in \mathbb{N}$. Trivially, (3.68) holds for $k = 1$ (see (3.64) and recall that $\mathcal{L}_1(V) = V$). When $k = 2$ we have

$$P^*([v_1, v_2]) \overset{(3.67)}{=} [P^*(v_1), v_2] + [v_1, P^*(v_2)] \overset{(3.64)}{=} [v_1, v_2] + [v_1, v_2] = 2\,[v_1, v_2].$$

We now suppose that (3.68) holds for a fixed k and we prove it for $k + 1$:

$$\begin{aligned}
&P^*\big([v_1, [v_2 \ldots [v_k, v_{k+1}] \ldots]]\big) \\
&\overset{(3.67)}{=} [P^*(v_1), [v_2 \ldots [v_k, v_{k+1}] \ldots]] + [v_1, P^*([v_2 \ldots [v_k, v_{k+1}] \ldots])]
\end{aligned}$$

(use the inductive hypothesis)

$$\begin{aligned}
&= [v_1, [v_2 \ldots [v_k, v_{k+1}] \ldots]] + [v_1, k\,[v_2 \ldots [v_k, v_{k+1}] \ldots]] \\
&= (k+1)\,[v_1, [v_2 \ldots [v_k, v_{k+1}] \ldots]].
\end{aligned}$$

STEP III.v We are finally ready to prove (3.61). Due to $\mathcal{L} = \bigoplus_{k \geq 1} \mathcal{L}_k(V)$, it is not restrictive to suppose that $\ell \in \mathcal{L}_k$ for some $k \in \mathbb{N}$. We then have

$$P(\ell) \overset{(3.65)}{=} k^{-1} P^*(\ell) \overset{(3.68)}{=} k^{-1} k \ell = \ell$$

This completes the proof.   □

Here we employed the following simple result.

Lemma 3.28. *Let $\mathfrak{g}$ be a Lie algebra. Let $\mathrm{End}(\mathfrak{g})$ (the vector space of the endomorphisms of $\mathfrak{g}$) be equipped with its commutator-algebra structure, related to its UAA structure coming from the composition $\circ$ of maps.*

Then $\mathrm{ad} : \mathfrak{g} \to \mathrm{End}(\mathfrak{g})$ is a Lie algebra morphism.

Proof. Let us fix $a, b \in \mathfrak{g}$. We have to prove that

$$\mathrm{ad}\left([a, b]_{\mathfrak{g}}\right) = [\mathrm{ad}\,(a), \mathrm{ad}\,(b)]_{\circ}.$$

Since the bracket on $\mathrm{End}(\mathfrak{g})$ is given by

$$[A, B]_{\circ} := A \circ B - B \circ A, \quad \text{for every } A, B \in \mathrm{End}(\mathfrak{g}),$$

all we have to prove is that

$$[[a, b]_{\mathfrak{g}}, c]_{\mathfrak{g}} = (\mathrm{ad}\,(a) \circ \mathrm{ad}\,(b))(c) - (\mathrm{ad}\,(b) \circ \mathrm{ad}\,(a))(c), \quad \forall\, c \in \mathfrak{g}.$$

In its turn this is equivalent to $[[a, b], c] = [a, [b, c]] - [b, [a, c]]$, which is a consequence of antisymmetry and the Jacoby identity for $[\cdot, \cdot]_{\mathfrak{g}}$.   □

Remark 3.29. Consider the unique linear map $S : \mathscr{T}(V) \to \mathscr{T}(V)$ such that

$$\begin{aligned}
S(1) &= 0, \\
S(v) &= v, \\
S(v_1 \otimes \cdots \otimes v_k) &= k\, v_1 \otimes \cdots \otimes v_k, \quad \forall\, k \geq 2
\end{aligned} \tag{3.69}$$

for any $v, v_1, \ldots, v_k \in V$. Then S *is a derivation of the UA algebra $\mathscr{T}(V)$.*

Indeed, by linearity, it is obvious that

$$S(t) = k\,t, \quad \text{for every } t \in \mathscr{T}_k(V) \text{ and every } k \geq 0. \tag{3.70}$$

Next, we prove that

$$S(t \cdot t') = S(t) \cdot t' + t \cdot S(t'), \quad \forall\, t, t' \in \mathscr{T}(V). \tag{3.71}$$

It holds $t = (t_n)_n$ and $t' = (t'_n)_n$ with $t_n, t'_n \in \mathscr{T}_n(V)$ for every $n \geq 0$, and t_n, t'_n are 0 for n large enough. Then we have

$$S(t \cdot t') = S\Big(\Big(\sum_{i+j=n} t_i \cdot t'_j\Big)_{n \geq 0}\Big) = \Big(n \sum_{i+j=n} t_i \cdot t'_j\Big)_{n \geq 0}$$

$$= \Big(\sum_{i+j=n} (i+j)\, t_i \cdot t'_j\Big)_{n \geq 0} = \Big(\sum_{i+j=n} (i\, t_i) \cdot t'_j\Big)_n + \Big(\sum_{i+j=n} t_i \cdot (j\, t'_j)\Big)_n$$

$$= S(t) \cdot t' + t \cdot S(t').$$

Thanks to Remark 2.18 on page 62, S is also a derivation of the commutator-algebra related to $\mathscr{T}(V)$. Furthermore, in view of (3.70), the restriction of S to $\mathcal{L}_k(V)$ coincides with $k\,\mathrm{Id}_{\mathcal{L}_k(V)}$. Hence the restriction of S to $\mathcal{L}(V)$ is an endomorphism of $\mathcal{L}(V)$. Actually, $S|_{\mathcal{L}(V)}$ is the unique linear map from $\mathcal{L}(V)$ to $\mathcal{L}(V)$ such that

$$\begin{aligned} S(v) &= v, \\ S\big([v_1, \ldots [v_{k-1}, v_k]\ldots]\big) &= k\,[v_1, \ldots [v_{k-1}, v_k]\ldots], \quad \forall\, k \geq 2. \end{aligned} \tag{3.72}$$

As a consequence of the above remarks *the restriction of S to $\mathcal{L}(V)$ is a derivation of the Lie algebra $\mathcal{L}(V)$*.

Now, in STEP III.iii of the proof of Lemma 3.26 we constructed a derivation of the Lie algebra $\mathcal{L}(V)$ whose restriction to V is the identity of V: namely this derivation is the map $P^*|_{\mathcal{L}(V)}$, where P^* is as in (3.64).

Since the above $S|_{\mathcal{L}(V)}$ is also a derivation of $\mathcal{L}(V)$ whose restriction to V is Id_V, since V Lie-generates $\mathcal{L}(V)$ and since two derivations coinciding on a set of Lie-generators coincide throughout $\mathcal{L}(V)$, this proves that

$$P^*|_{\mathcal{L}(V)} \equiv S|_{\mathcal{L}(V)}. \tag{3.73}$$

Incidentally, via (3.70), this gives another proof of (3.68).

3.3.2 Dynkin's Formula

Thanks to the Lemma of Dynkin, Specht and Wever in Sect. 3.3.1, we now easily get the following important characterization of the operation $\blacklozenge$.

Theorem 3.30 (Dynkin). *Let V be a vector space and let $P : \mathscr{T}(V) \to \mathcal{L}(V)$ be the linear map of Lemma 3.26. Then for every $u, v \in \mathcal{L}(V)$,*

$$\mathrm{Log}\,(\mathrm{Exp}\,u \cdot \mathrm{Exp}\,v) = \sum_{n=1}^{\infty} \frac{(-1)^{n+1}}{n} \sum_{(h,k) \in \mathbb{N}_n} \frac{P\big(u^{h_1} \cdot v^{k_1} \cdots u^{h_n} \cdot v^{k_n}\big)}{h_1! \cdots h_n!\, k_1! \cdots k_n!}.$$

Proof. If P is as in Lemma 3.26, then $P(\mathscr{T}_k(V)) = \mathcal{L}_k(V) \subset \mathscr{T}_k(V)$, for every $k \in \mathbb{N}$. As a consequence,

$$P\left(\bigoplus_{k\geq n} \mathscr{T}_k(V)\right) \subseteq \bigoplus_{k\geq n} \mathscr{T}_k(V), \qquad \forall\, n \geq 0.$$

Hence we are entitled to apply Lemma 2.79 (see page 103; take $k_n := n$ in (2.88)). As a consequence, considering $\mathscr{T}(V)$ and $\mathcal{L}(V)$ as subsets of the metric space $\widehat{\mathscr{T}}(V)$, P is uniformly continuous, so that it admits a unique linear continuous prolongation $\widehat{P} : \widehat{\mathscr{T}}(V) \to \widehat{\mathscr{T}}(V)$. Actually we have $\widehat{P}(\widehat{\mathscr{T}}(V)) \subseteq \overline{\mathcal{L}(V)}$, since $P(\mathscr{T}(V)) = \mathcal{L}(V)$ (recall that $\widehat{P}$ is the continuous prolongation of P and that $\mathscr{T}(V)$ is dense in $\widehat{\mathscr{T}}(V)$).

It is easily seen that $\widehat{P}$ is the identity on $\overline{\mathcal{L}(V)}$, since P is the identity on $\mathcal{L}(V)$. Hence, as $u \diamond v \in \overline{\mathcal{L}(V)}$ for $u, v \in \mathcal{L}(V)$ (by Theorem 3.20), we obtain (see (3.17))

$$u \diamond v = \widehat{P}(u \diamond v) = \widehat{P} \sum_{n=1}^{\infty} \frac{(-1)^{n+1}}{n} \sum_{(h_1,k_1),\ldots,(h_n,k_n)\neq(0,0)} \frac{u^{h_1} \cdot v^{k_1} \cdots u^{h_n} \cdot v^{k_n}}{h_1! \cdots h_n! \, k_1! \cdots k_n!}.$$

The thesis of the theorem now follows from the continuity – and the linearity – of $\widehat{P}$ (we can thus interchange $\widehat{P}$ with the summation operations in both series) and the fact that $\widehat{P} \equiv P$ on $\mathscr{T}(V)$. (Finally recall the notation for $\mathbb{N}_n$ in (3.20)). This ends the proof. $\square$

In particular, if $u, v \in V$, then $u^{h_1} \cdot v^{k_1} \cdots u^{h_n} \cdot v^{k_n}$ is an elementary tensor, homogeneous of degree $\sum_{i=1}^{n}(h_i + k_i)$, so that Theorem 3.30 and the definition of P give

$$u \diamond v = \sum_{n=1}^{\infty} \frac{(-1)^{n+1}}{n} \sum_{(h,k)\in\mathbb{N}_n} \frac{\left[u^{h_1}v^{k_1} \cdots u^{h_n}v^{k_n}\right]_\otimes}{h_1! \cdots h_n! \, k_1! \cdots k_n! \left(\sum_{i=1}^{n}(h_i + k_i)\right)}$$

$$= \sum_{j=1}^{\infty} \sum_{n=1}^{j} \frac{(-1)^{n+1}}{n} \sum_{\substack{(h,k)\in\mathbb{N}_n \\ |h|+|k|=j}} \frac{\left[u^{h_1}v^{k_1} \cdots u^{h_n}v^{k_n}\right]_\otimes}{h! \, k! \, (|h| + |k|)}.$$

Thus we have proved that

$$u \diamond v = \sum_{j=1}^{\infty} \sum_{n=1}^{j} \frac{(-1)^{n+1}}{n} \sum_{\substack{(h,k)\in\mathbb{N}_n \\ |h|+|k|=j}} \frac{\left[u^{h_1}v^{k_1} \cdots u^{h_n}v^{k_n}\right]_\otimes}{h! \, k! \, (|h| + |k|)}, \qquad u, v \in V.$$

(3.74)

[This is a partial version of our CBHD Theorem! What remains for us to do is to pass from $u, v \in V$ to $u, v \in \widehat{\mathscr{T}_+}(V)$.]

Remark 3.31. Another way to derive (3.74) (which plays – as we shall see below – the key rôle in deriving the general form of the BCHD Theorem) is described here. Let $u, v \in V$ be fixed. Let us reorder the series in (3.17) as follows:

$$u \blacklozenge v = \sum_{n=1}^{\infty} \frac{(-1)^{n+1}}{n} \sum_{j=1}^{\infty} \sum_{\substack{(h,k) \in \mathcal{N}_n \\ |h|+|k|=j}} \frac{u^{h_1} \cdot v^{k_1} \cdots u^{h_n} \cdot v^{k_n}}{h_1! \cdots h_n! \, k_1! \cdots k_n!}$$

(interchange the sums recalling that, in the j-sum, $n \leq |h| + |k| = j$)

$$\sum_{j=1}^{\infty} \sum_{n=1}^{j} \frac{(-1)^{n+1}}{n} \underbrace{\sum_{\substack{(h,k) \in \mathcal{N}_n \\ |h|+|k|=j}} \frac{u^{h_1} \cdot v^{k_1} \cdots u^{h_n} \cdot v^{k_n}}{h_1! \cdots h_n! \, k_1! \cdots k_n!}}_{=: \gamma_j}.$$

Note that $\gamma_j \in \mathscr{T}_j(V)$ for every $j \in \mathbb{N}$, since $u, v \in \mathscr{T}_1(V)$. Moreover, by Theorem 3.20, we have $\sum_{j=1}^{\infty} \gamma_j = u \blacklozenge v \in \overline{\mathcal{L}(V)}$ (since $V \subset \mathcal{L}(V)$). Thus we are in a position to apply Corollary 3.16 and infer that $\gamma_j \in \mathcal{L}_j(V)$ for every $j \in \mathbb{N}$. Consequently, by Lemma 3.26, we have $\gamma_j = P(\gamma_j)$ for every $j \in \mathbb{N}$. Hence we get

$$u \blacklozenge v = \sum_{j=1}^{\infty} P(\gamma_j) \quad \text{(by the linearity of } P\text{)}$$

$$= \sum_{j=1}^{\infty} \sum_{n=1}^{j} \frac{(-1)^{n+1}}{n} \sum_{\substack{(h,k) \in \mathcal{N}_n \\ |h|+|k|=j}} \frac{P\left(u^{h_1} \cdot v^{k_1} \cdots u^{h_n} \cdot v^{k_n}\right)}{h_1! \cdots h_n! \, k_1! \cdots k_n!}$$

$$= \sum_{j=1}^{\infty} \sum_{n=1}^{j} \frac{(-1)^{n+1}}{n} \sum_{\substack{(h,k) \in \mathcal{N}_n \\ |h|+|k|=j}} \frac{\left[u^{h_1} v^{k_1} \cdots u^{h_n} v^{k_n}\right]_{\otimes}}{h! \, k! \, (|h| + |k|)}.$$

In the last equality, we invoked the very definition of P in (3.59), noticing that $u^{h_1} \cdot v^{k_1} \cdots u^{h_n} \cdot v^{k_n} = u^{\otimes h_1} \otimes v^{\otimes k_1} \otimes \cdots \otimes u^{\otimes h_n} \otimes v^{\otimes k_n}$ is an elementary tensor of degree $|h| + |k|$. So, we re-derived (3.74).　　　$\square$

By using the map $\widehat{P}$ introduced in the proof of Theorem 3.30, we have another characterization of $\overline{\mathcal{L}(V)}$:

Theorem 3.32. *Let V be a vector space. Let $\widehat{P} : \widehat{\mathscr{T}}(V) \to \widehat{\mathscr{T}}(V)$ be the (unique) continuous prolongation of the map P in (3.59), that is,*

$\widehat{P} : \widehat{\mathscr{T}}(V) \to \widehat{\mathscr{T}}(V)$ *is such that*

$$P(1) = 0, \quad P(v) = v, \quad \text{and} \tag{3.75}$$

$$P(v_1 \otimes \cdots \otimes v_k) = k^{-1} [v_1, \ldots [v_{k-1}, v_k] \ldots], \quad \forall \, k \geq 2,$$

for any $v, v_1, \ldots, v_k \in V$. *Then we have*

$$\overline{\mathcal{L}(V)} = \{ t \in \widehat{\mathscr{T}}(V) \mid \widehat{P}(t) = t \}. \tag{3.76}$$

Proof. Since $\widehat{\mathscr{T}}(V) = \prod_{k=0}^{\infty} \mathscr{T}_k(V)$ and $\mathcal{L}(V) = \prod_{k=1}^{\infty} \mathcal{L}_k(V)$ and since $\widehat{P}(\mathscr{T}_k(V)) = \mathcal{L}_k(V)$, it is immediately seen that $\widehat{P}(\widehat{\mathscr{T}}(V)) = \overline{\mathcal{L}(V)}$. Hence, if $t \in \widehat{\mathscr{T}}$ is such that $t = \widehat{P}(t)$, then $t \in \overline{\mathcal{L}(V)}$. Conversely, if $t \in \overline{\mathcal{L}(V)}$, we have $t = \sum_{k=1}^{\infty} \ell_k$, with $\ell_k \in \mathcal{L}_k(V)$ for every $k \geq 1$. Then by the continuity of $\widehat{P}$ and by (3.60) we deduce

$$\widehat{P}(t) = \widehat{P}\left(\sum_{k=1}^{\infty} \ell_k \right) = \sum_{k=1}^{\infty} P(\ell_k) = \sum_{k=1}^{\infty} \ell_k = t.$$

This ends the proof. $\square$

3.3.3 The Final Proof of the CBHD Theorem

Let us now turn to the derivation of an explicit formula for $u \blacklozenge v$ for general $u, v \in \widehat{\mathscr{T}}_+(V)$. Let us recall that, for every $u, v \in \widehat{\mathscr{T}}_+(V)$, we introduced the notation

$$u \diamond v = \sum_{j=1}^{\infty} Z_j(u, v), \tag{3.77}$$

where $Z_j(u, v)$ is as in (3.15), i.e., with the now shorter notation:

$$Z_j(u, v) := \sum_{n=1}^{j} \frac{(-1)^{n+1}}{n} \sum_{\substack{(h,k) \in \mathcal{N}_n \\ |h| + |k| = j}} \frac{\left[u^{h_1} v^{k_1} \cdots u^{h_n} v^{k_n} \right]_{\otimes}}{h! \, k! \, (|h| + |k|)}. \tag{3.78}$$

By collecting together (3.74), the definition of Z_j in (3.78) and the definition of the $\diamond$ operation in (3.77), we get

$$u \blacklozenge v = u \diamond v, \quad \text{for every } u, v \in V. \tag{3.79}$$

To derive the CBHD Theorem 3.8, which is nothing but the identity

$$u \blacklozenge v = u \diamond v, \quad \text{for every } u, v \in \widehat{\mathscr{T}}_+(V), \tag{3.80}$$

we only have to write down (3.79) when $V = \mathbb{K}\langle x, y \rangle$, $u = x$, $v = y$, as $\mathbb{K}\langle x, y \rangle$ is the free vector space on two (non-commuting) indeterminates x, y. Then we shall go back (via a substitution argument) to any pair of $u, v \in \widehat{\mathscr{T}}_+(V)$, where V is an arbitrary vector space. A continuity argument is also needed. This is the essence of the following proof.

Proof. (of the CBHD *Theorem.)* Let V be any vector space over $\mathbb{K}$ and let us fix arbitrary $u, v \in \widehat{\mathscr{T}}_+(V)$. Equation (3.14) is then uniquely solved by $Z(u, v) = \mathrm{Log}(\mathrm{Exp} u \cdot \mathrm{Exp} v) = u \blacklozenge v$. By the universal property of the free vector space $\mathbb{K}\langle x, y \rangle$ (where $x \neq y$), there exists a unique linear map $\varphi : \mathbb{K}\langle x, y \rangle \to \widehat{\mathscr{T}}(V)$ such that $\varphi(x) = u$, $\varphi(y) = v$. Furthermore, by the universal property of the tensor algebra, there exists a unique UAA morphism $\Phi : \mathscr{T}(\mathbb{K}\langle x, y \rangle) \to \widehat{\mathscr{T}}(V)$ extending φ. We claim that Φ is also uniformly continuous. (As usual the tensor algebra $\mathscr{T}$ is considered as a subspace of the associated metric space $\widehat{\mathscr{T}}$.) Indeed, it is easily seen that the following property holds:

$$\Phi\left(\bigoplus_{k \geq N} \mathscr{T}_k(\mathbb{K}\langle x, y \rangle)\right) \subseteq \prod_{k \geq N} \mathscr{T}_k(V). \tag{3.81}$$

By arguing as in the proof of Lemma 2.79 (see page 430 in Chap. 7), this proves our claim. Hence there exists a unique *continuous* prolongation of Φ, say $\widehat{\Phi} : \widehat{\mathscr{T}}(\mathbb{K}\langle x, y \rangle) \to \widehat{\mathscr{T}}(V)$, which is also a UAA morphism (see again the proof of the cited Lemma 2.79).

Now, identity (3.79) holds when V is replaced by $\mathbb{K}\langle x, y \rangle$, thus providing an identity in $\widehat{\mathscr{T}}(\mathbb{K}\langle x, y \rangle)$, holding true for any two elements in $\mathbb{K}\langle x, y \rangle$. As a particular case, we can apply it to the pair $x, y \in \mathbb{K}\langle x, y \rangle$, thus obtaining $x \blacklozenge y = x \diamond y$. We next apply $\widehat{\Phi}$ to this identity, thus obtaining $\widehat{\Phi}(x \blacklozenge y) = \widehat{\Phi}(x \diamond y)$. Our final task is to show that this is precisely $u \blacklozenge v = u \diamond v$, thus proving (3.80) (for the arbitrariness of $u, v \in \widehat{\mathscr{T}}_+(V)$). Indeed, the precise argument is the following one:

$$u \blacklozenge v \overset{(a)}{=} \widehat{\Phi}(x \blacklozenge y) = \widehat{\Phi}(x \diamond y) \overset{(b)}{=} u \diamond v.$$

This is a consequence of the following two facts:

a. $\widehat{\Phi}(x \blacklozenge y) = u \blacklozenge v$; here we used the fact that $\widehat{\Phi}$ is a continuous UAA morphism, together with an application of (3.8) in Theorem 3.6 (note that (3.7) is fulfilled, as can easily be proved starting from (3.81)).
b. $\widehat{\Phi}(x \diamond y) = u \diamond v$; here we exploited again the cited properties of $\widehat{\Phi}$, the universal definition of the functions Z_j both on $\widehat{\mathscr{T}}(\mathbb{K}\langle x, y \rangle)$ and on $\widehat{\mathscr{T}}(V)$, and the fact that, $\widehat{\Phi}$ being a prolongation of Φ, one has

$$\widehat{\Phi}\left(\left[x^{h_1} y^{k_1} \cdots x^{h_n} y^{k_n} \right]_{\widehat{\mathscr{T}}(\mathbb{K}\langle x,y \rangle)} \right)$$

$$= \left[\overbrace{u \cdots [u}^{h_1 \text{ times}} \; \left[\overbrace{v \cdots [v}^{k_1 \text{ times}} \; \cdots \left[\overbrace{u \cdots [u}^{h_n \text{ times}} \; \left[\overbrace{v [\cdots v}^{k_n \text{ times}} \; \right]]]]]]]] \right. \right. \right]_{\widehat{\mathscr{T}}(V)}$$

$$= \left[u^{h_1} v^{k_1} \cdots u^{h_n} v^{k_n} \right]_{\otimes}.$$

This ends the proof of the CBHD Theorem.   □

Note that (3.80) is an improvement of (3.79), which cannot be deduced immediately from Theorem 3.30 (not even by replacing P with $\widehat{P}$), because

$$\widehat{P}(u^{h_1} v^{k_1} \cdots u^{h_n} v^{k_n})$$

cannot be explicitly written, for general $u, v \in \widehat{\mathscr{T}}_+(V)$.

As a consequence of the CBHD Theorem 3.8, we get the following result.

Corollary 3.33. *Let V be a vector space over the field $\mathbb{K}$ (of characteristic zero). Let $\diamond$ be the operation on $\widehat{\mathscr{T}}_+(V)$ defined by the series introduced in (3.24). Then $\diamond$ is associative and moreover $(\widehat{\mathscr{T}}_+(V), \diamond)$ is a group. Furthermore, we have the identity*

$$\mathrm{Exp}_{\otimes}(u \diamond v) = \mathrm{Exp}_{\otimes}(u) \cdot \mathrm{Exp}_{\otimes}(v), \qquad \text{for every } u, v \in \widehat{\mathscr{T}}_+(V). \tag{3.82}$$

Proof. By the CBHD Theorem 3.8, we know that $\diamond$ coincides with $\blacklozenge$ on $\widehat{\mathscr{T}}_+(V)$. Hence the corollary is proved by invoking the fact that $(\widehat{\mathscr{T}}_+(V), \blacklozenge)$ is a group (see Proposition 3.10) and by (3.19) (which gives (3.33)).   □

3.3.4 Some "Finite" Identities Arising from the Equality Between $\blacklozenge$ and $\diamond$

Let $(A, *)$ be any UA algebra. For every $(h, k) \in \mathbb{N}_n$ we set

$$D^*_{(h,k)} : A \times A \to A, \quad \text{where} \quad D^*_{(h,k)}(a, b)$$

$$= \left[\overbrace{a, \cdots [a}^{h_1 \text{ times}}, \; \overbrace{[b, \cdots [b}^{k_1 \text{ times}}, \cdots \overbrace{[a, \cdots [a}^{h_n \text{ times}}, \; \overbrace{[b, [\cdots, b}^{k_n \text{ times}} \;]_*]_* \cdots]_* \cdots]_* \cdots]_*]_* \cdots]_*.$$

Here as usual $[\cdot, \cdot]_*$ denotes the commutator $[a, b]_* = a * b - b * a$.

The fundamental identity (3.25) in the CBHD Theorem asserting the equality of the operations $\blacklozenge$ and $\diamond$ on $\widehat{\mathscr{T}}_+(V)$, holds in the particular case

$$V = \mathbb{K}\langle x, y \rangle, \qquad \text{with } x \neq y.$$

In this case (3.25) becomes explicitly

$$\sum_{n=1}^{\infty} \frac{(-1)^{n+1}}{n} \sum_{(h,k)\in\mathcal{N}_n} \frac{x^{h_1} \cdot y^{k_1} \cdots x^{h_n} \cdot y^{k_n}}{h!\,k!}$$

$$= \sum_{n=1}^{\infty} \frac{(-1)^{n+1}}{n} \sum_{(h,k)\in\mathcal{N}_n} \frac{D^{\cdot}_{(h,k)}(x,y)}{h!\,k!\,(|h|+|k|)}. \tag{3.83}$$

By projecting this identity on $\mathscr{T}_r(\mathbb{K}\langle x,y\rangle)$, we get the family of identities

$$\sum_{n=1}^{r} \frac{(-1)^{n+1}}{n} \sum_{\substack{(h,k)\in\mathcal{N}_n \\ |h|+|k|=r}} \frac{x^{h_1} \cdot y^{k_1} \cdots x^{h_n} \cdot y^{k_n}}{h!\,k!}$$

$$= \sum_{n=1}^{r} \frac{(-1)^{n+1}}{n} \sum_{\substack{(h,k)\in\mathcal{N}_n \\ |h|+|k|=r}} \frac{D^{\cdot}_{(h,k)}(x,y)}{h!\,k!\,(|h|+|k|)}, \quad \text{for every } r \in \mathbb{N}. \tag{3.84}$$

Alternatively, if we set, for any fixed $i,j \in \mathbb{N} \cup \{0\}$,

$$H_{i,j} := \mathrm{span}\big\{ x^{h_1} \cdot y^{k_1} \cdots x^{h_n} \cdot y^{k_n} \; : \; |h|=i, \; |k|=j \big\},$$

then it obviously holds that $\widehat{\mathscr{T}}(\mathbb{K}\langle x,y\rangle) = \prod_{i,j\geq 0} H_{i,j}$. Hence, starting from (3.83) we also obtain the identities

$$\sum_{n=1}^{i+j} \frac{(-1)^{n+1}}{n} \sum_{\substack{(h,k)\in\mathcal{N}_n \\ |h|=i,\; |k|=j}} \frac{x^{h_1} \cdot y^{k_1} \cdots x^{h_n} \cdot y^{k_n}}{h!\,k!}$$

$$= \sum_{n=1}^{i+j} \frac{(-1)^{n+1}}{n} \sum_{\substack{(h,k)\in\mathcal{N}_n \\ |h|=i,\; |k|=j}} \frac{D^{\cdot}_{(h,k)}(x,y)}{h!\,k!\,(|h|+|k|)}, \quad \text{for every } r \in \mathbb{N}. \tag{3.85}$$

valid for every fixed $i,j \geq 0$.

Since the above are all identities in the free associative algebra $\mathscr{T}(\mathbb{K}\langle x,y\rangle)$, by Theorem 2.85 on page 107 they specialize to every UA algebra. This proves the following result.

Theorem 3.34. *Let $(A,*)$ be any UA algebra over a field of characteristic zero. For every $r \in \mathbb{N}$ and every $a,b \in A$, we have*

$$\sum_{n=1}^{r} \frac{(-1)^{n+1}}{n} \sum_{\substack{(h,k)\in\mathcal{N}_n \\ |h|+|k|=r}} \frac{a^{*\,h_1} * b^{*\,k_1} * \cdots * a^{*\,h_n} * b^{*\,k_n}}{h!\,k!}$$

$$= \sum_{n=1}^{r} \frac{(-1)^{n+1}}{n} \sum_{\substack{(h,k)\in\mathcal{N}_n \\ |h|+|k|=r}} \frac{D^*_{(h,k)}(a,b)}{h!\,k!\,(|h|+|k|)}. \tag{3.86}$$

Summing up for $r = 1,\ldots,N$ (where $N \in \mathbb{N}$ is arbitrarily given) and interchanging the sums we also obtain the identity

$$\sum_{n=1}^{N} \frac{(-1)^{n+1}}{n} \sum_{\substack{(h,k)\in\mathcal{N}_n \\ |h|+|k|\leq N}} \frac{a^{*\,h_1} * b^{*\,k_1} * \cdots * a^{*\,h_n} * b^{*\,k_n}}{h!\,k!}$$

$$= \sum_{n=1}^{N} \frac{(-1)^{n+1}}{n} \sum_{\substack{(h,k)\in\mathcal{N}_n \\ |h|+|k|\leq N}} \frac{D^*_{(h,k)}(a,b)}{h!\,k!\,(|h|+|k|)}. \tag{3.87}$$

Analogously, for every fixed $i,j \in \mathbb{N} \cup \{0\}$, $(i,j) \neq (0,0)$ and every $a,b \in A$

$$\sum_{n=1}^{i+j} \frac{(-1)^{n+1}}{n} \sum_{\substack{(h,k)\in\mathcal{N}_n \\ |h|=i,\ |k|=j}} \frac{a^{*\,h_1} * b^{*\,k_1} * \cdots * a^{*\,h_n} * b^{*\,k_n}}{h!\,k!}$$

$$= \sum_{n=1}^{i+j} \frac{(-1)^{n+1}}{n} \sum_{\substack{(h,k)\in\mathcal{N}_n \\ |h|=i,\ |k|=j}} \frac{D^*_{(h,k)}(a,b)}{h!\,k!\,(|h|+|k|)}. \tag{3.88}$$

We shall return to look more closely at such type of identities in Sect. 4.1.

3.4 Résumé: The "Spine" of the Proof of the CBHD Theorem

To close the sections devoted to the first proof of the CBHD presented in this Book, we summarize the "backbone" of the arguments used throughout:

1. We introduce the unique UAA morphism $\delta : \mathscr{T} \to \mathscr{T} \otimes \mathscr{T}$ such that $\delta(v) = v \otimes 1 + 1 \otimes v$, for all $v \in V$; we denote by $\hat{\delta} : \widehat{\mathscr{T}} \to \widehat{\mathscr{T}\otimes\mathscr{T}}$ its continuous prolongation.
2. Having introduced the set $\Gamma(V) = \{x \in 1 + \widehat{\mathscr{T}_+} : \hat{\delta}(x) = x \otimes x\}$, one easily proves that it is a group, the Hausdorff group.

3. By means of the crucial characterization of Lie elements, due to Friedrichs, $\overline{\mathcal{L}(V)} = \{t \in \widehat{\mathcal{T}} : \widehat{\delta}(t) = t \otimes 1 + 1 \otimes t\}$, one shows that

$$\mathrm{Exp}(\overline{\mathcal{L}(V)}) = \Gamma(V).$$

4. This proves the Campbell-Baker-Hausdorff Theorem

$$\mathrm{Log}(\mathrm{Exp}\, u \cdot \mathrm{Exp}\, v) \in \overline{\mathcal{L}(V)}, \quad \text{for every } u, v \in \overline{\mathcal{L}(V)}.$$

   Indeed, if $u, v \in \overline{\mathcal{L}}$, then $\mathrm{Exp}\, u, \mathrm{Exp}\, v \in \Gamma$, so that (as Γ is a group) $\mathrm{Exp}\, u \cdot \mathrm{Exp}\, v \in \Gamma$; as a consequence, the logarithm of this product belongs to $\overline{\mathcal{L}}$, as $\mathrm{Exp}(\overline{\mathcal{L}}) = \Gamma$.

5. We introduce the unique linear map $P : \mathcal{T} \to \mathcal{T}$ such that $P(1) = 0$, and $P(v_1 \otimes \cdots \otimes v_k) = k^{-1} [v_1, \ldots [v_{k-1}, v_k] \ldots]$, for $k \geq 1$; also, $\widehat{P} : \widehat{\mathcal{T}} \to \widehat{\mathcal{T}}$ denotes the continuous prolongation of P.

6. The crucial Lemma of Dynkin-Specht-Wever holds, yielding another characterization of Lie elements as $\overline{\mathcal{L}(V)} = \{t \in \widehat{\mathcal{T}} : \widehat{P}(t) = t\}$.

7. For every $u, v \in \overline{\mathcal{L}}$, we know from the Campbell-Baker-Hausdorff Theorem that the element $\mathrm{Log}(\mathrm{Exp}\, u \cdot \mathrm{Exp}\, v)$ belongs to $\overline{\mathcal{L}(V)}$; hence $\widehat{P}$ leaves it unchanged and the Dynkin Theorem follows.

8. By applying Dynkin's Theorem for the special case $V = \mathbb{K}\langle x, y \rangle$ and $u = x$, $v = y$ one derives that

$$\mathrm{Log}(\mathrm{Exp}\, x \cdot \mathrm{Exp}\, y) = \sum_{j=1}^{\infty} \sum_{n=1}^{j} \frac{(-1)^{n+1}}{n} \sum_{\substack{(h,k) \in \mathbb{N}_n \\ |h|+|k|=j}} \frac{\left[x^{h_1} y^{k_1} \cdots x^{h_n} y^{k_n} \right]}{h! \, k! \, (|h| + |k|)}.$$

A substitution argument finally provides Dynkin's representation for $\mathrm{Log}(\mathrm{Exp}\, u \cdot \mathrm{Exp}\, v)$ for arbitrary $u, v \in \widehat{\mathcal{T}}(V)$ and any vector space V, over the field of null characteristic $\mathbb{K}$.
The CBHD Theorem is completely proved.

3.5 A Few Summands of the Dynkin Series

In this section we furnish a few summands of the Dynkin series. They could be computed directly from Dynkin's formula, but with a sizeable amount of computations, since this formula obviously does not take into account that the same commutator may stem from different choices of the summation indices, nor does it take into account possible cancellations resulting from skew-symmetry or the Jacobi identity. For example, in the explicit Dynkin representation (3.78) for the $Z_j(u, v)$, the summands for $j = 2$ are 7, of which 3 are non-vanishing:

$$n = 1 \qquad (h_1, k_1) \qquad \begin{cases} (2, 0): & \frac{1}{4}[x, x] = 0 \\ (1, 1): & \frac{1}{2}[x, y] \\ (0, 2): & \frac{1}{4}[y, y] = 0 \end{cases}$$

$$n = 2 \qquad (h_1, k_1),\ (h_2, k_2) \qquad \begin{cases} (1, 0),\ (1, 0): & -\frac{1}{4}[x, x] = 0 \\ (1, 0),\ (0, 1): & -\frac{1}{4}[x, y] \\ (0, 1),\ (1, 0): & -\frac{1}{4}[y, x] \\ (0, 1),\ (0, 1): & -\frac{1}{4}[y, y] = 0 \end{cases}$$

summing up to produce the well known $\frac{1}{2}[x, y]$, resulting from

$$\tfrac{1}{2}[x, y] - \tfrac{1}{4}[x, y] - \tfrac{1}{4}[y, x].$$

For $j = 3$ we have 24 summands, of which 10 are non-vanishing and (using skew symmetry) they sum up to produce

$$\tfrac{1}{12}[x[x, y]] + \tfrac{1}{12}[y[y, x]].$$

For $j = 4$ we have 82 summands (!!), of which 34 are non-vanishing and (after many cancellations and using skew symmetry) they sum up to produce

$$\tfrac{1}{48}[y[x[y, x]]] - \tfrac{1}{48}[x[y[x, y]]],$$

which, this time *in view of the Jacobi identity*, is indeed equal to $-\frac{1}{24}[x[y[x, y]]]$.

As it appears from these computations, it is not easy to handle the Dynkin series for explicit calculations by hand. Other recursion formulas may be of use for this purpose. For instance, by exploiting his formulas (1.70) and (1.71) (see page 30 of our Chap. 1), Hausdorff succeeded in a two-page calculation to write the expansion up to $j = 5$.

We here exhibit the expansion up to $j = 8$. We follow J. A. Oteo [134, Table III], who has provided a simplified explicit formula up to order 8 with the minimal number of commutators required. [In the formula below, the first four homogeneous summands are grouped in the first line, whereas the homogeneous summands of orders $5, 6, 7, 8$ are separated by a blank line]:

$$x \diamond y = x + y + \frac{1}{2}[x, y] + \frac{1}{12}\left([x[x, y]] + [y[y, x]]\right) - \frac{1}{24}[x[y[x, y]]]$$

$$+ \frac{1}{120}\left([x[y[x[x, y]]]] + [y[x[y[y, x]]]]\right)$$

$$- \frac{1}{360}\left([y[x[x[x, y]]]] + [x[y[y[y, x]]]]\right)$$

$$-\frac{1}{720}\left([x[x[x[x,y]]]]+[y[y[y[y,x]]]]\right)$$

$$+\frac{1}{1440}\,[x[y[y[y[x,y]]]]]-\frac{1}{720}\,[x[x[y[y[x,y]]]]]$$

$$+\frac{1}{240}\,[x[y[y[x[x,y]]]]]+\frac{1}{1440}\,[y[x[x[x[x,y]]]]]$$

$$+\frac{1}{30240}\left([x[x[x[x[x[x,y]]]]]]+[y[y[y[y[y[y,x]]]]]]\right)$$

$$-\frac{1}{10080}\left([y[x[x[x[x[x,y]]]]]]+[x[y[y[y[y[y,x]]]]]]\right)$$

$$+\frac{1}{2520}\left([x[x[y[x[x[x,y]]]]]]+[y[y[x[y[y[y,x]]]]]]\right)$$

$$+\frac{1}{10080}\left([y[y[y[y[x[x,y]]]]]]+[x[x[x[x[y[y,x]]]]]]\right)$$

$$-\frac{1}{1680}\left([y[x[y[y[x[x,y]]]]]]+[x[y[y[y[y[y,x]]]]]]\right)$$

$$-\frac{1}{3360}\left([y[y[x[y[x[x,y]]]]]]+[x[x[y[x[y[y,x]]]]]]\right)$$

$$+\frac{1}{7560}\left([y[y[y[x[x[x,y]]]]]]+[x[x[x[y[y[y,x]]]]]]\right)$$

$$-\frac{1}{1260}\left([y[x[y[x[x[x,y]]]]]]+[x[y[x[y[y[y,x]]]]]]\right)$$

$$+\frac{1}{3360}\left([y[y[x[x[x[x,y]]]]]]+[x[x[y[y[y[y,x]]]]]]\right)$$

$$-\frac{1}{60480}\,[x[y[y[y[y[y[x,y]]]]]]]$$

$$+\frac{1}{20160}\,[x[x[y[y[y[y[x,y]]]]]]]-\frac{1}{5040}\,[x[y[y[x[y[y[x,y]]]]]]]$$

$$+\frac{1}{20160}\,[x[y[y[y[y[x[x,y]]]]]]]-\frac{1}{60480}\,[y[x[x[x[x[x[x,y]]]]]]]$$

$$-\frac{1}{20160}\,[y[y[x[x[x[x[x,y]]]]]]]+\frac{1}{5040}\,[y[x[x[y[x[x[x,y]]]]]]]$$

$$-\frac{1}{20160}\,[y[x[x[x[y[x[x,y]]]]]]]+\frac{1}{15120}\,[x[y[y[y[x[x[x,y]]]]]]]$$

$$-\frac{1}{6720}\,[x[y[y[x[y[x[x,y]]]]]]]-\frac{1}{3360}\,[x[y[x[y[y[x[x,y]]]]]]]$$

$$+\frac{1}{120960}\,[x[x[x[y[y[y[x,y]]]]]]]-\frac{1}{5040}\,[x[x[y[y[y[x[x,y]]]]]]]$$

$$+\left\{\text{brackets of heights}\ge 9\right\}+\cdots$$

3.6   Further Reading: Hopf Algebras

The objective of this section is to give the definition of a Hopf algebra and to overview some very basic related facts, with the aim only to show that some results presented in this chapter can be properly restated (and proved) within the setting of Hopf algebras. A comprehensive exposition on Hopf algebras is definitely beyond our scope and the interested Reader is referred to introductory treatises (see e.g., [2, 34, 45, 120, 125, 165]). Here, we shall content ourselves with highlighting the fact that a great part of the formalism behind the proof of the CBHD Theorem presented in the foregoing sections has a deep connection with Hopf algebra theory.

First we introduce a new notation: for $i = 1, 2$, let V_i, W_i be vector spaces (over a field $\mathbb{K}$) and $\varphi_i : V_i \to W_i$ a linear map. Since the map

$$V_1 \times V_2 \ni (v_1, v_2) \mapsto \varphi_1(v_1) \otimes \varphi_2(v_2) \in W_1 \otimes W_2$$

is bilinear, there exists a unique linear map $\varphi_1 \otimes \varphi_2$ such that

$$(\varphi_1 \otimes \varphi_2) : V_1 \otimes V_2 \to W_1 \otimes W_2, \quad \varphi_1 \otimes \varphi_2(v_1 \otimes v_2) = \varphi_1(v_1) \otimes \varphi_2(v_2).$$

In what follows, as usual, the map id_V denotes the identity map on the set V. If V is a $\mathbb{K}$-vector space, we recall that we have natural isomorphisms $\mathbb{K} \otimes V \simeq V$ and $V \otimes \mathbb{K} \simeq V$ given by the unique linear maps[7] acting on elementary tensors as follows

$$\mathbb{K} \otimes V \longrightarrow V \qquad\qquad V \otimes \mathbb{K} \longrightarrow V$$
$$k \otimes v \mapsto k\,v, \qquad\qquad v \otimes k \mapsto k\,v,$$

where $k \in \mathbb{K}$ and $v \in V$ are arbitrary.

Remark 3.35. An equivalent way, useful for the topic of this section, of giving the axioms defining a unital associative algebra is the following one: A unital associative algebra (over a field $\mathbb{K}$) is a triple (A, M, u), where A is a vector space, $M : A \otimes A \to A$ and $u : \mathbb{K} \to A$ are linear maps, such that the following diagrams are commutative:

---

[7]These linear maps do exist since the functions

$$\mathbb{K} \times V \ni (k, v) \mapsto k\,v \in V, \qquad V \times \mathbb{K} \ni (v, k) \mapsto k\,v \in V$$

are bilinear (then we use the universal property of the tensor product). We leave it to the Reader to prove that the associated linear maps are isomorphisms of vector spaces.

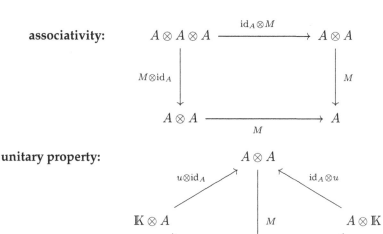

associativity:

$$A \otimes A \otimes A \xrightarrow{\text{id}_A \otimes M} A \otimes A$$

with vertical maps $M \otimes \text{id}_A$ and M, and

$$A \otimes A \xrightarrow{M} A$$

unitary property:

$$\mathbb{K} \otimes A \xrightarrow{u \otimes \text{id}_A} A \otimes A \xleftarrow{\text{id}_A \otimes u} A \otimes \mathbb{K}$$

$$A$$

with map M.

The commutativity of the first diagram is equivalent to

$$M(a \otimes M(a' \otimes a'')) = M(M(a \otimes a') \otimes a''), \qquad \forall \, a, a', a'' \in A.$$

The commutativity of the second diagram is equivalent to

$$M(u(k) \otimes a) = k\, a = M(a \otimes u(k)), \qquad \forall \, k \in \mathbb{K}, \ a \in A.$$

Hence, if (A, M, u) is as above, the map

$$* : A \times A \to A, \qquad a * a' := M(a \otimes a'), \quad a, a' \in A$$

endows A with the structure of an associative algebra, and the element $u(1_{\mathbb{K}})$ of A is a unit element for $*$. Vice versa, let $(A, *)$ be a UA algebra, according to the definition used so far in this Book. Then, since $A \times A \ni (a, a') \mapsto a * a' \in A$ is bilinear, there exists a unique linear map $M : A \otimes A \to A$ such that $M(a \otimes a') = a * a'$ for every $a, a' \in A$. If we further consider the map $u : \mathbb{K} \to A$ defined by $u(k) := k\, 1_A$ (where 1_A is the unit of A), then it is easily seen that the triple (A, M, u) satisfies the requirements in Remark 3.35.

Following Sweedler's words in his treatise on Hopf algebras [165, page 4], "dualizing" the above diagrams, that is, "turning all the arrows around", one obtains in a natural way the definition of coalgebra.

A *counital and coassociative coalgebra* (over the field $\mathbb{K}$) is a triple (C, Δ, ε), where C is a vector space, $\Delta : C \to C \otimes C$ and $\varepsilon : C \to \mathbb{K}$ are linear maps, such that the following diagrams are commutative:

coassociativity: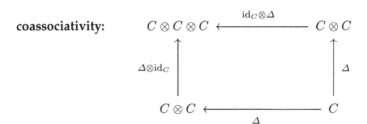

counitary property: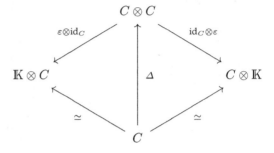

The map Δ is called *coproduct* and ε is called *counit*. In the sequel, by *co-UA coalgebra* we mean a counital and coassociative coalgebra.

Remark 3.36. While we derived the definition of co-UA coalgebra in a purely formal fashion, the structure of co-UA coalgebra carries along a significant property, which we now recall. All the details can be found in [26, III §11.2].

Let (C, Δ, ε) be a co-UA coalgebra over $\mathbb{K}$. Let $(A, *)$ be any UA algebra over $\mathbb{K}$. Note that, by the universal property of the tensor product, there exists a unique linear map

$$m_* : A \otimes A \longrightarrow A \quad \text{such that } m_*(a \otimes a') = a * a', \qquad (3.89)$$

for every $a, a' \in A$. We now aim to equip $\mathrm{Hom}(C, A)$ (the vector space of the $\mathbb{K}$-linear maps from C to A) with a composition law

$$\mu : \mathrm{Hom}(C, A) \times \mathrm{Hom}(C, A) \longrightarrow \mathrm{Hom}(C, A).$$

Let $u, v \in \mathrm{Hom}(C, A)$. Then, collecting all the notations given in this section, we can define a linear map $\mu(u, v) \in \mathrm{Hom}(C, A)$ by considering the following composition of maps:

$$\mu(u, v) : C \xrightarrow{\;\;\Delta\;\;} C \otimes C \xrightarrow{\;\;u \otimes v\;\;} A \otimes A \xrightarrow{\;\;m_*\;\;} A.$$

Then one can prove without difficulty that $(\text{Hom}(C, A), \mu)$ *is a unital associative algebra if* (and only if) C *is a co-UA algebra*. The unit element of $(\text{Hom}(C, A), \mu)$ is the linear mapping defined by

$$c \mapsto \varepsilon(c)\, 1_A, \quad \forall c \in C,$$

where 1_A is the unit element of A.

For example, notice that, for $u, v, w \in \text{Hom}(C, A)$, $\mu(\mu(u, v), w)$ and $\mu(u, \mu(v, w))$ are respectively given by the maps in the following diagrams:

$$C \xrightarrow{\;\Delta\;} C \otimes C \xrightarrow{\;\Delta \otimes \text{id}_C\;} C \otimes C \otimes C \xrightarrow{\;u \otimes v \otimes w\;} A \otimes A \otimes A \longrightarrow A$$

$$C \xrightarrow{\;\Delta\;} C \otimes C \xrightarrow{\;\text{id}_C \otimes \Delta\;} C \otimes C \otimes C \xrightarrow{\;u \otimes v \otimes w\;} A \otimes A \otimes A \longrightarrow A,$$

where the last arrow in each diagram describes the unique linear function mapping $a \otimes a' \otimes a''$ into $a * a' * a''$, for $a, a', a'' \in A$. Hence, if coassociativity holds, one gets $\mu(\mu(u, v), w) = \mu(u, \mu(v, w))$.

In order to define Hopf algebras, we need another definition.

A *bialgebra* (over the field $\mathbb{K}$) is a 5-tuple $(A, *, 1_A, \Delta, \varepsilon)$ where $(A, *, 1_A)$ is a unital associative algebra (with unit 1_A) and (A, Δ, ε) is a counital coassociative coalgebra (both structures are over $\mathbb{K}$) such that the following compatibility assumptions hold:

1. the coproduct $\Delta : A \to A \otimes A$ is a UAA morphism,[8]
2. the counit $\varepsilon : \mathbb{K} \to A$ is a UAA morphism.[9]

Let us introduce the following notation[10]

$$\Delta(a) = \sum_i f_{i,1}(a) \otimes f_{i,2}(a) \qquad (\text{for some } f_{i,1}(a), \, f_{i,2}(a) \text{ in } A), \qquad (3.90)$$

---

[8] We recall that, according to Proposition 2.41, page 81, $A \otimes A$ can be naturally equipped with the structure of UA algebra $(A \otimes A, \bullet)$ where

$$(a \otimes b) \bullet (a' \otimes b') = (a * a') \otimes (b * b'), \qquad \forall\, a, b, a', b' \in A.$$

[9] Here, $\mathbb{K}$ is endowed with the trivial UA algebra structure given by multiplication.

[10] The $f_{i,j}$ in (3.90) *do not* refer to any function on A. The meaning of (3.90) is the following: For every $a \in A$ there exists a finite set $\mathcal{F}(a) \subset \mathbb{N}$ and some there exist elements $f_{i,j}(a)$ of A (for $j = 1, 2$ and $i \in \mathcal{F}(a)$) such that

$$\Delta(a) = \sum_{i \in \mathcal{F}(a)} f_{i,1}(a) \otimes f_{i,2}(a).$$

The $f_{i,j}(a)$ are not uniquely defined, possibly.

the sum being finite. Using this notation, the compatibility condition (1) is equivalent to

$$\begin{cases} \Delta(1_A) = 1_A \otimes 1_A, \\ \Delta(a * a') = \sum_{i,i'} \big(f_{i,1}(a) * f_{i',1}(a') \big) \otimes \big(f_{i,2}(a) * f_{i',2}(a') \big), \end{cases}$$

for all $a, a' \in A$, whilst the compatibility condition (2) is equivalent to

$$\varepsilon(1_A) = 1_{\mathbb{K}}, \qquad \varepsilon(a * a') = \varepsilon(a)\, \varepsilon(a'),$$

for every $a, a' \in A$. Alternatively, the compatibility condition (1) is equivalent to commutativity of the following two diagrams:

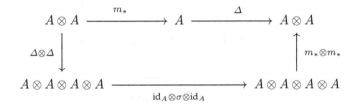

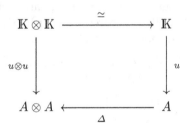

Here, m_* is as in (3.89), $u : \mathbb{K} \to A$ is the linear map defined by $u(k) = k\,1_A$ (for $k \in \mathbb{K}$), whilst $\sigma : A \otimes A \to A \otimes A$ is the unique linear map such that $\sigma(a \otimes a') = a' \otimes a$, for every $a, a' \in A$.

With the same notation, the compatibility condition (2) is equivalent to the commutativity of the following two diagrams:

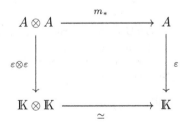

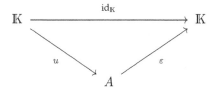

If $(A, *, 1_A, \Delta, \varepsilon)$ is a bialgebra, an element $a \in A$ is called:

- *primitive* (in A) if $\Delta(a) = a \otimes 1_A + 1_A \otimes a$.
- *grouplike* (in A) if $a \neq 0$ and $\Delta(a) = a \otimes a$.

The set $G(A) \subset A$ of the grouplike elements contains, for example, 1_A. Also, if $a \in G(A)$, then $\varepsilon(a) = 1_{\mathbb{K}}$ (as follows easily by using the counitary property). Moreover, $G(A)$ is closed under $*$-multiplication. This follows from the fact that Δ is a UAA morphism: indeed, if $x, y \in G(A)$ one has

$$\Delta(x * y) = \Delta(x) \bullet \Delta(y) = (x \otimes x) \bullet (y \otimes y) = (x * y) \otimes (x * y).$$

Note that a similar computation appeared in Part I of the proof of Theorem 3.23 (page 143), when showing that $\Gamma(V)$ is a group.

The set $P(A) \subseteq A$ of the primitive elements in A is a Lie subalgebra of the commutator algebra of A, as the following computation shows: given $p_1, p_2 \in P(A)$, one has

$$\Delta([p_1, p_2]_*) = \Delta(p_1 * p_2 - p_2 * p_1) = \Delta(p_1) \bullet \Delta(p_2) - \Delta(p_2) \bullet \Delta(p_1)$$

$$= (p_1 \otimes 1_A + 1_A \otimes p_1) \bullet (p_2 \otimes 1_A + 1_A \otimes p_2)$$

$$- (p_2 \otimes 1_A + 1_A \otimes p_2) \bullet (p_1 \otimes 1_A + 1_A \otimes p_1)$$

$$= (p_1 * p_2) \otimes 1_A + p_1 \otimes p_2 + p_2 \otimes p_1 + 1_A \otimes (p_1 * p_2) + \qquad (3.91)$$

$$- (p_2 * p_1) \otimes 1_A - p_2 \otimes p_1 - p_1 \otimes p_2 - 1_A \otimes (p_2 * p_1)$$

$$= (p_1 * p_2 - p_2 * p_1) \otimes 1_A + 1_A \otimes (p_1 * p_2 - p_2 * p_1)$$

$$= ([p_1, p_2]_*) \otimes 1_A + 1_A \otimes ([p_1, p_2]_*).$$

In the second equality we used the fact that the coproduct Δ is a UAA morphism of $(A, *)$ into $(A \otimes A, \bullet)$.

We remark that an analogous computation, with $\mathscr{T}(V)$ replacing A, arose in Friedrichs's Theorem 3.13, page 133.

A bialgebra A is said to be *primitively generated* if it is generated, as an algebra (that is, with respect to $*$), by the set of its primitive elements $P(A)$ (together with 1_A). A primitively generated bialgebra A is *cocommutative*, that is, the following diagram commutes:

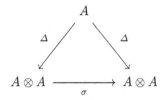

Here, as before, σ is the unique linear map such that $\sigma(a \otimes a') = a' \otimes a$, for every $a, a' \in A$ (note that σ is actually a UAA morphism). To prove the commutativity of this diagram for primitively generated bialgebras, it suffices to recall that Δ and σ are UAA morphisms, that A is generated, as an algebra, by $\{1_A\} \cup P(A)$, and to notice that, on this set of generators, we have

$$\sigma(\Delta(1_A)) = \sigma(1_A \otimes 1_A) = 1_A \otimes 1_A = \Delta(1_A),$$

$$\sigma(\Delta(p)) = \sigma(p \otimes 1_A + 1_A \otimes p) = 1_A \otimes p + p \otimes 1_A = \Delta(p),$$

for every $p \in P(A)$.

At last, we are ready for the main definition of this section.

Definition 3.37 (Hopf Algebra). A *Hopf algebra* (over the field $\mathbb{K}$) is a 6-tuple $(A, *, 1_A, \Delta, \varepsilon, S)$, where $(A, *, 1_A, \Delta, \varepsilon)$ is a bialgebra, and $S : A \to A$ is a linear map (called *the antipode*) such that the following is a commutative diagram:

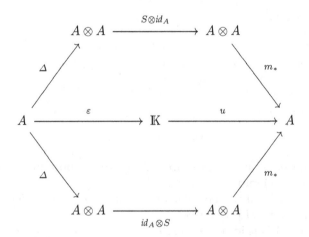

Following the notation in (3.90), this is equivalent to the requirement

$$\sum_i S(f_{i,1}(a)) * f_{i,2}(a) = \sum_i f_{i,1}(a) * S(f_{i,2}(a)) = \varepsilon(a) 1_A, \qquad (3.92)$$

for every $a \in A$.

In a Hopf algebra, the set $G(A)$ of the grouplike elements is indeed a group with respect to the multiplication of A, as we show below.

Remark 3.38. Let $(A, *, 1_A, \Delta, \varepsilon, S)$ be a Hopf algebra. The set of the grouplike elements $G(A)$ is a multiplicative subgroup of $(A, *)$. More precisely,

$$S(x) \in G(A) \quad \text{and} \quad S(x) * x = x * S(x) = 1_A, \quad \text{for every } x \in G(A).$$

Indeed, if $x \in G(A)$, we know that $\varepsilon(x) = 1_{\mathbb{K}}$, so that (as a direct application of the axioms of a Hopf algebra) we have

$$1_A = 1_{\mathbb{K}} 1_A = \varepsilon(x) 1_A = u(\varepsilon(x)) \qquad \text{(by definition of Hopf algebra)}$$
$$= (m_* \circ (S \otimes \mathrm{id}_A) \circ \Delta)(x) = m_* \circ (S \otimes \mathrm{id}_A)(x \otimes x)$$
$$= m_*(S(x) \otimes x) = S(x) * x.$$

The other identity $x * S(x) = 1_A$ is proved similarly. Finally, we obtain that $S(x) \in G(A)$ from the following arguments: First note that

$$(x \otimes x) \bullet (S(x) \otimes S(x)) = 1_A \otimes 1_A, \tag{3.93}$$

since $(x \otimes x) \bullet (S(x) \otimes S(x)) = (x * S(x)) \otimes (x * S(x)) = 1_A \otimes 1_A$. We also have

$$1_A \otimes 1_A = \Delta(1_A) = \Delta(S(x) \otimes x) = \Delta(S(x)) \bullet \Delta(x) = \Delta(S(x)) \bullet (x \otimes x),$$

whence

$$1_A \otimes 1_A = \Delta(S(x)) \bullet (x \otimes x). \tag{3.94}$$

If we multiply (w.r.t. $\bullet$) both sides of (3.93) by $\Delta(S(x))$, we get

$$\Delta(S(x)) = \Delta(S(x)) \bullet (1_A \otimes 1_A) \stackrel{(3.93)}{=} \Delta(S(x)) \bullet (x \otimes x) \bullet (S(x) \otimes S(x))$$
$$\stackrel{(3.94)}{=} 1_A \otimes 1_A \bullet (S(x) \otimes S(x)) = S(x) \otimes S(x).$$

This gives $\Delta(S(x)) = S(x) \otimes S(x)$, that is, $S(x) \in G(A)$.

We remark that a similar argument applied in Part II of the proof of Theorem 3.23 (page 143), when showing that the Hausdorff group is closed under inversion.

As our main example of a Hopf algebra, we give the following one.

Example 3.39. Let $\mathfrak{g}$ be a Lie algebra over $\mathbb{K}$ and let $\mathscr{U}(\mathfrak{g})$ be its universal enveloping algebra. We begin by claiming that *the map*

$$\delta : \mathfrak{g} \to \mathscr{U}(\mathfrak{g}) \otimes \mathscr{U}(\mathfrak{g}), \qquad \delta(x) := \mu(x) \otimes 1 + 1 \otimes \mu(x), \quad x \in \mathfrak{g}, \tag{3.95}$$

is u Lie algebra morphism, where $\mathscr{U}(\mathfrak{g}) \otimes \mathscr{U}(\mathfrak{g})$ is equipped with the Lie algebra structure resulting from the UAA structure given by the multiplication

$$(a \otimes b) \bullet (a' \otimes b') = (a \, a') \otimes (b \, b'), \qquad \forall \, a, b, a', b' \in \mathscr{U}(\mathfrak{g}).$$

As usual, simple juxtaposition denotes the multiplication in $\mathscr{U}(\mathfrak{g})$.

Indeed, to prove the above claim, we use the following computation:

$$\delta\big([x, y]_{\mathfrak{g}}\big) = \mu\big([x, y]_{\mathfrak{g}}\big) \otimes 1 + 1 \otimes \mu\big([x, y]_{\mathfrak{g}}\big) \qquad \text{(see (2.104))}$$
$$= [\mu(x), \mu(y)]_{\mathscr{U}} \otimes 1 + 1 \otimes [\mu(x), \mu(y)]_{\mathscr{U}}$$
$$= \big(\mu(x) \, \mu(y) - \mu(y) \, \mu(x)\big) \otimes 1 + 1 \otimes \big(\mu(x) \, \mu(y) - \mu(y) \, \mu(x)\big).$$

On the other hand, by arguing exactly as in (3.91), one proves that

$$[\delta(x), \delta(y)]_{\bullet} = \big(\mu(x) \, \mu(y) - \mu(y) \, \mu(x)\big) \otimes 1 + 1 \otimes \big(\mu(x) \, \mu(y) - \mu(y) \, \mu(x)\big).$$

This proves that δ is an LA morphism.

Let $A := \mathscr{U}(\mathfrak{g})^{\mathrm{op}}$ denote the opposite algebra of $\mathscr{U}(\mathfrak{g})$, that is, A has the same vector space structure as $\mathscr{U}(\mathfrak{g})$, whilst the multiplication in $\mathscr{U}(\mathfrak{g})^{\mathrm{op}}$, denoted by $\cdot_{\mathrm{op}}$, is defined by

$$a \cdot_{\mathrm{op}} b := b \, a, \qquad \forall \, a, b \in \mathscr{U}(\mathfrak{g}).$$

Clearly, $(A, \cdot_{\mathrm{op}})$ is a UA algebra, with the same unit element as that of $\mathscr{U}(\mathfrak{g})$. Consider the map

$$f : \mathfrak{g} \to A, \quad f(x) := -\mu(x) \in \mathscr{U}(\mathfrak{g}) = A.$$

We claim that f is a Lie algebra morphism of $\mathfrak{g}$ in the commutator Lie algebra of A. Indeed, for every $x, y \in \mathfrak{g}$, we have

$$[f(x), f(y)]_A = f(x) \cdot_{\mathrm{op}} f(y) - f(y) \cdot_{\mathrm{op}} f(x) = f(y)f(x) - f(x)f(y)$$
$$= (-\mu(y))(-\mu(x)) - (-\mu(x))(-\mu(y)) = \mu(y) \, \mu(x) - \mu(x) \, \mu(y)$$
$$= [\mu(y), \mu(x)]_{\mathscr{U}} = -[\mu(x), \mu(y)]_{\mathscr{U}} \overset{(2.104)}{=} -\mu([x, y]_{\mathfrak{g}}) = f([x, y]_{\mathfrak{g}}).$$

Hence, by the universal property in Theorem 2.92-(i), there exists a unique UAA morphism $f^{\mu} : \mathscr{U}(\mathfrak{g}) \to A$ such that $f^{\mu}(\mu(x)) = f(x)$, for every $x \in \mathfrak{g}$. Since the underlying vector space structure of A is precisely that of $\mathscr{U}(\mathfrak{g})$, this defines a linear map

$$S : \mathscr{U}(\mathfrak{g}) \longrightarrow \mathscr{U}(\mathfrak{g}), \qquad S(t) := f^{\mu}(t) \quad (t \in \mathscr{U}(\mathfrak{g})). \tag{3.96}$$

Obviously, S is not, in general, a UAA morphism, but it satisfies

$$S(t\,t') = S(t')\,S(t), \quad \text{for every } t, t' \in \mathcal{U}(\mathfrak{g}),$$

that is, S is an algebra antihomomorphism of $\mathcal{U}(\mathfrak{g})$.

Next, we consider the following maps:

(1) $\Delta : \mathcal{U}(\mathfrak{g}) \to \mathcal{U}(\mathfrak{g}) \otimes \mathcal{U}(\mathfrak{g})$ is the unique UAA morphism associated to the LA morphism δ in (3.95). In other words, following the notation in Theorem 2.92-(i), $\Delta = \delta^{\mu}$, that is, $\Delta : \mathcal{U}(\mathfrak{g}) \to \mathcal{U}(\mathfrak{g}) \otimes \mathcal{U}(\mathfrak{g})$ *is the unique UAA morphism such that* $\Delta(\mu(x)) = \mu(x) \otimes 1 + 1 \otimes \mu(x)$, *for every* $x \in \mathfrak{g}$.

(2) $\varepsilon : \mathcal{U}(\mathfrak{g}) \to \mathbb{K}$ is the unique UAA morphism such that $\varepsilon(\mu(x)) = 0$ for every $x \in \mathfrak{g}$.

(3) $S : \mathcal{U}(\mathfrak{g}) \to \mathcal{U}(\mathfrak{g})$ is the linear map defined in (3.96).

It can be proved that, *with the above coproduct Δ, counit ε and antipode S, $\mathcal{U}(\mathfrak{g})$ (together with its UAA algebra structure) is a Hopf algebra.* Furthermore, $\mathcal{U}(\mathfrak{g})$ is primitively generated (hence cocommutative), since $\{1\} \cup \mu(\mathfrak{g})$ is a set of algebra-generators for $\mathcal{U}(\mathfrak{g})$ (note that $\mu(\mathfrak{g}) \subseteq P(\mathcal{U}(\mathfrak{g}))$, by the very definition of Δ).

Remark 3.40. With the structures considered in Example 3.39, we claim that, *if $\mathbb{K}$ has characteristic zero, the set $P(\mathcal{U}(\mathfrak{g}))$ of the primitive elements of the bialgebra $\mathcal{U}(g)$ coincides with $\mu(\mathfrak{g}) \simeq \mathfrak{g}$.*

This can be proved *by rerunning the arguments in the proof of Friedrichs's Theorem 3.13* (see Cartier [34, Theorem 3.6.1]); note that Poincaré-Birkhoff-Witt's Theorem 2.94 is also needed. Indeed it suffices to replace, in the arguments on pages 134–135, $\mathcal{T}(V)$ with $\mathcal{U}(\mathfrak{g})$ and δ with the coproduct Δ.

So far we have pointed out many circumstances where the material presented in this chapter may be properly restated in terms of Hopf algebra theory: for instance, this happens in Friedrichs's Theorems 3.13 and 3.14 (the characterizations of $\mathcal{L}(V)$ and $\overline{\mathcal{L}(V)}$ as the set of primitive elements w.r.t. δ and $\widehat{\delta}$, respectively), or when dealing with the Hausdorff group $\Gamma(V)$ (the set of grouplike elements w.r.t. $\widehat{\delta}$, see Theorem 3.23).

As we have already highlighted, the core of the proof of the CBH Theorem furnished here relies on establishing the bijection between $\overline{\mathcal{L}(V)}$ and $\Gamma(V)$ via Exp (see Theorem 3.24), which is well pictured in Fig. 3.1 (page 146), where $\widehat{\delta}$-primitive elements and $\widehat{\delta}$-grouplike elements are involved. This definitely exhibits the likely existence of *a deep connection between the formalism behind the CBHD Theorem and Hopf algebra theory.*

For an in-depth analysis of the Hopf algebra structure of $\mathcal{U}(\mathfrak{g})$, the Reader is referred to the pioneering work by Milnor and Moore [120]. For example, Theorem 5.18 in [120, page 244] characterizes (in terms of suitable functors) the universal enveloping algebras $\mathcal{U}(\mathfrak{g})$ of the Lie algebras $\mathfrak{g}$ (over a field of characteristic zero) as the *primitively generated Hopf algebras*. For further characterizations, see Cartier [34, Section 3.8]. We also remark that, within the Hopf algebra theory presented in [120], the Poincaré-Birkhoff-Witt Theorem also assumes a natural place (see [120, Theorem 5.15, page 243]). Since this theorem plays a crucial rôle in the proof of the CBHD Theorem (as we will show with full particulars in Chap. 6), we have another piece of evidence for a connection between Hopf algebra theory and the topics of this Book.

Chapter 4
Some "Short" Proofs of the CBHD Theorem

THE aim of this chapter is to give all the details of five other proofs (besides the one given in Chap. 3) of the Campbell, Baker, Hausdorff Theorem, stating that $x \blacklozenge y := \mathrm{Log}(\mathrm{Exp}(x) \cdot \mathrm{Exp}(y))$ is a series of Lie polynomials in x, y. As we showed in Chap. 3, this is the "qualitative" part of the CBHD Theorem, and the actual formula expressing $x \blacklozenge y$ as an explicit series (that is, Dynkin's Formula) can be quite easily derived from this qualitative counterpart as exhibited in Sect. 3.3.

These proofs are – in the order presented here – respectively, by M. Eichler [59], D. Ž. Djoković [48], V. S. Varadarajan [171], C. Reutenauer [144], P. Cartier [33]. They are not presented in the chronological order in which they appeared in the literature, but rather we have followed a criterion adapted to our exposition: Eichler's proof requires the least prerequisites; Djoković's proof allows us to introduce all the machinery needed to make precise the manipulation of formal power series (and of series of endomorphisms) also used in the subsequent proofs; Varadarajan's proof may be viewed as the correct and most natural continuation of Djoković's; Reutenauer's proof has in common with Djoković's some core computations; Cartier's proof makes use of slightly more algebraic prerequisites and it leads us back to the ideas of Chap. 3.

The actual chronological order is:

Author	Paper, Book	Date
Cartier	[33]	1956
Eichler	[59]	1968
Varadarajan	[171] (book, 1st edition)	1974
Djoković	[48]	1975
Reutenauer	[144] (book)	1993

We decided to term the proofs presented here as "short" proofs of the CBHD Theorem (actually – we should have said – of the Campbell, Baker,

A. Bonfiglioli and R. Fulci, *Topics in Noncommutative Algebra*, Lecture Notes in Mathematics 2034, DOI 10.1007/978-3-642-22597-0_4, © Springer-Verlag Berlin Heidelberg 2012

Hausdorff Theorem). Indeed, an undoubted quality of compendiousness (if compared to the approach of the Bourbakist proof in Chap. 3) is evident for all these proofs; nonetheless we mitigated the adjective *short* by putting it in quotes, since these five proofs necessitate a certain amount of background prerequisites, not completely perspicuous from the original proofs. In fact, this Book should have convinced the Reader that a three-page proof (as in Eichler [59] or in Djoković [48]) of the Campbell-Baker-Hausdorff Theorem would be asking too much!

Before presenting the proofs, we would like to underline a few of their characteristic features, tracing some parallels and differences.

The first due remark is devoted to Varadarajan's argument. Indeed, whereas the other four proofs are in a purely algebraic setting, Varadarajan's original investigation of the Baker, Campbell, Hausdorff Formula in [171, Section 2.15] is concerned with the version of this theorem related to the Lie group setting. Notwithstanding this, the argument in [171] is so fitting with Djoković's, that we felt forced to think that Djoković's proof was somewhat incomplete without the remarkable recursion formula given by Varadarajan (this formula having been implicit in Hausdorff, see [78, eq. (29), page 31]). Moreover, the argument in [171] is so easily generalized to the abstract algebraic setting, that we also felt free to adapt it to our purpose, even though its original exposition is another one.

As we said, among the five proofs, Eichler's is undoubtedly the most devoid of prerequisites, apart from the existence of free Lie algebras (over two or three indeterminates). Even if this is a clear advantage (why did we not present Eichler's proof of the CBHD Theorem in the first Chapter of this Book, without the need of a foregoing – long – chapter of backgrounds?), it must be said that this ingenious tricky proof has the disadvantage of concealing the nature of the CBHD Theorem as being a result crossing the realms of Algebra, Analysis and Geometry. Moreover, Eichler's argument says nothing about possible recursive relations between the coefficients of the CBHD series, a crucial fact which is instead close at hand in Djoković's and in Reutenauer's proofs and completely transparent in Varadarajan's.

The contribution of Djoković's proof to the understanding of the CBHD Theorem is a palpable use of formal power series, in such a way that the rôle of some core computations (tracing back to Baker and Hausdorff) is made clearer than it appears from the Bourbakist approach. For instance, Djoković's computations recover – in a direct fashion – the milestone result that (roughly)

$$Z(t) := \mathrm{Log}\big(\mathrm{Exp}(t\,x) \cdot \mathrm{Exp}(t\,y)\big)$$

satisfies an "ordinary differential equation": solving this formal ODE (by the power series *Ansatz* method) allows us to discover that the coefficients of the CBHD series satisfy recursive relations, revealing their Lie-polynomial nature. The reverse side of the medal when it comes to formal handling of power series is that some arguments are not set into their proper algebraic context: For instance, what is the rôle of t in $Z(t)$? Is it a real parameter? Is it a free indeterminate? And then, what are the rôles of x, y? And what about convergence? [For example, these rôles are all clear in Varadarajan's original context: x, y are vector fields in the Lie algebra of a Lie group, t is a real parameter, and $Z(t)$ satisfies a genuine ODE on the Lie algebra.] To set things straight, we preferred to add to Djoković's original arguments (at the risk of straining the Reader's patience) all the needed algebraic background about power series of endomorphisms and formal power series in one indeterminate t, with coefficients in $\mathscr{T}(\mathbb{Q}\langle x, y \rangle)$ (the free associative $\mathbb{Q}$-algebra of the polynomials in two non-commuting indeterminates).

As already remarked, Varadarajan's explicit computations on the cited recursive relations furnish a sort of "prolongation" of Djoković's proof, both giving a manifest demonstration of the Lie-nature of $x \blacklozenge y$ and providing a recursion formula extremely powerful for handling convergence matters (a contribution which is not given by any of the other proofs of the Campbell, Baker, Hausdorff Theorem presented in this Book).

Next, Reutenauer's proof has in common with Djoković's a core computation, proving that (whenever D is a derivation) we have

$$D(\operatorname{Exp} x) * \operatorname{Exp}(-x) = \sum_{n=1}^{\infty} \frac{1}{n!} (\operatorname{ad} x)^{n-1} (D(x)).$$

This is the crucial part where brackets step in and (if properly generalized and if used for other suitable derivations as in Reutenauer's catchy argument) it allows us to recover and formalize the original ideas by Baker and Hausdorff themselves.

Finally, Cartier's proof – if compared to the others – is certainly the nearest to the approach given in Chap. 3, hence preserving the natural multi-faceted algebraic/analytic/geometric flavor of the CBHD Theorem. This proof furnishes a balanced *summa* of some explicit computations (such as those cited in Djoković's argument) together with the idea of obtaining suitable characterizations of Lie-elements (as in Friedrichs's Theorem from Chap. 3), having an interest in their own. Moreover, some remarkable algebraic arguments (such as those involved in the Lemma of Dynkin, Specht, Wever) are further exploited and clarified.

To end this introduction, we provide an overview of the chapter.

Warning: First, a general warning is due. In this chapter, we shall encounter some specific graded algebras $(A, *)$, with the associated metric space of the formal power series $(\widehat{A}, \widehat{*})$, the relevant exponential and logarithmic maps Exp_*, Log_*, and the relevant $\blacklozenge$ operation:

$$u \blacklozenge v = \operatorname{Log}_*(\operatorname{Exp}_*(u) \,\widehat{*}\, \operatorname{Exp}_*(v)).$$

At the risk of appearing tedious, we shall safeguard different notations according to the different algebras we shall encounter: this is intended to preserve the specificity of the contexts and of each CBHD Formula associated to each of these contexts. Indeed, if the use of a unified notation $\mathrm{Log}(\mathrm{Exp}(u) \cdot \mathrm{Exp}(v))$ has the clear advantage of avoiding a proliferation of symbols, the reverse of the medal is to induce the Reader to believe that there exists a single CBHD Theorem. Instead, we trust it is an interesting fact to establish what kind of CBHD formulas allow us to recover the CBHD Theorem 3.8 for the general algebra $\widehat{\mathscr{T}}(V)$, where V is any vector space over a field of null characteristic. This will be an item of investigation too.

Here is the plan of the chapter:

- To begin with, we furnish the precise statement of the CBHD Theorem in the context of formal power series in two free indeterminates over $\mathbb{Q}$, that is, the algebra $\widehat{\mathscr{T}}(\mathbb{Q}\langle x, y \rangle)$, and we show how to derive the general CBHD Theorem 3.8 from this (Sect. 4.1).

- Section 4.2 contains, without drawing breath, Eichler's proof of the Campbell, Baker, Hausdorff Theorem. This is formulated in the above context of $\widehat{\mathscr{T}}(\mathbb{Q}\langle x, y \rangle)$ and it says that, written

$$\log\left(e^x \mathbin{\widehat{}} e^y\right) = \sum_{j=1}^{\infty} \mathbf{F}_j(x, y),$$

where $\mathbf{F}_j(x, y)$ is a homogeneous polynomial in x, y of degree j, then $\mathbf{F}_j(x, y)$ is in fact a Lie-polynomial in x, y. The proof rests on an ingenious use of the identity

$$\left(e^A \mathbin{\widehat{}} e^B\right) \mathbin{\widehat{}} e^C = e^A \mathbin{\widehat{}} \left(e^B \mathbin{\widehat{}} e^C\right),$$

where A, B, C are three free non-commuting indeterminates. This fact has an independent interest for it illuminates the fact that the CBH Theorem is somewhat implicit in the *associativity* of the composition law, a fact that has many echoes in Lie group theory.

- Section 4.3 contains a detailed version of Djokovic's proof of the Campbell, Baker, Hausdorff Theorem. First, we need to introduce the relevant setting (see Sect. 4.3.1): Given a UA algebra A, we consider the algebra $A[t]$ of polynomials in the indeterminate t with coefficients in A; this is a graded algebra whose completion is $A[[t]]$, the algebra of formal power series in t and coefficients in A. We introduce in $A[[t]]$ the operator ∂_t (the formal derivative with respect to t), and we establish several of its remarkable properties. Then, through Sect. 4.3.2 we get closer and closer to Djokovic's original argument by making the choice of A: we take $A = \mathscr{T}(\mathbb{Q}\langle x, y \rangle)$ and we consider the element of $A[[t]]$ defined by

$$Z := \log(\exp(x\, t) * \exp(y\, t)).$$

We then discover that the actual expression of Z as a series in t is

$$Z = \sum_{j=1}^{\infty} \mathbf{F}_j(x, y)\, t^j,$$

where the $\mathbf{F}_j(x, y)$ are the same as in Eichler's proof. We further give all the details for making a precise use of formal series of endomorphisms of $A[[t]]$. We are thus ready for Djokovic's proof (Sect. 4.3.3): with all the former machinery at hands, we discover that Z satisfies a formal ODE, roughly writable as

$$\partial_t Z = \frac{\operatorname{ad} Z}{e^{\operatorname{ad} Z} - 1}\left(x + e^{\operatorname{ad} Z}(y)\right).$$

By unraveling this identity with respect to the coefficients $\mathbf{F}_j(x, y)$ of Z, we obtain a (qualitative) argument ensuring that $\mathbf{F}_{j+1}(x, y)$ is a Lie-polynomial, once this is true of $\mathbf{F}_1(x, y), \ldots, \mathbf{F}_j(x, y)$. Since $\mathbf{F}_1(x, y) = x + y$, we are done.

- Once Djokovic's differential equation for Z is established, in Sect. 4.5 we present an adaptation of Varadarajan's argument in [171, Section 2.15] to write this equation in the following more symmetric form (note the presence of $x \pm y$ instead of x and y)

$$\partial_t Z = \frac{\frac{1}{2}\operatorname{ad} Z}{\tanh\left(\frac{1}{2}\operatorname{ad} Z\right)}(x + y) - \tfrac{1}{2}(\operatorname{ad} Z)(x - y).$$

By substituting the formal power series for Z, this gives an explicit recursion formula for the coefficients $\mathbf{F}_j$, also involving the remarkable Bernoulli numbers. Besides its usefulness in deriving the Lie-polynomial nature of the $\mathbf{F}_j$, this formula will appear as a rewarding tool for the study of *convergence* matters (this will be done in Sect. 5.2.3). In Sect. 4.5.2, we exhibit another compact form of writing the "ODE" for Z, following G. Czichowski [44]:

$$\partial_t Z = \frac{\operatorname{ad} Z}{e^{\operatorname{ad} Z} - 1}(x) + \frac{-\operatorname{ad} Z}{e^{-\operatorname{ad} Z} - 1}(y).$$

This will allow us to write another (more compact) recursion formula for the coefficients $\mathbf{F}_j$.

- If we replace the above operator ∂_t with some "partial differential operators" (actually, derivations of the algebra $\widehat{\mathscr{T}}(\mathbb{Q}\langle x, y\rangle)$) of the form

$$D_x = H_1^y \frac{\partial}{\partial x} \quad \text{or} \quad D_y = H_1^x \frac{\partial}{\partial y}$$

(all will be clarified in Sect. 4.6), we discover that

$$H(x, y) := \mathrm{Log}(\mathrm{Exp}\, x \cdot \mathrm{Exp}\, y)$$

satisfies some sort of PDE identities looking like

$$H(x, y) = \exp(D_x)(x) \quad \text{or} \quad H(x, y) = \exp(D_y)(y).$$

This is, roughly, the leading idea in Reutenauer's proof, presented in Sect. 4.6. The rôle of the series H_1^x and H_1^y (which are obtained by gathering the summands in the series of $H(x, y)$ containing, respectively, x and y with degree 1) is paramount in Reutenauer's approach. Since this is true also of the early proofs by Pascal, Baker and Hausdorff, it follows that Reutenauer's argument gives us the chance of understanding these early approaches, getting to the core of the CBHD Theorem both mathematically and historically.

- Finally, Sect. 4.7 contains the details of Cartier's proof of the Campbell, Baker, Hausdorff Theorem. This rests on a new *ad hoc* characterization of $\overline{\mathcal{L}(V)}$ playing a rôle similar to that played by Friedrichs's characterization in the proof from Chap. 3. First, it is necessary to establish the well-behaved properties of some remarkable maps (Sect. 4.7.1), similar to those involved in the proof of the Lemma of Dynkin, Specht, Wever (see Sect. 3.3.1). By means of these maps, Cartier gives a characterization of $\mathrm{Exp}(\overline{\mathcal{L}(V)})$ (see Sect. 4.7.2), that is, of the so called Hausdorff group related to V (see Sect. 3.2.3). We remark that Cartier's proof directly applies to the general case of $\widehat{\mathscr{T}}(V)$, without the need to pass through the formal power series in two indeterminates, as in the other proofs presented in this chapter.

4.1 Statement of the CBHD Theorem for Formal Power Series in Two Indeterminates

Let us begin by collecting the due notation and definitions. The well known notation in (3.20), i.e.,

$$\mathcal{N}_n := \Big\{ (h, k) \,\big|\, h, k \in (\mathbb{N} \cup \{0\})^n, \ (h_1, k_1), \ldots, (h_n, k_n) \neq (0, 0) \Big\},$$

$$c_n := \frac{(-1)^{n+1}}{n}, \quad \mathbf{c}(h, k) := \frac{1}{h!\, k!\, (|h| + |k|)},$$

will apply throughout. We also give the following useful definition.

Definition 4.1. Let $(A, *)$ be an associative algebra (over a field of characteristic zero). Let $j \in \mathbb{N}$ be fixed. We set[1]

$$\mathbf{F}_j^A : A \times A \longrightarrow A$$

$$\mathbf{F}_j^A(u, v) := \sum_{n=1}^{j} c_n \sum_{(h,k) \in \mathcal{N}_n : |h| + |k| = j} \frac{u^{*h_1} * v^{*k_1} * \cdots * u^{*h_n} * v^{*k_n}}{h_1! \cdots h_n! \, k_1! \cdots k_n!}. \tag{4.1}$$

When there is no possibility of confusion, we shall also use the shorter notation

$$\mathbf{F}_j^* := \mathbf{F}_j^A. \tag{4.2}$$

The Reader has certainly recognized the similarity between $\mathbf{F}_j^A(u, v)$ and the term in parentheses in (3.23) on page 128, the latter intervening in the definition of the important $\blacklozenge$ operation. Indeed, if V is any vector space over a field of characteristic zero, if A is $\widehat{\mathscr{T}}(V)$, i.e., the algebra of the formal power series equipped with the usual Cauchy operation in (2.91) (page 105), and if finally $\blacklozenge$ is the operation introduced in (3.16) (page 126), then we have

$$\sum_{j=1}^{\infty} \mathbf{F}_j^{\widehat{\mathscr{T}}(V)}(u, v) = u \blacklozenge v, \qquad \forall\, u, v \in \widehat{\mathscr{T}}_+(V).$$

The maps $\mathbf{F}_j$ in Definition 4.1 have a sort of "universal" property, stated in the following:

Proposition 4.2. *Let A, B be two associative algebras and suppose $\varphi : A \to B$ is an algebra morphism. Then*

$$\varphi(\mathbf{F}_j^A(u, v)) = \mathbf{F}_j^B(\varphi(u), \varphi(v)), \quad \text{for every } u, v \in A \text{ and every } j \in \mathbb{N}. \tag{4.3}$$

In particular, if A is a subalgebra of B, we have

$$\mathbf{F}_j^A(u, v) = \mathbf{F}_j^B(u, v), \quad \text{for every } u, v \in A \text{ and every } j \in \mathbb{N}. \tag{4.4}$$

Proof. It follows immediately from the very definition (4.1) of the maps $\mathbf{F}_j$ and the fact that φ is an algebra morphism. $\qquad\square$

A direct computation gives:

$$\mathbf{F}_1^A(u, v) = u + v, \tag{4.5a}$$

---

[1]Since A may not contain an identity element, we need to clarify what we mean for u^{*h} if $h = 0$: When some of the h_i or k_i in (4.1) are null, we understand that the relevant factors u^{*h_i} and v^{*k_i} simply do not appear in the formula.

$$\mathbf{F}_2^A(u, v) = \frac{1}{2}\, [u, v]_*. \tag{4.5b}$$

Indeed, (4.5a) is trivial, whereas for (4.5b) we have

$$\mathbf{F}_2^A(u, v) = \sum_{(h_1, k_1):\, |h_1|+|k_1|=2} \frac{u^{*h_1} * v^{*k_1}}{h_1!\, k_1!} +$$

$$- \frac{1}{2} \sum_{\substack{(h_1,k_1),\,(h_2,k_2)\neq(0,0) \\ h_1+h_2+k_1+k_2=2}} \frac{u^{*h_1} * v^{*k_1} * u^{*h_2} * v^{*k_2}}{h_1!h_2!\, k_1!k_2!}$$

$$= \frac{u^{*2}}{2} + u * v + \frac{v^{*2}}{2} - \frac{1}{2}\big(u^{*2} + u * v + v * u + v^{*2}\big)$$

$$= \frac{u * v}{2} - \frac{v * u}{2} = \frac{1}{2}\, [u, v]_*.$$

This simple computation offers a glimpse of a fundamental fact concerning the maps $\mathbf{F}_j$ (a fact which we have already familiarized with in Chap. 3), that is, $\mathbf{F}_j^A(u, v)$ *is in fact a homogeneous Lie-polynomial of degree j in u, v, in the commutator-algebra associated to* $(A, *)$. We shall soon prove this fact explicitly.

We introduce the algebra we shall be primarily dealing with in this chapter.

Let $\{x, y\}$ be a set of cardinality 2. Let $\mathbb{Q}$ denote the field of rational numbers.[2] As usual,

$$\mathscr{T}(\mathbb{Q}\langle x, y\rangle)$$

denotes the tensor algebra of the free vector space $\mathbb{Q}\langle x, y\rangle$. [We know from Theorem 2.40 on page 79 that $\mathscr{T}(\mathbb{Q}\langle x, y\rangle)$ is isomorphic to the free associative algebra over the set $\{x, y\}$ and can be thus thought of as the associative algebra of words in x, y.] Then we consider the *algebra over $\mathbb{Q}$ of the formal power series in the two non-commuting indeterminates x, y*, namely (recalling the notation in Sect. 2.3.4, page 106)

$$\widehat{\mathscr{T}}(\mathbb{Q}\langle x, y\rangle).$$

As usual, following the general exposition in Sect. 3.1.1 (page 119) about exponentials and logarithms, since $\widehat{\mathscr{T}}(\mathbb{Q}\langle x, y\rangle)$ is the algebra of formal power series related to the *graded* algebra

---

[2]Without the possibility of confusion, if $\mathbb{K}$ is any field of characteristic zero, we shall also denote by $\mathbb{Q}$ the field of the rational numbers in $\mathbb{K}$, i.e., the least subfield of $\mathbb{K}$ containing $1_{\mathbb{K}}$.

$$\mathscr{T}(\mathbb{Q}\langle x, y\rangle) = \bigoplus_{j\geq 0} \mathscr{T}_j(\mathbb{Q}\langle x, y\rangle),$$

we are entitled to consider the relevant exponential function Exp.

To underline the specificity of the present setting, we shall resume the older notation $\widehat{}$ for the Cauchy operation on $\widehat{\mathscr{T}}(\mathbb{Q}\langle x, y\rangle)$ (see (2.82) in Definition 2.73, page 101) and we shall use the notation $z \mapsto \mathbf{e}^z$ for the relevant exponential function $z \mapsto \mathrm{Exp}(z)$, i.e.,

$$\mathbf{e}^z := \sum_{k=0}^{\infty} \frac{z^{\widehat{}k}}{k!}, \qquad \forall\, z \in \widehat{\mathscr{T}}_+(\mathbb{Q}\langle x, y\rangle).$$

The inverse of

$$\widehat{\mathscr{T}}_+(\mathbb{Q}\langle x, y\rangle) \ni z \mapsto \mathbf{e}^z \in 1 + \widehat{\mathscr{T}}_+(\mathbb{Q}\langle x, y\rangle)$$

is denoted by **log**. Finally, any element in

$$\mathcal{L}(\mathbb{Q}\langle x, y\rangle)$$

(that is, the free Lie algebra generated by the vector space $\mathbb{Q}\langle x, y\rangle$) will be called *a Lie polynomial in* x, y. In particular, any element of $\mathcal{L}_j(\mathbb{Q}\langle x, y\rangle)$ will be called a *homogeneous* Lie polynomial in x, y of (joint) degree j. (See Definition 2.46, page 85, for the relevant definitions and see (2.49) for an explanation of the $\mathcal{L}_j$ notation.)

The closure of $\mathcal{L}(\mathbb{Q}\langle x, y\rangle)$ in the topological space $\widehat{\mathscr{T}}(\mathbb{Q}\langle x, y\rangle)$ is denoted

$$\overline{\mathcal{L}(\mathbb{Q}\langle x, y\rangle)}.$$

With all the above notations at hand, the aim of this chapter is to provide short proofs of the following theorem, which we shall refer to as *the CBHD Theorem for formal power series in two non-commuting indeterminates.*

Theorem 4.3 (CBHD for $\widehat{\mathscr{T}}(\mathbb{Q}\langle x, y\rangle)$). *Using the above notation, we have:*

$$\mathbf{F}_j^{\mathscr{T}(\mathbb{Q}\langle x,y\rangle)}(x, y) \in \mathcal{L}_j(\mathbb{Q}\langle x, y\rangle), \quad \text{for every } j \in \mathbb{N}; \tag{4.6}$$

$$\mathbf{F}(x, y) := \sum_{j=1}^{\infty} \mathbf{F}_j^{\mathscr{T}(\mathbb{Q}\langle x,y\rangle)}(x, y) \quad \text{belongs to } \overline{\mathcal{L}(\mathbb{Q}\langle x, y\rangle)}; \tag{4.7}$$

$$\mathbf{F}(x, y) = \mathbf{log}(\mathbf{e}^x \,\widehat{}\, \mathbf{e}^y); \tag{4.8}$$

$$\mathbf{e}^{\mathbf{F}(x,y)} = \mathbf{e}^x \,\widehat{}\, \mathbf{e}^y. \tag{4.9}$$

Remark 4.4. We show that it is easy to prove the following implications:

$$(4.6) \implies (4.7) \implies (4.8) \implies (4.9).$$

Obviously, (4.9) follows from (4.8). Moreover, (4.7) is a consequence of (4.6) jointly with (see (3.45) at page 138)

$$\overline{\mathcal{L}(\mathbb{Q}\langle x, y \rangle)} = \prod_{j=1}^{\infty} \mathcal{L}_j(\mathbb{Q}\langle x, y \rangle).$$

Furthermore, (4.8) is a consequence of the definition of $\mathbf{F}$ in (4.7) (and the very definition of $\mathbf{F}_j$ in (4.1)). Indeed, we derive (4.8) by a straightforward computation (see e.g., (3.18) page 127):

$$\log(e^x \hat{\frown} e^y) = \sum_{n=1}^{\infty} \frac{(-1)^{n+1}}{n} \left(\sum_{(h,k) \neq (0,0)} \frac{x^{\hat{\frown} h} \hat{\frown} y^{\hat{\frown} k}}{h!\, k!} \right)^n$$

$$= \sum_{n=1}^{\infty} c_n \sum_{(h,k) \in \mathcal{N}_n} \frac{x^{\hat{\frown} h_1} \hat{\frown} y^{\hat{\frown} k_1} \hat{\frown} \cdots \hat{\frown} x^{\hat{\frown} h_n} \hat{\frown} y^{\hat{\frown} k_n}}{h_1! \cdots h_n!\, k_1! \cdots k_n!}$$

(by reordering the sum) $\qquad\qquad\qquad\qquad\qquad\qquad$ (4.10)

$$= \sum_{j=1}^{\infty} \left(\sum_{n=1}^{j} c_n \sum_{(h,k) \in \mathcal{N}_n : |h|+|k|=j} \frac{x^{\hat{\frown} h_1} \hat{\frown} y^{\hat{\frown} k_1} \hat{\frown} \cdots \hat{\frown} x^{\hat{\frown} h_n} \hat{\frown} y^{\hat{\frown} k_n}}{h_1! \cdots h_n!\, k_1! \cdots k_n!} \right)$$

$$= \sum_{j=1}^{\infty} \left(\sum_{n=1}^{j} c_n \sum_{(h,k) \in \mathcal{N}_n : |h|+|k|=j} \frac{x^{\cdot h_1} \cdot y^{\cdot k_1} \cdot \cdots \cdot x^{\cdot h_n} \cdot y^{\cdot k_n}}{h_1! \cdots h_n!\, k_1! \cdots k_n!} \right)$$

$$\overset{(4.1)}{=} \sum_{j=1}^{\infty} \mathbf{F}_j^{\mathcal{T}(\mathbb{Q}\langle x, y \rangle)}(x, y) = \mathbf{F}(x, y), \quad \text{by the definition of } \mathbf{F}(x, y)$$

in (4.7).

Hence, *the main task in Theorem 4.3 is to prove* (4.6).

The proof of (4.6) is the topic of Sects. 4.2, 4.3, 4.5, 4.7 below. But before proceeding, we would like to show that Theorem 4.3 – though it is formulated in the very special setting of $\widehat{\mathcal{T}}(\mathbb{Q}\langle x, y \rangle)$ – gives the Campbell, Baker, Hausdorff Theorem 3.20, in the following even stronger form.

Corollary 4.5. *Let V be a vector space (over a field $\mathbb{K}$ of characteristic zero). Consider the usual Exponential/Logarithm maps*

$$\widehat{\mathcal{T}}_+(V) \underset{\text{Log}}{\overset{\text{Exp}}{\rightleftarrows}} 1 + \widehat{\mathcal{T}}_+(V).$$ $\qquad\qquad\qquad$ (4.11)

Then, for every $u, v \in \widehat{\mathscr{T}_+}(V)$, the unique solution $Z(u,v)$ of

$$\mathrm{Exp}(Z(u,v)) = \mathrm{Exp}\, u \cdot \mathrm{Exp}\, v \qquad (4.12)$$

is given by (see also Definition 4.1)

$$Z(u,v) = \mathrm{Log}(\mathrm{Exp}\, u \cdot \mathrm{Exp}\, v) = \sum_{j=1}^{\infty} \mathbf{F}_j^{\widehat{\mathscr{T}}(V)}(u,v). \qquad (4.13)$$

Moreover, thanks to (4.6) we infer that, for every $j \in \mathbb{N}$, the general summand $\mathbf{F}_j^{\widehat{\mathscr{T}}(V)}(u,v)$ of the above series belongs to $\mathrm{Lie}\{u,v\}$ (the least Lie subalgebra of the commutator-algebra of $\widehat{\mathscr{T}}(V)$ containing u, v and can be expressed by a "universal" linear combination (with rational coefficients) of iterated Lie brackets of length j in u, v (jointly). Consequently,

$$\mathrm{Log}(\mathrm{Exp}\, u \cdot \mathrm{Exp}\, v) \in \overline{\mathrm{Lie}\{u,v\}}, \qquad (4.14)$$

the closure being taken in $\widehat{\mathscr{T}}(V)$ with the usual topology so that, as a particular case, (3.51) in the Campbell, Baker, Hausdorff Theorem 3.20 holds true.

Proof. Let $u, v \in \widehat{\mathscr{T}_+}(V)$. Since the maps in (4.11) are inverse to each other and $\mathrm{Exp}\, u \cdot \mathrm{Exp}\, v$ belongs to $1 + \widehat{\mathscr{T}_+}$, then (4.12) has the unique solution $Z(u,v) := \mathrm{Log}(\mathrm{Exp}\, u \cdot \mathrm{Exp}\, v)$. By Theorem 2.85-(1a), there exists a unique UAA morphism $\Phi_{u,v} : \mathscr{T}(\mathbb{K}\langle x,y\rangle) \to \widehat{\mathscr{T}}(V)$ such that

$$(\bigstar 1) \qquad\qquad \Phi_{u,v}(x) = u \quad \text{and} \quad \Phi_{u,v}(y) = v.$$

It is easily seen that, for every $k \in \mathbb{N}$, one has

$$(\bigstar 2) \qquad\qquad \Phi_{u,v}\left(\bigoplus_{j \geq k} \mathscr{T}_j(\mathbb{K}\langle x,y\rangle) \right) \subseteq \prod_{j \geq k} \mathscr{T}_j(V).$$

Hence, arguing as in the proof of Theorem 2.79 (page 103), we derive from $(\bigstar 2)$ that $\Phi_{u,v}$ is uniformly continuous and can be prolonged to a continuous UAA morphism

$$\widehat{\Phi}_{u,v} : \widehat{\mathscr{T}}(\mathbb{K}\langle x,y\rangle) \to \widehat{\mathscr{T}}(V).$$

We next consider the identity

$$(\bigstar 3) \qquad\qquad \log\left(e^x \widehat{\cdot}\, e^y\right) = \sum_{j=1}^{\infty} \mathbf{F}_j^{\mathscr{T}(\mathbb{Q}\langle x,y\rangle)}(x,y),$$

deriving from (4.7) and (4.8) in Theorem 4.3. We can apply $\widehat{\Phi}_{u,v}$ to both sides of ($\bigstar$3) since $\widehat{\mathscr{T}}(\mathbb{Q}\langle x,y\rangle)$ can be viewed as a subspace of $\widehat{\mathscr{T}}(\mathbb{K}\langle x,y\rangle)$. The fact that $\widehat{\Phi}_{u,v}$ is a continuous UAA morphism satisfying ($\bigstar$1) (and an application of Proposition 4.2) ensures that, by applying $\widehat{\Phi}_{u,v}$, we produce the identity

$$(\bigstar 4) \qquad \mathrm{Log}\big(\mathrm{Exp}(u)\cdot\mathrm{Exp}(v)\big) = \sum_{j=1}^{\infty} \mathbf{F}_j^{\widehat{\mathscr{T}}(V)}(u,v),$$

and (4.13) is completely proved.

If we consider (4.6) and the fact that $\Phi_{u,v}$ is a Lie algebra morphism of the commutator-algebras of $\mathscr{T}(\mathbb{K}\langle x,y\rangle)$ and $\widehat{\mathscr{T}}(V)$, we get

$$\mathbf{F}_j^{\widehat{\mathscr{T}}(V)}(u,v) \overset{(4.3)}{=} \Phi_{u,v}\big(\mathbf{F}_j^{\widehat{\mathscr{T}}(\mathbb{Q}\langle x,y\rangle)}(x,y)\big) = \Phi_{u,v}\big(\mathbf{F}_j^{\mathscr{T}(\mathbb{Q}\langle x,y\rangle)}(x,y)\big)$$

$$\in \Phi_{u,v}\big(\mathcal{L}(\mathbb{Q}\langle x,y\rangle)\big) \subseteq \mathrm{Lie}\{u,v\},$$

for every $j \in \mathbb{N}$. Summing up for $j \in \mathbb{N}$, we get

$$\mathrm{Log}\big(\mathrm{Exp}(u)\cdot\mathrm{Exp}(v)\big) \overset{(\bigstar 4)}{=} \sum_{j=1}^{\infty} \mathbf{F}_j^{\widehat{\mathscr{T}}(V)}(u,v) \in \overline{\mathrm{Lie}\{u,v\}},$$

and (4.14) follows.

Since, for every $j \in \mathbb{N}$, (4.6) states that $\mathbf{F}_j^{\mathscr{T}(\mathbb{Q}\langle x,y\rangle)}(x,y) \in \mathcal{L}_j(\mathbb{Q}\langle x,y\rangle)$, then the identity (proved above)

$$\mathbf{F}_j^{\widehat{\mathscr{T}}(V)}(u,v) = \Phi_{u,v}\big(\mathbf{F}_j^{\mathscr{T}(\mathbb{Q}\langle x,y\rangle)}(x,y)\big) \qquad (4.15)$$

guarantees that $\mathbf{F}_j^{\widehat{\mathscr{T}}(V)}(u,v)$ can be expressed by a "universal" linear combination – with rational coefficients – of iterated Lie brackets of length j in u,v (jointly), this universal linear combination being the one related to the Lie element $\mathbf{F}_j^{\mathscr{T}(\mathbb{Q}\langle x,y\rangle)}(x,y)$ (recall (4.6)!).

Finally, we derive (3.51) in the Campbell, Baker, Hausdorff Theorem 3.20. Indeed, if we take $u,v \in \overline{\mathcal{L}(V)}$, we obtain

$$\mathrm{Log}(\mathrm{Exp}\,u \cdot \mathrm{Exp}\,v) \in \overline{\mathrm{Lie}\{u,v\}} \in \overline{\mathrm{Lie}\{\overline{\mathcal{L}(V)}\}} = \overline{\mathcal{L}(V)}.$$

Here we applied the fact that $\mathrm{Lie}\{\overline{\mathcal{L}(V)}\} = \overline{\mathcal{L}(V)}$, since $\overline{\mathcal{L}(V)}$ is a Lie subalgebra of $\widehat{\mathscr{T}}(V)$ (see Remark 3.17). The proof is thus complete. $\qquad\square$

Another remarkable consequence of Theorem 4.3 (and of the Dynkin, Specht, Wever Lemma 3.26) is the following:

Corollary 4.6. *Let $(A, *)$ be an arbitrary associative algebra (over a field $\mathbb{K}$ of null characteristic). Then*

$$\mathbf{F}_j^A(u, v) \in \mathrm{Lie}\{u, v\}, \quad \text{for every } u, v \in A \text{ and every } j \in \mathbb{N}. \tag{4.16}$$

Here, $\mathrm{Lie}\{u, v\}$ denotes the least Lie subalgebra of the commutator-algebra of A containing u, v. More precisely, $F_j^A(u, v)$ is a linear combination (with rational coefficients) of iterated $*$-brackets of u, v with joint degree j.

Finally, $\mathbf{F}_j^A(u, v)$ can be expressed by a "universal" expression in the following precise sense: If $\Phi_{u,v} : \mathscr{T}_+(\mathbb{K}\langle x, y \rangle) \to A$ is the (unique) algebra morphism mapping x into u and y into v, then

$$\mathbf{F}_j^A(u, v) = \Phi_{u,v}\left(\mathbf{F}_j^{\mathscr{T}(\mathbb{K}\langle x, y \rangle)}(x, y)\right). \tag{4.17}$$

Hence, thanks to the Dynkin, Specht, Wever Lemma 3.26 (and by (4.6)), we have the explicit Dynkin expression

$$\mathbf{F}_j^A(u, v) = \sum_{n=1}^{j} \frac{(-1)^{n+1}}{n} \sum_{\substack{(h,k) \in \mathcal{N}_n \\ |h|+|k|=j}} \frac{\left[u^{h_1} v^{k_1} \cdots u^{h_n} v^{k_n}\right]_*}{h! \, k! \, (|h| + |k|)}. \tag{4.18}$$

Recall that $\left[u^{h_1} v^{k_1} \cdots u^{h_n} v^{k_n}\right]_*$ denotes the following nested iterated bracket (in the commutator-algebra of A):

$$\overbrace{[u, \cdots [u,}^{h_1 \text{ times}} \overbrace{[v, \cdots [v,}^{k_1 \text{ times}} \cdots \overbrace{[u, \cdots [u,}^{h_n \text{ times}} \overbrace{[v, [\cdots, v}^{k_n \text{ times}}]_*]_* \cdots]_* \cdots]_* \cdots]_*]_* \cdots]_*.$$

More explicitly, this gives the "associative-to-Lie" identity

$$
\begin{aligned}
\sum_{n=1}^{j} \frac{(-1)^{n+1}}{n} \sum_{\substack{(h,k) \in \mathcal{N}_n \\ |h|+|k|=j}} & \frac{u^{*h_1} * v^{*k_1} * \cdots * u^{*h_n} * v^{*k_n}}{h! \, k!} = \\
& = \sum_{n=1}^{j} \frac{(-1)^{n+1}}{n} \sum_{\substack{(h,k) \in \mathcal{N}_n \\ |h|+|k|=j}} \frac{\left[u^{h_1} v^{k_1} \cdots u^{h_n} v^{k_n}\right]_*}{h! \, k! \, (|h| + |k|)},
\end{aligned} \tag{4.19}
$$

valid for every $j \in \mathbb{N}$ and every $u, v \in A$.

Proof. By arguing as in the comments preceding Remark 5.6 (page 273), we can consider a *unital* associative algebra A_1 containing A as a subalgebra. Let us fix $u, v \in A$. By Theorem 2.85-(1a), there exists a unique UAA morphism $\Phi_{u,v} : \mathscr{T}(\mathbb{K}\langle x, y \rangle) \to A_1$ such that

$$\Phi_{u,v}(x) = u \quad \text{and} \quad \Phi_{u,v}(y) = v. \tag{4.20}$$

It is immediately seen that the restriction of $\Phi_{u,v}$ to $\mathscr{T}_+(\mathbb{K}\langle x,y\rangle)$ maps this latter set to A. For the sake of brevity, we denote this restriction again by $\Phi_{u,v}$. Hence,

$$\Phi_{u,v} : \mathscr{T}_+(\mathbb{K}\langle x,y\rangle) \to A$$

is an algebra morphism satisfying (4.20) (and it is the unique morphism with this property).

With the morphism $\Phi_{u,v}$ at hand, we are able to prove (4.17) via the following direct computation:

$$\Phi_{u,v}\left(\mathbf{F}_j^{\mathscr{T}(\mathbb{K}\langle x,y\rangle)}(x,y)\right)$$

$$\overset{(4.1)}{=} \Phi_{u,v}\left(\sum_{n=1}^{j} c_n \sum_{(h,k)\in\mathcal{N}_n : |h|+|k|=j} \frac{x^{\otimes h_1} \otimes y^{\otimes k_1} \otimes \cdots \otimes x^{\otimes h_n} \otimes y^{\otimes k_n}}{h_1!\cdots h_n! \, k_1!\cdots k_n!}\right)$$

($\Phi_{u,v}$ is an algebra morphism satisfying (4.20))

$$= \sum_{n=1}^{j} c_n \sum_{(h,k)\in\mathcal{N}_n : |h|+|k|=j} \frac{u^{*h_1} * v^{*k_1} * \cdots * u^{*h_n} * v^{*k_n}}{h_1!\cdots h_n! \, k_1!\cdots k_n!}$$

$$\overset{(4.1)}{=} \mathbf{F}_j^A(u,v).$$

To complete the proof, we first remark that it suffices to prove (4.18), for it implies (4.16) *a fortiori*. To this aim, let us take $V = \mathbb{K}\langle x,y\rangle$ in the Dynkin, Specht, Wever Lemma 3.26 (page 145). We thus obtain the (unique) linear map $P:\mathscr{T}(\mathbb{K}\langle x,y\rangle) \to \mathcal{L}(\mathbb{K}\langle x,y\rangle)$ such that

$$\begin{aligned}
&P(1) = 0, \\
&P(x) = x, \quad P(y) = y, \\
&P(z_1 \otimes \cdots \otimes z_k) = \tfrac{1}{k}[z_1, \ldots [z_{k-1}, z_k]_\otimes]_\otimes, \quad \forall\, k \geq 2,
\end{aligned} \tag{4.21}$$

for any choice of $z_1, \ldots, z_k$ in $\{x,y\}$. By the cited Lemma 3.26, we know that P is the identity map on $\mathcal{L}(\mathbb{K}\langle x,y\rangle)$. By (4.6) in Theorem 4.3, we have

$$\mathbf{F}_j^{\mathscr{T}(\mathbb{Q}\langle x,y\rangle)}(x,y) \in \mathcal{L}_j(\mathbb{Q}\langle x,y\rangle), \quad \text{for every } j \in \mathbb{N}.$$

Obviously, we can replace $\mathbf{F}_j^{\mathscr{T}(\mathbb{Q}\langle x,y\rangle)}(x,y)$ by $\mathbf{F}_j^{\mathscr{T}(\mathbb{K}\langle x,y\rangle)}(x,y)$, since

$$\mathscr{T}(\mathbb{Q}\langle x,y\rangle) \subseteq \mathscr{T}(\mathbb{K}\langle x,y\rangle).$$

Hence we have, in particular,

$$\mathbf{F}_j^{\mathscr{I}(\mathbb{K}\langle x,y\rangle)}(x,y) \in \mathcal{L}(\mathbb{K}\langle x,y\rangle), \quad \text{for every } j \in \mathbb{N}. \tag{4.22}$$

As a consequence, P leaves $\mathbf{F}_j^{\mathscr{I}(\mathbb{K}\langle x,y\rangle)}(x,y)$ unchanged, so that

$$\mathbf{F}_j^{\mathscr{I}(\mathbb{K}\langle x,y\rangle)}(x,y) = P\big(\mathbf{F}_j^{\mathscr{I}(\mathbb{K}\langle x,y\rangle)}(x,y)\big)$$

$$\overset{(4.1)}{=} P\bigg(\sum_{n=1}^{j} c_n \sum_{(h,k)\in\mathcal{N}_n:\,|h|+|k|=j} \frac{x^{\otimes h_1} \otimes y^{\otimes k_1} \otimes \cdots \otimes x^{\otimes h_n} \otimes y^{\otimes k_n}}{h_1!\cdots h_n!\,k_1!\cdots k_n!}\bigg)$$

$$\overset{(4.21)}{=} \sum_{n=1}^{j} c_n \sum_{(h,k)\in\mathcal{N}_n:\,|h|+|k|=j} \frac{\big[x^{h_1}y^{k_1}\cdots x^{h_n}y^{k_n}\big]_{\otimes}}{h!\,k!\,(|h|+|k|)}.$$

This gives the explicit formula

$$\mathbf{F}_j^{\mathscr{I}(\mathbb{K}\langle x,y\rangle)}(x,y) = \sum_{n=1}^{j} c_n \sum_{(h,k)\in\mathcal{N}_n:\,|h|+|k|=j} \frac{\big[x^{h_1}y^{k_1}\cdots x^{h_n}y^{k_n}\big]_{\otimes}}{h!\,k!\,(|h|+|k|)}. \tag{4.23}$$

Finally, we get

$$\mathbf{F}_j^A(u,v) \overset{(4.17)}{=} \Phi_{u,v}\big(\mathbf{F}_j^{\mathscr{I}(\mathbb{K}\langle x,y\rangle)}(x,y)\big)$$

$$\overset{(4.23)}{=} \Phi_{u,v}\bigg(\sum_{n=1}^{j} c_n \sum_{(h,k)\in\mathcal{N}_n:\,|h|+|k|=j} \frac{\big[x^{h_1}y^{k_1}\cdots x^{h_n}y^{k_n}\big]_{\otimes}}{h!\,k!\,(|h|+|k|)}\bigg)$$

($\Phi_{u,v}$ is also a commutator-algebra morphism satisfying (4.20))

$$= \sum_{n=1}^{j} c_n \sum_{(h,k)\in\mathcal{N}_n:\,|h|+|k|=j} \frac{\big[u^{h_1}v^{k_1}\cdots u^{h_n}v^{k_n}\big]_{*}}{h!\,k!\,(|h|+|k|)}.$$

This proves (4.18) and the proof is complete.   □

4.2 Eichler's Proof

In this section, we provide the proof of the CBHD Theorem 4.3 for formal power series in two indeterminates, as given by M. Eichler in [59]. If compared to Eichler's arguments in [59], the following exposition aims to present *all* the details behind the original proof. In particular, we shall use the universal properties of the UA algebras of the polynomials in two and three indeterminates. The notations of Sect. 4.1 will be used.

Let us consider the algebra $(\widehat{\mathscr{T}}(\mathbb{Q}\langle x, y\rangle), \widehat{\cdot})$ of the formal power series in two non-commuting indeterminates x, y. As we discussed in Sect. 4.1, there is only one element $\mathbf{F}(x, y)$ of $\widehat{\mathscr{T}}(\mathbb{Q}\langle x, y\rangle)$ satisfying the identity

$$\mathbf{e}^x \mathbin{\widehat{\cdot}} \mathbf{e}^y = \mathbf{e}^{\mathbf{F}(x,y)},$$

and this element is $\mathbf{F}(x, y) = \mathbf{log}(\mathbf{e}^x \mathbin{\widehat{\cdot}} \mathbf{e}^y)$. As an element of

$$\widehat{\mathscr{T}}(\mathbb{Q}\langle x, y\rangle) = \prod\nolimits_{n \geq 0} \mathscr{T}_n(\mathbb{Q}\langle x, y\rangle),$$

$\mathbf{F}(x, y)$ can be decomposed in a unique way as a power series $\sum_{n=0}^{\infty} \mathbf{F}_n(x, y)$, with $\mathbf{F}_n(x, y) \in \mathscr{T}_n(\mathbb{Q}\langle x, y\rangle)$ for every $n \geq 0$.

On the other hand, by the computation in (4.10) we discover that, if the individual maps $\mathbf{F}_n^{\mathscr{T}(\mathbb{Q}\langle x,y\rangle)}$ are defined as in (4.1), then $\sum_{n=1}^{\infty} \mathbf{F}_n^{\mathscr{T}(\mathbb{Q}\langle x,y\rangle)}(x, y)$ equals $\mathbf{log}(\mathbf{e}^x \mathbin{\widehat{\cdot}} \mathbf{e}^y)$. Hence $\mathbf{F}_0(x, y) = 0$ and $\mathbf{F}_n(x, y) = \mathbf{F}_n^{\mathscr{T}(\mathbb{Q}\langle x,y\rangle)}(x, y)$, for every $n \geq 1$. Hence, in facing the proof of Theorem 4.3, we are left to demonstrate the following result:

Theorem 4.7 (Eichler). *With the above notation, we have that*

$$\mathbf{F}_n^{\mathscr{T}(\mathbb{Q}\langle x,y\rangle)}(x, y) \text{ is a Lie element, for every } n \in \mathbb{N}. \qquad (4.24)$$

Note that (4.24) will give (4.6), since $\mathbf{F}_n^{\mathscr{T}(\mathbb{Q}\langle x,y\rangle)}(x, y)$ is clearly a homogeneous polynomial of degree n in x, y jointly.

We next turn to prove Theorem 4.7. To this aim, by exploiting the explicit expression of $\mathbf{F}_n$, we have proved in (4.5a), (4.5b) that one has

$$\mathbf{F}_1^{\mathscr{T}(\mathbb{Q}\langle x,y\rangle)}(x, y) = x + y,$$

$$\mathbf{F}_2^{\mathscr{T}(\mathbb{Q}\langle x,y\rangle)}(x, y) = \frac{1}{2}[x, y].$$

Hence, the claimed (4.24) is proved for $n = 1$ and $n = 2$. We now aim to argue by induction on n. To this end, we make the relevant

Inductive Hypothesis:

> *We suppose that* $\mathbf{F}_1^{\mathscr{T}(\mathbb{Q}\langle x,y\rangle)}(x, y), \ldots, \mathbf{F}_{n-1}^{\mathscr{T}(\mathbb{Q}\langle x,y\rangle)}(x, y)$ *are Lie-polynomials.*

We can suppose that $n \geq 3$. To complete the induction argument, we aim to prove that $\mathbf{F}_n^{\mathscr{T}(\mathbb{Q}\langle x,y\rangle)}(x, y)$ is a Lie polynomial too. To this end, the crucial device is to perform a "jump" in the context, passing from two indeterminates x, y to three indeterminates A, B, C: we shall perform a certain amount of (tricky) computations in this last context and we shall eventually go back to the x, y-context. This is performed below, splitting the proof in several steps.

4.2.1 Eichler's Inductive Argument

Step I. A general consequence of the inductive hypothesis. First of all, as a consequence of the inductive hypothesis, we claim that the following fact holds:

for every UA algebra A and every $u, v \in A$, then

$$\mathbf{F}_j^A(u, v) \quad (j = 1, 2, \ldots, n - 1) \tag{4.25}$$

belongs to $\mathrm{Lie}\{u, v\}$,

where $\mathrm{Lie}\{u, v\}$ is the least Lie subalgebra of the commutator-algebra of A which contains u, v.

Indeed, by the universal property of $\mathscr{T}(\mathbb{Q}\langle x, y\rangle)$ in Theorem 2.85-(1a), for every UA algebra A and every pair of elements $u, v \in A$, there exists a UAA morphism $\Phi_{u,v} : \mathscr{T}(\mathbb{K}\langle x, y\rangle) \to A$ such that $\Phi_{u,v}(x) = u$ and $\Phi_{u,v}(y) = v$. Then (see Proposition 4.2)

$$\Phi_{u,v}\big(\mathbf{F}_j^{\mathscr{T}(\mathbb{Q}\langle x,y\rangle)}(x, y)\big) \overset{(4.4)}{=} \Phi_{u,v}\big(\mathbf{F}_j^{\mathscr{T}(\mathbb{K}\langle x,y\rangle)}(x, y)\big)$$

$$\overset{(4.3)}{=} \mathbf{F}_j^A(\Phi_{u,v}(x), \Phi_{u,v}(y)) = \mathbf{F}_j^A(u, v). \tag{4.26}$$

If $\mathbf{F}_j^{\mathscr{T}(\mathbb{Q}\langle x,y\rangle)}(x, y)$ is a Lie polynomial, then it can be expressed by a linear combination of higher order brackets involving x and y. Hence, as a consequence of (4.26) (and the fact that any UAA morphism is also a commutator-LA morphism), we can express the far right-hand side of (4.26) by a linear combination of higher order brackets involving u and v. This proves (4.25).

Step II. The "Abelian" version of the CBHD Formula. Since the cited computation in (4.10) rests on the sole definitions of the Exponential and Logarithm maps, it can be immediately generalized to the following general context. If V is a vector space (over a field of characteristic zero $\mathbb{K}$), and if for every $n \in \mathbb{N}$ we consider the functions

$$\mathbf{F}_n^{\widehat{\mathscr{T}}(V)} : \widehat{\mathscr{T}}(V) \times \widehat{\mathscr{T}}(V) \to \widehat{\mathscr{T}}(V)$$

defined in (4.1) for the algebra $(\widehat{\mathscr{T}}(V), \cdot)$, then one has

$$\mathrm{Exp}(u) \cdot \mathrm{Exp}(v) = \mathrm{Exp}\left(\sum_{n=1}^{\infty} \mathbf{F}_n^{\widehat{\mathscr{T}}(V)}(u, v) \right), \qquad \text{for every } u, v \in \widehat{\mathscr{T}}_+(V).$$

(4.27)

We now prove the following lemma, which is an "Abelian" version of the CBHD Formula.

Lemma 4.8. *If* $u, v \in \widehat{\mathscr{T}}_+(V)$ *commute (i.e.,* $u \cdot v - v \cdot u = 0$*), then*

$$\mathrm{Exp}(u) \cdot \mathrm{Exp}(v) = \mathrm{Exp}(u + v). \tag{4.28}$$

Proof. Let $u, v \in \widehat{\mathscr{T}}_+(V)$ be such that $u \cdot v - v \cdot u = 0$. We expand both sides of (4.28), showing that they are indeed equal. On the one hand we have

$$\mathrm{Exp}(u + v) = \sum_{n=0}^{\infty} \frac{(u + v)^n}{n!} \qquad \text{(by Newton's binomial formula)}$$

$$= \sum_{n=0}^{\infty} \frac{1}{n!} \sum_{k=0}^{n} \binom{n}{k} u^k \cdot v^{n-k} = \sum_{n=0}^{\infty} \sum_{k=0}^{n} \frac{u^k \cdot v^{n-k}}{(n-k)!\, k!}.$$

On the other hand, we have

$$\mathrm{Exp}(u) \cdot \mathrm{Exp}(v) = \left(\lim_{I \to \infty} \sum_{i=0}^{I} \frac{u^i}{i!} \right) \cdot \left(\lim_{J \to \infty} \sum_{j=0}^{J} \frac{v^j}{j!} \right)$$

(by the continuity of the $\cdot$ operation)

$$= \lim_{I, J \to \infty} \sum_{0 \le i \le I,\, 0 \le j \le J} \frac{u^i \cdot v^j}{i!\, j!}$$

(since the double limit exists, then the same holds for the limit

"along the squares" and the two limits coincide)

$$= \lim_{N \to \infty} \sum_{0 \le i, j \le N} \frac{u^i \cdot v^j}{i!\, j!} = \lim_{N \to \infty} \left(\sum_{0 \le i+j \le N} \frac{u^i \cdot v^j}{i!\, j!} + \sum_{\substack{0 \le i, j \le N \\ i+j \ge N+1}} \frac{u^i \cdot v^j}{i!\, j!} \right)$$

$$\left(\text{the second sum vanishes as } N \to \infty \text{ for it belongs to } \textstyle\prod_{n \ge N+1} \mathscr{T}_n(V) \right)$$

$$= \lim_{N \to \infty} \sum_{0 \le i+j \le N} \frac{u^i \cdot v^j}{i!\, j!} = \lim_{N \to \infty} \sum_{n=0}^{N} \sum_{i+j=n} \frac{u^i \cdot v^j}{i!\, j!} = \lim_{N \to \infty} \sum_{n=0}^{N} \sum_{k=0}^{n} \frac{u^k \cdot v^{n-k}}{(n-k)!\, k!}$$

(we applied the associative and commutative laws of $+$)

$$= \sum_{n=0}^{\infty} \sum_{k=0}^{n} \frac{u^{k} \cdot v^{n-k}}{(n-k)!\,k!}.$$

Consequently (4.28) is proved. □

As a consequence of (4.27) and (4.28), by the injectivity of Exp, by

$$\mathbf{F}_{n}^{\mathscr{T}(V)}(u,v) = \mathbf{F}_{n}^{\widehat{\mathscr{T}}(V)}(u,v) \in \mathscr{T}_{n}(V), \quad \text{for every } u,v \in V \text{ and every } n \in \mathbb{N},$$

we get, for any pair of commuting elements $u,v \in V$:

$$\mathbf{F}_{n}^{\mathscr{T}(V)}(u,v) = \begin{cases} u+v, & \text{if } n=1, \\ 0, & \text{if } n \geq 2. \end{cases}$$

In particular, this gives

$$\mathbf{F}_{n}^{\mathscr{T}(V)}(\alpha\,u, \beta\,u) = 0, \qquad \begin{array}{l} \text{for every } n \geq 2, \text{ every } \alpha, \beta \in \mathbb{K}, \\ \text{every } \mathbb{K}\text{-vector space } V \text{ and every } u \in V. \end{array} \qquad (4.29)$$

Step III. An identity involving the coefficients of the CBHD Formula in three indeterminates. The crucial device is to consider the algebra of the formal power series related to the vector space

$$W := \mathbb{Q}\langle A, B, C \rangle,$$

where A, B, C are *three* non-commuting indeterminates. By the associativity of the relevant $\widehat{}$ operation, we get

$$\left(e^{A} \,\widehat{}\, e^{B}\right) \widehat{}\, e^{C} = e^{A} \,\widehat{}\, \left(e^{B} \,\widehat{}\, e^{C}\right), \qquad (4.30)$$

which is an equality in $\widehat{\mathscr{T}}(W)$.

By using (4.27) twice (with the vector space V replaced by the above W), the left-hand side of (4.30) becomes

$$\mathrm{Exp}\left(\sum_{j=1}^{\infty} \mathbf{F}_{j}^{\widehat{\mathscr{T}}(W)}(A,B)\right) \widehat{}\, \mathrm{Exp}(C)$$

$$= \mathrm{Exp}\left(\sum_{i=1}^{\infty} \mathbf{F}_{i}^{\widehat{\mathscr{T}}(W)}\left(\sum_{j=1}^{\infty} \mathbf{F}_{j}^{\widehat{\mathscr{T}}(W)}(A,B), C\right)\right).$$

Analogously, the right-hand side of (4.30) is

$$\mathrm{Exp}(A) \curvearrowright \mathrm{Exp}\left(\sum_{j=1}^{\infty} \mathbf{F}_j^{\widehat{\mathscr{T}}(W)}(B,C) \right)$$

$$= \mathrm{Exp}\left(\sum_{i=1}^{\infty} \mathbf{F}_i^{\widehat{\mathscr{T}}(W)}\left(A, \sum_{j=1}^{\infty} \mathbf{F}_j^{\widehat{\mathscr{T}}(W)}(B,C) \right) \right).$$

Hence, by taking into account the injectivity of Exp, (4.30) is equivalent to the following equality in $\widehat{\mathscr{T}}(\mathbb{Q}\langle A,B,C\rangle)$:

$$\sum_{i=1}^{\infty} \mathbf{F}_i^{\widehat{\mathscr{T}}(W)}\left(\textstyle\sum_{j=1}^{\infty} \mathbf{F}_j^{\widehat{\mathscr{T}}(W)}(A,B), C \right)$$

$$= \sum_{i=1}^{\infty} \mathbf{F}_i^{\widehat{\mathscr{T}}(W)}\left(A, \textstyle\sum_{j=1}^{\infty} \mathbf{F}_j^{\widehat{\mathscr{T}}(W)}(B,C) \right). \tag{4.31}$$

For the sake of brevity, we shall drop the superscript $\widehat{\mathscr{T}}(W)$ in the computations below. For any fixed $n \in \mathbb{N} \cup \{0\}$, we denote by

$$p_n : \widehat{\mathscr{T}}(W) \to \mathscr{T}_n(W)$$

the projection of $\widehat{\mathscr{T}}(W) = \prod_{n \geq 0} \mathscr{T}_n(W)$ onto the n-th factor $\mathscr{T}_n(W)$. We now apply p_n to both sides of (4.31). Since

$$\textstyle\sum_{j=1}^{\infty} \mathbf{F}_j(A,B), \;\; C, \;\; A, \;\; \textstyle\sum_{j=1}^{\infty} \mathbf{F}_j(B,C)$$

all belong to $\prod_{n \geq 1} \mathscr{T}_n(W)$, by applying p_n to (4.31) we can bound the sums over i to $i \leq n$, and so we get

$$p_n\left(\sum_{i=1}^{n} \mathbf{F}_i\left(\sum_{j=1}^{\infty} \mathbf{F}_j(A,B), C \right) \right) = p_n\left(\sum_{i=1}^{n} \mathbf{F}_i\left(A, \sum_{j=1}^{\infty} \mathbf{F}_j(B,C) \right) \right). \tag{4.32}$$

We now prove that the sums over j can also be truncated to length n. Indeed, consider the following computation

$$\mathbf{F}_i\left(\textstyle\sum_{j=1}^{\infty} \mathbf{F}_j(A,B), C \right)$$

$$= \sum_{r=1}^{i} c_r \sum_{\substack{(0,k)\in\mathcal{N}_r \\ |k|=i,\, h\neq 0}} \frac{C^i}{k!} + \sum_{r=1}^{i} c_r \sum_{\substack{(h,k)\in\mathcal{N}_r \\ |h|+|k|=i,\, h\neq 0}} \frac{1}{h!\, k!}$$

$$\times \left(\sum_{j=1}^{\infty} \mathbf{F}_j(A,B) \right)^{h_1} \curvearrowright C^{k_1} \cdots \left(\sum_{j=1}^{\infty} \mathbf{F}_j(A,B) \right)^{h_r} \curvearrowright C^{k_r} =: (\star).$$

Now take any nonzero exponent, say h_i, among the above $h_1, \ldots, h_r$: Note that, if in $\left(\sum_{j=1}^{\infty} \cdots\right)^{h_i}$ at least one of the summands over j has index $j > n$, then the contribution for the whole $(\star)$ turns out to be an element in $\mathscr{T}_\alpha(W)$ with $\alpha > n + |h| - 1 + |k| \geq n$ (as $h_i \geq 1$). This means that $(\star)$ is congruent, modulo $\prod_{\alpha > n} \mathscr{T}_\alpha(W)$, to

$$
\sum_{r=1}^{i} c_r \sum_{\substack{(0,k) \in \mathcal{N}_r \\ |k|=i,\, h \neq 0}} \frac{C^i}{k!} + \sum_{r=1}^{i} c_r \sum_{\substack{(h,k) \in \mathcal{N}_r \\ |h|+|k|=i,\, h \neq 0}} \frac{1}{h!\, k!}
$$

$$
\times \left(\sum_{j=1}^{n} \mathbf{F}_j(A, B)\right)^{h_1} \widehat{C^{k_1}} \cdots \left(\sum_{j=1}^{n} \mathbf{F}_j(A, B)\right)^{h_r} \widehat{C^{k_r}} \tag{4.33}
$$

$$
= \mathbf{F}_i\left(\sum_{j=1}^{n} \mathbf{F}_j(A, B), C\right).
$$

By the definition of p_n, this is equivalent to

$$
p_n\left(\sum_{i=1}^{n} \mathbf{F}_i\left(\sum_{j=1}^{\infty} \mathbf{F}_j(A, B), C\right)\right) = p_n\left(\sum_{i=1}^{n} \mathbf{F}_i\left(\sum_{j=1}^{n} \mathbf{F}_j(A, B), C\right)\right).
$$

We can argue analogously for the right-hand side of (4.32), so that (4.32) gives

$$
p_n\left(\sum_{i=1}^{n} \mathbf{F}_i^{\widehat{\mathscr{T}}(W)}\left(\sum_{j=1}^{n} \mathbf{F}_j^{\widehat{\mathscr{T}}(W)}(A, B), C\right)\right)
$$

$$
= p_n\left(\sum_{i=1}^{n} \mathbf{F}_i^{\widehat{\mathscr{T}}(W)}\left(A, \sum_{j=1}^{n} \mathbf{F}_j^{\widehat{\mathscr{T}}(W)}(B, C)\right)\right). \tag{4.34}
$$

Note that this is *de facto* an equality in $\mathscr{T}(\mathbb{Q}\langle A, B, C\rangle)$ and (4.34) holds replacing everywhere $\widehat{\mathscr{T}}(W)$ with $\mathscr{T}(W)$. Thus, the (tacit!) superscript $\widehat{\mathscr{T}}(W)$ will be meant henceforth as a (tacit) superscript $\mathscr{T}(W)$.

We now investigate (4.34) yet further, starting with its left-hand side (the right-hand being analogous). For any $n \geq 2$, we have

$$
p_n\left(\sum_{i=1}^{n} \mathbf{F}_i\left(\sum_{j=1}^{n} \mathbf{F}_j(A, B), C\right)\right) = p_n\left(\mathbf{F}_1\left(\sum_{j=1}^{n} \mathbf{F}_j(A, B), C\right)\right)
$$

$$
+ p_n\left(\mathbf{F}_n\left(\sum_{j=1}^{n} \mathbf{F}_j(A, B), C\right)\right) +
$$

$$
+ p_n\left(\sum_{1 \lessgtr i \lessgtr n} \mathbf{F}_i\left(\sum_{j=1}^{n} \mathbf{F}_j(A, B), C\right)\right) =: \text{I} + \text{II} + \text{III}.
$$

Thanks to (4.5a), we have (recalling that here, in the inductive step, $n > 1$)

$$I = p_n\left(\sum_{j=1}^n \mathbf{F}_j(A,B) + C\right) = p_n\left(\mathbf{F}_n(A,B) + C\right) = \mathbf{F}_n(A,B);$$

$$II = p_n\left(\mathbf{F}_n(\mathbf{F}_1(A,B),C)\right) = p_n\left(\mathbf{F}_n(A+B,C)\right) = \mathbf{F}_n(A+B,C). \qquad (4.35)$$

[Indeed, note that – by the explicit definition of $\mathbf{F}_j$ in (4.1) – $\mathbf{F}_n(A,B)$ and $\mathbf{F}_n(A+B,C)$ belong to $\mathscr{T}_n(W)$, since $A,B,C \in \mathscr{T}_1(W)$.] We now claim that

$$III = p_n\left(\sum_{1 \lneq i \lneq n} \mathbf{F}_i\left(\sum_{1 \le j \lneq n} \mathbf{F}_j(A,B), C\right)\right) \qquad (4.36)$$

Indeed, taking into account (4.33) and the notation therein, if $i > 1$, we can express $\mathbf{F}_i(\sum_{j=1}^n \mathbf{F}_j(A,B), C)$ as a linear combination of tensors in $\mathscr{T}_\alpha(W)$, with (here $r \in \{1,\dots,i\}$, $(h,k) \in \mathcal{N}_r$ and $|h| + |k| = i$)

$$\alpha = j_1^1 + \cdots + j_{h_1}^1 + k_1 + \cdots + j_1^r + \cdots + j_{h_r}^r + k_r$$

$$= j_1^1 + \cdots + j_{h_1}^1 - h_1 + \cdots + j_1^r + \cdots + j_{h_r}^r - h_r + i,$$

where the indices j are in $\{1,\dots,n\}$. If one of the indices j equals n (say, for simplicity, j_1^1) then the relevant α is equal to

$$n + \underbrace{j_2^1 + \cdots + j_{h_1}^1 - h_1}_{\ge h_1 - 1} + \underbrace{j_1^2 + \cdots + j_{h_2}^2 - h_2}_{\ge 0} + \cdots + \underbrace{j_1^r + \cdots + j_{h_r}^r - h_r}_{\ge 0} + i$$

$$\ge n + h_1 - 1 - h_1 + i = n + i - 1 > n.$$

This then proves that the sum

$$\sum_{1 \lneq i \lneq n} \mathbf{F}_i\left(\sum_{j=1}^n \mathbf{F}_j(A,B), C\right)$$

is congruent, modulo $\prod_{\alpha > n} \mathscr{T}_\alpha(W)$, to an analogous sum omitting the contribution, from the inner sum, from $j = n$. This proves the claimed (4.36). Summing up, and gathering (4.35)–(4.36), we get

$$p_n\left(\sum_{i=1}^n \mathbf{F}_i\left(\sum_{j=1}^n \mathbf{F}_j(A,B), C\right)\right)$$

$$= \mathbf{F}_n(A,B) + \mathbf{F}_n(A+B,C) + p_n\left\{\sum_{1 \lneq i \lneq n} \mathbf{F}_i\left(\sum_{1 \le j \lneq n} \mathbf{F}_j(A,B), C\right)\right\}. \qquad (4.37)$$

We get an analogous identity for the right-hand side of (4.34):

$$p_n\left(\sum_{i=1}^{n}\mathbf{F}_i\left(A,\sum_{j=1}^{n}\mathbf{F}_j(B,C)\right)\right)$$

$$= \mathbf{F}_n(B,C) + \mathbf{F}_n(A, B+C) + p_n\left\{\sum_{1\lneqq i\lneqq n}\mathbf{F}_i\left(A, \sum_{1\leq j\lneqq n}\mathbf{F}_j(B,C)\right)\right\}.$$

$$(4.38)$$

Gathering together (4.37) and (4.38), (4.34) becomes

$$\mathbf{F}_n(A,B) + \mathbf{F}_n(A+B,C) + p_n\left\{\sum_{1\lneqq i\lneqq n}\mathbf{F}_i\left(\sum_{1\leq j\lneqq n}\mathbf{F}_j(A,B),C\right)\right\}$$

$$= \mathbf{F}_n(B,C) + \mathbf{F}_n(A, B+C) + p_n\left\{\sum_{1\lneqq i\lneqq n}\mathbf{F}_i\left(A, \sum_{1\leq j\lneqq n}\mathbf{F}_j(B,C)\right)\right\}.$$

$$(4.39)$$

We recall that this is an identity in $\mathscr{T}(\mathbb{Q}\langle A, B, C\rangle)$.

Step IV. The identity (4.39) *revisited, modulo Lie polynomials.* We next prove that, as a consequence of the inductive hypothesis, the summands $p_n\{\cdots\}$ in (4.39) are Lie polynomials. Indeed, we can take $A = \mathscr{T}(W)$ and $u, v \in \mathcal{L}(W)$ in (4.25). Now, since $\mathrm{Lie}\{u,v\} \subseteq \mathcal{L}(W)$ whenever $u, v \in \mathcal{L}(W)$, then (4.25) also proves that

$$\mathbf{F}_j^{\mathscr{T}(W)}(u,v) \in \mathcal{L}(W), \qquad \forall\, u, v \in \mathcal{L}(W), \quad \forall\, j = 1, \ldots, n-1. \qquad (4.40)$$

We now introduce the equivalence relation $\sim$ on $\mathscr{T}(W)$ modulo the vector subspace $\mathcal{L}(W)$, i.e., $p_1 \sim p_2$ whenever $p_1 - p_2$ is a Lie-polynomial. We are then ready to make a crucial remark: Since the indices i and j in (4.39) are strictly less than n, then (4.40) ensures that the sums in curly braces on both lines of (4.39) do belong to $\mathcal{L}(W)$ (since $A, B, C \in \mathcal{L}(W)$). Consequently (since the components in each individual $\mathscr{T}_n$ of an element of $\mathcal{L}(W)$ also belong to $\mathcal{L}(W)$), by means of the $\sim$ relation, (4.39) yields

$$\mathbf{F}_n(A,B) + \mathbf{F}_n(A+B,C) \;\sim\; \mathbf{F}_n(B,C) + \mathbf{F}_n(A, B+C). \qquad (4.41)$$

Step V. Tricky computations. Now, we show that (4.41) and (4.29) suffice to prove that $\mathbf{F}_n(A,B) \sim 0$, which will turn out to be the main goal to complete the induction. First, as a particular case of (4.29) when $V = \mathbb{Q}\langle A, B, C\rangle$, we get

$$\mathbf{F}_n(\alpha\,u, \beta\,u) = 0, \qquad \begin{array}{l} \text{for every } \alpha, \beta \in \mathbb{Q} \\ \text{and every } u \in \{A, B, C\}. \end{array} \qquad (4.42)$$

Second, we need to show rigorously how we can replace A, B, C, in (4.41), with any of their linear combinations. This is accomplished in the following:

Lemma 4.9. *Let ℓ_1, ℓ_2, ℓ_3 be $\mathbb{Q}$-linear combinations of A, B, C. Suppose*

$$\sum_{h=1}^{N} \mathbf{F}_n^{\mathscr{T}(W)}\big(\alpha_{h,1}A + \alpha_{h,2}B + \alpha_{h,3}C,\, \beta_{h,1}A + \beta_{h,2}B + \beta_{h,3}C\big) \sim 0,$$

where $N \in \mathbb{N}$ and the coefficients α and β are in $\mathbb{Q}$. Then it also holds

$$\sum_{h=1}^{N} \mathbf{F}_n^{\mathscr{T}(W)}\big(\alpha_{h,1}\ell_1 + \alpha_{h,2}\ell_2 + \alpha_{h,3}\ell_3,\, \beta_{h,1}\ell_1 + \beta_{h,2}\ell_2 + \beta_{h,3}\ell_3\big) \sim 0.$$

Proof. Let $\Phi : \mathscr{T}(W) \to \mathscr{T}(W)$ be the unique UAA morphism mapping A, B, C respectively to ℓ_1, ℓ_2, ℓ_3 (see Theorem 2.85-(2a)). Since Φ is also a (commutator-) LA morphism, we have $\Phi(\mathcal{L}(W)) \subseteq \mathcal{L}(W)$. Then the assertion of the lemma is a direct consequence of (4.3). $\square$

The above lemma guarantees that we can replace, coherently on both sides of (4.41), A, B, C with any $\mathbb{Q}$-linear combinations of A, B, C themselves. In the following computations, we are going to apply this result repeatedly, and we shall refer to it as a "change of variable" or a "substitution".

To begin with, the substitution $C := -B$ in (4.41) yields

$$\mathbf{F}_n(A, B) \sim -\mathbf{F}_n(A + B, -B) + \mathbf{F}_n(A, 0) + \mathbf{F}_n(B, -B)$$
$$= -\mathbf{F}_n(A + B, -B)$$

(here we have also applied (4.42) twice) i.e.,

$$\mathbf{F}_n(A, B) \sim -\mathbf{F}_n(A + B, -B). \qquad (4.43)$$

Analogously, the substitution $A := -B$ in (4.41) yields

$$\mathbf{F}_n(B, C) \sim \mathbf{F}_n(-B, B) + \mathbf{F}_n(0, C) - \mathbf{F}_n(-B, B + C)$$
$$\overset{(4.42)}{=} -\mathbf{F}_n(-B, B + C),$$

and, via the change of variables $B \mapsto A, C \mapsto B$:

$$\mathbf{F}_n(A, B) \sim -\mathbf{F}_n(-A, A + B). \qquad (4.44)$$

Applying (4.44), (4.43) (with the substitutions $A \mapsto -A$, $B \mapsto A + B$) and again (4.44) (with the substitutions $A \mapsto B$, $B \mapsto -A - B$), we get

$$\mathbf{F}_n(A, B) \sim -\mathbf{F}_n(-A, A + B) \sim -\big(-\mathbf{F}_n(-A + A + B, -A - B)\big)$$
$$= \mathbf{F}_n(B, -A - B) \sim -\mathbf{F}_n(-B, B - A - B) = -\mathbf{F}_n(-B, -A)$$
$$= -(-1)^n \, \mathbf{F}_n(B, A).$$

The last equality is a consequence of the fact that $\mathbf{F}_n$ is a *homogeneous* polynomial of degree n. This gives

$$\mathbf{F}_n(A, B) \sim (-1)^{n+1} \, \mathbf{F}_n(B, A). \tag{4.45}$$

Now, we substitute $C = -\tfrac{1}{2} B$ in (4.41). This gives

$$\mathbf{F}_n(A, B) \sim -\mathbf{F}_n(A + B, -\tfrac{1}{2} B) + \mathbf{F}_n(A, B - \tfrac{1}{2} B) + \mathbf{F}_n(B, -\tfrac{1}{2} B)$$
$$\overset{(4.42)}{=} -\mathbf{F}_n(A + B, -\tfrac{1}{2} B) + \mathbf{F}_n(A, \tfrac{1}{2} B),$$

that is,

$$\mathbf{F}_n(A, B) \sim \mathbf{F}_n(A, \tfrac{1}{2} B) - \mathbf{F}_n(A + B, -\tfrac{1}{2} B). \tag{4.46}$$

Analogously, by substituting $A := -\tfrac{1}{2} B$ in (4.41), we derive

$$0 \sim \mathbf{F}_n(-\tfrac{1}{2} B, B) + \mathbf{F}_n(-\tfrac{1}{2} B + B, C) - \mathbf{F}_n(-\tfrac{1}{2} B, B + C) - \mathbf{F}_n(B, C)$$
$$\overset{(4.42)}{=} \mathbf{F}_n(\tfrac{1}{2} B, C) - \mathbf{F}_n(-\tfrac{1}{2} B, B + C) - \mathbf{F}_n(B, C);$$

in its turn, by the change of variables $B \mapsto A$, $C \mapsto B$, this gives

$$\mathbf{F}_n(A, B) \sim \mathbf{F}_n(\tfrac{1}{2} A, B) - \mathbf{F}_n(-\tfrac{1}{2} A, A + B). \tag{4.47}$$

Starting from (4.47) and by applying (4.46) to both of its right-hand summands yields

$$\mathbf{F}_n(A, B) \sim \mathbf{F}_n(\tfrac{1}{2} A, \tfrac{1}{2} B) - \mathbf{F}_n(\tfrac{1}{2} A + B, -\tfrac{1}{2} B) +$$
$$- \Big(\mathbf{F}_n(-\tfrac{1}{2} A, \tfrac{1}{2} A + \tfrac{1}{2} B) - \mathbf{F}_n(-\tfrac{1}{2} A + A + B, -\tfrac{1}{2} A - \tfrac{1}{2} B) \Big)$$
$$= \mathbf{F}_n(\tfrac{1}{2} A, \tfrac{1}{2} B) - \mathbf{F}_n(\tfrac{1}{2} A + B, -\tfrac{1}{2} B) +$$
$$- \mathbf{F}_n(-\tfrac{1}{2} A, \tfrac{1}{2} A + \tfrac{1}{2} B) + \mathbf{F}_n(\tfrac{1}{2} A + B, -\tfrac{1}{2} A - \tfrac{1}{2} B).$$

Now, we apply (4.44) to the third summand in the far right-hand of the above expression and (4.43) to the second and fourth summands. We thus get

$$\mathbf{F}_n(A, B) \sim \mathbf{F}_n(\tfrac{1}{2} A, \tfrac{1}{2} B) + \mathbf{F}_n(\tfrac{1}{2} A + \tfrac{1}{2} B, \tfrac{1}{2} B) +$$

$$+ \mathbf{F}_n(\tfrac{1}{2} A, \tfrac{1}{2} B) - \mathbf{F}_n(\tfrac{1}{2} B, \tfrac{1}{2} A + \tfrac{1}{2} B)$$

(recall that $\mathbf{F}_n$ is homogeneous of degree n)

$$= 2^{1-n} \mathbf{F}_n(A, B) + 2^{-n} \mathbf{F}_n(A + B, B) - 2^{-n} \mathbf{F}_n(B, A + B)$$

$$\overset{(4.45)}{\sim} 2^{1-n} \mathbf{F}_n(A, B) + 2^{-n}(1 + (-1)^n) \mathbf{F}_n(A + B, B).$$

This gives

$$(1 - 2^{1-n}) \mathbf{F}_n(A, B) \sim 2^{-n}(1 + (-1)^n) \mathbf{F}_n(A + B, B), \qquad (4.48)$$

which proves that $\mathbf{F}_n(A, B) \sim 0$ whenever $n \neq 1$ is odd. As for the case when n is even, we argue as follows. We substitute $A - B$ in place of A in (4.48):

$$(1 - 2^{1-n}) \mathbf{F}_n(A - B, B) \sim 2^{-n}(1 + (-1)^n) \mathbf{F}_n(A, B),$$

and we use (4.43) in the above left-hand side, getting (after multiplication by $-(1 - 2^{1-n})^{-1}$)

$$\mathbf{F}_n(A, -B) \sim -(1 - 2^{1-n})^{-1} 2^{-n}(1 + (-1)^n) \mathbf{F}_n(A, B). \qquad (4.49)$$

We substitute $-B$ instead of B in (4.49):

$$\mathbf{F}_n(A, B) \sim -(1 - 2^{1-n})^{-1} 2^{-n}(1 + (-1)^n) \mathbf{F}_n(A, -B),$$

and we finally apply (4.49) once again, obtaining

$$\mathbf{F}_n(A, B) \sim (1 - 2^{1-n})^{-2} 2^{-2n}(1 + (-1)^n)^2 \mathbf{F}_n(A, B). \qquad (4.50)$$

When n is even and $n \neq 2$ (say, $n = 2k$, $k \geq 2$), the scalar coefficient in the right-hand side of (4.50) equals

$$\frac{4}{(4^k - 2)^2},$$

which is different from 1 since $k \neq 1$. Consequently, (4.50) can hold only if

$$\mathbf{F}_n(A, B) \sim 0. \qquad (4.51)$$

Step VI. From three indeterminates to two. We claim that (4.51) ends the proof. Indeed, (4.51) rewrites as

$$\mathbf{F}_n^{\mathscr{T}(\mathbb{Q}\langle A,B,C\rangle)}(A,B) \in \mathcal{L}(\mathbb{Q}\langle A,B,C\rangle). \tag{4.52}$$

Let $\varphi : \mathscr{T}(\mathbb{Q}\langle A,B,C\rangle) \to \mathscr{T}(\mathbb{Q}\langle x,y\rangle)$ be the UAA morphism such that

$$\varphi(A) = x, \quad \varphi(B) = y, \quad \varphi(C) = 0.$$

The existence of φ follows from Theorem 2.85-(2a), page 107. Since a UAA morphism is also a (commutator-) LA morphism, it is immediately seen that

$$\varphi(\mathcal{L}(\mathbb{Q}\langle A,B,C\rangle)) = \mathcal{L}(\mathbb{Q}\langle x,y\rangle),$$

so that, by applying φ to (4.52) and by exploiting (4.3), we finally obtain

$$\mathbf{F}_n^{\mathscr{T}(\mathbb{Q}\langle x,y\rangle)}(x,y) \in \mathcal{L}(\mathbb{Q}\langle x,y\rangle).$$

This proves (4.24) by induction, and the proof is complete.   □

4.3 Djokovic's Proof

The aim of this section is to present the arguments by D. Ž. Djoković in [48] (provided with more details) for another "short" proof of the CBHD theorem for formal power series in two non-commuting indeterminates. As we shall see, the arguments are very different from those presented in Sect. 4.2 and the present proof can be considered (if only from the point of view of the length) somewhat "longer" than Eichler's. Nonetheless, Djoković's proof is undoubtedly more transparent and probably more natural, lacking any tricky steps such as those in Eichler's. More important, many of the ideas in [48] go back to some of the original computations made on the CBHD Theorem, by Baker and Hausdorff themselves, selecting only the very core calculations. Thus we consider the following proof as a significant milestone.

First, we need to make precise the algebraic structures we shall be dealing with. (This part is missing in [48].) We shall also provide suitable prerequisite lemmas. This is accomplished in the following section.

4.3.1 *Polynomials and Series in t over a UA Algebra*

Let $(A, \star)$ be a UA algebra over the field $\mathbb{K}$. We now define $A[t]$, namely *the A-module of the polynomials in one indeterminate t with coefficients in A*. The rigorous definition is the following one. We set

$$A[t] := \bigoplus_{k \in \mathbb{N} \cup \{0\}} A_k, \qquad \text{where } A_k := A \text{ for every } k \geq 0. \qquad (4.53)$$

Obviously, $A[t]$ is a vector space which becomes an associative algebra, if equipped with the usual Cauchy product:

$$(a_j)_{j \geq 0} * (b_j)_{j \geq 0} := \left(\sum_{i=0}^{j} a_i \star b_{j-i} \right)_{j \geq 0}. \qquad (4.54)$$

Also, $A[t]$ is a unital algebra whose unit is $(1_A, 0, 0, \ldots)$. Furthermore, $A[t]$ is a graded algebra, since the direct sum decomposition (4.53) also furnishes a grading, since if we set (for every $k \geq 0$)

$$A_k[t] := \underbrace{\{0\} \oplus \cdots \oplus \{0\}}_{k \text{ times}} \oplus A \oplus \{0\} \oplus \{0\} \oplus \cdots \subset A[t], \qquad (4.55)$$

then we have

$$A[t] = \bigoplus_{k \in \mathbb{N} \cup \{0\}} A_k[t], \qquad \text{and} \quad A_i[t] * A_j[t] \subseteq A_{i+j}[t] \quad \forall \, i, j \geq 0. \qquad (4.56)$$

We now turn to writing the elements of $A[t]$ as polynomials in one indeterminate, say t, with coefficients in A: to this end we set

$$t := (0, 1_A, 0, \ldots),$$

$$t^n := t^{*n} = (\underbrace{0, \ldots, 0}_{n \text{ times}}, 1_A, 0, \ldots)$$

(the last equality can be directly proved by induction). We agree to identify any element $a \in A$ with $(a, 0, 0, \ldots) \in A[t]$, thus embedding A in $A[t]$. Note that, via this identification, for every $a, b \in A$, we have

$$a * b \equiv (a, 0, 0, \ldots) * (b, 0, 0, \ldots) = (a \star b, 0, 0, \ldots) \equiv a \star b, \qquad (4.57)$$

i.e., the inclusion $A \hookrightarrow A[t]$ is indeed a UAA isomorphism of A onto its "copy" in $A[t]$. With this notation at hand, we can rewrite the generic element $(a_j)_{j \geq 0}$ of $A[t]$ as

$$(a_j)_{j \geq 0} = a_0 + a_1 * t^1 + a_2 * t^2 + \cdots = \sum_{j \geq 0} a_j * t^j$$

(the sum being finite), since one can directly prove by induction that

$$a_n * t^n = (a, 0, \ldots) * \big(\underbrace{0, \ldots, 0, 1_A, 0, \ldots}_{n \text{ times}}\big) = \big(\underbrace{0, \ldots, 0, a_n, 0, \ldots}_{n \text{ times}}\big).$$

To lighten the notation, we shall mainly drop the $*$ sign in front of the t^j and use the equivalent notation

$$(a_j)_{j \geq 0} = a_0 + a_1 \, t^1 + a_2 \, t^2 + \cdots = \sum_{j \geq 0} a_j \, t^j \qquad (4.58)$$

for the generic element of $A[t]$. *The representation of an element of $A[t]$ as in the right-hand side of* (4.58) *is obviously unique.*

As we know very well from Sect. 2.3.3 (page 101), we can consider the topological algebra $\widehat{A[t]}$ of the formal power series related to the graded algebra $A[t]$. We shall set

$$A[[t]] := \widehat{A[t]} \quad \big(= \textstyle\prod_{k \geq 0} A_k[t]\big),$$

and following the above convention (4.58), the elements of $A[[t]]$ will be denoted exactly as in (4.58), *this time the sum possibly being infinite*. As usual, for $N \in \mathbb{N} \cup \{0\}$, we set

$$\widehat{U}_N := \prod_{k \geq N} A_k[t] = \Big\{ \sum_{j \geq N} a_j \, t^j \in A[[t]] \,\Big|\, a_j \in A \,\forall j \geq N \Big\}. \qquad (4.59)$$

We recall that, from the results of Sect. 2.3, we know the following facts:

1. $A[[t]]$ is an algebra, when equipped with a Cauchy operation as in (4.54) (which we still denote by $*$); this operation is written as

$$\Big(\sum_{j \geq 0} a_j \, t^j\Big) * \Big(\sum_{j \geq 0} b_j \, t^j\Big) = \sum_{j \geq 0} \Big(\sum_{i=0}^{j} a_i \star b_{j-i}\Big) t^j. \qquad (4.60)$$

2. $A[[t]]$ is a topological algebra which is also a metric space with distance

$$d\Big(\sum_{j \geq 0} a_j \, t^j, \sum_{j \geq 0} b_j \, t^j\Big) := \begin{cases} 0, & \text{if } a_j = b_j \text{ for every } j \geq 0, \\ \exp\Big(-\min\{j \geq 0 \,:\, a_j \neq b_j\}\Big), & \text{otherwise.} \end{cases}$$

3. $A[t]$ is dense in $A[[t]]$ and $(A[[t]], d)$ is an isometric completion of $A[t]$ (with the induced metric).
4. The commutator in $A[[t]]$ has a particularly well-behaved form:

$$\left[\sum_{j\geq 0} a_j\, t^j, \sum_{j\geq 0} b_j\, t^j\right]_* = \sum_{j\geq 0}\left(\sum_{i=0}^{j}[a_i, b_{j-i}]_*\right)t^j, \qquad (4.61)$$

as follows easily from (4.60).

Following the general investigation of Exponentials/Logarithms on a graded algebra as in Sect. 3.1.1 (page 119), we know that the following maps are well posed and are inverse to each other (see also $\widehat{U}_1$ in (4.59)):

$$\exp : \widehat{U}_1 \longrightarrow 1 + \widehat{U}_1, \quad \exp(p) := \sum_{k=0}^{\infty}\frac{1}{k!}\, p^{*\,k},$$

$$\log : 1 + \widehat{U}_1 \longrightarrow \widehat{U}_1, \quad \log(1+p) = \sum_{k=1}^{\infty}\frac{(-1)^{k+1}}{k}\, p^{*\,k}.$$

Beware: Here, we decided to denote by exp and log the Exp and Log maps related to the completion of the graded algebra $A[t]$. With this choice (hoping not to ...drive the Reader crazy with the proliferation of the exp/log notations), we intend to separate the more specific context of the series in one indeterminate over A to the more general context of the series related to the tensor algebra of a vector space V. Soon we shall state and prove a CBHD Theorem (with its own exp and log maps) in the former more restrictive context of $A[[t]]$ (with $A = \mathscr{T}(\mathbb{Q}\langle x, y\rangle)$) and we will show how to derive from it the general CBHD Theorem (with its Exp and Log maps) for $\widehat{\mathscr{T}}(V)$: then we have preferred to use different notations in order to distinguish the two cases.

Remark 4.10. (1) The following equation holds:

$$\exp(a\, t) = \sum_{k=0}^{\infty}\frac{a^{*\,k}}{k!}\, t^k \quad \text{for every } a \in A. \qquad (4.62)$$

This follows immediately from the definition of exp and by the following obvious computation

$$(a\, t)^{*\,k} \overset{(4.58)}{=} (0, a, 0, \dots)^{*\,k} = \Big(\underbrace{0, \dots, 0}_{k\ \text{times}}, a^{*\,k}, 0, \dots\Big) \overset{(4.58)}{=} a^{*\,k}\, t^k. \qquad (4.63)$$

(2) For every $p \in \widehat{U}_1$, it holds that

$$(\exp p) * (\exp(-p)) = 1. \qquad (4.64)$$

This follows from the trivial "Abelian" version of the CBHD Theorem as in Proposition 3.5, page 121 (indeed, we can apply (3.6) since p and $-p$ commute w.r.t. $*$).

(3). Any power t^k (with $k \in \mathbb{N} \cup \{0\}$) belongs to the center of $A[[t]]$. Indeed, for every $p = (a_j)_j \in A[[t]]$ and every $k \geq 0$, we have

$$t^k * p = \underbrace{(0, \ldots, 0, 1_A, 0, \ldots)}_{k \text{ times}} * (a_0, a_1, \ldots) = \underbrace{(0, \ldots, 0, a_0, a_1, \ldots)}_{k \text{ times}}$$

$$= (a_0, a_1, \ldots) * \underbrace{(0, \ldots, 0, 1_A, 0, \ldots)}_{k \text{ times}} = p * t^k.$$

We now introduce a crucial definition for Djokovic's proof:

Definition 4.11 (∂_t in $A[[t]]$). Following the above notation, we introduce the map

$$\partial_t : A[[t]] \longrightarrow A[[t]]$$

$$\sum_{j \geq 0} a_j \, t^j \mapsto \sum_{j \geq 0} (j+1) \, a_{j+1} \, t^j. \tag{4.65}$$

We recognize the usual derivative operator

$$\partial_t \Big(\sum_{j \geq 0} a_j \, t^j \Big) = \sum_{j \geq 1} j \, a_j \, t^{j-1}.$$

In the proof presented in Sect. 4.3.3, we shall make use of the following results concerning the operator ∂_t.

Proposition 4.12. *Following Definition 4.11, we have the following facts:*

(i) *∂_t is a derivation of the algebra $(A[[t]], *)$.*

(ii) *For every $p \in A[[t]]$ and every $m \in \mathbb{N}$, we have*

$$\partial_t(p^{*m}) = \sum_{k=0}^{m-1} p^{*k} * (\partial_t p) * p^{*m-k-1}. \tag{4.66}$$

(iii) *∂_t is a continuous map.*

(iv) *For every $a \in A$, we have*

$$\partial_t \big(\exp(a\,t) \big) = a * \exp(a\,t) = \exp(a\,t) * a. \tag{4.67}$$

Proof. See page 445 for the proof. □

Remark 4.13. Obviously, $A[[t]]$ is a noncommutative algebra (if A is not); hence the so-called "evaluation maps" (which we introduce below) are not necessarily algebra morphisms. Such is the case, however, *when we evaluate at an element in the center of A.*

Indeed, let $(A, \star)$ be a UA algebra and let a belong to the center of A (i.e., $a \star b = b \star a$ for every $b \in A$). We consider the map

$$\mathrm{ev}_a : A[t] \longrightarrow A$$

$$\sum_{j \geq 0} b_j\, t^j \mapsto \sum_{j \geq 0} b_j \star a^{\star j}.$$

This map is obviously well posed since any element of $A[t]$ can be written in a unique way in the form $\sum_{j \geq 0} b_j * t^j$, where the sum is *finite* and any b_j belongs to A. We claim that:

If a belongs to the center of A, then ev_a is a UAA morphism.

Indeed, first we recognize that ev_a is linear (by the bilinearity of $\star$). Second,

$$\mathrm{ev}_a(1_{A[t]}) = \mathrm{ev}_a(1_A) = \mathrm{ev}_a(1_A * t^0) = 1_A \star a^0 = 1_A.$$

Finally, for every $p, q \in A[t]$, represented respectively by $p = \sum_{j \geq 0} a_j\, t^j$, $q = \sum_{j \geq 0} b_j\, t^j$, we have (here we agree that the coefficients a_j and b_j are defined for every $j \in \mathbb{N}$, possibly by setting them to be 0, for sufficiently large j)

$$\mathrm{ev}_a(p * q) = \sum_{j \geq 0} \left(\sum_{i=0}^{j} a_i \star b_{j-i} \right) \star a^{\star j}$$

(since a belongs to the center of A then any power of a does, too)

$$= \sum_{j \geq 0} \sum_{i=0}^{j} a_i \star a^{\star i} \star b_{j-i} \star a^{\star j-i}.$$

On the other hand,

$$\mathrm{ev}_a(p) \star \mathrm{ev}_a(q) = \left(\sum_{h \geq 0} a_h \star a^{\star h} \right) \star \left(\sum_{k \geq 0} b_k \star a^{\star k} \right)$$

(by the bilinearity and the associativity of $\star$)

$$= \sum_{h,k \geq 0} a_h \star a^{\star h} \star b_k \star a^{\star k}$$

(by the commutativity and the associativity *of the sum of A*)

$$= \sum_{r \geq 0} \sum_{h+k=r} a_h \star a^{\star h} \star b_k \star a^{\star k}$$

(by renaming $r := j$, $h := i$ so that $k = j - i$)

$$= \sum_{j \geq 0} \sum_{i=0}^{j} (a_i \star a^{\star\, i}) \star (b_{j-i} \star a^{\star\, j-i}).$$

This ends the proof that ev_a is a UAA morphism.   □

4.3.2   Background of Djoković's Proof

We begin by proving a preliminary result which will play a paramount rôle in the computations concerning the CBHD Formula. It can be stated in many different contexts and it roughly states that *conjugation composed with exponentiation is a Lie series of* ad *operators*. The most general of such contexts which is at our disposal in this Book (amply sufficing for our purposes) is the following one. Let A be a graded algebra and let $(\widehat{A}, *)$ be the relevant algebra of formal power series (see Sect. 2.3.3, page 101). Let also $\mathrm{Exp} : \widehat{A}_+ \to 1_A + \widehat{A}_+$ be the relevant Exponential function (see Sect. 3.1.1, page 119).

Theorem 4.14. *With the above notation, for every $u \in \widehat{A}_+$ and $z \in \widehat{A}$, we have*

$$\mathrm{Exp}(u) * z * \mathrm{Exp}(-u) = \sum_{h=0}^{\infty} \frac{1}{h!} \underbrace{[u \cdots [u, z]_* \cdots]_*}_{h \; times}.$$

In compact form, by using the usual adjoint map

$$\mathrm{ad}\, u : \widehat{A} \to \widehat{A}, \qquad (\mathrm{ad}\, u)(v) := [u, v]_*$$

(related to the commutator-algebra of $\widehat{A}$), this rewrites as:

$$\mathrm{Exp}(u) * z * \mathrm{Exp}(-u) = \sum_{h=0}^{\infty} \frac{1}{h!} (\mathrm{ad}\, u)^{\circ\, h}(z), \quad \text{for every } u \in \widehat{A}_+, \; z \in \widehat{A}. \quad (4.68)$$

Here, $(\mathrm{ad}\, u)^{\circ\, h}$ denotes the h-fold iteration of the endomorphism $\mathrm{ad}\, u$.

Note that (4.68) can be suggestively rewritten as follows, anticipating a notation (to come very soon) on formal series of endomorphisms:

$$\mathrm{Exp}(u) * z * \mathrm{Exp}(-u) = \exp(\mathrm{ad}\, u)(z).$$

Proof. For any $a, b \in \widehat{A}$, we set

$$
\begin{aligned}
L_a &: \widehat{A} \longrightarrow \widehat{A} & R_b &: \widehat{A} \longrightarrow \widehat{A} \\
L_a(x) &:= a * x & R_b(x) &:= x * b.
\end{aligned}
\qquad (4.69)
$$

In other words, L_a is the left multiplication on $(A, *)$ related to a and R_b is the right multiplication on $(A, *)$ related to b. We obviously have:

$$\text{Since } * \text{ is associative, } L_a \text{ and } R_b \text{ commute, for every } a, b \in A. \qquad (4.70)$$

Let now $u \in \widehat{A}_+$ and $z \in \widehat{A}$ be fixed. The following computation applies (each equality is explained below):

$$\text{Exp}(u) * z * \text{Exp}(-u) = \left(\sum_{i \geq 0} \frac{u^{*i}}{i!} \right) * z * \left(\sum_{j \geq 0} \frac{(-u)^{*j}}{j!} \right)$$

$$\overset{(1)}{=} \sum_{i,j \geq 0} \frac{1}{i!\,j!} u^{*i} * z * (-u)^{*j} \overset{(2)}{=} \sum_{h=0}^{\infty} \sum_{i+j=h} \frac{1}{i!\,j!} u^{*i} * z * (-u)^{*j}$$

$$\overset{(3)}{=} \sum_{h=0}^{\infty} \sum_{j=0}^{h} \frac{u^{*h-j} * z * (-u)^{*j}}{(h-j)!\,j!} = \sum_{h=0}^{\infty} \frac{1}{h!} \sum_{j=0}^{h} \binom{h}{j} u^{*h-j} * z * (-u)^{*j}$$

$$= \sum_{h=0}^{\infty} \frac{1}{h!} \left(\sum_{j=0}^{h} \binom{h}{j} (L_u)^{\circ h-j} \circ (R_{-u})^{\circ j} \right)(z)$$

$$\overset{(4)}{=} \sum_{h=0}^{\infty} \frac{1}{h!} \left(L_u + R_{-u} \right)^{\circ h}(z) \overset{(5)}{=} \sum_{h=0}^{\infty} \frac{1}{h!} \left(L_u - R_u \right)^{\circ h}(z)$$

$$\overset{(6)}{=} \sum_{h=0}^{\infty} \frac{1}{h!} (\text{ad } u)^{\circ h}(z).$$

Here we used the following results:

(1): $(\widehat{A}, *)$ is a topological algebra.
(2): The sum has been rearranged, by an argument similar to that applied within the proof of Proposition 3.5, page 121.
(3): By the commutativity of the sum.
(4): An application of Newton's binomial formula jointly with (4.70).
(5): By the bilinearity of $*$, we obviously have $R_{-u} \equiv -R_u$.
(6): $L_u - R_u = \text{ad } u$, indeed

$$(L_u - R_u)(v) = u * v - v * u = [u, v]_* = (\text{ad } u)(v).$$

This completes the proof of (4.68). $\square$

We denote by

$$A := \mathscr{T}(\mathbb{Q}\langle x, y \rangle)$$

the tensor algebra related to the free $\mathbb{Q}$-vector space over the set $\{x, y\}$, $x \neq y$. The elements of A will also be referred to as polynomials (over $\mathbb{Q}$) in two non-commuting indeterminates x, y. The elements of $\mathcal{L}(\mathbb{Q}\langle x, y\rangle) \subset A$ will be referred to as Lie polynomials in A (the elements of $\mathcal{L}_j(\mathbb{Q}\langle x, y\rangle)$, $j \geq 0$, being called homogeneous of degree j). The operation on A is denoted by the $\cdot$ symbol or simply by juxtaposition, whereas the operation on $\widehat{A} = \widehat{\mathcal{F}}(\mathbb{Q}\langle x, y\rangle)$ is denoted by the $\widehat{}$ symbol.

We then consider, following Sect. 4.3.1, the algebras

$$A[t] \quad \text{and} \quad A[[t]]$$

of polynomials (respectively, series) in the indeterminate t with coefficients in A. Both operations on these algebras will be denoted by $*$. The other definitions of $\widehat{U}_N$, exp, log, ∂_t from Sect. 4.3.1 will apply as well.

We consider the elements of $A[t]$ defined by

$$X := xt, \quad Y := yt. \tag{4.71}$$

Since both X and Y belong to $\widehat{U}_1$,

$$Z(t) := \log\left(\exp(X) * \exp(Y)\right) \tag{4.72}$$

is well posed. Obviously, this is an element of $A[[t]]$ and the notation "$Z(t)$" *does not refer to a function of t*. We have, in fact, $Z(t) \in \widehat{U}_1$.

By the very definitions of exp and log, we get

$$Z(t) = \sum_{n=1}^{\infty} \frac{(-1)^{n+1}}{n} \left(\sum_{(h,k) \neq (0,0)} \frac{X^{*h} * Y^{*k}}{h! \, k!} \right)^n$$

$$\overset{(3.20)}{=} \sum_{n=1}^{\infty} c_n \sum_{(h,k) \in \mathcal{N}_n} \frac{X^{*h_1} * Y^{*k_1} * \cdots * X^{*h_n} * Y^{*k_n}}{h_1! \cdots h_n! \, k_1! \cdots k_n!}$$

$$\overset{(\star)}{=} \sum_{n=1}^{\infty} c_n \sum_{(h,k) \in \mathcal{N}_n} \frac{x^{\cdot h_1} \cdot y^{\cdot k_1} \cdots x^{\cdot h_n} \cdot y^{\cdot k_n}}{h_1! \cdots h_n! \, k_1! \cdots k_n!} t^{|h|+|k|}$$

$$\overset{(\text{reorder})}{=} \sum_{j=1}^{\infty} \left(\sum_{n=1}^{j} c_n \sum_{(h,k) \in \mathcal{N}_n: \, |h|+|k|=j} \frac{x^{\cdot h_1} \cdot y^{\cdot k_1} \cdots x^{\cdot h_n} \cdot y^{\cdot k_n}}{h_1! \cdots h_n! \, k_1! \cdots k_n!} \right) t^j$$

In the starred equality we used the definitions of X, Y in (4.71), the fact that (see (4.63)) $X^{*h} = x^{\cdot h} t^h$ and $Y^{*k} = y^{\cdot k} t^k$ and, finally, the fact that (see Remark 4.10-(3)) the powers of t are in the center of $A[[t]]$. As a consequence (dropping the "$\cdot$" notation for simplicity), if we set

$$Z_j(x,y) := \sum_{n=1}^{j} c_n \sum_{(h,k)\in\mathcal{N}_n:\ |h|+|k|=j} \frac{x^{h_1} y^{k_1} \cdots x^{h_n} y^{k_n}}{h_1! \cdots h_n! \, k_1! \cdots k_n!}, \qquad (4.73)$$

we have provided the equality

$$Z(t) = \sum_{j=1}^{\infty} Z_j(x,y)\, t^j, \qquad (4.74)$$

which is also the decomposition of $Z(t)$ in $A[[t]] = \bigoplus_{j\geq 0} A_j[t]$ (see also (4.55)) since $Z_j(x,y) \in A$. With the notation of Definition 4.1, we recognize that

$$Z_j(x,y) = \mathbf{F}_j^A(x,y), \quad \forall\, j \geq 1. \qquad (4.75)$$

As a consequence of the above computations, we have proved that

$$\exp\Big(\sum_{j=1}^{\infty} Z_j(x,y)\, t^j \Big) = \exp(x\,t) * \exp(y\,t) \quad \text{in } A[[t]]. \qquad (4.76)$$

Our aim is to prove the following result, whose complete proof will be provided in the next section, after we have completed the due preliminary work:

Theorem 4.15 (Djoković). *With the above notation, we have*

$$Z_j(x,y) \in \mathcal{L}(\mathbb{Q}\langle x,y\rangle), \quad \text{for every } j \geq 1. \qquad (4.77)$$

Before passing to Djoković's argument, we would like to show that Theorem 4.15 furnishes the CBHD Theorem for formal power series in two indeterminates as stated in Theorem 4.3.

Indeed, once we have proved (4.77), the identity in (4.75) will yield

$$\mathbf{F}_j^{\mathcal{T}(\mathbb{Q}\langle x,y\rangle)}(x,y) \in \mathcal{L}(\mathbb{Q}\langle x,y\rangle), \quad \forall\, j \geq 1.$$

Since $Z_j(x,y)$ clearly belongs to $\mathcal{T}_j(\mathbb{Q}\langle x,y\rangle)$ (*de visu*, from (4.73)), this proves (4.6), which – as we already discussed in Remark 4.4 – is the key statement in Theorem 4.3.

Remark 4.16. On the other hand, it is possible to derive (4.9) by an "evaluation argument" (roughly speaking, by the "substitution $t = 1$") from (4.76); an argument which has some interest in its own right. This is given in the appendix, see page 447.

The last prerequisite that we need before proceeding with the proof of Theorem 4.15 is the following one.

Definition 4.17. Let E be an endomorphism of $A[[t]]$ with the following property:

$$E(\widehat{U}_N) \subseteq \widehat{U}_{N+1}, \quad \text{for all } N \geq 0. \tag{4.78}$$

The set of all the endomorphisms of $A[[t]]$ satisfying (4.78) will be denoted by $\mathcal{H}$. Let $f = \sum_{k \geq 0} a_k z^k$ be an element of $\mathbb{K}[[z]]$, i.e., a formal power series in one indeterminate z over the field $\mathbb{K}$.

Then, the formal expression

$$f_E := \sum_{k \geq 0} a_k E^k \tag{4.79}$$

defines unambiguously an endomorphism of $A[[t]]$, in the following precise sense:

$$f_E : A[[t]] \longrightarrow A[[t]]$$

$$p \mapsto f_E(p) := \sum_{k=0}^{\infty} a_k E^{\circ k}(p). \tag{4.80}$$

Here, $E^{\circ k}$ denotes the k-fold composition of the endomorphism E with itself.

The well posedness of f_E follows from the fact that condition (4.78) ensures that $E^{\circ k}(p) \in \widehat{U}_k$, for every $k \geq 0$ and every $p \in A[[t]]$: thus, $\lim_{k \to \infty} a_k E^{\circ k}(p) = 0$ in $A[[t]]$ so that the series in (4.80) converges (recall e.g., Remark 2.76-3). Moreover, the linearity of f_E follows from the linearity of E. In the sequel we shall write E^k instead of $E^{\circ k}$.

Obviously, for fixed $E \in \mathcal{H}$, the map sending f to f_E is linear, from $\mathbb{K}[[z]]$ to the vector space of the endomorphisms of $A[[t]]$, i.e., we have

$$(\alpha f + \beta g)_E \equiv \alpha f_E + \beta g_E, \tag{4.81}$$

for every $\alpha, \beta \in \mathbb{K}$, every $f, g \in \mathbb{K}[[z]]$ and every $E \in \mathcal{H}$. Moreover, if 1 denotes the formal power series $1 + 0z + 0z^2 + \cdots$, then

$$1_E \text{ is the identity map of } A[[t]], \text{ for every } E \in \mathcal{H}. \tag{4.82}$$

In the next section, we shall make a crucial use of the following result.

Lemma 4.18. *Let the notation in Definition (4.17) hold. For every $f, g \in \mathbb{K}[[z]]$ and every $E \in \mathcal{H}$, we have*

$$f_E \circ g_E \equiv (f \cdot g)_E. \tag{4.83}$$

Here, the $\circ$ on the above left-hand side is the composition of endomorphisms of $A[[t]]$, whereas the $\cdot$ on the right-hand side is the usual Cauchy product of formal power series in one indeterminate over $\mathbb{K}$.

In particular, if $f = \sum_{k \geq 0} a_k \, z^k \in \mathbb{K}[[z]]$ is such that $a_0 \neq 0$ and if $g \in \mathbb{K}[[z]]$ is the reciprocal of f then, for every $E \in \mathcal{H}$, the endomorphisms f_E and g_E are inverse to each other.

Proof. The last part of the assertion is obviously a consequence of the former. Indeed, if $f = \sum_{k \geq 0} a_k \, z^k \in \mathbb{K}[[z]]$ is such that $a_0 \neq 0$, then (see Chap. 9) there exists a (unique) formal power series g such that $f \cdot g = 1_\mathbb{K} = g \cdot f$; hence, by (4.83) we have (for every $E \in \mathcal{H}$)

$$f_E \circ g_E \equiv (f \cdot g)_E \equiv (1_\mathbb{K})_E \equiv \mathrm{Id}_{A[[t]]} \equiv (1_\mathbb{K})_E \equiv (g \cdot f)_E \equiv g_E \circ f_E.$$

We then turn to prove the first part of the assertion. Let $f, g \in \mathbb{K}[[z]]$ and $E \in \mathcal{H}$ be fixed. If $f = \sum_{k \geq 0} a_k \, z^k$, $g = \sum_{k \geq 0} b_k \, z^k$ and $p \in A[[t]]$, we have

$$(f_E \circ g_E)(p) \overset{(1)}{=} \sum_{k=0}^{\infty} a_k \, E^k \Big(\sum_{h=0}^{\infty} b_h \, E^h(p) \Big) \overset{(2)}{=} \sum_{k,h=0}^{\infty} a_k \, b_h \, E^{k+h}(p)$$

$$\overset{(3)}{=} \sum_{j=0}^{\infty} \Big(\sum_{k+h=j} a_k \, b_h \Big) E^j(p) \overset{(4)}{=} \sum_{j=0}^{\infty} (f \cdot g)_j E^j(p) \overset{(5)}{=} (f \cdot g)_E(p).$$

Here we have used the following facts:

1. We applied definition (4.80) twice.
2. E^k is (uniformly) continuous since one has $E^k(\widehat{U}_N) \subseteq \widehat{U}_{N+k}$ for every $N, k \geq 0$, in view of $E \in \mathcal{H}$.
3. A reordering argument.
4. The definition of Cauchy product in $\mathbb{K}[[z]]$.
5. Again definition (4.80).

This ends the proof.   □

4.3.3 Djoković's Argument

We are ready to give *all the details* of Djoković's proof in [48]. We follow all the notations introduced in Sect. 4.3.2.

As in (4.71) and (4.72), we consider

$$X := x\,t, \quad Y := y\,t, \quad Z(t) := \log\big(\exp(X) * \exp(Y)\big). \tag{4.84}$$

By the definition of $Z(t)$ it holds that

$$\exp(Z(t)) = \exp(x\,t) * \exp(y\,t) \quad \text{in } A[[t]]. \tag{4.85}$$

For the sake of brevity, we shall also write Z instead of $Z(t)$. We apply the derivation ∂_t introduced in Definition 4.11 to both sides of (4.85), getting

$$\partial_t(\exp(Z)) \overset{(4.85)}{=} \partial_t\big(\exp(x\,t) * \exp(y\,t)\big)$$

$$= \partial_t(\exp(x\,t)) * \exp(y\,t) + \exp(x\,t) * \partial_t(\exp(y\,t))$$

$$\overset{(4.67)}{=} x * \exp(x\,t) * \exp(y\,t) + \exp(x\,t) * \exp(y\,t) * y$$

$$\overset{(4.85)}{=} x * \exp(Z) + \exp(Z) * y.$$

This yields

$$\partial_t\big(\exp(Z)\big) = x * \exp(Z) + \exp(Z) * y.$$

Multiplying on the right by $\exp(-Z)$ (and using (4.64)) we infer

$$\partial_t\big(\exp(Z)\big) * \exp(-Z) = x + \exp(Z) * y * \exp(-Z).$$

An application of Theorem 4.14 to the second summand on the above right-hand side (recalling also that Z has no zero-degree term in t) thus produces the identity:

$$\partial_t\big(\exp(Z)\big) * \exp(-Z) = x + \sum_{h=0}^{\infty} \frac{1}{h!}\,(\operatorname{ad} Z)^{\circ\,h}(y). \tag{4.86}$$

On the other hand, a direct computation on the left-hand side of (4.86) yields (recall that ∂_t is continuous by Proposition 4.12-(iii))

$$\partial_t\big(\exp(Z)\big) * \exp(-Z) = \partial_t\Big(\sum_{m=0}^{\infty} \frac{1}{m!}\,Z^m\Big) * \Big(\sum_{k=0}^{\infty} \frac{1}{k!}(-Z)^k\Big)$$

$$= \Big(\sum_{m=1}^{\infty} \frac{1}{m!}\,\partial_t(Z^m)\Big) * \Big(\sum_{k=0}^{\infty} \frac{(-1)^k}{k!}\,Z^k\Big) \quad \text{(reordering)}$$

$$= \sum_{n=1}^{\infty} \frac{1}{n!} \sum_{m=1}^{n} (-1)^{n-m} \binom{n}{m} \partial_t(Z^m) * Z^{n-m}$$

$$\overset{(4.66)}{=} \sum_{n=1}^{\infty} \frac{1}{n!} \sum_{m=1}^{n} \sum_{k=0}^{m-1} (-1)^{n-m} \binom{n}{m} Z^k * (\partial_t Z) * Z^{m-k-1} * Z^{n-m}$$

$$\Big(\text{interchanging the inner sums: } \textstyle\sum_{m=1}^{n}\sum_{k=0}^{m-1} = \sum_{k=0}^{n-1}\sum_{m=k+1}^{n}\Big)$$

$$= \sum_{n=1}^{\infty} \frac{1}{n!} \sum_{k=0}^{n-1} \Big(\sum_{m=k+1}^{n} (-1)^{n-m} \binom{n}{m}\Big) Z^k * (\partial_t Z) * Z^{n-k-1}$$

$$\left(\text{we set } r := n - m \text{ and we use } \binom{n}{n-r} = \binom{n}{r} \right)$$

$$= \sum_{n=1}^{\infty} \frac{1}{n!} \sum_{k=0}^{n-1} \left(\sum_{r=0}^{n-k-1} (-1)^r \binom{n}{r} \right) Z^k * (\partial_t Z) * Z^{n-k-1}$$

$$\left(\text{see Lemma 4.19 below and use } \binom{n-1}{n-k-1} = \binom{n-1}{k} \right)$$

$$= \sum_{n=1}^{\infty} \frac{1}{n!} \sum_{k=0}^{n-1} (-1)^{n-k-1} \binom{n-1}{k} Z^k * (\partial_t Z) * Z^{n-k-1}$$

$$= \sum_{n=1}^{\infty} \frac{1}{n!} \sum_{k=0}^{n-1} \binom{n-1}{k} Z^k * (\partial_t Z) * (-Z)^{n-k-1}$$

$$\overset{(4.69)}{=} \sum_{n=1}^{\infty} \frac{1}{n!} \underbrace{\left(\sum_{k=0}^{n-1} \binom{n-1}{k} (L_Z)^{\circ k} \circ (R_{-Z})^{\circ n-k-1} \right)}_{= (L_Z + R_{-Z})^{\circ n-1} = (\mathrm{ad}\, Z)^{\circ n-1}} (\partial_t Z)$$

$$= \sum_{n=1}^{\infty} \frac{1}{n!} (\mathrm{ad}\, Z)^{\circ n-1} (\partial_t Z).$$

Note that the above computation holds for every $Z \in A[[t]]$. So we have provided the remarkable formula for the derivative of an exponential:

$$\partial_t(\exp(a)) = \exp(a) * \sum_{n=1}^{\infty} \frac{1}{n!} (\mathrm{ad}\, a)^{\circ n-1} (\partial_t a), \quad \forall a \in A[[t]]_+. \tag{4.87}$$

Summing up, going back to (4.86) (whose left-hand side we have just made explicit) we have proved the identity

$$\sum_{n=1}^{\infty} \frac{1}{n!} (\mathrm{ad}\, Z)^{\circ n-1} (\partial_t Z) = x + \sum_{h=0}^{\infty} \frac{1}{h!} (\mathrm{ad}\, Z)^{\circ h} (y), \tag{4.88}$$

having also used the following lemma, which we turn to prove:

Lemma 4.19. *For every $n \in \mathbb{N}$ it holds that*

$$\sum_{r=0}^{m} (-1)^r \binom{n}{r} = (-1)^m \binom{n-1}{m}, \quad 0 \le m \le n - 1. \tag{4.89}$$

Proof. The proof (though boiling down to a calculation on binomials) is more delicate than it seems. We argue fixing an arbitrary $n \in \mathbb{N}$ and proving (4.89) *by induction on m.* For $m = 0$, (4.89) is trivial. We next fix $k \geq 0$ such that $k + 1 \leq n - 1$ and we show that, supposing the truth of

$$\sum_{r=0}^{k}(-1)^r \binom{n}{r} = (-1)^k \binom{n-1}{k}, \tag{4.90}$$

we obtain the validity of

$$\sum_{r=0}^{k+1}(-1)^r \binom{n}{r} = (-1)^{k+1} \binom{n-1}{k+1}. \tag{4.91}$$

Indeed, we have

$$\sum_{r=0}^{k+1}(-1)^r \binom{n}{r} \overset{(4.90)}{=} (-1)^k \binom{n-1}{k} + (-1)^{k+1} \binom{n}{k+1}$$

$$= (-1)^k \left(\frac{(n-1)!}{k!\,(n-1-k)!} - \frac{n!}{(k+1)!\,(n-k-1)!} \right)$$

$$= (-1)^k \frac{(n-1)!(k+1-n)}{(k+1)!\,(n-k-1)!} = (-1)^{k+1} \frac{(n-1)!}{(k+1)!\,(n-k-2)!} = (-1)^{k+1} \binom{n-1}{k+1}$$

This proves (4.91). $\qquad\qquad\square$

The rest of the proof consists in deriving from (4.88) a sort of *"differential equation"* on $Z(t)$, which – solved by the method of power series – produces inductive relations on the coefficients of $Z(t)$ disclosing the fact that these are Lie elements.

To this aim, let us first fix some notation: we set

$$h, f \in \mathbb{Q}[[z]], \qquad h := \sum_{j=0}^{\infty} \frac{z^j}{j!}, \qquad f := \sum_{j=0}^{\infty} \frac{z^j}{(j+1)!}. \tag{4.92}$$

[Note that h, f furnish, respectively, the Maclaurin series of e^z and of $\varphi_1(z) := (e^z - 1)/z$, which are both entire functions on $\mathbb{C}$.]

Moreover, since the zero-degree term of f is non-vanishing, there exists a reciprocal series of f, say $g := \sum_{j=0}^{\infty} b_j z^j$, characterized by the identity

$$1 = \left(\sum_{j=0}^{\infty} b_j z^j \right) \cdot \left(\sum_{j=0}^{\infty} \frac{z^j}{(j+1)!} \right) \quad \text{in } \mathbb{Q}[[z]]. \tag{4.93}$$

Obviously, all the numbers b_j in (4.93) are in $\mathbb{Q}$. [Note that g furnishes the Maclaurin series of $\psi_1(z) := z/(e^z - 1)$, which – when $z \in \mathbb{R}$ – is a real analytic function whose Maclaurin series converges only on the real interval $(-2\pi, 2\pi)$; indeed – when $z \in \mathbb{C}$ – the poles of $\psi_1(z)$ nearest to the origin are $\pm 2\pi i$. The function $\psi_1(-z)$ is sometimes referred to as *Todd function*. The coefficients b_j in (4.93) can be expressed by means of the Bernoulli numbers, see Sect. 4.5.]

We now make a crucial remark:

ad (Z) *belongs to the set of endomorphisms* $\mathcal{K}$ *introduced in Definition 4.17.*

To this aim, we have to prove that (4.78) holds for $E = \mathrm{ad}\, Z$: this is obviously true since $Z \in \widehat{U}_1$, so that

$$(\mathrm{ad}\, Z)(u) = Z * u - u * Z \in \widehat{U}_{N+1},$$

for every $u \in \widehat{U}_N$ and every $N \geq 0$.

As a consequence, we are entitled to consider the endomorphisms of $A[[t]]$ defined by $h_{\mathrm{ad}\, Z}$, $f_{\mathrm{ad}\, Z}$ and $g_{\mathrm{ad}\, Z}$, according to the cited Definition 4.17. With these endomorphisms at hand, we recognize that (4.88) can be compactly rewritten as:

$$f_{\mathrm{ad}\, Z}(\partial_t Z) = x + h_{\mathrm{ad}\, Z}(y). \tag{4.94}$$

We now aim to "release" $\partial_t Z$ in the above equation: this is possible since $f_{\mathrm{ad}\, Z}$ is an invertible endomorphism. Indeed, as $g \cdot f = 1$, (4.83) in Lemma 4.18 ensures that $f_{\mathrm{ad}\, Z}$ and $g_{\mathrm{ad}\, Z}$ are inverse maps, so that (by applying $g_{\mathrm{ad}\, Z}$ to both sides of (4.94)) we get

$$\partial_t Z(t) = g_{\mathrm{ad}\, Z}\big(x + h_{\mathrm{ad}\, Z}(y)\big). \tag{4.95}$$

We finally show that this identity (obtained by Djoković in [48, eq. (8) page 210]) can be used to obtain some inductive relations on the coefficients of

$$Z = \sum_{j=1}^{\infty} Z_j(x,y)\, t^j,$$

ensuring that the coefficients Z_j are Lie elements (once we know that the first of them is a Lie element, which is true *de visu*, for $Z_1(x,y) = x + y$).

To this aim, we shall apply a very "qualitative" argument. Instead, in the next Sect. 4.5, by exploiting some ideas by Varadarajan [171, §2.15], we shall ameliorate identity (4.95) (rewriting it in an even more compact and more "symmetric" form) thus obtaining a quantitative *recursion formula* on the coefficients Z_j. This recursion formula will undoubtedly reveal the Lie-polynomial nature of any $Z_j(x,y)$.

Let us rerun Djoković's original argument. By the definition (4.65) of ∂_t, the left-hand side of (4.95) equals

$$\partial_t Z(t) \overset{(4.74)}{=} \sum_{r=0}^{\infty} (r+1)\, Z_{r+1}(x,y)\, t^r. \tag{4.96}$$

On the other hand, the right-hand side of (4.95) can be expanded as follows. We write $Z = \sum_{m=1}^{\infty} Z_m\, t^m$ and we notice that, for every $p \in A[[t]]$, one has

$$\operatorname{ad} Z(p) = \left[\sum_{m=1}^{\infty} Z_m \, t^m, p \right] = \sum_{m=1}^{\infty} [Z_m \, t^m, p]$$

$$= \sum_{m=1}^{\infty} (Z_m \, t^m * p - p * Z_m \, t^m) = \sum_{m=1}^{\infty} (Z_m * p - p * Z_m) \, t^m$$

$$= \sum_{m=1}^{\infty} (\operatorname{ad} Z_m)(p) \, t^m \quad \text{(this last adjoint being performed in } A\text{).}$$

Here, we applied the fact that the bracket operation is continuous on both of its arguments (recall that $A[[t]]$ is a topological algebra!) and the fact that any t^m is in the center of $A[[t]]$ (see Remark 4.10-(3)). We then have

$$g_{\operatorname{ad} z}\{x + h_{\operatorname{ad} Z}(y)\} = \left(\operatorname{Id} + \sum_{j=1}^{\infty} b_j \, (\operatorname{ad} Z)^j \right) \left\{ x + \left(\operatorname{Id} + \sum_{n=1}^{\infty} \frac{(\operatorname{ad} Z)^n}{n!} \right)(y) \right\}$$

$$= x + y + \sum_{r=1}^{\infty} \left(\sum_{n=1}^{\infty} \frac{1}{n!} \sum_{m_1 + \cdots + m_n = r} \operatorname{ad} Z_{m_1} \circ \cdots \circ \operatorname{ad} Z_{m_n}(y) \right) t^r$$

$$+ \sum_{k=1}^{\infty} \left\{ \sum_{j=1}^{\infty} b_j \sum_{m_1 + \cdots + m_j = k} \operatorname{ad} Z_{m_1} \circ \cdots \circ \operatorname{ad} Z_{m_j} \left(x + y \right. \right.$$

$$+ \sum_{l=1}^{\infty} \sum_{n=1}^{\infty} \frac{1}{n!} \sum_{h_1 + \cdots + h_n = l} \operatorname{ad} Z_{h_1} \circ \cdots \circ \operatorname{ad} Z_{h_n}(y) \, t^l \Bigg) \Bigg\} t^k$$

$$= x + y + \sum_{r=1}^{\infty} \left(\sum_{n=1}^{\infty} \frac{1}{n!} \sum_{m_1 + \cdots + m_n = r} \operatorname{ad} Z_{m_1} \circ \cdots \circ \operatorname{ad} Z_{m_n}(y) \right) t^r$$

$$+ \sum_{r=1}^{\infty} \left(\sum_{j=1}^{\infty} b_j \sum_{m_1 + \cdots + m_j = r} \operatorname{ad} Z_{m_1} \circ \cdots \circ \operatorname{ad} Z_{m_j}(x + y) \right) t^r$$

$$+ \sum_{r=1}^{\infty} \sum_{\substack{k,l \geq 1: \\ k+l=r}} \sum_{j,n=1}^{\infty} \frac{b_j}{n!} \sum_{\substack{m_1 + \cdots + m_j = k \\ h_1 + \cdots + h_n = l}}$$

$$\times \operatorname{ad} Z_{m_1} \circ \cdots \circ \operatorname{ad} Z_{m_j} \circ \operatorname{ad} Z_{h_1} \circ \cdots \circ \operatorname{ad} Z_{h_n}(y) \, t^r.$$

It has to be noticed that the coefficient of t^r in this last expression is a $\mathbb{Q}$-linear combination of terms of the following type:

$$\operatorname{ad} Z_{\alpha_1} \circ \cdots \circ \operatorname{ad} Z_{\alpha_n}(x) \quad \text{and} \quad \operatorname{ad} Z_{\alpha_1} \circ \cdots \circ \operatorname{ad} Z_{\alpha_n}(y)$$

with $\alpha_1 + \cdots + \alpha_n = r$.

In particular, by equating the coefficients of t^r from both sides of (4.95) (also taking into account (4.96)), one has:

- First of all, $Z_1(x, y) = x + y$;
- Then $Z_2(x, y)$ is a $\mathbb{Q}$-linear combination of commutators of height 2 of $Z_1(x, y)$ with x and with y;
- Moreover $Z_3(x, y)$ is a $\mathbb{Q}$-linear combination of commutators of height 3 of $Z_2(x, y)$ (or of $Z_1(x, y)$) and x or y, and so on...

An inductive argument now proves that $Z_r(x, y)$ is a $\mathbb{Q}$-linear combination of commutators of height r of x and y. In particular, $Z_r(x, y) \in \mathrm{Lie}\{x, y\}$ for every $r \in \mathbb{N}$. This actually demonstrates (4.77) and the proof of Theorem 4.15 is complete.   □

4.4   The "Spine" of the Proof

A deeper insight into Djoković's proof shows that the crucial steps are played by Theorem 4.14, providing the formula

$$\mathrm{Exp}(u) * z * \mathrm{Exp}(-u) = \sum_{n=0}^{\infty} \frac{1}{n!} (\mathrm{ad}\, u)^n(z), \quad u \in \widehat{A}_+, \; z \in \widehat{A}, \qquad (4.97)$$

and by the computations which led to the formula (4.87) for the derivative of an exponential. In compact notation, the above formula rewrites as

$$e^u * z * e^{-u} = e^{\mathrm{ad}\, u}(z), \qquad u \in \widehat{A}_+, \; z \in \widehat{A}.$$

Formula (4.97) will play a central rôle also in the proofs by Varadarajan, by Reutenauer and by Cartier; hence this must be considered by full right the keystone in the proof of the Campbell-Baker-Hausdorff Theorem. The aim of this section is to furnish a general formula for the action of a derivation on an exponential in the more general setting of graded algebras.

To begin with, we start with two lemmas having a (computational) interest in their own right.

Lemma 4.20. *Let* $(A, *)$ *be a UA algebra. Then*

$$(\mathrm{ad}\, b)^n(a) = \sum_{i=0}^{n} (-1)^i \binom{n}{i} b^{n-i} * a * b^i, \qquad (4.98)$$

for every $n \in \mathbb{N} \cup \{0\}$ *and every* $a, b \in A$.

Proof. The proof can be easily performed by induction on n, or by the following direct argument. If L_c, R_c denote left and right multiplication on A by c, we have (recall that right and left multiplications commute thanks to associativity!)

$$\sum_{i=0}^{n}(-1)^i \binom{n}{i} b^{n-i} * a * b^i = \sum_{i=0}^{n} \binom{n}{i} b^{n-i} * a * (-b)^i$$
$$= \sum_{i=0}^{n} \binom{n}{i} (L_b)^{\circ n-i} \circ (R_{-b})^i(a) = (L_b + R_{-b})^{\circ n}(a) = (\operatorname{ad} b)^{\circ n}(a).$$

This ends the proof. □

Lemma 4.21. *Let $(A, *)$ be a UA algebra and let D be a derivation of A. Then*

$$D(u^n) = \sum_{k=0}^{n-1} \binom{n}{k+1} u^{n-1-k} * (-\operatorname{ad} u)^k (Du) \tag{4.99}$$

$$= \sum_{k=0}^{n-1} \binom{n}{k+1} (\operatorname{ad} u)^k (Du) * u^{n-1-k}, \tag{4.100}$$

for every $n \in \mathbb{N}$ and every $u \in A$.

Proof. By induction on n. For $n = 1$ the thesis is trivial. We next suppose that (4.99) holds and we prove it when n is replaced by $n+1$. First, we notice that (by the identity $a * b = b * a - [b, a]_*$) we obtain

$$(-\operatorname{ad} u)^k(Du) * u = u * (-\operatorname{ad} u)^k(Du) - [u, (-\operatorname{ad} u)^k(Du)]_*$$
$$= u * (-\operatorname{ad} u)^k(Du) + (-\operatorname{ad} u)^{k+1}(Du).$$

For the sake of brevity, we set $U := -\operatorname{ad} u$. Note that the above equality gives

$$U^k(Du) * u = u * U^k(Du) + U^{k+1}(Du). \tag{4.101}$$

As a consequence, the following computation applies

$$D(u^{n+1}) = D(u^n * u) = D(u^n) * u + u^n * (Du) \quad \text{(inductive hypothesis)}$$

$$= u^n * (Du) + \sum_{k=0}^{n-1} \binom{n}{k+1} u^{n-1-k} * U^k(Du) * u$$

$$\stackrel{(4.101)}{=} u^n * Du + \sum_{k=0}^{n-1} \binom{n}{k+1} u^{n-k} * U^k(Du) + \sum_{k=0}^{n-1} \binom{n}{k+1} u^{n-1-k} * U^{k+1}(Du)$$

(rename the index in the second sum: $k + 1 = j$)

$$= u^n * Du + \sum_{k=0}^{n-1} \binom{n}{k+1} u^{n-k} * U^k(Du) + \sum_{j=1}^{n} \binom{n}{j} u^{n-j} * U^j(Du)$$

(we isolate the summands $k = 0$ and $j = n$ from the two sums)

$$= u^n * Du + n\, u^n * Du + U^n(Du) + \sum_{j=1}^{n-1} \left(\binom{n}{j} + \binom{n}{j+1} \right) u^{n-j} * U^j(Du)$$

$$= \sum_{j=0}^{n} \binom{n+1}{j+1} u^{n-j} * U^j(Du).$$

In the last equality we used the simple fact $\binom{n}{j} + \binom{n}{j+1} = \binom{n+1}{j+1}$. This proves the first identity in (4.99) and the second can be proved analogously. $\qquad\square$

In the sequel, $A = \bigoplus_{j \geq 0} A_j$ is a graded algebra, $\widehat{A} = \prod_{j \geq 0} A_j$ denotes its completion (thought of as the algebra of the formal power series on A) and, as usual, $\widehat{A}_+ = \prod_{j \geq 1} A_j$. Finally, $\mathrm{Exp} : \widehat{A}_+ \to 1 + \widehat{A}_+$ is the relevant exponential function. We denote by $*$ the operation both on A and on $\widehat{A}$.

We are ready for the following central result.

Theorem 4.22 (Differential of the Exponential). *Let $\widehat{A}$ be as above and let D be a continuous derivation of $\widehat{A}$. Then*

$$D(\mathrm{Exp}(u)) = \mathrm{Exp}(u) * \sum_{k=1}^{\infty} \frac{1}{k!} (-\mathrm{ad}\, u)^{k-1}(Du) \qquad (4.102)$$

$$= \sum_{k=1}^{\infty} \frac{1}{k!} (\mathrm{ad}\, u)^{k-1}(Du) * \mathrm{Exp}(u), \qquad (4.103)$$

for every $u \in \widehat{A}_+$, or, with compact notation,

$$D(e^u) = e^u * \frac{1 - e^{-\mathrm{ad}\, u}}{\mathrm{ad}\, u}(Du), \qquad D(e^u) = \frac{e^{\mathrm{ad}\, u} - 1}{\mathrm{ad}\, u}(Du) * e^u.$$

Proof. Since u is in $\widehat{A}_+$, $\mathrm{Exp}(u)$ is well-posed. D being continuous, we can pass D under the summation symbol, getting (note that $D(1) = 0$ since D is a derivation)

$$D(\mathrm{Exp}(u)) = \sum_{n=1}^{\infty} \frac{1}{n!} D(u^n) \overset{(4.99)}{=} \sum_{n=1}^{\infty} \frac{1}{n!} \sum_{k=0}^{n-1} \binom{n}{k+1} u^{n-1-k} * (-\mathrm{ad}\, u)^k(Du)$$

$$\left(\text{use } \tfrac{1}{n!} \binom{n}{k+1} = \tfrac{1}{(k+1)!(n-1-k)!}, \text{ interchange sums: } \sum_{n=1}^{\infty} \sum_{k=0}^{n-1} = \sum_{k=0}^{\infty} \sum_{n=k+1}^{\infty} \right.$$

$$\left. \text{and rename the dummy index: } m := n - 1 - k \right)$$

$$= \sum_{k=0}^{\infty} \sum_{m=0}^{\infty} \frac{u^m * (-\operatorname{ad} u)^k (Du)}{(k+1)!\, m!} \quad \text{(the iterated series split into a product)}$$

$$= \operatorname{Exp}(u) * \left(\sum_{k=0}^{\infty} \frac{(-\operatorname{ad} u)^k (Du)}{(k+1)!} \right) = \operatorname{Exp}(u) * \sum_{k=1}^{\infty} \frac{1}{k!} (-\operatorname{ad} u)^{k-1}(Du).$$

This ends the proof of the first identity in (4.102), whereas the second can be proved by an analogous computation (this time making use of (4.100) instead of (4.99)). □

If $\widehat{A}$ is as above, note that Lemma 4.20 gives back in a two-line-proof the above identity (4.97): for every $u \in \widehat{A}_+$ and every $z \in \widehat{A}$ we have

$$\sum_{n=0}^{\infty} \frac{1}{n!} (\operatorname{ad} u)^n (z) \overset{(4.98)}{=} \sum_{n=0}^{\infty} \frac{1}{n!} \sum_{i=0}^{n} (-1)^i \binom{n}{i} u^{n-i} * z * u^i$$

(interchange sums and rename the inner index $m := n - i$)

$$= \sum_{i=0}^{\infty} \sum_{m=0}^{\infty} \frac{1}{i!\, m!} u^m * z * (-u)^i = \operatorname{Exp}(u) * z * \operatorname{Exp}(-u).$$

Collecting together (4.97) and (4.102) we obtain the crucial "ODE" for $\operatorname{Exp}(Z)$, when Z solves $\operatorname{Exp}(Z) = \operatorname{Exp}(X) * \operatorname{Exp}(Y)$.

Indeed, if $X, Y, Z \in \widehat{A}_+$ are such that

$$\operatorname{Exp}(Z) = \operatorname{Exp}(X) * \operatorname{Exp}(Y), \tag{4.104}$$

and if D is any continuous derivation of $\widehat{A}$ satisfying

$$D(\operatorname{Exp}(X)) = x * \operatorname{Exp}(X), \quad D(\operatorname{Exp}(Y)) = \operatorname{Exp}(Y) * y, \tag{4.105}$$

for some $x, y \in \widehat{A}$, we immediately get

$$\sum_{k=1}^{\infty} \frac{1}{k!} (-\operatorname{ad} Z)^{k-1}(D Z) \overset{(4.102)}{=} \operatorname{Exp}(-Z) * D(\operatorname{Exp}(Z))$$

$$\overset{(4.104)}{=} \operatorname{Exp}(-Z) * D\big(\operatorname{Exp}(X) * \operatorname{Exp}(Y)\big)$$

$$= \operatorname{Exp}(-Z) * D(\operatorname{Exp}(X)) * \operatorname{Exp}(Y)$$

$$\quad + \operatorname{Exp}(-Z) * \operatorname{Exp}(X) * D(\operatorname{Exp}(Y))$$

$$\overset{(4.105)}{=} \operatorname{Exp}(-Z) * x * \operatorname{Exp}(X) * \operatorname{Exp}(Y)$$

$$\quad + \operatorname{Exp}(-Z) * \operatorname{Exp}(X) * \operatorname{Exp}(Y) * y$$

$$\overset{(4.104)}{=} \mathrm{Exp}(-Z) * x * \mathrm{Exp}(Z) + y$$

$$\overset{(4.97)}{=} \sum_{n=0}^{\infty} \frac{1}{n!} (-\mathrm{ad}\, Z)^n (x) + y.$$

If we introduce the usual formalism for power series of endomorphisms, this rewrites as

$$\frac{1 - e^{-\mathrm{ad}\, Z}}{\mathrm{ad}\, Z} (DZ) = e^{-\mathrm{ad}\, Z}(x) + y,$$

which, as we have seen, is the starting point for the proof of Djoković, based on the "differential equation"

$$DZ = \frac{\mathrm{ad}\, Z}{e^{\mathrm{ad}\, Z} - 1} \big(x + e^{\mathrm{ad}\, Z}(y)\big).$$

Note that the above formal ODE can be rewritten as follows (as in [44])

$$DZ = \frac{\mathrm{ad}\, Z}{e^{\mathrm{ad}\, Z} - 1}(x) + \frac{-\mathrm{ad}\, Z}{e^{-\mathrm{ad}\, Z} - 1}(y). \tag{4.106}$$

4.4.1 Yet Another Proof with Formal Power Series

In this short section, we describe an argument which allows us to prove the Campbell, Baker, Hausdorff Theorem and to derive an analogue of Dynkin's Formula *without the use of the Dynkin, Specht, Wever Lemma 3.26*. This is an adaptation of the argument used by Duistermaat and Kolk in the setting of Lie groups (see [52, Section 1.7]). We only give a formal sketch of the proof, leaving the details to the Reader (a rigorous argument needing the machinery on formal power series used in the previous sections).

As usual, we consider the algebra $A[[t]]$ of the formal power series in t over $A = \mathscr{T}(\mathbb{Q}\langle x, y \rangle)$. Set $Z(t) = \log(\exp(x\,t) * \exp(y\,t))$, the obvious identity $\exp(Z(t)) = \exp(x\,t) * \exp(y\,t)$ gives (by applying Theorem 4.14 three times)

$$e^{\mathrm{ad}\, Z(t)}(z) = \exp(Z(t)) * z * \exp(-Z(t))$$

$$= \exp(x\,t) * \exp(y\,t) * z * \exp(-y\,t) * \exp(-x\,t)$$

$$= \exp(x\,t) * e^{t\,\mathrm{ad}\, y}(z) * \exp(-x\,t) = e^{t\,\mathrm{ad}\, x}\big(e^{t\,\mathrm{ad}\, y}(z)\big),$$

holding true for every $z \in A[[t]]$. This means precisely

$$e^{\mathrm{ad}\, Z(t)} = e^{t\,\mathrm{ad}\, x} \circ e^{t\,\mathrm{ad}\, y}, \quad \text{or} \quad e^{-\mathrm{ad}\, Z(t)} = e^{-t\,\mathrm{ad}\, y} \circ e^{-t\,\mathrm{ad}\, x}. \tag{4.107}$$

We now exploit a useful identity on power series in one indeterminate z. From $\log(1+w)/w = \sum_{n=1}^{\infty} \frac{(-1)^{n+1}}{k} w^n$ we get (via the substitution $w = e^z - 1$)

$$\frac{z}{e^z - 1} = \frac{\log(1 + (e^z - 1))}{e^z - 1} = \sum_{n=1}^{\infty} \frac{(-1)^{n+1}}{n} (e^z - 1)^{n-1}.$$

The crucial trick here is to write $z/(e^z - 1)$ as a function of $e^z - 1$ itself.
Thus, we derive the analogous expansion (valid for every $w \in A[[t]]_+$)

$$\frac{\operatorname{ad} w}{e^{\operatorname{ad} w} - 1} = \sum_{n=1}^{\infty} \frac{(-1)^{n+1}}{n} (e^{\operatorname{ad} w} - 1)^{n-1}. \qquad (4.108)$$

Inserting this in the ODE-like identity (4.106) (when $w = \pm Z(t)$), one obtains (recall our notation $c_n = \frac{(-1)^{n+1}}{n}$)

$$Z' \overset{(4.106)}{=} \frac{\operatorname{ad} Z}{e^{\operatorname{ad} Z} - 1}(x) + \frac{-\operatorname{ad} Z}{e^{-\operatorname{ad} Z} - 1}(y)$$

$$\overset{(4.108)}{=} \sum_{n=1}^{\infty} c_n \, (e^{\operatorname{ad} Z} - 1)^{n-1}(x) + \sum_{n=1}^{\infty} c_n \, (e^{-\operatorname{ad} Z} - 1)^{n-1}(y)$$

$$\overset{(4.107)}{=} \sum_{n=1}^{\infty} c_n \, (e^{t \operatorname{ad} x} e^{t \operatorname{ad} y} - 1)^{n-1}(x) + \sum_{n=1}^{\infty} c_n \, (e^{-t \operatorname{ad} y} e^{-t \operatorname{ad} x} - 1)^{n-1}(y).$$

Now, by the explicit expansion

$$e^{t \operatorname{ad} x} e^{t \operatorname{ad} y} - 1 = \sum_{(h,k) \neq (0,0)} \frac{t^{h+k}}{h! \, k!} (\operatorname{ad} x)^h (\operatorname{ad} y)^k,$$

and after some computations, we easily derive

$$Z'(t) = x + y + \sum_{\substack{n \geq 2 \\ (h,k) \in \mathcal{N}_{n-1}}} \frac{c_n \, t^{|h|+|k|}}{h! \, k!}$$

$$\times \Big\{ (\operatorname{ad} x)^{h_1} (\operatorname{ad} y)^{k_1} \cdots (\operatorname{ad} x)^{h_{n-1}} (\operatorname{ad} y)^{k_{n-1}}(x)$$

$$+ (-1)^{|h|+|k|} (\operatorname{ad} y)^{h_1} (\operatorname{ad} x)^{k_1} \cdots (\operatorname{ad} y)^{h_{n-1}} (\operatorname{ad} x)^{k_{n-1}}(y) \Big\}.$$

By the following identity (with the obvious algebraic meaning of the "integral" and of $Z(0)$)

$$Z(1) = Z(0) + \int_0^1 Z'(t) \, dt,$$

and by the above expansion of $Z'(t)$, an "integration by series" gives at once

$$\log(\exp(x) * \exp(y)) = x + y + \sum_{\substack{n \geq 1 \\ (h,k) \in \mathcal{N}_n}} \frac{(-1)^n}{n+1} \frac{1}{h! \, k! \, (|h| + |k| + 1)}$$

$$\times \left\{ (\operatorname{ad} x)^{h_1} (\operatorname{ad} y)^{k_1} \cdots (\operatorname{ad} x)^{h_n} (\operatorname{ad} y)^{k_n} (x) \right.$$

$$\left. + (-1)^{|h|+|k|} (\operatorname{ad} y)^{h_1} (\operatorname{ad} x)^{k_1} \cdots (\operatorname{ad} y)^{h_n} (\operatorname{ad} x)^{k_n} (y) \right\}.$$

Note that this is an alternative form of Dynkin's Formula:

$$\log(\exp(x) * \exp(y)) = x + y$$

$$+ \sum_{\substack{n \geq 2 \\ (r,s) \in \mathcal{N}_n}} \frac{(-1)^{n+1}}{n \, r! \, s! \, (|r| + |s|)} (\operatorname{ad} x)^{r_1} (\operatorname{ad} y)^{s_1} \cdots (\operatorname{ad} x)^{r_n} (\operatorname{ad} y)^{s_n - 1}(y).$$

[Here, when $s_n = 0$ the term "$(\operatorname{ad} y)^{s_n - 1}(y)$" is suppressed and substituted by $(\operatorname{ad} x)^{r_n - 1}(x)$.] Another representation of $\log(e^x * e^y)$ can be obtained by rerunning the above arguments with $W(t) := \log(e^x * e^{ty})$. This time one has the "Cauchy problem"

$$W' = \frac{-\operatorname{ad} W}{e^{-\operatorname{ad} W} - 1}(y), \qquad W(0) = x,$$

so that the relevant "integral" representation gives

$$\log(\exp(x) * \exp(y)) = x + y + \sum_{\substack{n \geq 1 \\ (h,k) \in \mathcal{N}_n}} \frac{(-1)^n \, (-1)^{|h|}}{(n+1) \, h! \, k! \, (|h| + 1)} \times$$

$$\times (\operatorname{ad} y)^{h_1} (\operatorname{ad} x)^{k_1} \cdots (\operatorname{ad} y)^{h_n} (\operatorname{ad} x)^{k_n} (y).$$

Yet another formula can be derived with the choice $U(t) := \log(e^{tx} e^y)$. This time one has

$$U' = \frac{\operatorname{ad} U}{e^{\operatorname{ad} U} - 1}(x), \qquad U(0) = y,$$

so that the corresponding "integral" representation gives

$$\log(\exp(x) * \exp(y)) = x + y + \sum_{\substack{n \geq 1 \\ (h,k) \in \mathcal{N}_n}} \frac{(-1)^n}{(n+1)\, h!\, k!\, (|h|+1)}$$

$$\times (\operatorname{ad} x)^{h_1} (\operatorname{ad} y)^{k_1} \cdots (\operatorname{ad} x)^{h_n} (\operatorname{ad} y)^{k_n} (x).$$

4.5 Varadarajan's Proof

The aim of this section is to furnish a very explicit argument which provides, starting from equation (4.95), recursive relations on the coefficients $Z_j(x, y)$ (introduced in (4.73)), revealing the Lie-polynomial nature of any Z_j. These recursion formulas (see Theorem 4.23 below) have an interest in their own right, besides providing a proof of the CBH Theorem.

We shall follow the exposition of Varadarajan, [171, §2.15]. It has to be noticed that the setting of [171] is different from ours, in that it is concerned with Lie groups: actually, (4.95) is – in that setting – a "true" differential equation, t being a real parameter and x, y being vector fields (in a neighborhood of the origin) in the Lie algebra of a Lie group. Nonetheless, the arguments in [171, §2.15] perfectly adapt to our context of the formal power series in one indeterminate t over the algebra of the polynomials in two non-commuting indeterminates x, y. We follow all the notation introduced in the previous sections.

Let us consider the following formal power series in $\mathbb{Q}[[z]]$:

$$i = z,$$

$$f = \sum_{j=0}^{\infty} \frac{1}{(j+1)!}\, z^j,$$

$$g = \sum_{j=0}^{\infty} K_j\, z^j, \quad K_j := \frac{B_j}{j!},$$

$$h = \sum_{j=0}^{\infty} \frac{1}{j!}\, z^j, \tag{4.109}$$

$$k = \sum_{j=0}^{\infty} K_{2j}\, z^{2j},$$

where the coefficients B_j are the so called *Bernoulli numbers*, defined e.g., by the recursion formulas:

$$B_0 := 1, \qquad B_n := -n! \sum_{k=0}^{n-1} \frac{B_k}{k! \, (n+1-k)!} \qquad (n \geq 1). \tag{4.110}$$

In Chap. 9, we shall prove that the following identities hold in $\mathbb{Q}[[z]]$:

$$g + \tfrac{1}{2} i = k, \tag{4.111a}$$

$$f \cdot g = 1, \tag{4.111b}$$

$$g \cdot (h - 1) = i. \tag{4.111c}$$

We remark that the above formal power series respectively furnish the expansions of the following complex functions (with a small abuse of notation, we denote the function related to the series f by $f(z)$ and so on):

$$i(z) = z, \quad z \in \mathbb{C},$$

$$f(z) = \frac{e^z - 1}{z}, \quad z \in \mathbb{C},$$

$$g(z) = \frac{z}{e^z - 1}, \quad z \in \mathbb{C}: \ |z| < 2\pi, \tag{4.112}$$

$$h(z) = e^z, \quad z \in \mathbb{C},$$

$$k(z) = \frac{z/2}{\sinh(z/2)} \cosh(z/2), \quad z \in \mathbb{C}: \ |z| < 2\pi.$$

[Once (4.112) is known, it is immediate[3] to prove (4.111a)–(4.111c).] We now proceed with a chain of equalities starting from (4.95) (note that the notation in Definition 4.17 is also used):

$$\partial_t Z(t) \overset{(4.95)}{=} g_{\mathrm{ad}\,Z} \big(x + h_{\mathrm{ad}\,Z}(y) \big) = g_{\mathrm{ad}\,Z}(x + y) + g_{\mathrm{ad}\,Z} \big(-y + h_{\mathrm{ad}\,Z}(y) \big)$$

(here we use (4.81) and (4.82))

$$= (g + \tfrac{1}{2} i)_{\mathrm{ad}\,Z}(x+y) - \tfrac{1}{2} i_{\mathrm{ad}\,Z}(x+y) + g_{\mathrm{ad}\,Z} \big(-1_{\mathrm{ad}\,Z} + h_{\mathrm{ad}\,Z} \big)(y)$$

$$\overset{(4.111a)}{=} k_{\mathrm{ad}\,Z}(x+y) - \tfrac{1}{2} \,\mathrm{ad}\, Z(x+y) + g_{\mathrm{ad}\,Z} \circ (-1 + h)_{\mathrm{ad}\,Z}(y)$$

---

[3]Indeed, (4.111b) and (4.111c) are obvious; as for (4.111a) we have:

$$g(z) + \frac{1}{2} z = \frac{z}{e^z - 1} + \frac{1}{2} z = \frac{e^z + 1}{e^z - 1} \frac{z}{2} = \frac{e^{z/2} + e^{-z/2}}{e^{z/2} - e^{-z/2}} \frac{z}{2} = \frac{\cosh(z/2)}{\sinh(z/2)} \frac{z}{2} = k(z).$$

Once these identities hold – as they do! – in a neighborhood of the origin, analogous identities between the relevant Maclaurin series hold by elementary Calculus.

$$\overset{(4.83)}{=} k_{\mathrm{ad}\,Z}(x+y) - \tfrac{1}{2}\,[Z, x+y] + \big(g \cdot (h-1)\big)_{\mathrm{ad}\,Z}(y)$$

$$\overset{(4.111c)}{=} k_{\mathrm{ad}\,Z}(x+y) - \tfrac{1}{2}\,[Z, x+y] + i_{\mathrm{ad}\,Z}(y)$$

$$= k_{\mathrm{ad}\,Z}(x+y) - \tfrac{1}{2}\,[Z, x+y] + \mathrm{ad}\,Z(y)$$

$$= k_{\mathrm{ad}\,Z}(x+y) + \tfrac{1}{2}\,[Z, y-x] = k_{\mathrm{ad}\,Z}(x+y) + \tfrac{1}{2}\,[x-y, Z].$$

Summing up, we have proved

$$\partial_t Z = k_{\mathrm{ad}\,Z}(x+y) + \tfrac{1}{2}\,[x-y, Z], \tag{4.113}$$

where the bracket in the right-hand side is the commutator of $A[[t]]$. [Compare to [171, eq. (2.15.11)], where this identity is derived as a differential equation in the Lie algebra of a Lie group.]

We now aim to substitute for Z its expansion as a series in t over the algebra $A = \mathscr{T}(\mathbb{Q}\langle x, y\rangle)$ and then to equate, from both sides, the coefficients of t^n, thus deriving a family of identities in A.

To this aim, we recall that (see (4.74))

$$Z = \sum_{j=1}^{\infty} Z_j(x, y)\, t^j, \tag{4.114}$$

where the coefficients $Z_j(x, y)$ are as in (4.73), that is,

$$Z_j(x, y) = \sum_{n=1}^{j} \frac{(-1)^{n+1}}{n} \sum_{(h,k)\in\mathbb{N}_n:\,|h|+|k|=j} \frac{x^{h_1} y^{k_1} \cdots x^{h_n} y^{k_n}}{h_1! \cdots h_n!\, k_1! \cdots k_n!}.$$

Consequently, the left-hand side of (4.113) is given by (recalling the definition of ∂_t in (4.74))

$$\partial_t Z = \sum_{j=0}^{\infty} (j+1)\, Z_{j+1}(x, y)\, t^j. \tag{4.115}$$

We now turn to the two summands in the right-hand side of (4.113). First of all, by (4.114) and (4.61), the second summand is equal to

$$\tfrac{1}{2}\,[x-y, Z]_{A[[t]]} = \sum_{j=1}^{\infty} \tfrac{1}{2}\,[x-y, Z_j(x, y)]_A\, t^j. \tag{4.116}$$

Moreover, for any $p \in \mathbb{N}$, we have the following computation (we denote $Z_j(x, y)$ simply by Z_j):

$$(\operatorname{ad} Z)^{\circ 2p}(x+y) = \underbrace{[Z \cdots [Z, x+y]]}_{2p}{}_{A[[t]]}$$

$$= \left[\sum_{j_1=1}^{\infty} Z_{j_1} t^{j_1} \cdots \left[\sum_{j_{2p}=1}^{\infty} Z_{j_{2p}} t^{j_{2p}}, x+y \right] \right]_{A[[t]]}$$

$$\overset{(4.61)}{=} \sum_{j_1,\ldots,j_{2p} \geq 1} [Z_{j_1} \cdots [Z_{j_{2p}}, x+y]]_A \, t^{j_1+\cdots+j_{2p}}$$

$$= \sum_{j \geq 2p} \left(\sum_{\substack{k_1,\ldots,k_{2p} \geq 1 \\ k_1+\cdots+k_{2p}=j}} [Z_{k_1} \cdots [Z_{k_{2p}}, x+y]]_A \right) t^j.$$

Hence, by (4.80) and by the explicit expression of k in (4.109), we have

$$k_{\operatorname{ad} Z}(x+y) = x+y+\sum_{p=1}^{\infty} K_{2p}(\operatorname{ad} Z)^{\circ 2p}(x+y)$$

(see the computations above)

$$= x+y+\sum_{p=1}^{\infty} K_{2p} \sum_{j \geq 2p} \left(\sum_{\substack{k_1,\ldots,k_{2p} \geq 1 \\ k_1+\cdots+k_{2p}=j}} [Z_{k_1} \cdots [Z_{k_{2p}}, x+y]]_A \right) t^j$$

$$= x+y+\sum_{j=2}^{\infty} \left(\sum_{p=1}^{[h/2]} K_{2p} \sum_{\substack{k_1,\ldots,k_{2p} \geq 1 \\ k_1+\cdots+k_{2p}=j}} [Z_{k_1} \cdots [Z_{k_{2p}}, x+y]]_A \right) t^j.$$

Here $[h/2]$ denotes the integer part of $h/2$, i.e., the largest integer $\leq h/2$. Consequently, the first summand in the right-hand side of (4.113) is equal to

$$k_{\operatorname{ad} Z}(x+y) = x+y+\sum_{j=2}^{\infty} \left(\sum_{\substack{p \geq 1,\ 2p \leq j \\ k_1,\ldots,k_{2p} \geq 1 \\ k_1+\cdots+k_{2p}=j}} K_{2p} [Z_{k_1} \cdots [Z_{k_{2p}}, x+y]]_A \right) t^j.$$

$$(4.117)$$

Finally, summing up (4.115), (4.116) and (4.117), identity (4.113) rewrites as:

$$\sum_{j=0}^{\infty} (j+1) Z_{j+1}(x,y) t^j = x+y+\sum_{j=1}^{\infty} \tfrac{1}{2} [x-y, Z_j(x,y)]_A t^j$$

$$+\sum_{j=2}^{\infty} \left(\sum_{\substack{p \geq 1,\ 2p \leq j \\ k_1,\ldots,k_{2p} \geq 1 \\ k_1+\cdots+k_{2p}=j}} K_{2p} [Z_{k_1}(x,y) \cdots [Z_{k_{2p}}(x,y), x+y]]_A \right) t^j.$$

4.5.1 A Recursion Formula for the CBHD Series

As $A[[t]]$ is the product of all the spaces $A_k[t]$ (where $A_k[t] = \{a*t^k : a \in A\}$), by equating the coefficients of t^j from the above identity, we derive the proof of the recursion formula in the following remarkable result (compare to [171, eq. (2.15.15)]), where we also seize the opportunity to summarize the results obtained so far.

Theorem 4.23 (Varadarajan). *Let the above notation be fixed. Then the following recursion formula holds true:*

$$
\begin{cases}
Z_1(x,y) = x + y, \\[2mm]
Z_2(x,y) = \frac{1}{4}[x - y, Z_1(x,y)] \qquad \text{and, for } j \geq 2, \\[4mm]
Z_{j+1}(x,y) = \frac{1}{2(j+1)} [x - y, Z_j(x,y)] + \\[4mm]
\quad + \displaystyle\sum_{\substack{p\geq 1,\, 2p\leqslant j \\ k_1,\ldots,k_{2p}\geqslant 1 \\ k_1+\cdots+k_{2p}=j}} \frac{K_{2p}}{j+1} [Z_{k_1}(x,y)\cdots[Z_{k_{2p}}(x,y), x+y]\cdots].
\end{cases}
\tag{4.118}
$$

We explicitly summarize the direct implications of the above theorem.

Corollary 4.24. *As a consequence of Theorem 4.23, the following facts hold.*
   Formula (4.118) provides identities in $\mathcal{T}(\mathbb{Q}\langle x, y\rangle)$, the algebra over $\mathbb{Q}$ of the polynomials in two (non-commuting) indeterminates x, y. Indeed, $Z_j(x,y)$ is the polynomial defined by

$$
Z_j(x,y) := \sum_{n=1}^{j} \frac{(-1)^{n+1}}{n} \sum_{\substack{(h_1,k_1),\ldots,(h_n,k_n)\neq(0,0) \\ |h|+|k|=j}} \frac{x^{h_1}y^{k_1}\cdots x^{h_n}y^{k_n}}{h_1!\cdots h_n!\, k_1!\cdots k_n!},
$$

so that $Z(x,y) = \sum_{j=1}^{\infty} Z_j(x,y)$ is the (unique) element of $(\widehat{\mathcal{T}}(\mathbb{Q}\langle x,y\rangle), \widehat{\frown})$ (the algebra over $\mathbb{Q}$ of the formal power series in x, y) such that

$$
\exp(Z(x,y)) = \exp(x) \frown \exp(y).
$$

Consequently, formula (4.118) shows that $Z(x,y) = \log\big(\exp(x) \frown \exp(y)\big)$ is actually an element of the closure of $\mathcal{L}(\mathbb{Q}\langle x, y\rangle)$, the latter being the free Lie algebra generated by $\{x, y\}$.
   Equivalently, (4.118) proves that $Z_j(x,y)$ is a Lie polynomial (homogeneous of degree j) in x, y over $\mathbb{Q}$, that is, an element of $\mathcal{L}_j(\mathbb{Q}\langle x, y\rangle)$. As a final consequence, the CBHD Theorem 4.3 follows.

Proof. Since $Z_1(x,y) = x + y \in \mathcal{L}_1(\mathbb{Q}\langle x,y\rangle)$ it follows that

$$Z_2(x,y) = \tfrac{1}{4}[x - y, Z_1(x,y)] \in \mathcal{L}_2(\mathbb{Q}\langle x,y\rangle).$$

Let $j \in \mathbb{N}$, $j \geq 2$, suppose that $Z_1(x,y), \ldots, Z_j(x,y)$ are Lie polynomials, homogeneous of degrees $1, \ldots, j$ respectively. With the notation of (4.118), since $k_1, \ldots, k_{2p}$ in the right-hand side of the third identity in (4.118) are all $< j$, a simple inductive argument shows that $Z_{j+1}(x,y) \in \mathcal{L}_{j+1}(\mathbb{Q}\langle x,y\rangle)$. This ends the proof. $\qquad\square$

Here we have a ready-to-use consequence of the above results. This can be stated on general Lie/associative algebras over a field of characteristic zero.

Corollary 4.25. *For every $j \in \mathbb{N} \cup \{0\}$, we set $K_j := B_j/j!$, where the coefficients B_j are the Bernoulli numbers in (4.110).*

1. *Let $(\mathfrak{g}, [\cdot, \cdot]_\mathfrak{g})$ be any Lie algebra over a field of null characteristic. For every $j \in \mathbb{N}$ and every $u, v \in \mathfrak{g}$ we set*

$$Z_j^\mathfrak{g}(u,v) = \sum_{n=1}^{j} \frac{(-1)^{n+1}}{n} \sum_{\substack{(h,k)\in\mathcal{N}_n \\ |h|+|k|=j}} \frac{\left[u^{h_1}v^{k_1}\cdots u^{h_n}v^{k_n}\right]_\mathfrak{g}}{h!\,k!\,(|h|+|k|)}. \qquad (4.119)$$

Then the following recursion formula holds:

$$\begin{cases} Z_1^\mathfrak{g}(u,v) = u + v, \\[4pt] Z_2^\mathfrak{g}(u,v) = \tfrac{1}{4}[u - v, Z_1^\mathfrak{g}(u,v)]_\mathfrak{g} \\[10pt] Z_{j+1}^\mathfrak{g}(u,v) = \dfrac{1}{2(j+1)}\,[u - v, Z_j^\mathfrak{g}(u,v)]_\mathfrak{g} + \\[8pt] \quad + \displaystyle\sum_{\substack{p\geq 1,\ 2p\leq j \\ k_1,\ldots,k_{2p}\geq 1 \\ k_1+\cdots+k_{2p}=j}} \dfrac{K_{2p}}{j+1}\,[Z_{k_1}^\mathfrak{g}(u,v)\cdots[Z_{k_{2p}}^\mathfrak{g}(u,v), u+v]_\mathfrak{g}\cdots]_\mathfrak{g}, \end{cases}$$

for every $j \in \mathbb{N}$ and every $u, v \in \mathfrak{g}$.

2. *Let $(A, *)$ be any associative algebra over a field of null characteristic. For every $j \in \mathbb{N}$ and every $u, v \in A$, we set*

$$F_j^A(u,v) = \sum_{n=1}^{j} \frac{(-1)^{n+1}}{n} \sum_{\substack{(h,k)\in\mathcal{N}_n \\ |h|+|k|=j}} \frac{u^{*h_1} * y^{*k_1} * \cdots * x^{*h_n} * y^{*k_n}}{h_1!\cdots h_n!\,k_1!\cdots k_n!}.$$

Then, if Z_j^* is as in (4.119) relatively to the commutator-algebra $\mathfrak{g} := (A, [\cdot, \cdot]_*)$ (that is, $[u, v]_* = u * v - v * u$) then we have

$$F_j^A(u,v) = Z_j^*(u,v), \quad \text{for every } j \in \mathbb{N} \text{ and every } u,v \in A,$$

so that the following recursion formula holds:

$$
\begin{cases}
F_1^A(u,v) = u+v, \\[2mm]
F_2^A(u,v) = \frac{1}{4}[u-v, F_1^A(u,v)]_* \\[4mm]
F_{j+1}^A(u,v) = \frac{1}{2(j+1)} [u-v, F_j^A(u,v)]_* + \\[2mm]
\qquad + \displaystyle\sum_{\substack{p\geq 1,\, 2p\leqslant j \\ k_1,\ldots,k_{2p}\geqslant 1 \\ k_1+\cdots+k_{2p}=j}} \frac{K_{2p}}{j+1} [F_{k_1}^A(u,v)\cdots [F_{k_{2p}}^A(u,v), u+v]_* \cdots]_*,
\end{cases}
$$

for every $j \in \mathbb{N}$ and every $u, v \in A$.

Proof. This follows by collecting together Theorem 4.23, (4.17), (4.18) and by simple substitution arguments, based on Theorem 2.85 on page 107. □

Besides giving a proof of the CBHD Theorem, formula (4.118) also provides a tool for studying the *convergence* of the so-called CBHD series. Indeed, by rerunning the remarkable arguments by Varadarajan on the CBHD formula for Lie groups, it is possible to derive from (4.118) an extremely accurate estimate of the coefficients Z_j (see the proof of Theorem 2.15.4 in [171]), in the setting of finite dimensional Lie algebras or of normed Banach algebras. This will be done in details in Chap. 5 (precisely in Sect. 5.2.3).

4.5.2 Another Recursion Formula

Yet another more compact way of writing the "differential equation" solved by $Z = \sum_{j=1}^{\infty} Z_j(x,y)\, t^j$ is given by G. Czichowski in [44]. It is then possible to obtain another recursion formula for $Z_j(x,y)$. We follow the exposition in [44, page 88, 89].

Indeed, as a continuation of Djoković's computation, from formula (4.95) on page 214 we get

$$\partial_t Z \overset{(4.95)}{=} g_{\mathrm{ad}\, Z}\big(x + h_{\mathrm{ad}\, Z}(y)\big)$$

$$= g_{\mathrm{ad}\, Z}(x) + \big(g_{\mathrm{ad}\, Z} \circ h_{\mathrm{ad}\, Z}\big)(y).$$

By Lemma 4.18 on page 209 we have

$$g_{\mathrm{ad}\, Z} \circ h_{\mathrm{ad}\, Z} = (g \cdot h)_{\mathrm{ad}\, Z} = g_{-\mathrm{ad}\, Z}. \tag{4.120}$$

Indeed, in the last equality we used a simple equality involving the complex functions $g(z)$ and $h(z)$ associated to the formal power series $g = \sum_{j=0}^{\infty} K_j\, z^j$ and $h = \sum_{j=0}^{\infty} z^j / j!$, namely

$$g(z)\, h(z) = \frac{z}{e^z - 1}\, e^z = \frac{z}{1 - e^{-z}} = \frac{-z}{e^{-z} - 1} = g(-z).$$

Since this equality holds true for $|z| < 2\pi$, we derive an analogous identity for the related formal power series:

$$\left(\sum_{j=0}^{\infty} K_j\, z^j \right) \cdot \left(\sum_{j=0}^{\infty} \frac{z^j}{j!} \right) = \sum_{j=0}^{\infty} (-1)^j\, K_j\, z^j,$$

so that we are entitled to to derive the last equality in (4.120). Thus we obtain the more symmetric-looking equation for $\partial_t Z$:

$$\partial_t Z = g_{\mathrm{ad}\, Z}(x) + g_{-\mathrm{ad}\, Z}(y). \tag{4.121}$$

By inserting in (4.121) the actual expression of the formal power series for g, this is equivalent to

$$\partial_t Z = \sum_{j=0}^{\infty} K_j\, (\mathrm{ad}\, Z)^{\circ j} \big(x + (-1)^j\, y \big). \tag{4.122}$$

Incidentally, by the identity (4.111a), which is equivalent to

$$K_1 = -\tfrac{1}{2}, \qquad K_{2p+1} = 0 \quad \forall\, p \geq 1,$$

formula (4.122) gives back the identity (4.113), indeed

$$\partial_t Z \overset{(4.122)}{=} K_1\, (\mathrm{ad}\, Z)(x - y) + \sum_{p=0}^{\infty} K_{2p}\, (\mathrm{ad}\, Z)^{\circ\, 2p} \big(x + (-1)^{2p}\, y \big)$$

$$= -\tfrac{1}{2}\, [Z, x - y] + \sum_{p=0}^{\infty} K_{2p}\, (\mathrm{ad}\, Z)^{\circ\, 2p}(x + y)$$

$$= \tfrac{1}{2}\, [x - y, Z] + k_{\mathrm{ad}\, Z}(x + y).$$

By inserting in (4.122) the power series expression for Z and by equating the coefficients of t^j from both sides, we obtain the following recursion formula (which is obviously equivalent to that obtained in Corollary 4.25):

$$
\begin{cases}
Z_1(x,y) = x + y, \\[2mm]
Z_2(x,y) = \dfrac{K_1}{2}\,[Z_1(x,y), x - y] = \dfrac{1}{2}\,[x,y] \\[4mm]
Z_{n+1}(x,y) = \dfrac{1}{n+1} \sum_{\substack{1 \leqslant j \leqslant n \\ i_1,\dots,i_j \geqslant 1 \\ i_1+\cdots+i_j=n}} K_j\,[Z_{i_1}(x,y)\cdots[Z_{i_j}(x,y), x + (-1)^j y]\cdots],
\end{cases}
$$

holding true for every $n \in \mathbb{N}$.

Remark 4.26. Exactly as in the statement of Corollary 4.25, from the above recursive relations on the Lie polynomials $Z_j(x,y)$ we can obtain recursion formulas for $Z_j^{\mathfrak{g}}(u,v)$ on any Lie algebra $\mathfrak{g}$ and for the $F_j^A(u,v)$ on any associative algebra A. We leave it to Reader.

4.6 Reutenauer's Proof

The aim of this section is to provide all the details of the proof by Reutenauer of the Campbell-Baker-Hausdorff Theorem. We mainly follow [144, Section 3.4], adding to the exposition therein a discussion of all convergence conditions (always within the setting of the completion of graded algebras). To this aim, we will be forced to introduce some lemmas which are not given in [144] (see Lemma 4.34 and Corollary 4.35 below).

One of the most evident contributions of Reutenauer's proof is its focusing on some crucial computations about the CBHD Theorem – tracing back to Pascal, Campbell, Baker and Hausdorff – in a clear and concise way. In particular, the rôles of H_1^x and H_1^y (respectively, the series of the summands coming from $\mathrm{Log}(\mathrm{Exp}\,x \cdot \mathrm{Exp}\,y)$ containing x and y precisely once) will be clarified. The crucial part played by these series (evident in many of the original early arguments about the Campbell-Baker-Hausdorff Theorem) does not appear in the other proofs we presented so far. Finally, some analogies with the computations in Djoković's proof will be highlighted in due course.

Reutenauer's proof is based on a catchy argument involving derivation and exponentiation. Before looking at it, we first need a new definition.

Definition 4.27 (φ-Derivation). Let $(A, *)$ be an associative algebra. Let $\varphi : A \to A$ be an algebra morphism. We say that a map $D : A \to A$ is a φ-*derivation* if D is linear and the following condition holds

$$
D(a * b) = D(a) * \varphi(b) + \varphi(a) * D(b), \quad \text{for every } a, b \in A. \tag{4.123}
$$

The notation D_φ will also be used for φ-derivations.

It can be proved that φ-derivations have a particularly nice property when they are applied to formal power series (see Theorem 3.22 in [144]). All we need here is to consider the case of the exponential power series, as in the following lemma. The Reader will recognize the similarities with the computations in Djoković's proof (see Sect. 4.3.3).

Lemma 4.28. *Let $\mathbb{K}$ be a field of characteristic zero. Let $A = \bigoplus_{j=0}^\infty A_j$ be a graded UA algebra (with $A_0 = \mathbb{K}$) and let $\widehat{A} = \prod_{j=0}^\infty A_j$ denote, as usual, the topological UA algebra of the formal power series on A (see Sect. 2.3.3). Let $*$ denote both the operation on A and on $\widehat{A}$. Finally, let $\mathrm{Exp} : \widehat{A}_+ \to 1 + \widehat{A}_+$ be the relevant exponential function (as in Definition 3.2), where $\widehat{A}_+ = \prod_{j=1}^\infty A_j$.*

Suppose that $\varphi : \widehat{A} \to \widehat{A}$ is a UAA morphism and that $D_\varphi : \widehat{A} \to \widehat{A}$ is a continuous φ-derivation (according to Definition 4.27). Then, we have

$$D_\varphi(\mathrm{Exp}\, x) * \mathrm{Exp}(-\varphi(x)) = \sum_{n=1}^\infty \frac{1}{n!} \big(\mathrm{ad}\, \varphi(x)\big)^{n-1}(D_\varphi(x)), \qquad (4.124)$$

for every $x \in \widehat{A}_+$ such that $\varphi(x) \in \widehat{A}_+$.

[Here, given $a \in \widehat{A}$, we denote by $\mathrm{ad}\,(a) : \widehat{A} \to \widehat{A}$ the usual adjoint map with respect to the commutator of $\widehat{A}$, that is, $\mathrm{ad}\,(a)(b) = a * b - b * a$.]

Note that (4.124) can be rewritten as

$$D_\varphi(\mathrm{Exp}\, x) * \mathrm{Exp}(-\varphi(x)) = f(\mathrm{ad}\,\varphi(x))(D_\varphi(x)), \qquad (4.125)$$

where $f(z) = \sum_{n=1}^\infty z^{n-1}/n!$ is the Maclaurin expansion of the entire function $f(z) := (e^z - 1)/z$.

Proof. Let $x \in \widehat{A}_+$ be fixed. By hypothesis, $x, \varphi(x) \in \widehat{A}_+$ so that $\mathrm{Exp}(x)$ and $\mathrm{Exp}(-\varphi(x))$ are both well posed. The fact that $\varphi(x) \in \widehat{A}_+$ also ensures that

$$a_k := \big(\mathrm{ad}\,(\varphi(x))\big)^{k-1}(D_\varphi(x)) \in \prod_{j=k-1}^\infty A_j.$$

This yields $\lim_{k\to\infty} a_k = 0$ in $\widehat{A}$, whence the series in the right-hand side of (4.124) converges (making use, for example, of part 3 of Remark 2.76 on page 102).

We now prove a useful fact: given $x \in \widehat{A}$ and $m \in \mathbb{N}$, we have

$$D_\varphi(x^m) = \sum_{k=0}^{m-1} (\varphi(x))^k * (D_\varphi x) * (\varphi(x))^{m-k-1}. \qquad (4.126)$$

We prove (4.126) by induction on m. The case $m = 1$ is trivial. Assuming (4.126) to hold for $1, 2, \ldots, m$, we prove it for $m + 1$:

$$D_\varphi(x^{m+1}) = D_\varphi(x^m * x) \stackrel{(4.123)}{=} D_\varphi(x^m) * \varphi(x) + \varphi(x^m) * D_\varphi x$$

(by the inductive hypothesis, by the bilinearity and associativity of $*$

and the fact that φ is a UAA morphism so that $\varphi(x^m) = (\varphi(x))^m$)

$$= \sum_{k=0}^{m-1} (\varphi(x))^k * (D_\varphi x) * (\varphi(x))^{m-k} + (\varphi(x))^m * D_\varphi x$$

$$= \sum_{k=0}^{m} (\varphi(x))^k * (D_\varphi x) * (\varphi(x))^{m-k}.$$

We are now in a position to complete the proof. Indeed, the following computation applies:

$$D_\varphi(\mathrm{Exp}\, x) * \mathrm{Exp}(-\varphi(x)) = D_\varphi\left(\sum_{m=0}^{\infty} \frac{x^m}{m!} \right) * \left(\sum_{k=0}^{\infty} \frac{(-\varphi(x))^k}{k!} \right)$$

(D_φ is continuous by hypothesis)

$$= \left(\sum_{m=1}^{\infty} \frac{1}{m!} D_\varphi(x^m) \right) * \left(\sum_{k=0}^{\infty} \frac{(-1)^k}{k!} (\varphi(x))^k \right) \quad \text{(reordering)}$$

$$= \sum_{n=1}^{\infty} \frac{1}{n!} \sum_{m=1}^{n} (-1)^{n-m} \binom{n}{m} D_\varphi(x^m) * (\varphi(x))^{n-m}$$

$$\stackrel{(4.126)}{=} \sum_{n=1}^{\infty} \frac{1}{n!} \sum_{m=1}^{n} \sum_{k=0}^{m-1} (-1)^{n-m} \binom{n}{m} (\varphi(x))^k * (D_\varphi x) * (\varphi(x))^{n-k-1}$$

(interchanging the inner sums: $\sum_{m=1}^{n} \sum_{k=0}^{m-1} = \sum_{k=0}^{n-1} \sum_{m=k+1}^{n}$)

$$= \sum_{n=1}^{\infty} \frac{1}{n!} \sum_{k=0}^{n-1} \left(\sum_{m=k+1}^{n} (-1)^{n-m} \binom{n}{m} \right) (\varphi(x))^k * (D_\varphi x) * (\varphi(x))^{n-k-1}$$

(we rename $r := n - m$ and we use $\binom{n}{n-r} = \binom{n}{r}$)

$$= \sum_{n=1}^{\infty} \frac{1}{n!} \sum_{k=0}^{n-1} \left(\sum_{r=0}^{n-k-1} (-1)^r \binom{n}{r} \right) (\varphi(x))^k * (D_\varphi x) * (\varphi(x))^{n-k-1}$$

(see Lemma 4.19 on page 212 and use $\binom{n-1}{n-k-1} = \binom{n-1}{k}$)

$$= \sum_{n=1}^{\infty} \frac{1}{n!} \sum_{k=0}^{n-1} (-1)^{n-k-1} \binom{n-1}{k} (\varphi(x))^k * (D_\varphi x) * (\varphi(x))^{n-k-1}$$

$$= \sum_{n=1}^{\infty} \frac{1}{n!} \sum_{k=0}^{n-1} \binom{n-1}{k} (\varphi(x))^k * (D_\varphi x) * (-\varphi(x))^{n-k-1}$$

(we denote by L_a, R_a respectively the right and the left multiplications by $a \in \widehat{A}$ on the UA algebra $(\widehat{A}, *)$)

$$= \sum_{n=1}^{\infty} \frac{1}{n!} \underbrace{\left(\sum_{k=0}^{n-1} \binom{n-1}{k} (L_{\varphi(x)})^k \circ (R_{-\varphi(x)})^{n-1-k} \right)}_{= (L_{\varphi(x)} + R_{-\varphi(x)})^{n-1} = (\operatorname{ad} \varphi(x))^{n-1}} (D_\varphi x)$$

$$= \sum_{n=1}^{\infty} \frac{1}{n!} (\operatorname{ad} \varphi(x))^{n-1} (D_\varphi x).$$

This completes the proof of the lemma.   □

From now on, we take a set $\{x, y\}$, with $x \neq y$, and we consider the tensor algebra $\mathscr{T}(\mathbb{Q}\langle x, y \rangle)$ of the free vector space $\mathbb{Q}\langle x, y \rangle$ over the rationals, which we denote briefly by $\mathscr{T}$. We know that $\mathscr{T}$ is nothing but the free unital associative $\mathbb{Q}$-algebra over the set $\{x, y\}$ (see Theorem 2.40 on page 79). We also know that $\mathscr{T}$ is graded, with grading

$$\mathscr{T} = \bigoplus_{j=0}^{\infty} \mathscr{T}_j(\mathbb{Q}\langle x, y \rangle),$$

and its completion is

$$\widehat{\mathscr{T}}(\mathbb{Q}\langle x, y \rangle) = \prod_{j=0}^{\infty} \mathscr{T}_j(\mathbb{Q}\langle x, y \rangle),$$

which is simply the topological unital associative algebra of the formal power series in two non-commuting indeterminates x, y. We use the short notations $\mathscr{T}_j$ and $\widehat{\mathscr{T}}$ with clear meanings. We set

$$H(x, y) := \operatorname{Log}(\operatorname{Exp} x \cdot \operatorname{Exp} y).$$

By its very definition,

$$\operatorname{Exp}(H(x, y)) = \operatorname{Exp} x \cdot \operatorname{Exp} y. \tag{4.127}$$

In our former notation, $H(x, y)$ is precisely $x \blacklozenge y$ (see Sect. 3.1.3 on page 126). As is well known, this can be written as

$$H(x, y) = \sum_{n=1}^{\infty} \frac{(-1)^{n+1}}{n} \sum_{(h_1, k_1), \ldots, (h_n, k_n) \neq (0,0)} \frac{x^{h_1} y^{k_1} \cdots x^{h_n} y^{k_n}}{h_1! \cdots h_n! k_1! \cdots k_n!}.$$

We now reorder this series with respect to the increasing number of times the indeterminate x appears. More precisely, we set

$$H(x,y) = \sum_{j=0}^{\infty} H_j^x, \quad \text{where, for any } j \in \mathbb{N} \cup \{0\},$$

$$H_j^x = \sum_{n=1}^{\infty} \frac{(-1)^{n+1}}{n} \sum_{\substack{(h_1,k_1),\ldots,(h_n,k_n) \neq (0,0) \\ h_1+\cdots+h_n=j}} \frac{x^{h_1} y^{k_1} \cdots x^{h_n} y^{k_n}}{h_1! \cdots h_n! k_1! \cdots k_n!}. \tag{4.128}$$

In other words, H_j^x is the series of summands (chosen from the summands of $H(x,y)$) which contain x precisely j times. With our usual notation in this Book (see e.g., page 127) H_j^x may be written as

$$H_j^x = \sum_{n=1}^{\infty} c_n \sum_{(h,k) \in \mathcal{N}_n : |h|=j} \frac{x^{h_1} y^{k_1} \cdots x^{h_n} y^{k_n}}{h! \, k!}, \quad j \geq 0.$$

For example, we have

$$H_0^x = y. \tag{4.129}$$

Indeed the following computation applies

$$H_0^x = \sum_{n=1}^{\infty} \frac{(-1)^{n+1}}{n} \sum_{k_1,\ldots,k_n \neq 0} \frac{y^{k_1+\cdots+k_n}}{k_1! \cdots k_n!}$$

$$= \sum_{n=1}^{\infty} \frac{(-1)^{n+1}}{n} \left(\sum_{k \neq 0} \frac{y^k}{k!} \right)^n = \sum_{n=1}^{\infty} \frac{(-1)^{n+1}}{n} (\text{Exp}(y) - 1)^n$$

$$= \text{Log}(1 + (\text{Exp}(y) - 1)) = \text{Log}(\text{Exp}\, y) = y.$$

By means of Lemma 4.28, we can prove the following fundamental result (see [144, Corollary 3.24]).

Theorem 4.29. *With the above notation, we have*

$$H_1^x = x + \tfrac{1}{2}[x,y] + \sum_{p=1}^{\infty} \frac{B_{2p}}{(2p)!} (\text{ad}\, y)^{2p}(x), \tag{4.130}$$

where the coefficients B_n are the Bernoulli numbers (see e.g., (4.110)).

We recall that, although in the cited formula (4.110) we gave an explicit recursive definition of B_n, it is more convenient to anticipate a fact (proved in Chap. 9) about the Bernoulli numbers, namely that they are defined by the following generating function (see (9.39) on page 496):

$$\frac{x}{e^x - 1} = \sum_{n=0}^{\infty} \frac{B_n}{n!} x^n \qquad \text{if } |x| < 2\pi. \tag{4.131}$$

This immediately shows that $B_0 = 1$, $B_1 = -1/2$ and $B_{2k+1} = 0$ for every $k \geq 1$ (since $(e^x - 1)/x + x/2$ is an even function). As a consequence, (4.130) compactly rewrites as

$$H_1^x = \sum_{n=0}^{\infty} \frac{B_n}{n!} (\operatorname{ad} y)^n (x). \tag{4.132}$$

Proof (of Theorem 4.29). Let $\varphi : \mathscr{T} \to \mathscr{T}$ be the UAA morphism such that

$$\varphi(x) = 0 \quad \text{and} \quad \varphi(y) = y.$$

[This exists thanks to part 2 of Theorem 2.40 on page 79.] By the very definition of φ, we infer

$$\varphi\left(x^{h_1} y^{k_1} \cdots x^{h_n} y^{k_n}\right) = \begin{cases} \text{(if } |h| = 0) & x^{h_1} y^{k_1} \cdots x^{h_n} y^{k_n} = y^{|k|}, \\ \text{(if } |h| \neq 0) & 0, \end{cases} \tag{4.133}$$

for all nonnegative h and k. This immediately yields

$$\varphi\left(\bigoplus_{j \geq k} \mathscr{T}_j\right) \subseteq \bigoplus_{j \geq k} \mathscr{T}_j, \quad \text{for every } k \geq 0. \tag{4.134}$$

By means of Lemma 2.79 on page 103, (4.134) ensures that there exists a unique continuous prolongation of φ, say $\widehat{\varphi} : \widehat{\mathscr{T}} \to \widehat{\mathscr{T}}$ which is also a UAA morphism.

Let now $D : \mathscr{T} \to \mathscr{T}$ be the unique linear map such that

$$D(1) = 0, \quad D(x) = x, \quad D(y) = 0,$$

and such that

$$D\left(x^{h_1} y^{k_1} \cdots x^{h_n} y^{k_n}\right) = \begin{cases} \text{(if } |h| = 1) & x^{h_1} y^{k_1} \cdots x^{h_n} y^{k_n}, \\ \text{(if } |h| \neq 1) & 0, \end{cases} \tag{4.135}$$

for all nonnegative h and k. The existence of D is simply verified: it suffices to define it according to (4.135) on the basis of $\mathscr{T}$ given by

$$\{1\} \cup \left\{x^{h_1} y^{k_1} \cdots x^{h_n} y^{k_n} \mid n \in \mathbb{N}, \ (h_1, k_1), \ldots, (h_n, k_n) \in \{(1,0), (0,1)\}\right\},$$

and then to check that it fulfils (4.135) throughout. Note that D is the linear map which kills all the elementary monomials containing more than one x and preserving those containing x precisely once.

It is easily seen that D fulfills the same property as φ in (4.134). Hence, by the above cited lemma on continuous prolongations, there exists a unique linear continuous prolongation of D, say

$$D_{\widehat{\varphi}} : \widehat{\mathscr{T}} \to \widehat{\mathscr{T}}.$$

We claim that $D_{\widehat{\varphi}}$ is a $\widehat{\varphi}$-derivation of $\widehat{\mathscr{T}}$, according to Definition 4.27. As $D_{\widehat{\varphi}}$ is linear, all we have to prove is that

$$D_{\widehat{\varphi}}(a \cdot b) = D_{\widehat{\varphi}}(a) \cdot \widehat{\varphi}(b) + \widehat{\varphi}(a) \cdot D_{\widehat{\varphi}}(b), \tag{4.136}$$

when a, b are elementary monomials, say

$$a = x^{h_1} y^{k_1} \cdots x^{h_n} y^{k_n}, \quad b = x^{r_1} y^{s_1} \cdots x^{r_m} y^{s_m}.$$

The following computation holds:

$$D_{\widehat{\varphi}}(a \cdot b) = D_{\widehat{\varphi}}\big(x^{h_1} y^{k_1} \cdots x^{h_n} y^{k_n} x^{r_1} y^{s_1} \cdots x^{r_m} y^{s_m}\big)$$

$$= \begin{cases} a \cdot b, & \text{if } |h| + |r| = 1, \\ 0, & \text{otherwise.} \end{cases}$$

On the other hand we have

$$D_{\widehat{\varphi}}(a) \cdot \widehat{\varphi}(b) + \widehat{\varphi}(a) \cdot D_{\widehat{\varphi}}(b)$$

$$= \begin{cases} \text{if } |h| = 1 & a \cdot \begin{cases} b, & \text{if } |r| = 0, \\ 0, & \text{if } |r| \neq 0, \end{cases} \\ \text{if } |h| \neq 1 & 0 \end{cases} + \begin{cases} \text{if } |h| = 0 & a \cdot \begin{cases} b, & \text{if } |r| = 1, \\ 0, & \text{if } |r| \neq 1 \end{cases} \\ \text{if } |h| \neq 0 & 0 \end{cases}$$

$$= \begin{cases} a \cdot b, & \text{if } |h| = 1 \text{ and } |r| = 0 \\ 0 & \text{otherwise} \end{cases} + \begin{cases} a \cdot b, & \text{if } |h| = 0 \text{ and } |r| = 1 \\ 0 & \text{otherwise} \end{cases}$$

$$= \begin{cases} a \cdot b, & \text{if } |h| + |r| = 1 \\ 0 & \text{otherwise.} \end{cases}$$

De visu, this proves (4.136). We next remark that

$$D_{\widehat{\varphi}}(H(x,y)) = H_1^x. \tag{4.137}$$

Indeed, by the continuity of $D_{\widehat{\varphi}}$, the fact that $D_{\widehat{\varphi}}$ prolongs D and that D fulfills (4.135), we have

$$D_{\widehat{\varphi}}(H(x,y)) = \sum_{n=1}^{\infty} c_n \sum_{(h,k)\in N_n} D\big(x^{h_1}y^{k_1}\cdots x^{h_n}y^{k_n}\big)/(h!\,k!)$$

$$= \sum_{n=1}^{\infty} c_n \sum_{(h,k)\in N_n:\ |h|=1} x^{h_1}y^{k_1}\cdots x^{h_n}y^{k_n}/(h!\,k!) = H_1^x.$$

Moreover, we claim that

$$D_{\widehat{\varphi}}\big(\operatorname{Exp} x \cdot \operatorname{Exp} y\big) = x \cdot \operatorname{Exp} y. \tag{4.138}$$

Indeed, by the same arguments as above, one has

$$D_{\widehat{\varphi}}\big(\operatorname{Exp} x \cdot \operatorname{Exp} y\big) = D_{\widehat{\varphi}} \sum_{h,k\geq 0} x^h y^k/(h!\,k!)$$

$$= \sum_{h,k\geq 0} D(x^h y^k)/(h!\,k!) = \sum_{k\geq 0} x\,y^k/k! = x\operatorname{Exp} y.$$

Furthermore, we claim that

$$\widehat{\varphi}(H(x,y)) = y. \tag{4.139}$$

Indeed, the following computation applies:

$$\widehat{\varphi}(H(x,y)) = \widehat{\varphi} \sum_{n=1}^{\infty} \frac{(-1)^{n+1}}{n} \sum_{(h_1,k_1),\ldots,(h_n,k_n)\neq(0,0)} \frac{x^{h_1}y^{k_1}\cdots x^{h_n}y^{k_n}}{h_1!\cdots h_n!\,k_1!\cdots k_n!}$$

$$(\widehat{\varphi} \text{ is continuous and prolongs } \varphi)$$

$$= \sum_{n=1}^{\infty} \frac{(-1)^{n+1}}{n} \sum_{(h_1,k_1),\ldots,(h_n,k_n)\neq(0,0)} \frac{\varphi\big(x^{h_1}y^{k_1}\cdots x^{h_n}y^{k_n}\big)}{h_1!\cdots h_n!\,k_1!\cdots k_n!}$$

$$\overset{(4.133)}{=} \sum_{n=1}^{\infty} \frac{(-1)^{n+1}}{n} \sum_{k_1,\ldots,k_n\neq 0} \frac{y^{k_1+\cdots+k_n}}{k_1!\cdots k_n!} = H_0^x \overset{(4.129)}{=} y.$$

Finally, we are in a position to apply Lemma 4.28: Indeed, $A := \mathscr{T}$ has all the properties needed in that lemma, $\widehat{\varphi}$ is a UAA morphism of $\widehat{\mathscr{T}}$ and $D_{\widehat{\varphi}}$ is a continuous $\widehat{\varphi}$-derivation of $\widehat{\mathscr{T}}$. Hence (notice that $H(x,y) \in \widehat{\mathscr{T}}_+$ and $\widehat{\varphi}(H(x,y)) = y \in \widehat{\mathscr{T}}_+$) (4.124) gives the key step of the following chain of equalities:

$$x \cdot \operatorname{Exp}(y) \overset{(4.138)}{=} D_{\widehat{\varphi}}\big(\operatorname{Exp} x \cdot \operatorname{Exp} y\big) \overset{(4.127)}{=} D_{\widehat{\varphi}}\big(\operatorname{Exp}(H(x,y))\big)$$

$$\overset{(4.124)}{=} \sum_{n=1}^{\infty} \frac{1}{n!}\big(\operatorname{ad}\widehat{\varphi}(H(x,y))\big)^{n-1}\big(D_{\widehat{\varphi}}(H(x,y))\big) \cdot \operatorname{Exp}(\widehat{\varphi}(H(x,y)))$$

$$\left(\text{we now use the identities } \widehat{\varphi}(H(x,y)) \overset{(4.139)}{=} y \text{ and } D_{\widehat{\varphi}}(H(x,y)) \overset{(4.137)}{=} H_1^x\right)$$

$$= \sum_{n=1}^{\infty} \frac{1}{n!} (\operatorname{ad} y)^{n-1}(H_1^x) \cdot \operatorname{Exp}(y).$$

By canceling $\operatorname{Exp}(y)$ from both the far sides, we get

$$x = \sum_{n=1}^{\infty} \frac{1}{n!} (\operatorname{ad} y)^{n-1}(H_1^x). \tag{4.140}$$

Now, we notice that $\operatorname{ad}(y)$ maps $\prod_{j \geq N} \mathscr{T}_j$ into $\prod_{j \geq N+1} \mathscr{T}_j$, for every $N \geq 0$. Hence, arguing as in Lemma 4.18 on page 209, we infer that the endomorphism of $\widehat{\mathscr{T}}$ defined by

$$f(\operatorname{ad} y) := \sum_{n=1}^{\infty} \frac{1}{n!} (\operatorname{ad} y)^{n-1}$$

is well posed and that it is invertible, with inverse given by

$$g(\operatorname{ad} y) := \sum_{n=0}^{\infty} \frac{B_n}{n!} (\operatorname{ad} y)^n.$$

Indeed, this follows from the fact that the formal power series in $\mathbb{Q}[[z]]$

$$f(z) := \sum_{n=1}^{\infty} \frac{z^{n-1}}{n!} \quad \text{and} \quad g(z) := \sum_{n=0}^{\infty} \frac{B_n}{n!} z^n$$

are reciprocal to one another, for they are the Maclaurin series respectively of

$$\frac{e^z - 1}{z} \quad \text{and} \quad \frac{z}{e^z - 1} \qquad \text{(by recalling (4.131)).}$$

Hence (4.140) rewrites as

$$x = f(\operatorname{ad} y)(H_1^x).$$

By applying $g(\operatorname{ad} y)$ to both sides of this identity we get

$$g(\operatorname{ad} y)(x) = H_1^x.$$

Thanks to the definition of g, this is precisely (4.132), which (as we already remarked) is equivalent to (4.130). This ends the proof. $\qquad \square$

To proceed with our details behind Reutenauer's proof of the Campbell-Baker-Hausdorff Theorem, we need a preparatory lemma.

Lemma 4.30. *Let X be any nonempty set. Let $\{S_x\}_{x \in X}$ be some given family of elements of $\mathscr{T}(\mathbb{K}\langle X \rangle)$. Then there exists a unique continuous derivation D of the algebra $\widehat{\mathscr{T}}(\mathbb{K}\langle X \rangle)$ such that*

$$D(x) = S_x, \quad \text{for every } x \in X. \tag{4.141}$$

Proof. See page 449.   □

The previous lemma proves that the following definition is well posed.

Definition 4.31 (The Operator $S\,(\partial/\partial y)$). Let $S \in \widehat{\mathscr{T}}(\mathbb{Q}\langle x, y \rangle)$ be given. We denote by $S\,\frac{\partial}{\partial y}$ the unique continuous derivation of $\widehat{\mathscr{T}}(\mathbb{Q}\langle x, y \rangle)$ mapping x into 0 and y into S (see Lemma 4.30).

Let us note that the use of operators like $S\,(\partial/\partial y)$ just introduced goes back to the original papers by Baker [8] and by Hausdorff [78].

Hereafter, we set for brevity

$$\widehat{\mathscr{T}} := \widehat{\mathscr{T}}(\mathbb{Q}\langle x, y \rangle), \qquad \mathscr{T}_j := \mathscr{T}_j(\mathbb{Q}\langle x, y \rangle) \quad (j \geq 0),$$

and we resume from Sect. 2.3.3 the notation

$$\widehat{U}_n := \prod_{j \geq n} \mathscr{T}_j(\mathbb{Q}\langle x, y \rangle) \quad (n \geq 0).$$

Remark 4.32. At this point, we are forced to add some extra results to Reutenauer's arguments.[4] Within a few lines, we will meet a sort of exponential operator "$\exp(D)$", where D is a derivation of an associative algebra A over $\mathbb{Q}$. Unfortunately, it is not possible to define $\exp(D)$ unambiguously for general topological (or graded) algebras. For instance, as in Djoković's proof, we had to require D to satisfy a stronger hypothesis[5] in order to define a formal power series in D.

For example, consider $A = \mathbb{Q}[[t]]$ (the algebra of formal power series in one indeterminate t over $\mathbb{Q}$) and the derivation $D = \partial_t$ (see Definition 4.11). Then, if x is the element of $\mathbb{Q}[[t]]$ given by $x = \sum_{k=0}^{\infty} t^k/k!$, the series

$$\sum_{n=0}^{\infty} \frac{1}{n!} D^n(x)$$

does not converge in the topology of (the usual metric space) $\mathbb{Q}[[t]]$. Indeed (see e.g., Remark 2.76 on page 102), since $\mathbb{Q}[[t]]$ is an ultrametric space, a series $\sum_n a_n$ converges if and only if a_n tends to zero as $n \to \infty$. Now, in the above example we have

---

[4]See [144], page 78, when $\exp(D)$ is introduced.
[5]See the class $\mathcal{H}$ of endomorphisms in Definition 4.17 on page 209.

$$\partial_t x = x, \quad \text{so that} \quad (\partial_t)^n x = x \ \text{for every } n \geq 0.$$

Consequently, $D^n(x)/n! = x/n!$ does not converge to zero as $n \to \infty$ (recall that the usual Euclidean topology is by no means involved here!) so that $\sum_{n=0}^{\infty} D^n(x)/n!$ does not make sense at all in $\mathbb{Q}[[t]]$.

Another example of non-convergence can be given in the very context we are interested in. Indeed, take for instance

$$D : \widehat{\mathscr{T}} \to \widehat{\mathscr{T}}, \quad D := y \frac{\partial}{\partial y}.$$

Then, trivially, $D^n(y) = y$ for every $n \geq 0$. This proves that $\sum_{n \geq 0} D^n(y)/n!$ does not converge, for $y/n!$ does not vanish, as $n \to \infty$.

In the rest of Reutenauer's proof, the derivation $D = H_1^x \frac{\partial}{\partial y}$ of $\widehat{\mathscr{T}}$ is concerned. Even in this case, D does not belong *a priori* to the class $\mathcal{H}$ of endomorphisms introduced in Definition 4.17. Indeed, for example, $y \in \widehat{U}_1$ but

$$D(y) = H_1^x \frac{\partial}{\partial y}(y) = H_1^x \notin \widehat{U}_2.$$

Nonetheless, *due to certain very special properties of H_1^x*, we claim that that the series

$$\exp(D)(z) := \sum_{n=0}^{\infty} \frac{1}{n!} \left(H_1^x \frac{\partial}{\partial y} \right)^n (z)$$

converges in $\widehat{\mathscr{T}}$ *for every $z \in \widehat{\mathscr{T}}$*. We hope that our proof of the above claim will be welcomed. Indeed, without this convergence result the last part of Reutenauer's proof is only formal, whereas (once it is known that $\exp(D)$ defines a genuine map) the proof is completely justified. Also, this highlights how the properties of H_1^x intervene in making $\exp(D)$ a well-defined map.

Remark 4.33. Before proceeding we show that, once D fulfills a well-behaved convergence assumption, then $\exp(D)$ turns out to be a well-defined UAA morphism. Indeed, let $A, \widehat{A}$ be as in Lemma 4.28 and suppose that $D : \widehat{A} \to \widehat{A}$ is a derivation of $\widehat{A}$ satisfying the following hypothesis

$$D^n(\widehat{A}) \subseteq \widehat{U}_n, \quad \text{for every } n \geq 0, \tag{4.142}$$

where we have set $\widehat{U}_n := \prod_{j \geq n} A_j$ (for every $n \geq 0$). As a consequence of (4.142), for every $a \in \widehat{A}$, $D^n(a)/n! \longrightarrow 0$ as $n \to \infty$ in $\widehat{A}$, so that (by part 3 of Remark 2.76 on page 102) the series $\sum_{n=0}^{\infty} D^n(a)/n!$ converges in $\widehat{A}$ and the formula

$$\exp(D) : \widehat{A} \to \widehat{A}, \quad a \mapsto \sum_{n=0}^{\infty} \frac{D^n(a)}{n!}$$

defines an endomorphism of $\widehat{A}$. *We claim that $\exp(D)$ is a UAA morphism.*

Indeed, to begin with, it is easily seen by induction that the following Leibnitz formula is valid (see Proposition 4.40):

$$D^n(a * b) = \sum_{i=0}^{n} \binom{n}{i} D^i(a) * D^{n-i}(b), \quad \text{for every } a, b \in \widehat{A}, n \geq 0. \quad (4.143)$$

As a consequence, if $a, b \in \widehat{A}$, we have

$$\exp(D)(a * b) = \sum_{n=0}^{\infty} \frac{D^n(a * b)}{n!} = \sum_{n=0}^{\infty} \frac{1}{n!} \left(\sum_{i=0}^{n} \binom{n}{i} D^i(a) * D^{n-i}(b) \right)$$

$$= \sum_{n=0}^{\infty} \left(\sum_{i+j=n} \frac{D^i(a) * D^j(b)}{i!\, j!} \right) = \left(\sum_{i=0}^{\infty} \frac{D^i(a)}{i!} \right) * \left(\sum_{j=0}^{\infty} \frac{D^j(b)}{j!} \right)$$

$$= \exp(D)(a) * \exp(D)(b).$$

Observe that in the fourth equality we used again hypothesis (4.142), which allows us to reorder the series. The above computation shows that $\exp(D)$ is an algebra morphism. Finally, note that $\exp(D)$ is unital, for $\exp(D)(1_{\mathbb{K}}) = 1_{\mathbb{K}}$ (since $D^n(1_{\mathbb{K}}) = 0$ for every $n \geq 1$, D being a derivation). $\qquad\square$

We aim to prove the following preparatory lemmas, which will allow us to formalize Reutenauer's final argument for the proof of the CBH Theorem.

Lemma 4.34. *Let H_1^x be as in (4.128). Let $H_1^x \frac{\partial}{\partial y}$ be the continuous derivation of the algebra $\widehat{\mathscr{T}}(\mathbb{Q}\langle x, y \rangle)$, as in Definition 4.31. Then we have*

$$D^n(y) \in \widehat{U}_n, \quad \text{for every } n \geq 0. \quad (4.144)$$

Proof. Let $D := H_1^x \frac{\partial}{\partial y}$. Note that

$$D(y) = H_1^x. \quad (4.145)$$

We are entitled to apply Lemma 4.28 with $\varphi = \mathrm{Id}_{\widehat{\mathscr{T}}}$ since D is a continuous derivation of $\widehat{\mathscr{T}}$. Then (4.124) gives (note that $y \in \widehat{\mathscr{T}}_+$)

$$D(\mathrm{Exp}\, y) * \mathrm{Exp}(-y) = \sum_{n=1}^{\infty} \frac{1}{n!} (\mathrm{ad}\, y)^{n-1} (D(y))$$

$$\overset{(4.145)}{=} \sum_{n=1}^{\infty} \frac{1}{n!} (\mathrm{ad}\, y)^{n-1} (H_1^x) \overset{(4.140)}{=} x.$$

This gives

$$D(\mathrm{Exp}\, y) = x \cdot \mathrm{Exp}\, y. \quad (4.146)$$

By applying D to both sides (and recalling that D is a derivation and that $D(x) = 0$) we get

$$D^2(\operatorname{Exp} y) = D(x \cdot \operatorname{Exp} y) = x \cdot D(\operatorname{Exp} y) \overset{(4.146)}{=} x^2 \cdot \operatorname{Exp} y.$$

Arguing inductively (and using $D(x^n) = 0$ for every $n \geq 1$), we obtain

$$D^n(\operatorname{Exp} y) = x^n \cdot \operatorname{Exp} y, \quad \text{for every } n \geq 0. \tag{4.147}$$

In particular this gives

$$D^n(\operatorname{Exp} y) \in \widehat{U}_n, \quad \text{for every } n \geq 0. \tag{4.148}$$

We now claim that, by means of a delicate inductive estimate, (4.148) will allow us to prove the claimed (4.144).

First we notice that we have

$$D(\widehat{U}_k) \subseteq \widehat{U}_k, \quad \text{for every } k \geq 0. \tag{4.149}$$

In order to prove (4.149), it is sufficient to prove that, for every $k \geq 0$, D maps $\mathscr{T}_k$ into $\widehat{U}_k$. Indeed, we have

$$D(1) = 0, \quad D(x) = 0, D(y) = H_1^x \in \widehat{U}_1,$$

so that, for every $k \geq 2$ and any fixed $z_1, \ldots, z_k \in \{x, y\}$

$$D(z_1 \cdots z_k) = D(z_1) z_2 \cdots z_k + \cdots + z_1 \cdots z_{k-1} D(z_k)$$
$$\in \widehat{U}_1 \cdot \mathscr{T}_{k-1} + \mathscr{T}_1 \cdot \widehat{U}_1 \cdot \mathscr{T}_{k-2} + \cdots + \mathscr{T}_{k-1} \cdot \widehat{U}_1 \subseteq \widehat{U}_k.$$

From (4.149) we get at once

$$D^n(\widehat{U}_k) \subseteq \widehat{U}_k, \quad \text{for every } n, k \geq 0. \tag{4.150}$$

We are now ready to prove (4.144) by induction on n. Before embarking with the proof, we show explicitly the steps up to $n = 2, 3$ in order to make transparent the ideas involved. First of all, let us recall a general formula (holding true for any derivation of an associative algebra) which can be proved by a simple inductive argument (see Proposition 4.40 at the end of the section):

$$D^n(a_1 \cdots a_k) = \sum_{\substack{0 \leq i_1, \ldots, i_k \leq n \\ i_1 + \cdots + i_k = n}} \frac{n!}{i_1! \cdots i_k!} D^{i_1} a_1 \cdots D^{i_k} a_k. \tag{4.151}$$

The validity of (4.144) for $n = 0, 1$ is obvious since

$$D^0(y) = y \in \widehat{U}_1 \subset \widehat{U}_0, \qquad D(y) = H_1^x \in \widehat{U}_1.$$

We now consider the case $n = 2$: From (4.148), from the continuity of D and from (4.150), we infer

$$\widehat{U}_2 \ni D^2(\operatorname{Exp} y) = D^2\Big(\sum_{k=1}^{\infty} \frac{y^k}{k!}\Big) = D^2(1) + D^2(y) + D^2\Big(\sum_{k=2}^{\infty} \frac{y^k}{k!}\Big)$$

$$\overset{(4.150)}{=} D^2(y) + \{\text{an element of } \widehat{U}_2\}.$$

This proves that $D^2(y) \in \widehat{U}_2$. We next consider the case $n = 3$ (thus showing that the inductive argument becomes more complicated): Arguing as above, and this time using the acquired information $D^2(y) \in \widehat{U}_2$, we infer

$$\widehat{U}_3 \ni D^3(\operatorname{Exp} y) = D^3(1) + D^3(y) + \tfrac{1}{2} D^3(y^2) + D^3\Big(\sum_{k=3}^{\infty} \frac{y^k}{k!}\Big)$$

$$\overset{(4.150)}{=} D^3(y) + \tfrac{1}{2} D^3(y^2) + \{\text{an element of } \widehat{U}_3\}.$$

Thus, by the aid of this fact and by (4.151), we get

$$\widehat{U}_3 \ni D^3(y) + \tfrac{1}{2} D^3(y^2)$$

$$= D^3(y) + \tfrac{1}{2}\Big(D^3(y)\,y + 3\,D^2(y)\,D(y) + 3\,D(y)\,D^2(y) + y\,D^3(y)\Big)$$

$$\in D^3(y) + \tfrac{1}{2} D^3(y)\,y + \tfrac{1}{2}\,y\,D^3(y) + \underbrace{\widehat{U}_2 \cdot \widehat{U}_1 + \widehat{U}_1 \cdot \widehat{U}_2}_{\subseteq \widehat{U}_3}.$$

This gives

$$D^3(y) + \tfrac{1}{2} D^3(y)\,y + \tfrac{1}{2}\,y\,D^3(y) \in \widehat{U}_3.$$

By writing $D^3(y) = \sum_{j=0}^{\infty} a_j$ with $a_j \in \mathscr{T}_j$, this yields

$$\sum_{j=0}^{\infty} a_j + \tfrac{1}{2} \sum_{j=0}^{\infty} a_j\,y + \tfrac{1}{2} \sum_{j=0}^{\infty} y\,a_j \in \widehat{U}_3 = \prod_{j\geq 3} \mathscr{T}_j.$$

Since $a_j\,y$ and $y\,a_j$ belong to $\mathscr{T}_{j+1}$, this is equivalent to

$$\underbrace{a_0}_{\in \mathscr{T}_0} + \underbrace{(a_1 + \tfrac{1}{2} a_0\,y + \tfrac{1}{2}\,y\,a_0)}_{\in \mathscr{T}_1} + \underbrace{(a_2 + \tfrac{1}{2} a_1\,y + \tfrac{1}{2}\,y\,a_1)}_{\in \mathscr{T}_2} = 0$$

This is possible if and only if

$$
\begin{cases}
a_0 & = 0 \\
a_1 + \frac{1}{2}\,a_0\,y + \frac{1}{2}\,y\,a_0 = 0 \\
a_2 + \frac{1}{2}\,a_1\,y + \frac{1}{2}\,y\,a_1 = 0.
\end{cases}
$$

Solving this system from top to bottom, we derive $a_0 = a_1 = a_2 = 0$, whence $D^3(y) = \sum_{j=3}^{\infty} a_j$. This proves that $D^3(y) \in \widehat{U}_3$, as $a_j \in \mathcal{T}_j$ for every j.

We are now ready for the inductive step: we suppose we already know that $D^j(y) \in \widehat{U}_j$ for $j = 0, 1, \ldots, n-1$ and we prove $D^n(y) \in \widehat{U}_n$. From (4.148) we know that $D^n(\mathrm{Exp}\, y) \in \widehat{U}_n$, whence

$$
D^n\Big(\sum_{k=0}^{n-1}\frac{y^k}{k!}\Big) = D^n(\mathrm{Exp}\, y) - \underbrace{D^n\Big(\sum_{k=n}^{\infty}\frac{y^k}{k!}\Big)}_{\in \widehat{U}_n} \in \widehat{U}_n + D^n(\widehat{U}_n) \subseteq \widehat{U}_n,
$$

where we have also applied (4.149). This proves (since $D^n(1) = 0$)

$$
D^n\Big(\sum_{k=1}^{n-1}\frac{y^k}{k!}\Big) \in \widehat{U}_n.
$$

This gives, by the aid of identity (4.151) (applied for $a_1, \ldots, a_k = y$),

$$
\widehat{U}_n \ni \sum_{k=1}^{n-1}\frac{D^n(y^k)}{k!} = \sum_{k=1}^{n-1}\frac{1}{k!}\sum_{\substack{0 \le i_1, \ldots, i_k \le n \\ i_1 + \cdots + i_k = n}} \frac{n!}{i_1! \cdots i_k!}\, D^{i_1} y \cdots D^{i_k} y
$$

$$
= \sum_{k=1}^{n-1}\frac{1}{k!}\big(D^n(y)\, y^{k-1} + y\, D^n(y)\, y^{k-2} + \cdots + y^{k-1}\, D^n(y)\big) + \tag{4.152}
$$

$$
+ \Bigg\{\sum_{k=1}^{n-1}\frac{n!}{k!}\sum_{\substack{0 \le i_1, \ldots, i_k < n \\ i_1 + \cdots + i_k = n}} \frac{D^{i_1} y \cdots D^{i_k} y}{i_1! \cdots i_k!}\Bigg\}.
$$

Note that in the sum in curly braces there appear powers of D with exponent strictly less than n. We are then allowed to apply the inductive hypothesis and derive that each of its summands belongs to

$$
\widehat{U}_{i_1} \cdots \widehat{U}_{i_k} \subseteq \widehat{U}_{i_1 + \cdots + i_k}, \quad \text{with } i_1 + \cdots + i_k = n.
$$

Hence the sum in braces belongs to $\widehat{U}_n$, so that (4.152) gives

$$\sum_{k=1}^{n-1} \frac{1}{k!} \left(D^n(y) \, y^{k-1} + y \, D^n(y) \, y^{k-2} + \cdots + y^{k-1} \, D^n(y) \right) \in \widehat{U}_n. \qquad (4.153)$$

We decompose $D^n(y) \in \widehat{\mathscr{T}} = \prod_{j \geq 0} \mathscr{T}_j$ into its components:

$$D^n(y) = \sum_{j=0}^{\infty} a_y, \quad \text{with } a_j \in \mathscr{T}_j \text{ for every } j \geq 0. \qquad (4.154)$$

Thus (4.153) rewrites as

$$\sum_{j=0}^{\infty} \sum_{k=1}^{n-1} \frac{1}{k!} \left(a_j \, y^{k-1} + y \, a_j \, y^{k-2} + \cdots + y^{k-1} \, a_j \right) \in \widehat{U}_n.$$

By reordering the above double-sum, we get

$$\sum_{s=1}^{\infty} \sum_{1 \leq k \leq n-1, \, j \geq 0: \, j+k=s} \frac{1}{k!} \left(a_j \, y^{k-1} + \cdots + y^{k-1} \, a_j \right) \in \widehat{U}_n. \qquad (4.155)$$

Note that the sum in parentheses belongs to

$$\mathscr{T}_j \cdot \mathscr{T}_{k-1} + \mathscr{T}_1 \cdot \mathscr{T}_j \cdot \mathscr{T}_{k-2} + \cdots + \mathscr{T}_{k-1} \cdot \mathscr{T}_j \subseteq \mathscr{T}_{j+k-1} = \mathscr{T}_{s-1}.$$

As a consequence, (4.155) can hold if and only if

$$\sum_{\substack{1 \leq k \leq n-1, \, j \geq 0 \\ j+k=s}} \frac{1}{k!} \left(a_j \, y^{k-1} + \cdots + y^{k-1} \, a_j \right) = 0, \quad \text{for every } s = 1, \ldots, n.$$

$$\qquad (4.156)$$

When $s = 1$, (4.156) gives (notice that $k \geq 1$ in the sum) $a_j = 0$. Thus we can delete from (4.156) the index $j = 0$. In particular, when $s = 2$, (4.156) gives

$$0 = \sum_{\substack{1 \leq k \leq n-1, \, j \geq 1 \\ j+k=2}} \frac{1}{k!} \left(a_j \, y^{k-1} + \cdots + y^{k-1} \, a_j \right) = a_1.$$

Again, this proves that we can erase from (4.156) the indices $j = 0, 1$. After finitely many steps we are then able to prove that $a_0 = a_1 = \cdots = a_{n-1} = 0$. As a consequence, (4.154) gives

$$D^n(y) = \sum_{j=n}^{\infty} a_y \in \widehat{U}_n,$$

which is the desired equation (4.144). This completes the proof. $\qquad \square$

With Lemma 4.34 at hand, we are able to prove the following result.

Corollary 4.35. *Let the notation in Lemma 4.34 hold. Then we have*

$$D^n(z) \in \widehat{U}_n, \quad \text{for every } n \geq 0 \text{ and every } z \in \widehat{\mathscr{T}}. \tag{4.157}$$

As a consequence, for every $z \in \widehat{\mathscr{T}}$, the series $\sum_{n \geq 0} D^n(z)/n!$ converges in $\widehat{\mathscr{T}}$ and, by posing

$$\exp(D) : \widehat{\mathscr{T}} \to \widehat{\mathscr{T}}, \quad \exp(D)(z) := \sum_{n=0}^{\infty} \frac{1}{n!} \left(H_1^x \frac{\partial}{\partial y} \right)^n (z), \tag{4.158}$$

we define an endomorphism of $\widehat{\mathscr{T}}$. Actually, $\exp(D)$ is a continuous UAA morphism of the algebra $\widehat{\mathscr{T}}(\mathbb{Q}\langle x, y \rangle)$.

Proof. D being continuous (whence D^n also), in order to prove (4.157) it suffices to show that

$$D^n(\mathscr{T}_k) \subseteq \widehat{U}_{\max\{n,k\}}, \quad \text{for every } n, k \geq 0. \tag{4.159}$$

When n or k is 0 this is trivial, so we can assume $n, k \geq 1$. Moreover, the inclusion

$$D^n(\mathscr{T}_k) \subseteq \widehat{U}_k, \quad \text{for every } n, k \geq 0, \tag{4.160}$$

follows from (4.150). So (4.159) will follow if we demonstrate

$$D^n(\mathscr{T}_k) \subseteq \widehat{U}_n, \quad \text{for every } n, k \geq 0. \tag{4.161}$$

To prove this, first note that (4.144) together with $D^n(x) = 0$ gives

$$D^n(z) \in \widehat{U}_n, \quad \text{for every } n \geq 0 \text{ and every } z \in \{x, y\}. \tag{4.162}$$

Let now $z = z_1 \cdots z_k$ with $z_1, \ldots, z_k \in \{x, y\}$. From (4.151) we get

$$D^n(z_1 \cdots z_k) = \sum_{\substack{0 \leq i_1, \ldots, i_k \leq n \\ i_1 + \cdots + i_k = n}} \frac{n!}{i_1! \cdots i_k!} D^{i_1} z_1 \cdots D^{i_k} z_k$$

$$\underset{\in}{\overset{(4.162)}{}} \sum_{\substack{0 \leq i_1, \ldots, i_k \leq n \\ i_1 + \cdots + i_k = n}} \widehat{U}_{i_1} \cdots \widehat{U}_{i_k} \subseteq \widehat{U}_n.$$

This proves (4.161), which – together with (4.160) – gives at once (4.159) (recall that $\{\widehat{U}_N\}_N$ is a decreasing sequence of sets).

For every $z \in \widehat{\mathscr{T}}$, (4.157) ensures that $\lim_{n \to \infty} D^n(z)/n! = 0$ so that (see Remark 2.76-3) the map $\exp(D)$ in (4.158) is well posed and it is clearly an endomorphism of $\widehat{\mathscr{T}}$. Furthermore, (4.157) gives $D^n(\widehat{\mathscr{T}}) \subseteq \widehat{U}_n$, so that Remark 4.33 ensures that $\exp(D)$ is a UAA morphism. Finally, the continuity of $\exp(D)$ is a consequence of

$$\exp(D)(\widehat{U}_k) \subseteq \widehat{U}_k, \quad \forall\, k \geq 0, \tag{4.163}$$

which follows at once from (4.150) and the convergence of the series expressing $\exp(D)$: indeed, if $z \in \widehat{U}_k$, then

$$\exp(D)(z) = \sum_{n=0}^{\infty} \frac{1}{n!} \underbrace{D^n(z)}_{\in D^n(\widehat{U}_k) \subseteq \widehat{U}_k} \in \overline{\widehat{U}_k} = \widehat{U}_k.$$

This completes the proof.   □

We are ready to complete the proof of the following result. This is the rigorous formalization by Reutenauer of a celebrated result due separately to Baker and to Hausdorff (see also Chap. 1 for historical references).

Theorem 4.36 (Reutenauer). *Let H_1^x be as in (4.128). Let $H_1^x \frac{\partial}{\partial y}$ be the continuous derivation of the algebra $\widehat{\mathscr{T}}(\mathbb{Q}\langle x, y\rangle)$ as in Definition 4.31. Then, the following equality holds*

$$H(x, y) = \sum_{n=0}^{\infty} \frac{1}{n!} \left(H_1^x \frac{\partial}{\partial y} \right)^n (y), \tag{4.164}$$

where $H(x, y) = \mathrm{Log}(\mathrm{Exp}\, x \cdot \mathrm{Exp}\, y)$. Furthermore, if H_n^x is as in (4.128), we have

$$H_n^x = \frac{1}{n!} \left(H_1^x \frac{\partial}{\partial y} \right)^n (y), \quad \text{for every } n \geq 0. \tag{4.165}$$

Proof. Set $D := H_1^x \frac{\partial}{\partial y}$. By means of Corollary 4.35, the series in the right-hand side of (4.164) converges in $\widehat{\mathscr{T}}$. In the course of the proof of Lemma 4.34, we also proved that the following explicit equality holds (see (4.147)):

$$D^n(\mathrm{Exp}\, y) = x^n \cdot \mathrm{Exp}\, y, \quad \text{for every } n \geq 0. \tag{4.166}$$

Hence we have

$$\sum_{n=0}^{\infty} \frac{D^n(\mathrm{Exp}\, y)}{n!} \overset{(4.166)}{=} \sum_{n=0}^{\infty} \frac{x^n \cdot \mathrm{Exp}\, y}{n!} = \mathrm{Exp}(x) \cdot \mathrm{Exp}(y) \overset{(4.127)}{=} \mathrm{Exp}(H(x, y)).$$

By definition of $\exp(D)$ in (4.158), this yields

$$\exp(D)(\mathrm{Exp}\, y) = \mathrm{Exp}(H(x, y)). \tag{4.167}$$

On the other hand, since $\exp(D)$ is a continuous UAA morphism of $\widehat{\mathscr{T}}$ (recall Corollary 4.35), we also have

$$\exp(D)(\mathrm{Exp}\, y) = \sum_{k=0}^{\infty} \frac{1}{k!} \exp(D)(y^k) = \sum_{k=0}^{\infty} \frac{1}{k!} \left(\exp(D)(y) \right)^k$$

$$= \mathrm{Exp}\left(\exp(D)(y) \right),$$

that is,

$$\exp(D)(\mathrm{Exp}\, y) = \mathrm{Exp}\left(\exp(D)(y) \right). \tag{4.168}$$

Note that the above right-hand side is well defined, since we have

$$\exp(D)(y) \in \exp(D)\big(\widehat{U}_1\big) \stackrel{(4.163)}{\in} \widehat{U}_1.$$

Collecting together (4.167) and (4.168), we infer

$$\mathrm{Exp}(H(x,y)) = \mathrm{Exp}\left(\exp(D)(y) \right),$$

and the injectivity of Exp finally gives $H(x,y) = \exp(D)(y)$. This is precisely (4.164). In order to complete the proof, we need only to show (4.165), that is,

$$H_n^x = \frac{D^n(y)}{n!}, \quad \forall\, n \geq 0. \tag{4.169}$$

Recalling the very definition of H_n^x in (4.128), H_x^n collects the summands, out of the series for $H(x,y)$, containing x precisely n times. Since we have proved (4.164), which is nothing but

$$H(x,y) = \sum_{n=0}^{\infty} \frac{D^n(y)}{n!},$$

(4.169) will follow if we show that $D^n(y)/n!$ is a sum of words containing x precisely n times.

This is true of $D^0(y) = y$ and $D^1(y) = H_1^x$. We now argue by induction supposing that, for $j = 0, \ldots, n-1$, $D^j(y)$ can be expressed as a (convergent) series of words containing x precisely j times: we then prove this for $j = n$. To this aim, using $D^n(y) = D(D^{n-1}(y))$, the inductive hypothesis (and the continuity of D) shows that it is sufficient to prove that, if w is an elementary word in x,y containing x exactly $n-1$ times, then $D(w)$ is a series containing x exactly n times. Any such word w can be written as follows (unless it does not contain y, in which case it is killed by D and the thesis holds)

$$w = x^{h_1}\, y\, x^{h_2}\, y \cdots x^{h_i}\, y\, x^{h_{i+1}},$$

with $h_i, \ldots, h_{i+1} \geq 0$ and $h_1 + \cdots + h_{i+1} = n - 1$. Since D is a derivation which kills any power of x, we have

$$D(w) = x^{h_1} D(y) x^{h_2} y \cdots x^{h_i} y x^{h_{i+1}} + x^{h_1} y x^{h_2} D(y) \cdots x^{h_i} y x^{h_{i+1}} +$$

$$\cdots + x^{h_1} y x^{h_2} y \cdots x^{h_i} D(y) x^{h_{i+1}}$$

$$= x^{h_1} H_1^x x^{h_2} y \cdots x^{h_i} y x^{h_{i+1}} + \cdots + x^{h_1} y x^{h_2} y \cdots x^{h_i} H_1^x x^{h_{i+1}}.$$

Since H_1^x is a series of summands containing x exactly once, it is evident by the above computation that the number of times x occurs in the summands expressing $D(w)$ is incremented by 1 with respect to w, as we aimed to prove. This ends the proof of the theorem. $\qquad\square$

By means of Reutenauer's Theorem 4.36, we are able to give the fourth proof of the Campbell, Baker, Hausdorff Theorem for this chapter.

Corollary 4.37 (Campbell, Baker, Hausdorff). *Let H_j^x be as in (4.128). Then*

$$H_j^x \in \overline{\mathcal{L}(\mathbb{Q}\langle x, y \rangle)}, \quad \text{for every } j \geq 0. \tag{4.170}$$

As a consequence $\mathrm{Log}(\mathrm{Exp}\, x \cdot \mathrm{Exp}\, y)$ belongs to $\overline{\mathcal{L}(\mathbb{Q}\langle x, y \rangle)}$ too, that is, it is a series of Lie polynomials in x, y.

Proof. Clearly, it suffices to prove (4.170), for the rest of the proof will follow from (4.170), since

$$\mathrm{Log}(\mathrm{Exp}\, x \cdot \mathrm{Exp}\, y) = H(x, y) = \sum_{j=0}^{\infty} H_j^x.$$

We prove (4.170) by induction on j. First we have $H_0^x = y$ (see (4.129)) and $H_1^x \in \overline{\mathcal{L}(\mathbb{Q}\langle x, y \rangle)}$ (in view of (4.130)). We now suppose that H_{j-1}^x is a Lie series and we prove it for H_j^x. To this aim, set $D = H_1^x (\partial/\partial y)$, it holds that

$$H_j^x \overset{(4.165)}{=} \frac{D^j(y)}{j!} = \frac{1}{j} D\left(\frac{D^{j-1}(y)}{(j-1)!}\right) \overset{(4.165)}{=} \frac{1}{j} D(H_{j-1}^x).$$

Hence, we are done if we show that D maps Lie series into Lie series. Since D is a derivation of the associative algebra $\widehat{\mathscr{T}}$, it is a derivation of the commutator-algebra associated to $\widehat{\mathscr{T}}$ (see Remark 2.18 on page 62). In particular, D maps $\mathcal{L} := \mathcal{L}(\mathbb{Q}\langle x, y \rangle)$ into $\overline{\mathcal{L}} := \overline{\mathcal{L}(\mathbb{Q}\langle x, y \rangle)}$. This fact, together with the continuity of D, with (4.149) and the fact that any element of $\overline{\mathcal{L}}$ is a series of elements of $\mathcal{L}$, proves that $D(\overline{\mathcal{L}}) \subseteq \overline{\mathcal{L}}$. This completes the proof. $\qquad\square$

Remark 4.38. Obviously, we can also rewrite the series of $H(x, y)$ as

$$\mathrm{Log}(\mathrm{Exp}\, x \cdot \mathrm{Exp}\, y) = \sum_{j=0}^{\infty} H_j^y,$$

where H_j^y groups together the summands containing y precisely j times. All the results of this section then have a dual version:

- Lemma 4.28 has the following analogue (the hypotheses and notation are the same): *Suppose that* $\varphi : \widehat{A} \to \widehat{A}$ *is a UAA morphism and that* $D_\varphi : \widehat{A} \to \widehat{A}$ *is a continuous φ-derivation. Then, we have*

$$\mathrm{Exp}(-\varphi(x)) * D_\varphi(\mathrm{Exp}\, x) = \sum_{n=1}^{\infty} \frac{1}{n!} \big(- \,\mathrm{ad}\,\varphi(x)\big)^{n-1}(D_\varphi(x)), \qquad (4.171)$$

for every $x \in \widehat{A}_+$ such that $\varphi(x) \in \widehat{A}_+$. This means that

$$\mathrm{Exp}(-\varphi(x)) * D_\varphi(\mathrm{Exp}\, x) = f(-\mathrm{ad}\,\varphi(x))(D_\varphi(x))$$
$$= \widetilde{f}(\mathrm{ad}\,\varphi(x))(D_\varphi(x)), \tag{4.172}$$

where $f(z) = \sum_{n=1}^{\infty} z^{n-1}/n!$ is the Maclaurin expansion of the entire function $f(z) := (e^z - 1)/z$, and $\widetilde{f}(z) = \sum_{n=1}^{\infty} (-1)^n z^{n-1}/n!$ is the Maclaurin expansion of the entire function $\widetilde{f}(z) := (1 - e^{-z})/z$.
- The analogue of Theorem 4.29 states that

$$H_1^y = y + \tfrac{1}{2}[x, y] + \sum_{p=1}^{\infty} \frac{B_{2p}}{(2\,p)!} (\mathrm{ad}\, x)^{2\,p}(y)$$

$$= \sum_{n=0}^{\infty} \frac{(-1)^n\, B_n}{n!} (\mathrm{ad}\, x)^n(y) \tag{4.173}$$

$$= g(-\mathrm{ad}\, x)(y) = \widetilde{g}(\mathrm{ad}\, x)(y),$$

where the B_n are the Bernoulli numbers, whereas $g(z) = \sum_{n=0}^{\infty} B_n\, z^n/n!$ is the Maclaurin expansion of the function

$$g(z) := \frac{z}{e^z - 1} = 1/f(z),$$

and $\widetilde{g}(z) = \sum_{n=0}^{\infty} (-1)^n\, B_n\, z^n/n!$ is the Maclaurin expansion of

$$\widetilde{g}(z) := \frac{z}{1 - e^{-z}} = 1/\widetilde{f}(z).$$

– Let $D = H_1^y \frac{\partial}{\partial x}$ be the unique continuous derivation of $\widehat{\mathcal{F}}(\mathbb{Q}\langle x, y\rangle)$ such that $D(x) = H_1^y$ and $D(y) = 0$. Then, an analogue of Corollary 4.35 ensures that: $D^n(z) \in \widehat{U}_n$ for every $n \geq 0$ and every $z \in \widehat{\mathcal{F}}$. As a consequence, the series $\sum_{n\geq 0} D^n(z)/n!$ converges in $\widehat{\mathcal{F}}$ and the formula

$$\exp(D) : \widehat{\mathcal{F}} \to \widehat{\mathcal{F}}, \quad \exp(D)(z) := \sum_{n=0}^{\infty} \frac{1}{n!} \left(H_1^y \frac{\partial}{\partial x} \right)^n (z)$$

defines an endomorphism of $\widehat{\mathcal{F}}$. Actually, $\exp(D)$ is a continuous UAA morphism of the algebra $\widehat{\mathcal{F}}$.

– Finally, the analogue of Reutenauer's Theorem 4.36 ensures that

$$H(x, y) = \exp(D)(x) = \sum_{n=0}^{\infty} \frac{1}{n!} \left(H_1^y \frac{\partial}{\partial x} \right)^n (x) = \exp\left(H_1^y \frac{\partial}{\partial x} \right)(x),$$

(4.174)

where $H(x, y) = \mathrm{Log}(\mathrm{Exp}\, x \cdot \mathrm{Exp}\, y)$, and furthermore

$$H_n^y = \frac{1}{n!} \left(H_1^y \frac{\partial}{\partial x} \right)^n (x), \quad \text{for every } n \geq 0.$$

(4.175)

Here, we have set (as stated above)

$$H(x, y) = \sum_{j=0}^{\infty} H_j^y, \quad \text{where, for any } j \in \mathbb{N} \cup \{0\},$$

$$H_j^y = \sum_{n=1}^{\infty} \frac{(-1)^{n+1}}{n} \sum_{\substack{(h_1, k_1), \ldots, (h_n, k_n) \neq (0,0) \\ k_1 + \cdots + k_n = j}} \frac{x^{h_1} y^{k_1} \cdots x^{h_n} y^{k_n}}{h_1! \cdots h_n! k_1! \cdots k_n!}.$$

(4.176)

Remark 4.39. As a consequence of the above reasonings, the CBHD operation $\diamond$ can be written as follows

$$x \diamond y = \sum_{n=0}^{\infty} \frac{1}{n!} \left(H_1^y \frac{\partial}{\partial x} \right)^n (x).$$

(4.177)

In a certain quantitative sense, this states that H_1^y (i.e., the series of the summands in $x \diamond y$ containing y precisely once) completely determines $x \diamond y$.

We think that it is not inappropriate to draw a parallel between this fact and what happens in Lie groups. To explain the analogy we have in mind, we make a simple example supposing that $\mathbb{R}^N$ is endowed with a Lie group structure by a map $(x, y) \mapsto x * y$ (everything works the same in any abstract Lie group). If e is the identity of $\mathbb{G} = (\mathbb{R}^N, *)$, it is known that the Jacobian matrix at $y = e$ of the map $\tau_x(y) := x * y$ has the following remarkable property: its N columns determine N vector fields which constitute a basis for the Lie algebra of the Lie group $\mathbb{G}$. Note that the cited Jacobian matrix $\mathcal{J}_{\tau_x}(e)$ is completely determined by the terms in the Maclaurin expansion of $y \mapsto x * y$ containing y with degree 1. Hence $\mathcal{J}_{\tau_x}(e)$ plays the same rôle as H_1^y.

Now, by general results of Lie group theory, it is well known that the Lie algebra of a (simply connected) Lie group completely determines (up to isomorphism) the group itself. Even more explicitly, it can be proved that the integral curves of the left-invariant vector fields allow us to reconstruct the operation $x * y$ of the group, at least for (x, y) in a neighborhood of (e, e).

The stated parallel is now evident: as H_1^y determines $x \diamond y$ by formula (4.177), the Jacobian matrix $\mathcal{J}_{\tau_x}(e)$ determines the operation $x * y$ via the integral curves of its column vector fields. Even more closely, paralleling the identity

$$x \diamond y = \exp(D)(x), \tag{4.178}$$

the cited integral curves are, by sheer chance, of "exponential type": indeed, if X is a left-invariant vector field, the integral curve $\gamma(t)$ of X starting at x has the Maclaurin expansion

$$\gamma(t) \sim \sum_{k=0}^{\infty} \frac{X^k(x)}{k!} t^k =: \exp(tX)(x).$$

If we further recall the formula (see e.g., [21, Proposition 1.2.29])

$$x * y = \exp(\text{Log } y)(x) \quad (\text{for } x, y \text{ in a neighborhood of } e),$$

then the analogy with (4.178) becomes completely palpable.   □

In this section we made use of the following result.

Proposition 4.40. *Let $(A, *)$ be an associative algebra and let D be a derivation of A. Then*

$$D^n(a_1 * \cdots * a_k) = \sum_{\substack{0 \leq i_1, \ldots, i_k \leq n \\ i_1 + \cdots + i_k = n}} \frac{n!}{i_1! \cdots i_k!} D^{i_1} a_1 * \cdots * D^{i_k} a_k, \tag{4.179}$$

for every $n, k \geq 1$ and every $a_1, \ldots, a_k \in A$.
Proof. See page 451.   □

4.7 Cartier's Proof

The aim of this section is to give a detailed exposition of Cartier's proof of the CBHD Theorem, as given in the paper [33]. We shall adapt the notations of the cited paper to those of the present Book.

For instance, with some slight modifications[6] if compared to [33], we suppose we are given a set X and a field $\mathbb{K}$ of characteristic zero, and we consider the free vector space

$$V := \mathbb{K}\langle X \rangle,$$

its tensor algebra $\mathscr{T}(V)$ and, as a subset of the latter, the free Lie algebra $\mathcal{L}(V)$. We further consider the subspace of $\mathscr{T}(V)$ whose elements have vanishing component of degree zero: this is the usual $\mathscr{T}_+(V)$. Furthermore, we consider the algebra of the formal power series related to $\mathscr{T}(V)$, that is, the algebra $\widehat{\mathscr{T}}(V)$, endowed with the usual structure of metric space. The closure of $\mathcal{L}(V)$ in $\widehat{\mathscr{T}}(V)$ is denoted by $\overline{\mathcal{L}(V)}$ and $\widehat{\mathscr{T}_+}(V)$ is the ideal of all formal power series having vanishing degree-zero term.

4.7.1 Some Important Maps

Next we have to introduce some important maps. We consider the unique linear map (it is easy to verify that this is well defined)

$$g : \mathscr{T}(V) \longrightarrow \mathscr{T}(V) \quad \text{such that}$$

$$\begin{cases} g(1) = 0, \quad g(v_1) = v_1, \\ g(v_1 \otimes \cdots \otimes v_k) = [v_1 \cdots [v_{k-1}, v_k] \cdots], \end{cases} \tag{4.180}$$

for every $v_1, \ldots, v_k \in V$ and every $k \geq 2$. Furthermore, we consider the unique endomorphism D of $\mathscr{T}(V) = \bigoplus_{k=0}^{\infty} \mathscr{T}_k(V)$ whose restriction to $\mathscr{T}_k(V)$ is k times the identity map of $\mathscr{T}_k(V)$. More explicitly, D is the unique linear map

$$D : \mathscr{T}(V) \longrightarrow \mathscr{T}(V) \quad \text{such that}$$

$$\begin{cases} D(1) = 0, \quad D(v_1) = v_1, \\ D(v_1 \otimes \cdots \otimes v_k) = k\, v_1 \otimes \cdots \otimes v_k, \end{cases} \tag{4.181}$$

for every $v_1, \ldots, v_k \in V$ and every $k \geq 2$. It is immediately seen that D is a derivation of $\mathscr{T}(V)$:

---

[6]Actually, in [33], instead of the arbitrary set X, finite sets $\{x_1, \ldots, x_n\}$ are considered and the Lie algebra $\mathcal{L}(V)$ is substituted by Lie(X), the free Lie algebra over $\mathbb{Q}$ generated by X, as we defined it in (2.56) on page 90, but eventually (see [33, page 243]) this Lie algebra becomes identified with $\mathcal{L}(\mathbb{Q}\langle x_1, \ldots, x_n \rangle)$. Since we have previously proved that these algebras are isomorphic (see Theorem 2.56 on page 91) we fix this identification from the beginning.

$$D(u \cdot w) = D\big(\sum_{k=0}^{\infty}(\sum_{i+j=k} u_i \otimes w_j)\big) = \sum_{k=0}^{\infty} k(\sum_{i+j=k} u_i \otimes w_j)$$
$$= \sum_{k=0}^{\infty}(\sum_{i+j=k}(i+j)\, u_i \otimes w_j)$$
$$= \sum_{k=0}^{\infty}(\sum_{i+j=k}(i\, u_i) \otimes w_j) + \sum_{k=0}^{\infty}(\sum_{i+j=k} u_i \otimes (j\, w_j))$$
$$= (Du) \cdot w + u \cdot (Dw).$$

$$(4.182)$$

Here $u = \sum_{k=0}^{\infty} u_k$, $w = \sum_{k=0}^{\infty} w_k$ and $u_k, w_k \in \mathscr{T}_k(V)$ for every $k \geq 0$ (and u_k, w_k are different to 0 only for finitely many k).

We remark that both g and D map $\mathscr{T}_k(V)$ to itself. This is enough to prove that g, D are uniformly continuous linear maps, when $\mathscr{T}(V)$ is considered as a subspace of the usual metric space $\widehat{\mathscr{T}}(V)$ (see Lemma 2.79 on page 103: take $k_n := n$ in hypothesis (2.88)). As a consequence, g and D can be uniquely prolonged to continuous maps $\widehat{g}, \widehat{D}$ respectively, which are also endomorphisms of $\widehat{\mathscr{T}}(V)$. Taking into account the fact that

$$\widehat{\mathscr{T}}(V) = \prod_{k=0}^{\infty} \mathscr{T}_k(V),$$

the explicit actions of $\widehat{g}$ and $\widehat{D}$ on elementary elements $v_1 \otimes \cdots \otimes v_k$ of $\widehat{\mathscr{T}}(V)$ are exactly as in (4.180) and (4.181) respectively:

$$\widehat{g}, \widehat{D} : \widehat{\mathscr{T}}(V) \longrightarrow \widehat{\mathscr{T}}(V) \quad \text{are such that}$$

$$(4.183)$$

$$\begin{cases} \widehat{g}(1) = 0, \quad \widehat{g}(v_1) = v_1, \\ \widehat{g}(v_1 \otimes \cdots \otimes v_k) = [v_1 \cdots [v_{k-1}, v_k] \cdots], \end{cases}$$

and $\quad \begin{cases} \widehat{D}(1) = 0, \quad \widehat{D}(v_1) = v_1, \\ \widehat{D}(v_1 \otimes \cdots \otimes v_k) = k\, v_1 \otimes \cdots \otimes v_k, \end{cases}$

for every $v_1, \ldots, v_k \in V$ and every $k \geq 2$. Moreover, the same computation as in (4.182) proves that $\widehat{D}$ is a derivation of $\widehat{\mathscr{T}}(V)$.

Remark 4.41. We observe that the restrictions of $\widehat{g}$ and $\widehat{D}$ to $\widehat{\mathscr{T}}_+$ are endomorphisms of $\widehat{\mathscr{T}}_+$. Moreover

$$\widehat{D}\big|_{\widehat{\mathscr{T}}_+} : \widehat{\mathscr{T}}_+(V) \to \widehat{\mathscr{T}}_+(V)$$

is bijective and its inverse, say $\widehat{d}$, is the unique linear map

$$\widehat{d} : \widehat{\mathscr{T}}_+(V) \longrightarrow \widehat{\mathscr{T}}_+(V) \quad \text{such that}$$

$$(4.184)$$

$$\begin{cases} \widehat{d}(v_1) = v_1, \\ \widehat{d}(v_1 \otimes \cdots \otimes v_k) = \frac{1}{k}\, v_1 \otimes \cdots \otimes v_k, \end{cases}$$

for every $v_1, \ldots, v_k \in V$ and every $k \geq 2$. As a consequence, it is immediately seen that the composition $\widehat{g} \circ \widehat{d}$ coincides with the restriction to $\widehat{\mathscr{T}}_+$ of the map $\widehat{P}$ in Theorem 3.32 on page 153. Now, since any iterated bracket $[v_1 \cdots [v_{k-1}, v_k] \cdots]$ is a linear combination of elementary tensors of order k on V, it is straightforwardly shown that

$$\widehat{g} \circ \widehat{d} = \widehat{d} \circ \widehat{g} \quad \text{on } \widehat{\mathscr{T}}_+(V). \tag{4.185}$$

Thus, recalling that $\widehat{d} = (\widehat{D})^{-1}$ on $\widehat{\mathscr{T}}_+$, by the chain of equivalences

$$\widehat{P}(t) = t \Leftrightarrow \widehat{g}(\widehat{d}(t)) = t \Leftrightarrow \widehat{d}(\widehat{g}(t)) = t \Leftrightarrow (\widehat{D})^{-1}(\widehat{g}(t)) = t \Leftrightarrow \widehat{g}(t) = \widehat{D}(t),$$

and by (3.76) on page 154, we get the further characterization of $\overline{\mathcal{L}(V)}$

$$\overline{\mathcal{L}(V)} = \{t \in \widehat{\mathscr{T}}_+(V) \mid \widehat{g}(t) = \widehat{D}(t)\}. \tag{4.186}$$

This can be viewed as a restatement of the Dynkin, Specht, Wever Lemma, in terms of the maps $\widehat{g}, \widehat{D}$.

Finally, we need to construct a further map, whose existence is more delicate. We claim that *there exists a morphism of UA algebras $\widehat{\theta}$*

$$\widehat{\theta} : \widehat{\mathscr{T}}(V) \longrightarrow \mathrm{End}(\widehat{\mathscr{T}}_+(V)) \quad \text{such that}$$

$$\begin{cases} \widehat{\theta}(1) = \mathrm{Id}_{\widehat{\mathscr{T}}_+(V)}, \quad \widehat{\theta}(v_1) = \mathrm{ad}\,(v_1), \\ \widehat{\theta}(v_1 \otimes \cdots \otimes v_k) = \mathrm{ad}\,(v_1) \circ \cdots \circ \mathrm{ad}\,(v_k), \end{cases} \tag{4.187}$$

for every $v_1, \ldots, v_k \in V$ and every $k \geq 2$. In (4.187), given $v \in V$, we are considering the usual adjoint map $\mathrm{ad}\,(v)$ as an endomorphism of $\widehat{\mathscr{T}}_+(V)$:

$$\mathrm{ad}\,(v) : \widehat{\mathscr{T}}_+(V) \to \widehat{\mathscr{T}}_+(V), \quad w \mapsto \mathrm{ad}\,(v)(w) = v \cdot w - w \cdot v;$$

note that this is actually an endomorphism of $\widehat{\mathscr{T}}_+(V)$, since $\mathscr{T} \cdot \widehat{\mathscr{T}}_+ \subseteq \widehat{\mathscr{T}} \cdot \widehat{\mathscr{T}}_+ \subseteq \widehat{\mathscr{T}}_+$ and analogously $\widehat{\mathscr{T}}_+ \cdot \mathscr{T} \subseteq \widehat{\mathscr{T}}_+$. We next prove our claim for $\widehat{\theta}$ in the following lemma.

Lemma 4.42. *There exists a UAA morphism $\widehat{\theta}$ as in (4.187). Also, $\widehat{\theta}(t)$ is a continuous endomorphism of $\widehat{\mathscr{T}}_+(V)$, for every $t \in \widehat{\mathscr{T}}(V)$. Furthermore, if $\{\gamma_k\}_k$ is any sequence in $\widehat{\mathscr{T}}(V)$ such that $t := \sum_{k=0}^{\infty} \gamma_k$ is convergent in $\widehat{\mathscr{T}}(V)$, then we have*

$$\widehat{\theta}(t)(\tau) = \sum_{k=0}^{\infty} \widehat{\theta}(\gamma_k)(\tau), \quad \forall\, \tau \in \widehat{\mathscr{T}}_+(V), \tag{4.188}$$

the series on the right-hand side being convergent in the metric space $\widehat{\mathscr{T}}_+(V)$.
Finally,

$$\widehat{\theta}(\ell) = \mathrm{ad}\,(\ell), \quad \text{for every } \ell \in \overline{\mathcal{L}(V)}, \tag{4.189}$$

both sides being meant as endomorphisms of $\widehat{\mathscr{T}}_+(V)$.

Proof. The proof is simple – though tedious. For completeness, we give it in the appendix, page 452. □

The maps $\widehat{g}$ and $\widehat{\theta}$ are related by the following result.

Lemma 4.43. *With the above notation, we have*

$$\widehat{g}(x \cdot y) = \widehat{\theta}(x)(\widehat{g}(y)), \quad \text{for every } x \in \widehat{\mathscr{T}}(V) \text{ and every } y \in \widehat{\mathscr{T}}_+(V). \tag{4.190}$$

Proof. First note that $\widehat{g}(y)$ is in $\widehat{\mathscr{T}}_+$, thanks to (4.183) and the fact that $y \in \widehat{\mathscr{T}}_+(V)$.

To begin with (4.190), we imitate the proof of Lemma 8.5, on page 466, in proving that

$$g(x \cdot y) = \theta(x)(g(y)), \quad \text{for every } x \in \mathscr{T}(V), y \in \mathscr{T}_+(V), \tag{4.191}$$

where g is as in (4.180) and θ is as in (7.65).

Proof of (4.191): If $x = k \in \mathscr{T}_0(V)$, (4.191) is trivially true, indeed we have $g(k \cdot y) = g(k\,y) = k\,g(y)$ (since g is linear) and (recalling that θ is a UAA morphism, hence unital)

$$\theta(k)(g(y)) = k\,\mathrm{Id}_{\mathscr{T}_+}(g(y)) = k\,g(y).$$

Thus we are left to prove (4.191) when both x, y belong to $\mathscr{T}_+(V)$; moreover, by linearity, we can assume without loss of generality that

$$x = v_1 \otimes \cdots \otimes v_k \quad \text{and} \quad y = w_1 \otimes \cdots \otimes w_h,$$

with $h, k \geq 1$ and the vectors v and w are elements of V:

$$
\begin{aligned}
g(x \cdot y) &= g\big(v_1 \otimes \cdots \otimes v_k \otimes w_1 \otimes \cdots \otimes w_h\big) \\
&\overset{(4.180)}{=} [v_1, \ldots [v_k, [w_1, \ldots [w_{h-1}, w_h] \ldots]] \ldots] \\
&= \mathrm{ad}\,(v_1) \circ \cdots \circ \mathrm{ad}\,(v_k)\big([w_1, \ldots [w_{h-1}, w_h] \ldots]\big) \\
&\quad \text{(by (7.65) and (4.180), the cases } h = 1 \text{ and } h > 1 \text{ being analogous)} \\
&= \theta(v_1 \otimes \cdots \otimes v_k)(g(w_1 \otimes \cdots \otimes w_h)) = \theta(x)(g(y)).
\end{aligned}
$$

We finally prove (4.190). Let $x = (x_k)_k \in \widehat{\mathscr{T}}$ and $y = (y_k)_k \in \widehat{\mathscr{T}}_+$. We have (recall that $\widehat{g}$ is the continuous prolongation of g)

$$\widehat{g}(y) = \sum_{k=0}^{\infty} g(y_k), \quad \text{and } g(y_k) \in \mathscr{T}_k(V) \text{ for every } k \geq 1.$$

As a consequence, the definition (7.66) of $\widehat{\theta}$ gives

$$\widehat{\theta}(x)(\widehat{g}(y)) = \sum_{h,k=0}^{\infty} \theta(x_h)(g(y_k)) \qquad \text{(we can apply (4.191) for } y_0 = 0\text{)}$$

$$= \sum_{h,k=0}^{\infty} g(x_h \cdot y_k) \qquad \text{(reordering)}$$

$$= \sum_{j=0}^{\infty} \sum_{h+k=j} g(x_h \cdot y_k) = \sum_{j=0}^{\infty} g\left(\sum_{h+k=j} x_h \cdot y_k\right)$$

$$= \sum_{j=0}^{\infty} g\big((x \cdot y)_j\big) = \widehat{g}(x \cdot y).$$

This completes the proof of (4.190).   □

4.7.2 A New Characterization of Lie Elements

With the results in Sect. 4.7.1 at hand, it is possible to give a new characterization of the elements of $\overline{\mathcal{L}(V)}$.

It is easily seen that the formula

$$x \blacklozenge y := \mathrm{Log}(\mathrm{Exp}(x) \cdot \mathrm{Exp}(y)), \quad x, y \in \widehat{\mathscr{T}}_+(V) \tag{4.192}$$

defines on $\widehat{\mathscr{T}}_+(V)$ a *group structure*. Indeed, we proved this in Proposition 3.10 (page 127) as a consequence of the fact that $1 + \widehat{\mathscr{T}}_+(V)$ is a multiplicative subgroup of $\widehat{\mathscr{T}}(V)$ (see Lemma 3.1 on page 118) together with the fact that Exp and Log are inverse to each other.

We next make the following definition

$$\mathbf{H} := \big\{ u \in 1 + \widehat{\mathscr{T}}_+(V) \,\big|\, \mathrm{Log}(u) \in \overline{\mathcal{L}(V)} \big\} = \mathrm{Exp}(\overline{\mathcal{L}(V)}). \tag{4.193}$$

Note that u^{-1} is well posed for every $u \in \mathbf{H}$, since $\mathbf{H}$ is a subset of $1 + \widehat{\mathscr{T}}_+(V)$, which is – as we recalled above – a multiplicative subgroup of $\widehat{\mathscr{T}}(V)$.

Following Cartier's [33, Lemme 2], we prove the following result.

Theorem 4.44 (Cartier). *Let $\mathbf{H}$ be the set in (4.193). Let also $\widehat{g}, \widehat{D}$ be the linear maps in (4.183) and let $\widehat{\theta}$ be the UAA morphism constructed in Lemma 4.42.*

Then, an element $u \in 1 + \widehat{\mathscr{T}}_+(V)$ belongs to $\mathbf{H}$ if and only if the following two conditions hold:

$$\hat{g}(u-1) = \hat{D}(u) \cdot u^{-1}, \tag{4.194}$$

$$\hat{\theta}(u)(x) = u \cdot x \cdot u^{-1}, \quad \text{for every } x \in \widehat{\mathscr{T}_+}(V). \tag{4.195}$$

Proof. Let $u \in 1 + \widehat{\mathscr{T}_+}(V)$ and set $z := \mathrm{Log}(u)$, that is, $u = \mathrm{Exp}(z)$. Note that $z \in \widehat{\mathscr{T}_+}(V)$. Let us denote by L_z and R_z, respectively, the endomorphisms of $\widehat{\mathscr{T}_+}(V)$ given by

$$L_z(x) := z \cdot x, \quad R_z(x) := x \cdot z, \quad \forall\, x \in \widehat{\mathscr{T}_+}(V).$$

For every $x \in \widehat{\mathscr{T}_+}(V)$, thanks to Theorem 4.14 on page 205, it holds that

$$u \cdot x \cdot u^{-1} = \mathrm{Exp}(z) \cdot x \cdot \mathrm{Exp}(-z) = \sum_{h=0}^{\infty} \frac{1}{h!} (\mathrm{ad}\, z)^{\circ\, h}(x).$$

Noting that $\mathrm{ad}\,(z) = L_z - R_z$ and using the same formalism as in Definition 4.17 on page 209, this can be rewritten as

$$u \cdot x \cdot u^{-1} = e^{L_z - R_z}(x). \tag{4.196}$$

We are indeed entitled to apply the cited formalism, for $\mathrm{ad}\,(z)$ is an endomorphism of $\widehat{\mathscr{T}_+}(V)$ mapping $\hat{U}_N := \prod_{k=N}^{\infty} \mathscr{T}_k(V)$ into $\hat{U}_{N+1} = \prod_{k=N+1}^{\infty} \mathscr{T}_k(V)$, for every $N \geq 0$ (this follows from the fact that $z \in \widehat{\mathscr{T}_+}(V)$). We claim that, with the same formalism,

$$\hat{\theta}(u) = e^{\hat{\theta}(z)}. \tag{4.197}$$

To begin with, we note that $\hat{\theta}(z)$ maps $\hat{U}_N$ into $\hat{U}_{N+1}$ for every $N \geq 0$, so that the right-hand side of (4.197) makes sense. Indeed, if $\tau \in \hat{U}_N$, we have $\tau = (\tau_k)_k$ with $\tau_k = 0$ for every $k = 0, \ldots, N-1$. Then, it holds that

$$\hat{\theta}(z)(\tau) \overset{(7.66)}{=} \sum_{h,k=0}^{\infty} \theta(z_h)(\tau_k) = \sum_{h \geq 1,\ k \geq N} \theta(z_h)(\tau_k) \in \prod_{k=N+1}^{\infty} \mathscr{T}_k(V),$$

since $z_0 = 0$ being $z \in \widehat{\mathscr{T}_+}(V)$. Next, we prove (4.197). First we remark that, if $z = (z_k)_k \in \widehat{\mathscr{T}_+}$ (with $z_k \in \mathscr{T}_k(V)$ for every $k \geq 0$ and $z_0 = 0$), then

$$\mathrm{Exp}(z) = 1 + \sum_{j=1}^{\infty} \left(\sum_{k=1}^{j} \frac{1}{k!} \sum_{\alpha_1 + \cdots + \alpha_k = j} z_{\alpha_1} \cdots z_{\alpha_1} \right).$$

Note that the last sum in parentheses belongs to $\mathscr{T}_j(V)$. Hence, by applying the definition of $\hat{\theta}$ in (7.66), we get

$$\widehat{\theta}(u) = \widehat{\theta}(\mathrm{Exp}(z)) = \mathrm{Id} + \sum_{j=1}^{\infty} \theta\Big(\sum_{k=1}^{j} \frac{1}{k!} \sum_{\alpha_1+\cdots+\alpha_k=j} z_{\alpha_1}\cdots z_{\alpha_1}\Big)$$

(recall that θ is a UAA morphism)

$$= \mathrm{Id} + \sum_{j=1}^{\infty} \Big(\sum_{k=1}^{j} \frac{1}{k!} \sum_{\alpha_1+\cdots+\alpha_k=j} \theta(z_{\alpha_1})\circ\cdots\circ\theta(z_{\alpha_1})\Big)$$

$$= \sum_{k=0}^{\infty} \frac{1}{k!}\Big(\sum_{h=1}^{\infty}\theta(z_h)\Big)^{\circ k} \overset{(4.188)}{=} \sum_{k=0}^{\infty} \frac{1}{k!}\big(\widehat{\theta}(z)\big)^{\circ k} = e^{\widehat{\theta}(z)}.$$

This proves (4.197). Collecting together (4.196) and (4.197), we see that

$$(4.195)\ holds\quad if\ and\ only\ if\quad e^{\widehat{\theta}(z)} \equiv e^{L_z-R_z}. \qquad (4.198)$$

We claim that the condition $e^{\widehat{\theta}(z)} \equiv e^{L_z-R_z}$ is equivalent to $\widehat{\theta}(z) = L_z - R_z$. We have to prove a sort of injectivity condition, the argument being not obvious since it is known that the exp maps are not injective in any context where they have sense. A possible way to formulate this is the following one: Let us consider the class of endomorphisms

$$\mathcal{H} := \Big\{ E \in \mathrm{End}(\widehat{\mathscr{T}_+}(V)) \Big| E\Big(\prod_{k=N}^{\infty}\mathscr{T}_k(V)\Big) \subseteq \prod_{k=N+1}^{\infty}\mathscr{T}_k(V),\ \forall\ N \geq 0 \Big\}.$$

Given a formal power series $\sum_{k=0}^{\infty} a_k z^k \in \mathbb{K}[[z]]$, we have already discussed that $\sum_{k=0}^{\infty} a_k E^k$ is well defined as an endomorphism of $\widehat{\mathscr{T}_+}$, provided that $E \in \mathcal{H}$. We next remark that

$$\sum_{k=0}^{\infty} a_k E^k\ is\ an\ element\ of\ \mathcal{H},\ whenever\ E \in \mathcal{H}\ and\ if\ a_0 = 0.$$

Indeed, if $a_0 = 0$ and $E \in \mathcal{H}$ we have

$$\sum_{k=1}^{\infty} a_k \underbrace{E^k(\tau)}_{\in\prod_{k=N+k}^{\infty}\mathscr{T}_k(V)} \in \prod_{k=N+1}^{\infty}\mathscr{T}_k(V),$$

for every $\tau \in \prod_{k=N}^{\infty}\mathscr{T}_k(V)$. For example, considering the formal power series of $e^z - 1$, that is, $\sum_{k=1}^{\infty} z^k/k!$, the following map $\mathbf{E}$ is well defined

$$\mathcal{H} \ni E \overset{\mathbf{E}}{\mapsto} e^E - \mathrm{Id} := \sum_{k=1}^{\infty} \frac{1}{k!} E^k \in \mathcal{H}.$$

Now, by simple arguments, it is easily seen that the above map is invertible and its inverse is precisely

$$\mathcal{H} \ni E \overset{\mathbf{L}}{\mapsto} \log(E + \mathrm{Id}) := \sum_{k=1}^{\infty} \frac{(-1)^{k+1}}{k} E^k \in \mathcal{H}.$$

As a consequence, if $A, B \in \mathcal{H}$ are such that $e^A = e^B$ then $e^A - \mathrm{Id} = e^B - \mathrm{Id}$ (which belong to $\mathcal{H}$), that is, $\mathbf{E}(A) = \mathbf{E}(B)$ so that, by applying the above map $\mathbf{L}$ to this identity we get $A = B$.

All these arguments can be applied to the identity on the far right-hand of (4.198), since $\widehat{\theta}(z)$ and $L_z - R_z = \mathrm{ad}\,(z)$ belong to the above class $\mathcal{H}$ (as $z \in \widehat{\mathscr{T}_+}$). We have thus proved that

$$(4.195)\ holds\quad if\ and\ only\ if\quad \widehat{\theta}(z) \equiv \mathrm{ad}\,(z). \tag{4.199}$$

Now, by rerunning a computation as in Djoković's proof and also in Reutenauer's (see page 211 for the details which we omit here), the fact that $\widehat{D}$ is a derivation of $\widehat{\mathscr{T}}(V)$ (plus some properties of the binomial coefficients) gives

$$\widehat{D}(u) \cdot u^{-1} = \sum_{n=1}^{\infty} \frac{1}{n!} (\mathrm{ad}\, z)^{\circ\, n-1}(\widehat{D}(z)).$$

Note that, by (4.183), the restriction of $\widehat{D}$ to $\widehat{\mathscr{T}_+}$ is a derivation of $\widehat{\mathscr{T}_+}$, so that $\widehat{D}(z) \in \widehat{\mathscr{T}_+}(V)$. If we introduce the formal power series

$$\varphi(z) := \frac{e^z - 1}{z} \sim \sum_{n=1}^{\infty} \frac{z^{n-1}}{n!},$$

then the above computation can be rewritten as

$$\widehat{D}(u) \cdot u^{-1} = \varphi(\mathrm{ad}\,(z))(\widehat{D}(z)). \tag{4.200}$$

We next consider the left-hand side of (4.194). For this, we have the following chain of equalities:

$$\widehat{g}(u - 1) = \widehat{g}(\mathrm{Exp}(z) - 1) = \widehat{g}\left(\sum_{n=1}^{\infty} \frac{z^n}{n!} \right) = \widehat{g}\left(\left(\sum_{n=1}^{\infty} \frac{z^{n-1}}{n!} \right) \cdot z \right)$$

$$\overset{(4.190)}{=} \widehat{\theta}\left(\sum_{n=1}^{\infty} \frac{z^{n-1}}{n!} \right)(\widehat{g}(z)) \overset{(4.188)}{=} \sum_{n=1}^{\infty} \frac{\widehat{\theta}(z^{n-1})}{n!}(\widehat{g}(z))$$

$$(\widehat{\theta}\ \text{is a UAA morphism})$$

$$= \sum_{n=1}^{\infty} \frac{1}{n!} \left(\widehat{\theta}(z) \right)^{\circ\, n-1}(\widehat{g}(z)) = \varphi\left(\widehat{\theta}(z) \right)(\widehat{g}(z)).$$

We have thus proved that

$$\widehat{g}(u-1) = \varphi\big(\widehat{\theta}(z)\big)(\widehat{g}(z)).   \tag{4.201}$$

Thus, collecting together (4.200) and (4.201), we see that

(4.194) *holds   if and only if* $\varphi\big(\widehat{\theta}(z)\big)(\widehat{g}(z)) = \varphi(\mathrm{ad}\,(z))(\widehat{D}(z)).$ (4.202)

From a result analogous to the one stated in the last part of Lemma 4.18 (page 209), we recognize that $\varphi(E)$ is invertible, for every $E \in \mathcal{H}$ (since the formal power series for φ admits a reciprocal). Then, *if* (4.195) *holds*, we see from (4.199) that $\widehat{\theta}(z) \equiv \mathrm{ad}\,(z)$ so that the far right-hand of (4.202) becomes $\varphi(\mathrm{ad}\,(z))(\widehat{g}(z)) = \varphi(\mathrm{ad}\,(z))(\widehat{D}(z))$, that is (by the above remarks), $\widehat{g}(z) = \widehat{D}(z)$.

  By all the above results, we easily see that

Conditions (4.194) *and* (4.195) *hold if and only if the following conditions hold:*

$$\widehat{\theta}(z) \equiv \mathrm{ad}\,(z),   \tag{4.203}$$

$$\widehat{g}(z) = \widehat{D}(z).   \tag{4.204}$$

Indeed, the following implications hold:

$$(4.195) \overset{\text{see (4.199)}}{\Longleftrightarrow} (4.203);$$

$$(4.194)\ \&\ (4.195) \overset{\text{see (4.199)}}{\Longrightarrow} (4.204);$$

$$(4.203)\ \&\ (4.204) \overset{\text{see (4.202)}}{\Longrightarrow} (4.194);$$

so that   (4.194) & (4.195) $\Longleftrightarrow$ (4.203) & (4.204).

The final step is to prove that (4.203) is actually contained in (4.204). This will prove that

(4.194) & (4.195) *hold if and only if* (4.204) *holds*.

This will end the proof, for, in view of (4.186), condition (4.204), that is,

$$\widehat{g}(\mathrm{Log}(u)) = \widehat{D}(\mathrm{Log}(u)),$$

is equivalent to $\mathrm{Log}(u) \in \overline{\mathcal{L}(V)}$, that is, $u \in \mathbf{H}$.

We are thus left to prove the implication

$$(4.204) \quad \Longrightarrow \quad (4.203).$$

If $z \in \widehat{\mathscr{T}_+}$ is such that $\widehat{g}(z) = \widehat{D}(z)$, then by (4.186) it holds that $z \in \overline{\mathcal{L}(V)}$. As a consequence we are entitled to apply (4.189), thus getting $\widehat{\theta}(z) = \mathrm{ad}\,(z)$. This is precisely (4.203). This completes the proof. $\qquad\qquad\qquad\square$

The above theorem gives at once the following result:

Corollary 4.45. *Let* **H** *be the set introduced in* (4.193). *Then* **H** *is a multiplicative subgroup of* $1 + \widehat{\mathscr{T}_+}(V)$.

Proof. We already know that $1 + \widehat{\mathscr{T}_+}(V)$ is a multiplicative subgroup of $\widehat{\mathscr{T}}(V)$. Clearly **H** is closed under multiplicative inversion: indeed if $u \in$ **H**, there exists $\ell \in \overline{\mathcal{L}(V)}$ such that $u = \mathrm{Exp}(\ell)$ so that

$$u^{-1} = (\mathrm{Exp}(\ell))^{-1} = \mathrm{Exp}(-\ell) \in \mathrm{Exp}(\overline{\mathcal{L}(V)}) = \mathbf{H}.$$

Hence, we have to prove that

$$u \cdot v \in \mathbf{H}, \quad \text{for every } u, v \in \mathbf{H}.$$

Set $w := u \cdot v$. By means of Theorem 4.44, all we have to prove is that w satisfies conditions (4.194) and (4.195), knowing that u, v satisfy them as well. We start with (4.195): let $x \in \widehat{\mathscr{T}_+}(V)$, then we have

$$\begin{aligned}
\widehat{\theta}(w)(x) &= \widehat{\theta}(u \cdot v)(x) \quad (\widehat{\theta} \text{ is a UAA morphism}) \\
&= \widehat{\theta}(u) \circ \widehat{\theta}(v)(x) \quad (v \text{ satisfies } (4.195)) \\
&= \widehat{\theta}(u)(v \cdot x \cdot v^{-1}) \quad (u \text{ satisfies } (4.195)) \\
&= u \cdot v \cdot x \cdot v^{-1} \cdot u^{-1} = w \cdot x \cdot w^{-1}.
\end{aligned}$$

We next prove (4.194):

$$\begin{aligned}
\widehat{D}(w) \cdot w^{-1} &= \widehat{D}(u \cdot v) \cdot v^{-1} \cdot u^{-1} \quad (\widehat{D} \text{ is a derivation}) \\
&= \widehat{D}(u) \cdot v \cdot v^{-1} \cdot u^{-1} + u \cdot \widehat{D}(v) \cdot v^{-1} \cdot u^{-1} \\
&= \widehat{D}(u) \cdot u^{-1} + u \cdot \widehat{D}(v) \cdot v^{-1} \cdot u^{-1} \\
&\quad (u \text{ and } v \text{ satisfy } (4.194)) \\
&= \widehat{g}(u - 1) + u \cdot \widehat{g}(v - 1) \cdot u^{-1} \quad (u \text{ satisfies } (4.195))
\end{aligned}$$

$$= \widehat{g}(u-1) + \widehat{\theta}(u)(\widehat{g}(v-1)) \overset{(4.190)}{=} \widehat{g}(u-1) + \widehat{g}(u \cdot (v-1))$$
$$= \widehat{g}(u-1+u \cdot v - u) = \widehat{g}(u \cdot v - 1) = \widehat{g}(w-1).$$

(Here we also applied the fact that $v - 1 \in \widehat{\mathscr{T}_+}$ for $v \in \mathbf{H} \subseteq 1 + \widehat{\mathscr{T}_+}$.) This completes the proof of the corollary.   □

As a consequence of the above corollary, we can prove the following:

Corollary 4.46 (Campbell, Baker, Hausdorff). *If V is a vector space over a field of characteristic zero, then $(\overline{\mathcal{L}(V)}, \blacklozenge)$ is a group, where $\blacklozenge$ is as in (4.192).*

In particular we have

$$\mathrm{Log}(\mathrm{Exp}(x) \cdot \mathrm{Exp}(y)) \in \overline{\mathcal{L}(V)}, \quad \text{for every } x, y \in \overline{\mathcal{L}(V)}. \tag{4.205}$$

Proof. Being $\mathbf{H} = \mathrm{Exp}(\overline{\mathcal{L}(V)})$, the map $\mathrm{Exp}|_{\overline{\mathcal{L}(V)}} : \overline{\mathcal{L}(V)} \to \mathbf{H}$ is clearly a bijection, with inverse $\mathrm{Log}|_{\mathbf{H}} : \mathbf{H} \to \overline{\mathcal{L}(V)}$. Hence, in view of the very definition of $\blacklozenge$ in (4.192), the fact that $(\overline{\mathcal{L}(V)}, \blacklozenge)$ is a group follows from the fact that $(\mathbf{H}, \cdot)$ is a group (see Corollary 4.45). This ends the proof.   □

Starting from the above Corollary 4.46, we can further derive the CBHD Theorem itself, proceeding as in Sects. 3.3.2 and 3.3.3.

Chapter 5
Convergence of the CBHD Series and Associativity of the CBHD Operation

THE aim of this chapter is twofold. On the one hand, we aim to study the convergence of the Dynkin series

$$u \diamond v := \sum_{j=1}^{\infty} \left(\sum_{n=1}^{j} \frac{(-1)^{n+1}}{n} \sum_{\substack{(h_1,k_1),\ldots,(h_n,k_n) \neq (0,0) \\ h_1+k_1+\cdots+h_n+k_n=j}} \right.$$
$$\left. \times \frac{(\mathrm{ad}\, u)^{h_1} (\mathrm{ad}\, v)^{k_1} \cdots (\mathrm{ad}\, u)^{h_n} (\mathrm{ad}\, v)^{k_n-1}(v)}{h_1! \cdots h_n! \, k_1! \cdots k_n! \left(\sum_{i=1}^{n}(h_i + k_i) \right)} \right),$$

in various contexts. For instance, this series can be investigated in any nilpotent Lie algebra (over a field of characteristic zero) where it is actually a finite sum, or in any finite dimensional real or complex Lie algebra and, more generally, its convergence can be studied in any *normed Banach-Lie algebra* (over $\mathbb{R}$ or $\mathbb{C}$). For example, the case of the normed Banach algebras (becoming normed Banach-Lie algebras if equipped with the associated commutator) will be extensively considered here.

On the other hand, once the well-posedness of the "operation" $\diamond$ has been established (at least in a neighborhood of the origin), the problem of its "associativity" can be considered. For instance, we shall obtain a local result, providing the identity

$$a \diamond (b \diamond c) = (a \diamond b) \diamond c,$$

at least when a, b, c belong to a neighborhood of the origin of any normed Banach-Lie algebra. Also, this identity turns out to be global when we are dealing with nilpotent Lie algebras, a fact which is frequently considered as a folklore consequence of the CBHD Theorem, but which deserves – in our opinion – a rigorous derivation.

What is more, in the context of finite-dimensional nilpotent Lie algebras $\mathfrak{n}$, we are able to solve the problem of finding a (connected and simply

A. Bonfiglioli and R. Fulci, *Topics in Noncommutative Algebra*, Lecture Notes in Mathematics 2034, DOI 10.1007/978-3-642-22597-0_5, © Springer-Verlag Berlin Heidelberg 2012

connected) Lie group whose Lie algebra is isomorphic to $\mathfrak{n}$: namely, $(\mathfrak{n}, \diamond)$ solves this problem. By invoking only very basic facts on Lie groups, we are thus in a position to prove the so-called *Third Fundamental Theorem of Lie, in its global form,* for $\mathfrak{n}$. This gives a very significant application of the CBHD operation for Lie group theory (actually, the original context where the CBHD Theorem was born).

In dealing with these topics, we shall make use of as much information as possible deriving from the identities implicit in the general CBHD Theorem: these identities are the following ones (the relevant notation has been introduced in Chap. 3):

$$x \diamond y = x \blacklozenge y, \qquad \mathrm{Exp}(u \diamond v) = \mathrm{Exp}(u) \cdot \mathrm{Exp}(v), \qquad a \diamond (b \diamond c) = (a \diamond b) \diamond c.$$

As a matter of fact, these are identities between formal power series of the tensor algebra of a vector space (over a field of characteristic zero), so that their applicability to different settings cannot be forced and, if we want to use them in other contexts, some work must be accomplished. But this work is worthwhile, since it yields useful identities on any associative or Lie algebra.

Whereas the first identity has been already investigated in this Book (see e.g., Theorem 3.34 on page 157 or identity (4.19) on page 185), it is within the scope of the present chapter to study the other two. For example, when suitably "truncated", we can obtain from the second identity a family of equalities valid in any associative algebra and serving, for example, as a starting point for plenty of applications in Analysis (e.g. of some partial differential operators) or in Differential Geometry (e.g. of Lie groups). Obviously, by "truncating" the third of the above identities, we are able to provide the starting point for the associativity investigations of this chapter.

An alternative approach to the above topics may be obtained by a systematical use of analytic functions on a domain in a Banach space: once it is known how to deal with functions (locally) admitting power series expansions, many results (including the associativity of $\diamond$, since the map $(a, b) \mapsto a \diamond b$ turns out to be analytical!) can be derived at once by identities between formal power series. For this alternative approach, we refer the Reader to the very effective exposition given by Hofmann and Morris [91, Chapter 5], allowing ourselves to furnish only a brief sketch (see Sect. 5.5.1). By making use of a "unique continuation" result proved in the setting of analytic functions between Banach spaces, we will be able to exhibit in full details an example of *non-convergence* of the CBHD series in a Banach algebra (namely, that of the real 2×2 matrices). This is done in Sect. 5.6.

As the problem of finding the largest domain of convergence of the CBHD series is still an open question, it is beyond the scope of this chapter to provide final results. We shall instead furnish an overview of references at the end of the chapter (see Sect. 5.7).

The exposition is organized as follows:

- Section 5.1 provides identities -for a general associative algebra- resulting from the CBHD identity $\mathrm{Exp}(u \diamond v) = \mathrm{Exp}(u) \cdot \mathrm{Exp}(v)$.
- Section 5.2 collects results on the convergence of the Dynkin series $u \diamond v$ in various contexts: first (as an introductory section) on finite dimensional Lie algebras, then – more generally – on Banach-Lie algebras. In the former setting, a result concerning the real analyticity of $\diamond$ will also be given; in the latter context, a result on the rate of convergence of the Dynkin series is provided (see Theorem 5.31). Finally, Sect. 5.2.3 furnishes an adaptation of a remarkable argument by Varadarajan on an improved domain of convergence for the cited series, this argument gathering together the algebraic recursion formula found in the previous chapter (see Sect. 4.5.1) plus an interesting technique from the Theory of ODEs.
- In Sect. 5.3 we study the associativity property of the CBHD operation $(x, y) \mapsto x \diamond y$. As considered in the previous sections, the right context in which this infinite sum makes sense is that of Banach-Lie algebras, where $\diamond$ is a priori defined only in a neighborhood of the origin. We shall then prove (in Sect. 5.3.2) that $\diamond$ defines a *local group*, so that in particular it is associative (in a suitable neighborhood of the origin).
- Another case of interest is that of the nilpotent Lie algebras, which we take up in Sect. 5.4: in this case $\diamond$ is globally defined (since the associated series becomes in fact a finite sum) and it defines a group on the whole algebra (see Sect. 5.4.1). Furthermore, we shall prove that once $\mathfrak{n}$ is finite-dimensional – besides being nilpotent – then $(\mathfrak{n}, \diamond)$ is a *Lie group*, whose Lie algebra is isomorphic to $\mathfrak{n}$ itself. This solves the Third Fundamental Theorem of Lie, in *global* form, for finite-dimensional nilpotent Lie algebras, thus furnishing a remarkable application of the CBHD operation (see Sect. 5.4.2 for the details).
- Section 5.5 is devoted to the CBHD formula for normed Banach algebras $(A, *, \|\cdot\|)$. Indeed, in this context it is possible to define the exponential function Exp (think, for instance, of the algebra of square matrices) and to consider a further problem, besides that of the convergence of the series expressing $u \diamond v$: namely, we are interested in deriving the identity

$$\mathrm{Exp}(u \diamond v) = \mathrm{Exp}(u) * \mathrm{Exp}(v).$$

This is done in Theorem 5.56. In Sect. 5.5.1, we define analytic functions between Banach spaces and we prove some basic facts, including a Unique Continuation Theorem.
- Section 5.6 is devoted to exhibiting an example of *failure of convergence for the CBHD series*. We use the results on analytic functions from Sect. 5.5.1. We also establish a result of independent interest (see Theorem 5.67):

If the CBHD series expressing $a \diamond b$ is convergent for some a, b in a Banach algebra $(A, *)$, then the equality $\mathrm{Exp}(a) * \mathrm{Exp}(b) = \mathrm{Exp}(a \diamond b)$ necessarily holds.

- Section 5.7 collects some references on the literature on convergence of the CBHD series and on related topics.

5.1 "Finite" Identities Obtained from the CBHD Theorem

One of the most useful applications of the CBHD Formula is that it provides, as a byproduct, identities in any associative algebra (without the requirement of any topology) simply by taking the projections onto the respective subspaces $\mathscr{T}_k$ w.r.t. $\mathscr{T} = \bigoplus_k \mathscr{T}_k$. We describe in this section how this can be accomplished.

Throughout, V *will denote a fixed vector space over a field $\mathbb{K}$ of characteristic zero.* (A hypothesis which will not be recalled in the sequel.)

Let $N \in \mathbb{N}$ be fixed. We set

$$H_N := \Big\{ (u_0, u_1, \ldots, u_N, 0, 0, \ldots) \in \mathscr{T}(V) \,\big|\, u_j \in \mathscr{T}_j(V),\ \forall j \in \mathbb{N} \cup \{0\} \Big\}. \tag{5.1}$$

We shall consider H_N both as a subspace of $\mathscr{T}(V)$ or of $\widehat{\mathscr{T}}(V)$, depending on the occasion. We obviously have (see the notation in Remark 2.80, page 104)

$$H_N \simeq \mathscr{T}(V)/U_{N+1} \simeq \widehat{\mathscr{T}}(V)/\widehat{U}_{N+1}. \tag{5.2}$$

We shall denote by π_N the endomorphism of $\widehat{\mathscr{T}}(V)$ acting as follows:

$$\pi_N : \widehat{\mathscr{T}}(V) \to \widehat{\mathscr{T}}(V), \qquad (u_0, u_1, \ldots) \mapsto (u_0, u_1, \ldots, u_N, 0, 0, \ldots).$$

In practice (see (5.2)) π_N is the natural projection of $\widehat{\mathscr{T}}(V)$ onto the quotient $H_N \simeq \widehat{\mathscr{T}}(V)/\widehat{U}_{N+1}$.

Remark 5.1. The following simple facts are obvious from the definition of π_N:

1. $\pi_N(\mathscr{T}(V)) = \pi_N(\widehat{\mathscr{T}}(V)) = H_N$.
2. one has $\pi_N(\mathscr{T}_+(V)) \subset \mathscr{T}_+(V)$, and $\pi_N(\widehat{\mathscr{T}}_+(V)) \subset \widehat{\mathscr{T}}_+(V)$, or equivalently, $\pi_N(U_1) \subset U_1$, and $\pi_N(\widehat{U}_1) \subset \widehat{U}_1$.
3. $\pi_N(u) = \pi_N(v)$ if and only if the homogeneous components of degrees $0, 1, \ldots, N$ of u and of v coincide.
4. $\ker(\pi_N) = \widehat{U}_{N+1}$.
5. π_N is the identity on H_N, whence on $\mathscr{T}_0(V), \mathscr{T}_1(V), \ldots, \mathscr{T}_N(V)$ too.

Unfortunately π_N is not an algebra morphism,[1] though it possesses the following "quasi-morphism" property:

$$\pi_N(u \cdot v) = \pi_N\big(\pi_N(u) \cdot \pi_N(v)\big), \qquad \text{for every } u, v \in \widehat{\mathscr{T}}(V). \tag{5.3}$$

Indeed, for every $u = (u_j)_j$, $v = (v_j)_j$ in $\widehat{\mathscr{T}}(V)$ we have

$$\pi_N(u \cdot v) = \pi_N\left(\left(\sum_{k=0}^{j} u_{j-k} \otimes v_k\right)_{j \geq 0}\right)$$

$$= \left(u_0 \otimes v_0, u_1 \otimes v_0 + u_0 \otimes v_1, \ldots, \sum_{k=0}^{N} u_{N-k} \otimes v_k, 0, \ldots\right).$$

On the other hand, we also have

$$\pi_N(u) \cdot \pi_N(v) = (u_0, u_1, \ldots, u_N, 0, \ldots) \cdot (v_0, v_1, \ldots, v_N, 0, \ldots)$$

$$= \left(\sum_{k=0}^{j} \overline{u}_{j-k} \otimes \overline{v}_k\right)_{j \geq 0},$$

where we have set

$$\overline{u}_j := \begin{cases} u_j, & \text{if } 0 \leq j \leq N, \\ 0, & \text{if } j \geq N+1, \end{cases} \quad \text{and, analogously,} \quad \overline{v}_j := \begin{cases} v_j, & \text{if } 0 \leq j \leq N, \\ 0, & \text{if } j \geq N+1. \end{cases}$$

Now, if $0 \leq j \leq N$, we obviously have

$$\sum_{k=0}^{j} \overline{u}_{j-k} \otimes \overline{v}_k = \sum_{k=0}^{j} u_{j-k} \otimes v_k,$$

and (5.3) follows from Remark 5.1-3.   □

The following result will help us in applying the π_N map on both sides of the "exponential" identity $\mathrm{Exp}(u \diamond v) = \mathrm{Exp}(u) \cdot \mathrm{Exp}(v)$ (resulting from the CBHD Theorem, see Corollary 3.33, page 156).

Lemma 5.2. *Let* $\mathrm{Exp} : \widehat{\mathscr{T}}_+(V) \to \widehat{\mathscr{T}}(V)$ *be the usual exponential function. Then*

$$\pi_N(\mathrm{Exp}(u)) = \pi_N(\mathrm{Exp}(\pi_N(u))), \tag{5.4}$$

for every $u \in \widehat{\mathscr{T}}_+(V)$ *and every* $N \in \mathbb{N}$. *In other words,* $\mathrm{Exp}(u)$ *and* $\mathrm{Exp}(\pi_N(u))$ *have the same components in* $\widehat{\mathscr{T}}(V) = \prod_{j=0}^{\infty} \mathscr{T}_j(V)$ *up to degree* N.

---

[1]Indeed, for example we have

$$\pi_1(u \cdot v) = \big(u_0 \otimes v_0, u_0 \otimes v_1 + u_1 \otimes v_0, 0, 0, \ldots\big)$$

$$\neq \pi_1(u) \cdot \pi_1(v) = (u_0, u_1, 0, \ldots) \cdot (v_0, v_1, 0, \ldots)$$

$$= \big(u_0 \otimes v_0, u_0 \otimes v_1 + u_1 \otimes v_0, u_1 \otimes v_1, 0, \ldots\big).$$

Proof. First of all we have (recall that π_N is linear)

$$\pi_N(\text{Exp}(u)) = \pi_N\left(\sum_{n=0}^{N} u^n/n!\right) + \pi_N\left(\underbrace{\sum_{n=N+1}^{\infty} u^n/n!}_{\in \hat{U}_{N+1} \text{ for } u \in \hat{U}_1}\right)$$

(the second summand vanishes, for $\pi_N \equiv 0$ on $\hat{U}_{N+1}$)

$$= \sum_{n=0}^{N} \frac{\pi_N(u^n)}{n!} \qquad \text{(use (5.3) and induction)}$$

$$= \sum_{n=0}^{N} \frac{\pi_N\left((\pi_N(u))^n\right)}{n!} = \pi_N\left(\sum_{n=0}^{N} \frac{(\pi_N(u))^n}{n!}\right)$$

(argue as above, recalling that $\pi_N(\hat{U}_1) = \hat{U}_1$)

$$= \pi_N\left(\sum_{n=0}^{N} \frac{(\pi_N(u))^n}{n!} + \sum_{n=N+1}^{\infty} \frac{(\pi_N(u))^n}{n!}\right)$$

$$= \pi_N(\text{Exp}(\pi_N(u))).$$

Here we have used twice the following fact: if $v \in \hat{U}_1$ then $v^n \in \hat{U}_n$, so that $v^n \in \hat{U}_{N+1}$ for every $n \geq N+1$, whence

$$\sum_{n=N+1}^{\infty} \frac{v^n}{n!} \in \overline{\hat{U}_{N+1}} = \hat{U}_{N+1}.$$

The proof of (5.4) is complete.   □

From the proof of Lemma 5.2 we immediately derive the following result.

Lemma 5.3. *With the notation of Lemma 5.2, we have*

$$\pi_N(\text{Exp}(u)) = \sum_{n=0}^{N} \frac{u^n}{n!} \qquad \text{for every } u \in V \text{ and } N \in \mathbb{N}. \qquad (5.5)$$

Proof. From the computations in Lemma (5.2), we have

$$\pi_N(\text{Exp}(u)) = \sum_{n=0}^{N} \pi_N(u^n)/n!.$$

Then we immediately get (5.5) by noticing that, if $u \in V$ then

$$u^n = u^{\otimes n} \in \mathscr{T}_N(V),$$

so that $\pi_N(u^n) = u^n$ for every $n = 0, 1, \ldots, N$.   □

From Lemma 5.3, we get:

Lemma 5.4. *With the notation of Lemma 5.2, we have*

$$\pi_N\big(\mathrm{Exp}(u) \cdot \mathrm{Exp}(v)\big) = \sum_{0 \leq i+j \leq N} \frac{u^i \cdot v^j}{i!\,j!}, \qquad (5.6)$$

for every $u, v \in V$ and $N \in \mathbb{N}$.

Proof. Let $u, v \in V$. The following computation applies:

$$\pi_N\big(\mathrm{Exp}(u) \cdot \mathrm{Exp}(v)\big) \overset{(5.3)}{=} \pi_N\Big(\pi_N(\mathrm{Exp}\,u) \cdot \pi_N(\mathrm{Exp}\,v)\Big)$$

$$\overset{(5.5)}{=} \pi_N\bigg(\sum_{i,j=0}^{N} \frac{u^i \cdot v^j}{i!\,j!}\bigg) = \sum_{0 \leq i+j \leq N} \frac{u^i \cdot v^j}{i!\,j!}.$$

In the last equality we used the fact that $u, v \in V = \mathscr{T}_1(V)$ implies $u^i \cdot v^j \in \mathscr{T}_{i+j}(V)$, and we further exploited the fact that π_N is the identity on $\mathscr{T}_0, \dots, \mathscr{T}_N$, whereas it vanishes on $\mathscr{T}_{N+1}, \mathscr{T}_{N+2}, \dots$ $\qquad\square$

We now aim to apply the map π_N to the identity

$$\mathrm{Exp}(u \diamond v) = \mathrm{Exp}(u) \cdot \mathrm{Exp}(v), \qquad \forall\, u, v \in V,$$

resulting from (3.82) at page 156. We get

$$\pi_N\big(\mathrm{Exp}(u \diamond v)\big) = \pi_N\big(\mathrm{Exp}(u) \cdot \mathrm{Exp}(v)\big), \qquad \forall\, u, v \in V,\, N \in \mathbb{N}. \qquad (5.7)$$

The right-hand side of (5.7) has been computed in (5.6), whilst the left-hand side equals

$$\pi_N\Big(\mathrm{Exp}\big(\pi_N(u \diamond v)\big)\Big),$$

thanks to (5.4) (recall that $u \diamond v \in \widehat{\mathscr{T}_+}(V)$ whenever $u, v \in \widehat{\mathscr{T}_+}(V)$). In its turn, $\pi_N(u \diamond v)$ can be easily computed as follows (see (3.27), page 130):

$$\pi_N(u \diamond v) = \pi_N\bigg(\sum_{n=1}^{\infty}\bigg(c_n \sum_{(h,k)\in\mathcal{N}_n} \mathbf{c}(h,k)\,\big[u^{h_1}v^{k_1}\cdots u^{h_n}v^{k_n}\big]_\otimes\bigg)\bigg)$$

$$\bigg(\text{notice that } \big[u^{h_1}v^{k_1}\cdots u^{h_n}v^{k_n}\big]_\otimes \in \mathscr{T}_{|h|+|k|} \text{ since } u, v \in V,$$

$$\text{and recall that } |h| + |k| \geq n \text{ for every } (h,k) \in \mathcal{N}_n\bigg)$$

$$= \sum_{n=1}^{N} \left(c_n \sum_{(h,k)\in\mathcal{N}_n:\ |h|+|k|\leq N} \mathbf{c}(h,k) \left[u^{h_1} v^{k_1} \cdots u^{h_n} v^{k_n} \right]_{\otimes} \right)$$

$$= \eta_N(u,v),$$

where $\eta_N(u,v)$ was introduced in (3.30), page 131. Hence we have proved the following equality

$$\pi_N(u \diamond v) = \eta_N(u,v), \qquad \forall\, u,v \in V,\ N \in \mathbb{N}. \tag{5.8}$$

Collecting together all the above facts, we have derived the identity

$$\pi_N\Big(\mathrm{Exp}\big(\eta_N(u,v)\big) \Big) = \sum_{0\leq i+j\leq N} \frac{u^i \cdot v^j}{i!\, j!}, \tag{5.9}$$

valid for every $u,v \in V$ and every $N \in \mathbb{N}$. By the definition of π_N, this means that

$$\mathrm{Exp}\big(\eta_N(u,v)\big) \equiv \sum_{0\leq i+j\leq N} \frac{u^i \cdot v^j}{i!\, j!} \quad \text{modulo } \widehat{U}_{N+1}, \tag{5.10}$$

again for every $u,v \in V$ and every $N \in \mathbb{N}$.

We now aim to expand $\mathrm{Exp}\big(\eta_N(u,v)\big)$, to derive further information from (5.9) and (5.10). To this end, we have

$$\mathrm{Exp}\big(\eta_N(u,v)\big) = \sum_{s=0}^{N} \frac{(\eta_N(u,v))^s}{s!} + \sum_{s=N+1}^{\infty} \frac{(\eta_N(u,v))^s}{s!}.$$

Now notice that the second sum in the above right-hand side belongs to $\widehat{U}_{N+1}$. Indeed, by the very definition of $\eta_N(u,v)$, for every $u,v \in V$ we have

$$\eta_N(u,v) \in \mathscr{T}_1(V) \oplus \cdots \oplus \mathscr{T}_N(V) \subset \widehat{U}_1,$$

so that $(\eta_N(u,v))^s \in \widehat{U}_{N+1}$ for every $s \geq N+1$. Consequently

$$\mathrm{Exp}\big(\eta_N(u,v)\big) \equiv \sum_{s=0}^{N} \frac{(\eta_N(u,v))^s}{s!} \quad \text{modulo } \widehat{U}_{N+1}, \tag{5.11}$$

for every $u,v \in V$ and every $N \in \mathbb{N}$. From (5.10) and (5.11), we get an important "finite" identity:

$$\sum_{s=0}^{N} \frac{(\eta_N(u,v))^s}{s!} \equiv \sum_{0\leq i+j\leq N} \frac{u^i \cdot v^j}{i!\, j!} \quad \text{modulo } \widehat{U}_{N+1}, \tag{5.12}$$

for every $u, v \in V$ and every $N \in \mathbb{N}$. Notice that "modulo $\widehat{U}_{N+1}$" in the above (5.12) can be replaced by

$$\text{"modulo } \mathscr{T}_{N+1} \oplus \mathscr{T}_{N+2} \oplus \cdots \oplus \mathscr{T}_{N^2}\text{"}$$

since the left-hand side of (5.12) belongs to $\mathscr{T}_0 \oplus \mathscr{T}_1 \oplus \cdots \oplus \mathscr{T}_{N^2}$. We are thus in a position to derive our main result for this section, a theorem giving a "finite" version of the CBHD Formula $\mathrm{Exp}(u \diamond v) = \mathrm{Exp}(u) \cdot \mathrm{Exp}(v)$ (which is an identity in the space of formal power series $\widehat{\mathscr{T}}(V)$).

Theorem 5.5. *Let V be a vector space over a field of characteristic zero. Let $N \in \mathbb{N}$. Then there exists a function*

$$\mathcal{R}_{N+1} : V \times V \to \bigoplus_{n=N+1}^{N^2} \mathscr{T}_n(V)$$

such that the following identity in $\mathscr{T}(V)$ holds for any choice of $u, v \in V$:

$$\sum_{s=0}^{N} \frac{1}{s!} \left(\sum_{n=1}^{N} \frac{(-1)^{n+1}}{n} \sum_{\substack{(h,k) \in \mathcal{N}_n \\ |h|+|k| \leq N}} \frac{\left[u^{h_1} v^{k_1} \cdots u^{h_n} v^{k_n} \right]_{\otimes}}{h!\, k!\, (|h| + |k|)} \right)^s$$

$$= \sum_{0 \leq i+j \leq N} \frac{u^{\otimes i} \otimes v^{\otimes j}}{i!\, j!} + \mathcal{R}_{N+1}(u, v). \tag{5.13}$$

Obviously, the "remainder" function $\mathcal{R}_{N+1}$ in the above statement is expressed by the "universal" expression

$$\mathcal{R}_{N+1}(u, v) := \sum_{s=0}^{N} \frac{1}{s!} \left(\sum_{n=1}^{N} \frac{(-1)^{n+1}}{n} \sum_{\substack{(h,k) \in \mathcal{N}_n \\ |h|+|k| \leq N}} \frac{\left[u^{h_1} v^{k_1} \cdots u^{h_n} v^{k_n} \right]_{\otimes}}{h!\, k!\, (|h| + |k|)} \right)^s$$

$$- \sum_{0 \leq i+j \leq N} \frac{u^{\otimes i} \otimes v^{\otimes j}}{i!\, j!}. \tag{5.14}$$

The important (and nontrivial!) fact about $\mathcal{R}_{N+1}$ is that the right-hand side of (5.14) belongs to $\mathscr{T}_{N+1}(V) \oplus \mathscr{T}_{N+2}(V) \oplus \cdots \oplus \mathscr{T}_{N^2}(V)$.

Since Theorem 5.5 establishes an identity in the tensor algebra of an arbitrary vector space (on a field of 0 characteristic $\mathbb{K}$), we can derive an analogous result *on an arbitrary associative algebra* (over $\mathbb{K}$), as described below.

Remark 5.6. Let $(A, *)$ be an associative algebra (over the field $\mathbb{K}$). In case A is not unital (or even if it is), we can "add" an element to A so that A becomes (isomorphic to) a subalgebra of a UAA algebra A_1, in the following way. Let us equip the vector space $A_1 := \mathbb{K} \times A$ with the operation

$$(k_1, a_1) \star (k_2, a_2) := (k_1 k_2, k_1 a_2 + k_2 a_1 + a_1 * a_2), \qquad (5.15)$$

for every $k_1, k_2 \in \mathbb{K}$ and every $a_1, a_2 \in A$. Then it is easily seen that $(A_1, \star)$ is a UA algebra with unit $(1_\mathbb{K}, 0_A)$ and A is isomorphic (as an associative algebra) to $\{0\} \times A$, via the algebra isomorphism

$$A \ni a \overset{\Psi}{\mapsto} (0, a) \in \{0\} \times A \subset A_1.$$

By identifying A with $\Psi(A)$, we may say that *any associative algebra is the subalgebra of a unital associative algebra*. $\qquad\qquad \square$

Now let $a, b \in A$ and let $\{x, y\}$ be a set of cardinality 2. Since a, b are elements of A_1 too, and since A_1 is a UA algebra, by Theorem 2.85-(1a) there exists a unique UAA morphism $\Phi_{a,b} : \mathscr{T}(\mathbb{K}\langle x, y \rangle) \to A_1$ such that

$$\Phi_{a,b}(x) = a \quad \text{and} \quad \Phi_{a,b}(y) = b. \qquad (5.16)$$

We can apply Theorem 5.5 when $V = \mathbb{K}\langle x, y \rangle$ and $u = x$, $v = y$, so that we get the following identity in $\mathscr{T}(\mathbb{K}\langle x, y \rangle)$:

$$\sum_{s=0}^{N} \frac{1}{s!} \left(\sum_{n=1}^{N} \frac{(-1)^{n+1}}{n} \sum_{\substack{(h,k) \in \mathbb{N}_n \\ |h|+|k| \leq N}} \frac{\left[x^{h_1} y^{k_1} \cdots x^{h_n} y^{k_n}\right]_\otimes}{h!\, k!\, (|h| + |k|)} \right)^s$$

$$- \sum_{0 \leq i+j \leq N} \frac{x^{\otimes i} \otimes y^{\otimes j}}{i!\, j!} = \mathcal{R}_{N+1}(x, y). \qquad (5.17)$$

Here the symbol "$\otimes$" (and the s-power) obviously refers to the algebraic structure on the tensor algebra $\mathscr{T}(\mathbb{K}\langle x, y \rangle)$. We are certainly entitled to apply the UAA morphism $\Phi_{a,b}$ to the identity (5.17). Since we have $\Phi_{a,b}(x^{\otimes i} \otimes y^{\otimes j}) = a^{*i} * b^{*j}$ and

$$\Phi_{a,b}\left(\left[x^{h_1} y^{k_1} \cdots x^{h_n} y^{k_n}\right]_\otimes\right) = \left[a^{h_1} b^{k_1} \cdots a^{h_n} b^{k_n}\right]_*$$

(indeed, recall that any morphism of associative algebras is also a Lie algebra morphism of the associated commutator-algebras), we get

$$\sum_{s=0}^{N} \frac{1}{s!} \left(\sum_{n=1}^{N} \frac{(-1)^{n+1}}{n} \sum_{\substack{(h,k) \in \mathbb{N}_n \\ |h|+|k| \leq N}} \frac{\left[a^{h_1} b^{k_1} \cdots a^{h_n} b^{k_n}\right]_*}{h!\, k!\, (|h| + |k|)} \right)^{*s}$$

$$= \sum_{0 \le i+j \le N} \frac{a^{*i} * b^{*j}}{i! \, j!} + \mathcal{R}^*_{N+1}(a, b).$$

Here, $\mathcal{R}^*_{N+1}(a, b)$ denotes the element of A obtained by formally substituting u with a and v with b in (5.14) and the $\otimes$ operation by $*$. What is remarkable here is to observe that $\mathcal{R}^*_{N+1}(a, b)$ is a sum of $*$-products where a, b jointly appear at least $N + 1$ times (and at most N^2 times). By means of (5.14), we could also provide a bound for the number of such summands in $\mathcal{R}^*_{N+1}(a, b)$.

All the above arguments lead to the following remarkable (and "ready-to-use") result, a consequence of the CBHD Formula for an arbitrary associative algebra. In stating this result, we use the following notation: If A is an associative algebra and $n \in \mathbb{N}$, we set

$$A^n := \mathrm{span}\{a_1 * \cdots * a_n \mid a_1, \ldots, a_n \in A\}. \tag{5.18}$$

Note that $A = A^1 \supseteq A^2 \supseteq A^3 \supseteq \cdots$.

Theorem 5.7. *Let $(A, *)$ be an associative algebra over a field of characteristic zero $\mathbb{K}$. Let $N \in \mathbb{N}$. Then there exists a function*

$$\mathcal{R}^*_{N+1} : A \times A \to A^{N+1}$$

such that the following identity in A holds for any choice of $a, b \in A$:

$$\sum_{s=0}^{N} \frac{1}{s!} \left(\sum_{n=1}^{N} \frac{(-1)^{n+1}}{n} \sum_{\substack{(h,k) \in \mathcal{N}_n \\ |h|+|k| \le N}} \frac{\left[a^{h_1} b^{k_1} \cdots a^{h_n} b^{k_n} \right]_*}{h! \, k! \, (|h| + |k|)} \right)^{*s}$$

$$= \sum_{0 \le i+j \le N} \frac{a^{*i} * b^{*j}}{i! \, j!} + \mathcal{R}^*_{N+1}(a, b). \tag{5.19}$$

More precisely, $\mathcal{R}^*_{N+1}(a, b)$ is obtained by substituting x, y with a, b respectively, in (5.17) and by replacing the $\otimes$ operation of $\mathcal{T}(\mathbb{K}\langle x, y \rangle)$ by $*$. In particular $\mathcal{R}^*_{N+1}(a, b)$ is a sum of elements of $A^{N+1}, A^{N+2}, \ldots, A^{N^2}$, since $\mathcal{R}_{N+1}(x, y)$ is expressed by a "universal" polynomial in $\bigoplus_{n=N+1}^{N^2} \mathcal{T}_n(\mathbb{K}\langle x, y \rangle)$ (see Theorem 5.5). In (5.19) we exploited our usual notation

$$\left[a^{h_1} b^{k_1} \cdots a^{h_n} b^{k_n} \right]_*$$

$$= \overbrace{[a, \cdots [a,}^{h_1 \text{ times}} \overbrace{[b, \cdots [b,}^{k_1 \text{ times}} \cdots \overbrace{[a, \cdots [a,}^{h_n \text{ times}} \overbrace{[b, [\cdots, b}^{k_n \text{ times}}]_*]_* * \cdots]_* \cdots]_* \cdots]_*]_* * \cdots]_* \tag{5.20}$$

where $[\alpha, \beta]_* = \alpha * \beta - \beta * \alpha$, for every $\alpha, \beta \in A$.

Just to give an idea of the scope of our applications of the above theorem, in the forthcoming Part II of this Book we will apply (5.19) when $(A, *)$ is, for example, the algebra of linear partial differential operators with smooth real coefficients on $\mathbb{R}^N$ (equipped with the operation $\circ$ of composition of operators) and a, b are tX, tY, where X, Y are vector fields (i.e., linear PDO's of order 1) and t is a real parameter. Our precise knowledge of the remainder term will allow us to estimate $\mathcal{R}^*_{N+1}(tX, tY)$, simply by factoring t^{N+1} throughout. The identity resulting from (5.19) in this context has a remarkable meaning in the theory of ODEs as well as in Lie group theory.

We can obtain a further "closed" identity deriving from the main CBHD identity

$$\mathrm{Exp}(x) \cdot \mathrm{Exp}(y) = \mathrm{Exp}\Big(\textstyle\sum_{n=1}^{\infty} Z_n(x,y) \Big), \quad \text{in } \widehat{\mathscr{T}}(\mathbb{K}\langle x, y\rangle).$$

Indeed, by expanding both sides we immediately get

$$\sum_{h,k \geq 0} \frac{x^h y^k}{h!\, k!} = 1 + \sum_{j=1}^{\infty} \frac{1}{j!} \sum_{\alpha_1,\ldots,\alpha_j \in \mathbb{N}} Z_{\alpha_1}(x,y) \cdots Z_{\alpha_j}(x,y).$$

This last identity can be easily projected on $\mathscr{T}_N(\mathbb{K}\langle x, y\rangle)$, recalling that (for every $\alpha \in \mathbb{N}$)

$$x, y \in \mathscr{T}_1(\mathbb{K}\langle \{x, y\}\rangle), \qquad Z_\alpha(x,y) \in \mathscr{T}_\alpha(\mathbb{K}\langle x, y\rangle).$$

We derive, for every $N \in \mathbb{N}$,

$$\sum_{\substack{h,k \in \mathbb{N} \cup \{0\} \\ h+k=N}} \frac{x^h y^k}{h!\, k!} = \sum_{j=1}^{N} \frac{1}{j!} \sum_{\substack{\alpha_1,\ldots,\alpha_j \in \mathbb{N} \\ \alpha_1 + \cdots + \alpha_j = N}} Z_{\alpha_1}(x,y) \cdots Z_{\alpha_j}(x,y).$$

Letting N vary in $\mathbb{N}$, we have a family of identities in the *tensor algebra* $\mathscr{T}(\mathbb{K}\langle x, y\rangle)$. By the universal property of the latter, we can then obtain an analogous family of identities in any associative algebra:

Theorem 5.8. *Let $(A, *)$ be an associative algebra over a field of characteristic zero. Then the following identity holds for every $N \in \mathbb{N}$ and all choices of $a, b \in A$:*

$$\sum_{\substack{h,k \in \mathbb{N} \cup \{0\} \\ h+k=N}} \frac{a^{*h} * b^{*k}}{h!\, k!} = \sum_{j=1}^{N} \frac{1}{j!} \sum_{\substack{\alpha_1,\ldots,\alpha_j \in \mathbb{N} \\ \alpha_1 + \cdots + \alpha_j = N}} Z^*_{\alpha_1}(a,b) * \cdots * Z^*_{\alpha_j}(a,b), \quad (5.21)$$

$$\text{where} \quad Z_\alpha^*(a,b) = \sum_{n=1}^\alpha \frac{(-1)^{n+1}}{n} \sum_{\substack{(h,k)\in\mathcal{N}_n \\ |h|+|k|=\alpha}} \frac{\left[a^{h_1}b^{k_1}\cdots a^{h_n}b^{k_n}\right]_*}{h!\,k!\,(|h|+|k|)}.$$

Here, we exploited our usual notation (5.20) for $[a^{h_1}\cdots b^{k_n}]_*$.
Moreover, by summing up for $1 \le N \le R$, we obtain the following identity

$$\sum_{\substack{h,k\in\mathbb{N}\cup\{0\} \\ h+k\le R}} \frac{a^{*h}*b^{*k}}{h!\,k!} = 1 + \sum_{j=1}^R \frac{1}{j!} \sum_{\substack{\alpha_1,\dots,\alpha_j\in\mathbb{N} \\ \alpha_1+\cdots+\alpha_j\le R}} Z_{\alpha_1}^*(a,b)*\cdots*Z_{\alpha_j}^*(a,b),$$

$$(5.22)$$

valid for every $R \in \mathbb{N}$ and every $a,b \in A$.

For example, (5.21) becomes, for $N = 2$ and $N = 3$ (we temporarily drop the $*$ notation):

$$N = 2: \quad \frac{a^2}{2} + ab + \frac{b^2}{2} = \tfrac{1}{2}[a,b] + \tfrac{1}{2}(a+b)(a+b)$$

$$N = 3: \quad \frac{a^3}{6} + \frac{a^2b}{2} + \frac{ab^2}{2} + \frac{b^3}{6} = \tfrac{1}{12}[a,[a,b]] + \tfrac{1}{12}[b,[b,a]]$$

$$+ \tfrac{1}{6}(a+b)(a+b)(a+b) + \tfrac{1}{2}\left((a+b)\tfrac{1}{2}[a,b] + \tfrac{1}{2}[a,b](a+b)\right).$$

5.2 Convergence of the CBHD Series

The main topic of the CBHD Theorem is that it expresses $\mathrm{Log}(\mathrm{Exp}(u) \cdot \mathrm{Exp}(v))$ (an object which can be defined only in a UA algebra, for it involves powers and multiplications) as a formal power series, say $u \diamond v$, in the *Lie algebra* generated by $\{u, v\}$. But $u \diamond v$ makes sense (being a series of Lie polynomials), *mutatis mutandis*, even in an abstract Lie algebra equipped with a topology: for example in any finite dimensional Lie algebra or in any so-called *Banach-Lie algebra*. The aim of this section is to study the convergence of the series $u \diamond v$ in these contexts.

For the sake of simplicity, we first consider finite dimensional Lie algebras and then, in Sect. 5.2.2, we generalize our results to the case of Banach-Lie algebras. It is beyond our scope here to embark on an investigation of the *best* domain of convergence (a topic of very recent study in literature): the interested Reader will be referred to appropriate sources at the end of the chapter (see Sect. 5.7).

Throughout this chapter, we use the following definition.

Definition 5.9 (Normally Convergent Series of Functions). Let A be any set and let $(X, \|\cdot\|)$ be a normed space (over $\mathbb{R}$ or $\mathbb{C}$). Let $f_n : A \to X$ ($n \in \mathbb{N}$) be a sequence of functions. We say that the series of functions $\sum_{n=1}^{\infty} f_n$ *converges normally on A* if the real-valued series $\sum_{n=1}^{\infty} \sup_{a \in A} \|f_n(a)\|$ is convergent.

Moreover, with the same notation as above, we say that $\sum_{n=1}^{\infty} f_n$ is *absolutely convergent at a* if a is an element of A and if it holds that $\sum_{n=1}^{\infty} \|f_n(a)\| < \infty$. Obviously, if $\sum_{n=1}^{\infty} f_n$ converges normally on A, then it converges absolutely at every $a \in A$. Also, it is trivially seen that, if X is a Banach space and if $\sum_{n=1}^{\infty} f_n$ is absolutely convergent at $a \in A$, then the X-valued sequence $\left\{ \sum_{n=1}^{N} f_n(a) \right\}_{N \in \mathbb{N}}$ converges in X (a simple consequence of the triangle inequality and of the completeness of X); in this case we obviously set $\sum_{n=1}^{\infty} f_n(a) := \lim_{N \to \infty} \sum_{n=1}^{N} f_n(a)$.

Finally, as usual, we say that the series of functions $\sum_{n=1}^{\infty} f_n$ *converges uniformly on A to the function $f : A \to X$*, if the following fact holds: For every $\varepsilon > 0$, there exists $N_\varepsilon \in \mathbb{N}$ such that $\left\| f(a) - \sum_{n=1}^{N} f_n(a) \right\| < \varepsilon$, for every $a \in A$ and every $N \geq N_\varepsilon$.

Following the above notation, we recall the well-known result of Analysis stating that, if X is a *Banach* space, any series of functions $\sum_{n=1}^{\infty} f_n$ which is normally convergent on A is also uniformly convergent on A, to the (well defined) function $A \ni a \mapsto f(a) := \sum_{n=1}^{\infty} f_n(a) \in X$.

5.2.1 The Case of Finite Dimensional Lie Algebras

Let $\mathfrak{g}$ be a Lie algebra over a field of characteristic zero $\mathbb{K}$. We recall some definitions and notation coming from several parts of this Book. If $[\cdot, \cdot]_\mathfrak{g}$ (or simply $[\cdot, \cdot]$) denotes the Lie bracket on $\mathfrak{g}$, we set as usual

$$\left[a^{h_1} b^{k_1} \cdots a^{h_n} b^{k_n} \right]_\mathfrak{g}$$

$$:= [\underbrace{a, \cdots [a}_{h_1 \text{ times}}, \underbrace{[b, \cdots [b}_{k_1 \text{ times}}, \cdots \underbrace{[a, \cdots [a}_{h_n \text{ times}}, [\underbrace{b, [\cdots, b}_{k_n \text{ times}}]_\mathfrak{g}]_\mathfrak{g}]_\mathfrak{g} \cdots]_\mathfrak{g} \cdots]_\mathfrak{g} \cdots]_\mathfrak{g} \cdots]_\mathfrak{g} \tag{5.23}$$

for any choices of $h_1, \ldots, h_n, k_1, \ldots, k_n$ in $\mathbb{N} \cup \{0\}$ (not all vanishing simultaneously). The Reader will take care not to confuse the power-like notation "a^{h_1}" (and similar) as an effective power coming from some associative algebra: this is just a notation to mean the right-hand side of (5.23).

If $h = (h_1, \ldots, h_n)$ and $k = (k_1, \ldots, k_n)$ are multi-indices from $(\mathbb{N} \cup \{0\})^n$, with $(h, k) \neq (0, 0)$, we also set, briefly,

$$D_{(h,k)}^\mathfrak{g}(a, b) := \left[a^{h_1} b^{k_1} \cdots a^{h_n} b^{k_n} \right]_\mathfrak{g}, \qquad a, b \in \mathfrak{g}. \tag{5.24}$$

The notation $D_{(h,k)}(a, b)$ will sometimes apply as well and we shall also write

$$(h, k) = (h_1, \ldots, h_n, k_1, \ldots, k_n),$$

suppressing redundant parentheses. For example, we have (note the "intertwining" of the coordinates of h and k)

$$D_{(0,3,2,4)}(a, b) = [b[b[a[a[a[b[b[b, b]]]]]]]] = 0,$$
$$D_{(2,0,0,1)}(a, b) = [a[a, b]] = D_{(1,1,0,1)}(a, b).$$

We also recall that we introduced the useful notation

$$\mathcal{N}_n := \left\{ (h, k) \mid h, k \in (\mathbb{N} \cup \{0\})^n, \ (h_1, k_1), \ldots, (h_n, k_n) \neq (0, 0) \right\}. \quad (5.25)$$

Moreover, given $n \in \mathbb{N}$ and $(h, k) \in \mathcal{N}_n$, we also set

$$c_n := \frac{(-1)^{n+1}}{n}, \qquad \mathbf{c}(h, k) := \frac{1}{h! \, k! \, (|h| + |k|)}, \quad (5.26)$$

with $|h| = |(h_1, \ldots, h_n)| := h_1 + \ldots + h_n$, and $|k|$ analogously. The rational numbers in (5.26) have a precise sense in $\mathbb{K}$ too (for instance, $c_n = (-1)^{n+1} (n \cdot 1_{\mathbb{K}})^{-1}$ and so on).

We are thus now in a position to define a sequence of functions $\{\eta_N^{\mathfrak{g}}\}_N$ on $\mathfrak{g}$ as follows: given $N \in \mathbb{N}$, we set

$$\eta_N^{\mathfrak{g}} : \mathfrak{g} \times \mathfrak{g} \longrightarrow \mathfrak{g}$$

$$\eta_N^{\mathfrak{g}}(a, b) := \sum_{n=1}^{N} c_n \sum_{(h,k) \in \mathcal{N}_n : |h|+|k| \leq N} \mathbf{c}(h, k) \, D_{(h,k)}^{\mathfrak{g}}(a, b). \quad (5.27)$$

By reordering the summands expressing $\eta_N^{\mathfrak{g}}$ as in (3.32) on page 131, we have

$$\eta_N^{\mathfrak{g}}(a, b) = \sum_{j=1}^{N} Z_j^{\mathfrak{g}}(a, b) \quad \text{for every } a, b \in \mathfrak{g},$$

$$\text{where} \quad Z_j^{\mathfrak{g}}(a, b) := \sum_{n=1}^{j} c_n \sum_{(h,k) \in \mathcal{N}_n : |h|+|k|=j} \mathbf{c}(h, k) \, D_{(h,k)}^{\mathfrak{g}}(a, b) \quad (5.28)$$

(where terms have been grouped in "homogeneous-like" summands Z_j). In other words, $\eta_N^{\mathfrak{g}}(a, b)$ is the N-th partial sum of the series related to the summands $Z_j^{\mathfrak{g}}(a, b)$. We thus recognize that the series appearing in the CBHD Theorem is nothing but $\lim_{N \to \infty} \eta_N^{\mathfrak{g}}(a, b)$. *The problem here is to give sufficient conditions on $\mathfrak{g}$ and on $a, b \in \mathfrak{g}$ ensuring that that this limit exists in $\mathfrak{g}$.*

To this end, for the rest of this introductory section, $\mathfrak{g}$ will denote a fixed *finite dimensional Lie algebra over* $\mathbb{R}$. [Essentially, all the results can be extended to the complex case. Furthermore, in the next sections, we shall largely generalize the finite dimensional case, in considering Banach-Lie algebras.]

We denote by $m \in \mathbb{N}$ the dimension of $\mathfrak{g}$. Given a (linear) basis $\mathcal{E} = \{e_1, \ldots, e_m\}$ for $\mathfrak{g}$, we set

$$\|a\|_{\mathcal{E}} := \sqrt{(a_1)^2 + \cdots + (a_m)^2}, \qquad \begin{array}{l} \text{where } a_1, \ldots, a_m \in \mathbb{R} \text{ are such} \\ \text{that } a = a_1\, e_1 + \cdots + a_m\, e_m. \end{array} \qquad (5.29)$$

[That is, $\|\cdot\|_{\mathcal{E}}$ is the standard Euclidean norm on $\mathfrak{g}$ when this is identified with $\mathbb{R}^m$ via coordinates w.r.t. $\mathcal{E}$.] We have the following:

Lemma 5.10. *With the above notation, there exists a basis $\mathcal{E}$ for $\mathfrak{g}$ such that*

$$\big\|[a,b]_{\mathfrak{g}}\big\|_{\mathcal{E}} \leq \|a\|_{\mathcal{E}} \cdot \|b\|_{\mathcal{E}}, \qquad \forall\, a, b \in \mathfrak{g}. \qquad (5.30)$$

More precisely, given an arbitrary basis $\mathcal{M} = \{\mu_1, \ldots, \mu_m\}$ for $\mathfrak{g}$, the basis $\mathcal{E} = \{e_1, \ldots, e_m\}$ can be chosen in the following way:

$$e_k := \varepsilon\, \mu_k, \qquad k = 1, \ldots, m, \qquad (5.31)$$

where $\varepsilon > 0$ is a structural constant only depending on $\mathfrak{g}$ and $\mathcal{M}$.

Proof. [Roughly, this follows from the continuity of $(a,b) \mapsto [a,b]_{\mathfrak{g}}$ (as a bilinear map on a finite dimensional vector space!), together with a "magnification" argument. We provide a more constructive proof as follows.]

Let $\mathcal{M} = \{\mu_1, \ldots, \mu_m\}$ be any basis for $\mathfrak{g}$. There exist structural scalars $c_{i,j}^k$ such that

$$[\mu_i, \mu_j]_{\mathfrak{g}} = \textstyle\sum_{k=1}^m c_{i,j}^k\, \mu_k, \qquad \forall\, i, j \in \{1, \ldots, m\}.$$

Introducing the (skew-symmetric) matrices $C^k := (c_{i,j}^k)_{i,j \leq m}$, and denoting by $|\cdot|_{\mathrm{Eu}}$ the standard Euclidean norm on $\mathbb{R}^m$, let us denote by

$$\||C^k\|| := \max\Big\{ \big|C^k\, x\big|_{\mathrm{Eu}} : x \in \mathbb{R}^m,\ |x|_{\mathrm{Eu}} = 1 \Big\},$$

which is the usual operator norm of the matrix C^k. Then we have

$$\big\|[a,b]_{\mathfrak{g}}\big\|_{\mathcal{M}} = \bigg\| \sum_{i,j,k=1}^m a_i\, b_j\, c_{i,j}^k\, \mu_k \bigg\|_{\mathcal{M}} = \bigg(\sum_{k=1}^m \big(\textstyle\sum_{i,j=1}^m a_i\, b_j\, c_{i,j}^k\big)^2 \bigg)^{1/2}$$

$$= \bigg(\sum_{k=1}^m \big|(a_1, \ldots, a_m) \cdot C^k \cdot (b_1, \ldots, b_m)^T\big|^2 \bigg)^{1/2}$$

$$\leq \left(\sum_{k=1}^{m} |(a_1, \ldots, a_m)|_{\text{Eu}}^2 \cdot \||C^k\||^2 \cdot |(b_1, \ldots, b_m)|_{\text{Eu}}^2 \right)^{1/2}$$

$$= K \|a\|_{\mathcal{E}} \|b\|_{\mathcal{E}}, \quad \text{where} \quad K := \left(\sum_{k=1}^{m} \||C^k\||^2 \right)^{1/2},$$

where we have also set $a = \sum_{i=1}^{m} a_i \, \mu_i$, $b = \sum_{j=1}^{m} a_j \, \mu_j$.

We have thus proved the existence of $K \geq 0$ such that

$$\|[a, b]_{\mathfrak{g}}\|_{\mathcal{M}} \leq K \|a\|_{\mathcal{M}} \|b\|_{\mathcal{M}}, \qquad \forall \, a, b \in \mathfrak{g}. \tag{5.32}$$

[We can suppose that $K > 0$ since the case $K = 0$ occurs iff $\mathfrak{g}$ is Abelian, in which case the assertion of the present lemma is obvious.] Let $\mathcal{E} = \{e_1, \ldots, e_m\}$ be the basis of $\mathfrak{g}$ as in (5.31), with the choice $\varepsilon = 1/K$. We obviously have

$$\|a\|_{\mathcal{E}} = \frac{1}{\varepsilon} \|a\|_{\mathcal{M}}, \quad \forall \, a \in \mathfrak{g}. \tag{5.33}$$

Hence, by (5.32), we get

$$\|[a, b]_{\mathfrak{g}}\|_{\mathcal{E}} \overset{(5.33)}{=} \frac{1}{\varepsilon} \|[a, b]_{\mathfrak{g}}\|_{\mathcal{M}} \overset{(5.32)}{\leq} \frac{1}{\varepsilon} K \|a\|_{\mathcal{E}} \|b\|_{\mathcal{E}} \quad (\text{recall that } K = 1/\varepsilon)$$

$$= \frac{1}{\varepsilon} \|a\|_{\mathcal{M}} \cdot \frac{1}{\varepsilon} \|b\|_{\mathcal{M}} \overset{(5.33)}{=} \|a\|_{\mathcal{E}} \cdot \|b\|_{\mathcal{E}},$$

and (5.30) follows. $\qquad\qquad\square$

We henceforth fix any basis $\mathcal{E}$ for $\mathfrak{g}$ such that (5.30) holds. The existence of at least one basis with this property follows from the above lemma. We denote the associated norm $\| \cdot \|_{\mathcal{E}}$ simply by $\| \cdot \|$. We thus have

$$\|[a, b]_{\mathfrak{g}}\| \leq \|a\| \cdot \|b\|, \qquad \text{for every } a, b \in \mathfrak{g}. \tag{5.34}$$

By the definition of $D^{\mathfrak{g}}$ in (5.23)–(5.24), an inductive argument *based only on* (5.34) proves that

$$\|D_{(h,k)}^{\mathfrak{g}}(a, b)\| \leq \|a\|^{|h|} \cdot \|b\|^{|k|}, \qquad \begin{matrix} \text{for every } a, b \in \mathfrak{g} \\ \text{and every } (h, k) \in \mathcal{N}_n. \end{matrix} \tag{5.35}$$

We are thus in a position to prove a fundamental estimate concerning the CBHD series.

Theorem 5.11 (Fundamental Estimate). *Let* $\mathfrak{g}$ *be a real Lie algebra of finite dimension. Let* $\| \cdot \|$ *be a norm on* $\mathfrak{g}$ *satisfying* (5.34).

Then, for every $N \in \mathbb{N}$ and every $a, b \in \mathfrak{g}$, we have the estimate

$$\sum_{n=1}^{N} |c_n| \sum_{(h,k) \in \mathcal{N}_n : |h|+|k| \leq N} \mathbf{c}(h,k) \|D^{\mathfrak{g}}_{(h,k)}(a,b)\|$$

$$\leq \sum_{n=1}^{N} \frac{1}{n} \left(e^{\|a\|} e^{\|b\|} - 1 \right)^n. \tag{5.36}$$

In view of the definitions of $\eta_N^{\mathfrak{g}}$ and of $Z_j^{\mathfrak{g}}$ in (5.27) and (5.28) respectively, the above theorem immediately implies the following corollary.

Corollary 5.12. *Let $\mathfrak{g}$ be a real Lie algebra of finite dimension. Let $\| \cdot \|$ be a norm on $\mathfrak{g}$ satisfying (5.34). Finally, let $\eta_N^{\mathfrak{g}}(a,b)$ and $Z_j^{\mathfrak{g}}(a,b)$ be as in (5.27) and (5.28). Then, for every $N \in \mathbb{N}$ and every $a, b \in \mathfrak{g}$, we have the estimates*

$$\left.\begin{array}{r}\|\eta_N^{\mathfrak{g}}(a,b)\| \\[1em] \sum_{j=1}^{N} \|Z_j^{\mathfrak{g}}(a,b)\|\end{array}\right\} \leq \sum_{n=1}^{N} \frac{1}{n} \left(e^{\|a\|} e^{\|b\|} - 1 \right)^n. \tag{5.37}$$

Remark 5.13. The hypothesis of finite dimensionality of $\mathfrak{g}$ in the above Theorem 5.11 and Corollary 5.12 are only temporary: they will be dropped in the next Sect. 5.2.2, provided a norm as in (5.34) exists.

Remark 5.14. For future repeated references, we prove the following identity:

$$\sum_{(h,k) \in \mathcal{N}_n} \frac{A^{|h|} B^{|k|}}{h! \, k!} = (e^{A+B} - 1)^n, \tag{5.38}$$

for every $n \in \mathbb{N}$ and every $A, B \in \mathbb{R}$.

Indeed, it holds that

$$\sum_{(h,k) \in \mathcal{N}_n} \frac{A^{|h|} B^{|k|}}{h! \, k!} = \left(\sum_{i,j \in \mathbb{N} \cup \{0\} : (i,j) \neq (0,0)} \frac{A^i B^j}{i! \, j!} \right)^n$$

$$= \left(\sum_{i \geq 0} \frac{A^i}{i!} \cdot \sum_{j \geq 0} \frac{B^j}{j!} - 1 \right)^n = \left(e^A e^B - 1 \right)^n.$$

Proof (of Theorem 5.11). In view of (5.35), the left-hand side of (5.36) is bounded above by the sum

$$\sum_{n=1}^{N} |c_n| \sum_{(h,k) \in \mathcal{N}_n : |h|+|k| \leq N} \mathbf{c}(h,k) \|a\|^{|h|} \cdot \|b\|^{|k|}$$

$$\overset{(5.26)}{=} \sum_{n=1}^{N} \frac{1}{n} \sum_{(h,k)\in\mathbb{N}_n \,:\, |h|+|k|\leq N} \frac{\|a\|^{|h|} \cdot \|b\|^{|k|}}{h!\,k!\,(|h|+|k|)}$$

$\Big($ we obtain an upper estimate, by erasing the denominator $|h| + |k| \geq 1$

and by dropping the condition "$|h| + |k| \leq N$" in the inner sum $\Big)$

$$\leq \sum_{n=1}^{N} \frac{1}{n} \sum_{(h,k)\in\mathbb{N}_n} \frac{\|a\|^{|h|} \cdot \|b\|^{|k|}}{h!\,k!} \overset{(5.38)}{=} \sum_{n=1}^{N} \frac{1}{n} \left(e^{\|a\|}\, e^{\|b\|} - 1\right)^n.$$

This completes the proof. □

Remark 5.15. Let $\mathfrak{g}$ be a finite dimensional real Lie algebra. Let $\|\cdot\|_*$ be any norm on $\mathfrak{g}$. Let $\|\cdot\|$ be a norm on $\mathfrak{g}$ such that (5.34) holds. Since all norms on a finite dimensional real vector space are equivalent,[2] it follows from (5.34) that there exists $M > 0$ such that

$$\left\|[a,b]_{\mathfrak{g}}\right\|_* \leq M \, \|a\|_* \cdot \|b\|_*, \qquad \text{for every } a, b \in \mathfrak{g}. \tag{5.39}$$

(Equivalently, (5.39) follows from the continuity of the bilinear map $(a, b) \mapsto [a, b]_{\mathfrak{g}}$). Then, by multiplying both sides of (5.39) times M, we see that the norm

$$\|a\|_\star := M \, \|a\|_*, \qquad a \in \mathfrak{g}$$

---

[2]To prove the equivalence of all norms on a finite-dimensional real vector space, it suffices to prove that, given a norm $\|\cdot\|_*$ on $\mathfrak{g}$, there exist constants $\alpha, \beta > 0$ such that

$$\alpha \, \|a\|_\varepsilon \leq \|a\|_* \leq \beta \, \|a\|_\varepsilon, \qquad \text{for every } a, b \in \mathfrak{g},$$

where $\|\cdot\|_\varepsilon$ is the norm in (5.29), related to some fixed basis $\mathcal{E} = \{e_1, \ldots, e_m\}$ for $\mathfrak{g}$. In turn, since a norm is homogeneous, the above inequalities are equivalent to

$$\alpha \leq \|\xi\|_* \leq \beta, \qquad \text{for every } \xi \in \mathfrak{g} \text{ such that } \|\xi\|_\varepsilon = 1.$$

Let $\mathcal{T}$ be the topology on $\mathfrak{g}$ induced by $\|\cdot\|_\varepsilon$. Since $K := \{\xi \in \mathfrak{g} : \|\xi\|_\varepsilon = 1\}$ is obviously a compact subset of $\mathfrak{g}$ w.r.t. $\mathcal{T}$ (indeed $\mathfrak{g} \simeq \mathbb{R}^m$ via $\mathcal{E}$, and $\mathcal{T}$ is the Euclidean norm related to $\mathcal{E}$), and $0 \notin K$, the last inequalities follow from an application of the Weierstrass Theorem to the $\mathcal{T}$-continuous function $\|\cdot\|_*$ which is strictly positive away from the origin. The cited $\mathcal{T}$-continuity of $\|\cdot\|_*$ derives from the following computation:

$$\|x^n - x^0\|_* = \left\|(x_1^n - x_1^0)\,e_1 + \cdots + (x_m^n - x_m^0)\,e_m\right\|_*$$

$$\leq |x_1^n - x_1^0| \cdot \|e_1\|_* + \cdots + |x_m^n - x_m^0| \cdot \|e_m\|_* \xrightarrow{n\to\infty} 0,$$

whenever $x^n \to x^0$ with respect to $\|\cdot\|_\varepsilon$ (which means in fact $|x_i^n - x_i^0| \xrightarrow{n\to\infty} 0$ for every $i = 1, \ldots, m$). □

satisfies (5.34) too. Hence, Theorem 5.11 and Corollary 5.12 hold by replacing $\|\cdot\|$ with $\|\cdot\|_*$. This results in:

Corollary 5.16. *Let $\mathfrak{g}$ be a real Lie algebra of finite dimension. Let $\|\cdot\|_*$ be any norm on $\mathfrak{g}$ and let M be a positive constant satisfying (5.39). Then, for every $N \in \mathbb{N}$ and every $a, b \in \mathfrak{g}$, the quantities*

$$\|\eta_N^{\mathfrak{g}}(a,b)\|_*, \qquad \sum_{j=1}^N \|Z_j^{\mathfrak{g}}(a,b)\|_*,$$

$$\sum_{n=1}^N |c_n| \sum_{(h,k)\in\mathcal{N}_n \,:\, |h|+|k|\leq N} \mathbf{c}(h,k) \|D_{(h,k)}^{\mathfrak{g}}(a,b)\|_* \tag{5.40}$$

are all bounded above by

$$M^{-1} \sum_{n=1}^N \frac{1}{n} \left(e^{M(\|a\|_* + \|b\|_*)} - 1\right)^n.$$

Here, as usual, $\eta_N^{\mathfrak{g}}(a,b)$, $Z_j^{\mathfrak{g}}(a,b)$ and $D_{(h,k)}^{\mathfrak{g}}(a,b)$ are as in (5.27), (5.28) and (5.24).

We can now derive from Theorem 5.11 a result on the convergence of the CBHD series $\sum_{j=1}^\infty Z_j^{\mathfrak{g}}(a,b)$.

This is based on either of the following simple estimates[3]

$$\sum_{n=1}^N \frac{1}{n} \left(e^A e^B - 1\right)^n \leq \begin{cases} \displaystyle\sum_{n=1}^\infty \frac{1}{n} \left(e^A e^B - 1\right)^n = -\log(2 - e^{A+B}) \\[4mm] \displaystyle\sum_{n=1}^\infty \left(e^A e^B - 1\right)^n = \dfrac{e^{A+B}-1}{2-e^{A+B}} \end{cases} \tag{5.41}$$

holding true for real numbers A, B such that $|e^{A+B} - 1| < 1$, that is, when $A + B < \log 2$. Obviously, the former estimate gives a sharper bound, for $\log(1/(1-x)) \leq x/(1-x)$ for every $x \in [-1, 1]$.

To derive the cited result on the convergence of the CBHD series, suppose that $a, b \in \mathfrak{g}$ are such that

$$\|a\| + \|b\| < \log 2,$$

where $\|\cdot\|$ is any norm on $\mathfrak{g}$ satisfying (5.34). Then, by (5.41), the series

---

[3]Here, we used the following well-known Maclaurin expansions: $\sum_{n=1}^\infty \frac{x^n}{n} = -\log(1-x)$ (valid for $-1 \leq x < 1$), and $\sum_{n=1}^\infty x^n = \frac{x}{1-x}$ (valid for $-1 < x < 1$).

$$\sum_{n=1}^{\infty} \frac{1}{n} \left(e^{\|a\|} e^{\|b\|} - 1 \right)^n$$

converges, and in fact it equals $-\log(2 - e^{\|a\|+\|b\|})$. Hence, by (5.37) in Corollary 5.12, we are easily able to show that the sequence $\sum_{j=1}^{N} Z_j^{\mathfrak{g}}(a,b) = \eta_N^{\mathfrak{g}}(a,b)$ is Cauchy. Since $\mathfrak{g}$ equipped with the norm $\|\cdot\|$ is a *Banach space*, i.e., a complete metric space (recall that $\mathfrak{g}$ is a finite dimensional real vector space and all norms are equivalent on $\mathfrak{g}$), this condition ensures the convergence in $(\mathfrak{g}, \|\cdot\|)$ of the series $\sum_{j=1}^{\infty} Z_j^{\mathfrak{g}}(a,b)$.

Indeed, we demonstrate that $\{\eta_N^{\mathfrak{g}}(a,b)\}_N$ is a Cauchy sequence: we have

$$\left\| \eta_{n+p}^{\mathfrak{g}}(a,b) - \eta_n^{\mathfrak{g}}(a,b) \right\| = \left\| \sum_{j=1}^{n+p} Z_j^{\mathfrak{g}}(a,b) - \sum_{j=1}^{n} Z_j^{\mathfrak{g}}(a,b) \right\| \tag{5.42}$$

$$= \left\| \sum_{j=n+1}^{n+p} Z_j^{\mathfrak{g}}(a,b) \right\| \leq \sum_{j=n+1}^{n+p} \left\| Z_j^{\mathfrak{g}}(a,b) \right\| \leq \sum_{j=n+1}^{\infty} \left\| Z_j^{\mathfrak{g}}(a,b) \right\|,$$

and the far right-hand side vanishes as $n \to \infty$, since it is the n-th remainder of a convergent series.

This proves that for every $a, b \in \mathfrak{g}$ with $\|a\| + \|b\| < \log 2$, there exists

$$a \diamond b := \lim_{N \to \infty} \eta_N^{\mathfrak{g}}(a,b) = \sum_{j=1}^{\infty} Z_j^{\mathfrak{g}}(a,b).$$

Moreover, we have the estimate

$$\|a \diamond b\| = \left\| \sum_{j=1}^{\infty} Z_j^{\mathfrak{g}}(a,b) \right\| \leq \sum_{j=1}^{\infty} \left\| Z_j^{\mathfrak{g}}(a,b) \right\|$$

$$\text{(thanks to (5.37) and (5.41))}$$

$$\leq \log \left(\frac{1}{2 - e^{\|a\|+\|b\|}} \right).$$

Collecting all the above facts, we have proved the following result.

Theorem 5.17. (Convergence of the CBHD series for finite dimensional Lie algebras). *Let $\mathfrak{g}$ be a finite dimensional real Lie algebra. Let $\|\cdot\|$ be a norm on $\mathfrak{g}$ satisfying (5.34). If $Z_j^{\mathfrak{g}}(a,b)$ is as in (5.28), let us set*

$$a \diamond b := \sum_{j=1}^{\infty} Z_j^{\mathfrak{g}}(a,b), \quad \text{whenever this series converges in } \mathfrak{g}, \tag{5.43}$$

$\mathfrak{g}$ *being equipped with the Banach space structure related to the norm $\|\cdot\|$ (or to any other norm on $\mathfrak{g}$, since all norms on $\mathfrak{g}$ are equivalent).*

Then a sufficient condition for the existence of $a \diamond b$ is that the couple (a, b) belongs to the "diagonal square"

$$D := \left\{ (a, b) \in \mathfrak{g} \times \mathfrak{g} : \|a\| + \|b\| < \log 2 \right\}. \tag{5.44}$$

For example, this is the case if a, b belong to the "disc" centred at the origin

$$Q := \left\{ a \in \mathfrak{g} : \|a\| < \tfrac{1}{2} \log 2 \right\}, \tag{5.45}$$

since $Q \times Q \subset D$. Moreover, the following inequality holds

$$\|a \diamond b\| \leq \log \left(\frac{1}{2 - e^{\|a\| + \|b\|}} \right), \qquad \text{for every } (a, b) \in D. \tag{5.46}$$

Finally, the series $\sum_{j=1}^{\infty} Z_j^{\mathfrak{g}}(a, b)$ converges normally[4] on every set of the type

$$\left\{ (a, b) \in \mathfrak{g} \times \mathfrak{g} : \|a\| + \|b\| \leq \delta \right\}, \qquad \text{with } \delta < \log 2.$$

Note that, modulo the "compatibility" of the norm $\| \cdot \|$ with the Lie algebra structure of $\mathfrak{g}$ (this just means that (5.34) holds), the disc Q with centre 0 (on which we have been able to prove that the CBHD series converges) has the "universal" radius

$$\tfrac{1}{2} \log 2 \approx 0.3465735902 \ldots$$

Remark 5.18. Obviously, the condition $(a, b) \in D$ is by no means necessary for the convergence of the series expressing $a \diamond b$. Indeed, if $a \in \mathfrak{g}$ is arbitrary and $b = 0$ we have $Z_1^{\mathfrak{g}}(a, 0) = a + 0$ and $Z_j^{\mathfrak{g}}(a, 0) = 0$ for every $j \geq 2$, so that $\sum_{j=1}^{\infty} Z_j^{\mathfrak{g}}(a, 0) = a$ converges. Analogously, if $b = -a$ we have $Z_1^{\mathfrak{g}}(a, -a) = a - a$ and $Z_j^{\mathfrak{g}}(a, -a) = 0$ for every $j \geq 2$, so that $\sum_{j=1}^{\infty} Z_j^{\mathfrak{g}}(a, -a) = 0$ converges too, whatever the choice of $a \in \mathfrak{g}$.

Remark 5.19. As argued in Remark 5.15, if $\mathfrak{g}$ is as in the above theorem and if $\| \cdot \|_*$ is any norm on $\mathfrak{g}$, the above results hold true by replacing D and Q respectively by the sets

$$D_* = \left\{ (a, b) \in \mathfrak{g} \times \mathfrak{g} : \|a\| + \|b\| < \frac{\log 2}{M} \right\},$$

$$Q_* = \left\{ a \in \mathfrak{g} : \|a\|_* < \frac{\log 2}{2M} \right\},$$

where $M > 0$ is a constant satisfying (5.39). Moreover, the estimate (5.46) holds in the following form

---

[4]Recall Definition 5.9, page 278.

$$\|a \diamond b\|_* \leq \frac{1}{M} \log\left(\frac{1}{2 - \exp\left(M\|a\|_* + M\|b\|_*\right)}\right), \qquad \text{for every } (a,b) \in D_*.$$

Remark 5.20. The estimate (5.46) has two important consequences:

1. *The function $D \ni (a,b) \mapsto a \diamond b \in \mathfrak{g}$ is continuous at $(0,0)$.* Indeed, we have

$$\|a \diamond b - 0 \diamond 0\| = \|a \diamond b\| \leq \log\left(\frac{1}{2 - e^{\|a\| + \|b\|}}\right)$$

$$\xrightarrow{(a,b) \to (0,0)} \log\left(\frac{1}{2 - 1}\right) = 0.$$

Obviously, more is true: the above function is continuous on the whole of D, since it is the sum of a series (of continuous functions) which converges *normally* (hence uniformly) on every set of the type

$$\left\{(a,b) \in \mathfrak{g} \times \mathfrak{g} : \|a\| + \|b\| \leq \delta\right\},$$

where $\delta < \log 2$. Note that the partial sums $\eta_N^{\mathfrak{g}}(a,b) = \sum_{j=1}^{N} Z_j^{\mathfrak{g}}(a,b)$ are continuous functions on $\mathfrak{g} \times \mathfrak{g}$, for the $Z_j^{\mathfrak{g}}$ are Lie polynomials.

As we shall see in the next theorem, $\diamond$ is much more than continuous: it is indeed real analytic in a neighborhood of $(0,0)$.

2. *By shrinking Q, $\diamond$ can be iterated.* More precisely, if a, b belong to

$$\widetilde{Q} := \left\{a \in \mathfrak{g} : \|a\| < \tfrac{1}{2} \log\left(2 - 1/\sqrt{2}\right)\right\},$$

then $a \diamond b \in Q$. Indeed, if $a, b \in \widetilde{Q}$, we have (notice that $\widetilde{Q} \subset Q \subset D$)

$$\|a \diamond b\| \overset{(5.46)}{\leq} \log\left(\frac{1}{2 - e^{\|a\| + \|b\|}}\right)$$

$$\leq \log\left(\frac{1}{2 - \exp\log\left(2 - 1/\sqrt{2}\right)}\right) = \frac{\log 2}{2},$$

that is, by definition of Q, $a \diamond b \in Q$. As a consequence we deduce that $(a \diamond b) \diamond c$ and $a \diamond (b \diamond c)$ make sense, whenever $a, b, c \in \widetilde{Q}$. The comparison of these latter two elements of $\mathfrak{g}$, i.e., the study of the associativity of the "local operation" $\diamond$ will be considered in Sect. 5.3.

Note that, just as for Q, the disc $\widetilde{Q}$ has the "universal" radius

$$\tfrac{1}{2}\log\left(2 - \frac{1}{\sqrt{2}}\right) \approx 0.1284412561\ldots$$

We have the following remarkable result. Actually, the hypothesis of finite dimensionality of $\mathfrak{g}$ is not necessary, as we will show in Sect. 5.5.1.

Theorem 5.21 (Real Analyticity of the $\diamond$ Function). *Suppose that the hypotheses of Theorem 5.17 hold true. Let Q be as in (5.45). Then there exists an open neighborhood $\mathfrak{Q}$ of $0 \in \mathfrak{g}$ contained in Q such that the map*

$$\mathfrak{Q} \times \mathfrak{Q} \ni (a, b) \mapsto a \diamond b \in \mathfrak{g}$$

is real analytic.

Here in referring to real-analyticity, by the aid of any basis for $\mathfrak{g}$ (recall that $m := \dim \mathfrak{g} < \infty$), we are identifying $\mathfrak{g} \times \mathfrak{g}$ with $\mathbb{R}^{2m}$ and $\mathfrak{g}$ with $\mathbb{R}^m$ and the term "real-analytic" inherits the obvious meaning.

Before embarking on the proof of the theorem, we need to recall what is meant by an analytic function on an open set $\mathcal{U} \subseteq \mathbb{R}^{2m}$: This is a function $f(x, y)$ in $C^\infty(\mathcal{U}, \mathbb{R})$ (x, y denote coordinates in $\mathbb{R}^m$) such that for every $(a^0, b^0) \in \mathcal{U}$ there exists $r_0 > 0$ such that the sum of the series

$$\sum_{N=0}^{\infty} \sum_{\substack{\alpha,\beta \in (\mathbb{N}\cup\{0\})^m \\ \alpha_1+\cdots+\alpha_m+\beta_1+\cdots+\beta_m=N}} \left| \frac{\partial^N f(a^0, b^0)}{\partial x_1^{\alpha_1} \cdots \partial x_m^{\alpha_m} \partial y_1^{\beta_1} \cdots \partial y_m^{\beta_m}} \right|$$

$$\times |a_1 - a_1^0|^{\alpha_1} \cdots |a_m - a_m^0|^{\alpha_m} \cdot |b_1 - b_1^0|^{\beta_1} \cdots |b_m - b_m^0|^{\beta_m}$$

is finite (the way summands are arranged is immaterial, since the summands are nonnegative) and it holds that

$$f(a, b) = \sum_{N=0}^{\infty} \sum_{\substack{\alpha,\beta \in (\mathbb{N}\cup\{0\})^m \\ \alpha_1+\cdots+\alpha_m+\beta_1+\cdots+\beta_m=N}} \frac{\partial^N f(a^0, b^0)}{\partial x_1^{\alpha_1} \cdots \partial x_m^{\alpha_m} \partial y_1^{\beta_1} \cdots \partial y_m^{\beta_m}}$$

$$\times (a_1 - a_1^0)^{\alpha_1} \cdots (a_m - a_m^0)^{\alpha_m} \cdot (b_1 - b_1^0)^{\beta_1} \cdots (b_m - b_m^0)^{\beta_m},$$

for every $(a, b) \in \mathcal{U}$ satisfying $\sum_{j=1}^m (|a_j - a_j^0| + |b_j - b_j^0|) < r_0$.

A sufficient condition for real analyticity is the following one: there exists $\varepsilon > 0$ such that, if I is the real interval $(-\varepsilon, \varepsilon)$, we have

$$f(a, b) = \sum_{N=0}^{\infty} \sum_{\substack{\alpha,\beta \in (\mathbb{N}\cup\{0\})^m \\ \alpha_1+\cdots+\alpha_m+\beta_1+\cdots+\beta_m=N}} C_{\alpha,\beta}\, a_1^{\alpha_1} \cdots a_m^{\alpha_m}\, b_1^{\beta_1} \cdots b_m^{\beta_m},$$

for every $a, b \in I^m$ (the m-fold Cartesian product of I with itself), where the $C_{\alpha,\beta}$ are real constants such that

$$\sum_{N=0}^{\infty} \sum_{\substack{\alpha,\beta \in (\mathbb{N} \cup \{0\})^m \\ \alpha_1 + \cdots + \alpha_m + \beta_1 + \cdots + \beta_m = N}} |C_{\alpha,\beta}| \, \varepsilon^N < \infty.$$

In this case f is real analytic on I^{2m} (the $2m$-fold Cartesian product of I with itself). This is the case for example if f is defined as the series related to a sequence of functions $\{z_j(a,b)\}_{j \geq 1}$ which are polynomials in $a, b \in \mathbb{R}^m$ of common degree j, with

$$z_j(a,b) = \sum_{\substack{(\alpha,\beta) \in (\mathbb{N} \cup \{0\})^{2m} \\ \alpha_1 + \cdots + \alpha_m + \beta_1 + \cdots + \beta_m = j}} C_{\alpha,\beta} \, a_1^{\alpha_1} \cdots a_m^{\alpha_m} b_1^{\beta_1} \cdots b_m^{\beta_m} \tag{5.47}$$

and the constants $C_{\alpha,\beta}$ satisfy

$$\sum_{j=1}^{\infty} \sum_{\substack{\alpha,\beta \in (\mathbb{N} \cup \{0\})^m \\ \alpha_1 + \cdots + \alpha_m + \beta_1 + \cdots + \beta_m = j}} |C_{\alpha,\beta}| \, \varepsilon^j < \infty. \tag{5.48}$$

Proof (of Theorem 5.21). Let us fix a basis $\mathcal{E} = \{e_1, \ldots, e_m\}$ as in Lemma 5.10. We denote $\|\cdot\|_{\mathcal{E}}$ simply by $\|\cdot\|$ and, for every fixed $i = 1, \ldots, m$, we introduce the projection (onto the i-th component w.r.t. $\mathcal{E}$)

$$\Pi_i : \mathfrak{g} \longrightarrow \mathbb{R}, \qquad \Pi_i(a_1 \, e_1 + \cdots + a_m \, e_m) := a_i \quad (a_1, \ldots, a_m \in \mathbb{R}).$$

We also identify $\mathfrak{g}$ with $\mathbb{R}^m$ via the map

$$\Pi : \mathfrak{g} \longrightarrow \mathbb{R}^m, \qquad \Pi(g) := \big(\Pi_1(g), \ldots, \Pi_m(g)\big),$$

and $\mathfrak{g} \times \mathfrak{g}$ with $\mathbb{R}^{2m}$, accordingly.

We have to find an open neighborhood $\mathcal{Q}$ of $0 \in \mathfrak{g}$ such that for $\ell = 1, \ldots, m$ all the functions

$$\mathbb{R}^{2m} \ni (a,b) \mapsto P_\ell(a,b) := \Pi_\ell\big((\Pi^{-1}(a)) \diamond (\Pi^{-1}(b))\big) \in \mathbb{R}$$

are real analytic on $\Pi(\mathcal{Q}) \times \Pi(\mathcal{Q}) \subset \mathbb{R}^{2m}$.

Since $\mathfrak{g}$ is a real Lie algebra, there exist real numbers $c_{i,j}^k$ such that

$$[e_i, e_j]_{\mathfrak{g}} = \sum_{k=1}^{m} c_{i,j}^k \, e_k, \quad \forall \, i, j \in \{1, \ldots, m\}.$$

We denote by c the positive constant

$$\mathbf{c} := 1 + \max\left\{|c_{i,j}^k| \;:\; i,j,k = 1,\ldots,m\right\}. \tag{5.49}$$

[Note that c depends on $\mathfrak{g}$ and $\mathcal{E}$.] An inductive argument shows that one has

$$[v^N, [v^{N-1} \cdots [v^2, v^1] \cdots]]_{\mathfrak{g}} = \sum_{\ell=1}^m e_\ell$$

$$\times \left(\sum_{s_1,\ldots,s_{N-2}=1}^m \; \sum_{i_1,\ldots,i_N=1}^m v_{i_N}^N \cdots v_{i_1}^1 c_{i_N,s_{N-2}}^\ell c_{i_{N-1},s_{N-3}}^{s_{N-2}} \cdots c_{i_3,s_1}^{s_2} c_{i_2,i_1}^{s_1} \right) \tag{5.50}$$

for every choice of $v^1,\ldots,v^N$ in $\mathfrak{g}$; here we have set $v^k = v_1^k e_1 + \ldots + v_m^k e_m$. If $a, b \in \mathbb{R}^m$ and $\bar{a} := \Pi^{-1}(a)$, $\bar{b} := \Pi^{-1}(b)$, we have (for every $(h,k) \in \mathcal{N}_n$)

$$D_{(h,k)}^{\mathfrak{g}}(\bar{a},\bar{b}) = \left[\underbrace{\bar{a}\cdots\bar{a}}_{h_1} \underbrace{\bar{b}\cdots\bar{b}}_{k_1} \cdots \underbrace{\bar{a}\cdots\bar{a}}_{h_n} \underbrace{\bar{b}\cdots\bar{b}}_{k_n} \right]_{\mathfrak{g}}$$

$$= \sum_{\ell=1}^m e_\ell \sum_{1\le is,\,js\le m} a_{i_1^1} \cdots a_{i_{h_1}^1} b_{j_1^1} \cdots b_{j_{k_1}^1} \cdots a_{i_1^n} \cdots a_{i_{h_n}^n} b_{j_1^n} \cdots b_{j_{k_n}^n}$$

$$\times \sum_{1\le s_1,\ldots,s_{|h|+|k|-2}\le m} \left\{ \begin{array}{c} \text{a certain product} \\ \text{(indexed over } \ell \text{ and over the } i,j,s) \\ \text{of } |h|+|k|-1 \text{ structure constants of type } c \end{array} \right\}.$$

This gives a representation of $\Pi_\ell\big(Z_j^{\mathfrak{g}}(a,b)\big)$ of the following type:

$$\sum_{n=1}^j c_n \sum_{(h,k)\in\mathcal{N}_n\,:\,|h|+|k|=j} \mathbf{c}(h,k) \sum_{\substack{1\le i_1^1,\ldots,i_{h_1}^1,\ldots,i_1^n,\ldots,i_{h_n}^n \\ j_1^1,\ldots,j_{k_1}^1,\ldots,j_1^n,\ldots,j_{k_n}^n}\le m}$$

$$\times a_{i_1^1} \cdots a_{i_{h_1}^1} b_{j_1^1} \cdots b_{j_{k_1}^1} \cdots a_{i_1^n} \cdots a_{i_{h_n}^n} b_{j_1^n} \cdots b_{j_{k_n}^n} \times$$

$$\times \sum_{1\le s_1,\ldots,s_{|h|+|k|-2}\le m} \left\{ \begin{array}{c} \text{a certain product} \\ \text{(indexed over } \ell \text{ and over the } i,j,s) \\ \text{of } |h|+|k|-1 \text{ structure constants of type } c \end{array} \right\}.$$

This is actually a decomposition of $z_j(a,b) := \Pi_\ell\big(Z_j^{\mathfrak{g}}(a,b)\big)$ in sums of monomials in the (a,b)-coordinates where the same monomial may appear several times. We group monomials together to produce a representation of $z_j(a,b)$ as in (5.47). Hence, in view of the triangle inequality for the absolute value in $\mathbb{R}$, we certainly provide an upper bound for the series in (5.48), if we estimate the following series of nonnegative summands

$$\sum_{j=1}^{\infty}\sum_{n=1}^{j}|c_n| \sum_{(h,k)\in\mathcal{N}_n \,:\, |h|+|k|=j} \mathbf{c}(h,k) \sum_{1\le \frac{i_1^1,\dots,i_{h_1}^1,\cdots,i_1^n,\dots,i_{h_n}^n}{j_1^1,\dots,j_{k_1}^1,\cdots,j_1^n,\dots,j_{k_n}^n}\le m} \varepsilon^j$$

$$\times \sum_{1\le s_1,\dots,s_{|h|+|k|-2}\le m} \left| \begin{array}{c} \text{a certain product} \\ \text{(indexed over } \ell \text{ and over the } i,\,j,\,s) \\ \text{of } |h|+|k|-1 \text{ structure constants of type } c \end{array} \right| .$$

Taking into account the constant $\mathbf{c}$ in (5.49), the above series is bounded above by

$$\sum_{j=1}^{\infty}\sum_{n=1}^{j}\frac{1}{n} \sum_{(h,k)\in\mathcal{N}_n \,:\, |h|+|k|=j} \frac{1}{h!\,k!}$$

$$\times \sum_{1\le \frac{i_1^1,\dots,i_{h_1}^1,\cdots,i_1^n,\dots,i_{h_n}^n}{j_1^1,\dots,j_{k_1}^1,\cdots,j_1^n,\dots,j_{k_n}^n}\le m} \varepsilon^j \sum_{1\le s_1,\dots,s_{|h|+|k|-2}\le m} c^{|h|+|k|-1}$$

$$\le \sum_{j=1}^{\infty}\sum_{n=1}^{j}\frac{1}{n} \sum_{(h,k)\in\mathcal{N}_n \,:\, |h|+|k|=j} \frac{1}{h!\,k!}\, m^{|h|+|k|}\, \varepsilon^j\, m^{|h|+|k|-2}\, c^{|h|+|k|-1}$$

$$= \frac{1}{\mathbf{c}\,m^2}\sum_{j=1}^{\infty}\sum_{n=1}^{j}\frac{1}{n} \sum_{(h,k)\in\mathcal{N}_n \,:\, |h|+|k|=j} \frac{1}{h!\,k!}\,(m^2\,\varepsilon\,\mathbf{c})^{|h|+|k|}$$

(interchange the j and n sums and recall that $|h|+|k|\ge n$ for $(h,k)\in\mathcal{N}_n$)

$$= \frac{1}{\mathbf{c}\,m^2}\sum_{n=1}^{\infty}\frac{1}{n} \sum_{(h,k)\in\mathcal{N}_n} \frac{1}{h!\,k!}\,(m^2\,\varepsilon\,\mathbf{c})^{|h|+|k|}$$

$$\overset{(5.38)}{=} \frac{1}{\mathbf{c}\,m^2}\sum_{n=1}^{\infty}\frac{1}{n}\Big(\exp(2\,m^2\,\varepsilon\,\mathbf{c})-1\Big)^{n}.$$

The far right-hand series converges provided $|\exp(2\,m^2\,\varepsilon\,\mathbf{c})-1|<1$, that is, by choosing

$$\varepsilon < \frac{\log 2}{2\,\mathbf{c}\,m^2}.$$

With this choice of ε, by the remarks on real analyticity preceding this proof, we infer that $P_\ell(a,b)$ is real analytic on the $2m$-fold Cartesian product of the interval $(-\varepsilon, \varepsilon)$. As a consequence, the assertion of Theorem 5.21 follows by choosing

$$\mathfrak{Q} = \left\{ g \in \mathfrak{g} \,\middle|\, g = a_1\, e_1 + \cdots + a_m\, e_m \; : \; |a_1|, \ldots, |a_m| < \varepsilon \right\}.$$

This ends the proof.   □

5.2.2 The Case of Banach-Lie Algebras

In this section we generalize the results of Sect. 5.2.1 (except for those on real analyticity) to the wider setting of so-called Banach-Lie algebras. To this end we need some definitions which will be used also in the next sections. We declare the following convention:

Convention: *Throughout this section, $\mathbb{K}$ denotes the field of real numbers $\mathbb{R}$ or of complex numbers $\mathbb{C}$.*

Definition 5.22. The following definitions are given:

Normed algebra: A triple $(A, *, \|\cdot\|)$ is a *normed algebra* if $(A, *)$ is a unital associative algebra over $\mathbb{K}$ and if $(A, \|\cdot\|)$ is a normed vector space over $\mathbb{K}$ with the following property: there exists $M > 0$ such that

$$\|x * y\| \le M \, \|x\| \cdot \|y\|, \qquad \text{for every } x, y \in A. \tag{5.51a}$$

If the constant in (5.51a) can be chosen equal to 1, that is, if

$$\|x * y\| \le \|x\| \cdot \|y\|, \qquad \text{for every } x, y \in A, \tag{5.51b}$$

then $\|\cdot\|$ is said to be *compatible* with the multiplication $*$ of A (or just $\|\cdot\|$ is *compatible with A*).

Banach algebra: A normed algebra $(A, *, \|\cdot\|)$ is called a *Banach algebra*, if the normed space $(A, \|\cdot\|)$ is complete.

Normed Lie algebra: A triple $(\mathfrak{g}, [\cdot,\cdot]_\mathfrak{g}, \|\cdot\|)$ is a *normed Lie algebra* if $(\mathfrak{g}, [\cdot,\cdot]_\mathfrak{g})$ is a Lie algebra over $\mathbb{K}$ and if $(\mathfrak{g}, \|\cdot\|)$ is a normed vector space over $\mathbb{K}$ with the following property: there exists $M > 0$ such that

$$\left\|[x, y]_\mathfrak{g}\right\| \le M \, \|x\| \cdot \|y\|, \qquad \text{for every } x, y \in \mathfrak{g}. \tag{5.51c}$$

If the constant in (5.51c) can be chosen equal to 1, that is, if

$$\left\|[x, y]_\mathfrak{g}\right\| \le \|x\| \cdot \|y\|, \qquad \text{for every } x, y \in \mathfrak{g}, \tag{5.51d}$$

then $\| \cdot \|$ is said to be *compatible* with the Lie bracket of $\mathfrak{g}$ (or just $\| \cdot \|$ is *compatible with* $\mathfrak{g}$).

Banach-Lie algebra: A normed Lie algebra $(\mathfrak{g}, [\cdot, \cdot]_\mathfrak{g}, \| \cdot \|)$ is called a *Banach-Lie algebra*, if $(\mathfrak{g}, \| \cdot \|)$ is a complete normed space.

It is possible to give equivalent definitions of normed (Lie) algebras, by making use of the following remarks. Throughout, if $(V, \| \cdot \|)$ is a real or complex normed vector space, we equip V with the topology induced by its norm $\| \cdot \|$ and we denote by $B(x, r)$ the $\| \cdot \|$-ball about $x \in V$ of radius $r > 0$ (so that a basis for the topology is the family $\{B(x, r) \mid x \in V, \ r > 0\}$).

Remark 5.23. Let $(V, \| \cdot \|)$ be a real or complex normed vector space. Let $k \in \mathbb{N}$ be fixed and let also

$$F : \underbrace{V \times \cdots \times V}_{k \text{ times}} \longrightarrow V$$

be a k-linear map. Then F is continuous if and only if there exists a positive constant M such that

$$\|F(x_1, \ldots, x_k)\| \leq M \|x_1\| \cdots \|x_k\|, \quad \forall \, x_1, \ldots, x_k \in V. \tag{5.52}$$

Indeed, if (5.52) holds, then F is continuous since we have

$$\|F(x_1, \ldots, x_k) - F(\xi_1, \ldots, \xi_k)\| = \left\| \sum_i F(\xi_1, \ldots, \xi_{i-1}, x_i - \xi_i, x_{i+1}, \ldots, x_k) \right\|$$

$$\leq M \sum_i \|\xi_1\| \cdots \|\xi_{i-1}\| \cdot \|x_i - \xi_i\| \cdot \|x_{i+1}\| \cdots \|x_k\| \longrightarrow 0,$$

as $x_i \to \xi_i$ (for every $i = 1, \ldots, k$).

Conversely, suppose that F is continuous. In particular, given the open neighborhood $B(0, 1)$ of $F(0, \ldots, 0) = 0 \in V$, due to the continuity of F at $(0, \ldots, 0)$, there exists $\varepsilon > 0$ such that

$$F\left(\underbrace{B(0, \varepsilon) \times \cdots \times B(0, \varepsilon)}_{k \text{ times}} \right) \subseteq B(0, 1). \tag{5.53}$$

Then for every $x_1, \ldots, x_k \in V \setminus \{0\}$ we have (thanks to the multi-linearity of F and the homogeneity property of the norm)

$$\|F(x_1, \ldots, x_k)\| = \left(\frac{2}{\varepsilon} \right)^k \|x_1\| \cdots \|x_k\| \cdot \left\| F\left(\frac{\varepsilon}{2} \frac{x_1}{\|x_1\|}, \ldots, \frac{\varepsilon}{2} \frac{x_k}{\|x_k\|} \right) \right\|$$

$$\leq \left(\frac{2}{\varepsilon} \right)^k \|x_1\| \cdots \|x_k\| \cdot 1.$$

In deriving the above "$\leq$" sign we used (5.53), by noticing that

$$\frac{\varepsilon}{2} \cdot \frac{x_i}{\|x_i\|} \in B(0, \varepsilon), \quad \text{for every } i = 1, \ldots, k.$$

Hence (5.52) holds with the choice $M = 2^k/\varepsilon^k$, when all the x_i are different from 0. Since (5.52) trivially holds when at least one of the x_i equals 0, (5.52) is completely proved. $\square$

Remark 5.24. Let $(A, *)$ be a UA algebra and suppose $\|\cdot\|$ is a norm on A. Since $A \times A \ni (x, y) \mapsto x * y \in A$ is a bilinear map, then (thanks to Remark 5.23) *condition* (5.51a) *is equivalent to the continuity of* $*$.

Analogously, let $\mathfrak{g}$ be a Lie algebra and suppose $\|\cdot\|$ is a norm on $\mathfrak{g}$. Since $\mathfrak{g} \times \mathfrak{g} \ni (x, y) \mapsto [x, y] \in \mathfrak{g}$ is a bilinear map, then (thanks to Remark 5.23) *condition* (5.51c) *is equivalent to the continuity of the Lie bracket operation.*

Remark 5.25. Let V be a real or complex vector space and suppose $\|\cdot\|$ is a norm on V. *Then the maps*

$$V \times V \to V, \quad (x, y) \mapsto x + y, \qquad\qquad \mathbb{K} \times V \to V, \quad (k, y) \mapsto k\,y$$

are continuous (with respect to the associated topologies, V being endowed with the topology induced by the norm and $\mathbb{K} = \mathbb{R}, \mathbb{C}$ of the standard Euclidean topology).

As a consequence, by taking into account Remark 5.24, we see that the following definitions are equivalent:

- $(A, *, \|\cdot\|)$ *is a normed algebra iff – with its operations of vector space and of UA algebra – A is a topological algebra w.r.t. the topology induced by* $\|\cdot\|$.
- $(\mathfrak{g}, [\cdot, \cdot]_\mathfrak{g}, \|\cdot\|)$ *is a normed Lie algebra iff – with its operations of vector space and of Lie algebra – $\mathfrak{g}$ is a topological Lie algebra w.r.t. the topology induced by* $\|\cdot\|$.

Remark 5.26. *Any normed algebra* $(A, *, \|\cdot\|)$ *is a normed Lie algebra,* with the commutator Lie-bracket $[x, y]_* = x * y - y * x$. Indeed, if M is as in (5.51a), one has

$$\big\|[x, y]_*\big\| = \|x * y - y * x\| \leq \|x * y\| + \|y * x\| \leq 2M\,\|x\| \cdot \|y\|,$$

for every $x, y \in A$, so that inequality (5.51c) holds by replacing M with $2M$.

Remark 5.27. Let $(A, *, \|\cdot\|)$ be a normed algebra. *Then, by a "magnification" process, we can obtain from* $\|\cdot\|$ *a compatible norm.* Indeed, let M be a positive constant as in (5.51a) related to $\|\cdot\|$. Then the norm

$$\|x\|_\star := M\,\|x\|, \qquad x \in A, \tag{5.54}$$

is compatible with $*$. Indeed, we have

$$\|x * y\|_\star \overset{(5.51a)}{=} M \|x * y\| \leq M M \|x\| \cdot \|y\| = M \|x\| \cdot M \|y\| = \|x\|_\star \cdot \|y\|_\star,$$

for every $x, y \in A$, so that (5.51b) holds.

The same reasoning works for a normed Lie algebra $\mathfrak{g}$: if M is a positive constant as in (5.51c), then the norm in $\| \cdot \|_\star = M \| \cdot \|$ is compatible with $\mathfrak{g}$.

Remark 5.28. Some authors do not include in the definition of normed algebra the condition that A be *unital*. Our definition is by no means restrictive, for the following reason: Let $(A, *)$ be a real or complex associative algebra *not necessarily unital*, and suppose that $\| \cdot \|$ is a norm on A such that $*$ is continuous (that is, (5.51a) holds). Let $(A_1, \star)$ be as in Remark 5.6 on page 273. We define a norm $\| \cdot \|_\star$ on A_1 as follows:

$$\|(k, a)\|_\star := |k| + \|a\|, \quad k \in \mathbb{K}, \quad a \in A. \tag{5.55}$$

(Here $|\cdot|$ denotes the usual absolute value in $\mathbb{K} = \mathbb{R}, \mathbb{C}$.) It is easily seen that $\| \cdot \|_\star$ is actually a norm on A_1 which coincides with $\| \cdot \|$ on A (thought of as a subalgebra of A_1): hence the map

$$A \ni a \overset{\Psi}{\mapsto} (0, a) \in \{0\} \times A \subset A_1$$

is not only a UAA isomorphism, but also an isomorphism of the normed spaces $(A, \| \cdot \|)$ and $(\Psi(A), \| \cdot \|_\star)$. We claim that $(A_1, \star, \| \cdot \|_\star)$ is a normed algebra and that $\| \cdot \|_\star$ is compatible with $\star$ if $\| \cdot \|$ is compatible with $*$. Indeed, one has (recall (5.15) on page 274)

$$\left\| (k_1, a_1) \star (k_2, a_2) \right\|_\star = \left\| (k_1 k_2, k_1 a_2 + k_2 a_1 + a_1 * a_2) \right\|_\star$$

$$\overset{(5.55)}{=} |k_1 k_2| + \left\| k_1 a_2 + k_2 a_1 + a_1 * a_2 \right\|$$

(by the triangle inequality and the homogeneity of $\| \cdot \|$)

$$\leq |k_1| \cdot |k_2| + |k_1| \cdot \|a_2\| + |k_2| \cdot \|a_1\| + \|a_1 * a_2\|$$

$$\overset{(5.51a)}{\leq} |k_1| \cdot |k_2| + |k_1| \cdot \|a_2\| + |k_2| \cdot \|a_1\| + M \|a_1\| \cdot \|a_2\|$$

$$\leq \max\{1, M\} \cdot \left(|k_1| \cdot |k_2| + |k_1| \cdot \|a_2\| + |k_2| \cdot \|a_1\| + \|a_1\| \cdot \|a_2\| \right)$$

$$= \max\{1, M\} \cdot \|(k_1, a_1)\|_\star \cdot \|(k_2, a_2)\|_\star.$$

As a consequence, (5.51a) holds for $\| \cdot \|_\star$ too, with the constant M replaced by $\max\{1, M\}$ and the claimed facts are proved. □

We are now ready to state several results, whose proofs follow *verbatim* by rerunning the related proofs in Sect. 5.2.1.

Theorem 5.29 (Fundamental Estimate). *Let $\mathfrak{g}$ be a real or complex normed Lie algebra and let $\| \cdot \|$ be a norm on $\mathfrak{g}$ compatible with the Lie bracket. Finally let $\eta_N^{\mathfrak{g}}(a,b)$, $Z_j^{\mathfrak{g}}(a,b)$ and $D_{(h,k)}^{\mathfrak{g}}(a,b)$ be as in (5.27), (5.28), (5.24) respectively.*

Then, for every $N \in \mathbb{N}$ and every $a, b \in \mathfrak{g}$, the sum

$$\sum_{n=1}^{N} \frac{1}{n} \left(e^{\|a\|+\|b\|} - 1 \right)^n$$

furnishes an upper bound for any of the following:

$$\|\eta_N^{\mathfrak{g}}(a,b)\|, \qquad \textstyle\sum_{j=1}^{N} \|Z_j^{\mathfrak{g}}(a,b)\|,$$

$$\sum_{n=1}^{N} |c_n| \sum_{(h,k) \in \mathcal{N}_n \,:\, |h|+|k| \leq N} \mathbf{c}(h,k)\, \|D_{(h,k)}^{\mathfrak{g}}(a,b)\|.$$

Proof. Verbatim as in the proofs of Theorem 5.11 and Corollary 5.12. A key rôle is played by the estimate

$$\left\| D_{(h,k)}^{\mathfrak{g}}(a,b) \right\| \leq \|a\|^{|h|} \cdot \|b\|^{|k|}, \qquad \begin{array}{l} \text{for every } a, b \in \mathfrak{g} \\ \text{and every } (h,k) \in \mathcal{N}_n, \end{array}$$

which immediately follows by compatibility of the norm $\| \cdot \|$ with the Lie bracket of $\mathfrak{g}$. $\square$

Theorem 5.30 (Convergence of the CBHD series for Banach-Lie algebras). *Let $\mathfrak{g}$ be a Banach-Lie algebra. Let $\| \cdot \|$ be a norm on $\mathfrak{g}$ compatible with the Lie bracket. If $Z_j^{\mathfrak{g}}(a,b)$ is as in (5.28), let us set*

$$a \diamond b := \sum_{j=1}^{\infty} Z_j^{\mathfrak{g}}(a,b), \quad \text{whenever this series converges in } \mathfrak{g}, \tag{5.56}$$

$\mathfrak{g}$ *being equipped with its given Banach space structure.*

Then a sufficient condition for the existence of $a \diamond b$ is that the couple (a,b) belongs to the set

$$D := \left\{ (a,b) \in \mathfrak{g} \times \mathfrak{g} \,:\, \|a\| + \|b\| < \log 2 \right\}. \tag{5.57}$$

For example, this is the case if a and b belong to the $\| \cdot \|$-disc centred at the origin

$$Q := \left\{ a \in \mathfrak{g} \,:\, \|a\| < \tfrac{1}{2} \log 2 \right\}, \tag{5.58}$$

since $Q \times Q \subset D$. Moreover, the following inequality holds

$$\|a \diamond b\| \leq \log \left(\frac{1}{2 - e^{\|a\| + \|b\|}} \right), \qquad \text{for every } (a, b) \in D. \qquad (5.59)$$

Finally, the series of functions $\sum_{j=1}^{\infty} Z_j^{\mathfrak{g}}(a, b)$ converges normally[5] (hence uniformly) on every set of the type

$$D_\delta := \{ (a, b) \in \mathfrak{g} \times \mathfrak{g} : \|a\| + \|b\| \leq \delta \}, \qquad \text{with } \delta < \log 2. \qquad (5.60)$$

Actually, the above result of normal convergence and the bound (5.59) are also valid for the "majorizing" series

$$\sum_{j=1}^{\infty} \sum_{n=1}^{j} |c_n| \sum_{\substack{(h,k) \in \mathcal{N}_n \\ |h| + |k| = j}} \mathbf{c}(h, k) \left\| D_{(h,k)}^{\mathfrak{g}}(a, b) \right\|.$$

Proof. Verbatim as in the proof of Theorem 5.17. We reproduce the main element of the computation. Let $0 < \delta < \log 2$ be fixed arbitrarily. Suppose that $a, b \in \mathfrak{g}$ are such that $\|a\| + \|b\| < \delta$. We claim that in this case $\{\eta_N^{\mathfrak{g}}(a, b)\}_N$ is a Cauchy sequence in $\mathfrak{g}$: Indeed, we have

$$\left\| \eta_{n+p}^{\mathfrak{g}}(a, b) - \eta_n^{\mathfrak{g}}(a, b) \right\| = \left\| \sum_{j=1}^{n+p} Z_j^{\mathfrak{g}}(a, b) - \sum_{j=1}^{n} Z_j^{\mathfrak{g}}(a, b) \right\|$$

$$= \left\| \sum_{j=n+1}^{n+p} Z_j^{\mathfrak{g}}(a, b) \right\| \leq \sum_{j=n+1}^{n+p} \left\| Z_j^{\mathfrak{g}}(a, b) \right\| \leq \sum_{j=n+1}^{\infty} \left\| Z_j^{\mathfrak{g}}(a, b) \right\|.$$

Now, by the estimate in Theorem 5.29, the above far right-hand side vanishes as $n \to \infty$, since it is the n-th remainder of the convergent series:

$$\sum_{j=1}^{\infty} \left\| Z_j^{\mathfrak{g}}(a, b) \right\| \leq \sum_{j=1}^{\infty} \sum_{n=1}^{j} |c_n| \sum_{\substack{(h,k) \in \mathcal{N}_n \\ |h| + |k| = j}} \mathbf{c}(h, k) \left\| D_{(h,k)}^{\mathfrak{g}}(a, b) \right\|$$

$$\leq \sum_{n=1}^{\infty} \frac{1}{n} \left(e^{\|a\| + \|b\|} - 1 \right)^n \leq \sum_{n=1}^{\infty} \frac{1}{n} \left(e^\delta - 1 \right)^n = -\log(2 - e^\delta).$$

Since $(\mathfrak{g}, \|\cdot\|)$ is a Banach space, this proves that the series $\sum_{j=1}^{N} Z_j^{\mathfrak{g}}(a, b) = \eta_N^{\mathfrak{g}}(a, b)$ converges in $\mathfrak{g}$, normally on the set $\{\|a\| + \|b\| \leq \delta\}$. By the

---

[5]Recall Definition 5.9, page 278.

arbitrariness of $\delta < \log 2$, we infer the existence of $a \diamond b = \sum_{j=1}^{\infty} Z_j^{\mathfrak{g}}(a, b)$, for every $(a, b) \in D$. Moreover, the above computations also yield (5.59).

Finally, the claimed total convergence on the set D_δ in (5.60) follows from the estimate (proved above)

$$\sum_{j=1}^{\infty} \sup_{(a,b)\in D_\delta} \|Z_j^{\mathfrak{g}}(a, b)\|$$

$$\leq \sum_{j=1}^{\infty} \sum_{n=1}^{j} |c_n| \sum_{\substack{(h,k)\in N_n \\ |h|+|k|=j}} c(h, k) \sup_{(a,b)\in D_\delta} \left\|D_{(h,k)}^{\mathfrak{g}}(a, b)\right\|$$

$$\leq \sum_{n=1}^{\infty} \frac{1}{n} \left(e^\delta - 1\right)^n = -\log(2 - e^\delta) < \infty,$$

the equality holding true for $\delta < \log 2$.   □

With some additional work, we are able to provide an estimate of the rate of convergence of $\sum_{j=1}^{N} Z_j^{\mathfrak{g}}(a, b)$ to $a \diamond b$:

Theorem 5.31 (Rate of Convergence of the CBHD series). *Let $\mathfrak{g}$ be a Banach-Lie algebra. Let $\| \cdot \|$ be a norm on $\mathfrak{g}$ compatible with the Lie bracket. Let $Z_j^{\mathfrak{g}}(a, b)$ be as in (5.28) and let $\diamond$ be as in (5.56).*

Then, for every $a, b \in \mathfrak{g}$ such that $\|a\| + \|b\| < \log 2$, we have

$$\left\|a \diamond b - \sum_{j=1}^{N} Z_j^{\mathfrak{g}}(a, b)\right\| \leq \frac{(e^{\|a\|+\|b\|} - 1)^{N+1}}{(N + 1)\left(2 - e^{\|a\|+\|b\|}\right)} +$$

$$+ \sum_{n=1}^{N} (e^{\|a\|+\|b\|} - 1)^{n-1} \left(\sum_{i \geq \frac{N+1}{n}} \frac{(\|a\| + \|b\|)^i}{i!}\right). \tag{5.61}$$

[An application of the Lebesgue dominated convergence Theorem will prove that the right-hand side of (5.61) vanishes as $N \to \infty$, see Remark 5.32 below.]

Proof. Let us fix $a, b \in \mathfrak{g}$ such that $\|a\| + \|b\| < \log 2$. Then by (5.56), we have $a \diamond b = \sum_{j=1}^{\infty} Z_j^{\mathfrak{g}}(a, b)$, the series being convergent. Hence

$$\left\|a \diamond b - \sum_{j=1}^{N} Z_j^{\mathfrak{g}}(a, b)\right\| = \left\|\sum_{j=N+1}^{\infty} Z_j^{\mathfrak{g}}(a, b)\right\|$$

$$= \left\|\sum_{j=N+1}^{\infty} \sum_{n=1}^{j} c_n \sum_{\substack{(h,k)\in N_n \\ |h|+|k|=j}} c(h, k)\, D_{(h,k)}^{\mathfrak{g}}(a, b)\right\|$$

$$\overset{(5.35)}{\leq} \sum_{j=N+1}^{\infty} \sum_{n=1}^{j} \frac{1}{n} \sum_{\substack{(h,k)\in\mathcal{N}_n \\ |h|+|k|=j}} \frac{\|a\|^{|h|}\,\|b\|^{|k|}}{h!\,k!}$$

(we interchange the sums in n and in j)

$$= \sum_{n=1}^{N} \sum_{j=N+1}^{\infty} \sum_{\substack{(h,k)\in\mathcal{N}_n \\ |h|+|k|=j}} \{\cdots\} + \sum_{n=N+1}^{\infty} \sum_{j=n}^{\infty} \sum_{\substack{(h,k)\in\mathcal{N}_n \\ |h|+|k|=j}} \{\cdots\} =: \mathrm{I} + \mathrm{II}.$$

We estimate I and II separately:

$$\mathrm{II} = \sum_{n=N+1}^{\infty} \sum_{j=n}^{\infty} \sum_{\substack{(h,k)\in\mathcal{N}_n \\ |h|+|k|=j}} \frac{\|a\|^{|h|}\,\|b\|^{|k|}}{n\,h!\,k!} = \sum_{n=N+1}^{\infty} \sum_{\substack{(h,k)\in\mathcal{N}_n \\ |h|+|k|\geq n}} \frac{\|a\|^{|h|}\,\|b\|^{|k|}}{n\,h!\,k!}$$

$\Big($recall that it is always true that $|h| + |k| \geq n$ for every $(h,k) \in \mathcal{N}_n\Big)$

$$= \sum_{n=N+1}^{\infty} \frac{1}{n} \sum_{(h,k)\in\mathcal{N}_n} \frac{\|a\|^{|h|}\,\|b\|^{|k|}}{h!\,k!} \overset{(5.38)}{=} \sum_{n=N+1}^{\infty} \frac{1}{n}\Big(e^{\|a\|}\,e^{\|b\|} - 1\Big)^{n}$$

$$\leq \frac{1}{N+1} \sum_{n=N+1}^{\infty} \Big(e^{\|a\|}\,e^{\|b\|} - 1\Big)^{n} = \frac{(e^{\|a\|+\|b\|} - 1)^{N+1}}{(N+1)\,(2 - e^{\|a\|+\|b\|})},$$

where in the last equality we used the fact that $|e^{\|a\|+\|b\|} - 1| < 1$ together with the well-known formula $\sum_{n=N+1}^{\infty} q^n = q^{N+1}/(1-q)$ if $|q| < 1$. This gives the first summand of (5.61). Moreover

$$\mathrm{I} = \sum_{n=1}^{N} \sum_{j=N+1}^{\infty} \sum_{\substack{(h,k)\in\mathcal{N}_n \\ |h|+|k|=j}} \frac{\|a\|^{|h|}\,\|b\|^{|k|}}{n\,h!\,k!} = \sum_{n=1}^{N} \sum_{\substack{(h,k)\in\mathcal{N}_n \\ |h|+|k|\geq N+1}} \frac{\|a\|^{|h|}\,\|b\|^{|k|}}{n\,h!\,k!} =: (\star).$$

Now note that if $(h,k) \in \mathcal{N}_n$ is such that $|h| + |k| \geq N+1$ then *one at least* among $h_1 + k_1, \ldots, h_n + k_n$ is greater than $(N+1)/n$. This gives

$$(\star) \leq \sum_{n=1}^{N} \frac{1}{n} \Bigg\{ \sum_{\substack{h_1+k_1\geq \frac{N+1}{n} \\ (h_2,k_2),\ldots,(h_n,k_n)\neq(0,0)}} + \cdots + \sum_{\substack{h_n+k_n\geq \frac{N+1}{n} \\ (h_1,k_1),\ldots,(h_{n-1},k_{n-1})\neq(0,0)}} \Bigg\} \times$$

$$\times \frac{\|a\|^{|h|}\,\|b\|^{|k|}}{h!\,k!} = \sum_{n=1}^{N} \frac{1}{n}\cdot n \sum_{i+j\geq \frac{N+1}{n}} \frac{\|a\|^{i}\,\|b\|^{j}}{i!\,j!} \cdot \sum_{(h,k)\in\mathcal{N}_{n-1}} \frac{\|a\|^{|h|}\,\|b\|^{|k|}}{h!\,k!}$$

$$= \sum_{n=1}^{N} \Bigg(\sum_{i\geq \frac{N+1}{n}} \frac{(\|a\| + \|b\|)^{i}}{i!} \Bigg) \cdot (e^{\|a\|+\|b\|} - 1)^{n-1}.$$

This gives the second summand of (5.61). To produce the last equality above we used (5.38) and the following computation: for every $H \in \mathbb{N}$ and every $A, B \in \mathbb{R}$,

$$\sum_{i \geq H} \frac{(A+B)^i}{i!} = \sum_{i \geq H} \frac{1}{i!} \sum_{j=0}^{i} \binom{i}{j} A^j B^{i-j} = \sum_{i \geq H} \sum_{j=0}^{i} \frac{A^j B^{i-j}}{j!\,(i-j)!}$$

$$= \sum_{i \geq H} \sum_{r+s=i} \frac{A^r B^s}{r!\,s!} = \sum_{r+s \geq H} \frac{A^r B^s}{r!\,s!}.$$

This completes the proof.   □

Remark 5.32. Let $a, b \in \mathfrak{g}$ be such that $\|a\| + \|b\| < \log 2$. We prove that the right-hand side of (5.61) vanishes as $N \to \infty$. For brevity, we set $\sigma := \|a\| + \|b\|$. As $0 \leq \sigma < \log 2$, the first summand of (5.61) vanishes as $N \to \infty$ (since $|e^\sigma - 1| < 1$). The second summand is bounded above by

$$\gamma_N := \sum_{n=1}^{\infty} (e^\sigma - 1)^{n-1} \left(\sum_{i \geq \frac{N+1}{n}} \frac{\sigma^i}{i!} \right).$$

We claim that $\lim_{N \to \infty} \gamma_N = 0$. Let us set, for every $n, N \in \mathbb{N}$,

$$\alpha_n := (e^\sigma - 1)^{n-1}, \quad \beta_{n,N} := \sum_{i \geq \frac{N+1}{n}} \frac{\sigma^i}{i!},$$

so that $\gamma_N = \sum_{n=1}^{\infty} \alpha_n \beta_{n,N}$. The proof will be complete if we show that we can pass the limit across the series sign:

$$\lim_{N \to \infty} \sum_{n=1}^{\infty} \alpha_n \beta_{n,N} = \sum_{n=1}^{\infty} \alpha_n \lim_{N \to \infty} \beta_{n,N} = 0,$$

since, for every fixed $n \in \mathbb{N}$, $\lim_{N \to \infty} \beta_{n,N} = 0$ because $\sum_{i \geq 0} \frac{\sigma^i}{i!}$ is a convergent series. We then have to show that we can interchange the limit and the series signs. This follows for example by an application of the Lebesgue dominated convergence theorem (think of the series as an integral with respect to the counting measure!): all we have to provide is a nonnegative sequence g_n (independent of N) such that

$$|\alpha_n \beta_{n,N}| \leq g_n \quad \text{and such that} \quad \sum_{n=1}^{\infty} g_n < \infty.$$

The choice $g_n := e^\sigma \alpha_n$ does the job:

- We have $\sum\limits_{n=1}^{\infty} g_n = \sum\limits_{n=1}^{\infty} e^\sigma \alpha_n = e^\sigma \sum\limits_{n=1}^{\infty} (e^\sigma - 1)^{n-1} < \infty$ for $|e^\sigma - 1| < 1$.
- For every $n, N \geq 1$,

$$|\alpha_n \beta_{n,N}| \leq (e^\sigma - 1)^{n-1} \sum_{i \geq 0} \frac{\sigma^i}{i!} = (e^\sigma - 1)^{n-1} e^\sigma = g_n.$$

This ends the argument.   □

5.2.3 An Improved Domain of Convergence

In Theorem 5.30, as a byproduct of a domain of convergence of

$$\sum_{j=1}^{\infty} \sum_{n=1}^{j} |c_n| \sum_{\substack{(h,k) \in \mathcal{N}_n \\ |h|+|k|=j}} \mathbf{c}(h,k) \left\| D^{\mathfrak{g}}_{(h,k)}(a,b) \right\|, \tag{5.62}$$

we furnished a domain of convergence

$$D = \left\{ (a,b) \in \mathfrak{g} \times \mathfrak{g} : \|a\| + \|b\| < \log 2 \right\} \tag{5.63}$$

for the series $\sum_{j=1}^{\infty} \|Z_j^{\mathfrak{g}}(a,b)\|$, the latter being bounded by the series in (5.62). However, there may exist larger domains of convergence for $\sum_{j=1}^{\infty} \|Z_j^{\mathfrak{g}}(a,b)\|$, not necessarily working for (5.62). Roughly speaking, this is due to the fact that "cancellations" may occur in $\|Z_j^{\mathfrak{g}}(a,b)\|$ *before* the bounding series (5.62) is produced. It is then not unexpected that there exists an improved domain for the total convergence of $\sum_{j=1}^{\infty} Z_j^{\mathfrak{g}}(a,b)$, and this we shall provide in the following theorem.

We closely follow the ideas in the remarkable proof by Varadarajan in [171, Section 2.15, p.118–120] (which immediately adapt from the case of the Lie algebra of a Lie group to our context of Banach-Lie algebras). To this end, we shall make use of the recursion formula for Z_j proved in Sect. 4.5 on page 223, and a catchy argument from the theory of ODEs.

Theorem 5.33 (Improved Convergence of the CBHD Series). *Let $\mathfrak{g}$ be a Banach-Lie algebra. Let $\|\cdot\|$ be a norm on $\mathfrak{g}$ compatible with the Lie bracket. If $Z_j^{\mathfrak{g}}(a,b)$ is as in (5.28), the series $\sum_{j=1}^{\infty} Z_j^{\mathfrak{g}}(a,b)$ converges normally (hence uniformly) on every set of the type*

$$\widehat{D}_\rho := \{ (a,b) \in \mathfrak{g} \times \mathfrak{g} : \|a\| + \|b\| \leq \rho \}, \tag{5.64}$$

where $\rho > 0$ is strictly less than the absolute constant

$$\delta = \int_0^{2\pi} \frac{d\,s}{2 + \frac{1}{2}s - \frac{1}{2}s \cot\left(\frac{1}{2}s\right)} \approx 2.173\ldots \tag{5.65}$$

In particular, $\sum_{j=1}^{\infty} Z_j^{\mathfrak{g}}(a,b)$ converges (absolutely) for every fixed (a,b) on the whole set

$$\widehat{D} := \left\{(a,b) \in \mathfrak{g} \times \mathfrak{g} : \|a\| + \|b\| < \delta\right\}. \tag{5.66}$$

Note that the set D in (5.63) is properly contained in $\widehat{D}$, for $\log 2 < \delta$.

Proof. Let the constants K_{2p} be as in (4.109) on page 223, that is (see also (9.43) on page 496), the following complex Maclaurin expansion holds

$$\frac{z}{e^z - 1} = -\frac{z}{2} + 1 + \sum_{p=1}^{\infty} K_{2p}\, z^{2p}, \quad |z| < 2\,\pi. \tag{5.67}$$

As $\frac{z}{e^z-1} + \frac{z}{2}$ is an even function, we also have

$$\frac{z}{1 - e^{-z}} = \frac{z}{2} + 1 + \sum_{p=1}^{\infty} K_{2p}\, z^{2p}, \quad |z| < 2\,\pi. \tag{5.68}$$

We set (z still denoting a complex variable)

$$F(z) := 1 + \sum_{p=1}^{\infty} |K_{2p}|\, z^{2p}, \quad \text{whenever } |z| < 2\,\pi. \tag{5.69}$$

Obviously, the radius of convergence of this last series is the same as that of the series in (5.67), i.e., $2\,\pi$, whence (5.69) is well posed. Actually, taking into account the alternating signs of the numbers K_{2p}, it is easily proved that[6] (see Newman, So, Thompson [131, page 305])

$$F(z) = 2 - \frac{z}{2}\frac{\cos(z/2)}{\sin(z/2)}. \tag{5.70}$$

---

[6]Indeed, one has

$$F(z) = 1 + \sum_{p=1}^{\infty} |K_{2p}|\, z^{2p} = 1 + \sum_{p=1}^{\infty} (-1)^{p-1} K_{2p}\, z^{2p} = 2 - \left(1 + \sum_{p=1}^{\infty} K_{2p}\,(i\,z)^{2p}\right)$$

$$\overset{(5.67)}{=} 2 - \left(\frac{i\,z}{e^{i\,z} - 1} + \frac{i\,z}{2}\right) = 2 - \frac{z}{2}\frac{\cos(z/2)}{\sin(z/2)}.$$

Let us consider the complex ODE (see also [171, eq. (2.15.20)])

$$y' = \tfrac{1}{2} y + F(y), \quad y(0) = 0, \tag{5.71}$$

that is, equivalently (see the above (5.70))

$$y' = 2 + \tfrac{1}{2} y - \tfrac{1}{2} y \cot\left(\tfrac{1}{2} y\right), \quad y(0) = 0. \tag{5.72}$$

By the general theory of ordinary Cauchy problems, there exists $\delta > 0$ and a solution $y(z)$ to (5.71) which is holomorphic in the disc centered at the origin with radius δ. A remarkable result by Newman, So, Thompson [131] proves that the actual value of δ is the following one

$$\delta = \int_0^{2\pi} \frac{d\,s}{2 + \tfrac{1}{2} s - \tfrac{1}{2} s \cot\left(\tfrac{1}{2} s\right)} \approx 2.173\ldots$$

[For this result, see [131, Section 5]. We will not have occasion to invoke the actual value of δ, but – rather – the fact that it is larger than $\log 2$ so that the present proof in fact furnishes an improvement of our previous Theorem 5.30]. If we set (recall that $y(0) = 0$)

$$y(z) = \sum_{n=1}^{\infty} \gamma_n\, z^n, \quad |z| < \delta, \tag{5.73}$$

then a recursion formula for γ_n can be straightforwardly derived by exploiting (5.69) and (5.71). Indeed, by inserting the expansions of y and of F in (5.71), we have

$$\sum_{n=0}^{\infty} (n+1)\,\gamma_{n+1}\, z^n = \sum_{n=1}^{\infty} \frac{\gamma_n}{2}\, z^n + 1 + \sum_{p=1}^{\infty} |K_{2p}| \left(\sum_{n=1}^{\infty} \gamma_n\, z^n \right)^{2p}$$

$$= 1 + \sum_{n=1}^{\infty} \frac{\gamma_n}{2}\, z^n + \sum_{p=1}^{\infty} |K_{2p}| \sum_{k_1,\ldots,k_{2p} \geq 1} \gamma_{k_1} \cdots \gamma_{k_{2p}}\, z^{k_1 + \cdots + k_{2p}}.$$

Then by equating the coefficients of z^n (for $n \geq 0$) from the above far right/left-hand sides, we get the following recursion formula:

$$\begin{cases} \gamma_1 = 1, \quad 2\gamma_2 = \tfrac{1}{2}\gamma_1, \qquad \text{and, for } n \geq 2, \\[2mm] \gamma_{n+1} = \frac{1}{2\,(n+1)}\,\gamma_n + \displaystyle\sum_{\substack{p \geq 1,\ 2p \leq n \\ k_1,\ldots,k_{2p} \geq 1,\ k_1 + \cdots + k_{2p} = n}} \frac{|K_{2p}|}{n+1}\,\gamma_{k_1} \cdots \gamma_{k_{2p}}. \end{cases} \tag{5.74}$$

We observe that, by (5.74),

$$\gamma_n > 0 \text{ for every } n \in \mathbb{N}. \tag{5.75}$$

Now, since the complex power series $\sum_{n=1}^{\infty} \gamma_n z^n$ is convergent for $|z| < \delta$, we have (by invoking classical results on the absolute convergence of power series)

$$\sum_{n=1}^{\infty} \gamma_n |z|^n \overset{(5.75)}{=} \sum_{n=1}^{\infty} |\gamma_n z^n| < \infty, \quad \text{whenever } |z| < \delta,$$

so that

$$\sum_{n=1}^{\infty} \gamma_n \rho^n < \infty, \quad \text{for every } \rho \in (-\delta, \delta). \tag{5.76}$$

Let us now go back to the recursion formula for $Z_j^{\mathfrak{g}}$, derived in Corollary 4.25 on page 228, which we here rewrite, for convenience of reading:

$$
\begin{cases}
Z_1^{\mathfrak{g}}(u,v) = u + v, \\[2mm]
Z_2^{\mathfrak{g}}(u,v) = \frac{1}{4}[u - v, Z_1^{\mathfrak{g}}(u,v)]_{\mathfrak{g}} \\[2mm]
Z_{n+1}^{\mathfrak{g}}(u,v) = \frac{1}{2(n+1)} [u - v, Z_n^{\mathfrak{g}}(u,v)]_{\mathfrak{g}} + \\[2mm]
\quad + \displaystyle\sum_{\substack{p \geq 1,\, 2p \leqslant n \\ k_1,\dots,k_{2p} \geqslant 1 \\ k_1 + \cdots + k_{2p} = n}} \frac{K_{2p}}{n+1} [Z_{k_1}^{\mathfrak{g}}(u,v) \cdots [Z_{k_{2p}}^{\mathfrak{g}}(u,v), u+v]_{\mathfrak{g}} \cdots]_{\mathfrak{g}},
\end{cases}
\tag{5.77}
$$

valid for every $j \in \mathbb{N}$ and every $u, v \in \mathfrak{g}$. Let us consider the norm $\|\cdot\|$ compatible with $\mathfrak{g}$ and let us derive some estimates starting from the above recursion formula. We take any two elements $u, v \in \mathfrak{g}$ and we set

$$d := \|u\| + \|v\|.$$

Note that $\|u \pm v\| \leq \|u\| + \|v\| = d$. Let the numbers γ_n be as in (5.74). Starting from the first two identities in (5.77), we have

$$\|Z_1^{\mathfrak{g}}(u,v)\| \leq \|u\| + \|v\| = d = d\,\gamma_1$$
$$\|Z_2^{\mathfrak{g}}(u,v)\| \leq \tfrac{1}{4} \|[u-v, Z_1^{\mathfrak{g}}(u,v)]_{\mathfrak{g}}\| \leq \tfrac{1}{4} d \|Z_1^{\mathfrak{g}}(u,v)\| \leq \tfrac{1}{4} d^2 \gamma_1 = d^2 \gamma_2.$$

We claim that

$$\|Z_n^{\mathfrak{g}}(u,v)\| \leq d^n \gamma_n, \quad \forall\, n \in \mathbb{N}. \tag{5.78}$$

We prove this by induction on n, noting that the cases $n = 1, 2$ have just been checked. Supposing (5.78) to hold up to a fixed $n \in \mathbb{N}$, we prove it for

$n + 1$. Indeed, by applying (5.77), we have

$$\left\| Z_{n+1}^{\mathfrak{g}}(u,v) \right\| \leq \tfrac{1}{2(n+1)} \, \|u - v\| \cdot \| Z_n^{\mathfrak{g}}(u,v) \|$$

$$+ \sum_{\substack{p \geq 1,\ 2p \leqslant n \\ k_1,\ldots,k_{2p} \geqslant 1 \\ k_1 + \cdots + k_{2p} = n}} \frac{|K_{2p}|}{n+1} \, \| Z_{k_1}^{\mathfrak{g}}(u,v) \| \cdots \| Z_{k_{2p}}^{\mathfrak{g}}(u,v) \| \cdot \|u + v\|$$

(use the induction hypothesis together with $\|u \pm v\| \leq d$)

$$\leq \tfrac{1}{2(n+1)} \, d \cdot d^n \, \gamma_n + \sum_{\substack{p \geq 1,\ 2p \leqslant n \\ k_1,\ldots,k_{2p} \geqslant 1 \\ k_1 + \cdots + k_{2p} = n}} \frac{|K_{2p}|}{n+1} \, d^{k_1} \, \gamma_{k_1} \cdots d^{k_{2p}} \, \gamma_{k_{2p}} \cdot d$$

$$= d^{n+1} \cdot \left(\tfrac{1}{2(n+1)} \, \gamma_n + \sum_{\substack{p \geq 1,\ 2p \leqslant n \\ k_1,\ldots,k_{2p} \geqslant 1 \\ k_1 + \cdots + k_{2p} = n}} \frac{|K_{2p}|}{n+1} \, \gamma_{k_1} \cdots \gamma_{k_{2p}} \right)$$

$$= d^{n+1} \, \gamma_{n+1} \quad \text{in view of (5.74)}.$$

So (5.78) is proved by induction. Note that it can be rewritten as the interesting estimate

$$\| Z_n^{\mathfrak{g}}(u,v) \| \leq (\|u\| + \|v\|)^n \, \gamma_n, \quad \forall \, n \in \mathbb{N}, \quad \forall \, u, v \in \mathfrak{g}. \tag{5.79}$$

Let now $\rho > 0$ be such that $\rho < \delta$, where δ is as in (5.65) and let us consider the set $\widehat{D}_\rho$ in (5.64). Then we have (as $\rho < \delta$)

$$\sum_{n=1}^{\infty} \sup_{(u,v) \in \widehat{D}_\rho} \| Z_n^{\mathfrak{g}}(u,v) \| \overset{(5.79)}{\leq} \sum_{n=1}^{\infty} \gamma_n \, \rho^n \overset{(5.76)}{<} \infty.$$

This proves that the series $\sum_{n=1}^{\infty} Z_n^{\mathfrak{g}}(u,v)$ converges normally, hence uniformly, for $(u,v) \in \widehat{D}_\rho$. This ends the proof. $\square$

5.3 Associativity of the CBHD Operation

The aim of this section is to study the *associativity* property of the CBHD operation obtained by considering the map $(x,y) \mapsto x \diamond y = \sum_{j=1}^{\infty} Z_j(x,y)$. The first task is to discover the right setting where this infinite sum makes sense. This happens for instance, as we saw in Sects. 5.2.1 and 5.2.2, when we are dealing with a finite dimensional Lie algebra or, more generally, with a Banach-Lie algebra: in these cases, $\diamond$ is a priori defined only in a neighborhood of the origin. We shall then prove (in Sect. 5.3.2) that $\diamond$ defines a *local group*, so that in particular $\diamond$ is associative (in a suitable neighborhood of the origin). The other case of interest is that of the *nilpotent Lie algebras* $\mathfrak{n}$: in this case $\diamond$ is globally defined (since the associated series becomes in fact a finite sum) and it defines a group on the whole of $\mathfrak{n}$ (see Sect. 5.4.1).

5.3.1 *"Finite" Identities from the Associativity of* $\diamond$

We first need to provide ready-to-use identities (for general associative algebras) encoded in the associativity of the CBHD operations $\blacklozenge$ and $\diamond$ on the completion $\widehat{\mathscr{T}}(V)$ of the tensor algebra of a vector space V.

To this end, we fix henceforth a vector space V over a field $\mathbb{K}$ of characteristic zero. The notation of Sect. 5.1 is also used:

$$H_N, \quad \pi_N, \quad \widehat{U}_N, \quad (\widehat{\mathscr{T}}(V), \cdot), \quad \ldots$$

along with other well-known notations (used throughout the Book)

$$[\cdots]_\otimes, \quad N_n, \quad \mathbf{c}_n, \quad \mathbf{c}(h,k), \quad Z_j, \quad \ldots$$

We recall that we introduced an operation $\diamond$ on $\widehat{U}_1 = \widehat{\mathscr{T}}_+(V)$ as follows

$$u \diamond v = \sum_{n=1}^{\infty} \left(c_n \sum_{(h,k) \in N_n} \mathbf{c}(h,k) \left[u^{h_1} v^{k_1} \cdots u^{h_n} v^{k_n} \right]_\otimes \right), \quad u,v \in \widehat{U}_1.$$

We have the following lemma:

Lemma 5.34. *1. For every* $u, v \in \widehat{\mathscr{T}}(V)$ *and every* $N \in \mathbb{N}$

$$\pi_N([u,v]_\otimes) = \pi_N([\pi_N(u), \pi_N(v)]_\otimes). \tag{5.80a}$$

2. For every $k \geq 2$, *every* $t_1, \ldots, t_k \in \widehat{\mathscr{T}}(V)$ *and every* $N \in \mathbb{N}$

$$\pi_N\left([t_1, [t_2, [\cdots, t_k]\cdots]]_\otimes\right) = \pi_N\left([\pi_N t_1, [\pi_N t_2, [\cdots, \pi_N t_k]\cdots]]_\otimes\right). \tag{5.80b}$$

For every $u, v \in \widehat{\mathscr{T}}(V)$ *and every* $N \in \mathbb{N}$

$$\pi_N(u \diamond v) = \pi_N((\pi_N u) \diamond (\pi_N v)). \tag{5.81}$$

3. Proof. (1). This follows from (5.3) together with our former definition of $[u,v]_\otimes$ *as* $u \cdot v - v \cdot u$.

(2). This follows from an induction argument based on (5.80a) (recall also that $\pi_N \circ \pi_N = \pi_N$). For example, the case $k = 3$ reads

$$\pi_N([t_1, [t_2, t_3]]) = \pi_N([\pi_N t_1, \pi_N[t_2, t_3]]) = \pi_N([\pi_N t_1, \pi_N[\pi_N t_2, \pi_N t_3]])$$
$$= \pi_N([\pi_N(\pi_N t_1), \pi_N[\pi_N t_2, \pi_N t_3]])$$
$$= \pi_N([\pi_N t_1, [\pi_N t_2, \pi_N t_3]]),$$

where in the last equality we used (5.80a) (read from right to left!).
(3). Let $u, v \in \widehat{U}_1$. Then we have

$$\pi_N(u \diamond v) = \pi_N \left(\sum_{n=1}^{N} \left(c_n \sum_{\substack{(h,k) \in \mathcal{N}_n \\ |h|+|k| \leq N}} \mathbf{c}(h, k) \left[u^{h_1} v^{k_1} \cdots u^{h_n} v^{k_n} \right]_\otimes \right. \right.$$

$$\left. + \sum_{n=N+1}^{\infty} c_n \sum_{|h|+|k| \leq N} \{\cdots\} + \sum_{n=1}^{\infty} c_n \sum_{|h|+|k| \geq N+1} \{\cdots\} \right)$$
$$\underbrace{\phantom{+ \sum_{n=N+1}^{\infty} c_n \sum_{|h|+|k| \leq N} \{\cdots\} + \sum_{n=1}^{\infty} c_n \sum_{|h|+|k| \geq N+1} \{\cdots\}}}_{\in \widehat{U}_{N+1}}$$

$$= \sum_{n=1}^{N} \left(c_n \sum_{\substack{(h,k) \in \mathcal{N}_n \\ |h|+|k| \leq N}} \mathbf{c}(h, k) \, \pi_N \left[u^{h_1} v^{k_1} \cdots u^{h_n} v^{k_n} \right]_\otimes \right)$$

$$\overset{(5.80b)}{=} \sum_{n=1}^{N} \left(c_n \sum_{\substack{(h,k) \in \mathcal{N}_n \\ |h|+|k| \leq N}} \mathbf{c}(h, k) \, \pi_N \left[(\pi_N u)^{h_1} (\pi_N v)^{k_1} \cdots \right]_\otimes \right)$$

(argue exactly as above, by using $\pi_N(\widehat{U}_1) \subseteq \widehat{U}_1$)

$$= \pi_N((\pi_N u) \diamond (\pi_N v)).$$

This ends the proof. □

Incidentally, in the above computations we have also proved that

$$\pi_N(u \diamond v) = \pi_N \sum_{n=1}^{N} c_n \sum_{\substack{(h,k) \in \mathcal{N}_n \\ |h|+|k| \leq N}} \mathbf{c}(h, k) \left[u^{h_1} v^{k_1} \cdots u^{h_n} v^{k_n} \right]_\otimes, \quad \forall\, u, v \in \widehat{\mathcal{T}}_+(V).$$

$$(5.82)$$

Let now $V = \mathbb{K}\langle x, y, z \rangle$, where $\{x, y, z\}$ is a set of cardinality three. We know from Corollary 3.33 on page 156 that $\diamond$ is associative on $\widehat{\mathcal{T}}_+(V)$, so that in particular we have (recalling that $x, y, z \in V = \mathcal{T}_1 \subset \widehat{\mathcal{T}}_+$)

$$x \diamond (y \diamond z) = (x \diamond y) \diamond z.$$

We fix $N \in \mathbb{N}$. By applying π_N to both sides of the above equality (and noticing that x, y, z are left unchanged by π_N for they belong to $\mathscr{T}_1$), we get

$$\pi_N(x \diamond \pi_N(y \diamond z)) = \pi_N(\pi_N(x \diamond y) \diamond z). \tag{5.83}$$

Now, from (5.8) on page 272, we have

$$\pi_N(x \diamond y) = \eta_N(x, y) = \sum_{n=1}^{N} c_n \sum_{\substack{(h,k) \in \mathcal{N}_n \\ |h|+|k| \leq N}} \mathbf{c}(h, k) \left[x^{h_1} y^{k_1} \cdots x^{h_n} y^{k_n} \right]_\otimes,$$

and an analogous formula holds for $\pi_N(y \diamond z)$. As a consequence, (5.83) can be rewritten as follows:

$$\pi_N \left(x \diamond \left(\sum_{n=1}^{N} c_n \sum_{\substack{(h,k) \in \mathcal{N}_n \\ |h|+|k| \leq N}} \mathbf{c}(h, k) \, D^\otimes_{(h,k)}(y, z) \right) \right)$$

$$= \pi_N \left(\left(\sum_{n=1}^{N} c_n \sum_{\substack{(h,k) \in \mathcal{N}_n \\ |h|+|k| \leq N}} \mathbf{c}(h, k) \, D^\otimes_{(h,k)}(x, y) \right) \diamond z \right), \tag{5.84}$$

where we have invoked the usual notation

$$D^\otimes_{(h,k)}(y, z) := \left[x^{h_1} y^{k_1} \cdots x^{h_n} y^{k_n} \right]_\otimes, \qquad (h, k) \in \mathcal{N}_n.$$

If we now make use of (5.82), we can take the identity (5.84) even further: indeed, recalling that two elements of $\widehat{\mathscr{T}}(V)$ have the same π_N-image if and only if their difference belongs to $\widehat{U}_{N+1}$, (5.84) becomes

$$\sum_{n=1}^{N} c_n \sum_{\substack{(h,k) \in \mathcal{N}_n \\ |h|+|k| \leq N}} \mathbf{c}(h, k) \, D^\otimes_{(h,k)} \left(x, \sum_{m=1}^{N} c_m \sum_{\substack{(\alpha,\beta) \in \mathcal{N}_m \\ |\alpha|+|\beta| \leq N}} \mathbf{c}(\alpha, \beta) \, D^\otimes_{(\alpha,\beta)}(y, z) \right)$$

$$- \sum_{n=1}^{N} c_n \sum_{\substack{(h,k) \in \mathcal{N}_n \\ |h|+|k| \leq N}} \mathbf{c}(h, k) \, D^\otimes_{(h,k)} \left(\sum_{m=1}^{N} c_m \sum_{\substack{(\alpha,\beta) \in \mathcal{N}_m \\ |\alpha|+|\beta| \leq N}} \mathbf{c}(\alpha, \beta) \, D^\otimes_{(\alpha,\beta)}(x, y), z \right)$$

$$=: \mathcal{R}_N(x, y, z) \in \widehat{\mathscr{T}}_{N+1}(V). \tag{5.85}$$

Actually, much more is true:

1. Since $D^{\otimes}_{(h,k)}(a,b)$ is a Lie polynomial in a, b, we immediately recognize that $\mathcal{R}_N(x,y,z)$ in (5.85) is a Lie polynomial in x, y, z, that is, an element of the free Lie algebra $\mathcal{L}(\mathbb{K}\langle x,y,z \rangle)$.
2. Moreover, the fact that (see (5.85)), $\mathcal{R}_N(x,y,z)$ belongs to $\widehat{\mathcal{T}}_{N+1}(V)$, also ensures that $\mathcal{R}_N(x,y,z)$ belongs to $\bigoplus_{j \geq N+1} \mathcal{L}_j(\mathbb{K}\langle x,y,z \rangle)$.
3. Furthermore, by analyzing closely the expression of $\mathcal{R}_N(x,y,z)$ in (5.85), we recognize that the maximal height of its summands (iterated brackets in x, y, z) does not exceed N^2, whence

$$\mathcal{R}_N(x,y,z) \in \bigoplus_{j=N+1}^{N^2} \mathcal{L}_j(\mathbb{K}\langle x,y,z \rangle).$$

4. Finally, (5.85) can be seen as an identity in the free associative algebra over three indeterminates $\mathcal{T}(\mathbb{K}\langle x,y,z \rangle)$ or as an identity in the free Lie algebra over three indeterminates $\mathcal{L}(\mathbb{K}\langle x,y,z \rangle)$.
5. An explicit (though awesome!) expression of $\mathcal{R}_N(x,y,z)$ is given by the following formula: recalling that

$$\sum_{m=1}^{N} c_m \sum_{\substack{(\alpha,\beta)\in\mathcal{N}_m \\ |\alpha|+|\beta|\leq N}} \mathbf{c}(\alpha,\beta)\, D^{\otimes}_{(\alpha,\beta)}(y,z) = \sum_{j=1}^{N} Z^{\otimes}_j(y,z),$$

and that $\quad Z^{\otimes}_j(y,z) \in \mathcal{T}_j(\mathbb{K}\langle x,y,z \rangle),$

one has:

$$\mathcal{R}_N(x,y,z) \tag{5.86}$$

$$= \sum_{n=1}^{N} c_n \sum_{\substack{(h,k)\in\mathcal{N}_n \\ |h|+|k|\leq N}} \mathbf{c}(h,k) \sum_{\substack{1\leq j_1^1,\dots,j_{k_1}^1,\dots,j_1^n,\dots,j_{k_n}^n \leq N \\ h_1+\dots+h_n+j_1^1+\dots+j_{k_1}^1+\dots+j_1^n+\dots+j_{k_n}^n \geq N+1}}$$

$$\times \left[\underbrace{x\cdots x}_{h_1} Z^{\otimes}_{j_1^1}(y,z)\cdots Z^{\otimes}_{j_{k_1}^1}(y,z) \cdots \underbrace{x\cdots x}_{h_n} Z^{\otimes}_{j_1^n}(y,z)\cdots Z^{\otimes}_{j_{k_n}^n}(y,z) \right]_{\otimes}$$

$$- \sum_{n=1}^{N} c_n \sum_{\substack{(h,k)\in\mathcal{N}_n \\ |h|+|k|\leq N}} \mathbf{c}(h,k) \sum_{\substack{1\leq j_1^1,\dots,j_{h_1}^1,\dots,j_1^n,\dots,j_{h_n}^n \leq N \\ k_1+\dots+k_n+j_1^1+\dots+j_{h_1}^1+\dots+j_1^n+\dots+j_{h_n}^n \geq N+1}}$$

$$\times \left[Z^{\otimes}_{j_1^1}(x,y)\cdots Z^{\otimes}_{j_{h_1}^1}(x,y)\underbrace{z\cdots z}_{k_1} \cdots Z^{\otimes}_{j_1^n}(x,y)\cdots Z^{\otimes}_{j_{h_n}^n}(x,y)\underbrace{z\cdots z}_{k_n} \right]_{\otimes}.$$

By all the above remarks and by the universal property of the free associative algebra and of the free Lie algebra over three indeterminates (see Theorem 2.85 on page 107), identity (5.85) proves the following two theorems. These theorems can be seen as "finite" versions of the associativity information encoded in the $\diamond$ operation.

Theorem 5.35 (Finite Associativity of $\diamond$ on an Associative Algebra). *Let N $\in \mathbb{N}$. Let $(A, *)$ be an associative algebra over a field of characteristic zero. Then there exists a function*

$$\mathcal{R}_N^* : A \times A \times A \longrightarrow \underbrace{A * \cdots * A}_{N + 1 \text{ times}}$$

such that the following identity holds for every choice of $a, b, c \in A$:

$$\sum_{n=1}^{N} c_n \sum_{\substack{(h,k)\in\mathcal{N}_n \\ |h|+|k|\leq N}} \mathbf{c}(h, k)\, D_{(h,k)}^* \left(a, \sum_{m=1}^{N} c_m \sum_{\substack{(\alpha,\beta)\in\mathcal{N}_m \\ |\alpha|+|\beta|\leq N}} \mathbf{c}(\alpha, \beta)\, D_{(\alpha,\beta)}^*(b, c) \right)$$

$$- \sum_{n=1}^{N} c_n \sum_{\substack{(h,k)\in\mathcal{N}_n \\ |h|+|k|\leq N}} \mathbf{c}(h, k)\, D_{(h,k)}^* \left(\sum_{m=1}^{N} c_m \sum_{\substack{(\alpha,\beta)\in\mathcal{N}_m \\ |\alpha|+|\beta|\leq N}} \mathbf{c}(\alpha, \beta)\, D_{(\alpha,\beta)}^*(a, b), c \right)$$

$$= \mathcal{R}_N^*(a, b, c). \tag{5.87}$$

More precisely, $\mathcal{R}_N^*(a, b, c)$ can be obtained by the replacements

$$x \mapsto a, \quad y \mapsto b, \quad z \mapsto c, \qquad \otimes \mapsto *$$

in $\mathcal{R}_N(x, y, z)$ introduced in (5.85) – and explicitly written in (5.86) – which is a "universal" polynomial in three indeterminates belonging to

$$\bigoplus_{j=N+1}^{N^2} \mathcal{L}_j(\mathbb{K}\langle x, y, z\rangle) \subset \bigoplus_{j=N+1}^{N^2} \mathcal{T}_j(\mathbb{K}\langle x, y, z\rangle).$$

In particular, $\mathcal{R}_N^*(a, b, c)$ can be expressed as a linear combination of $*$-commutator Lie brackets (and of $*$-products) where a, b, c appear at least $(N + 1)$-times (and no more than N^2 times).

Here, along with the usual notation in (5.25) and (5.26) for $\mathcal{N}_n, c_n, \mathbf{c}(h, k)$, we have used the notations $[a, b]_* = a * b - b * a$ and

$$D_{(h,k)}^*(a, b)$$

$$= [\overbrace{a, \cdots [a}^{h_1 \text{ times}}, \overbrace{[b, \cdots [b}^{k_1 \text{ times}}, \cdots \overbrace{[a, \cdots [a}^{h_n \text{ times}}, \overbrace{[b, [\cdots, b}^{k_n \text{ times}}]_*]_*]_* \cdots]_* \cdots]_* \cdots]_*]_* \cdots]_*.$$

Theorem 5.36 (Finite Associativity of $\diamond$ on a Lie Algebra). *Let $N \in \mathbb{N}$. Let $\mathfrak{g}$ be a Lie algebra over a field of characteristic zero. Then there exists a function*

$$\mathcal{R}_N^{\mathfrak{g}} : \mathfrak{g} \times \mathfrak{g} \times \mathfrak{g} \longrightarrow \underbrace{[\mathfrak{g}, [\cdots, \mathfrak{g}] \cdots]}_{N+1 \text{ times}}$$

such that the following identity holds for every choice of $a, b, c \in \mathfrak{g}$:

$$\sum_{n=1}^{N} c_n \sum_{\substack{(h,k) \in \mathcal{N}_n \\ |h|+|k| \leq N}} \mathbf{c}(h,k) \, D_{(h,k)}^{\mathfrak{g}} \left(a, \sum_{m=1}^{N} c_m \sum_{\substack{(\alpha,\beta) \in \mathcal{N}_m \\ |\alpha|+|\beta| \leq N}} \mathbf{c}(\alpha,\beta) \, D_{(\alpha,\beta)}^{\mathfrak{g}}(b,c) \right)$$

$$- \sum_{n=1}^{N} c_n \sum_{\substack{(h,k) \in \mathcal{N}_n \\ |h|+|k| \leq N}} \mathbf{c}(h,k) \, D_{(h,k)}^{\mathfrak{g}} \left(\sum_{m=1}^{N} c_m \sum_{\substack{(\alpha,\beta) \in \mathcal{N}_m \\ |\alpha|+|\beta| \leq N}} \mathbf{c}(\alpha,\beta) \, D_{(\alpha,\beta)}^{\mathfrak{g}}(a,b), c \right)$$

$$= \mathcal{R}_N^{\mathfrak{g}}(a,b,c). \tag{5.88}$$

More precisely, $\mathcal{R}_N^{\mathfrak{g}}(a,b,c)$ can be obtained by the replacements

$$x \mapsto a, \quad y \mapsto b, \quad z \mapsto c, \qquad [\cdot,\cdot]_{\otimes} \mapsto [\cdot,\cdot]_{\mathfrak{g}}$$

in $\mathcal{R}_N(x,y,z)$ introduced in (5.85) – and explicitly written in (5.86) – which is a "universal" Lie polynomial in three indeterminates belonging to

$$\bigoplus_{j=N+1}^{N^2} \mathcal{L}_j(\mathbb{K}\langle x, y, z \rangle).$$

In particular, $\mathcal{R}_N^{\mathfrak{g}}(a,b,c)$ can be expressed as a linear combination of iterated $\mathfrak{g}$-brackets in a, b, c of heights at least $(N+1)$ (and at most N^2 times).

Here, along with the usual notation in (5.25) and (5.26) for $\mathcal{N}_n, c_n, \mathbf{c}(h,k)$, we have used the notation

$$D_{(h,k)}^{\mathfrak{g}}(a,b)$$

$$= \overbrace{[a, \cdots [a}^{h_1 \text{ times}}, \overbrace{[b, \cdots [b}^{k_1 \text{ times}}, \cdots \overbrace{[a, \cdots [a}^{h_n \text{ times}}, \overbrace{[b, [\cdots, b}^{k_n \text{ times}}]_{\mathfrak{g}}]_{\mathfrak{g}} \cdots]_{\mathfrak{g}} \cdots]_{\mathfrak{g}} \cdots]_{\mathfrak{g}}]_{\mathfrak{g}} \cdots]_{\mathfrak{g}}.$$

Remark 5.37. Let $\mathfrak{g}$ be a Lie algebra. If $Z_j^{\mathfrak{g}}$ denotes as usual the map

$$Z_j^{\mathfrak{g}} : \mathfrak{g} \times \mathfrak{g} \to \mathfrak{g}, \quad Z_j^{\mathfrak{g}}(a,b) := \sum_{n=1}^{j} c_n \sum_{(h,k) \in \mathcal{N}_n : |h|+|k|=j} \mathbf{c}(h,k) \, D_{(h,k)}^{\mathfrak{g}}(a,b),$$

there is an *explicit* (although awesome!) way of writing (5.88) in a closed form (which amounts in projecting the identity $x \diamond (y \diamond z) = (x \diamond y) \diamond z$ on $\mathcal{L}_N(\mathbb{K}\langle x, y, z \rangle)$): *For every $N \in \mathbb{N}$ it holds that*

$$
\sum_{n=1}^{N} c_n \sum_{\substack{(h,k) \in \mathcal{N}_n \\ |h|+|k| \leq N}} \mathbf{c}(h,k) \sum_{\substack{1 \leq r_1^1, \ldots, r_{k_1}^1, \cdots, r_1^n, \ldots, r_{k_n}^n \leq N \\ h_1 + \cdots + h_n + r_1^1 + \cdots + r_{k_1}^1 + \cdots + r_1^n + \ldots + r_{k_n}^n = N}}
$$

$$
\times \left[\underbrace{x \cdots x}_{h_1 \text{ times}} Z_{r_1^1}^{\mathfrak{g}}(y,z) \cdots Z_{r_{k_1}^1}^{\mathfrak{g}}(y,z) \cdots \underbrace{x \cdots x}_{h_n \text{ times}} Z_{r_1^n}^{\mathfrak{g}}(y,z) \cdots Z_{r_{k_n}^n}^{\mathfrak{g}}(y,z) \right]_{\mathfrak{g}}
$$

$$
= \sum_{n=1}^{N} c_n \sum_{\substack{(h,k) \in \mathcal{N}_n \\ |h|+|k| \leq N}} \mathbf{c}(h,k) \sum_{\substack{1 \leq r_1^1, \ldots, r_{h_1}^1, \cdots, r_1^n, \ldots, r_{h_n}^n \leq N \\ k_1 + \cdots + k_n + r_1^1 + \cdots + r_{h_1}^1 + \cdots + r_1^n + \ldots + r_{h_n}^n = N}}
$$

$$
\times \left[Z_{r_1^1}^{\mathfrak{g}}(x,y) \cdots Z_{r_{h_1}^1}^{\mathfrak{g}}(x,y) \underbrace{z \cdots z}_{k_1 \text{ times}} \cdots Z_{r_1^n}^{\mathfrak{g}}(x,y) \cdots Z_{r_{h_n}^n}^{\mathfrak{g}}(x,y) \underbrace{z \cdots z}_{k_n \text{ times}} \right]_{\mathfrak{g}}.
$$

5.3.2 Associativity for Banach-Lie Algebras

Suppose $\mathbb{K} = \mathbb{R}$ or $\mathbb{K} = \mathbb{C}$ and $(\mathfrak{g}, [\cdot, \cdot]_{\mathfrak{g}}, \| \cdot \|)$ is a Banach-Lie algebra over $\mathbb{K}$ and that $\| \cdot \|$ is compatible with the Lie bracket of $\mathfrak{g}$ (see Definition 5.22). We use the notations introduced at the beginning of Sect. 5.2.1:

$$
D_{(h,k)}^{\mathfrak{g}}, \quad \mathcal{N}_n, \quad c_n, \quad \mathbf{c}(h,k), \quad \eta_N^{\mathfrak{g}}, \quad Z_j^{\mathfrak{g}}, \quad \ldots
$$

For the sake of brevity, the superscript "$\mathfrak{g}$" will be frequently omitted; analogously we shall denote the Lie bracket on $\mathfrak{g}$ simply by $[\cdot, \cdot]$.

Thanks to Theorem 5.30, we know that the map $(a, b) \mapsto a \diamond b$ defined by

$$
a \diamond b := \sum_{j=1}^{\infty} Z_j^{\mathfrak{g}}(a, b)
$$

is well posed for every a, b belonging to the $\| \cdot \|$-disc centred at the origin

$$
Q := \left\{ a \in \mathfrak{g} \, : \, \|a\| < \tfrac{1}{2} \log 2 \right\}.
$$

Making use of the estimate in (5.59) and arguing as in Remark 5.20-2, if furthermore a, b belong to the disc

$$
\widetilde{Q} := \left\{ a \in \mathfrak{g} \, : \, \|a\| < \tfrac{1}{2} \log \left(2 - 1/\sqrt{2} \right) \right\}, \tag{5.89}
$$

then $a \diamond b \in Q$ and the series of functions $\sum_{j=1}^{\infty} Z_j^{\mathfrak{g}}(a, b)$ converges normally, hence uniformly, on every set of the form

$$D_\delta := \left\{(a, b) \in \mathfrak{g} \times \mathfrak{g} : \|a\| + \|b\| \leq \delta\right\}, \qquad \text{with } \delta < \log 2,$$

hence, for example, on $\widetilde{Q} \times \widetilde{Q}$, for

$$\widetilde{Q} \times \widetilde{Q} \subseteq \left\{(a, b) \in \mathfrak{g} \times \mathfrak{g} : \|a\| + \|b\| \leq \log(2 - 1/\sqrt{2})\right\},$$

and $\log(2 - 1/\sqrt{2}) < \log 2$. As a consequence

$$a \diamond b, \quad b \diamond c, \qquad (a \diamond b) \diamond c, \quad a \diamond (b \diamond c)$$

are well-posed for every $a, b, c \in \widetilde{Q}$. The aim of this section is to prove the next result.

Theorem 5.38. (Local Associativity of the CBHD Operation for Banach-Lie Algebras). *Let $(\mathfrak{g}, [\cdot, \cdot]_\mathfrak{g}, \|\cdot\|)$ be a real or complex Banach-Lie algebra. Let $\|\cdot\|$ be compatible with the Lie bracket of $\mathfrak{g}$. Let also $\widetilde{Q}$ be as in (5.89). Then we have*

$$(a \diamond b) \diamond c = a \diamond (b \diamond c), \quad \text{for every } a, b, c \in \widetilde{Q}. \tag{5.90}$$

Proof. By the remarks preceding this theorem we have

$$a \diamond b \in Q, \quad \text{for every } a, b \in \widetilde{Q}. \tag{5.91}$$

Moreover, by the results in Theorem 5.30, we know that

$$\eta_N(x, y) \xrightarrow{N \to \infty} x \diamond y, \qquad \text{for every } x, y \in Q, \tag{5.92}$$

and the sequence of functions $\{\eta_N(x, y)\}_N$ converges uniformly to $x \diamond y$ on

$$D_\delta := \left\{(x, y) \in \mathfrak{g} \times \mathfrak{g} : \|x\| + \|y\| \leq \delta\right\}, \qquad \text{with } \delta < \log 2.$$

Remember that this means precisely:

$$\lim_{N \to \infty} \sup_{(x,y) \in D_\delta} \left\|\eta_N(x, y) - x \diamond y\right\| = 0 \qquad (\text{with } \delta < \log 2). \tag{5.93}$$

Let us fix henceforth $a, b, c \in \widetilde{Q}$. Then there exists σ such that

$$\max\{\|a\|, \|b\|, \|c\|\} < \sigma < \tfrac{1}{2} \log\left(2 - 1/\sqrt{2}\right). \tag{5.94}$$

The claimed (5.90) will follow if we are able to prove the following facts

$$\lim_{N\to\infty} \eta_N(\eta_N(a,b),c) = (a \diamond b) \diamond c, \tag{5.95a}$$

$$\lim_{N\to\infty} \eta_N(a, \eta_N(b,c)) = a \diamond (b \diamond c), \tag{5.95b}$$

$$\lim_{N\to\infty} \eta_N(\eta_N(a,b),c) = \lim_{N\to\infty} \eta_N(a, \eta_N(b,c)). \tag{5.95c}$$

We split the proof in two steps.

Proof of (5.95a) *and* (5.95b). It suffices to demonstrate the former, the latter being analogous. We shall prove much more than (5.95a), in proving that

$$\lim_{N,M\to\infty} \eta_N(\eta_M(a,b),c) = (a \diamond b) \diamond c, \qquad a,b,c \in \widetilde{Q}, \tag{5.96}$$

in the sense of *double limits*.[7] In facing (5.96) we make use of a theorem on (interchanging) double limits in a complete metric space, Theorem 5.39 below. Indeed, we claim that

$$\lim_{N\to\infty} \eta_N(\eta_M(a,b),c) = \eta_M(a,b) \diamond c, \qquad \text{uniformly w.r.t. } M, \tag{5.97a}$$

$$\lim_{M\to\infty} \eta_N(\eta_M(a,b),c) = \eta_N(a \diamond b, c), \qquad \text{for every fixed } N \in \mathbb{N}. \tag{5.97b}$$

Then a direct application of Theorem 5.39 to the double sequence $\ell_N^M :=$ $\eta_N(\eta_M(a,b),c)$ in the *complete* metric space $\mathfrak{g}$ (recall that $(\mathfrak{g}, \|\cdot\|)$ is a Banach space by hypothesis!) proves the existence and the equality of the following three limits

$$\lim_{N\to\infty} \left(\lim_{M\to\infty} \eta_N(\eta_M(a,b),c) \right) = \lim_{M\to\infty} \left(\lim_{N\to\infty} \eta_N(\eta_M(a,b),c) \right)$$

$$= \lim_{N,M\to\infty} \eta_N(\eta_M(a,b),c). \tag{5.98}$$

---

[7]We recall that, given a double sequence $\{\ell_{N,M} : N, M \in \mathbb{N}\}$ valued in a metric space (X, d) and given $\ell \in X$, we write

$$\lim_{N,M\to\infty} \ell_{N,M} = \ell \quad \text{in } X,$$

iff for every $\varepsilon > 0$ there exists $N_\varepsilon \in \mathbb{N}$ such that

$$d(\ell_{N,M}, \ell) < \varepsilon, \quad \text{for every } N, M \geq N_\varepsilon.$$

If this holds, obviously the "diagonal" sequence $\{\ell_{N,N}\}_{N\in\mathbb{N}}$ has limit in X and this limit equals ℓ.

Then the following argument applies in showing (5.96):

$$\lim_{N,M\to\infty} \eta_N(\eta_M(a,b),c) \overset{(5.98)}{=} \lim_{N\to\infty}\left(\lim_{M\to\infty}\eta_N(\eta_M(a,b),c)\right)$$

$$\overset{(5.97b)}{=} \lim_{N\to\infty}\eta_N(a\diamond b,c)\overset{(5.92)}{=}(a\diamond b)\diamond c.$$

In the last equality, we are indeed entitled to invoke (5.92) (applied to $x :=$ $a\diamond b$ and $y := c$), for $\widetilde{Q}\subseteq Q$ and since $a\diamond b\in Q$ by (5.91).

We are thus left to prove the claimed (5.97a) and (5.97b):

(5.97a): Let us choose $\delta := \log\sqrt{4-\sqrt{2}}$. Note that $0 < \delta < \log 2$. Hence $\{\eta_N(x,y)\}_N$ converges uniformly on D_δ to $x\diamond y$, that is (see (5.93)), for every $\varepsilon > 0$ there exists $N_\varepsilon\in\mathbb{N}$ such that

$$\|\eta_N(x,y)-x\diamond y\| < \varepsilon \qquad \left(\begin{array}{c}\text{for every }N\geq N_\varepsilon\\ \text{and every }(x,y)\in D_\delta\end{array}\right). \qquad (5.99)$$

We claim that in (5.99) we can take $y := c$ and $x := \eta_M(a,b)$: this follows if we are able to prove that $\|\eta_M(a,b)\| + \|c\| \leq \delta$, which, in turn, derives from the computation:

$$\|\eta_M(a,b)\| + \|c\| \leq \text{ (by Theorem 5.29) } \sum_{n=1}^{M}\frac{\left(e^{\|a\|+\|b\|}-1\right)^n}{n}+\|c\|$$

$$\overset{(5.94)}{\leq} \sum_{n=1}^{\infty}\frac{1}{n}\left(e^{2\sigma}-1\right)^n+\sigma \overset{(5.41)}{=} \log\left(\frac{1}{2-e^{2\sigma}}\right)+\sigma$$

$$\overset{(5.94)}{<} \log\left(\frac{1}{2-\exp\log\left(2-1/\sqrt{2}\right)}\right)+\tfrac{1}{2}\log\left(2-\frac{1}{\sqrt{2}}\right)$$

$$= \log\sqrt{2}+\log\sqrt{2-\frac{1}{\sqrt{2}}} = \log\sqrt{4-\sqrt{2}} = \delta.$$

Taking $y = c$ and $x = \eta_M(a,b)$ in (5.99) gives

$$\|\eta_N(\eta_M(a,b),c)-\eta_M(a,b)\diamond c\| < \varepsilon \qquad \left(\begin{array}{c}\text{for every }N\geq N_\varepsilon\\ \text{and every }M\in\mathbb{N}\end{array}\right).$$

This is precisely (5.97a).

(5.97b): Let $N\in\mathbb{N}$ be fixed. The map

$$\mathfrak{g}\times\mathfrak{g}\ni(x,y)\mapsto\eta_N(x,y)\in\mathfrak{g}$$

is continuous, for this is a Lie polynomial (recall that $\mathfrak{g}$ is a Banach-Lie algebra, whence the map $[\cdot, \cdot]_{\mathfrak{g}}$ is continuous on $\mathfrak{g} \times \mathfrak{g}$ by Remark 5.24 on page 294). Then we have

$$\lim_{M \to \infty} \eta_N(\eta_M(a, b), c) = \eta_N\left(\lim_{M \to \infty} \eta_M(a, b), c\right).$$

On the other hand, the above right-hand side equals $\eta_N(a \diamond b, c)$, thanks to (5.92) (since $a, b \in \widetilde{Q} \subseteq Q$).

Proof of (5.95c). Let $a, b, c \in \widetilde{Q}$ be fixed. We have to prove that

$$\eta_N(\eta_N(a, b), c) - \eta_N(a, \eta_N(b, c))$$

vanishes, as $N \to \infty$. [Note that (5.88) on its own is not enough to end the proof: we need the more "quantitative" information given by formula (5.86).]

By Theorem 5.36, the difference $\eta_N(\eta_N(a, b), c) - \eta_N(a, \eta_N(b, c))$ equals $\mathcal{R}_N(a, b, c)$, where this is given by formula (5.86), replacing x, y, z with a, b, c respectively, and turning everywhere $[\cdot, \cdot]_{\otimes}$ and $Z_j^{\otimes}$ into $[\cdot, \cdot]_{\mathfrak{g}}$ and $Z_j^{\mathfrak{g}}$.

We thus have (recall that $\|[\xi, \eta]_{\mathfrak{g}}\| \leq \|\xi\| \cdot \|\eta\|$ for every $\xi, \eta \in \mathfrak{g}$)

$$\|\mathcal{R}_N(a, b, c)\|$$

$$\leq \sum_{n=1}^{N} \frac{1}{n} \sum_{\substack{(h,k) \in \mathcal{N}_n \\ |h|+|k| \leq N}} \frac{1}{h! \, k!} \sum_{\substack{1 \leq j_1^1, \ldots, j_{k_1}^1, \cdots, j_1^n, \ldots, j_{k_n}^n \leq N \\ h_1 + \cdots + h_n + j_1^1 + \cdots + j_{k_1}^1 + \cdots + j_1^n + \cdots + j_{k_n}^n \geq N+1}}$$

$$\times \|a\|^{|h|} \cdot \left\|Z_{j_1^1}^{\mathfrak{g}}(b, c)\right\| \cdots \left\|Z_{j_{k_1}^1}^{\mathfrak{g}}(b, c)\right\| \cdots \left\|Z_{j_1^n}^{\mathfrak{g}}(b, c)\right\| \cdots \left\|Z_{j_{k_n}^n}^{\mathfrak{g}}(b, c)\right\|$$

$$+ \sum_{n=1}^{N} \frac{1}{n} \sum_{\substack{(h,k) \in \mathcal{N}_n \\ |h|+|k| \leq N}} \frac{1}{h! \, k!} \sum_{\substack{1 \leq j_1^1, \ldots, j_{h_1}^1, \cdots, j_1^n, \ldots, j_{h_n}^n \leq N \\ k_1 + \cdots + k_n + j_1^1 + \cdots + j_{h_1}^1 + \cdots + j_1^n + \cdots + j_{h_n}^n \geq N+1}}$$

$$\times \|c\|^{|k|} \left\|Z_{j_1^1}^{\mathfrak{g}}(a, b)\right\| \cdots \left\|Z_{j_{h_1}^1}^{\mathfrak{g}}(a, b)\right\| \cdots \left\|Z_{j_1^n}^{\mathfrak{g}}(a, b)\right\| \cdots \left\|Z_{j_{h_n}^n}^{\mathfrak{g}}(a, b)\right\|.$$

Now note that for every $j \in \mathbb{N}$ and every $\xi, \eta \in \mathfrak{g}$

$$\|Z_j^{\mathfrak{g}}(\xi, \eta)\| \leq \sum_{n=1}^{j} \frac{1}{n} \sum_{(h,k) \in \mathcal{N}_n : |h|+|k|=j} \frac{1}{h! \, k!} \|D_{(h,k)}^{\mathfrak{g}}(\xi, \eta)\|$$

$$\leq \sum_{n=1}^{j} \frac{1}{n} \sum_{(h,k) \in \mathcal{N}_n : |h|+|k|=j} \frac{\|\xi\|^{|h|} \|\eta\|^{|k|}}{h! \, k!}$$

$$(\text{by } (5.94)) \quad \leq \sum_{n=1}^{j} \frac{1}{n} \sum_{(h,k)\in\mathcal{N}_n:\,|h|+|k|=j} \frac{\sigma^{|h|+|k|}}{h!\,k!}.$$

As a consequence $\|\mathcal{R}_N(a,b,c)\|$ is bounded above by two sums analogous to the following one:

$$\sum_{n=1}^{N} \frac{1}{n} \sum_{\substack{(h,k)\in\mathcal{N}_n \\ |h|+|k|\leq N}} \frac{1}{h!\,k!} \sum_{\substack{1\leq j_1^1,\ldots,j_{k_1}^1,\cdots,j_1^n,\ldots,j_{k_n}^n \leq N \\ |h|+j_1^1+\cdots+j_{k_1}^1+\cdots+j_1^n+\cdots+j_{k_n}^n \geq N+1}} \cdot\, \sigma^{|h|}$$

$$\times \sum_{m_1^1=1}^{j_1^1} \frac{1}{m_1^1} \sum_{\substack{(\alpha_1^1,\beta_1^1)\in\mathcal{N}_{m_1^1} \\ |\alpha_1^1|+|\beta_1^1|=j_1^1}} \frac{\sigma^{|\alpha_1^1|+|\beta_1^1|}}{\alpha_1^1!\,\beta_1^1!} \cdots \sum_{m_{k_1}^1=1}^{j_{k_1}^1} \frac{1}{m_{k_1}^1} \sum_{\substack{(\alpha_{k_1}^1,\beta_{k_1}^1)\in\mathcal{N}_{m_{k_1}^1} \\ |\alpha_{k_1}^1|+|\beta_{k_1}^1|=j_{k_1}^1}} \frac{\sigma^{|\alpha_{k_1}^1|+|\beta_{k_1}^1|}}{\alpha_{k_1}^1!\,\beta_{k_1}^1!} \cdots$$

$$\times \sum_{m_1^n=1}^{j_1^n} \frac{1}{m_1^n} \sum_{\substack{(\alpha_1^n,\beta_1^n)\in\mathcal{N}_{m_1^n} \\ |\alpha_1^n|+|\beta_1^n|=j_1^n}} \frac{\sigma^{|\alpha_1^n|+|\beta_1^n|}}{\alpha_1^n!\,\beta_1^n!} \cdots \sum_{m_{k_n}^n=1}^{j_{k_n}^n} \frac{1}{m_{k_n}^n} \sum_{\substack{(\alpha_{k_n}^n,\beta_{k_n}^n)\in\mathcal{N}_{m_{k_n}^n} \\ |\alpha_{k_n}^n|+|\beta_{k_n}^n|=j_{k_n}^n}} \frac{\sigma^{|\alpha_{k_n}^n|+|\beta_{k_n}^n|}}{\alpha_{k_n}^n!\,\beta_{k_n}^n!}.$$

This series is obtained by preserving only the powers of σ with exponent $\geq N+1$ from the following series:

$$\sum_{n=1}^{N} \frac{1}{n} \sum_{\substack{(h,k)\in\mathcal{N}_n \\ |h|+|k|\leq N}} \frac{1}{h!\,k!} \sum_{1\leq j_1^1,\ldots,j_{k_1}^1,\cdots,j_1^n,\ldots,j_{k_n}^n \leq N} \cdot\, \sigma^{|h|} \times \{\text{as above}\ldots\}.$$

In its turn this last series can be majorized by dropping the requirement that the indices j are $\leq N$ and then by using the identity

$$\sum_{j=1}^{\infty}\sum_{m=1}^{j} \frac{1}{m} \sum_{\substack{(\alpha,\beta)\in\mathcal{N}_m \\ |\alpha|+|\beta|=j}} \frac{\sigma^{|\alpha|+|\beta|}}{\alpha!\,\beta!} = \sum_{m=1}^{\infty} \frac{1}{m} \sum_{j=m}^{\infty} \sum_{\substack{(\alpha,\beta)\in\mathcal{N}_m \\ |\alpha|+|\beta|=j}} \frac{\sigma^{|\alpha|+|\beta|}}{\alpha!\,\beta!}$$

$$= \sum_{m=1}^{\infty} \frac{1}{m} \sum_{\substack{(\alpha,\beta)\in\mathcal{N}_m \\ |\alpha|+|\beta|\geq m}} \frac{\sigma^{|\alpha|+|\beta|}}{\alpha!\,\beta!} = \sum_{m=1}^{\infty} \frac{1}{m} \sum_{(\alpha,\beta)\in\mathcal{N}_m} \frac{\sigma^{|\alpha|+|\beta|}}{\alpha!\,\beta!}$$

$$\overset{(5.38)}{=} \sum_{m=1}^{\infty} \frac{1}{m}(e^{2\sigma}-1)^m \overset{(5.41)}{=} \log\left(\frac{1}{2-e^{2\sigma}}\right).$$

Thus $\|\mathcal{R}_N(a,b,c)\|$ is bounded above by the series obtained by preserving the summands σ^j with $j \geq N+1$ from expansion of the following series:

$$2\sum_{n=1}^{N}\frac{1}{n}\sum_{\substack{(h,k)\in\mathcal{N}_n\\|h|+|k|\le N}}\frac{1}{h!\,k!}\sigma^{|h|}\cdot\left(\log\left(\frac{1}{2-e^{2\sigma}}\right)\right)^{|k|}$$

$$\le 2\sum_{n=1}^{\infty}\frac{1}{n}\sum_{(h,k)\in\mathcal{N}_n}\frac{1}{h!\,k!}\sigma^{|h|}\cdot\left(\log\left(\frac{1}{2-e^{2\sigma}}\right)\right)^{|k|}$$

$$\stackrel{(5.38)}{=}2\sum_{n=1}^{\infty}\frac{1}{n}\left(e^{\sigma}\,(2-e^{2\sigma})^{-1}-1\right)^{n}\stackrel{(5.41)}{=}2\log\left(\frac{2(2-e^{2\sigma})}{e^{\sigma}}\right)=:F(\sigma).$$

[We are indeed entitled to apply (5.41) since $\sigma + \log((2-e^{2\sigma})^{-1}) < \log 2$ in view of (5.94).] Now note that $t \mapsto F(t)$ from the above far right-hand side is the function

$$(-\infty,\tfrac{\log 2}{2}) \ni t \mapsto F(t) := \log 4 - 2\,t + 2\log\left(1 + (1 - e^{2\,t})\right),$$

which is real analytic and which coincides with its Maclaurin expansion in its domain, since $|1 - e^{2\,t}| < 1$ whenever $t < \frac{\log 2}{2}$. We have thus proved

$$\|\mathcal{R}_N(a,b,c)\| \le \sum_{j=N+1}^{\infty}\frac{F^{(j)}(0)}{j!}\,\sigma^j,$$

and the right-hand side vanishes as $N \to \infty$, since the Maclaurin series of F converges at σ, and $\sigma < \frac{\log 2}{2}$ in view of (5.94). This ends the proof. □

In the previous proof we made use of the following result from Analysis.

Theorem 5.39 (Interchanging Double Limits). *Let (X,d) be a complete metric space. Let $\{\ell_n^k\}_{k,n\in\mathbb{N}}$ be a double sequence in X. Suppose that the following hypotheses hold:*

(i) *the limit $\lim_{n\to\infty}\ell_n^k$ exists, uniformly with respect to k;*
(ii) *for every fixed $n \in \mathbb{N}$, the limit $\lim_{k\to\infty}\ell_n^k$ exists.*

Then all the following limits exist and all are equal

$$\lim_{k\to\infty}\left(\lim_{n\to\infty}\ell_n^k\right) = \lim_{n\to\infty}\left(\lim_{k\to\infty}\ell_n^k\right) = \lim_{n,k\to\infty}\ell_n^k. \tag{5.100}$$

Proof. Let us set, for every $k, n \in \mathbb{N}$

$$\ell^k := \lim_{n\to\infty}\ell_n^k, \qquad \ell_n := \lim_{k\to\infty}\ell_n^k,$$

these limits existing in view of hypotheses (i) and (ii), which have the following precise meanings, respectively:

$$\forall \varepsilon > 0 \quad \exists \, N(\varepsilon) \in \mathbb{N}: \quad d(\ell^k, \ell_n^k) < \varepsilon \quad \forall \, n \geq N(\varepsilon), \ \forall \, k \in \mathbb{N};$$
$$\text{(5.101a)}$$

$$\forall \, n \in \mathbb{N}, \ \forall \, \varepsilon > 0 \quad \exists \, K(n, \varepsilon) \in \mathbb{N}: \quad d(\ell_n, \ell_n^k) < \varepsilon \quad \forall \, k \geq K(n, \varepsilon).$$
$$\text{(5.101b)}$$

We claim that the sequence $\{\ell^k\}_{k \in \mathbb{N}}$ is Cauchy in X. Indeed, given $\varepsilon > 0$ let us choose an $N(\varepsilon)$ as in (5.101a) and then let us choose a $K(N(\varepsilon), \varepsilon)$ as in (5.101b). Then for every $h, k \geq K(N(\varepsilon), \varepsilon)$ we have

$$d(\ell^h, \ell^k) \leq d(\ell^h, \ell_{N(\varepsilon)}^h) + d(\ell_{N(\varepsilon)}^h, \ell_{N(\varepsilon)}) + d(\ell_{N(\varepsilon)}, \ell_{N(\varepsilon)}^k) + d(\ell_{N(\varepsilon)}^k, \ell^k) \leq 4\,\varepsilon.$$

Since (X, d) is complete, the limit $\ell := \lim_{k \to \infty} \ell^k$ exists in X. This proves the existence of the first limit in (5.100).

We next claim that $\ell = \lim_{n \to \infty} \ell_n$ (thus proving the existence of the second limit in (5.100) and its equality with the former). Let $\varepsilon > 0$ be given. As $\ell = \lim_{k \to \infty} \ell^k$, there exists $k(\varepsilon) \in \mathbb{N}$ such that $d(\ell, \ell^k) < \varepsilon$ for every $k \geq k(\varepsilon)$. Moreover, if $N(\varepsilon)$ is as in (5.101a), for every $n \geq N(\varepsilon)$ there exists $K(n, \varepsilon)$ such that (5.101b) holds. Let us set $\nu(n, \varepsilon) := k(\varepsilon) + K(n, \varepsilon)$. Then for every $n \geq N(\varepsilon)$ we have

$$d(\ell, \ell_n) \leq d(\ell, \ell^{\nu(n, \varepsilon)}) + d(\ell^{\nu(n, \varepsilon)}, \ell_n^{\nu(n, \varepsilon)}) + d(\ell_n^{\nu(n, \varepsilon)}, \ell_n) < 3\,\varepsilon.$$

This proves the second claim. Finally, as for the third limit in (5.100) and its equality with the others, we argue as follows. Let $\varepsilon > 0$ be given and let $k(\varepsilon)$ be as above and $N(\varepsilon)$ be as in (5.101a): then, for every $k, n \geq k(\varepsilon) + N(\varepsilon)$,

$$d(\ell, \ell_n^k) \leq d(\ell, \ell^k) + d(\ell^k, \ell_n^k) < 2\,\varepsilon.$$

This completes the proof.   □

Remark 5.40. The hypothesis of completeness of X in Theorem 5.39 can be removed – it being understood that (i) and (ii) are still assumed – provided it is replaced by one of the following conditions:

(iii) The limit $\lim_{k \to \infty} \left(\lim_{n \to \infty} \ell_n^k \right)$ exists and is in X.
(iii)′ The limit $\lim_{n \to \infty} \left(\lim_{k \to \infty} \ell_n^k \right)$ exists and is in X.

Remark 5.41. Another possible (very compact) way of rewriting the CBHD operation on a Banach-Lie algebra $\mathfrak{g}$ is the following one:

$$x + \int_0^1 \Psi(e^{\mathrm{ad}\, x} \circ e^{t\,\mathrm{ad}\, y})(y) \, \mathrm{d}t,$$

where $\Psi(z) = z \log(z)/(z - 1)$ (which is analytic in the complex open disc about $1 \in \mathbb{C}$ of radius 1) and x, y are sufficiently close to $0 \in \mathfrak{g}$.

5.4   Nilpotent Lie Algebras and the Third Theorem of Lie

The aim of this section is to consider the CBHD operation $\diamond$ on a nilpotent
Lie algebras $\mathfrak{n}$: We shall prove that, in this context, $(\mathfrak{n}, \diamond)$ is a group, so
that $\diamond$ is in particular (globally) associative (Sect. 5.4.1). Furthermore, we
prove that, as long as $\mathfrak{n}$ is also finite-dimensional – besides being nilpotent
– then $(\mathfrak{n}, \diamond)$ is a *Lie group*, whose Lie algebra is isomorphic to $\mathfrak{n}$ itself. This
solves the so-called Third Fundamental Theorem of Lie (in its *global* form),
for finite-dimensional nilpotent Lie algebras, thus furnishing a remarkable
application of the CBHD operation (Sect. 5.4.2).

5.4.1   Associativity for Nilpotent Lie Algebras

Suppose $\mathbb{K}$ is a field of characteristic zero and that $\mathfrak{n}$ is a *nilpotent* Lie algebra
over $\mathbb{K}$, with step of nilpotency $r \in \mathbb{N}$. We recall that this has the following
meaning: Introducing the *descending central series* of $\mathfrak{n}$

$$\mathfrak{n}_1 := \mathfrak{n}, \quad \mathfrak{n}_{n+1} := [\mathfrak{n}, \mathfrak{n}_n] = \mathrm{span}\{[g, g_n] : g \in \mathfrak{n}, \ g_n \in \mathfrak{n}_n\} \quad (n \in \mathbb{N}), \tag{5.102}$$

then $\mathfrak{n}$ is nilpotent of step r iff $\mathfrak{n}_r \neq \{0\}$ and $\mathfrak{n}_{r+1} = \{0\}$.
    We use the notations introduced at the beginning of Sect. 5.2.1:

$$D_{(h,k)}^{\mathfrak{n}}, \quad \mathcal{N}_n, \quad c_n, \quad \mathbf{c}(h, k), \quad \eta_N^{\mathfrak{n}}, \quad Z_j^{\mathfrak{n}}, \quad \ldots$$

For the sake of brevity, the superscript "$\mathfrak{n}$" will frequently be omitted;
analogously we shall denote the Lie bracket on $\mathfrak{n}$ simply by $[\cdot, \cdot]$.
    Since $\mathfrak{n}$ is nilpotent of step r, the formal series which defines the operation
$\diamond$ on $\mathfrak{n}$ reduces to a finite sum. We thus set

$$\diamond_{\mathfrak{n}} : \mathfrak{n} \times \mathfrak{n} \longrightarrow \mathfrak{n}$$

$$\xi \diamond_{\mathfrak{n}} \eta := \sum_{n=1}^{r} \frac{(-1)^{n+1}}{n} \sum_{(h,k) \in \mathcal{N}_n \,:\, |h|+|k| \leq r} \frac{D_{(h,k)}^{\mathfrak{n}}(\xi, \eta)}{h! \, k! \, (|h| + |k|)}. \tag{5.103}$$

Our main result for this section is the following theorem (which is usually
treated in the literature as a folklore fact; it is our firm opinion that it
actually deserves a respectful proof). Note that *we make no hypothesis of finite-
dimensionality for $\mathfrak{n}$.*

Theorem 5.42 (The CBHD Operation on a Nilpotent Lie Algebra). *Let $\mathfrak{n}$ be
a nilpotent Lie algebra over a field of null characteristic. Let $r \in \mathbb{N}$ denote the step
of nilpotency of $\mathfrak{n}$. Let also $\diamond_{\mathfrak{n}}$ be the operation on $\mathfrak{n}$ defined in (5.103).*

Then $(\mathfrak{n}, \diamond_{\mathfrak{n}})$ is a group with identity 0 and inversion given by $\xi \mapsto -\xi$. In particular, $\diamond_{\mathfrak{n}}$ is associative on $\mathfrak{n}$.

Proof. Obviously, $a \diamond_{\mathfrak{n}} 0 = 0 \diamond_{\mathfrak{n}} a = a$ for every $a \in \mathfrak{n}$. Moreover,

$$D^{\mathfrak{n}}_{(h,k)}(\lambda\, a, \mu\, a) = 0, \qquad \text{for every} \qquad \left(\begin{array}{l} \lambda, \mu \in \mathbb{K}, \ a \in \mathfrak{n} \\ (h, k) \in \mathcal{N}_n \ : \ |h| + |k| \geq 2 \end{array} \right).$$

Consequently, this gives

$$(\lambda\, a) \diamond_{\mathfrak{n}} (\mu\, a) = D^{\mathfrak{n}}_{(1,0)}(\lambda\, a, \mu\, a) + D^{\mathfrak{n}}_{(0,1)}(\lambda\, a, \mu\, a) = \lambda\, a + \mu\, a,$$

so that $a \diamond_{\mathfrak{n}} (-a) = (-a) \diamond_{\mathfrak{n}} a = 0$, for all $a \in \mathfrak{n}$.

So we are left to prove the associativity of $\diamond_{\mathfrak{n}}$. We provide two proofs of this fact. The first one makes use of the identities concerning the associativity of $\diamond$ on a Lie algebra obtained in Theorem 5.36 (which actually required some hard work on "truncating" the $\diamond$ operation). The second proof, independent of this latter machinery, goes back directly to the associativity of the $\diamond$ operation on $\widehat{\mathscr{T}}(V)$, which is our original CBHD Theorem (and it also makes use of some general results on nilpotent Lie algebras of independent interest).

First Proof. Let $a, b, c \in \mathfrak{n}$ be given. We apply Theorem 5.36 when $\mathfrak{g}$ is our nilpotent Lie algebra $\mathfrak{n}$ and N is its step of nilpotency r. Note that the far left-hand side of (5.88) is exactly $a \diamond_{\mathfrak{n}} (b \diamond_{\mathfrak{n}} c) - (a \diamond_{\mathfrak{n}} b) \diamond_{\mathfrak{n}} c$. The statement of that theorem ensures that this difference belongs to

$$\underbrace{[\mathfrak{n}, [\cdots, \mathfrak{n}] \cdots]}_{r + 1 \text{ times}} = \mathfrak{n}_{r+1} = \{0\}.$$

Thus, associativity is proved.

Second Proof. Let $a, b, c \in \mathfrak{n}$ be given. Let $\{x, y, z\}$ be a set of cardinality 3. By the universal property of $W := \mathcal{L}(\mathbb{K}\langle x, y, z \rangle)$ in Theorem 2.85-(2b), there exists a unique LA morphism $\Phi_{a,b,c} : W \to \mathfrak{n}$ such that

$$\Phi_{a,b,c}(x) = a, \quad \Phi_{a,b,c}(y) = b, \quad \text{and} \quad \Phi_{a,b,c}(z) = c. \tag{5.104}$$

By Lemma 5.43 below, there exists an LA morphism $\overline{\Phi} : \overline{W} \longrightarrow \mathfrak{n}$ prolonging $\Phi_{a,b,c}$ and with the property:

$$\overline{\Phi}(\ell) = 0, \quad \text{for all } \ell \in \prod_{k=r+1}^{\infty} \mathcal{L}_k(\mathbb{K}\langle x, y, z \rangle). \tag{5.105}$$

Here $\overline{W}$ is the closure of W as a subset of the usual topological space $\widehat{\mathscr{T}}(\mathbb{K}\langle x, y, z \rangle)$ and $\mathcal{L}_k$ is defined in the usual way. We claim that

$$\overline{\Phi}(t \diamond t') = \overline{\Phi}(t) \diamond_n \overline{\Phi}(t'), \qquad \forall\, t, t' \in \overline{W}. \tag{5.106}$$

To prove this, first note that $t \diamond t' \in \overline{W}$ for every $t, t' \in \overline{W}$. Indeed, we have

$$t \diamond t' = \sum_{n=1}^{\infty} c_n \sum_{(h,k) \in \mathbb{N}_n} c(h, k) \underbrace{D^{\otimes}_{(h,k)}(t, t')}_{\in \overline{W}},$$

since $\overline{W}$ is a Lie subalgebra of $\widehat{\mathscr{T}}(\mathbb{K}\langle x, y, z \rangle)$ (see Remark 3.17 on page 139); hence $t \diamond t'$ itself belongs to $\overline{W}$, since it is the sum of a convergent series in $\overline{W}$. To prove (5.106), we argue as follows:

$$\overline{\Phi}(t \diamond t') = \overline{\Phi}\Big(\sum_{n=1}^{r} c_n \sum_{(h,k) \in \mathbb{N}_n\,:\,|h|+|k| \leq r} c(h, k)\, D^{\otimes}_{(h,k)}(t, t') +$$

$$+ \underbrace{\sum_{n=r+1}^{\infty} c_n \sum_{(h,k) \in \mathbb{N}_n} \{\cdots\} + \sum_{n=1}^{r} c_n \sum_{(h,k) \in \mathbb{N}_n\,:\,|h|+|k| \geq r+1} \{\cdots\}}_{\in \prod_{k=r+1}^{\infty} \mathcal{L}_k(\mathbb{K}\langle x,y,z \rangle)} \Big)$$

$$\overset{(5.105)}{=} \sum_{n=1}^{r} c_n \sum_{(h,k) \in \mathbb{N}_n\,:\,|h|+|k| \leq r} c(h, k)\, \overline{\Phi}\big(D^{\otimes}_{(h,k)}(t, t') \big)$$

($\overline{\Phi}$ is a Lie algebra morphism)

$$= \sum_{n=1}^{r} c_n \sum_{(h,k) \in \mathbb{N}_n\,:\,|h|+|k| \leq r} c(h, k)\, D^{\otimes}_{(h,k)}\big(\overline{\Phi}(t), \overline{\Phi}(t') \big) = \overline{\Phi}(t) \diamond_n \overline{\Phi}(t').$$

Since $\diamond$ is associative on $\overline{W}$, we have

$$(\bigstar) \qquad x \diamond (y \diamond z) = (x \diamond y) \diamond z.$$

Observing that

$$x, y, z, \quad y \diamond z,\ x \diamond y, \quad x \diamond (y \diamond z), \quad (x \diamond y) \diamond z$$

all belong to $\overline{W}$, we can apply $\overline{\Phi}$ to both sides of $(\bigstar)$ and we are entitled to make use of (5.106) twice on both sides, getting

$$\overline{\Phi}(x) \diamond_n (\overline{\Phi}(y) \diamond_n \overline{\Phi}(z)) = (\overline{\Phi}(x) \diamond_n \overline{\Phi}(y)) \diamond_n \overline{\Phi}(z).$$

Recalling that $\overline{\Phi}$ prolongs $\Phi_{a,b,c}$ and that (5.104) holds, this is equivalent to $a \diamond_n (b \diamond_n c) = (a \diamond_n b) \diamond_n c$ and the proof is complete. $\qquad \square$

Here we used the following result.

Lemma 5.43. *Let $\mathfrak{n}$ be a nilpotent Lie algebra over the (arbitrary) field $\mathbb{K}$. Let V be a $\mathbb{K}$-vector space and let $f : V \to \mathfrak{n}$ be any linear map. If $\mathcal{L}(V)$ is the free Lie algebra generated by V, we denote by $\overline{f} : \mathcal{L}(V) \to \mathfrak{n}$ the unique Lie algebra morphism prolonging f (this exists by Theorem 2.49-(i), on page 86).*

Then there exists a Lie algebra morphism $\overline{\overline{f}} : \overline{\mathcal{L}(V)} \to \mathfrak{n}$ prolonging $\overline{f}$, with the additional property

$$\overline{\overline{f}}(\ell) = 0, \quad \text{for all } \ell \in \prod_{k=r+1}^{\infty} \mathcal{L}_k(V), \tag{5.107}$$

where r is the step of nilpotency of $\mathfrak{n}$. Finally, $\overline{\overline{f}}$ is the unique Lie algebra morphism from $\overline{\mathcal{L}(V)}$ to $\mathfrak{n}$ prolonging f and satisfying (5.107).

Proof. First we prove uniqueness. Let $r \in \mathbb{N}$ denote the step of nilpotency of $\mathfrak{n}$. Let $F : \overline{\mathcal{L}(V)} \to \mathfrak{n}$ be an LA morphism prolonging f and null on $\prod_{k=r+1}^{\infty} \mathcal{L}_k(V)$. We first claim that F also prolongs $\overline{f} : \mathcal{L}(V) \to \mathfrak{n}$. This is equivalent to $F|_{\mathcal{L}(V)} \equiv \overline{f}$, which follows by the fact that $F|_{\mathcal{L}(V)}$ is an LA morphism from $\mathcal{L}(V)$ to $\mathfrak{n}$ prolonging f, a property which uniquely characterizes $\overline{f}$ (by the cited Theorem 2.49-(i)). Moreover, for every $\ell = (\ell_k)_{k \geq 1} \in \prod_{k=1}^{\infty} \mathcal{L}_k(V)$, the following computation applies:

$$F(\ell) = F(\ell_1, \ell_2, \ldots, \ell_r, 0, 0, \ldots) + F(0, 0, \ldots, 0, \ell_{r+1}, \ell_{r+2}, \ldots)$$

$$= F(\ell_1, \ell_2, \ldots, \ell_r, 0, 0, \ldots) = \sum_{k=1}^{r} F(\ell_k) = \sum_{k=1}^{r} \overline{f}(\ell_k).$$

In the last equality we used the fact that $\ell_k \in \mathcal{L}_k(V) \subseteq \mathcal{L}(V)$, for every $k \in \mathbb{N}$ together with the fact that F prolongs $\overline{f}$. The above argument proves that (since $\overline{f}$ is uniquely determined by f) F is uniquely determined by f.

We now prove the existence part. With the above notation, we set

$$F : \overline{\mathcal{L}(V)} \to \mathfrak{n}, \quad F\big((\ell_k)_{k \geq 1}\big) := \sum_{k=1}^{r} \overline{f}(\ell_k),$$

where $\ell_k \in \mathcal{L}_k(V)$ for every $k \in \mathbb{N}$. We show that $\overline{\overline{f}} := F$ has the properties claimed in the assertion of the lemma.

To begin with, F is obviously well-posed and linear. Moreover, it clearly prolongs $\overline{f}$ on $\mathcal{L}(V)$, for any element of $\mathcal{L}(V)$ has the form $(\ell_k)_{k \geq 1}$, where the ℓ_k are null for k large enough. Furthermore, property (5.107) immediately follows from the fact that any element of $\prod_{k=r+1}^{\infty} \mathcal{L}_k(V)$ has the form $(\ell_k)_{k \geq 1}$ with $\ell_k \in \mathcal{L}_k(V)$ for every $k \in \mathbb{N}$ and with $\ell_1 = \ell_2 = \cdots = \ell_r = 0$.

We are left to show that F is an LA morphism. This follows from the computation below: let $a = (a_k)_k$ and $b = (b_k)_k$ be arbitrary elements of

$\overline{\mathcal{L}(V)}$ (that is, $a_k, b_k \in \mathcal{L}_k(V)$ for every $k \in \mathbb{N}$); then we have

$$F([a,b]_\otimes) = F\Big(\big(\textstyle\sum_{i+j=k}[a_i,b_j]_\otimes\big)_{k\in\mathbb{N}}\Big) \overset{(1)}{=} \sum_{k=1}^{r} \overline{f}(\textstyle\sum_{i+j=k}[a_i,b_j]_\otimes)$$

$$\overset{(2)}{=} \sum_{k=1}^{r} \sum_{i+j=k} \big[\overline{f}(a_i), \overline{f}(b_j)\big]_\mathfrak{n} \overset{(3)}{=} \Big[\textstyle\sum_{i=1}^{r}\overline{f}(a_i), \sum_{j=1}^{r}\overline{f}(b_j)\Big]_\mathfrak{n}$$

$$\overset{(4)}{=} [F(a), F(b)]_\mathfrak{n}.$$

Here we used the following facts:

1. $\sum_{i+j=k}[a_i,b_j]_\otimes \in \mathcal{L}_k(V)$ for every $k \in \mathbb{N}$, together with the definition of F.
2. $\overline{f}$ is an LA morphism.
3. We used the r-step nilpotency of $\mathfrak{n}$ together with the following argument. Since $a_i \in \mathcal{L}_i(V)$ then a_i is a linear combination of $\otimes$-brackets of length i of elements of V, so that – by the LA morphism property of $\overline{f}$ – we deduce that $\overline{f}(a_i)$ is a linear combination of $\mathfrak{n}$-brackets of length i of elements of $\mathfrak{n}$, that is, an element of $\mathfrak{n}_i$ (see the notation in (5.102)). Thus, whenever $i + j \geq r + 1$ we have – by (2.11) on page 60 –

$$\big[\overline{f}(a_i), \overline{f}(b_j)\big]_\mathfrak{n} \in [\mathfrak{n}_i, \mathfrak{n}_j]_\mathfrak{n} \subseteq \mathfrak{n}_{i+j} = \{0\}.$$

4. The very definition of F.

This ends the proof.   □

Here is another remarkable property of the group $(\mathfrak{n}, \diamond_\mathfrak{n})$ in Theorem 5.42.

Lemma 5.44. *Let $\mathfrak{n}$ be a nilpotent Lie algebra over a field of null characteristic. Let $r \in \mathbb{N}$ denote the step of nilpotency of $\mathfrak{n}$. Let also $\diamond_\mathfrak{n}$ be the operation on $\mathfrak{n}$ defined in (5.103). Then the group $(\mathfrak{n}, \diamond_\mathfrak{n})$ is nilpotent of step r.*

Proof. We drop the notation $\diamond_\mathfrak{n}$ and replace it with $\diamond$. If we set

$$\alpha : \mathfrak{n} \times \mathfrak{n} \to \mathfrak{n}, \quad \alpha(X,Y) := X \diamond Y \diamond (-X) \diamond (-Y), \tag{5.108}$$

we need to show that

$$\alpha(X_{r+1}, \cdots \alpha(X_3, \alpha(X_2, X_1)) \cdots) = 0, \quad \text{for every } X_1, \ldots, X_{r+1} \in \mathfrak{n}, \tag{5.109}$$

and that there exist $X_1, \ldots, X_r \in \mathfrak{n}$ such that

$$\alpha(X_r, \cdots \alpha(X_3, \alpha(X_2, X_1)) \cdots) \neq 0. \tag{5.110}$$

We know that, as for the CBHD operation $\diamond$, we can write

$$X \diamond Y = X + Y + H_2(X, Y)$$
$$= X + Y + \tfrac{1}{2}[X, Y] + H_3(X, Y) \qquad \forall\, X, Y \in \mathfrak{n}, \tag{5.111}$$

where H_2 (respectively, H_3) is a Lie polynomial in $\mathfrak{n}$, sums of Lie monomials of heights in $\{2, \ldots, r\}$ (respectively, in $\{3, \ldots, r\}$). We know very well that these polynomials can be written in a "universal" way, suitable for all nilpotent Lie algebras of step r. So, more correctly, we can think of H_2 and H_3 as *functions* defined on $\mathfrak{n} \times \mathfrak{n}$ and taking values in $\mathfrak{n}$ such that $H_2(X, Y)$ and $H_3(X, Y)$ are obtained by substituting X for x and Y for y in two well determined Lie polynomials belonging to the free Lie algebra $\mathcal{L}(\mathbb{Q}\langle x, y \rangle)$ (here $\mathbb{Q}\langle x, y \rangle$ denotes the free vector space over $\mathbb{Q}$ on two non-commuting indeterminates x, y): more precisely, we have

$$H_2(x, y) \in \bigoplus_{k=2}^{r} \mathcal{L}_k(\mathbb{Q}\langle x, y \rangle), \qquad H_3(x, y) \in \bigoplus_{k=3}^{r} \mathcal{L}_k(\mathbb{Q}\langle x, y \rangle).$$

We begin with some enlightening computations which will explain the general background ideas.

- We claim that there exists $R_3(x, y) \in \bigoplus_{k=3}^{r} \mathcal{L}_k(\mathbb{Q}\langle x, y \rangle)$ such that

$$Y \diamond X \diamond (-Y) \diamond (-X) = [Y, X] + R_3(X, Y), \qquad \forall\, X, Y \in \mathfrak{n}. \tag{5.112}$$

(Recall that $\diamond$ is associative!) Indeed (5.111) gives

$$(Y \diamond X) \diamond ((-Y) \diamond (-X))$$
$$= Y \diamond X + (-Y) \diamond (-X) + \tfrac{1}{2}[Y \diamond X, (-Y) \diamond (-X)]$$
$$\quad + H_3(Y \diamond X, (-Y) \diamond (-X))$$
$$= (Y + X + \tfrac{1}{2}[Y, X] + H_3(Y, X)) + (-Y - X + \tfrac{1}{2}[Y, X] + H_3(-Y, -X))$$
$$\quad + \tfrac{1}{2}[Y + X + H_2(Y, X), -(Y + X) + H_2(-Y, -X)] + P_3(X, Y)$$
$$= [Y, X] + H_3(Y, X) + H_3(-Y, -X)$$
$$\quad + \tfrac{1}{2}([Y + X, -(Y + X)] + P_3'(X, Y)) + P_3(X, Y)$$
$$= [Y, X] + R_3(X, Y),$$

for some $P_3, P_3', R_3 \in \bigoplus_{k=3}^{r} \mathcal{L}_k(\mathbb{Q}\langle x, y \rangle)$. Hence (5.112) follows.

- We claim that there exists $R_4(x, y, z) \in \bigoplus_{k=4}^{r} \mathcal{L}_k(\mathbb{Q}\langle x, y, z \rangle)$ such that

$$Z \diamond Y \diamond X \diamond (-Y) \diamond (-X) \diamond (-Z) \diamond X \diamond Y \diamond (-X) \diamond (-Y)$$
$$= [Z, [Y, X]] + R_4(X, Y, Z), \qquad \text{for every } X, Y, Z \in \mathfrak{n}. \tag{5.113}$$

In order to prove (5.113), we start by recalling that $-W$ is the $\diamond$-inverse of W (for any $W \in \mathfrak{n}$). As a consequence we immediately get

$$X \diamond Y \diamond (-X) \diamond (-Y) = -(Y \diamond X \diamond (-Y) \diamond (-X)). \tag{5.114}$$

Recalling the notation in (5.108) note that (5.112) reads as

$$\alpha(Y, X) = [Y, X] + R_3(X, Y), \qquad \text{for every } X, Y \in \mathfrak{n}. \tag{5.115}$$

From (5.114) and (5.115) it follows that

$$
\begin{aligned}
Z &\diamond (Y \diamond X \diamond (-Y) \diamond (-X)) \diamond (-Z) \diamond (X \diamond Y \diamond (-X) \diamond (-Y)) \\
&= Z \diamond \alpha(Y, X) \diamond (-Z) \diamond (-\alpha(Y, X)) \\
&= \alpha(Z, \alpha(Y, X)) \\
&\overset{(5.115)}{=} [Z, \alpha(Y, X)] + R_3(\alpha(Y, X), Z) \\
&\overset{(5.115)}{=} [Z, [Y, X] + R_3(X, Y)] + R_3([Y, X] + R_3(X, Y), Z) \\
&= [Z, [Y, X]] + R_4(X, Y, Z),
\end{aligned}
$$

for some $R_4(x, y) \in \bigoplus_{k=4}^{r} \mathcal{L}_k(\mathbb{Q}\langle x, y \rangle)$. This proves (5.113). Note that the latter can be rewritten (again recalling the notation in (5.108)) as

$$\alpha(Z, \alpha(Y, X)) = [Z, [Y, X]] + R_4(X, Y, Z), \quad \text{for every } X, Y, Z \in \mathfrak{n}. \tag{5.116}$$

To go on with the proof of the present Lemma 5.44, we take the opportunity to state and prove a result which has importance in its own right.

Lemma 5.45. *Let $q \in \mathbb{N}$, $q \geq 2$. Let $\{x_1, \ldots, x_q\}$ denote a set of cardinality q. Then there exists a Lie polynomial*

$$R_{q+1}(x_1, \ldots, x_q) \in \bigoplus_{k=q+1}^{\infty} \mathcal{L}_k(\mathbb{Q}\langle x_1, \ldots, x_q \rangle)$$

such that, for every nilpotent Lie algebra $\mathfrak{n}$ (over a field of characteristic zero) and every choice of $X_1, \ldots, X_q \in \mathfrak{n}$

$$
\begin{aligned}
\alpha(X_q, &\cdots \alpha(X_3, \alpha(X_2, X_1)) \cdots) \\
&= [X_q, \cdots [X_3, [X_2, X_1]] \cdots] + R_{q+1}(X_1, \ldots, X_q),
\end{aligned}
\tag{5.117}
$$

where $\alpha(X, Y) = X \diamond Y \diamond (-X) \diamond (-Y)$ and $\diamond$ is the CBHD operation $\diamond_{\mathfrak{n}}$ introduced in (5.103).

Proof. We argue by induction on $q \geq 2$. The case $q = 2$ follows from (5.112). Let us now prove the statement for $q + 1$, assuming it to be true for q. The following computation applies:

$$\alpha(X_{q+1}, \alpha(X_q, \cdots \alpha(X_3, \alpha(X_2, X_1)) \cdots))$$

$$\overset{(5.115)}{=} \left[X_{q+1}, \alpha(X_q, \cdots \alpha(X_3, \alpha(X_2, X_1)) \cdots) \right]$$

$$+ R_3(\alpha(X_q, \cdots \alpha(X_3, \alpha(X_2, X_1)) \cdots), X_{q+1})$$

(by the inductive hypothesis)

$$= \left[X_{q+1}, [X_q, \cdots [X_3, [X_2, X_1]] \cdots] + R_{q+1}(X_1, \ldots, X_q) \right]$$

$$+ R_3 \Big([X_q, \cdots [X_3, [X_2, X_1]] \cdots] + R_{q+1}(X_1, \ldots, X_q), X_{q+1} \Big)$$

$$= [X_{q+1}, [X_q, \cdots [X_3, [X_2, X_1]] \cdots]] +$$

$$+ \Big\{ [X_{q+1}, R_{q+1}(X_1, \ldots, X_q)]$$

$$+ R_3 \big([X_q, \cdots [X_3, [X_2, X_1]] \cdots] + R_{q+1}(X_1, \ldots, X_q), X_{q+1} \big) \Big\}$$

$$= [X_{q+1}, [X_q, \cdots [X_3, [X_2, X_1]] \cdots]] + \{ R_{q+2}(X_1, \ldots, X_q, X_{q+1}) \},$$

for some $R_{q+2}(x_1, \ldots, x_q, x_{q+1}) \in \bigoplus_{k=q+2}^{\infty} \mathcal{L}_k(\mathbb{Q}\langle x_1, \ldots, x_q, x_{q+1} \rangle)$. $\square$

We are now able to end the proof of Lemma 5.44. To begin with, (5.109) immediately follows from (5.117), choosing $q = r + 1$. Indeed, in this case both

$$[X_{r+1}, \cdots [X_2, X_1] \cdots] \quad \text{and} \quad R_{r+2}(X_1, \ldots, X_{r+1})$$

vanish identically, since $\mathfrak{n}$ is nilpotent of step r, the latter vanishing in view of the fact that

$$R_{r+2} \in \bigoplus_{k=r+2}^{\infty} \mathcal{L}_k(\mathbb{Q}\langle x_1, \ldots, x_q \rangle).$$

The proof of (5.110) is even simpler: If we take $q = r$ in (5.117), we have the very precise identity

$$\alpha(X_r, \cdots \alpha(X_2, X_1) \cdots) = [X_r, \cdots [X_2, X_1] \cdots], \qquad (5.118)$$

holding true due to the fact that $R_{r+1}(X_1, \ldots, X_r) = 0$, since R_{r+1} is a sum of Lie monomials with length $\geq r + 1$ (recall that $\mathfrak{n}$ is nilpotent of step r). Now, there exists at least one r-tuple $X_1, \ldots, X_r$ in $\mathfrak{n}$ for which the right-hand side of (5.118) is non-vanishing, because $\mathfrak{n}$ has step of nilpotency equal to r. Thus the same is true of the left-hand side of (5.118). This ends the proof. $\square$

5.4.2 The Global Third Theorem of Lie for Nilpotent Lie Algebras

We are now ready to prove a central result, a significant application of the CBHD operation in the context of Lie groups. In the proofs below, we suppose that the Reader is familiar with basic notions of Lie groups.

Throughout this section, $\mathfrak{n}$ will denote a fixed *real* nilpotent Lie algebra. We denote by r its step of nilpotency. As usual, we denote by

$$\diamond_\mathfrak{n} : \mathfrak{n} \times \mathfrak{n} \longrightarrow \mathfrak{n}$$

$$\xi \diamond_\mathfrak{n} \eta := \sum_{j=1}^{r} \sum_{n=1}^{j} \frac{(-1)^{n+1}}{n} \sum_{(h,k)\in\mathcal{N}_n \,:\, |h|+|k|=j} \frac{1}{h! \, k! \, (|h|+|k|)} \tag{5.119}$$
$$\times (\operatorname{ad}\xi)^{h_1} (\operatorname{ad}\eta)^{k_1} \cdots (\operatorname{ad}\xi)^{h_n} (\operatorname{ad}\eta)^{k_n - 1}(\eta)$$

the CBHD operation on $\mathfrak{n}$. We have the following remarkable result.

Theorem 5.46. *Let $\mathfrak{n}$ be a finite-dimensional real nilpotent Lie algebra. Let r be its step of nilpotency. Let $\diamond_\mathfrak{n}$ be as in (5.119).*

Then $(\mathfrak{n}, \diamond_\mathfrak{n})$ is a Lie group whose Lie algebra is isomorphic to $\mathfrak{n}$.

More specifically, $(\mathfrak{n}, \diamond_\mathfrak{n})$ is nilpotent of step r and the underlying manifold is analytic, connected and simply connected. Indeed, via a global chart, we can identify $\mathfrak{n}$ with $\mathbb{R}^N$, where $N = \dim_\mathbb{R}(\mathfrak{n})$, and $\diamond_\mathfrak{n}$ can be expressed, in this global chart, as a polynomial function of the associated coordinates.

The proof of this theorem is postponed to page 334.

We recall that, given a finite-dimensional Lie algebra $\mathfrak{g}$, the existence of a Lie group whose Lie algebra is isomorphic to $\mathfrak{g}$ is known as the *global* version of the Third Fundamental Theorem of Lie[8] (see, e.g., [171, Theorem 3.15.1, page 230]). Thanks to Theorem 5.46, we are able to prove the global version of Lie's Third Theorem for finite dimensional real nilpotent Lie algebras, in a very direct and simple way: by making use of the CBHD operation.

Theorem 5.47 (Global Third Theorem of Lie for Nilpotent Lie Algebras). *Suppose $\mathfrak{n}$ is a finite-dimensional real nilpotent Lie algebra.*

Then there exists a simply connected analytic Lie group whose Lie algebra is isomorphic to $\mathfrak{n}$.

Proof. It suffices to take the group $(\mathfrak{n}, \diamond_\mathfrak{n})$ as in Theorem 5.46 above.   □

Remark 5.48. Actually, by analyzing our arguments below, we will prove much more. Even in absence of the hypothesis of nilpotency, it will turn

---

[8]Many authors simply call it "the Third Fundamental Theorem of Lie".

out that the "local group" defined by the CBHD series on a neighborhood of the identity of any finite-dimensional real Lie algebra $\mathfrak{g}$ (see Theorem 5.17) is such that, roughly speaking, the associated "locally left-invariant" vector fields form a Lie algebra isomorphic to $\mathfrak{g}$ itself. More precisely, by our proofs below we are able to prove the following fact:

Let $\mathfrak{g}$ be a finite-dimensional real Lie algebra which is endowed with a (globally defined) operation turning it into a Lie group and such that this operation coincides with the CBHD series $\diamond$ on a neighborhood of the identity 0. Then the Lie algebra of $\mathfrak{g}$ is isomorphic to $\mathfrak{g}$ itself.

In order to give the proof of Theorem 5.46 (see page 334), we first need to recall some simple facts about Lie groups (we shall assume the Reader to be sufficiently familiar with the basic definitions).

Let $(G, \cdot)$ be a (real, smooth) Lie group with Lie algebra $\mathfrak{g}$ (thought of as the set of smooth left-invariant vector fields on G). We denote by N the dimension of G (as a smooth manifold) and by e the identity of G. We know that $\mathfrak{g}$ is N-dimensional and can be identified with the tangent space to G at e, denoted henceforth by $T_e(G)$. The associated natural identification (which is an isomorphism of vector spaces) is

$$\alpha : \mathfrak{g} \to T_e(G), \quad \alpha(X) := X_e. \tag{5.120}$$

The inverse of α is the function which maps a given $\mathbf{v} \in T_e(G)$ into the (smooth and left-invariant) vector field X such that

$$X_x = \mathrm{d}_e \tau_x(\mathbf{v}), \qquad \text{for all } x \in G.$$

Here τ_x is the left-translation by x on G and $\mathrm{d}_e \tau_x$ denotes the differential of τ_x at the identity.

The natural Lie bracket $[\cdot, \cdot]_{\mathfrak{g}}$ on $\mathfrak{g}$ (the Lie bracket of vector fields on the manifold G, which is nothing but the commutator resulting from the enveloping algebra of G, i.e., the algebra of the smooth linear differential operators – of any order – on G, equipped with the usual operation of composition) is "pushed-forward" by α to an operation on $T_e(G)$, obviously endowing $T_e(G)$ with the structure of a *Lie algebra isomorphic to* $\mathfrak{g}$ and such that α is a Lie algebra isomorphism. We denote this operation by $[\cdot, \cdot]_e$: explicitly

$$[\mathbf{u}, \mathbf{v}]_e := \alpha\Big(\big[\alpha^{-1}(\mathbf{u}), \alpha^{-1}(\mathbf{v})\big]_{\mathfrak{g}} \Big), \qquad \text{for every } \mathbf{u}, \mathbf{v} \in T_e(G). \tag{5.121}$$

Let us fix a local chart (U, φ) centered at the identity e of G (this means that U is an open neighborhood of e in G and that φ is a homeomorphism of U onto an open subset $\varphi(U)$ of $\mathbb{R}^N$; the adjective *centered* means that we assume $\varphi(e)$ to equal $0 \in \mathbb{R}^N$). It is well-known that a basis of $T_e(G)$ is given by the following derivations at e (depending on the chart!)

$$\frac{\partial}{\partial x_1}\Big|_e, \ldots, \frac{\partial}{\partial x_N}\Big|_e, \tag{5.122}$$

defined by (for any $i = 1, \ldots, N$)

$$\frac{\partial}{\partial x_i}\Big|_e f := \frac{\partial}{\partial y_i}\Big|_0 \{(f \circ \varphi^{-1})(y_1, \ldots, y_N)\},$$

for every smooth function $f : G \to \mathbb{R}$ (the y_i in the above right-hand side denote the standard coordinates on $\mathbb{R}^N$). As a consequence, any element $\mathbf{u}$ of $T_e(G)$ has the form

$$\mathbf{u} = u_1 \frac{\partial}{\partial x_1}\Big|_e + \cdots + u_N \frac{\partial}{\partial x_N}\Big|_e, \tag{5.123}$$

for a uniquely determined N-tuple of real numbers $(u_1, \ldots, u_N)$. Fixing a chart (U, φ) as above, for any smooth function $f : G \to \mathbb{R}$ we denote by $\widehat{f}$ the expression of f in the corresponding local coordinates, that is,

$$\widehat{f} := f \circ \varphi^{-1} : \varphi(U) \to \mathbb{R}.$$

By an abuse of notation, we denote by $\widehat{m}$ the coordinate expression of the multiplication of G around e (by shrinking U if necessary, we here suppose that U is so small that, for every $x, y \in U$, $x \cdot y$ belongs to the domain of some chart on G centered at e on which the coordinate function φ is defined): more precisely we set

$$\widehat{m} : \varphi(U) \times \varphi(U) \to \mathbb{R}^N, \qquad \begin{array}{l} \widehat{m}(\alpha, \beta) := \varphi\big(\varphi^{-1}(\alpha) \cdot \varphi^{-1}(\beta)\big) \\ \text{for every } \alpha, \beta \in \varphi(U). \end{array} \tag{5.124}$$

Obviously, $\widehat{m}$ is a smooth map on an open neighborhood of $(0, 0) \in \mathbb{R}^N \times \mathbb{R}^N$. We also let $\widehat{m} = (\widehat{m}_1, \ldots, \widehat{m}_N)$ denote the component functions of $\widehat{m}$.

We are ready to prove a useful lemma, showing that the mixed partial derivatives in the Hessian matrix of $\widehat{m}$ at $(0, 0)$ suffice to determine (in a very precise way) the Lie algebra structure of $\mathfrak{g}$.

Lemma 5.49. *With all the above notation, we have*

$$\left[\sum_{i=1}^N u_i \frac{\partial}{\partial x_i}\Big|_e, \sum_{j=1}^N v_j \frac{\partial}{\partial x_j}\Big|_e \right] = \sum_{h=1}^N \left(\sum_{i,j=1}^N (u_i v_j - u_j v_i) \frac{\partial^2 \widehat{m}_h(0,0)}{\partial \alpha_i \partial \beta_j} \right) \frac{\partial}{\partial x_h}\Big|_e, \tag{5.125}$$

for every $u_i, v_i \in \mathbb{R}$ ($i = 1, \ldots, N$).

Equivalently, the structure constants of the Lie algebra $(T_e(G), [\cdot, \cdot]_e)$ *with respect to the basis in (5.122) are given by the formula:*

$$\left[\frac{\partial}{\partial x_i}\Big|_e, \frac{\partial}{\partial x_j}\Big|_e\right]_e = \sum_{h=1}^{N}\left(\frac{\partial^2 \widehat{m}_h(0,0)}{\partial \alpha_i \, \partial \beta_j} - \frac{\partial^2 \widehat{m}_h(0,0)}{\partial \alpha_j \, \partial \beta_i}\right)\frac{\partial}{\partial x_h}\Big|_e, \qquad (5.126)$$

for any $i, j \in \{1, \dots, N\}$.

Proof. Let the notation preceding the statement of the lemma be fixed. We have the following computation ($\mathbf{u}$ is as in (5.123) and for $\mathbf{v}$ we follow analogous notation; moreover $f : G \to \mathbb{R}$ is any smooth function):

$$[\mathbf{u}, \mathbf{v}]_e(f) \overset{(5.121)}{=} \left[\alpha^{-1}(\mathbf{u}), \alpha^{-1}(\mathbf{v})\right]_{\mathfrak{g}}(f)(e)$$

$$= (\alpha^{-1}(\mathbf{u}))\big|_e(\alpha^{-1}(\mathbf{v})(f)) - \{\text{analogous, interchange } \mathbf{u}, \mathbf{v}\}$$

$$= \mathbf{u}\big(x \mapsto \mathbf{v}(f \circ \tau_x)\big) - \{\text{analogous, interchange } \mathbf{u}, \mathbf{v}\}$$

$$= \left(\sum_{i=1}^{N} u_i \frac{\partial}{\partial x_i}\Big|_e\right)\left\{\left(\sum_{j=1}^{N} v_j \frac{\partial}{\partial y_j}\Big|_e\right)f(x \cdot y)\right\} - \{\text{analogous} \dots\}$$

$$= \sum_{i,j=1}^{N} u_i\, v_j \frac{\partial}{\partial x_i}\Big|_e \frac{\partial}{\partial y_j}\Big|_e \{f(x \cdot y)\} - \left\{\begin{array}{c}\text{analogous,}\\ \text{interchange } us \text{ with } vs\end{array}\right\}.$$

We have, by definition of $\partial/\partial x_i$ and of $\widehat{f}, \widehat{m}$,

$$\frac{\partial}{\partial x_i}\Big|_e \frac{\partial}{\partial y_j}\Big|_e \{f(x \cdot y)\} = \frac{\partial}{\partial \alpha_i}\Big|_0 \frac{\partial}{\partial \beta_j}\Big|_0 \{f(\varphi^{-1}(\alpha) \cdot \varphi^{-1}(\beta))\}$$

$$= \frac{\partial}{\partial \alpha_i}\Big|_0 \frac{\partial}{\partial \beta_j}\Big|_0 \{\widehat{f}(\widehat{m}(\alpha, \beta))\} = \quad \text{(by the chain rule)}$$

$$= \frac{\partial}{\partial \alpha_i}\Big|_0 \sum_{h=1}^{N} \frac{\partial \widehat{f}}{\partial x_h}(\widehat{m}(\alpha, 0))\frac{\partial \widehat{m}_h}{\partial \beta_j}(\alpha, 0)$$

(again by the chain rule, together with $\widehat{m}(\alpha, 0) = \alpha$)

$$= \sum_{h=1}^{N} \frac{\partial^2 \widehat{f}}{\partial x_i \, \partial x_h}(0)\frac{\partial \widehat{m}_h}{\partial \beta_j}(0,0) + \sum_{h=1}^{N} \frac{\partial \widehat{f}}{\partial x_h}(0)\frac{\partial^2 \widehat{m}_h}{\partial \alpha_i \, \partial \beta_j}(0,0).$$

Now note that one has (as $\widehat{m}_h(0, \beta) = \beta$)

$$\frac{\partial \widehat{m}_h}{\partial \beta_j}(0,0) = \frac{\partial}{\partial \beta_j}\{\widehat{m}_h(0, \beta)\} = \frac{\partial}{\partial \beta_j}\beta_h = \delta_{j,h},$$

where $\delta_{j,h}$ is the usual Kronecker symbol. As a consequence we have the following formula

$$\frac{\partial}{\partial x_i}\bigg|_e \frac{\partial}{\partial y_j}\bigg|_e \{f(x \cdot y)\} = \frac{\partial^2 \widehat{f}}{\partial x_i\, \partial x_j}(0) + \sum_{h=1}^{N} \frac{\partial \widehat{f}}{\partial x_h}(0)\, \frac{\partial^2 \widehat{m}_h}{\partial \alpha_i\, \partial \beta_j}(0,0).$$

Going back to the computation for $[\mathbf{u}, \mathbf{v}]_e(f)$, we obtain

$$[\mathbf{u}, \mathbf{v}]_e(f) = \sum_{i,j=1}^{N} u_i\, v_j\, \frac{\partial^2 \widehat{f}(0)}{\partial x_i\, \partial x_j} + \sum_{i,j,h=1}^{N} u_i\, v_j\, \frac{\partial \widehat{f}(0)}{\partial x_h}\, \frac{\partial^2 \widehat{m}_h(0,0)}{\partial \alpha_i\, \partial \beta_j}$$

$$- \{\text{analogous, interchange } us \text{ with } vs\}$$

$$= \sum_{i,j=1}^{N} u_i\, v_j\, \frac{\partial^2 \widehat{f}(0)}{\partial x_i\, \partial x_j} - \sum_{i,j=1}^{N} v_i\, u_j\, \frac{\partial^2 \widehat{f}(0)}{\partial x_i\, \partial x_j}$$

$$+ \sum_{i,j,h=1}^{N} u_i\, v_j\, \frac{\partial \widehat{f}(0)}{\partial x_h}\, \frac{\partial^2 \widehat{m}_h(0,0)}{\partial \alpha_i\, \partial \beta_j} - \sum_{i,j,h=1}^{N} v_i\, u_j\, \frac{\partial \widehat{f}(0)}{\partial x_h}\, \frac{\partial^2 \widehat{m}_h(0,0)}{\partial \alpha_i\, \partial \beta_j}$$

(the first two sums cancel each other, by Schwarz's Theorem!)

$$= \sum_{h=1}^{N} \left(\sum_{i,j=1}^{N} (u_i\, v_j - u_j\, v_i)\, \frac{\partial^2 \widehat{m}_h(0,0)}{\partial \alpha_i\, \partial \beta_j} \right) \frac{\partial \widehat{f}(0)}{\partial x_h}.$$

This gives the desired (5.125), by the very definition of $(\partial/\partial x_h)|_e$. Obviously, (5.126) is a particular case of (5.125).   □

In the next result, Proposition 5.50, we make use of the following definition.

Let $\mathfrak{g}$ be a Lie algebra. Let $\mathcal{B} = \{e_i\}_{i \in \mathfrak{I}}$ be a basis for $\mathfrak{g}$. Then there exist uniquely defined scalars $c_{i,j}^k$ such that

$$[e_i, e_j]_{\mathfrak{g}} = \sum_{k \in \mathfrak{I}} c_{i,j}^k\, e_k, \quad \text{for every } i, j \in \mathfrak{I}. \tag{5.127}$$

Obviously, the above sum is unambiguous, for it is finite for every $i, j \in \mathfrak{I}$ (by definition of a – linear – basis for a vector space). We say that the coefficients $c_{i,j}^k$ are the *structure constants* of the Lie algebra $\mathfrak{g}$ with respect to the basis $\mathcal{B}$.

Note that, by the skew-symmetry of the Lie bracket one has

$$c_{i,j}^k = -c_{j,i}^k, \quad \text{for every } i, j, k \in \mathfrak{I}. \tag{5.128}$$

Moreover, by the the Jacobi identity, we also obtain that

$$\sum_{r \in \mathcal{J}} (c_{i,j}^r \, c_{r,k}^s + c_{j,k}^r \, c_{r,i}^s + c_{k,i}^r \, c_{r,j}^s) = 0, \quad \text{for every } i, j, k, s \in \mathcal{J}. \tag{5.129}$$

Actually, if $\mathfrak{g}$ is a vector space over $\mathbb{K}$ with a basis $\mathcal{B} = \{e_i\}_{i \in \mathcal{J}}$ and $\{c_{i,j}^k\}_{i,j,k \in \mathcal{J}}$ is a family of scalars satisfying (5.128) and (5.129) (and such that, for every $i, j \in \mathcal{J}$, the $c_{i,j}^k$s are non-vanishing only for a finite – possibly empty – set of indices k in $\mathcal{J}$), then the unique bilinear operation $[\cdot, \cdot]_{\mathfrak{g}}$ on $\mathfrak{g}$ defined by (5.127) endows $\mathfrak{g}$ with the structure of a Lie algebra. We will have no occasion to apply this fact, though. Instead, we shall make use of the following well-known fact, stating that the structure constants completely determine the Lie algebra, up to isomorphism.

Proposition 5.50. *Let $\mathfrak{g}, \mathfrak{h}$ be two Lie algebras (over the same field). Then $\mathfrak{g}$ and $\mathfrak{h}$ are isomorphic as Lie algebras if and only if there exist a basis $\mathcal{G} = \{g_i\}_{i \in \mathcal{J}}$ for $\mathfrak{g}$ and a basis $\mathcal{H} = \{h_i\}_{i \in \mathcal{J}}$ for $\mathfrak{h}$ (indexed over the same set $\mathcal{J}$) such that the associated structure constants coincide.*

More precisely, this last condition means that

$$c_{i,j}^k = \gamma_{i,j}^k \quad \text{for every } i, j, k \in \mathcal{J}, \text{ where}$$

$$\begin{aligned} [g_i, g_j]_{\mathfrak{g}} &= \sum_{k \in \mathcal{J}} c_{i,j}^k \, g_k, \\ [h_i, h_j]_{\mathfrak{h}} &= \sum_{k \in \mathcal{J}} \gamma_{i,j}^k \, h_k, \end{aligned} \quad \text{for every } i, j \in \mathcal{J}. \tag{5.130}$$

Proof. We split the proof in two parts.

I. Suppose $\mathcal{G} = \{g_i\}_{i \in \mathcal{J}}$ is a basis for $\mathfrak{g}$ and $\mathcal{H} = \{h_i\}_{i \in \mathcal{J}}$ is a basis for $\mathfrak{h}$ such that (5.130) holds. We need to prove that $\mathfrak{g}, \mathfrak{h}$ are isomorphic Lie algebras. Let $\Psi : \mathfrak{g} \to \mathfrak{h}$ be the unique linear map mapping g_i into h_i, for every $i \in \mathcal{J}$. Obviously, Ψ is a vector space isomorphism (for its inverse is the unique linear map from $\mathfrak{h}$ to $\mathfrak{g}$ mapping h_i into g_i, for every $i \in \mathcal{J}$). We claim that Ψ is a Lie algebra isomorphism. Indeed, given $u, v \in \mathfrak{g}$ there exist scalars $\alpha_i(u), \alpha_i(v)$ (for every $i \in \mathcal{J}$) such that $\alpha_i(u) \neq 0$ and $\alpha_i(v) \neq 0$ for only a finite set (possibly empty) of indices i in $\mathcal{J}$ and such that

$$u = \sum_{i \in \mathcal{J}} \alpha_i(u) \, g_i, \quad v = \sum_{i \in \mathcal{J}} \alpha_i(v) \, g_i.$$

Then we have

$$\Psi([u,v]_{\mathfrak{g}}) = \Psi\left(\sum_{i,j \in \mathcal{J}} \alpha_i(u) \, \alpha_j(v) \, [g_i, g_j]_{\mathfrak{g}} \right)$$

$$= \Psi\left(\sum_{i,j,k \in \mathcal{J}} \alpha_i(u) \, \alpha_j(v) \, c_{i,j}^k \, g_k \right) = \sum_{i,j,k \in \mathcal{J}} \alpha_i(u) \, \alpha_j(v) \, c_{i,j}^k \, \Psi(g_k)$$

$$= \sum_{i,j,k \in \mathcal{I}} \alpha_i(u)\, \alpha_j(v)\, c_{i,j}^k\, h_k \overset{(5.130)}{=} \sum_{i,j,k \in \mathcal{I}} \alpha_i(u)\, \alpha_j(v)\, \gamma_{i,j}^k\, h_k$$

$$= \sum_{i,j \in \mathcal{I}} \alpha_i(u)\, \alpha_j(v)\, [h_i, h_j]_{\mathfrak{h}} = \sum_{i,j \in \mathcal{I}} \alpha_i(u)\, \alpha_j(v)\, [\Psi(g_i), \Psi(g_j)]_{\mathfrak{h}}$$

$$= \left[\Psi\Big(\sum_{i \in \mathcal{I}} \alpha_i(u)\, g_i \Big), \Psi\Big(\sum_{j \in \mathcal{I}} \alpha_j(v)\, g_i \Big) \right]_{\mathfrak{h}} = [\Psi(u), \Psi(v)]_{\mathfrak{h}}.$$

II. Vice versa, suppose that $\mathfrak{g}, \mathfrak{h}$ are isomorphic Lie algebras. Let $\Psi : \mathfrak{g} \to \mathfrak{h}$ be an isomorphism of Lie algebras. Let us fix any indexed basis $\mathcal{G} = \{g_i\}_{i \in \mathcal{I}}$ for $\mathfrak{g}$. Since an LA morphism is in particular a vector space isomorphism, then the system $\mathcal{H} := \{\Psi(g_i)\}_{i \in \mathcal{I}}$ is a basis for $\mathfrak{h}$. We set $h_i := \Psi(g_i)$ for every $i \in \mathcal{I}$. With obvious notation, we denote by $c_{i,j}^k$ the structure constants of $\mathfrak{g}$ w.r.t. $\mathcal{G}$ and by $\gamma_{i,j}^k$ the structure constants of $\mathfrak{h}$ w.r.t. $\mathcal{H}$. We aim to prove that $c_{i,j}^k = \gamma_{i,j}^k$ for every $i, j, k \in \mathcal{I}$. This derives from the following computation: Fixing any $i, j \in \mathcal{I}$, we have:

$$\sum_{k \in \mathcal{I}} \gamma_{i,j}^k\, h_k = [h_i, h_j]_{\mathfrak{h}} = [\Psi(g_i), \Psi(g_j)]_{\mathfrak{h}} = \Psi([g_i, g_j]_{\mathfrak{g}})$$

$$= \Psi\Big(\sum_{k \in \mathcal{I}} c_{i,j}^k\, g_k \Big) = \sum_{k \in \mathcal{I}} c_{i,j}^k\, \Psi(g_k) = \sum_{k \in \mathcal{I}} c_{i,j}^k\, h_k.$$

By equating left-hand side and right-hand side, we have

$$\sum_{k \in \mathcal{I}} (\gamma_{i,j}^k - c_{i,j}^k)\, h_k,$$

so that, by the linear independence of the vectors h_k we deduce that $\gamma_{i,j}^k - c_{i,j}^k = 0$ for every $k \in \mathcal{I}$. The arbitrariness of i, j proves the claimed equality of the structure constants. This ends the proof. $\square$

With the above lemmas at hand, we are ready to give the:

Proof (of Theorem 5.46.). Let $\mathfrak{n}$ be a finite-dimensional real nilpotent Lie algebra. Let r be its step of nilpotency. Let also $\diamond_{\mathfrak{n}}$ be as in (5.119). As $\mathfrak{n}$ is nilpotent, we know from Theorem 5.42 that $(\mathfrak{n}, \diamond_{\mathfrak{n}})$ is a group with identity 0 and inversion given by $\mathfrak{n} \ni \xi \mapsto -\xi \in \mathfrak{n}$. Since $\mathfrak{n}$ is nilpotent of step r, we know from Lemma 5.44 that the group $(\mathfrak{n}, \diamond_{\mathfrak{n}})$ is nilpotent of step r.

Let us set $N := \dim(\mathfrak{n})$. By hypothesis N is finite. Let us fix any basis $\mathcal{E} = \{E_1, \ldots, E_N\}$ for the underlying vector space of $\mathfrak{n}$. We denote by $\varphi : \mathfrak{n} \to \mathbb{R}^N$ the linear map defined by

$$\varphi(x_1 E_1 + \cdots + x_N E_N) := (x_1, \ldots, x_N), \quad \forall\ x_1, \ldots, x_N \in \mathbb{R}.$$

Trivially, this is an isomorphism of vector spaces and it also plays the rôle of a coordinate map for a global chart on $\mathfrak{n}$, when $\mathfrak{n}$ is endowed with the usual smooth structure of a finite-dimensional vector space (obviously, this smooth structure is independent of the basis $\mathcal{E}$). With respect to this fixed set of coordinates, the operation $\diamond_\mathfrak{n}$ in (5.119) has polynomial component functions and (by recalling the facts above) the group inversion map is simply $\mathbb{R}^N \ni x \mapsto -x \in \mathbb{R}^N$.

This proves that $(\mathfrak{n}, \diamond_\mathfrak{n})$ is also a Lie group of dimension N. In the rest of the proof, we denote this Lie group by G.

The main task is now to show that the Lie algebra of the Lie group G (denoted by $\mathfrak{g}$) is isomorphic (as a Lie algebra) to $\mathfrak{n}$ itself. Here, $\mathfrak{g}$ is endowed with the usual bracket *of vector fields* (denoted by $[\cdot, \cdot]_\mathfrak{g}$) whereas $\mathfrak{n}$ is endowed with its primordial Lie bracket $[\cdot, \cdot]_\mathfrak{n}$. From the facts recalled before Lemma 5.49, we know that the Lie algebra $(\mathfrak{g}, [\cdot, \cdot]_\mathfrak{g})$ is isomorphic to the Lie algebra $(T_0(G), [\cdot, \cdot]_0)$, where $T_0(G)$ is the tangent space to G at 0 (recall that 0 is the identity of G) and $[\cdot, \cdot]_0$ is the Lie bracket in (5.121) (see also (5.120)).

Hence, to complete the proof it suffices to show that the Lie algebras $(T_0(G), [\cdot, \cdot]_0)$ and $(\mathfrak{n}, [\cdot, \cdot]_\mathfrak{n})$ are isomorphic. Since they are both N-dimensional as vector spaces (recall that $T_0(G)$ has the same dimension as G, i.e., N), to prove the claimed LA isomorphism it suffices, in view of Proposition 5.50, to exhibit bases for $T_0(G)$ and for $\mathfrak{n}$ having the same structure constants. We denote the structure constants of $\mathfrak{n}$ with respect to the above basis $\mathcal{E}$ by $c_{i,j}^k$, which means that

$$[E_i, E_j]_\mathfrak{n} = \sum_{k=1}^N c_{i,j}^k E_k, \quad \text{for every } i, j = 1, \ldots, N. \tag{5.131}$$

We know that $T_0(G)$ admits the basis

$$\mathbf{B} := \left\{ \frac{\partial}{\partial x_1}\Big|_0, \ldots, \frac{\partial}{\partial x_N}\Big|_0 \right\},$$

where these partial derivatives are meant in the usual sense (of smooth manifolds) with respect to the given coordinate map φ as above. Now, identity (5.126) in Lemma 5.49 proves that the structure constants of $(T_0(G), [\cdot, \cdot]_0)$ with respect to the basis $\mathcal{B}$ are given by the numbers (indexed in the obvious way)

$$\frac{\partial^2 \widehat{m}_k(0,0)}{\partial \alpha_i \, \partial \beta_j} - \frac{\partial^2 \widehat{m}_k(0,0)}{\partial \alpha_j \, \partial \beta_i}, \quad i, j, k \in \{1, \ldots, N\}.$$

Here, according to (5.124), $\widehat{m}$ is the following function of $(\alpha, \beta) \in \mathbb{R}^N \times \mathbb{R}^N$

$$\widehat{m}(\alpha, \beta) = \varphi\left(\varphi^{-1}(\alpha) \diamond_\mathfrak{n} \varphi^{-1}(\beta)\right)$$

$$= \varphi\left((\alpha_1 E_1 + \cdots + \alpha_N E_N) \diamond_\mathfrak{n} (\beta_1 E_1 + \cdots + \beta_N E_N)\right).$$

Equivalently, the component functions $\widehat{m}_k$ of $\widehat{m}$ are given by

$$\left(\sum_{i=1}^{N} \alpha_i E_i\right) \diamond_n \left(\sum_{j=1}^{N} \beta_j E_j\right) = \sum_{k=1}^{N} \widehat{m}_k(\alpha, \beta) E_k. \tag{5.132}$$

We claim that *the structure constants of* $(T_0(G), [\cdot, \cdot]_0)$ *with respect to* $\mathcal{B}$ *coincide with those of* $(\mathfrak{n}, [\cdot, \cdot]_\mathfrak{n})$ *with respect to* $\mathcal{E}$. So, all we have to prove is that

$$c_{i,j}^k = \frac{\partial^2 \widehat{m}_k(0,0)}{\partial \alpha_i \, \partial \beta_j} - \frac{\partial^2 \widehat{m}_k(0,0)}{\partial \alpha_j \, \partial \beta_i}, \tag{5.133}$$

for every $i, j, k \in \{1, \ldots, N\}$, where the constants $c_{i,j}^k$ are as in (5.131) and $\widehat{m}$ is as in (5.132), that is, $\widehat{m}$ is the coordinate expression of the CBHD operation with respect to the basis $\mathcal{E}$. Let us explicitly remark that:

In the case of finite-dimensional nilpotent Lie algebras, we have reduced the (global version of) Lie's Third Theorem to an explicit computation on the Campbell, Baker, Hausdorff, Dynkin operation.

Now, the proof of (5.133) is a simple computation. Indeed, since we are interested only in the second derivatives of the components of $\widehat{m}$ at $(0,0)$, in view of (5.132) we can focuss on the summands expressing $\diamond_n$ which are brackets of height not exceeding 2. Recalling that we have

$$\xi \diamond_n \eta = \xi + \eta + \tfrac{1}{2}[\xi, \eta]_\mathfrak{n} + \{\text{brackets of heights} \geq 3 \text{ in } \xi, \eta\},$$

we infer

$$\sum_{k=1}^{N} \widehat{m}_k(\alpha, \beta) E_k = \left(\sum_{i=1}^{N} \alpha_i E_i\right) \diamond_n \left(\sum_{j=1}^{N} \beta_j E_j\right)$$

$$= \sum_{i=1}^{N} \alpha_i E_i + \sum_{j=1}^{N} \beta_j E_j + \frac{1}{2} \sum_{i,j=1}^{N} \alpha_i \beta_j [E_i, E_j]_\mathfrak{n} +$$

$$+ \left\{\text{brackets of heights} \geq 3 \text{ in } \sum_i \alpha_i E_i, \sum_j \beta_j E_j\right\}$$

$$\overset{(5.131)}{=} \sum_{k=1}^{N} \left(\alpha_k + \beta_k + \frac{1}{2} \sum_{i,j=1}^{N} \alpha_i \beta_j c_{i,j}^k\right) E_k + \{\cdots\}.$$

By comparing the far left/right-hand sides, we derive (for $k = 1, \ldots, N$)

$$\widehat{m}_k(\alpha, \beta) = \alpha_k + \beta_k + \frac{1}{2} \sum_{i,j=1}^{N} \alpha_i \beta_j c_{i,j}^k + O(\|(\alpha, \beta)\|^3), \tag{5.134}$$

as $(\alpha, \beta) \to (0,0)$. This gives

$$\frac{\partial^2 \widehat{m}_k(0,0)}{\partial \alpha_i \, \partial \beta_j} - \frac{\partial^2 \widehat{m}_k(0,0)}{\partial \alpha_j \, \partial \beta_i} = \frac{1}{2} c_{i,j}^k - \frac{1}{2} c_{j,i}^k \stackrel{(5.128)}{=} c_{i,j}^k,$$

and (5.133) follows. This ends the proof.   □

Remark 5.51. By analyzing the proofs of Proposition 5.50 and of Theorem 5.46, we have proved the following fact.

Given a finite-dimensional real nilpotent Lie algebra $\mathfrak{n}$, denoted by G the Lie group $(\mathfrak{n}, \diamond_\mathfrak{n})$ and by $\text{Lie}(G)$ the Lie algebra of G, a Lie algebra isomorphism $\Psi : \mathfrak{n} \to \text{Lie}(\mathbb{G})$ can be obtained as follows: Fixed any linear basis $E_1, \ldots, E_N$ of $\mathfrak{n}$, Ψ is the unique linear function mapping E_i (for any $i = 1, \ldots, N$) into the left-invariant vector field X_i such that

$$X_i(f)(x) = \frac{\mathrm{d}}{\mathrm{d}\, t}\Big|_{t=0} f(x \diamond_\mathfrak{n} (t \, E_i)),$$

for every $x \in G$ and every smooth $f : G \to \mathbb{R}$.

In a certain sense this shows that, in order to determine the Lie algebra $\text{Lie}(G)$, it suffices to consider, from the CBHD operation $x \diamond y$, the summands where y appears only with order 1. [A simpler formula can be provided for such summands. This formula has a long history, tracing back to Campbell, Baker and Pascal, as we showed in Chap. 1.]

5.5 The CBHD Operation and Series in Banach Algebras

Let $(A, *, \| \cdot \|)$ be a (real or complex) Banach algebra where $\| \cdot \|$ is compatible with $*$, according to Definition 5.22. We denote by $[\cdot, \cdot]_*$ (or simply by $[\cdot, \cdot]$) the commutator related to the algebra $(A, *)$. Hence $(A, [\cdot, \cdot]_*, \| \cdot \|)$ becomes a Banach Lie algebra (see Remark 5.26) and we have

$$\big\| [x, y]_* \big\| \leq 2 \, \|x\| \cdot \|y\|, \quad \text{for every } x, y \in A.$$

By Remark 5.27, we know that the norm $\| \cdot \|_\star = 2 \, \| \cdot \|$ is compatible with the Banach Lie algebra A. As a consequence, the results of Theorems 5.30 and 5.38 hold in the present context too, replacing $\| \cdot \|$ in their statements with the present norm $\| \cdot \|_\star$. We thus get (rewriting everything in terms of the present norm $\| \cdot \| = \frac{1}{2} \| \cdot \|_\star$) the following theorem.

This theorem gives us information on the convergence of the CBHD series relative to A and the local associativity of the CBHD operation on A.

Theorem 5.52. *Suppose $(A, *, \| \cdot \|)$ is a Banach algebra, $\| \cdot \|$ being a compatible norm on A. Let $Z_j^*(a, b)$ be as in (5.28), relative to the commutator-algebra of A,* *that is (see also (5.23)–(5.26) for the relevant notation),*

$$Z_j^*(a,b) = \sum_{n=1}^{j} c_n \sum_{(h,k)\in\mathcal{N}_n:\,|h|+|k|=j} \mathbf{c}(h,k)\, D_{(h,k)}^*(a,b), \qquad a,b \in A.$$

Let us set

$$a \diamond b := \sum_{j=1}^{\infty} Z_j^*(a,b), \quad \text{whenever this series converges in } A, \qquad (5.135)$$

A being equipped with its given Banach space structure.

Then a sufficient condition for the existence of $a \diamond b$ is that the couple (a,b) belongs to the set

$$\mathbf{D} = \left\{ (a,b) \in A \times A \,:\, \|a\| + \|b\| < \tfrac{1}{2} \log 2 \right\}.$$

This is the case, e.g., if a, b belong to the disc $\{a \in A \,:\, \|a\| < \tfrac{1}{4} \log 2\}$. Moreover,

$$\|a \diamond b\| \leq \tfrac{1}{2} \log \left(\frac{1}{2 - e^{2(\|a\|+\|b\|)}} \right), \qquad \text{for every } (a,b) \in \mathbf{D}.$$

Finally, the series of functions $\sum_{j=1}^{\infty} Z_j^*(a,b)$ converges normally (hence uniformly) on every set of the type

$$\{(a,b) \in A \times A \,:\, \|a\| + \|b\| \leq \delta\}, \qquad \text{with } \delta < \tfrac{1}{2} \log 2.$$

Let $\mathbf{Q} = \{a \in A \,:\, \|a\| < \tfrac{1}{4} \log(2 - 1/\sqrt{2})\}$, then we have

$$(a \diamond b) \diamond c = a \diamond (b \diamond c), \quad \text{for every } a,b,c \in \mathbf{Q}.$$

Unlike Banach Lie algebras, on a Banach algebra other interesting problems arise – also directly linked to the original CBHD Theorem – which we now describe. The exponential function

$$\mathrm{Exp} : A \to A, \qquad \mathrm{Exp}(a) := \sum_{k=0}^{\infty} \frac{a^{*k}}{k!}, \qquad a \in A$$

is well defined on the whole of A, for we have

$$\left\| \sum_{k=n}^{n+p} \frac{a^{*k}}{k!} \right\| \leq \sum_{k=n}^{n+p} \frac{\|a\|^k}{k!} \leq \sum_{k=n}^{\infty} \frac{\|a\|^k}{k!} \xrightarrow{n\to\infty} 0,$$

so that $\{\sum_{k=0}^{n} \frac{a^{*k}}{k!}\}_{n \in \mathbb{N}}$ is a Cauchy sequence on A, hence it is convergent (for A is complete). The problem which we aim to address is the following:

Determine a set $\mathbf{E} \subseteq A$ *such that* $a \diamond b$ *converges for every* $a, b \in \mathbf{E}$ *and*

$$\mathrm{Exp}(a \diamond b) = \mathrm{Exp}(a) * \mathrm{Exp}(b), \quad \text{for every } a, b \in \mathbf{E}. \tag{5.136}$$

This may be referred to as *the CBHD Theorem for Banach algebras.*

Incidentally, not only the $\diamond$ operation has a meaning on a Banach algebra, but so also does the analogue of the $\blacklozenge$ operation from Sect. 3.1.3 (page 126), that is,

$$a \blacklozenge b := \sum_{n=1}^{\infty} \frac{(-1)^{n+1}}{n} \sum_{(h,k) \in \mathcal{N}_n} \frac{a^{*h_1} * b^{*k_1} * \cdots * a^{*h_n} * b^{*k_n}}{h_1! \cdots h_n! k_1! \cdots k_n!}. \tag{5.137}$$

Some additional attention must be paid in the context of Banach algebras (if compared to the setting of $\widehat{\mathscr{T}}(V)$), for the above sum over $\mathcal{N}_n$ is infinite and it has to be properly interpreted.

Yet another couple of problems which we aim to address are the following ones: *Determine sets* $\mathbf{E}_i \subseteq A$ *(with* $i = 1, 2$*) such that* $a \blacklozenge b$ *converges for every* $a, b \in \mathbf{E}_i$ *and for which it holds that*

$$\mathrm{Exp}(a \blacklozenge b) = \mathrm{Exp}(a) * \mathrm{Exp}(b), \quad \text{for every } a, b \in \mathbf{E}_1; \tag{5.138}$$

$$a \diamond b = a \blacklozenge b, \quad \text{for every } a, b \in \mathbf{E}_2. \tag{5.139}$$

The solutions to these last two questions will give, as a byproduct, the above CBHD Theorem for Banach algebras.

We now face to solve all of these problems.

Lemma 5.53. *Let all the above notation apply. Let* $a, b \in A$. *Then, for every* $n \in \mathbb{N}$, *the infinite sum*

$$\varXi_n(a, b) := \sum_{(h,k) \in \mathcal{N}_n} \frac{a^{*h_1} * b^{*k_1} * \cdots * a^{*h_n} * b^{*k_n}}{h_1! \cdots h_n! k_1! \cdots k_n!} \tag{5.140}$$

makes sense in A *as the limit of ("polynomial") functions*

$$\varXi_n(a, b) = \lim_{N \to \infty} \sum_{\substack{(h,k) \in \mathcal{N}_n \\ |h|+|k| \leq N}} \frac{1}{h! \, k!} a^{*h_1} * b^{*k_1} * \cdots * a^{*h_n} * b^{*k_n}, \tag{5.141}$$

or as the sum of the series of ("homogeneous" polynomial) functions

$$\sum_{j=n}^{\infty} \left(\sum_{\substack{(h,k)\in\mathbb{N}_n \\ |h|+|k|=j}} \frac{1}{h!\,k!}\, a^{*\,h_1} * b^{*\,k_1} * \cdots * a^{*\,h_n} * b^{*\,k_n} \right). \tag{5.142}$$

Furthermore, these limits/series of functions converge uniformly on every disc centred at the origin $\{a \in A : \|a\| \leq R\}$, for any finite $R > 0$.

Actually, we have (if 1_A denotes the identity element of $(A, *)$)

$$\Xi_n(a, b) = \big(\mathrm{Exp}(a) * \mathrm{Exp}(b) - 1_A\big)^{*\,n}, \quad \text{for every } a, b \in A. \tag{5.143}$$

Finally, for every $n \in \mathbb{N}$ and every $a, b \in A$, we have the estimate

$$\|\Xi_n(a, b)\| \leq \sum_{(h,k)\in\mathbb{N}_n} \frac{1}{h!\,k!}\, \|a\|^{|h|}\, \|b\|^{|k|} = \big(e^{\|a\|+\|b\|} - 1\big)^n. \tag{5.144}$$

Proof. We fix henceforth $a, b \in A$, $n \in \mathbb{N}$. Given $N \geq n$, we denote by $\Psi_N(a, b)$ the sequence of functions whose limit appears in (5.141). The following computation applies (here $N \geq n$ and $P \in \mathbb{N}$):

$$\|\Psi_{N+P}(a, b) - \Psi_N(a, b)\|$$

$$= \left\| \sum_{\substack{(h,k)\in\mathbb{N}_n \\ N+1\leq|h|+|k|\leq N+P}} \frac{1}{h!\,k!}\, a^{*\,h_1} * b^{*\,k_1} * \cdots * a^{*\,h_n} * b^{*\,k_n} \right\|$$

$$\leq \sum_{\substack{(h,k)\in\mathbb{N}_n \\ N+1\leq|h|+|k|\leq N+P}} \frac{1}{h!\,k!}\, \|a\|^{|h|}\, \|b\|^{|k|} \quad (\text{we set } R := \max\{\|a\|, \|b\|\})$$

$$\leq \sum_{\substack{(h,k)\in\mathbb{N}_n \\ N+1\leq|h|+|k|}} \frac{1}{h!\,k!}\, R^{|h|+|k|} =: I_N.$$

Now, in view of identity (5.38), the above far right-hand side I_N is equal to the sum of the powers with exponent $\geq N + 1$ in the Maclaurin expansion of the function

$$\mathbb{R} \ni R \mapsto (e^{2R} - 1)^n \overset{(5.38)}{=} \sum_{(h,k)\in\mathbb{N}_n} \frac{R^{|h|}\, R^{|k|}}{h!\,k!}.$$

Since the above function is real analytic and it coincides throughout with its Maclaurin expansion, I_N vanishes as $N \to \infty$. This proves that $\{\Psi_N(a, b)\}_N$ is Cauchy in A, hence that it converges. The uniform convergence claimed in the statement of this lemma is a consequence of the above computations.

Furthermore, given $N \in \mathbb{N}$, $N \geq n$ we have (by reordering)

$$
\begin{aligned}
\Psi_N(a, b) &= \sum_{\substack{(h,k) \in \mathbb{N}_n \\ |h|+|k| \leq N}} \frac{1}{h!\, k!}\, a^{* h_1} * b^{* k_1} * \cdots * a^{* h_n} * b^{* k_n} \\
&= \sum_{j=n}^{N} \left(\sum_{\substack{(h,k) \in \mathbb{N}_n \\ |h|+|k| = j}} \frac{1}{h!\, k!}\, a^{* h_1} * b^{* k_1} * \cdots * a^{* h_n} * b^{* k_n} \right),
\end{aligned}
$$

so that $\Psi_N(a, b)$ is actually equal to the N-th partial sum of the series of functions in (5.142). Finally, by (5.38) we infer

$$
\left\| \Xi_n(a, b) \right\| \leq \sum_{(h,k) \in \mathbb{N}_n} \frac{1}{h!\, k!} \|a\|^{|h|}\, \|b\|^{|k|} = (e^{\|a\|+\|b\|} - 1)^n.
$$

Actually, as $(A, *)$ is a topological algebra, we have

$$
\left(\mathrm{Exp}(a) * \mathrm{Exp}(b) - 1_A \right)^{* n} = \left(\sum_{(h,k) \neq (0,0)} \frac{a^{* h} * b^{* k}}{h!\, k!} \right)^{* n}
$$

$$
= \sum_{(h_1, k_1), \ldots, (h_n, k_n) \neq (0,0)} \frac{a^{* h_1} * b^{* k_1} * \cdots * a^{* h_n} * b^{* k_n}}{h_1! \cdots h_n!\, k_1! \cdots k_n!} = \Xi_n(a, b).
$$

This proves (5.143), thus ending the proof. $\qquad\square$

Theorem 5.54 (Well Posedness of $\blacklozenge$ on a Banach Algebra). *Suppose $(A, *, \|\cdot\|)$ is a Banach algebra, $\|\cdot\|$ being a compatible norm on A.*
For every (a, b) belonging to the set

$$
\mathbf{E}_0 := \left\{ (a, b) \in A \times A \,:\, \|a\| + \|b\| < \log 2 \right\}, \tag{5.145}
$$

the series of functions

$$
a \blacklozenge b := \sum_{n=1}^{\infty} \frac{(-1)^{n+1}}{n} \left(\sum_{(h,k) \in \mathbb{N}_n} \frac{a^{* h_1} * b^{* k_1} * \cdots * a^{* h_n} * b^{* k_n}}{h_1! \cdots h_n!\, k_1! \cdots k_n!} \right) \tag{5.146}
$$

is (absolutely) convergent in A. For example, this holds for a, b in the disc

$$
\mathbf{E}_1 := \left\{ a \in A \,:\, \|a\| < \tfrac{1}{2} \log 2 \right\} \tag{5.147}
$$

since $\mathbf{E}_1 \times \mathbf{E}_1 \subset \mathbf{E}_0$.

The convergence of the above series is total (hence uniform) on any set of the type $\{(a, b) \in A \times A : \|a\| + \|b\| \leq \delta\}$, *with* $0 < \delta < \log 2$. *Furthermore, the same results of convergence apply for the majorizing series*

$$\sum_{n=1}^{\infty} \left| \frac{(-1)^{n+1}}{n} \right| \left(\sum_{(h,k) \in \mathcal{N}_n} \frac{\|a^{*h_1} * b^{*k_1} * \cdots * a^{*h_n} * b^{*k_n}\|}{h_1! \cdots h_n! k_1! \cdots k_n!} \right).$$

Finally, formula (5.138) holds, that is,

$$\mathrm{Exp}(a \blacklozenge b) = \mathrm{Exp}(a) * \mathrm{Exp}(b), \quad \text{for every } (a, b) \in \mathbf{E}_0, \tag{5.148}$$

hence in particular for every $(a, b) \in \mathbf{E}_1$.

Proof. Let $a, b \in A$. By Lemma (5.53) (see also the notation in (5.140)), $\Xi_n(a, b)$ makes sense for every $n \in \mathbb{N}$. By exploiting (5.144), we have, if $(a, b) \in \mathbf{E}_0$:

$$\sum_{n=1}^{\infty} \left\| \frac{(-1)^{n+1}}{n} \Xi_n(a, b) \right\|$$

$$\leq \sum_{n=1}^{\infty} \left| \frac{(-1)^{n+1}}{n} \right| \left(\sum_{(h,k) \in \mathcal{N}_n} \frac{\|a^{*h_1} * b^{*k_1} * \cdots * a^{*h_n} * b^{*k_n}\|}{h_1! \cdots h_n! k_1! \cdots k_n!} \right)$$

$$\leq \sum_{n=1}^{\infty} \frac{1}{n} \sum_{(h,k) \in \mathcal{N}_n} \frac{\|a\|^{|h|} * \|b\|^{|k|}}{h! \, k!} = \sum_{n=1}^{\infty} \frac{(e^{\|a\| + \|b\|} - 1)^n}{n}$$

$$\overset{(5.41)}{=} \log \left(\frac{1}{2 - \exp(\|a\| + \|b\|)} \right) < \infty.$$

So, if $(a, b) \in \mathbf{E}_0$ the series is convergent (thanks to the completeness of A). The same estimate as above proves the total convergence of the series (5.146) on any set of the type $\{(a, b) \in A \times A : \|a\| + \|b\| \leq \delta\}$, with $0 < \delta < \log 2$.

As a last task, we aim to prove (5.148). Given $(a, b) \in \mathbf{E}_0$, we have

$$\mathrm{Exp}(a \blacklozenge b) = \sum_{k=0}^{\infty} \frac{1}{k!} \left(\sum_{n=1}^{\infty} c_n \left(\mathrm{Exp}(a) * \mathrm{Exp}(b) - 1_A \right)^{*n} \right)^{*k}$$

$$= 1_A + \sum_{k=1}^{\infty} \frac{1}{k!} \sum_{n_1, \ldots, n_k \geq 1} c_{n_1} \cdots c_{n_k} \left(\mathrm{Exp}(a) * \mathrm{Exp}(b) - 1_A \right)^{*n_1 + \cdots + n_k}$$

$$= 1_A + \sum_{j=1}^{\infty} \left(\mathrm{Exp}(a) * \mathrm{Exp}(b) - 1_A \right)^{*j} \left(\sum_{k=1}^{j} \frac{1}{k!} \sum_{\substack{n_1, \ldots, n_k \geq 1 \\ n_1 + \cdots + n_k = j}} c_{n_1} \cdots c_{n_k} \right)$$

$$= 1_A + \mathrm{Exp}(a) * \mathrm{Exp}(b) - 1_A = \mathrm{Exp}(a) * \mathrm{Exp}(b).$$

In the second to last equality we applied (9.26) on page 491, ensuring that

$$\sum_{k=1}^{j} \frac{1}{k!} \sum_{\substack{n_1,\ldots,n_k \geq 1 \\ n_1+\cdots+n_k=j}} c_{n_1} \cdots c_{n_k} = 0, \quad \text{for every } j \geq 2.$$

This ends the proof. □

Theorem 5.55 (The Identity of $\diamond$ and $\blacklozenge$ on a Banach Algebra). *Let $(A, *, \|\cdot\|)$ be a Banach algebra, $\|\cdot\|$ being a compatible norm on A.*

For every (a, b) belonging to the set

$$\mathbf{D} := \Big\{ (a,b) \in A \times A \ : \ \|a\| + \|b\| < \tfrac{1}{2} \log 2 \Big\}, \tag{5.149}$$

the series of functions

$$a \diamond b := \sum_{j=1}^{\infty} \Big(\sum_{n=1}^{j} c_n \sum_{(h,k) \in \mathcal{N}_n : |h|+|k|=j} c(h,k) \, D^*_{(h,k)}(a,b) \Big) \tag{5.150}$$

is (absolutely) convergent in A. For example, this holds for a, b in the disc

$$\mathbf{E}_2 = \Big\{ a \in A \ : \ \|a\| < \tfrac{1}{4} \log 2 \Big\}, \tag{5.151}$$

since $\mathbf{E}_2 \times \mathbf{E}_2 \subset \mathbf{D}$.

The convergence of the above series is total (hence uniform) on any set of the type $\{(a,b) \in A \times A \ : \ \|a\| + \|b\| \leq \delta \}$, with $0 < \delta < \tfrac{1}{2} \log 2$.

Finally, formula (5.139) holds, that is, if $\blacklozenge$ is as in (5.146),

$$a \diamond b = a \blacklozenge b, \quad \text{for every } (a,b) \in \mathbf{D}, \tag{5.152}$$

hence in particular for every $a, b \in \mathbf{E}_2$.

Proof. The first part of the statement is already contained in Theorem 5.52, so we are left to prove (5.152).

Let $\mathbf{D}$ be as in (5.149). Let also $\Xi_n(a, b)$ be as in (5.140). Note that $\mathbf{D} \subset \mathbf{E}_0$, where the latter is the set introduced in (5.145). Then, in view of Theorem 5.54, we know that

$$\sum_{n=1}^{N} \frac{(-1)^{n+1}}{n} \Xi_n(a,b) \xrightarrow{N \to \infty} a \blacklozenge b, \tag{5.153}$$

uniformly for $(a, b) \in \mathbf{D}$ (actually, the series is normally convergent).

Let us fix any positive δ with $\delta < \frac{1}{2}\log 2$. We claim that

$$\sum_{n=1}^{N} \frac{(-1)^{n+1}}{n} \, \Xi_n(a,b) \xrightarrow{N\to\infty} a \diamond b,$$
(5.154)

uniformly on the set $A_\delta := \{(a,b) \in A \times A \,:\, \|a\| + \|b\| \leq \delta\}$.

Once this claim is proved, thanks to (5.153) the proof of the theorem is complete (note that we shall derive (5.152) by the arbitrariness of $\delta < \frac{1}{2}\log 2$).

For every $N \in \mathbb{N}$ the following computation holds (we drop the $*$ notation in the exponents):

$$\sum_{n=1}^{N} \frac{(-1)^{n+1}}{n} \, \Xi_n(a,b)$$

$$\overset{(5.142)}{=} \sum_{n=1}^{N} \frac{(-1)^{n+1}}{n} \sum_{j=n}^{\infty} \left(\sum_{\substack{(h,k)\in\mathcal{N}_n \\ |h|+|k|=j}} \frac{a^{h_1} * b^{k_1} * \cdots * a^{h_n} * b^{k_n}}{h!\,k!} \right)$$

(interchanging the sums over j and over n)

$$= \left\{ \sum_{j=1}^{N}\sum_{n=1}^{j} + \sum_{j=N+1}^{\infty}\sum_{n=1}^{N} \right\} \frac{(-1)^{n+1}}{n} \sum_{\substack{(h,k)\in\mathcal{N}_n \\ |h|+|k|=j}} \frac{a^{h_1} * \cdots * b^{k_n}}{h!\,k!}$$

$$=: \{L_N(a,b) + R_N(a,b)\}.$$

Here we have set

$$L_N(a,b) := \sum_{j=1}^{N}\sum_{n=1}^{j} \frac{(-1)^{n+1}}{n} \sum_{\substack{(h,k)\in\mathcal{N}_n \\ |h|+|k|=j}} \frac{a^{h_1} * b^{k_1} * \cdots * a^{h_n} * b^{k_n}}{h!\,k!},$$

$$R_N(a,b) := \sum_{j=N+1}^{\infty}\sum_{n=1}^{N} \frac{(-1)^{n+1}}{n} \sum_{\substack{(h,k)\in\mathcal{N}_n \\ |h|+|k|=j}} \frac{a^{h_1} * b^{k_1} * \cdots * a^{h_n} * b^{k_n}}{h!\,k!}.$$

From identity (4.18) in Corollary 4.6, we know that for every $j \in \mathbb{N}$ and every $a, b \in A$ it holds that

$$\sum_{n=1}^{j} \frac{(-1)^{n+1}}{n} \sum_{\substack{(h,k)\in\mathcal{N}_n \\ |h|+|k|=j}} \frac{a^{h_1} * b^{k_1} * \cdots * a^{h_n} * b^{k_n}}{h_1!\cdots h_n!\,k_1!\cdots k_n!}$$

$$= \sum_{n=1}^{j} \frac{(-1)^{n+1}}{n} \sum_{\substack{(h,k)\in\mathcal{N}_n \\ |h|+|k|=j}} \frac{D^*_{(h,k)}(a,b)}{h!\,k!\,(|h|+|k|)}. \tag{5.155}$$

As a consequence (see also Theorem 5.52)

$$L_N(a,b) \overset{(5.155)}{=} \sum_{j=1}^{N} Z_j^*(a,b) \xrightarrow{N\to\infty} \sum_{j=1}^{\infty} Z_j^*(a,b) = a \diamond b,$$

uniformly for $(a,b) \in A_\delta$. We are thus left to prove that the "remainder" term $R_N(a,b)$ vanishes as $N \to \infty$, uniformly for $(a,b) \in A_\delta$. Indeed, we have

$$\|R_N(a,b)\| \leq \sum_{j=N+1}^{\infty} \sum_{n=1}^{N} \frac{1}{n} \sum_{\substack{(h,k)\in\mathcal{N}_n \\ |h|+|k|=j}} \frac{1}{h!\,k!} \|a\|^{|h|} \|b\|^{|k|}$$

$$\leq \sum_{n=1}^{\infty} \frac{1}{n} \sum_{\substack{(h,k)\in\mathcal{N}_n \\ |h|+|k|\geq N+1}} \frac{1}{h!\,k!} \|a\|^{|h|} \|b\|^{|k|}$$

$$\left(\text{note that } \max\{\|a\|, \|b\|\} \leq \|a\| + \|b\| \leq \delta\right)$$

$$\leq \sum_{n=1}^{\infty} \frac{1}{n} \sum_{\substack{(h,k)\in\mathcal{N}_n \\ |h|+|k|\geq N+1}} \frac{1}{h!\,k!} \delta^{|h|+|k|}.$$

We claim that

$$\lim_{N\to\infty} \sum_{n=1}^{\infty} \frac{1}{n} \sum_{\substack{(h,k)\in\mathcal{N}_n \\ |h|+|k|\geq N+1}} \frac{\delta^{|h|+|k|}}{h!\,k!} = 0, \quad \text{whenever } |\delta| < \tfrac{1}{2}\log 2. \tag{5.156}$$

This follows by an application of Lebesgue's dominated convergence Theorem. Indeed, suppose we can pass the limit $N \to \infty$ under the sign of $\sum_{n=1}^{\infty}$. Then we have

$$\lim_{N\to\infty} \sum_{n=1}^{\infty} \frac{1}{n} \sum_{\substack{(h,k)\in\mathcal{N}_n \\ |h|+|k|\geq N+1}} \frac{\delta^{|h|+|k|}}{h!\,k!} = \sum_{n=1}^{\infty} \frac{1}{n} \left(\lim_{N\to\infty} \sum_{\substack{(h,k)\in\mathcal{N}_n \\ |h|+|k|\geq N+1}} \frac{\delta^{|h|+|k|}}{h!\,k!} \right) = 0.$$

Indeed, the limit in parentheses is zero *for every fixed n*, since the infinite sum

$$\sum_{(h,k)\in\mathcal{N}_n} \frac{\delta^{|h|+|k|}}{h!\,k!} \overset{(5.38)}{=} (e^{2\delta} - 1)^n \qquad (5.157)$$

is finite. To ensure that we can apply Lebesgue's theorem, it suffices to collect together the following facts:

1. We have the estimate (uniform w.r.t. N)

$$\left| \frac{1}{n} \sum_{\substack{(h,k)\in\mathcal{N}_n \\ |h|+|k|\geq N+1}} \frac{\delta^{|h|+|k|}}{h!\,k!} \right| \leq \frac{1}{n} \sum_{(h,k)\in\mathcal{N}_n} \frac{|\delta|^{|h|+|k|}}{h!\,k!} \overset{(5.38)}{=} \frac{1}{n} (e^{2|\delta|} - 1)^n;$$

2. The above uniform bounding sequence is "summable" in n, that is,

$$\sum_{n=1}^{\infty} \frac{1}{n} (e^{2|\delta|} - 1)^n \overset{(5.41)}{=} -\log(2 - e^{2|\delta|}) \in \mathbb{R},$$

this fact holding true because $2\,|\delta| < \log 2$.

From (5.156) and the above estimate of $\|R_N(a,b)\|$, we get $\lim_{N\to\infty} R_N(a,b) = 0$, uniformly for $(a,b) \in A_\delta$. Summing up,

$$\sum_{n=1}^{N} \frac{(-1)^{n+1}}{n} \Xi_n(a,b) = L_N(a,b) + R_N(a,b) \xrightarrow{N\to\infty} a \diamond b + 0 = a \diamond b,$$

uniformly for $(a,b) \in A_\delta$, so that (5.154) is proved. $\qquad\qquad\square$

Gathering together the results so far, we obtain the following

Theorem 5.56 (The CBHD Theorem for Banach Algebras). *Let $(A, *, \|\cdot\|)$ be a Banach algebra, $\|\cdot\|$ being a compatible norm on A.*
The following formula holds

$$\mathrm{Exp}(a \diamond b) = \mathrm{Exp}(a) * \mathrm{Exp}(b),$$
$$\text{for every } a, b \in A \text{ such that } \|a\| + \|b\| < \tfrac{1}{2} \log 2. \qquad (5.158)$$

In particular, (5.136) holds by choosing $\mathbf{E} = \mathbf{E}_2$, the latter set being as in (5.151).

Proof. Let $\mathbf{D}$ and $\mathbf{E}_0$ be as in (5.149) and (5.145), respectively. Note that $\mathbf{D} \subset \mathbf{E}_0$. Then for every $(a,b) \in \mathbf{D}$ we have

$$\mathrm{Exp}(a \diamond b) \overset{(5.152)}{=} \mathrm{Exp}(a \blacklozenge b) \overset{(5.148)}{=} \mathrm{Exp}(a) * \mathrm{Exp}(b).$$

This completes the proof. $\qquad\qquad\square$

5.5.1 An Alternative Approach Using Analytic Functions

In this section, we furnish the sketch of a possible alternative approach for attacking some of the topics of the previous sections. This approach uses the theory of *analytic functions* on a Banach space. For example, this approach is the one followed, very effectively, by Hofmann and Morris in [91, Chapter 5] for the study of the operation $\log(\exp x \exp y)$ on a Banach algebra.

Here we confine ourselves to giving the basic definitions and sketching some ideas. For an exhaustive exposition on how to prove, via analytic functions, the well-behaved properties of the operation ♦ and other interesting results, we refer the Reader to the cited reference [91, pages 111–130].

Analytic function theory provides a very powerful tool, despite some (slightly tedious) preliminary machinery. Nonetheless, we hope that our former exposition on how to attack associativity and convergence topics has some advantages too. For example, it can be adapted to any setting where analytic theory is not available (or is much more onerous), such as in the context of "exponential-like" formulas for smooth (and *non-smooth*) vector fields (see e.g., the context of the paper [124]).

Our exposition in this section is a bit more informal (and less detailed) than in the rest of the Book. We begin with the relevant definition.

Let E, F be Banach spaces (over the same field $\mathbb{K}$, where $\mathbb{K} = \mathbb{R}$ or $\mathbb{K} = \mathbb{C}$). Given $k \in \mathbb{N}$, we denote by $L_k(E^{\times k}, F)$ the set of the continuous k-linear maps $\varphi : E^{\times k} \to F$ (where $E^{\times k}$ denotes the k-fold Cartesian product of E). To say that $\varphi \in L_k(E^{\times k}, F)$ is continuous means that the real number

$$|||\varphi||| := \sup_{\|h_1\|_E, \dots, \|h_k\|_E \leq 1} \|\varphi(h_1, \dots, h_k)\|_F \qquad (5.159)$$

is finite. If $h \in E$, we denote by $h^{\times k}$ the k-tuple of $E^{\times k}$ with all entries equal to h. For $u \in E$ and $r > 0$ we let

$$B_r(u) = \{x \in E : \|x - u\|_E < r\}.$$

In the sequel, we denote by $\| \cdot \|$ the norm on any given Banach space (dropping subscripts like those in $\| \cdot \|_E$ and $\| \cdot \|_F$).

Definition 5.57 (Analytic Function). Let $f : \Omega \to F$, where $\Omega \subseteq E$ is an open set. We say that f is *analytic* in Ω if, for every $u \in \Omega$, there exist $r > 0$ and a family $\{\varphi_k\}_{k \in \mathbb{N} \cup \{0\}}$ such that:

1. $\varphi_0 \in F$ and $\varphi_k \in L_k(E^{\times k}, F)$ for every $k \in \mathbb{N}$;
2. The series of nonnegative real numbers $\sum_{k=1}^{\infty} |||\varphi_k||| \, r^k$ is convergent;
3. $B_r(u) \subseteq \Omega$ and $f(x) = \varphi_0 + \sum_{k=1}^{\infty} \varphi_k((x - u)^{\times k})$, for every $x \in B_r(u)$.

The expansion in point 3 above is called *the power series expansion of f at u* and the φ_k are called the *coefficients* of this expansion. Note that, under condition (2), the series $\sum_{k=1}^{\infty} \varphi_k((x-u)^{\times k})$ is a convergent series in the Banach space F, and the series is normally convergent on $\overline{B_r(u)}$ (recall Definition 5.9). Indeed, one has

$$\sum_{k \geq 1} \sup_{\|x-u\| \leq r} \left\| \varphi_k((x-u)^{\times k}) \right\| \leq \sum_{k \geq 1} \||\varphi_k\|| \, r^k < \infty.$$

Remark 5.58. If we have two power series expansions of f at u, with coefficients $\{\varphi_k\}_k$ and $\{\varphi'_k\}_k$ respectively, we cannot derive that $\varphi_k \equiv \varphi'_k$ for every k; instead, *the coefficients φ_k and φ'_k coincide on the diagonal of $E^{\times k}$*, i.e.,

$$\begin{cases} \varphi_0 = \varphi'_0, \\ \varphi_1 \equiv \varphi'_1 \quad \text{on } E, \\ \varphi_k(\omega^{\times k}) = \varphi'_k(\omega^{\times k}), \quad \text{for every } \omega \in E. \end{cases} \tag{5.160}$$

Indeed, given $\omega \in B_r(0)$ and $t \in \mathbb{K}$ with $|t| < 1$, we have

$$\varphi_0 + \sum_{k=1}^{\infty} t^k \varphi_k(\omega^{\times k}) = f(u + t\omega) = \varphi'_0 + \sum_{k=1}^{\infty} t^k \varphi'_k(\omega^{\times k}).$$

Taking $t = 0$ we get $\varphi_0 = \varphi'_0 = f(u)$. Canceling out $f(u)$, we obtain

$$t \sum_{k=1}^{\infty} t^{k-1} \varphi_k(\omega^{\times k}) = f(u + t\omega) - f(u) = t \sum_{k=1}^{\infty} t^{k-1} \varphi'_k(\omega^{\times k}).$$

Dividing by t and letting $t \to 0$, we obtain

$$\varphi_1(\omega) = \varphi'_1(\omega) = D_u f(\omega) := \lim_{t \to 0} \frac{f(u + t\omega) - f(u)}{t}.$$

Analogously (canceling $f(u) + t\, D_u f(\omega)$ and dividing by t^2) we get

$$\varphi_2(\omega, \omega) = \varphi'_2(\omega, \omega) = D_u^2 f(\omega) := \lim_{t \to 0} \frac{f(u + t\omega) - f(u) - t\, D_u f(\omega)}{t^2},$$

and so on. This proves (5.160) for small ω. The assertion for all values of ω then follows by the k-linearity of φ_k, φ'_k.

Remark 5.59 (Analyticity of Power Series). Let conditions (1) and (2) in Definition 5.57 hold, for some $r > 0$. Then the function $f(x) := \sum_{k=1}^{\infty} \varphi_k(x^{\times k})$ is analytic on $B_r(0)$. More precisely, given $u \in B_r(0)$, we are able to provide a power series expansion for f at u on the whole ball $B_{r-\|u\|}(u)$, *which is the largest ball centered at u contained in $B_r(0)$.*

Let us set $\rho := r - \|u\|$. We begin with the following central computation, holding true for any $w \in E$ such that $\|w\| < \rho$:

$$f(u+w) = \sum_{n\geq 1} \varphi_n((u+w)^{\times n}) \quad \text{(by the n-linearity of φ_n)}$$

$$= \sum_{n\geq 1} \sum_{z\in\{u,w\}^{\times n}} \varphi_n(z) = \sum_{n\geq 1} \sum_{k=0}^{n} \sum_{z\in\mathcal{H}_k^n} \varphi_n(z) = \sum_{k\geq 0} \sum_{n\geq k} \sum_{z\in\mathcal{H}_k^n} \varphi_n(z).$$

Here we have partitioned the Cartesian product $\{u,w\}^{\times n}$ into the disjoint union $\bigcup_{k=0}^{n} \mathcal{H}_k^n$, where $\mathcal{H}_k^n$ is the set of n-tuples containing w precisely k times (whence u appears $n-k$ times). Note that *the cardinality of $\mathcal{H}_k^n$ is equal to the binomial coefficient* $\binom{n}{k}$. The above computation leads us to seek for the coefficients of the expansion of f at u, say $\{\psi_k\}_k$, in such a way that

$$\psi_k(w^{\times k}) = \sum_{n\geq k} \sum_{z\in\mathcal{H}_k^n} \varphi_n(z), \quad \forall\, k \in \mathbb{N}.$$

Note that any element z of $\mathcal{H}_k^n$ has the form

$$z = \Big(u,\dots,u,w,u,\dots,u,w,\dots,u,\dots,u,w,u,\dots,u \Big),$$

where w occurs exactly k times and u occurs $n-k$ times (some of the above strings $u,\dots,u$ may not appear). This leads us to define ψ_k (as element of $L_k(E^{\times k}, F)$) as follows:

$$\psi_k(w_1,\dots,w_k) := \sum_{n\geq k} \sum_{z(w_1,\dots,w_k)} \varphi_n\big(z(w_1,\dots,w_k)\big), \quad w_1,\dots,w_k \in E,$$

where the inner sum runs over the elements of $E^{\times n}$ of the following form (with the obvious meaning)

$$z(w_1,\dots,w_k) = \Big(u,\dots,u,w_1,u,\dots,u,w_2,\dots,u,\dots,u,w_k,u,\dots,u \Big).$$

It is then not difficult to prove that one has

$$\big\|\psi_k(w_1,\dots,w_k)\big\| \leq \underbrace{\sum_{n\geq k} \binom{n}{k} \||\varphi_n\|| \, \|u\|^{n-k}}_{=:J_k(u)} \cdot \|w_1\| \cdots \|w_k\|.$$

Note that $J_k(u)$ is a *finite* real number since one has (for $\rho := r - \|u\|$)

$$\sum_{k \geq 0} J_k(u)\, \rho^k = \sum_{n \geq 0} \||\varphi_n\|| \sum_{k=0}^{n} \binom{n}{k} \|u\|^{n-k}\, \rho^k = \sum_{n \geq 0} \||\varphi_n\|| \, r^n < \infty,$$

in view of the convergence condition (2). This proves, all at once, that $\psi_k \in L_k(E^{\times k}, F)$ and that $\sum_{k \geq 0} \||\psi_k\|| \, \rho^k < \infty$. This is what we desired to prove.
□

Example 5.60. Let $(A, *, \| \cdot \|)$ be a Banach algebra with a norm compatible with the product. Suppose $\{a_k\}_{k \geq 0}$ is a sequence in A for which there exists $r > 0$ satisfying

$$\sum_{k=0}^{\infty} \|a_k\|\, r^k < \infty.$$

Then the function $f : B_r(0) \to A$, defined by

$$f(x) := \sum_{k=0}^{\infty} a_k * x^{*k} \tag{5.161}$$

is analytic in $B_r(0)$. Indeed, by Remark 5.59, it suffices to show that f admits a power series expansion (in the sense of Definition 5.57) at 0. To this end, it is immediately seen that the definitions $\varphi_0 := a_0$ and

$$\varphi_k : A^{\times k} \to A, \quad \varphi_k(\omega_1, \dots, \omega_k) := a_k * \omega_1 * \cdots * \omega_k$$

fulfil the axioms (1)–(3) in the cited definition, since $\||\varphi_k\|| \leq \|a_k\|$.

We remark that, for an arbitrary Banach algebra A, *not all* analytic functions (converging on some ball centered at 0) have the form (5.161). For example, if $(A, *)$ is the algebra of 2×2 real matrices (with the usual matrix product) the map

$$f \begin{pmatrix} x_{1,1} & x_{1,2} \\ x_{2,1} & x_{2,2} \end{pmatrix} = \begin{pmatrix} x_{2,2} & 0 \\ 0 & 0 \end{pmatrix}$$

is analytic on A but it does not take the form (5.161) on any ball centered at the origin. Also, the functions of the type (5.161) do not even form an algebra (with the multiplication of functions derived from the multiplication of A). Instead, let us consider the set $\mathcal{A}$ of the functions f of the form

$$f(x) = \sum_{k=0}^{\infty} a_k\, x^{*k}, \tag{5.162}$$

where $a_k \in \mathbb{K}$ for every $k \in \mathbb{N} \cup \{0\}$, and such that there exists $r > 0$ (depending on f) satisfying $\sum_{k=0}^{\infty} |a_k|\, r^k < \infty$. Then $\mathcal{A}$ is an algebra, with respect to the product of functions inherited from the $*$ product. Indeed, one has

$$\left(\textstyle\sum_{k=0}^{\infty} a_k\, x^{*k}\right) * \left(\textstyle\sum_{k=0}^{\infty} b_k\, x^{*k}\right) = \textstyle\sum_{k=0}^{\infty} c_k\, x^{*k},$$

with $c_k = \sum_{i=0}^{k} a_i\, b_{k-i}$ and, if $\sum_{k=0}^{\infty} |a_k|\, r_1^k < \infty$ and $\sum_{k=0}^{\infty} |b_k|\, r_2^k < \infty$, then $\sum_{k=0}^{\infty} |c_k|\, r^k < \infty$ with $r = \min\{r_1, r_2\}$.

Remark 5.61. Next, it is not difficult to prove that *the composition of analytic functions is analytic.* Indeed, let E, F, G be Banach spaces over $\mathbb{K}$. Let $g : V \to W$ and $f : W \to G$ be analytic functions, where $V \subseteq E$ and $W \subseteq F$ are open sets. Let $v \in V$ be fixed. Let $\varepsilon > 0$ be so small that the following conditions hold:

(a) g has a power series expansion $g(x) = \sum_{n \geq 0} G_n (x - v)^{\times n}$ on $B_\varepsilon(v) \subseteq V$;
(b) $g(B_\varepsilon(v))$ is contained in an open ball centered at $g(v)$, say $B_R(g(v)) \subseteq W$, where f has a power series expansion $f(y) = \sum_{i > 0} F_i (y - g(v))^{\times i}$;
(c) The sum of the real valued series $\sum_{n \geq 1} |||G_n|||\, \varepsilon^n$ is less than the above R.

It is then not difficult to prove that, for every $x \in B_\varepsilon(v)$, one has

$$f(g(x)) = \sum_{i \geq 0} F_i \left(\sum_{n \geq 1} G_n (x - v)^{\times n} \right)^{\times i}$$

$$= f(g(v)) + \sum_{j \geq 1} \sum_{\substack{i, n_1, \ldots, n_i \geq 1 \\ n_1 + \cdots + n_i = j}} F_i \left(G_{n_1} (x - v)^{\times n_1}, \ldots, G_{n_i} (x - v)^{\times n_i} \right).$$

Hence, we can define $H_j : E^{\times j} \to G$ by declaring that $H_j(\omega_1, \ldots, \omega_j)$ equals

$$\sum_{\substack{i, n_1, \ldots, n_i \geq 1 \\ n_1 + \cdots + n_i = j}} F_i \left(G_{n_1}(\omega_1, \ldots, \omega_{n_1}), G_{n_2}(\omega_{n_1+1}, \ldots, \omega_{n_1+n_2}), \ldots \right),$$

for every $\omega_1, \ldots, \omega_j \in E$. This clearly defines an element of $L_j(E^{\times j}, G)$, since the above sum is *finite*. Also, one has

$$\sum_{j \geq 1} |||H_j|||\, \varepsilon^j \leq \sum_{j \geq 1} \sum_{\substack{i, n_1, \ldots, n_i \geq 1 \\ n_1 + \cdots + n_i = j}} |||F_i||| \cdot |||G_{n_1}||| \cdots |||G_{n_i}|||\, \varepsilon^{n_1} \cdots \varepsilon^{n_i}$$

$$= \sum_{i \geq 1} |||F_i||| \left(\sum_{n \geq 1} |||G_n|||\, \varepsilon^n \right)^i \leq \sum_{i \geq 1} |||F_i|||\, R^i < \infty.$$

Here we also used condition (c) on ε. The finiteness of the far right-hand series is a consequence of condition (b) above (together with axiom (2) in Definition 5.57). Finally, our choice of H_j also fulfils the requirement $(f \circ g)(x) = f(g(v)) + \sum_{j \geq 1} H_j(x - v)^{\times j}$ on $B_\varepsilon(v)$, and we are done. □

Arguing as above, one proves the following fact.

Remark 5.62. Let $(A, *, \| \cdot \|)$ be a Banach algebra with a norm compatible with the product. Let $\mathcal{A}$ be the algebra of functions considered at the end of Example 5.60. Let f be as in (5.162) and let $g \in \mathcal{A}$ be of the form $g(x) = \sum_{k=1}^{\infty} b_k \, x^{*k}$. Then $h(x) := f(g(x))$ is well posed for $x \in B_\varepsilon(0)$ for a small $\varepsilon > 0$ and h belongs to $\mathcal{A}$ (the ball where h has an expansion as in (5.162) being possibly smaller than $B_\varepsilon(0)$).

The following result is probably one of the most important in the theory of analytic functions. We shall prove it as a simple consequence of the estimate of the radius of convergence contained in Remark 5.59.

Theorem 5.63 (Unique Analytic Continuation). *Suppose E, F are Banach spaces, Ω is an open and connected subset of E, and $f, g : \Omega \to F$ are analytic functions. If f and g coincide on an open (nonempty) subset of Ω, then f and g coincide throughout Ω.*

Proof. Let A denote the set of the points $u \in \Omega$ possessing an open neighborhood where f and g coincide. By the hypothesis that f and g coincide on a nonempty open subset of Ω, we deduce that A is not empty. The proof will be complete if we show that A is both open (which is trivial) and closed relatively to Ω.

To this end, let $\{x_k\}_k$ be a sequence in A and suppose that $x_k \to x_0$ as $k \to \infty$, with $x_0 \in \Omega$. We have to prove that $x_0 \in A$. Let $\varepsilon > 0$ be such that f, g admit convergent power series expansions on $B_\varepsilon(x_0)$. Since $x_k \to x_0$, there exists $k \in \mathbb{N}$ sufficiently large that $x_k \in B_{\varepsilon/2}(x_0)$. This ensures that the ball $B_k := B_{\varepsilon - \|x_k - x_0\|}(x_k)$ "captures" x_0. By the results in Remark 5.59, f and g do admit power series expansions at x_k, converging on the whole ball B_k, say

$$(\bigstar) \qquad \begin{cases} f(x) = \sum_{n \geq 0} F_n(x - u_k)^{\times n}, \\ g(x) = \sum_{n \geq 0} G_n(x - u_k)^{\times n}, \end{cases} \quad \text{for every } x \in B_k.$$

On the other hand, from $x_k \in A$ we deduce that f, g coincide on a neighborhood of x_k (not necessarily containing x_0). Hence, by (5.160), we obtain that (for every $n \geq 0$) F_n and G_n coincide on the diagonal of $E^{\times n}$. By $(\bigstar)$ above, this yields $f \equiv g$ on B_k. Since $x_0 \in B_k$, we derive that $x_0 \in A$. □

To test the power of Theorem 5.63 let us look at the following example (see the proof of [91, Proposition 5.3]).

Let $(A, *, \| \cdot \|)$ be a Banach algebra (the norm being compatible with the product). Let $f(x) := \sum_{k=0}^{\infty} \frac{1}{k!} x^{*k}$ and $g(x) := \sum_{k=1}^{\infty} \frac{(-1)^{k+1}}{k} x^{*k}$ (in other words, $f(x) = \mathrm{Exp}(x)$, $g(x) = \mathrm{Log}(1_A + x)$). Using Example 5.60, as $\sum_{k \geq 0} r^k / k! = e^r < \infty$ for every $r > 0$, and $\sum_{k \geq 1} r^k / k = -\log(1 - r) < \infty$

for every $r \in (0,1)$, it follows that f is analytic on A and g is analytic on $B_1(0)$. We prove that $f(g(x)) = 1 + x$ on $B_1(0)$.

Let us set $h := f \circ g$. By Remarks 5.59 and 5.61, h is analytic on $B_1(0)$. By Remark 5.62, h has a power series expansion of the special form $h(x) = \sum_{n=0}^{\infty} c_n x^{*n}$ (with $c_n \in \mathbb{K}$) on some ball $B_\varepsilon(0)$. Let us choose $x = \lambda 1_A$, with $\lambda \in \mathbb{K}$ (recall that $\mathbb{K}$ is $\mathbb{R}$ or $\mathbb{C}$). Then obviously $h(\lambda 1_A) = \left(\sum_{n=0}^{\infty} c_n \lambda^n \right) 1_A$ whenever $|\lambda| < \varepsilon/\|1_A\|$. On the other hand, directly from the definitions,

$$h(\lambda 1_A) = f(g(\lambda 1_A)) = f(\ln(1 + \lambda) 1_A) = e^{\ln(1+\lambda)} 1_A = (1 + \lambda) 1_A,$$

whenever $|\lambda| < 1/\|1_A\|$. Hence we have $\sum_{n=0}^{\infty} c_n \lambda^n = 1 + \lambda$, for λ near the origin. This gives $c_0 = c_1 = 1$ and $c_n = 0$ for every $n \geq 2$. Thus $h(x) = 1_A + x$ for $x \in B_\varepsilon(0)$. By Theorem 5.63, this identity is valid on the largest connected neighborhood of 0 where h is analytic, hence on $B_1(0)$.

Example 5.64. As for the topics of this Book, remarkable examples of analytic functions are provided by the CBHD series in a Banach algebra, or more generally, on a Banach-Lie algebra.

Indeed, let $(\mathfrak{g}, [\cdot, \cdot], \| \cdot \|_\mathfrak{g})$ be a Banach-Lie algebra (here the norm is assumed to be compatible with the bracket). For $j \in \mathbb{N}$, let $Z_j^\mathfrak{g}(a,b)$ be as in (5.28). Consider the usual function defined by the CBHD series

$$\Omega \ni (a,b) \mapsto f(a,b) := a \diamond b = \sum_{j=1}^{\infty} Z_j^\mathfrak{g}(a,b),$$

where $\Omega = \{(a,b) \in \mathfrak{g} \times \mathfrak{g} : \|a\|_\mathfrak{g} + \|b\|_\mathfrak{g} < \frac{1}{2} \log 2\}$. We claim that $f : \Omega \to \mathfrak{g}$ *is an analytic function on the open set* Ω *of the Banach space* $\mathfrak{g} \times \mathfrak{g}$. To prove this, let us consider $E := \mathfrak{g} \times \mathfrak{g}$ with its Banach space structure resulting from the norm

$$\|(a,b)\|_E := \|a\|_\mathfrak{g} + \|b\|_\mathfrak{g}, \qquad a, b \in \mathfrak{g}.$$

Note that Ω is the ball in E centred at $(0,0)$ with radius $\frac{1}{2} \log 2$. In view of Remark 5.59, it is enough to show that f admits a power series expansion at $(0,0) \in \mathfrak{g} \times \mathfrak{g}$, converging on every set of the form $\{\|a\|_\mathfrak{g} + \|b\|_\mathfrak{g} \leq r\}$ with $r < \frac{1}{2} \log 2$. We define the coefficients in this expansion as follows:

$\varphi_0 := 0;$

$\varphi_1 : E \to \mathfrak{g}, \quad \varphi_1(\omega) := a + b, \qquad\qquad\qquad$ where $\omega = (a,b);$

$\begin{cases} \varphi_2 : E \times E \to \mathfrak{g}, \\ \varphi_2(\omega_1, \omega_2) := \frac{1}{2} [a_1, b_2], \end{cases}$ where $\begin{aligned} \omega_1 &= (a_1, b_1), \\ \omega_2 &= (a_2, b_2); \end{aligned}$

$\begin{cases} \varphi_3 : E^{\times 3} \to \mathfrak{g}, \\ \varphi_3(\omega_1, \omega_2, \omega_3) := \frac{1}{12} \left([a_1, [a_2, b_3]] + [b_1, [b_2, a_3]] \right), \end{cases}$ where $\begin{aligned} \omega_i &= (a_i, b_i), \\ i &= 1, 2, 3; \end{aligned}$

$$\begin{cases} \varphi_4 : E^{\times 4} \to \mathfrak{g}, \\ \varphi_3(\omega_1, \omega_2, \omega_3, \omega_4) := -\frac{1}{24}[a_1, [b_2, [a_3, b_4]]], \end{cases} \quad \text{where} \quad \begin{matrix} \omega_i = (a_i, b_i), \\ i = 1, 2, 3, 4; \end{matrix}$$

and so on... Then, following the above idea, it is not difficult (although very tedious) to define φ_j in $L_j(E^{\times j}, \mathfrak{g})$ in such a way that

$$\varphi_j((a, b)^{\times j}) = Z_j^{\mathfrak{g}}(a, b), \quad \text{for every } j \in \mathbb{N} \text{ and } (a, b) \in E.$$

Also, by means of the fundamental estimate in Theorem 5.29, one proves that

$$\||\varphi_j\|| \le \sum_{n=1}^{j} |c_n| \sum_{(h,k) \in \mathbb{N}_n \,:\, |h|+|k|=j} \mathbf{c}(h, k),$$

so that (provided that $2\, r < \log 2$)

$$\sum_{j=1}^{\infty} \||\varphi_j\|| \, r^j \le \sum_{j=1}^{\infty} \sum_{n=1}^{j} |c_n| \sum_{(h,k) \in \mathbb{N}_n \,:\, |h|+|k|=j} \mathbf{c}(h, k) \, r^{|h|+|k|}$$

$$\le \sum_{n=1}^{\infty} \frac{1}{n} \sum_{(h,k) \in \mathbb{N}_n} \frac{r^{|h|} \, r^{|k|}}{h! \, k!} \overset{(5.38)}{=} \sum_{n=1}^{\infty} \frac{1}{n} \left(e^{2r} - 1\right)^n = -\log(2 - e^{2r}) < \infty.$$

Remark 5.65. Roughly speaking, the computations involving analytic functions can be performed at the level of the underlying *formal power series setting*, where – as we have seen – arguments are generally much easier, since all convergence matters are valid "for free". For example, one of the most significant applications of the theory of analytic functions to our area of interest is that, for example, one can prove associativity for $\diamond$ directly from the formal power series setting, that is, from the identities established in Sect. 5.3.1 (see Theorems 5.35 and 5.36 or the closed identity in Remark 5.37). We leave the details to the Reader.

5.6 An Example of Non-convergence of the CBHD Series

After having investigated extensively the topic of the convergence of the CBHD series, it is compulsory to provide an example of *non-convergence*. We exhibit such an example in one of the simplest contexts of non-commutative algebras: that of real square matrices with the usual matrix product. Throughout this section, $(\mathcal{M}, \cdot)$ denotes the UA algebra of 2×2 matrices with real entries. This is a Banach algebra, with the so-called *Frobenius norm*

$$\|A\|_F := \sqrt{\text{trace}(A^T \cdot A)} = \sqrt{\sum_{i,j=1}^{2}(a_{i,j})^2}, \quad A = (a_{i,j})_{i,j \le 2} \in \mathcal{M}.$$

This norm is compatible with the product of $\mathcal{M}$, as an application of the Cauchy-Schwarz inequality shows at once. We begin with the following example.

Example 5.66. This counterexample is due to Wei, [177].[9] Consider the matrices

$$A := \begin{pmatrix} 0 & -5\pi/4 \\ 5\pi/4 & 0 \end{pmatrix}, \qquad B := \begin{pmatrix} 0 & 1 \\ 0 & 0 \end{pmatrix}, \tag{5.163}$$

we claim that *there does not exist any $C \in \mathcal{M}$ such that $\exp(A) \cdot \exp(B) = \exp(C)$.*

To prove this, we first observe that, after simple calculations[10] one has

$$\exp(A) = \begin{pmatrix} -1/\sqrt{2} & 1/\sqrt{2} \\ -1/\sqrt{2} & -1/\sqrt{2} \end{pmatrix}, \qquad \exp(B) = \begin{pmatrix} 1 & 1 \\ 0 & 1 \end{pmatrix},$$

so that

$$\exp(A) \cdot \exp(B) = \begin{pmatrix} -1/\sqrt{2} & 0 \\ -1/\sqrt{2} & -\sqrt{2} \end{pmatrix} =: D.$$

We now argue by contradiction: we suppose there exists $C \in \mathcal{M}$ such that $\exp(C) = D$. Then $\exp(C/2)$ is a square root of D in $\mathcal{M}$. We reach a contradiction if we show that there does not exist any matrix in $\mathcal{M}$ whose square equals D. Indeed, if this were the case, one would have

$$\begin{pmatrix} -1/\sqrt{2} & 0 \\ -1/\sqrt{2} & -\sqrt{2} \end{pmatrix} = \begin{pmatrix} a & b \\ c & d \end{pmatrix}^2 = \begin{pmatrix} a^2 + bc & b(a+d) \\ c(a+d) & d^2 + bc \end{pmatrix},$$

for some real numbers a, b, c, d. Now, by equating the entries of place $(1, 2)$ of the far sides of the above equality, it must hold that $b = 0$ or $a + d = 0$; but $b = 0$ is in contradiction with the equalities of the entries of place $(1, 1)$ (there does not exist any *real* a such that $a^2 = -1/\sqrt{2}$), whilst $a + d = 0$ is in

---

[9]We take this opportunity to correct a misprint in the example by Wei, who declares the intention to use the matrix $2B$, but then makes the computation with B.

[10]Indeed, B is nilpotent of degree 2 and A can be easily diagonalized. Alternatively, we have the formula

$$\exp \begin{pmatrix} \alpha & -\beta \\ \beta & \alpha \end{pmatrix} = \begin{pmatrix} e^\alpha \cos\beta & -e^\alpha \sin\beta \\ e^\alpha \sin\beta & e^\alpha \cos\beta \end{pmatrix}, \qquad \alpha, \beta \in \mathbb{R},$$

which can be straightforwardly obtained by observing that the matrix $\begin{pmatrix} \alpha & -\beta \\ \beta & \alpha \end{pmatrix}$ is the matrix of the endomorphism of $\mathbb{R}^2 \equiv \mathbb{C}$ obtained by multiplication times $\alpha + i\beta \in \mathbb{C}$.

contradiction with the equalities of the entries of place $(2, 1)$. This ends the argument.[11]

In Example 5.66, we showed that there does not exist any matrix $C \in \mathcal{M}$ such that $\exp(A) \cdot \exp(B) = \exp(C)$. Does this prove that the CBHD series for A, B is not convergent? The answer to this (non-trivial) question is positive: this is a consequence of Theorem 5.67 below, applied to the case of the Banach algebra $(\mathcal{M}, \cdot, \| \cdot \|_F)$. Note that, in order to answer the above question, we make implicit use of the theory of analytic functions, since Theorem 5.67 will be proved by the theorem of unique continuation, Theorem 5.63.

The above arguments allow us to discover that, *if A and B are as in (5.163),
the CBHD series $\sum_{j=1}^{\infty} Z_j(A, B)$ is not convergent in the algebra of real 2×2
matrices.* Here, the Lie bracket defining Z_j is the commutator related to the usual associative algebra structure of $\mathcal{M}$. Note also that A, B belong to $\mathfrak{sl}_2(\mathbb{R}) = \{A \in \mathcal{M} : \operatorname{trace}(A) = 0\}$, which is a Lie subalgebra of $\mathcal{M}$ (indeed, for every $A, B \in \mathcal{M}$, $[A, B] = A \cdot B - B \cdot A$ is in $\mathfrak{sl}_2(\mathbb{R})$, since $\operatorname{trace}(A \cdot B) = \operatorname{trace}(B \cdot A)$). The algebra $\mathfrak{sl}_2(\mathbb{R})$ can be equipped with a norm compatible with the bracket, turning it into a Banach-Lie algebra (see [118]):

$$\|A\| := \sqrt{2} \, \|A\|_F, \quad A \in \mathfrak{sl}_2(\mathbb{R}).$$

As a consequence, the above counterexample also shows that the CBHD series $\sum_{j=1}^{\infty} Z_j(A, B)$ is not convergent *in the Banach-Lie algebra $\mathfrak{sl}_2(\mathbb{R})$.*

For other counterexamples, see e.g., [14, 35, 36, 122, 175, 177].

The following result, of independent interest, is a consequence of the theorem of unique continuation, and it is the main goal of this section.

Theorem 5.67. *Let $(A, *, \| \cdot \|)$ be a Banach algebra, the norm being compatible with the product. Let $a, b \in A$. If the CBHD series $a \diamond b = \sum_{j=1}^{\infty} Z_j^*(a, b)$ converges in A, then $\operatorname{Exp}(a) * \operatorname{Exp}(b) = \operatorname{Exp}(a \diamond b)$.*

We begin with another remarkable result.

Lemma 5.68 (Abel). *Let $(E, \| \cdot \|)$ be a (real or complex) Banach space. Suppose
$\{e_n\}_{n \geq 0}$ is a sequence in E such that the series $\sum_{n=0}^{\infty} e_n$ converges in E. Let*

$$f_n : [0, 1] \to E, \qquad f_n(t) := e_n \, t^n \quad (n \in \mathbb{N} \cup \{0\}).$$

Then the series of functions $\sum_{n=0}^{\infty} f_n(t)$ is uniformly convergent on $[0, 1]$.

Proof. To begin with, we observe that the series of functions $\sum_{n=0}^{\infty} f_n(t)$ is normally convergent on $[-r, r]$, for every $r \in (0, 1)$. Indeed, since $\sum_{n=0}^{\infty} e_n$ is convergent, the sequence $\{e_n\}_n$ vanishes as $n \to \infty$, hence it is bounded.

---

[11]Note that here the underlying field plays a major rôle. Indeed, as can be seen by using the Jordan normal form of a square matrix, *in the complex case* the map exp is onto the set of invertible matrices.

Let $M > 0$ be such that $\|e_n\| \leq M$ for every $n \geq 0$. Then we have

$$\sum_{n\geq 0} \sup_{t\in[-r,r]} \|f_n(t)\| \leq \sum_{n\geq 0} M\, r^n < \infty, \quad \text{since } r \in (0,1).$$

It follows that $\sum_{n\geq 1} f_n(t)$ is convergent for every $t \in (-1,1]$.

In order to demonstrate the uniform convergence on $[0,1]$, it is enough to show that the uniform Cauchy condition holds for the sequence of functions $s_n(t) := \sum_{k=0}^{n} f_k(t)$ (recall that E is Banach). That is, we must prove that for every $\varepsilon > 0$ there exists $n_\varepsilon \in \mathbb{N}$ such that, if $n \geq n_\varepsilon$, we have

$$\|s_{n+p}(t) - s_n(t)\| \leq \varepsilon, \quad \text{for every } t \in [0,1] \text{ and every } p \in \mathbb{N}.$$

To this end, let $\varepsilon > 0$ be fixed. The hypothesis that $\sum_{n\geq 0} e_n$ is convergent is equivalent to the fact that the sequence

$$R_n := \sum_{k=n}^{\infty} e_k \quad (n \in \mathbb{N} \cup \{0\})$$

vanishes as $n \to \infty$. Hence, there exists $n_\varepsilon \in \mathbb{N}$ such that

$$\|R_n\| \leq \varepsilon, \quad \text{for every } n \geq n_\varepsilon. \tag{5.164}$$

We now perform a "summation by parts": noticing that $e_k = R_k - R_{k+1}$,

$$s_{n+p}(t) - s_n(t) = \sum_{k=n+1}^{n+p} e_k\, t^k = \sum_{k=n+1}^{n+p} (R_k - R_{k+1})\, t^k$$

$$= \sum_{k=n+1}^{n+p} R_k\, t^k - \sum_{k=n+2}^{n+p+1} R_k\, t^{k-1}$$

$$= R_{n+1}\, t^{n+1} - R_{n+p+1}\, t^{n+p} + \sum_{k=n+2}^{n+p} R_k\, (t^k - t^{k-1}).$$

If $n \geq n_\varepsilon$, taking into account (5.164), the triangle inequality thus gives

$$\|s_{n+p}(t) - s_n(t)\|$$

$$\leq \|R_{n+1}\|\, |t^{n+1}| + \|R_{n+p+1}\|\, |t^{n+p}| + \sum_{k=n+2}^{n+p} \|R_k\|\, |t^k - t^{k-1}|$$

$$\leq 2\varepsilon + \varepsilon \sum_{k=n+2}^{n+p} |t^k - t^{k-1}| \leq 2\varepsilon + \varepsilon\, (t^{n+1} - t^{n+p}) \leq 4\varepsilon$$

(recall that $t \in [0,1]$ so that $|t^k - t^{k-1}| = t^{k-1} - t^k$).

Summing up, $\|s_{n+p}(t) - s_n(t)\| \leq 4\,\varepsilon$, whenever $n \geq n_\varepsilon$. Since this is true for every $t \in [0, 1]$ and every $p \in \mathbb{N}$, the proof is complete. $\square$

Proof (of Theorem 5.67). Let $a, b, Z_j(a, b)$ be as in the statement of the theorem. We begin by observing that the function

$$F : (-1, 1) \longrightarrow A, \quad F(t) := \sum_{n=1}^{\infty} Z_n(a, b)\, t^n$$

is analytic, in the sense of Definition 5.57 (here $(-1, 1)$ is a subset of the usual Banach space $\mathbb{R}$). For $n \geq 1$ we set

$$\varphi_n : \mathbb{R}^{\times n} \to A, \quad \varphi_n(\omega_1, \ldots, \omega_n) := Z_n(a, b)\, \omega_1 \cdots \omega_n,$$

for every $\omega_1, \ldots, \omega_n \in \mathbb{R}$. Obviously, $\varphi_n \in L_n(\mathbb{R}^{\times n}, A)$ and $\||\varphi_n\|| = \|Z_n(a, b)\|$. Thus, for every fixed $r \in\,]0, 1[$ we have $\sum_{n \geq 1} \||\varphi_n\||\, r^n < \infty$. This follows by arguing as in the proof of Lemma 5.68, taking into account the hypothesis that $\sum_{n \geq 1} Z_n(a, b)$ converges in A. By Remark 5.59, this proves that the restriction of F to $(-r, r)$ is analytic. Since $r \in (0, 1)$ is arbitrary, this proves that F is analytic on $(-1, 1)$. By Abel's Lemma 5.68, we obtain that $F(t)$ is continuous up to $t = 1$, since $\sum_{n=1}^{\infty} Z_n(a, b)\, t^n$ is the *sum of a uniformly convergent series of continuous functions on* $[0, 1]$.

Since $\mathrm{Exp} : A \to A$ is analytic (see Example 5.60) and since the composition of analytic functions is analytic (see Remark 5.61), all the above facts prove that the function $\mathrm{Exp} \circ F$ is analytic on $(-1, 1)$ and it is continuous up to $t = 1$. On the other hand, the function

$$G : \mathbb{R} \to A, \quad G(t) := \mathrm{Exp}(t\, a) * \mathrm{Exp}(t\, b)$$

is obviously analytic, again in the sense of Definition 5.57. This follows from the analyticity of Exp and the (evident!) analyticity of the function $A \times A \ni (a, b) \mapsto a * b \in A$; thus G is analytic, thanks to the results in Remark 5.61.

We claim that $\mathrm{Exp}(F(t))$ and $G(t)$ coincide on a neighborhood of $t = 0$. Once this claim is proved, *by the Unique Continuation Theorem 5.63*, we will be able to infer that

$$\mathrm{Exp}(F(t)) = G(t), \quad \text{for every } t \in (-1, 1).$$

Letting $t \to 1^-$ in the above identity (and recalling that $\mathrm{Exp} \circ F$ and G are continuous up to $t = 1$), we get $\mathrm{Exp}(F(1)) = G(1)$. This identity is equivalent to $\mathrm{Exp}(\sum_{n=1}^{\infty} F_n(a, b)) = \mathrm{Exp}(a) * \mathrm{Exp}(b)$, the thesis of Theorem 5.67.

Hence we are left to prove the above claim. First notice that, *as a consequence of the homogeneity of Z_n,* one has

$$F(t) = \sum_{n=1}^{\infty} Z_n(a, b)\, t^n = \sum_{n=1}^{\infty} Z_n(t\, a, t\, b) = (t\, a) \diamond (t\, b). \qquad (5.165)$$

By Theorem 5.56, we are thus able to derive that

$$\mathrm{Exp}\big((t\,a) \diamond (t\,b)\big) = \mathrm{Exp}(t\,a) * \mathrm{Exp}(t\,b), \qquad (5.166)$$

provided that $\|t\,a\| + \|t\,b\| < \frac{1}{2}\log 2$. By (5.165), identity (5.166) means that

$$\mathrm{Exp}(F(t)) = G(t) \quad \text{for every } t \in (-\varepsilon, \varepsilon),$$

where ε is sufficiently small that $\varepsilon\,(\|a\| + \|b\|) < \frac{1}{2}\log 2$. This proves that $\mathrm{Exp} \circ F$ and G coincide on a neighborhood of 0. The proof is complete. $\quad\square$

5.7 Further References

First, we would like to point out that the rôle of the CBHD Theorem is prominent not only for usual (finite-dimensional) Lie groups, but also within the context of *infinite dimensional* Lie groups (see Neeb [130] for a comprehensive survey on these groups). As for the topics of this Book, the notion of BCH-group (Baker-Campbell-Hausdorff group) is particularly significant.[12] See e.g., Birkhoff [12, 13]; Boseck, Czichowski and Rudolph [24]; Czyż [42] and [43, Chapter IV]; Djoković and Hofmann [49,88]; Dynkin [56]; Glöckner [66–68]; Glöckner and Neeb [69]; Gordina [73]; Hilgert, Hofmann [83]; Hofmann [86, 87]; Hofmann and Morris [92] (see also references therein); Hofmann and Neeb [93]; Neeb [130]; Omori [133]; Robart [147,148]; Schmid [152]; Van Est and Korthagen [170]; Vasilescu [173]; Wojtyński [182].

For example, an important class of (possibly) infinite dimensional Lie groups is that of the *locally exponential Lie groups* (see Section IV in Neeb's treatise [130]): this vast class comprises the so called *BCH-Lie groups* (those for which the CBHD series defines an analytic local multiplication on a neighborhood of the origin of the appropriate Lie algebra) and, in particular, it contains the Banach-Lie groups. In the setting of locally exponential Lie groups, many of the classical Lie theoretic results possess a proper analogue (see e.g., [130, Theorems IV.1.8, IV.1.18, IV.1.19, IV.2.8, IV.3.3, Proposition IV.2.7]; we explicitly remark that Theorem IV.2.8 in [130] contains a notable "universal property" of the CBHD series).

We now proceed to give references on the problems of convergence and "optimization" of the CBHD series. Let us begin with a quick descriptive overview of the related topics.

---

[12]The theory of infinite-dimensional Lie groups has grown rapidly in the last decade. A comprehensive list of references about infinite-dimensional Lie groups is beyond our scope here. We restrict ourselves to just citing a few references, having some intersection with the CBHD Theorem or with the BCH-groups.

It is not uncommon to meet in the literature unprecise statements concerning the CBHD series, asserting that this is given by

$$x + y + \tfrac{1}{2}[x,y] + \tfrac{1}{12}[x,[x,y]] + \tfrac{1}{12}[y,[y,x]] + \cdots \qquad \left(\begin{array}{c}\text{but what does}\\ \text{"}\cdots\text{" mean?}\end{array}\right)$$

We can accept this as some shorthand due to typographical reasons, but besides concealing the genuine formula for higher order terms, this formula lacks precision for more serious reasons: *Are the summands grouped together? IF they are, how are they grouped together?* And also: *How are summands ordered?*

These are not far-fetched questions, for it is well-known from elementary Analysis that sometimes by associating the summands from a non-convergent series we may turn it into a convergent one and, even if a series converges, the permutation law may fail as well (this is always the case for conditionally convergent series). The fact that the CBHD series arises from a logarithmic series (recall in particular the alternating factor $(-1)^{n+1}/n$) highlights that possible cancellations may play a prominent rôle (as we saw, e.g., in Theorem 5.33 on page 301).

A more subtle problem is the way the CBHD series itself is presented: For instance, is the "fourth order" term

$$-\tfrac{1}{24}[x,[y,[x,y]]] \quad \text{or is it} \quad -\tfrac{1}{48}[x,[y,[x,y]]] + \tfrac{1}{48}[y,[x,[y,x]]] ?$$

Actually (thanks to the Jacobi identity or – equivalently – as a direct calculation in the enveloping algebra shows, see Example 2.96 on page 113) these coincide in any Lie algebra, but the problem of a "minimal presentation" becomes paramount when we consider e.g., the absolute convergence problem.

For example, in the general context of formal power series over $\mathbb{Q}$ in two free indeterminates x, y, we know that $\mathrm{Log}(\mathrm{Exp}(x) \cdot \mathrm{Exp}(y))$ equals

$$x \bullet y := \sum_{n=1}^{\infty} \frac{(-1)^{n+1}}{n} \sum_{(h_1,k_1),\ldots,(h_n,k_n) \neq (0,0)} \frac{x^{h_1} y^{k_1} \cdots x^{h_n} y^{k_n}}{h_1! \cdots h_n! k_1! \cdots k_n!},$$

but the formal power series in the above right-hand side, as it is presented here, is far from being written in terms of some basis for $\mathscr{T}(\mathbb{Q}\langle x, y\rangle)$. That is, the same monomial may stem from different choices of (h, k) in the inner summation. For example, $x^2 y$ may come from

$$n = 1 : (h_1, k_1) = (2, 1) \quad \text{or} \quad n = 2 : (h_1, k_1) = (1, 0), \ (h_2, k_2) = (1, 1).$$

The so-called *Goldberg presentation* of $\mathrm{Log}(\mathrm{Exp}(x) \cdot \mathrm{Exp}(y))$ is actually a rewriting of the above series with respect to the basis for the associative algebra $\mathscr{T}(\mathbb{Q}\langle x, y\rangle)$ given by the elementary words

$$1, \quad x, \ y, \quad x^2, \ xy, \ yx, \ y^2,$$
$$x^3, \ x^2y, \ xyx, \ xy^2, \ yx^2, \ yxy, \ y^2x, \ y^3,$$
$$x^4, \ x^3y, \ x^2yx, \ x^2y^2, \ xyx^2, \ xyxy, \ xy^2x, \ xy^3,$$
$$yx^3, \ yx^2y, \ yxyx, \ yxy^2, \ y^2x^2, \ y^2xy, \ y^3x, \ y^4, \ \ldots \tag{5.167}$$

and so on. But now another interesting problem arises: we know very well from the CBHD Theorem that the above $x \blacklozenge y$ is equal to the series

$$x \diamond y := \sum_{j=1}^{\infty} \sum_{n=1}^{j} \frac{(-1)^{n+1}}{n} \sum_{(h,k) \in \mathcal{N}_n \, : \, |h|+|k|=j} \frac{1}{h! \, k! \, (|h| + |k|)}$$
$$\times \, (\mathrm{ad}\, x)^{h_1} (\mathrm{ad}\, y)^{k_1} \cdots (\mathrm{ad}\, x)^{h_n} (\mathrm{ad}\, y)^{k_n - 1}(y),$$

which is a formal Lie-series in the closure of $\mathcal{L}(\mathbb{Q}\langle x, y\rangle)$, called the *Dynkin presentation* of $\mathrm{Log}(\mathrm{Exp}(x) \cdot \mathrm{Exp}(y))$. The same problem of non-minimality of the presentation is now even more evident, in that infinite summands of this presentation are vanishing (take for instance $k_n \geq 2$) and since skew-symmetry and the Jacobi identity make possible cancellations even more thoroughly concealed. But Dynkin's is not the unique Lie presentation: for example, by using Goldberg presentation and the map P in the Dynkin, Specht, Wever Lemma, we can obtain another Lie presentation by dividing each word times its length and by substituting for each word its related right-nested Lie monomial. For instance, the independent words in (5.167) become (omitting all vanishing ones):

$$1, \quad x, \quad y, \quad \tfrac{1}{2}[x, y], \quad \tfrac{1}{2}[y, x],$$
$$\tfrac{1}{3}[x, [x, y]], \quad \tfrac{1}{3}[x, [y, x]], \quad \tfrac{1}{3}[y, [x, y]], \quad \tfrac{1}{3}[y, [y, x]],$$
$$\tfrac{1}{4}[x, [x, [x, y]]], \quad \tfrac{1}{4}[x, [x, [y, x]]], \quad \tfrac{1}{4}[x, [y, [x, y]]], \quad \tfrac{1}{4}[x, [y, [y, x]]],$$
$$\tfrac{1}{4}[y, [x, [x, y]]], \quad \tfrac{1}{4}[y, [x, [y, x]]], \quad \tfrac{1}{4}[y, [y, [x, y]]], \quad \tfrac{1}{4}[y, [y, [y, x]]], \quad \ldots$$

Unfortunately, *this is not a basis* for $\mathcal{L}(\mathbb{Q}\langle x, y\rangle)$, so the obtained Lie presentation is subject to further simplifications.

For each of the above different presentations, we may study the related problem of the convergence (when we are dealing with a normed Banach-Lie or Banach algebra) and further problems for the absolute convergence. For example, let us denote by

$$\sum_{j=1}^{\infty} W_j(x, y), \quad \text{with} \quad W_j(x, y) = \sum_{w \in \mathcal{W}_j} g(w) \, w, \tag{5.168}$$

the cited Goldberg's presentation, where

$$W_j := \left\{ x^{h_1} y^{k_1} \cdots x^{h_j} y^{k_j} \,\middle|\, (h_1, k_1), \ldots, (h_j, k_j) \in \{(1,0), (0,1)\} \right\}.$$

Then, as far as absolute convergence of Goldberg's presentation is concerned, i.e., the convergence of

$$\sum_{j=1}^{\infty} \|W_j(x, y)\|, \tag{5.169}$$

there are at least two interesting nonnegative series providing upper bounds for the series in (5.169).[13] On the one hand we have the majorizing series

$$\sum_{j=1}^{\infty} \sum_{w \in W_j} |g(w)| \, \|w\|, \tag{5.170}$$

and, on the other hand, we may consider the majorizing series

$$\sum_{j=1}^{\infty} \sum_{n=1}^{j} \left| \frac{(-1)^{n+1}}{n} \right| \sum_{\substack{(h,k) \in \mathbb{N}_n \\ |h|+|k|=j}} \frac{\|x^{h_1} y^{k_1} \cdots x^{h_n} y^{k_n}\|}{h_1! \cdots h_n! \, k_1! \cdots k_n!}. \tag{5.171}$$

Obviously, the convergence of (5.170) ensures the convergence of (5.169); in turn, it is easily seen that the convergence of (5.171) ensures that of (5.170). To visualize these facts more closely, let us notice that the terms resulting from $j = 2$ in the series in (5.169), (5.170) and (5.171) are respectively

(5.169) : $\left\| \frac{1}{2} xy - \frac{1}{2} yx \right\|,$

(5.170) : $\frac{1}{2} \|xy\| + \frac{1}{2} \|yx\|,$

(5.171) : $\frac{1}{2}\|x^2\| + \|xy\| + \frac{1}{2}\|y^2\| + \frac{1}{2} \left(\|x^2\| + \|xy\| + \|yx\| + \|y^2\| \right).$

Indeed, in (5.170) we have the further cancellations (if compared to (5.171)):

$$\|x^2\| \left| \frac{1}{2} - \frac{1}{2} \right| + \|xy\| \left| 1 - \frac{1}{2} \right| + \|yx\| \left| -\frac{1}{2} \right| + \|y^2\| \left| \frac{1}{2} - \frac{1}{2} \right|.$$

---

[13]Note that, by uniqueness of the expansion of $\log(e^x e^y)$ in the free associative algebra over x and y, $W_j(x, y)$ coincides with the usual Lie polynomial $Z_j(x, y)$ from the CBHD series (see e.g., (5.28)). We temporarily used different notations, because $W_j(x, y)$ does not make sense in an arbitrary Lie algebra, whereas $Z_j(x, y)$ does.

By these remarks, it is then not unexpected that the domain of convergence of (5.171) may be smaller than that of (5.170), which – in its turn – may be smaller than that of (5.169).

Furthermore, knowing that $W_j(x, y)$ is a Lie polynomial, if we apply the Dynkin-Specht-Wever Lemma, we get from (5.168) yet another presentation:

$$\log(e^x e^y) = \sum_{j=1}^{\infty} \left(\sum_{w \in \mathcal{W}_j} \frac{g(w)}{j} [w] \right), \tag{5.172}$$

called the *Goldberg commutator presentation*. Here, as usual, $[w]$ denotes the right-nested iterated commutator based on the word w. To take us even further, we mention that each Lie presentation has related convergence/absolute-convergence problems and that the use of norms compatible with the Lie-bracket is more suitable in this context; but this brings us to yet further problems, since a norm compatible with the Lie bracket does not come necessarily from a norm (or twice a norm) compatible with some underlying associative structure (even if the Lie algebra is obtained as the commutator-algebra of an associative algebra).

At this point, we believe we have convinced the Reader that the problem of convergence for the *many forms* of the BCHD series is quite rich and complicated. It is beyond the scope of this Book to collect all the known results so far about these problems (some being very recent). We confine ourselves to providing some references on related problems. (The references are listed in chronological order and we limit ourselves to mentioning some results – from each paper – *only involving the CBHD Theorem*.)

Magnus, 1950 [111]: Magnus solves a problem of group theory (a restricted version of *Burnside's problem*, see also Magnus, Karrass, Solitar [113] and Michel [117]) with the aid of the results of Baker [8] and Hausdorff [78], in particular by what is referred to (in this article) as *Baker-Hausdorff differentiation*.

Magnus, 1954 [112]: Magnus introduces his pioneering formula for $\Omega(t)$ (for the notation, see Sect. 1.2 of our Chap. 1), which he calls *the continuous analogue of the Baker-Hausdorff formula*. He also provides a proof of the Campbell-Baker-Hausdorff Theorem, making use of an equivalent "additive" formulation of Friedrichs's criterion for Lie polynomials: by means of [112, Theorem I, page 653], the proof of the CBH Theorem boils down to

$$\log(e^{x+x'} e^{y+y'}) = \log(e^x e^y) + \log(e^{x'} e^{y'}),$$

where x, y commute with x', y'. Still, the CBH formula is considered as a tool in the investigation of $\Omega(t)$, and – possibly – for a proof of an explicit formula for the expansion of $\Omega(t)$. The proof of the existence and Lie-algebraic nature of $\Omega(t)$ is very elegant and it relies only on the computations by Baker and Hausdorff and their *polarization operators* (the above mentioned Baker-Hausdorff differentiation).

Chen, 1957 [37]: The Campbell-Baker-Hausdorff Theorem is derived as a corollary of what is called by Chen the *generalized Baker-Hausdorff formula*, which is in close connection with Magnus's continuous analogue (roughly speaking, Chen's formula proves the existence of a Magnus-type expansion, without producing it in explicit form). The main tool is a suitable *iterated integration of paths* (and the early computations by Baker and Hausdorff are also used).

Goldberg, 1956 [71]: Goldberg discovers his (soon famous) recursive formula for the coefficients of the presentation of $\log(e^x e^y)$ in the basis consisting of monomials of the associative algebra of words in x, y: the so-called *Goldberg presentation*. Goldberg's procedure is effective for computer implementation, as suggested early on by the author himself.

Wei, 1963 [177]: Global validity of "the theorems of Baker-Hausdorff and of Magnus" is discussed and a list of several examples of *convergence failure* is also provided (see Sect. 5.6): the number π appears for the first time in relation to the seize of the domain of convergence for the CBHD series. The two cited theorems (say, the discrete and the continuous versions) are treated in a unified way and their joint relevance in mathematical physics is first neatly pointed out (quantum mechanical systems with a time-dependent Hamiltonian; linear stochastic motions). Together with the coeval [128, 178, 180], this is the first of a long series of papers concerning the CBHD (and the Magnus) Theorem in the *Journal of Mathematical Physics*.

Richtmyer, Greenspan, 1965 [146]: The very first computer implementation of the CBHD formula in its integral form

$$\log(e^x e^y) = x + \int_0^1 \Psi\Big(\exp(\mathrm{ad}\, x) \circ \exp(t \mathrm{ad}\, y) \Big)(y)\, \mathrm{d}t, \qquad (5.173)$$

where $\Psi(z) = (z \log z)/(z-1)$.

Eriksen, 1968 [61]: Various forms of the (therein called) "Baker-Hausdorff expansion" for $\log(e^x e^y)$ are obtained. Recurrence relations for the coefficients are also provided. In particular, the results of Goldberg [71] are exploited in order to get new *Lie polynomial* representations, anticipating an analogue of the Goldberg commutator representation, and giving new ones.

Mielnik, Plebański, 1970 [119]: With the advent of quantum theories, the CBHD formula steps down in favor of Magnus's continuous analogue. The passage "from the discrete to the continuous" is clear in this paper. The first part (pages 218–230) presents a comprehensive state-of-the-art (up to 1970) of the CBHD Theorem, whilst the rest of the paper is devoted to the Magnus expansion, including some new number-theoretical and combinatorial aspects (see also Bialynicki-Birula, Mielnik, Plebański [10]). There is furnished in [119, eq. (7.18), page 240] a new compact integral representation for $\Omega(t)$, in close connection with Dynkin's combinatorial proof of the CBH Theorem in [55].

Michel, 1974 [118]: For real Banach-Lie algebras (with a norm satisfying the condition $\|[a, b]\| \leq \|a\| \|b\|$), it is announced that the series $z(x, y) = \log(e^x e^y) = \sum_n z_n(x, y)$ (where $z_n(x, y)$ is homogeneous of degree n in x, y jointly) converges for

$$\|x\| < \int_{\|y\|}^{2\pi} \frac{1}{2 + \frac{t}{2}(1 - \cot \frac{t}{2})} \, dt,$$

and in the symmetric domain obtained by interchanging x and y. Michel credits M. Mérigot [115] for this result, as a consequence of the study of a suitable ODE satisfied by $z(x, y)$. Furthermore, it is also announced that, by considering the generating function given by Goldberg in [71] for the coefficient of $x^n y^m$, absolute non-convergence results hold in the domain $\|x\| + \|y\| \geq 2\pi$. Finally, by considering the special example of $\mathfrak{sl}_2(\mathbb{R})$, and the norm $\|X\| = \sqrt{2 \operatorname{trace}(X^T X)}$, the boundary of the convergence domain is explicitly parametrized and it is shown that the divergence domain can be larger than $\|x\| + \|y\| \geq 2\pi$. Expressibility of the CBHD series in terms of the Širšov basis is studied and the explicit expansion up to joint degree 10 is provided (see also [116] and [117]).

Suzuki, 1977 [162]: the domain of convergence $\|x\| + \|y\| < \log 2$ of the CBHD series is established in the general setting of Banach algebras, as a consequence of convergence estimates for the *Zassenhaus formula*.[14] See also Steinberg [164].

Macdonald, 1981 [109]: the coefficients of the expansion of $\log(e^x e^y)$ with respect to a Hall basis for the free Lie algebra $\mathcal{L}(\mathbb{Q}\langle x, y\rangle)$ are furnished.

Thompson, 1982 [166]: Fully rediscovering Goldberg's presentation after about 25 years of "silence" (apart from [61, 119]), Thompson remarks

---

[14]The Zassenhaus formula reads as follows: given non-commuting indeterminates A, B one has

$$e^{\lambda(A+B)} = e^{\lambda A} e^{\lambda B} e^{\lambda^2 C_2} e^{\lambda^3 C_3} \cdots,$$

where

$$C_2 = \frac{1}{2!} \frac{\partial^2}{\partial \lambda^2}\Big|_{\lambda=0} \left(e^{-\lambda B} e^{-\lambda A} e^{\lambda(A+B)}\right) = \frac{1}{2} [B, A],$$

$$C_3 = \frac{1}{3!} \frac{\partial^3}{\partial \lambda^3}\Big|_{\lambda=0} \left(e^{-\lambda^2 C_2} e^{-\lambda B} e^{-\lambda A} e^{\lambda(A+B)}\right) = \frac{1}{3} [C_2, A + 2B],$$

$$C_n = \frac{1}{n!} \frac{\partial^n}{\partial \lambda^n}\Big|_{\lambda=0} \left(e^{-\lambda^{n-1} C_{n-1}} \cdots e^{-\lambda^2 C_2} e^{-\lambda B} e^{-\lambda A} e^{\lambda(A+B)}\right).$$

It can be proved that C_n is a Lie polynomial in A, B homogeneous of degree n.

that a Lie presentation of $\exp(e^x e^y)$, based on Goldberg's, is immediately derived, say

$$\sum_{n=1}^{\infty} \sum_{|w|=n} g_w \, w \quad \xrightarrow{\text{Dynkin, Specht, Wever}} \quad \sum_{n=1}^{\infty} \sum_{|w|=n} \frac{g_w}{n} \, [w].$$

This is indeed what we formerly referred to as the Goldberg commutator presentation.

Newman, Thompson, 1987 [132]:   explicit computations of the Goldberg coefficients are given (up to length 20), by implementing Goldberg's algorithm.

Strichartz, 1987 [160]:   Strichartz reobtains Magnus's expansion (but [112] is not referenced), calling it *the generalized CBHD Formula*, as an explicit formulation of the results by Chen [37]. Applications to problems of sub-Riemannian geometry are given (Carnot-Carathéodory metrics), as well as improvements for the convergence of the CBHD series in the context of Banach algebras (say, $\|x\| + \|y\| < 1/2$). A simple yet evocative parallel is suggested:

> "The Magnus expansion is to the classical $u(t) = \exp\left(\int_0^t A(s)\,ds\right)$,
> as the CBHD Formula is to the classical $e^{x+y} = e^x e^y$".

It is interesting to notice that Strichartz's derivation of his explicit formula for $\Omega(t)$ makes use of Friedrichs's criterion $\delta(x) = x \otimes 1 + 1 \otimes x$ for Lie elements, plus ODE techniques.

Bose, 1989 [23]:   An algorithm providing the coefficients in Dynkin's series is furnished: Bose's procedure is not recursive, but it computes directly the coefficient of any bracket in Dynkin's presentation.

Thompson, 1989 [167]:   In a Banach (resp., a Banach-Lie) algebra, and with respect to a norm compatible with the associative (resp., Lie) multiplication, it is proved that the Goldberg presentations (the associative and the commutator ones, respectively) converge for $\max\{\|x\|, \|y\|\} < 1$.

Newman, So, Thompson, 1989 [131]:   The absolute convergence for different presentations of $\log(e^x e^y)$ is studied. Earlier results by Strichartz [160] and coeval ones by Thompson [167] are improved. For example, the actual value of $\delta/2 \approx 1.08686$ in Varadarajan's convergence estimate $\max\{\|x\|, \|y\|\} < \delta/2$ is provided; also, it is proved that the majorizing series in (5.170) diverges on the boundary of $\{\|x\| = 1\}$.

Day, So, Thompson, 1991 [46]:   Instead of the usual decomposition in joint homogeneous components in x, y, the decomposition

$$\log(e^x e^y) = \sum_{n \geq 0} z_n^y(x, y)$$

is used, ordered with respect to increasing powers of y. Accordingly, a recursion formula for z_n^y is used; one which had appeared in the early paper by Hausdorff [78].[15] The same ODE technique as in Varadarajan [171] and Newman, So, Thompson [131] is followed: it turns out that the y-expansion yields has a slightly improved domain of convergence $\max\{\|x\|, \|y\|\} < 1.23575$. Also, Sect. 2 of [46] makes use of some crucial computations on the differentiation of exponentials, which we met in the proofs by Djoković and by Varadarajan.

Oteo, 1991 [134]: An expansion of $\log(e^x e^y)$ up to joint degree 8 in x and y with a *minimal* number of brackets is given (whereas in some common bases – such as Lyndon's – the brackets required are more numerous). A comparison with the method of polar derivatives is also discussed. Richtmyer and Greenspan's formula (5.173) is proved by the well-established ODE technique, which can be traced back to Hausdorff: new Lie series representations for $z = \log(e^x e^y)$ are derived.

Vinokurov, 1991 [175]: In the context of Banach algebras, an an equivalent version of Magnus's formula is reobtained (but [160] is quoted, instead of [112]), in analogy with the results in [10, 119]). As a consequence, the "Hausdorff formula" for any Banach algebra and for any compatible norm is derived, with convergence $\|x\| + \|y\| < 1$. Vinokurov gives an example of non-convergence

$$x = \begin{pmatrix} 0 & \varepsilon \\ 0 & 0 \end{pmatrix}, \quad y = (1 + \varepsilon) \begin{pmatrix} 0 & -\pi \\ \pi & 0 \end{pmatrix}, \quad \varepsilon > 0,$$

so that the upper bound π reappears.

Kolsrud, 1993 [103]: An expansion is given up to (joint) degree 9 in x and y of the series $\log(e^x e^y)$ with a maximal reduction of the number of different multiple brackets (see also [134]) .

Blanes, Casas, Oteo, Ros, 1998 [15]: The attention is by now drowned by the continuous CBHD theorem. Starting from recursion formulas for the Magnus expansion (see Klarsfeld, Oteo [102]), a larger domain of convergence for the series of $\Omega(t)$ is provided (in the matrix context). As a byproduct, this gives the absolute convergence of $\sum_j Z_j(x, y)$, for

$$\|x\| + \|y\| < \delta/2 \approx 1.08686.$$

(Independently, this result was also obtained by Moan, 1998 [121].)

Reinsch, 2000 [145]: An extraordinarily simple formula for z_n in the expansion $\log(e^x e^y) = \sum_{n=1}^{\infty} z_n(x, y)$ in joint homogeneous components is given: it involves only a finite number of matrix multiplications and is easily implemented. No sums over multi-indices or partitions appear, and no noncommutative computations are required.

---

[15]Indeed, [46, Theorem I, pag 209] is Hausdorff's formula $\left(y \frac{\partial}{\partial y}\right) z = \chi(y, z)$, see equation (28) on page 31 of [78].

Moan, Oteo, 2001 [123]: Renouncing the use of the Lie representation of the Magnus expansion ("mainly due to the rather intricate nature of the Lie algebra bases", see [123, page 503]), another domain of convergence is obtained in the context of Banach algebras. Former results by Bialynicki-Birula, Mielnik, Plebański [10] and by Mielnik, Plebański [119] are used. When applied to obtain convergence of the $\log(e^x e^y)$ series, the domain of convergence gives back $\max\{\|x\|, \|y\|\} < 1$, see [131, 167, 175].

Blanes, Casas, 2004 [14]: A significant improvement of the convergence of the series expansion for $\log(e^x e^y)$ is given. For example, in a *Banach-Lie algebra* with a norm satisfying $\|[a, b]\| \leq \mu \|a\| \|b\|$, the new domain of convergence is expressed as

$$\left\{(x, y) : \ \|x\| < \frac{1}{\mu} \int_{\mu \|y\|}^{2\pi} h(t)\, dt\right\} \cup \left\{(x, y) : \ \|y\| < \frac{1}{\mu} \int_{\mu \|x\|}^{2\pi} h(t)\, dt\right\},$$

where $h(t) = 1/\left(2 + \frac{t}{2} - \frac{t}{2} \cot \frac{t}{2}\right)$. Note that this is the same domain announced by Michel [118] when $\mu = 1$ (the case of a general $\mu > 0$ followed by magnification). The above result is obtained by deriving an ODE for the function $t \mapsto \log(\exp(\varepsilon t\, x) \exp(\varepsilon\, y))$ and by expanding the solution as a power series in ε. Numerous examples of non-convergence are also given. For instance, examples are shown of matrices x, y *outside* the set defined by $\{\|x\| + \|y\| \leq \pi\}$, and arbitrarily approaching the boundary $\|x\| + \|y\| = \pi$, for which the usual CBHD series diverges (here $\|[a, b]\| \leq 2 \|a\| \|b\|$). By replacing $\|\cdot\|$ with $2\|\cdot\|$, this also furnishes an example of a norm satisfying $\|[a, b]\| \leq \|a\| \|b\|$ and elements x, y approaching from outside the boundary of $\{\|x\| + \|y\| < 2\pi\}$, for which the CBHD series diverges (in agreement with the results announced in [118]).

Theorem 4.1 of [14] recalls an (unpublished) article by B.S. Mityagin (announced in 1990), which contains the following result: In a Hilbert space of finite dimension > 2, the operator function $(x, y) \mapsto \log(e^x e^y)$ is analytic (hence well-posed) in $\{\|x\| + \|y\| < \pi\}$. This must hold in particular for the space of square matrices of order ≥ 2 with the Euclidean norm $\|(a_{ij})\|_E = (\sum_{i,j} |a_{ij}|^2)^{1/2}$. Note that this norm satisfies $\|[A, B]\|_E \leq 2 \|A\|_E \|B\|_E$, so that Blanes and Casas's examples of non-convergence show the "optimality" of the result announced by Mityagin.

Casas, 2007 [35]: The convergence of the Magnus expansion is considered. It is shown that improved results can be given in the setting of *Hilbert spaces* (see [35, Theorem 3.1, page 15006] for the precise statement); this is done by the use of some spectral estimates and properties of the unit sphere in a Hilbert space. In the finite dimensional case, this gives a proof of Mityagin's result; in the special case of square matrices, convergence is ensured in $\{\|X\| + \|Y\| < \pi\}$ (where $\|\cdot\|$ is the norm induced by the inner product: for example the above $\|\cdot\|_E$ is allowed). For complex square matrices, a suitable delicate spectral analysis gives further improvements, see [35, §4]. (For example, the matrix norm $\| \cdot \| := 2 \| \cdot \|_E$ satisfies $\|[A, B]\| \leq \|A\| \|B\|$, so that we have convergence in the set $\{\|X\| + \|Y\| < 2\pi\}$.)

Kurlin, 2007 [105]: If L is the free Lie algebra generated by two indeterminates x, y, a closed formula for $\log(e^x e^y)$ is given with respect to a basis for (the completion of) the free metabelian Lie algebra $L/[[L, L], [L, L]]$. Applications to exponential equations in Lie algebras are also given. See the rôle of the CBH Theorem in Kashiwara, Vergne [100] and in Alekseev, Meinrenken [4].

Moan, Niesen, 2008 [122]: For real square matrices, the convergence of the Magnus expansion is considered and the following domain of convergence is given (by using spectral properties):

$$\int_0^t \|A(s)\|_2 \, \mathrm{d}s < \pi.$$

Here $\|A\|_2 = \max_{|x| \leq 1} |Ax|$ and $|\cdot|$ is the usual Euclidean norm. This result was already obtained by Casas [35] in the general case of Hilbert spaces. As a byproduct, if $t = 2$ and $A(s) = \chi_{[0,1)}(s) X + \chi_{[1,2]}(s) Y$, we get a convergence result for the CBHD series, provided that $\|X\|_2 + \|Y\|_2 < \pi$. [Note that $\|\cdot\|_2$ satisfies $\|[A, B]\|_2 \leq 2 \|A\|_2 \|B\|_2$, so that by magnification times 2, we obtain a particular case of a Lie-submultiplicative norm $\| \cdot \| := 2 \| \cdot \|_2$ with convergence in the recurrent set $\{\|X\| + \|Y\| < 2\pi\}$.] Optimality of the bound π is ensured by providing explicit examples.

Blanes, Casas, Oteo, Ros, 2009 [16]: This is a complete and exhaustive report on the Magnus expansion and on the great variety of its applications (including the CBHD formula), with a comprehensive list of references on the subject, which we definitely recommend to the interested Reader. Historical and mathematical up-to-date results are exposed and a detailed comparison with earlier literature is given.

Casas, Murua, 2009 [36]: A new and efficient algorithm is given for generating the CBHD series in a *Hall-Viennot basis* for the free Lie algebra on two generators (see [144] for the relevant definitions). Improved results on convergence, in the matrix case, are also given (following the spectral techniques in [35]).

Chapter 6
Relationship Between the CBHD Theorem, the PBW Theorem and the Free Lie Algebras

$\mathbf{T}$HE aim of this chapter is to unravel the close relationship existing between the Theorems of CBHD and of Poincaré-Birkhoff-Witt ("PBW" for short) and to show how the existence of free Lie algebras intervenes. We have analyzed, in Chap. 3, how the PBW Theorem intervenes in the classical approach to the proof of CBHD, in that PBW can be used to prove in a simple way Friedrichs's characterization of $\mathcal{L}(V)$ (see Theorem 3.13 on page 133). Also, as for the proofs of CBHD in Chap. 4, the rôle of the free Lie algebras was broadly manifest.

Yet, we have not mentioned so far a quite surprising fact: the opposite path can be followed too, i.e. *the PBW Theorem can be proved by means of CBHD*.

To this end, in this chapter we shall investigate a result by Cartier [33]: Indeed in the cited paper, *as an application of the* CBHD *Theorem*, Cartier gives a proof of a result, referred to as "Théorème de Birkhoff-Witt" (see [33, Paragraph 6, page 247]), which in fact implies the classical PBW. We will here illustrate Cartier's argument, thus providing a self-contained proof of PBW involving CBHD.

Furthermore, the paramount rôle of free Lie algebras will be clarified and their intertwining with PBW and CBHD will be shown in details. Indeed, the proof of PBW involving CBHD also requires (see Theorem 6.5 below) two important results:

- The existence of the free Lie algebra $\mathrm{Lie}(X)$ related to a set X.
- The isomorphism $\mathrm{Lie}(X) \simeq \mathcal{L}(\mathbb{K}\langle X \rangle)$, the latter being the smallest Lie subalgebra – containing X – of the tensor algebra of the free vector space $\mathbb{K}\langle X \rangle$.

The fact that the proof of PBW – via CBHD – requires these two results turns out to be a rather delicate circumstance, because both results are classically

A. Bonfiglioli and R. Fulci, *Topics in Noncommutative Algebra*, Lecture Notes in Mathematics 2034, DOI 10.1007/978-3-642-22597-0_6,
© Springer-Verlag Berlin Heidelberg 2012

derived *from* PBW itself.[1] As our goal is to give an alternative proof of PBW, it is evident that we cannot make use of any of its corollaries. Consequently, the necessity of a proof of the existence of free Lie algebras independent of PBW becomes clear.

By sheer chance, an exposition which is *independent* of PBW of the existence of the free Lie algebra related to a set can be found in Chap. 2.[2]

As for the proof of the isomorphism $\mathrm{Lie}(X) \simeq \mathcal{L}(\mathbb{K}\langle X\rangle)$, it is provided in Sect. 8.1, from page 463. This isomorphism leads to the existence of a free Lie algebra related to a given set X *containing* (set-theoretically) X itself. This is furnished in Corollary 8.6.

The relationship of the present chapter with the rest of the Book can be visualized in the following diagram. The arrows suggest the path to be followed, but should *not* be considered as actual implications. Thus, the diagram must be read in the sequence I–II–III.

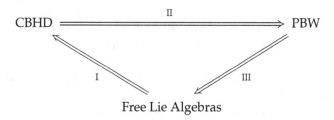

Free Lie Algebras

Path I corresponds to Eichler's argument (see Chap. 4, Sect. 4.2). Path II is the one this chapter is entirely devoted to. Path III, finally, has been given in that proof of Theorem 2.49 furnished on page 112.

At this point, it is clear that a profound intertwining between all three of PBW, CBHD and FLA (short for "free Lie algebras") occurs. We summarize it with the following result, which is a byproduct of the results in this chapter, plus some other results in this Book. The Reader is referred to Definition 2.50 on page 88 to recall the distinction between a free Lie algebra *related to* a set and *over* a set.

Theorem 6.1. *Let us consider the following statements (see also the notation in the PBW Theorem 2.94 on page 111 for statement* (a)*; here* $\{a, b, c\}$ *is a set of cardinality three and all linear structures are understood to be defined over a field of characteristic zero):*

(a) *The set* $\{X_i\}_{i\in I}$ *is independent in* $\mathcal{U}(\mathfrak{g})$.
(b) *Any Lie algebra* $\mathfrak{g}$ *can be embedded in its enveloping algebra* $\mathcal{U}(\mathfrak{g})$.

---

[1]See e.g. our proof at page 112. For a proof of the isomorphism $\mathrm{Lie}(X) \simeq \mathcal{L}(\mathbb{K}\langle X\rangle)$ which uses PBW, see also Reutenauer [144, Theorem 0.5] or Bourbaki [27, Chapitre II, §3, n.1].

[2]See Theorem 2.54, page 91, whose proof is at page 459.

(c) *For every set $X \neq \emptyset$, there exists a free Lie algebra over X.*
(d) *FLA: For every set $X \neq \emptyset$, the free Lie algebra $\mathrm{Lie}(X)$ related to X exists.*
(e) *The free Lie algebra over $\{a, b, c\}$ exists, and Theorem CBHD holds.*
(f) *Theorem PBW holds.*

Then these results can be proved each by another in the following circular sequence:

$$(a) \Rightarrow (b) \Rightarrow (c) \Rightarrow (d) \Rightarrow (e) \Rightarrow (f) \Rightarrow (a).$$

We observe that statement (c) *is proved in Chap. 8, without any prerequisite and without the aid of any of the other statements. Also, the isomorphism of the free Lie algebra related to X with the Lie algebra of the Lie polynomials in the letters of X can be proved independently of* (b) *and of* (f).

Proof. The following arguments apply.

(a) $\Rightarrow$ (b): this is obvious by the definition $X_i = \pi(x_i)$;
(b) $\Rightarrow$ (c): this is the modern approach to the proof of the existence of free Lie algebras, the one we followed in the proof of Theorem 2.49 given on page 112[3];
(c) $\Rightarrow$ (d): this is obvious from the relevant definitions;
(d) $\Rightarrow$ (e): FLA trivially implies the existence of $\mathrm{Lie}\{a, b, c\}$ which – in its turn – implies the existence of a free Lie algebra *over* $\{a, b, c\}$ as in Sect. 8.1.1 of this Book; also, the fact that FLA implies CBHD is contained in Eichler's proof, investigated in Sect. 4.2;
(e) $\Rightarrow$ (f): following Cartier's paper [33], this is accomplished in the present chapter;
(f) $\Rightarrow$ (a): this is obvious. □

Remark 6.2. Some remarks are in order:

– We note that the implications (d)$\Rightarrow$(e) and (e)$\Rightarrow$(f) actually require a lot of work. Though, the "circularity" of the above statements (see also Fig. 6.1) seems to have some theoretical interest in its own right, especially the "long" implication (a)$\Longrightarrow$(f), proving that PBW *can be derived by what is usually its very special corollary* (a).
– We remark that *the existence of free Lie algebras is not only a consequence of the* PBW *Theorem* (as it is commonly understood), *but the converse is also true*, this time by making use of the CBHD Theorem: this is the "long" implication (c)$\Longrightarrow$(f).
– Curiously, as highlighted by the "very long" implication (e)$\Longrightarrow$(d), we remark that just the *existence* of $\mathrm{Lie}\{a, b, c\}$ is in principle sufficient to

---

[3]Indeed, given a nonempty set X, we can construct the free vector space $V = \mathbb{K}\langle X \rangle$ and proceed as in the proof of Theorem 2.49 given at page 112. This gives a free Lie algebra *over* X since $\mathcal{L}(V)$ contains V which canonically contains X.

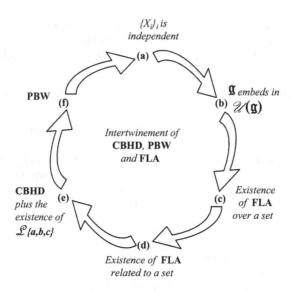

Fig. 6.1 Figure of Theorem 6.1

prove the existence of *all* free Lie algebras (but this time we require both CBHD and PBW to round off the implication.)

Bibliographical Note. The interdependence of the CBHD and PBW Theorems is neatly suggested by Cartier's investigations. More precisely:

(a) In [32, Exposé n. 1], it is announced that the Campbell-Hausdorff formula can be deduced by some special lemmas concerning the PBW Theorem (see part (2) of the Remarque on page 9 in [32]).
(b) In [32, Exposé n. 22], the Campbell-Hausdorff formula is proved by the aid of Friedrichs's characterization of Lie elements. The PBW Theorem is required here.
(c) In [33], as mentioned above, the implication CBHD $\Rightarrow$ PBW is studied; moreover, the CBHD formula is proved, by means of the existence of free Lie algebras, together with a new characterization of Lie polynomials (see Sect. 4.7 in Chap. 4).

As already remarked, the existence of free Lie algebras is usually proved as a consequence of the imbedding of $\mathfrak{g}$ in $\mathscr{U}(\mathfrak{g})$: See e.g. Hochschild [85, Chapter X, Section 2], Humphreys [95, Chapter V, Section 17.5], Jacobson [99, Chapter V, Section 4], Varadarajan [171, Section 3.2].

A proof of the existence of free Lie algebras *independent* of this cited fact seems to appear only in Reutenauer [144, Section 0.2] and in Bourbaki [27, Chapitre II, §2, n.2]. In these books Lie(X) is constructed as a quotient of the non-associative free magma generated by X. A proof of the isomorphism Lie(X) $\simeq$ $\mathcal{L}(\mathbb{K}\langle X \rangle)$ also appears in [27, Chapitre II, §3, n. 1] and in

[144, Theorem 0.5], but – respectively – the PBW Theorem[4] and the imbedding $\mathfrak{g} \hookrightarrow \mathscr{U}(\mathfrak{g})$ are used.

Some of the results contained in this chapter have been announced in [19].

6.1 Proving PBW by Means of CBHD

We briefly outline the track we will follow in this chapter, mainly based on Cartier's arguments in [33].

Let $\mathfrak{g}$ be a Lie algebra and consider the set $\mathscr{S}(\mathfrak{g}) \subset \mathscr{T}(\mathfrak{g})$ of its symmetric tensors (see Definition 10.18 on page 511).

- A bilinear map $F : \mathscr{S}(\mathfrak{g}) \times \mathscr{S}(\mathfrak{g}) \to \mathscr{T}(\mathfrak{g})$ is constructed, by making use of a sort of CBHD operation $(u,v) \mapsto Z^{\mathfrak{g}}(u,v)$, where (roughly) the $\otimes$-brackets $\left[u^{h_1} v^{k_1} \cdots u^{h_n} v^{k_n} \right]_\otimes$ (see (3.15), page 126) are replaced by the Lie brackets in $\mathfrak{g}$.
- It is shown that $u \star v := F(u,v)$ defines a *binary* and *associative* operation on $\mathscr{S}(\mathfrak{g})$.
- With the operation $\star$ at hand, a projection $\Pi : \mathscr{T}(\mathfrak{g}) \to \mathscr{S}(\mathfrak{g})$ is constructed, by setting $\Pi(a_1 \otimes \cdots \otimes a_n) = a_1 \star \cdots \star a_n$. It turns out that Π is the identity on $\mathscr{S}(\mathfrak{g})$ and that its kernel is $\mathscr{J}(\mathfrak{g})$ (the two-sided ideal introduced in (2.99), page 108).
- Finally, one infers $\mathscr{T}(\mathfrak{g}) = \mathscr{S}(\mathfrak{g}) \oplus \mathscr{J}(\mathfrak{g})$ ("Théorème de Birkhoff-Witt") which in fact implies PBW (see the proof on page 387).

To accomplish the proof, we first present some preliminary work (Sect. 6.1.1) and then rerun Cartier's original argument towards the derivation of PBW via CBHD (Sect. 6.1.2). In the former section, our approach is certainly much less elegant than Cartier's [33]; we hope it might be welcomed nonetheless for some explicit computations using Dynkin's series, in the spirit of the previous chapters.

6.1.1 Some Preliminaries

Let us now begin with the actual proof. The first goal is to introduce a suitable composition law, somehow inspired by the well-known identity

$$\mathrm{Exp}_\otimes(u) \cdot \mathrm{Exp}_\otimes(v) = \mathrm{Exp}_\otimes(Z(u,v))$$

---

[4]See [27, Chapitre II, §3, n.1, Théorème 1] where it is employed [25, Chapitre I, §2, n.7, Corollaire 3 du Théorème 1] which is the PBW Theorem.

from the CBHD Theorem 3.8. Let $\mathfrak{g}$ be a fixed Lie algebra. As usual, for every $j \in \mathbb{N}$, we define Lie polynomial functions on $\mathfrak{g}$ by setting[5]

$$Z_j^{\mathfrak{g}} : \mathfrak{g} \times \mathfrak{g} \to \mathfrak{g}, \qquad Z_j^{\mathfrak{g}}(a, b) := \sum_{n=1}^{j} c_n \sum_{\substack{(h,k) \in \mathcal{N}_n \\ |h|+|k|=j}} \mathbf{c}(h, k) \left[a^{h_1} b^{k_1} \cdots a^{h_n} b^{k_n} \right]_{\mathfrak{g}}.$$

(6.1)

For brevity, we also let – on occasion –

$$D_{(a,b)}^{\mathfrak{g}}(h, k) := \left[a^{h_1} b^{k_1} \cdots a^{h_n} b^{k_n} \right]_{\mathfrak{g}}, \qquad (6.2)$$

for every $(h, k) \in \mathcal{N}_n$ and every $a, b \in \mathfrak{g}$. Note that

$$Z_j^{\mathfrak{g}}(a, b), \ D_{(a,b)}^{\mathfrak{g}}(h, k) \in \mathfrak{g} \quad \text{for every } a, b \in \mathfrak{g} \text{ and every } (h, k) \in \mathcal{N}_n.$$

Then – roughly speaking – for $a, b \in \mathfrak{g}$, we consider the *formal* objects

$$Z^{\mathfrak{g}}(a, b) := \sum_{j \geq 1} Z_j^{\mathfrak{g}}(a, b), \qquad \exp(Z^{\mathfrak{g}}(a, b)) := \sum_{k \geq 0} \frac{1}{k!} \left(Z^{\mathfrak{g}}(a, b) \right)^{\cdot k},$$

where $\cdot$ denotes as usual the natural (Cauchy product) operation in the tensor algebra $\mathscr{T}(\mathfrak{g})$. Finally, we let $f_{i,j}(a, b)$ be the sum of terms in the expansion of $\exp(Z^{\mathfrak{g}}(a, b))$ where a and b appear, respectively, i and j times. Rigorously, as in [33], one can introduce the algebra of the formal power series in two commuting indeterminates S, T and coefficients in $\mathscr{T}(\mathfrak{g})$ and define the maps $f_{i,j}$ to be those functions resulting from the following identity

$$\sum_{i,j \geq 0} f_{i,j}(a, b)\, S^i T^j = \exp(Z(aS, bT)).$$

(This has also the advantage to make it unnecessary to use Dynkin's series.) Alternatively, one can introduce the following family of *explicit* functions: $f_{0,0}(a, b) := 1$ and, for $i, j \in \mathbb{N} \cup \{0\}$ with $(i, j) \neq (0, 0)$,

$$f_{i,j} : \mathfrak{g} \times \mathfrak{g} \to \mathscr{T}(\mathfrak{g}), \qquad\qquad\qquad\qquad (6.3)$$

$$f_{i,j}(a, b) := \sum_{\mathcal{A}_{i,j}} \frac{1}{s!}\, c_{n_1} \cdots c_{n_s}\, \mathbf{c}(h^{(1)}, k^{(1)}) \cdots \mathbf{c}(h^{(s)}, k^{(s)})$$

$$\times D_{(a,b)}^{\mathfrak{g}}(h^{(1)}, k^{(1)}) \otimes \cdots \otimes D_{(a,b)}^{\mathfrak{g}}(h^{(s)}, k^{(s)}),$$

---

[5]Following our customary notation, $[\cdot, \cdot]_{\mathfrak{g}}$ denotes the Lie bracket in $\mathfrak{g}$ and

$$\left[a^{h_1} b^{k_1} \cdots a^{h_n} b^{k_n} \right]_{\mathfrak{g}} := (\operatorname{ad} a)^{h_1} \circ (\operatorname{ad} b)^{k_1} \circ \cdots \circ (\operatorname{ad} a)^{h_n} \circ (\operatorname{ad} b)^{k_n - 1}(v).$$

Also, c_n and $\mathbf{c}(h, k)$ are as in (3.20) on page 127.

where the summation is over the set $\mathcal{A}_{i,j}$ defined by

$$s \geq 1, \ n_1, \ldots, n_s \geq 1, \ (h^{(1)}, k^{(1)}) \in N_{n_1}, \ldots, (h^{(s)}, k^{(s)}) \in N_{n_s}$$
$$|h^{(1)}| + \cdots + |h^{(s)}| = i, \quad |k^{(1)}| + \cdots + |k^{(s)}| = j. \tag{6.4}$$

We observe that $\mathcal{A}_{i,j}$ is finite, since $i + j \geq n^{(1)} + \cdots + n^{(s)} \geq s$. Thus,

$$f_{i,j}(a,b) \in \mathscr{T}_1(\mathfrak{g}) \oplus \cdots \oplus \mathscr{T}_{i+j}(\mathfrak{g}).$$

For example, an explicit calculation gives

$$f_{2,1}(a,b) = \tfrac{1}{12}[a,[a,b]] + \tfrac{1}{4}(a \otimes [a,b] + [a,b] \otimes a)$$
$$+ \tfrac{1}{6}(a^{\otimes 2} \otimes b + a \otimes b \otimes a + b \otimes a^{\otimes 2});$$

$$f_{3,1}(a,b) = \tfrac{1}{24} a \otimes [a,[a,b]] + \tfrac{1}{24}[a,[a,b]] \otimes a + \tfrac{1}{12} a^{\otimes 2} \otimes [a,b]$$
$$+ \tfrac{1}{12} a \otimes [a,b] \otimes a + \tfrac{1}{12}[a,b] \otimes a^{\otimes 2} + \tfrac{1}{24} a^{\otimes 3} \otimes b$$
$$+ \tfrac{1}{24} a^{\otimes 2} \otimes b \otimes a + \tfrac{1}{24} a \otimes b \otimes a^{\otimes 2} + \tfrac{1}{24} b \otimes a^{\otimes 3}.$$

This also shows that the functions $f_{i,j}$ are not homogeneous. We claim that, for every $i, j \in \mathbb{N} \cup \{0\}$ there exists a *bilinear map* $F_{i,j}$ such that

$$F_{i,j} : \mathscr{S}_i(\mathfrak{g}) \times \mathscr{S}_j(\mathfrak{g}) \to \mathscr{T}(\mathfrak{g}),$$
$$F_{i,j}\left(\tfrac{a^{\otimes i}}{i!}, \tfrac{b^{\otimes j}}{j!}\right) = f_{i,j}(a,b), \quad \forall \, a, b \in \mathfrak{g}. \tag{6.5}$$

First we set $F_{0,0} : \mathbb{K} \times \mathbb{K} \to \mathscr{T}(\mathfrak{g})$, $F_{0,0}(k_1, k_2) := k_1 k_2$ for every $k_1, k_2 \in \mathbb{K}$, which clearly satisfies (6.5), since $f_{0,0} \equiv 1$.

To prove (6.5) for $(i,j) \neq (0,0)$, we argue as follows. If $i \in \mathbb{N}$, let $\mathfrak{g}^i$ be the i-fold Cartesian product $\mathfrak{g} \times \cdots \times \mathfrak{g}$ and set $\mathfrak{g}^0 := \mathbb{K}$. Let us consider the function $\Xi_{i,j} : \mathfrak{g}^i \times \mathfrak{g}^j \to \mathscr{T}(\mathfrak{g})$ defined by

$$\Xi_{i,j}(z, w) := \sum_{\mathcal{A}_{i,j}} \frac{1}{s!} \, c_{n_1} \cdots c_{n_s} \, \mathbf{c}(h^{(1)}, k^{(1)}) \cdots \mathbf{c}(h^{(s)}, k^{(s)})$$

$$\times \Big[z_1 \ldots z_{h_1^{(1)}}, w_1 \ldots w_{k_1^{(1)}} \cdots$$

$$\cdots z_{h_1^{(1)} + \cdots + h_{n_1-1}^{(1)} + 1} \cdots z_{|h^{(1)}|}, w_{k_1^{(1)} + \cdots + k_{n_1-1}^{(1)} + 1} \cdots w_{|k^{(1)}|} \Big]_{\mathfrak{g}}$$

$$\otimes \Big[z_{|h^{(1)}| + 1} \cdots z_{|h^{(1)}| + h_1^{(2)}}, w_{|k^{(1)}| + 1} \cdots w_{|k^{(1)}| + k_1^{(2)}} \cdots$$

$$\cdots z_{|h^{(1)}| + h_1^{(2)} + \cdots + h_{n_2-1}^{(2)} + 1} \cdots z_{|h^{(1)}| + |h^{(2)}|},$$

$$w_{|k^{(1)}| + k_1^{(2)} + \cdots + k_{n_2-1}^{(2)} + 1} \cdots w_{|k^{(1)}| + |k^{(2)}|} \Big]_{\mathfrak{g}}$$

$$\otimes \cdots \otimes$$

$$\left[z_{|h^{(1)}|+\cdots+|h^{(s-1)}|+1} \cdots z_{|h^{(1)}|+\cdots+|h^{(s-1)}|+h_1^{(s)}}, \right.$$

$$w_{|k^{(1)}|+\cdots+|k^{(s-1)}|+1} \cdots w_{|k^{(1)}|+\cdots+|k^{(s-1)}|+k_1^{(s)}}, \quad \cdots$$

$$z_{|h^{(1)}|+\cdots+|h^{(s-1)}|+h_1^{(s)}+\cdots+h_{n_s-1}^{(s)}+1} \cdots z_{|h^{(1)}|+\cdots+|h^{(s-1)}|+|h^{(s)}|},$$

$$\left. w_{|k^{(1)}|+\cdots+|k^{(s-1)}|+k_1^{(s)}+\cdots+k_{n_s-1}^{(s)}+1} \cdots w_{|k^{(1)}|+\cdots+|k^{(s-1)}|+|k^{(s)}|} \right]_{\mathfrak{g}}.$$

A direct look at $\Xi_{i,j}$ shows that it is a multilinear map on $\mathfrak{g}^i \times \mathfrak{g}^j$. Hence, it defines a bilinear map from $\mathscr{T}_i(\mathfrak{g}) \times \mathscr{T}_j(\mathfrak{g})$ to $\mathscr{T}(\mathfrak{g})$, whose restriction to $\mathscr{S}_i(g) \times \mathscr{S}_j(g)$ we denote by $\widetilde{F}_{i,j}$. Finally, we set

$$F_{i,j}(a,b) := \widetilde{F}_{i,j}(i!\,a, j!\,b).$$

Obviously, $F_{i,j} : \mathscr{S}_i(\mathfrak{g}) \times \mathscr{S}_j(\mathfrak{g}) \to \mathscr{T}(\mathfrak{g})$ is bilinear. Also, the identity in the far right-hand side of (6.5) is a direct consequence of the definitions of $\Xi_{i,j}$ and $F_{i,j}$, as it results from the following computation:

$$F_{i,j}\Big(\frac{a^{\otimes i}}{i!}, \frac{b^{\otimes j}}{j!}\Big) = \widetilde{F}_{i,j}(a^{\otimes i}, b^{\otimes j}) = \Xi_{i,j}(\underbrace{a,\dots,a}_{i \text{ times}}, \underbrace{b,\dots,b}_{j \text{ times}}) = f_{i,j}(a,b).$$

We are now ready for the following definition.

Definition 6.3. With the above notation, we set

$$\star : \mathscr{S}(\mathfrak{g}) \times \mathscr{S}(\mathfrak{g}) \to \mathscr{T}(g), \qquad u \star v := \sum_{i,j \geq 0} F_{i,j}(u_i, v_j), \quad \forall\, u, v \in \mathscr{S}(\mathfrak{g}),$$

where $u = \sum_i u_i$, $v = \sum_i v_i$ (the sums being finite), with $u_i, v_i \in \mathscr{S}_i(\mathfrak{g})$ for every $i \geq 0$ (see e.g. the notation in (10.27), page 509).

For example, if $a, b \in \mathfrak{g} \hookrightarrow \mathscr{S}(\mathfrak{g})$, one has

$$a \star b = F_{1,1}(a,b) = \tfrac{1}{2}\,[a,b]_{\mathfrak{g}} + \tfrac{1}{2}(a \otimes b + b \otimes a). \tag{6.6}$$

Since the maps $F_{i,j}$ are bilinear on their domains, it follows that the same is true of $\star$. We are now in a position to prove the following fact.

Lemma 6.4. *Following Definition 6.3, we have*

$$u \star v \in \mathscr{S}(\mathfrak{g}), \quad \text{for every } u, v \in \mathscr{S}(\mathfrak{g}). \tag{6.7}$$

Hence, $(\mathscr{S}(\mathfrak{g}), \star)$ is an algebra.

Proof. It is enough to prove the assertion when $u \in \mathscr{S}_i$, $v \in \mathscr{S}_j$, for fixed $i, j \in \mathbb{N} \cup \{0\}$. Furthermore, in view of (10.28), it is also not restrictive to assume that $u = a^{\otimes i}/i!$, $v = b^{\otimes j}/j!$, for some $a, b \in \mathfrak{g}$. In this case we have $u \star v = F_{i,j}(u, v) = F_{i,j}(a^{\otimes i}/i!, b^{\otimes j}/j!) = f_{i,j}(a, b)$ (see (6.5)). Thus, (6.7) will follow if we prove

$$f_{i,j}(a, b) \in \mathscr{S}(\mathfrak{g}), \quad \text{for every } a, b \in \mathfrak{g} \text{ and every } i, j \in \mathbb{N} \cup \{0\}. \tag{6.8}$$

By a direct glimpse to the explicit formula (6.3), one can prove directly that, by permuting the indices of summation in $D^{\mathfrak{g}}_{(a,b)}(h^{(i)}, k^{(i)})$, the corresponding coefficients are left unchanged. Thus (6.8) follows from general characterizations of symmetric tensors. (Another direct proof of this fact is furnished on page 456 with the collateral aim to exhibit the single homogeneous components in $f_{i,j}$.) □

We next turn to prove that the algebra $(\mathscr{S}(\mathfrak{g}), \star)$ is indeed a *UA algebra*. Clearly $1 \in \mathbb{K} \equiv \mathscr{S}_0(\mathfrak{g})$ is a unit element for $\star$. Indeed, to begin with, we show that $1 \star v = v$ for every $v \in \mathscr{S}(\mathfrak{g})$. It suffices to prove it when $v = b^{\otimes j}/j!$ for $j \geq 1$ (the case $j = 0$ being trivial). The following computation holds

$$1 \star \frac{b^{\otimes j}}{j!} = F_{0,j}(1, b^{\otimes j}/j!) \overset{(6.5)}{=} f_{0,j}(1, b)$$

$$\overset{(6.3)}{=} \frac{1}{j!} c_1^j \, \mathbf{c}^j(0, 1) \underbrace{b \otimes \cdots \otimes b}_{j \text{ times}} = \frac{b^{\otimes j}}{j!}.$$

One analogously proves that $u \star 1 = u$ for every $u \in \mathscr{S}(\mathfrak{g})$. To prove that $(\mathscr{S}(\mathfrak{g}), \star)$ is a UA algebra, we are left to prove the associativity of $\star$, demonstrated in the following theorem. To this end, we make use of the associativity of the CBHD operation in the *free Lie algebra* on three non-commuting indeterminates, together with a substitution argument which permits us to obtain analogous identities in $\mathfrak{g}$.

Theorem 6.5. *Following Definition 6.3, $\star$ is an associative operation. Hence, the algebra $(\mathscr{S}(\mathfrak{g}), \star)$ is a unital associative algebra.*

Our proof is quite technical and the Reader interested in the proof of PBW can pass directly to Sect. 6.1.2.

Proof. The proof is divided into several steps.

STEP I. Since the powers $\{v^{\otimes n} : n \geq 0, v \in \mathfrak{g}\}$ span $\mathscr{S}(\mathfrak{g})$ (see Proposition 10.16 on page 510), it suffices to prove

$$\left(\frac{u^{\otimes i}}{i!} \star \frac{v^{\otimes j}}{j!} \right) \star \frac{w^{\otimes k}}{k!} = \frac{u^{\otimes i}}{i!} \star \left(\frac{v^{\otimes j}}{j!} \star \frac{w^{\otimes k}}{k!} \right), \tag{6.9}$$

for every $u, v, w \in \mathfrak{g}$ and every $i, j, k \in \mathbb{N} \cup \{0\}$.

The case when $ijk = 0$ is trivial (since 1 is the unit of $\star$), so we can assume $i, j, k \geq 1$. By the definition of $\star$ (together with (6.5)), (6.9) amounts to prove

$$f_{i,j}(u, v) \star \frac{w^{\otimes k}}{k!} = \frac{u^{\otimes i}}{i!} \star f_{j,k}(v, w). \tag{6.10}$$

In order to prove (6.10), we need some work on the CBHD formula in the free tensor algebra generated by two and three distinct noncommutative indeterminates, which is done in the next two steps.

STEP II. Let us set $\mathfrak{h} := \mathcal{L}(\mathbb{K}\langle x, y, z\rangle)$, the free Lie algebra generated by the set $S := \{x, y, z\}$ of cardinality three (see Definition 2.46 on page 85). Let $\varphi_1 : \mathbb{K}\langle S\rangle \to \mathfrak{g}$ be the unique linear function mapping x, y, z respectively into u, v, w. Hence, being $\mathfrak{h}$ a free Lie algebra over S (see Definition 2.50, and Theorem 2.56), there exists a unique Lie algebra morphism $\varphi_2 : \mathfrak{h} \to \mathfrak{g}$ extending φ_1. From the inclusion $\mathfrak{g} \hookrightarrow \mathscr{T}(\mathfrak{g})$, there exists a unique UAA morphism $\Phi : \mathscr{T}(\mathfrak{h}) \to \mathscr{T}(\mathfrak{g})$ extending φ_2.

Next, the definitions of $f_{i,j}, F_{i,j}, \star$ apply replacing $\mathfrak{g}$ with $\mathfrak{h}$: on this occasion, we denote the associated maps by $f_{i,j}^{\mathfrak{h}}, F_{i,j}^{\mathfrak{h}}, \star^{\mathfrak{h}}$. The notation $f_{i,j}^{\mathfrak{g}}, F_{i,j}^{\mathfrak{g}}, \star^{\mathfrak{g}}$ has the obvious analogous meaning. It is not difficult to prove that, thanks to the "universal" expression of $f_{i,j}$, the following facts hold:

$$\begin{aligned}
\Phi\big(f_{i,j}^{\mathfrak{h}}(a, b)\big) &= f_{i,j}^{\mathfrak{g}}\big(\Phi(a), \Phi(b)\big), \qquad \forall\, a, b \in \mathfrak{h}; \\
\Phi\big(U \star^{\mathfrak{h}} V\big) &= (\Phi(U)) \star^{\mathfrak{g}} (\Phi(V)), \qquad \forall\, U, V \in \mathscr{S}(\mathfrak{h}).
\end{aligned} \tag{6.11}$$

Thus, in order to obtain (6.10), it is enough to prove

$$f_{i,j}^{\mathfrak{h}}(x, y) \star^{\mathfrak{h}} \frac{z^{\otimes k}}{k!} = \frac{x^{\otimes i}}{i!} \star^{\mathfrak{h}} f_{j,k}^{\mathfrak{h}}(y, z). \tag{6.12}$$

Indeed, by applying Φ to both sides of (6.12) and by exploiting (6.11), one gets (6.10) precisely. Roughly speaking, we have fixed a particular algebraic context where to perform our computations, that of the free Lie algebra $\mathfrak{h} := \mathcal{L}(\mathbb{K}\langle x, y, z\rangle)$.

STEP III. If $S = \{x, y, z\}$ is as in Step II, let us set $V := \mathbb{K}\langle S\rangle$ and consider the completion $\widehat{\mathscr{T}}(V)$ of the tensor algebra of V. Let the notation in (3.14), (3.15) of the CBHD Theorem apply (see page 125). In particular, Z denotes the associated Dynkin series.

Since Z defines on $\widehat{\mathscr{T}}_+(V)$ an operation coinciding with the *associative* operation $u \blacklozenge v := \mathrm{Log}(\mathrm{Exp}\, u \cdot \mathrm{Exp}\, v)$, we have $Z(Z(x, y), z) = Z(x, Z(y, z))$. Recalling that $Z = \sum_{\alpha \geq 1} Z_\alpha$, this is equivalent to

$$\sum_{\alpha \geq 1} Z_\alpha\Big(\sum_{\beta \geq 1} Z_\beta(x, y), z\Big) = \sum_{\alpha \geq 1} Z_\alpha\Big(x, \sum_{\beta \geq 1} Z_\beta(y, z)\Big).$$

This is an equality of two elements in $\widehat{\mathscr{T}}_+(V)$. So we are allowed to take exponentials (relative to $\widehat{\mathscr{T}}_+(V)$) of both sides, getting

$$\sum_{s\geq 0}\frac{1}{s!}\left(\sum_{\alpha\geq 1}Z_\alpha\left(\sum_{\beta\geq 1}Z_\beta(x,y),z\right)\right)^{\otimes s}=\sum_{s\geq 0}\frac{1}{s!}\left(\sum_{\alpha\geq 1}Z_\alpha\left(x,\sum_{\beta\geq 1}Z_\beta(y,z)\right)\right)^{\otimes s}.$$

(6.13)

Let $W=\mathbb{K}\langle X,Y\rangle$, where $X\neq Y$. In $\widehat{\mathscr{T}}(W)$, the computation below holds:

$$\exp\left(\sum_{\alpha\geq 1}Z_\alpha(X,Y)\right)=\sum_{s=0}^{\infty}\frac{1}{s!}\left(\sum_{n\geq 1}\sum_{(h,k)\in\mathcal{N}_n}c_n\,\mathbf{c}(h,k)D^{\otimes}_{(X,Y)}(h,k)\right)^{\otimes s}$$

$$=\sum_{s=0}^{\infty}\frac{1}{s!}\sum_{n_1\geq 1}\sum_{(h^{(1)},k^{(1)})\in\mathcal{N}_{n_1}}c_{n_1}\,\mathbf{c}(h^{(1)},k^{(1)})D^{\otimes}_{(X,Y)}(h^{(1)},k^{(1)})\cdots$$

$$\cdots\sum_{n_s\geq 1}\sum_{(h^{(s)},k^{(s)})\in\mathcal{N}_{n_s}}c_{n_s}\,\mathbf{c}(h^{(s)},k^{(s)})D^{\otimes}_{(X,Y)}(h^{(s)},k^{(s)})$$

$$=\sum_{i,j\geq 0}f^{\otimes}_{i,j}(X,Y).$$

Here, we have applied (6.3), (6.4) together with the introduction of functions $f^{\otimes}_{i,j}:\widehat{\mathscr{T}}(W)\times\widehat{\mathscr{T}}(W)\to\widehat{\mathscr{T}}(W)$ completely analogous to the function $f_{i,j}$ in (6.3), where $\otimes$ is now the algebra operation in $\widehat{\mathscr{T}}(W)$ and $D^{\otimes}$ replaces $D^{\mathfrak{g}}$ when we are considering the Lie algebra structure of $\widehat{\mathscr{T}}(W)$.

We have thus derived

$$\sum_{i,j\geq 0}f^{\otimes}_{i,j}(X,Y)=\sum_{s=0}^{\infty}\frac{1}{s!}\left(\sum_{\alpha\geq 1}Z_\alpha(X,Y)\right)^{\otimes s},\qquad\text{in }\widehat{\mathscr{T}}(\mathbb{K}\langle X,Y\rangle).\quad(6.14)$$

From the universal property of the tensor algebra of the free vector space $\mathbb{K}\langle X,Y\rangle$ (plus an obvious argument of continuity), we are allowed to make substitutions of X,Y in (6.14). More precisely, the following equalities hold:

$$\sum_{i,j\geq 0}f^{\otimes}_{i,j}\left(\sum_{\beta\geq 1}Z_\beta(x,y),z\right)=\sum_{s\geq 0}\frac{1}{s!}\left(\sum_{\alpha\geq 1}Z_\alpha\left(\sum_{\beta\geq 1}Z_\beta(x,y),z\right)\right)^{\otimes s}$$

$$\overset{(6.13)}{=}\sum_{s\geq 0}\frac{1}{s!}\left(\sum_{\alpha\geq 1}Z_\alpha\left(x,\sum_{\beta\geq 1}Z_\beta(y,z)\right)\right)^{\otimes s}=\sum_{i,j\geq 0}f^{\otimes}_{i,j}\left(x,\sum_{\beta\geq 1}Z_\beta(y,z)\right).$$

Indeed, the first and third equalities follow from (6.14) by the choices:

first equality: $X = \sum_{\beta \geq 1} Z_\beta(x, y),$ $Y = z,$

second equality: $X = x,$ $Y = \sum_{\beta \geq 1} Z_\beta(y, z).$

As a consequence, we have proved the identity in $\widehat{\mathscr{T}}(\mathbb{K}\langle x, y, z \rangle)$:

$$\sum_{i,j \geq 0} f_{i,j}^\otimes \Big(\sum_{\beta \geq 1} Z_\beta(x, y), z \Big) = \sum_{i,j \geq 0} f_{i,j}^\otimes \Big(x, \sum_{\beta \geq 1} Z_\beta(y, z) \Big). \qquad (6.15)$$

Here, obviously, the maps $f_{i,j}^\otimes$ are now functions related to the algebra $\widehat{\mathscr{T}}(V)$. Let us next consider the following subspace of $\widehat{\mathscr{T}}(V)$:

$$W_{i,j,k} := \mathrm{span}\Big\{ x^{a_1} y^{b_1} z^{c_1} \cdots x^{a_n} y^{b_n} z^{c_n} \,\Big|\, n \in \mathbb{N}, a_1, b_1, c_1, \ldots, a_n, b_n, c_n \geq 0, a_1$$

$$+ \cdots + a_n = i, \ b_1 + \cdots + b_n = j, \ c_1 + \cdots + c_n = k \Big\}.$$

We have $W_{i,j,k} \subset \mathscr{T}(V)$, $\mathscr{T}(V) = \bigoplus_{i,j,k \geq 0} W_{i,j,k}$ and $\widehat{\mathscr{T}}(V) = \prod_{i,j,k \geq 0} W_{i,j,k}$.
Thus the natural projection $L_{i,j,k} : \widehat{\mathscr{T}}(V) \to W_{i,j,k}$ is well defined and we are entitled to apply $L_{i,j,k}$ to both sides of (6.15).

By the definition of $f_{i,j}$ (see (6.3), (6.4) and note that b "occurs" j times in $f_{i,j}(a, b)$), this gives

$$L_{i,j,k}\Big(\sum_{s \geq 0} f_{s,k}^\otimes \Big(\sum_{\beta \geq 1} Z_\beta(x, y), z \Big) \Big) = L_{i,j,k}\Big(\sum_{s \geq 0} f_{i,s}^\otimes \Big(x, \sum_{\beta \geq 1} Z_\beta(y, z) \Big) \Big). \qquad (6.16)$$

Moreover, in (6.16) all sums over s and β can be taken to be finite (say, $0 \leq s \leq i + j + k$, $1 \leq \beta \leq i + j + k$), and, for brevity, we shall do this without explicitly writing it.

STEP IV. We now make a crucial remark: *Equality (6.16) straightforwardly specializes to an equality of elements in the tensor algebra* $\mathscr{T}(\mathfrak{h})$. This follows from the fact that $\mathfrak{h}$ is a *free* Lie algebra over $\{x, y, z\}$ and by observing that $Z_\beta(x, y)$ and $Z_\beta(y, z)$ belong to $\mathfrak{h}$ for every $\beta \geq 1$. Hence, (6.16) yields an analogous identity in $\mathscr{T}(\mathfrak{h})$, replacing $\otimes$ with $\mathfrak{h}$ (see the notation introduced in Step II). This identity can be rewritten as follows:

$$L_{i,j,k} \sum_{s \geq 0} F_{s,k}^\mathfrak{h} \Big(\tfrac{1}{s!} \big(\sum_\beta Z_\beta(x, y) \big)^s, \tfrac{z^k}{k!} \Big) = L_{i,j,k} \sum_{s \geq 0} F_{i,s}^\mathfrak{h} \Big(\tfrac{x^i}{i!}, \tfrac{1}{s!} \big(\sum_\beta Z_\beta(y, z) \big)^s \Big).$$

By the definition of $\star^\flat$, this can be further rewritten as

$$L_{i,j,k}\left(\left(\sum_{s\geq 0}\tfrac{1}{s!}\Big(\sum_\beta Z_\beta(x,y)\Big)^s\right)\star^\flat \tfrac{z^k}{k!}\right)=L_{i,j,k}\left(\tfrac{x^i}{i!}\star^\flat\left(\sum_{s\geq 0}\tfrac{1}{s!}\Big(\sum_\beta Z_\beta(y,z)\Big)^s\right)\right).$$

Finally, a simple calculation based on the definition of $F_{i,j}^\flat$ shows that the above is equivalent to (6.12) and the proof is complete.   □

6.1.2   Cartier's Proof of PBW via CBHD

With the UA algebra $(\mathscr{S}(\mathfrak{g}),\star)$ at hand, we are ready to complete the proof of "CBHD⇒PBW", by recalling Cartier's argument (with all details) in the remainder of this chapter.

We begin with three lemmas. The first one is the key tool and the actual link between the "mixed" exponential $\exp(Z^\mathfrak{g}(a,b))$ and the exponential in the CBHD Theorem $\mathrm{Exp}(Z(a,b)) = \mathrm{Exp}(a)\cdot\mathrm{Exp}(b)$, as an identity in $\widehat{\mathscr{T}}(\mathfrak{g})$.

Lemma 6.6. *Let $\mathscr{J}(\mathfrak{g})$ be the two-sided ideal in $\mathscr{T}(\mathfrak{g})$ generated by the set*

$$\{x\otimes y-y\otimes x-[x,y]_\mathfrak{g} \; : \; x,y\in\mathfrak{g}\}.$$

Then for every $a,b\in\mathfrak{g}$, $(h,k)\in N_n$ and every $n\in\mathbb{N}$,

$$D_{(a,b)}^\mathfrak{g}(h,k) \equiv D_{(a,b)}^\otimes(h,k) \quad (\textit{modulo }\mathscr{J}(\mathfrak{g})). \tag{6.17}$$

Here, following (6.2), $D_{(a,b)}^\otimes(h,k)$ denotes the usual nested commutator in $\mathscr{T}(\mathfrak{g})$, where the latter is equipped with the Lie bracket $[\cdot,\cdot]_\otimes$ naturally related to its associative algebra structure.

Proof. This result follows by an inductive argument. Indeed, one has $[a,b]_\otimes = a\otimes b-b\otimes a \equiv [a,b]_\mathfrak{g}$ modulo $\mathscr{J}(\mathfrak{g})$, and, arguing by induction, we find

$$[a_1[a_2\ldots a_{k+1}]_\otimes]_\otimes \equiv [a_1[a_2\ldots a_{k+1}]_\mathfrak{g}]_\otimes \equiv [a_1[a_2\ldots a_{k+1}]_\mathfrak{g}]_\mathfrak{g} = [a_1\ldots a_{k+1}]_\mathfrak{g},$$

which completes the proof.   □

With (6.17) at hand, we next prove our second lemma, which at last gives us a justification for the definitions of $f_{i,j}$, $F_{i,j}$, and $\star$.

Lemma 6.7. *For every $a_1,\ldots,a_n\in\mathscr{S}(\mathfrak{g})$ and every $n\in\mathbb{N}$, we have*

$$a_1\star\cdots\star a_n \equiv a_1\cdot\ldots\cdot a_n \quad (\textit{modulo }\mathscr{J}(\mathfrak{g})), \tag{6.18}$$

Proof. Let us argue by induction. The case $n = 1$ is trivial. Let us turn to the case $n = 2$: we claim that

$$u \star v \equiv u \cdot v \ (\text{modulo } \mathscr{J}(\mathfrak{g})), \qquad \text{for every } u, v \in \mathscr{S}(\mathfrak{g}). \tag{6.19}$$

By bilinearity, it suffices to prove (6.19) when $u = a^{\otimes i}/i!$ and $v = b^{\otimes j}/j!$ for $a, b \in \mathfrak{g}$ and $i, j \in \mathbb{N}$. The following argument then applies

$$u \star v = F_{i,j}\left(\tfrac{a^{\otimes i}}{i!}, \tfrac{b^{\otimes j}}{j!}\right) \stackrel{(6.5)}{=} f_{i,j}(a, b)$$

$$\stackrel{(6.3)}{=} \sum_{\mathcal{A}_{i,j}} \frac{1}{s!} \left(\prod_{i=1}^{s} c_{n_i}\, \mathbf{c}(h^{(i)}, k^{(i)})\right) D^{\mathfrak{g}}_{(a,b)}(h^{(1)}, k^{(1)}) \otimes \cdots \otimes D^{\mathfrak{g}}_{(a,b)}(h^{(s)}, k^{(s)})$$

(by (6.17), modulo $\mathscr{J}(\mathfrak{g})$)

$$\equiv \sum_{\mathcal{A}_{i,j}} \frac{1}{s!} \left(\prod_{i=1}^{s} c_{n_i}\, \mathbf{c}(h^{(i)}, k^{(i)})\right) D^{\otimes}_{(a,b)}(h^{(1)}, k^{(1)}) \otimes \cdots \otimes D^{\otimes}_{(a,b)}(h^{(s)}, k^{(s)})$$

$$\stackrel{(6.4)}{=} \big\{\text{the summand in } \mathrm{Exp}(Z(a, b)) \text{ containing } i\text{-times } a, \text{ and } j\text{-times } b\big\}$$

(by the CBHD Theorem itself!)

$$= \big\{\text{the summand in } \mathrm{Exp}(a) \cdot \mathrm{Exp}(b) \text{ containing } i\text{-times } a, \text{ and } j\text{-times } b\big\}$$

$$= \frac{a^{\otimes i}}{i!} \cdot \frac{b^{\otimes j}}{j!} = u \cdot v.$$

We can now argue by induction

$$a_1 \star a_2 \star \cdots \star a_n = a_1 \star (a_2 \star \cdots \star a_n)$$

$$\equiv a_1 \star (a_2 \cdot \ldots \cdot a_n) \equiv a_1 \cdot a_2 \cdot \ldots \cdot a_n.$$

Note that we are using the fact that $(\mathscr{S}(\mathfrak{g}), \star)$ is an *associative* algebra. □

We next take up another step towards PBW, the third lemma:

Lemma 6.8. *For every $a \in \mathfrak{g}$ and every $i, j \in \mathbb{N} \cup \{0\}$,*

$$F_{i,j}\left(\frac{a^{\otimes i}}{i!}, \frac{a^{\otimes j}}{j!}\right) = \frac{a^{\otimes(i+j)}}{i!\, j!}. \tag{6.20}$$

Proof. This follows by collecting together (6.3), (6.5) and the identity

$$D^{\mathfrak{g}}_{(a,a)}(h,k) = \begin{cases} 0, & \text{if } |h| + |k| \geq 2, \\ a, & \text{if } |h| + |k| = 1. \end{cases}$$

This gives the following computation

$$F_{i,j}\left(\tfrac{a^{\otimes i}}{i!}, \tfrac{a^{\otimes j}}{j!}\right) = f_{i,j}(a,a)$$

$$= \frac{a^{\otimes(i+j)}}{(i+j)!} \sum_{\substack{(h^{(1)},k^{(1)}),\dots,(h^{(i+j)},k^{(i+j)})\in\{(1,0),(0,1)\} \\ h^{(1)}+\cdots+h^{(i+j)}=i,\ k^{(1)}+\cdots+k^{(i+j)}=j}} 1 = \frac{a^{\otimes(i+j)}}{i!\,j!}.$$

In the second identity, we used the following argument: the sum in (6.3) extends over $(h^{(r)}, k^{(r)}) \in \mathcal{N}_1$ with $|h^{(r)}| + |k^{(r)}| = 1$ and $\sum_{r=1}^{s} |h^{(r)}| = i$, $\sum_{r=1}^{s} |k^{(r)}| = j$ whence $i + j = \sum_{r=1}^{s}(|h^{(r)}| + |k^{(r)}|)$, which forces $s = i + j$ and $c_{n_r} = \mathbf{c}(h^{(r)}, k^{(r)}) = 1$. The last equality holds since the above sum equals, by an easy combinatorial argument, the binomial $\binom{i+j}{i}$.   □

Before stating the proposition which will decisively lead to PBW (see Theorem 6.10 below), we state a last remark:

Remark 6.9. For every $a \in \mathfrak{g}$ and every $n \in \mathbb{N}$,

$$\underbrace{a \star \cdots \star a}_{n \text{ times}} = a^{\otimes n}. \tag{6.21}$$

Proof. The proof is by induction. The case $n = 1$ is obvious. Supposing (6.21) to hold for n, we derive the $(n+1)$-case from the following computation:

$$\underbrace{a \star \cdots \star a}_{n+1 \text{ times}} = a^{\otimes n} \star a = n!\left(\frac{a^{\otimes n}}{n!} \star a\right) = n!\,F_{n,1}\left(\frac{a^{\otimes n}}{n!}, a\right)$$

$$\overset{(6.20)}{=} n!\,\frac{a^{\otimes(n+1)}}{n!\,1!} = a^{\otimes(n+1)},$$

and the proof is complete.   □

Let us now consider the unique linear map $\Pi : \mathscr{T}(\mathfrak{g}) \to \mathscr{S}(\mathfrak{g})$ such that $\Pi(1) = 1$ and

$$\Pi(a_1 \otimes \cdots \otimes a_n) = a_1 \star \cdots \star a_n, \qquad \left(\begin{array}{l} \text{for every } a_1, \dots, a_n \in \mathfrak{g} \\ \text{and every } n \in \mathbb{N} \end{array}\right). \tag{6.22}$$

Obviously, $\Pi : (\mathscr{T}(\mathfrak{g}), \cdot) \to (\mathscr{S}(\mathfrak{g}), \star)$ is the unique UAA morphism extending the inclusion $\mathfrak{g} \equiv \mathscr{S}_1(\mathfrak{g}) \hookrightarrow \mathscr{S}(\mathfrak{g})$.

Theorem 6.10 (Cartier). *With the above notation, the following facts hold:*

$$\Pi(t) \equiv t \quad (\text{modulo } \mathscr{J}(\mathfrak{g})), \qquad \text{for every } t \in \mathscr{T}(\mathfrak{g}); \qquad (6.23\text{a})$$

$$\Pi(v) = v, \qquad\qquad\qquad\quad \text{for every } v \in \mathscr{S}(\mathfrak{g}); \qquad (6.23\text{b})$$

$$\Pi(h) = 0, \qquad\qquad\qquad\quad \text{for every } h \in \mathscr{J}(\mathfrak{g}); \qquad (6.23\text{c})$$

$$\ker(\Pi) = \mathscr{J}(\mathfrak{g}). \qquad\qquad\qquad\qquad\qquad\qquad\qquad (6.23\text{d})$$

Proof. By linearity, we can prove (6.23a) by checking it when $t = a_1 \otimes \cdots \otimes a_n$, since $a_1, \ldots, a_n \in \mathfrak{g}$ and $n \in \mathbb{N}$ are arbitrary:

$$\Pi(a_1 \otimes \cdots \otimes a_n) = a_1 \star \cdots \star a_n \quad (\text{by (6.18), modulo } \mathscr{J}(\mathfrak{g}))$$

$$\equiv a_1 \cdot \ldots \cdot a_n = a_1 \otimes \cdots \otimes a_n.$$

Moreover, by linearity, we can prove (6.23b) by checking it when $v = a^{\otimes n}$, for $a \in \mathfrak{g}$ and $n \in \mathbb{N}$:

$$\Pi(a^{\otimes n}) \overset{(6.22)}{=} a^{\star n} \overset{(6.21)}{=} a^{\otimes n}.$$

Next, we turn to prove (6.23c): since the typical element of $\mathscr{J}(\mathfrak{g})$ is spanned by elements of the form

$$t \cdot (x \otimes y - y \otimes x - [x,y]_{\mathfrak{g}}) \cdot t', \qquad \text{for } x, y \in \mathfrak{g} \text{ and } t, t' \in \mathscr{T}(\mathfrak{g}),$$

the fact that Π is a UAA morphism shows that (6.23c) will follow if we prove that $\Pi(x \otimes y - y \otimes x - [x,y]_{\mathfrak{g}}) = 0$, for every $x, y \in \mathfrak{g}$. This latter fact is a consequence of the computation below:

$$\Pi(x \otimes y - y \otimes x - [x,y]_{\mathfrak{g}}) \overset{(6.22)}{=} x \star y - y \star x - \Pi([x,y]_{\mathfrak{g}})$$

$$= x \star y - y \star x - [x,y]_{\mathfrak{g}} \overset{(6.6)}{=} \tfrac{1}{2}[x,y]_{\mathfrak{g}} + \tfrac{1}{2}(x \otimes y + y \otimes x)$$

$$- \left(\tfrac{1}{2}[y,x]_{\mathfrak{g}} + \tfrac{1}{2}(y \otimes x + x \otimes y)\right) - [x,y]_{\mathfrak{g}} = 0.$$

The second equality comes from (6.23b) and the fact that $[x,y]_{\mathfrak{g}}$ belongs to $\mathfrak{g} \subset \mathscr{S}(\mathfrak{g})$.

Finally, from (6.23a) and (6.23c), we obtain (6.23d). Indeed, (6.23c) proves $\ker(\Pi) \supseteq \mathscr{J}(\mathfrak{g})$; conversely, if $t \in \ker(\Pi)$, we have $0 = \Pi(t) = t + h$ with $h \in \mathscr{J}(\mathfrak{g})$ (exploiting (6.23a)), so that $t = -h \in \mathscr{J}(\mathfrak{g})$, whence the reverse inclusion $\ker(\Pi) \subseteq \mathscr{J}(\mathfrak{g})$ follows. $\qquad\qquad\qquad\qquad\square$

Summing up, we infer that $\Pi : \mathscr{T}(\mathfrak{g}) \to \mathscr{S}(\mathfrak{g})$ is surjective, its restriction to $\mathscr{S}(\mathfrak{g})$ is the identity, i.e., Π is a projection onto $\mathscr{S}(\mathfrak{g})$; moreover its kernel is $\mathscr{J}(\mathfrak{g})$. As a consequence,

$$\mathscr{T}(\mathfrak{g}) = \mathscr{J}(\mathfrak{g}) \oplus \mathscr{S}(\mathfrak{g}). \qquad\qquad\qquad\qquad (6.24)$$

We are finally ready to present a proof of the PBW Theorem 2.94 – which is different from the one given in Chap. 8 – and which involves the CBHD Theorem 3.8.

Remark 6.11. By standard arguments, which we here recall, *the decomposition* (6.24) *implies PBW.*

For the sake of brevity, we omit $\mathfrak{g}$ in the notation. By (6.24), for every $t \in \mathscr{T}$ there exists a unique $s(t) \in \mathscr{S}$ such that $t - s(t) \in \mathscr{J}$. Hence, the following map is an isomorphism of vector spaces

$$s : \mathscr{U} \to \mathscr{S}, \qquad s(\pi(t)) := s(t) \quad \forall\, t \in \mathscr{T}. \tag{6.25}$$

(Recall that $\pi : \mathscr{T} \to \mathscr{U}$ is the canonical projection onto the quotient $\mathscr{U} = \mathscr{T}/\mathscr{J}$.) The inverse map s^{-1} is obviously given by $s^{-1} = \pi|_{\mathscr{S}}$. Hence, a basis for $\mathscr{U}$ can be obtained, via the linear isomorphism s^{-1}, from the following well known basis $\mathcal{B}$ for $\mathscr{S}$ (see the notation in Chap. 10, in particular (10.35) on page 511 for the map Q):

$$\mathcal{B} := \{1\} \cup \big\{ Q(e_{i_1} \otimes \cdots \otimes e_{i_n}) \,\big|\, n \in \mathbb{N},\ i_1, \ldots, i_n \in I,\ i_1 \preccurlyeq \ldots \preccurlyeq i_n \big\}.$$

Here (as in the statement of PBW), $\{e_i\}_{i \in I}$ denotes an indexed (linear) basis for $\mathfrak{g}$, where I is totally ordered by the relation $\preccurlyeq$. Consequently, $\mathcal{C} := s^{-1}(\mathcal{B})$ is a basis for $\mathscr{U}$, which we denote in the following way:

$$\mathcal{C} = \bigcup_{n \in \mathbb{N} \cup \{0\}} \mathcal{C}_n, \quad \text{where } \mathcal{C}_0 := \{\pi(1)\} \text{ and}$$
$$\mathcal{C}_n := \big\{ \pi\big(Q(e_{i_1} \otimes \cdots \otimes e_{i_n})\big) \,\big|\, i_1, \ldots, i_n \in I,\ i_1 \preccurlyeq \ldots \preccurlyeq i_n \big\}, \quad n \in \mathbb{N}. \tag{6.26}$$

We also set

$$\mathcal{W} = \bigcup_{n \in \mathbb{N} \cup \{0\}} \mathcal{W}_n, \quad \text{where } \mathcal{W}_0 := \{\pi(1)\} \text{ and}$$
$$\mathcal{W}_n := \big\{ \pi(e_{i_1} \otimes \cdots \otimes e_{i_n}) \,\big|\, i_1, \ldots, i_n \in I,\ i_1 \preccurlyeq \ldots \preccurlyeq i_n \big\}, \quad n \in \mathbb{N}.$$

The statement of PBW is precisely equivalent to the fact that $\mathcal{W}$ is a basis for $\mathscr{U}$. This latter fact is a consequence of the following claims:

$$\pi(\mathscr{T}_n(\mathfrak{g})) \text{ is spanned by } \mathcal{W}_1, \ldots, \mathcal{W}_n, \text{ for every } n \in \mathbb{N}; \tag{6.27a}$$

$$\pi\big(Q(e_{i_1} \otimes \cdots \otimes e_{i_n})\big) = \pi(e_{i_1} \otimes \cdots \otimes e_{i_n}) + \left\{ \begin{array}{c} \text{linear combination in} \\ \mathcal{W}_1 \cup \cdots \cup \mathcal{W}_{n-1} \end{array} \right\};$$
$$\tag{6.27b}$$

$$\mathcal{W}_n \text{ is spanned by } \mathcal{C}_1 \cup \cdots \cup \mathcal{C}_n, \text{ for every } n \in \mathbb{N}; \tag{6.27c}$$

$$\pi(e_{i_1} \otimes \cdots \otimes e_{i_n}) = \pi\big(Q(e_{i_1} \otimes \cdots \otimes e_{i_n})\big) + \left\{ \begin{array}{c} \text{linear combination in} \\ \mathcal{C}_1 \cup \cdots \cup \mathcal{C}_{n-1} \end{array} \right\}.$$

$$(6.27\text{d})$$

Here $n \in \mathbb{N}$ and $i_1, \ldots, i_n \in I$ are arbitrary. The claimed (6.27a) is a consequence of the fact that $\{e_{i_1} \otimes \cdots \otimes e_{i_n} \mid i_1, \ldots, i_n \in I\}$ is a basis for $\mathscr{T}_n(\mathfrak{g})$ together with an inductive argument based on the standard computation:

$$v_i \otimes \cdots \otimes (v_i \otimes v_{i+1}) \cdots \otimes v_n = v_i \otimes \cdots \big(v_{i+1} \otimes v_i$$

$$+ \{v_i \otimes v_{i+1} - v_{i+1} \otimes v_i - [v_i, v_{i+1}]\}$$

$$+ [v_i, v_{i+1}]\big) \cdots \otimes v_n {=} v_i \otimes \cdots v_{i+1} \otimes v_i \cdots \otimes v_n$$

$$+ \left\{ \begin{array}{c} \text{element in} \\ \mathscr{J}(\mathfrak{g}) \end{array} \right\} + \left[\begin{array}{c} \text{element in} \\ \mathscr{T}_{n-1}(\mathfrak{g}) \end{array} \right].$$

(6.27b) follows from the above computation, which gives:

$$\pi\big(Q(e_{i_1} \otimes \cdots \otimes e_{i_n})\big) = \frac{1}{n!} \sum_{\sigma \in \mathfrak{S}_n} \pi\big(e_{i_{\sigma(1)}} \otimes \cdots \otimes e_{i_{\sigma(n)}}\big)$$

$$= \frac{1}{n!}\big(n!\, \pi(e_{i_1} \otimes \cdots \otimes e_{i_n}) + \pi(j_n) + \pi(r_n)\big),$$

where $j_n \in \mathscr{J}$ and $r_n \in \mathscr{T}_1 \oplus \cdots \oplus \mathscr{T}_{n-1}$. If we apply $\pi(\mathscr{J}) = \{0\}$ and (6.27a), the above identities then give

$$\pi\big(Q(e_{i_1} \otimes \cdots \otimes e_{i_n})\big) = \pi(e_{i_1} \otimes \cdots \otimes e_{i_n})$$

$$+ \{ \text{linear combination in } \mathcal{W}_1 \cup \cdots \cup \mathcal{W}_{n-1} \}.$$

(6.27c) is proved by induction on $n \in \mathbb{N}$, by using (6.27b) and the fact that $\pi\big(Q(e_{i_1} \otimes \cdots \otimes e_{i_n})\big) \in \mathcal{C}_n$. Moreover, (6.27d) comes from (6.27b) and (6.27c).

We are now in a position to prove that $\mathcal{W}$ is a basis for $\mathscr{U}$:

- $\mathcal{W}$ *generates* $\mathscr{U}$: This follows from (6.27a).
- $\mathcal{W}$ *is independent*: Let H be a finite set of pairwise distinct n-tuples of ordered elements (w.r.t. $\preccurlyeq$) of I (with arbitrary n) and let $\lambda(i) \in \mathbb{K}$ for every $i \in H$. Then (6.27d) immediately gives

$$\sum_{i=(i_1,\ldots,i_n)\in H} \lambda(i)\, \pi(e_{i_1} \otimes \cdots \otimes e_{i_n})$$

$$= \sum_{i\in H} \lambda(i)\, \pi\big(Q(e_{i_1} \otimes \cdots \otimes e_{i_n})\big) + \left\{ \begin{array}{c} \text{linear combination in} \\ \mathcal{C}_1 \cup \cdots \cup \mathcal{C}_{n-1} \end{array} \right\}.$$

Since the sum $\sum_{i \in H}(\cdots)$ in the above far right-hand term is an element of $\mathcal{C}_n$ and since $\mathcal{C} = \bigcup_n \mathcal{C}_n$ is independent, the above left-hand term vanishes if and only if $\sum_{i \in H} \lambda(i) \, \pi\big(Q(e_{i_1} \otimes \cdots \otimes e_{i_n})\big) = 0$. In its turn, this is possible if and only if $\lambda(i) = 0$ for every $i \in H$ (see (6.26) and exploit the independence of the elements in $\mathcal{C}_n$). This proves the independence of W and the proof of PBW is complete. $\square$

Part II
Proofs of the Algebraic Prerequisites

Chapter 7
Proofs of the Algebraic Prerequisites

$\mathbf{T}$HE aim of this chapter is to collect all the missing proofs of the results in Chap. 2. The chapter is divided into several sections, corresponding to those of Chap. 2. Finally, Sect. 7.8 collects some proofs from Chaps. 4 and 6 too, considered as less crucial in favor of economy of presentation in the chapters they originally belonged to.

7.1 Proofs of Sect. 2.1.1

Proof (of Theorem 2.6, page 52). (i) It suffices to gather Proposition 2.1 and Remark 2.5. The actual definition of F^χ is

$$F^\chi \left(\sum_{j=1}^n \lambda_j \, \chi(v_j) \right) := \sum_{j=1}^n \lambda_j \, F(v_j),$$

for $n \in \mathbb{N}$, $\lambda_1, \ldots, \lambda_n \in \mathbb{K}$ and any pairwise distinct $v_1, \ldots, v_n \in S$.

(ii) Let V, φ be as in the statement, i.e., for every vector space X and every map $F : S \to X$ there exists a unique linear map making the following a commutative diagram:

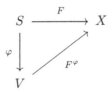

Let us choose $X := \mathbb{K}\langle S \rangle$ and $F := \chi$. We denote by χ^φ the linear map closing the following diagram

A. Bonfiglioli and R. Fulci, *Topics in Noncommutative Algebra*, Lecture Notes in Mathematics 2034, DOI 10.1007/978-3-642-22597-0_7,
© Springer-Verlag Berlin Heidelberg 2012

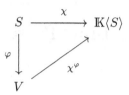

On the other hand, from what was shown in (1), there exists one and only one linear map φ^χ making the following diagram commute:

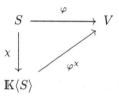

It suffices to prove that the two linear maps φ^χ and χ^φ are inverse to each other, thus giving the isomorphism desired between V and $\mathbb{K}\langle S\rangle$. First,

$$\chi^\varphi \circ \varphi^\chi = \mathrm{Id}_{\mathbb{K}\langle S\rangle}$$

can be easily verified: as

$$\chi^\varphi(\varphi^\chi(\chi(s))) = \chi^\varphi(\varphi(s)) = \chi(s) \qquad \forall\, s \in S,$$

the linear map $\chi^\varphi \circ \varphi^\chi$ plays the rle of χ^χ in the diagram

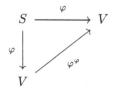

But the same diagram obviously admits $\mathrm{Id}_{\mathbb{K}\langle S\rangle}$ as a "closing" map, and, as such a closing map is unique, $\chi^\varphi \circ \varphi^\chi = \mathrm{Id}_{\mathbb{K}\langle S\rangle}$.

The second identity

$$\varphi^\chi(\chi^\varphi(v)) = v, \qquad \forall\, v \in V$$

can be proved analogously: We consider the diagram

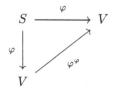

together with the chain of equalities

$$\varphi^{\chi}(\chi^{\varphi}(\varphi(s))) = \varphi^{\chi}(\chi(s)) = \varphi(s) \qquad \forall\, s \in S.$$

Hence, the two maps Id_V and $\chi^{\varphi} \circ \varphi^{\chi}$ both close the last diagram (that which is closed by φ^{φ}), so that they necessarily coincide. The maps χ^{φ} and φ^{χ} are thus inverse to each other and $\mathbb{K}\langle S \rangle \simeq V$ canonically.

We prove the injectivity of φ. Let us assume $\varphi(s) = \varphi(t)$. Then

$$\chi(s) = \chi^{\varphi}(\varphi(s)) = \chi^{\varphi}(\varphi(t)) = \chi(t),$$

but χ is injective, so $s = t$.

We prove that the set $\{\varphi(s)|\ s \in S\}$ is a basis for V. We begin to show the linear independence. Let $s_1, ..., s_n \in S$ be pairwise distinct and $\lambda_j \in \mathbb{K}$ with $\sum_j \lambda_j \varphi(s_j) = 0$. Applying the linear map χ^{φ}, we get $\sum_j \lambda_j \chi(s_j) = 0$ and so $\lambda_1 = \cdots = \lambda_n = 0$ (see Remark 2.5). Finally, we prove that the vectors $\{\varphi(s)|\ s \in S\}$ span V. To this aim, consider the trivial map

$$F : S \to \mathbb{K}, \quad s \mapsto 0$$

and the "open" diagram

$$
\begin{array}{ccc}
S & \xrightarrow{\ \ F\ \ } & \mathbb{K} \\[2pt]
{\scriptstyle \varphi}\big\downarrow & & \\[6pt]
V & &
\end{array}
\qquad\qquad (7.1)
$$

This diagram is obviously closed by the map $V \to \mathbb{K}$ which is identically 0. If $\{\varphi(s) : s \in S\}$ did not span V, we could fix an element

$$w \in V \setminus \mathrm{span}\{\varphi(s) : s \in S\}$$

and complete the set $\{\varphi(s) : s \in S\} \cup \{w\}$ to a basis for V, say $\mathcal{D}$. We may then define a (unique) linear map $\vartheta : V \to \mathbb{K}$ such that:

$$\vartheta(v) := \begin{cases} 1, & \text{if } v = w \\ 0, & \text{if } v \in \mathcal{D} \setminus \{w\}. \end{cases}$$

It is immediately seen that ϑ closes the diagram (7.1), thus contradicting the uniqueness of the closing map. $\square$

7.2   Proofs of Sect. 2.1.2

Proof (of Lemma 2.24, page 68). (i) Since M_{alg} coincides with $\mathbb{K}\langle M\rangle$, we can apply Theorem 2.6 (from which we also inherit the notation) to produce the (unique!) linear map $f^\chi : \mathbb{K}\langle M\rangle = M_{\text{alg}} \to A$ with property (2.16). The proof of (i) is accomplished if we show that f^χ is also a magma morphism (when f is). Denoted by $(M,.)$, $(M_{\text{alg}}, *)$, $(A, \star)$ the associated operations, we have[1]

$$f^\chi\Big(\big(\textstyle\sum_{i=1}^p \lambda_i\,\chi(m_i)\big) * \big(\sum_{j=1}^q \mu_j\,\chi(n_j)\big)\Big)$$

$$= f^\chi\Big(\sum_{1\le i\le p,\,1\le j\le q} \lambda_i\mu_j\,\chi(m_i.n_j)\Big) = \sum_{1\le i\le p,\,1\le j\le q} \lambda_i\mu_j\,f^\chi(\chi(m_i.n_j))$$

$$= \sum_{1\le i\le p,\,1\le j\le q} \lambda_i\mu_j\,f(m_i.n_j) = \sum_{1\le i\le p,\,1\le j\le q} \lambda_i\mu_j\,f(m_i) \star f(n_j)$$

$$= \sum_{1\le i\le p,\,1\le j\le q} \lambda_i\mu_j\,f^\chi(\chi(m_i)) \star f^\chi(\chi(n_j))$$

$$= f^\chi(\textstyle\sum_{i=1}^p \lambda_i\,\chi(m_i)) \star f^\chi(\sum_{j=1}^q \mu_j\,\chi(n_j)),$$

for any arbitrary $p, q \in \mathbb{N}$, $\lambda_1, \ldots, \lambda_p \in \mathbb{K}$, $\mu_1, \ldots, \mu_q \in \mathbb{K}$, $m_1, \ldots, m_p \in M$, $n_1, \ldots, n_q \in M$. This completes the proof of (i).

(ii) Argue as in the proof of Theorem 2.6, by also using the fact that χ is injective and that $\chi(M)$ is a basis of M_{alg}.

(iii) This follows from (i) and (ii) above, together with $f^\chi(\chi(e)) = f(e)$ (here e denotes the unit of the monoid M) and the fact that $f(e)$ is the unit of A since $f : M \to A$ is a monoid morphism.   □

7.3   Proofs of Sect. 2.1.3

Proof (of Theorem 2.30, page 74).
(i) *Uniqueness.* Any candidate linear map ϕ closing the diagram

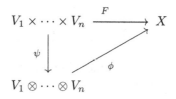

---

[1]The first equality comes from the definition of $*$, the second and sixth from the linearity of f^χ, the third and fifth from (2.16), the fourth from the fact that f is a magma morphism.

satisfies also

$$\phi(v_1 \otimes \cdots \otimes v_n) = \phi(\psi(v_1, \ldots, v_n)) = F(v_1, \ldots, v_n).$$

Now, as the set $\{v_1 \otimes \cdots \otimes v_n \mid v_i \in V_i\}$ spans $V_1 \otimes \cdots \otimes V_n$, ϕ is uniquely determined so $\phi = F^\psi$.

Existence. The map $\psi : V_1 \times \cdots \times V_n \to V_1 \otimes \cdots \otimes V_n$ is the composition of χ and π, as follows:

$$V_1 \times \cdots \times V_n$$

$$\chi \downarrow$$

$$\mathbb{K}\langle V_1 \times \cdots \times V_n \rangle$$

$$\pi \downarrow$$

$$V_1 \otimes \cdots \otimes V_n$$

First, thanks to the characteristic property of the free vector space, there exists a unique linear map F^χ such that the diagram

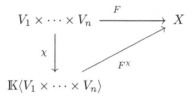

commutes. Furthermore, we claim that $\ker(F^\chi) \supseteq W$, where we have denoted by W (as done in Sect. 2.1.3) the subspace of $\mathbb{K}\langle V_1 \times \cdots \times V_n \rangle$ generated by the elements of the form

$$\chi(v_1, \ldots, a\, v_i, \ldots, v_n) - a\, \chi(v_1, \ldots, v_i, \ldots, v_n),$$

$$\chi(v_1, \ldots, v_i + v_i', \ldots, v_n) - \chi(v_1, \ldots, v_i, \ldots, v_n) - \chi(v_1, \ldots, v_i', \ldots, v_n),$$

where $v_i, v_i' \in V_i$, and $a \in \mathbb{K}$. The claim follows from a straightforward computation:

$$F^\chi(\chi(v_1, \ldots, a\, v_i, \ldots, v_n) - a\, \chi(v_1, \ldots, v_i, \ldots, v_n))$$

$$= F^\chi(\chi(v_1, \ldots, a\, v_i, \ldots, v_n)) - a\, F^\chi(\chi(v_1, \ldots, v_i, \ldots, v_n))$$

$$= F(v_1, \ldots, a\, v_i, \ldots, v_n) - a\, F(v_1, \ldots, v_n) = 0.$$

The fact that

$$\chi(v_1,\ldots,v_i+v_i',\ldots,v_n) - \chi(v_1,\ldots,v_i,\ldots,v_n) - \chi(v_1,\ldots,v_i',\ldots,v_n)$$

belongs to $\ker(F^\chi)$ is analogous and its proof is left to the Reader. It is then possible to apply Proposition 2.2-(i) to derive the existence of a (unique) linear map $\widetilde{F^\chi} : V_1 \otimes \cdots \otimes V_n \to X$ making the following diagram commute:

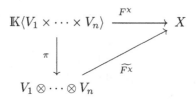

In particular,

$$\widetilde{F^\chi}(v_1 \otimes \cdots \otimes v_n) = \widetilde{F^\chi}(\pi(\chi(v_1,\ldots,v_n))) = F^\chi(\chi(v_1,\ldots v_n))) = F(v_1,\ldots,v_n),$$

so that the map $\widetilde{F^\chi}$ is precisely the map F^ψ we were looking for.

(ii) As $\psi : V_1 \times \cdots \times V_n \to V_1 \otimes \cdots \otimes V_n$ is an n-linear map, the assumptions of the hypothesis allow us to choose $X := V_1 \otimes \cdots \otimes V_n$ and $F := \psi$ and infer the existence of a unique ψ^φ making the following diagram commute:

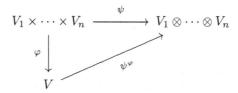

Analogously, for what was proved in (i), there exists one and only one linear map φ^ψ making the following diagram commute:

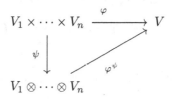

Now, as in the proof of Theorem 2.6, it suffices to show that ψ^φ and φ^ψ are inverse to each other. Indeed, on the one hand, we have

$$\psi^{\varphi}(\varphi^{\psi}(v_1 \otimes \cdots \otimes v_n)) = \psi^{\varphi}(\varphi^{\psi}(\psi(v_1, \ldots, v_n)))$$
$$= \psi^{\varphi}(\varphi(v_1, \ldots, v_n)) = \psi(v_1, \ldots, v_n) = v_1 \otimes \cdots \otimes v_n,$$

i.e., $\psi^{\varphi} \circ \varphi^{\psi} = \mathrm{Id}_{V_1 \otimes \cdots \otimes V_n}$ on a system of generators and consequently on $V_1 \otimes \cdots \otimes V_n$. On the other hand, it holds that

$$\varphi^{\psi}(\pi^{\varphi}(\varphi(v_1, \ldots, v_n))) = \varphi^{\psi}(\psi(v_1, \ldots, v_n)) = \varphi(v_1, \ldots, v_n).$$

So, in order to complete the proof, we need to show that V is generated by $\{\varphi(v_1, \ldots, v_n) | v_i \in V_i\}$. Consider again the diagram

$$
\begin{array}{ccc}
V_1 \times \cdots \times V_n & \xrightarrow{\;\;\psi\;\;} & V_1 \otimes \cdots \otimes V_n \\
{\scriptstyle \varphi} \downarrow & & \\
V & &
\end{array}
$$

which, as we have just seen, admits uniquely one linear closing map ψ^{φ}. If $\{\varphi(v_1, \ldots, v_n) | v_i \in V_i\}$ did not span V, it would be possible to find another linear closing map, simply by *arbitrarily* extending ψ^{φ} to a linear function defined on the whole of V. $\qquad\square$

Proof (of Theorem 2.31, page 74). It is trivial[2] to show that the set

$$\left\{ v_i \otimes w_k \right\}_{(i,k) \in \mathfrak{I} \times \mathfrak{K}}$$

generates $V \otimes W$. To prove the linear independence, we need the following three lemmas.

Lemma 7.1. *Let V, W be vector spaces and let $E \subseteq V$, $F \subseteq W$ be vector subspaces. Then the (abstract) tensor product $E \otimes F$ is isomorphic to the following subset of $V \otimes W$:*

$$\mathrm{span}\{e \otimes f \,:\, e \in E, \; f \in F\},$$

via the canonical map $E \otimes F \ni e \otimes f \mapsto e \otimes f \in V \otimes W$.

Proof. Let us denote by $\mathcal{V}$ the subset of $V \otimes W$ given by

$$\mathrm{span}\{e \otimes f \,:\, e \in E, \; f \in F\}.$$

---

[2] This is a consequence of the facts that $\{v_i\}_{i \in \mathfrak{I}}$ generates V, $\{w_k\}_{k \in \mathfrak{K}}$ generates W and that $\otimes$ is bilinear.

We will show that V with the map

$$\psi : E \times F \to V, \quad (e, f) \mapsto e \otimes f$$

has the universal property of the tensor product $E \otimes F$. To this end, let X be a vector space with a bilinear map $g : E \times F \to X$. We need to show that there exists a unique linear map g^ψ making the following diagram commute:

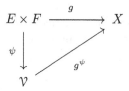

Let us consider any bilinear prolongation γ of g defined on the whole $V \times W$ (see Lemma 7.2 below). Given the map

$$\psi : V \times W \to V \otimes W$$

$$(v, w) \to v \otimes w,$$

there exists a unique γ^ψ making the following a commutative diagram:

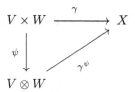

Actually, the restriction $\gamma^\psi|_V$ of γ to V verifies

$$\gamma^\psi|_V(e \otimes f) = \gamma^\psi(e \otimes f) = \gamma(e, f) = g(e, f),$$

thus it is the map g^ψ we aimed to exhibit.   □

Lemma 7.2. *Let V, W be vector spaces and let $E \subseteq V$, $F \subseteq W$ be vector subspaces. Let X be a vector space and let $g : E \times F \to X$ be a bilinear map. Then there exists $\tilde{g} : V \times W \to X$ bilinear, prolonging g.*

Proof. Let $\mathcal{E} := \{e_i\}_{i \in \mathcal{I}}$ and $\mathcal{J} := \{f_j\}_{j \in \mathcal{J}}$ be bases for E and F respectively. Let us extend them to bases $\mathcal{E}^* := \{e_i\}_{i \in \mathcal{I}^*}$, $\mathcal{J}^* := \{f_j\}_{j \in \mathcal{J}^*}$ of V and W (where $\mathcal{I}^* \supseteq \mathcal{I}$ and $\mathcal{J}^* \supseteq \mathcal{J}$). We define the function $\tilde{g} : V \times W \to X$ as follows:

$$\tilde{g}\left(\sum_{i \in \mathcal{I}^*} c_i e_i, \sum_{j \in \mathcal{J}^*} d_j f_j\right) := g\left(\sum_{i \in \mathcal{I}} c_i e_i, \sum_{j \in \mathcal{J}} d_j f_j\right).$$

Then $\widetilde{g}|_{E\times F} = g$ from the very definition of g; furthermore, $\widetilde{g}$ is bilinear:

$$\widetilde{g}(\alpha v + \beta v', w) = \widetilde{g}\left(\alpha \sum_{i\in\mathfrak{J}*} c_i e_i + \beta \sum_{i\in\mathfrak{J}*} c'_i e_i, \sum_{j\in\mathfrak{J}*} d_j f_j\right)$$
$$= g\left(\alpha \sum_{i\in\mathfrak{J}} c_i e_i + \beta \sum_{i\in\mathfrak{J}} c'_i e_i, \sum_{j\in\mathfrak{J}} d_j f_j\right)$$
$$= \alpha g\left(\sum_{i\in\mathfrak{J}} c_i e_i, \sum_{j\in\mathfrak{J}} d_j f_j\right) + \beta g\left(\sum_{i\in\mathfrak{J}} c'_i e_i, \sum_{j\in\mathfrak{J}} d_j f_j\right)$$
$$= \alpha \widetilde{g}(v, w) + \beta \widetilde{g}(v', w),$$

and analogously for the second variable. $\square$

Lemma 7.3. *Let V, W be finite-dimensional vector spaces with bases respectively $\{v_1, \ldots, v_p\}$ and $\{w_1, \ldots, w_q\}$. Then $\{v_i \otimes w_j : i = 1 \ldots, p, j = 1, \ldots, q\}$ is a basis for $V \otimes W$.*

Proof. Step 1. Let us begin with the construction of a suitable bilinear map

$$\Phi : V \times W \to \mathrm{Bil}(V, W),$$

where $\mathrm{Bil}(V, W)$ is the vector space of the bilinear functions from $V \times W$ to $\mathbb{K}$. The map Φ is defined as follows. Given

$$x = \sum_{i=1}^{p} x_i v_i \in V, \quad y = \sum_{j=1}^{q} y_i w_j \in W,$$

we define two linear maps $\mathrm{d}x : V \to \mathbb{K}$, $\mathrm{d}y : W \to \mathbb{K}$ as follows

$$\mathrm{d}x\left(\sum_{i=1}^{p} a_i v_i\right) := x_1 a_1 + \cdots + x_p a_p, \quad \mathrm{d}y\left(\sum_{j=1}^{p} b_j w_j\right) := y_1 b_1 + \cdots + y_q b_q.$$

As $\mathrm{d}x, \mathrm{d}y$ are linear, the map

$$\mathrm{d}x \times \mathrm{d}y : V \times W \to \mathbb{K}, \quad (a, b) \mapsto \mathrm{d}x(a) \cdot \mathrm{d}y(b)$$

is bilinear. We let

$$\Phi : V \times W \to \mathrm{Bil}(V, W), \quad (x, y) \mapsto \mathrm{d}x \times \mathrm{d}y.$$

It is immediately seen that Φ is (well posed and) bilinear in its turn.
    Hence there exists a unique $\Phi^\psi : V \otimes W \to \mathrm{Bil}(V, W)$ such that

$$\Phi^\psi(x \otimes y) = \mathrm{d}x \times \mathrm{d}y, \quad \forall\, x \in V, y \in W.$$

Step 2. We prove that the map Φ^ψ in Step 1 is a vector space isomorphism. Obviously, as i, j vary respectively in $\{1, \ldots, p\}$, $\{1, \ldots, q\}$, the elements of the form $\mathrm{d}v_i \times \mathrm{d}w_j$ generate $\mathrm{Bil}(V, W)$. On the other hand $\Phi^\psi(v_i \otimes w_j) = \mathrm{d}v_i \times \mathrm{d}w_j$, whence Φ^ψ is surjective. Furthermore, we claim that Φ^ψ admits an inverse, namely the following function

$$\Upsilon : \mathrm{Bil}(V, W) \longrightarrow V \otimes W$$

$$f \mapsto \sum_{i=1}^{p} \sum_{j=1}^{q} f(v_i, w_j)\, v_i \otimes w_j.$$

Indeed, given $\tau \in V \otimes W$, there exist scalars $\tau_{i,j} \in \mathbb{K}$ (whose possible non-uniqueness is presently immaterial) such that

$$\tau = \sum_{i=1}^{p} \sum_{j=1}^{q} \tau_{i,j}\, v_i \otimes w_j.$$

As a consequence,

$$(\Upsilon \circ \Phi^{\psi})(\tau) = \Upsilon\left(\sum_{i,j} \tau_{i,j}\, dv_i \times dw_j \right)$$

$$= \sum_{i=1}^{p} \sum_{j=1}^{q} \sum_{h,k} \tau_{h,k}\, (dv_h \times dw_k)(v_i, w_j)\, v_i \otimes w_j$$

$$= \sum_{i,j} \tau_{i,j}\, v_i \otimes w_j = \tau.$$

On the other hand, for every $f \in \mathrm{Bil}(V, W)$ we have

$$(\Phi^{\psi} \circ \Upsilon)(f) = \Phi^{\psi}\left(\sum_{i,j} f(v_i, w_j) v_i \otimes w_j \right) = \sum_{i,j} f(v_i, w_j)\, dv_i \times dw_j,$$

and the right-hand side coincides with f, as a direct computation shows (by expressing f as a linear combination of the generating elements $dv_i \times dw_j$). Finally, we deduce from the above facts that Φ^{ψ} is actually an isomorphism, with inverse Υ.

   Step 3. From what we have already proved of Theorem 2.31, the set $\{v_i \otimes w_j\}_{1 \le i \le p, 1 \le j \le q}$ spans $V \otimes W$. Now, in view of Step 2,

$$\dim(V \otimes W) = \dim(\mathrm{Bil}(V, W)) = pq \quad \text{(from basic Linear Algebra),}$$

and the latter coincides with the cardinality of

$$\{v_i \otimes w_j : i = 1 \ldots, p, j = 1, \ldots, q\}.$$

So these vectors are necessarily linear independent and thus give a basis for $V \otimes W$.   □

We now proceed with the proof of Theorem 2.31. We are left to prove that the vectors $\{v_i \otimes w_k\}_{(i,k) \in \mathfrak{J} \times \mathfrak{K}}$ are linearly independent. This will be the case iff, for every finite set $\mathcal{H} \subseteq \mathcal{J} \times \mathcal{K}$, the vectors $\{v_i \otimes w_k\}_{(i,k)\in\mathcal{H}}$ are linearly independent. We can assume, without loss of generality by simply "enlarging" $\mathcal{H}$ if necessary, that $\mathcal{H}$ has the following "rectangular" form:

$$\mathcal{H} := \mathcal{J}_p \times \mathcal{K}_q,$$

where $\mathcal{J}_p := \{i_1, \ldots, i_p\} \subseteq \mathcal{J}$ and $\mathcal{K}_q := \{k_1, \ldots, k_q\} \subseteq \mathcal{K}$. Let

$$E_p := \mathrm{span}\{v_i \mid i \in \mathcal{J}_p\}; \qquad F_q := \mathrm{span}\{w_k \mid k \in \mathcal{K}_q\}$$

By Lemma 7.1, we have

$$E_p \otimes F_q \cong \mathrm{span}\{e \otimes f \mid e \in E_p, f \in F_q\} =: \mathcal{V}_{p,q}.$$

By Lemma 7.3, the set $\{v_i \otimes w_k\}_{(i,k)\in\mathcal{J}_p \times \mathcal{K}_q}$ is a basis for $E_p \otimes F_q$. This means that, via the canonical map

$$E_p \otimes F_q \longleftrightarrow \mathcal{V}_{p,q} \subseteq V \otimes W$$
$$e \otimes f \leftrightarrow e \otimes f$$

the vectors $\{v_i \otimes w_k\}_{(i,k)\in\mathcal{J}_p \times \mathcal{K}_q}$ form a basis for $\mathcal{V}_{p,q}$ and are thus linearly independent. This argument holds for every choice of finite $\mathcal{J}_p \subseteq \mathcal{J}$ and finite $\mathcal{K}_q \subseteq \mathcal{K}$, so the theorem is completely proved.   □

Proof (of Theorem 2.32, page 75). Let

$$\mathcal{B}_k := \{v_i^k\}_{i\in I^k}, \qquad \mathcal{C}_k := \{w_j^k\}_{j\in J^k}$$

be bases, respectively, for V_k (with $k = 1, \ldots, n$) and for W_k (with $k = 1, \ldots, m$). When i_k varies in I^k (for any $k = 1, \ldots, n$) and when j_k varies in J^k (for any $k = 1, \ldots, m$), by Theorem 2.31, we deduce that the vectors

$$(v_{i_1}^1 \otimes \cdots \otimes v_{i_n}^n) \otimes (w_{j_1}^1 \otimes \cdots \otimes w_{j_m}^m)$$

constitute a basis for $(V_1 \otimes \cdots \otimes V_n) \otimes (W_1 \otimes \cdots \otimes W_m)$. Hence there is a unique linear map

$$L : (V_1 \otimes \cdots \otimes V_n) \otimes (W_1 \otimes \cdots \otimes W_m) \to V_1 \otimes \cdots \otimes V_n \otimes W_1 \otimes \cdots \otimes W_m$$

verifying (for every indices i and j)

$$L\left((v_{i_1}^1 \otimes \cdots \otimes v_{i_n}^n) \otimes (w_{j_1}^1 \otimes \cdots \otimes w_{j_m}^m)\right) = v_{i_1}^1 \otimes \cdots \otimes v_{i_n}^n \otimes w_{j_1}^1 \otimes \cdots \otimes w_{j_m}^m.$$

This is clearly a vector space isomorphism, because another application of Theorem 2.31 ensures that the vectors

$$v_{i_1}^1 \otimes \cdots \otimes v_{i_n}^n \otimes w_{j_1}^1 \otimes \cdots \otimes w_{j_m}^m$$

(when the indices i and j vary as above) constitute a basis for $V_1 \otimes \cdots \otimes V_n \otimes W_1 \otimes \cdots \otimes W_m$. Moreover, the inverse L^{-1} of L is clearly defined by

$$v_{i_1}^1 \otimes \cdots \otimes v_{i_n}^n \otimes w_{j_1}^1 \otimes \cdots \otimes w_{j_m}^m \mapsto (v_{i_1}^1 \otimes \cdots \otimes v_{i_n}^n) \otimes (w_{j_1}^1 \otimes \cdots \otimes w_{j_m}^m).$$

Using the linearity of L and the linearity of $\otimes$, it is easy to show that L acts on every elementary tensor of its domain as it acts on the elements of the basis:

$$L\Big((v^1 \otimes \cdots \otimes v^n) \otimes (w^1 \otimes \cdots \otimes w^m)\Big) = v^1 \otimes \cdots \otimes v^n \otimes w^1 \otimes \cdots \otimes w^m,$$

for every $v^i \in V_i$, $w^j \in W_j$. This ends the proof.   □

Proof (of Theorem 2.38, page 77). (i) Clearly, there exists at most one map $\overline{f}$ as in Theorem 2.38-(i), namely the map $\overline{f}$ as in (2.31).

To prove existence, we argue as follows. For every $k \in \mathbb{N}$, let us consider the map

$$f_k : \underbrace{V \times \cdots \times V}_{k \text{ times}} \to A$$

defined by $f_k(v_1, \ldots, v_k) := f(v_1) \star \cdots \star f(v_k)$, where $\star$ is the algebra operation of A. As f is linear, f_k is multilinear, and hence (see Theorem 2.30-(i)) there is a unique linear map

$$f_k^\psi : \underbrace{V \otimes \cdots \otimes V}_{k \text{ times}} \to A$$

such that $f_k^\psi(v_1 \otimes \cdots \otimes v_k) = f_k(v_1, \ldots, v_k) = f(v_1) \star \cdots \star f(v_k)$. As a consequence of Theorem 2.8, there exists a unique linear map

$$f_\Sigma : \bigoplus_{k \geq 1} \underbrace{V \otimes \cdots \otimes V}_{k \text{ times}} = \mathscr{T}_+(V) \to A$$

such that, for every $k \in \mathbb{N}$,

$$f_\Sigma(v_1 \otimes \cdots \otimes v_k) = f_k^\psi(v_1 \otimes \cdots \otimes v_k) = f(v_1) \star \cdots \star f(v_k).$$

This immediately proves that f_Σ is an algebra morphism and that (taking $k = 1$) $f_\Sigma|_V$ prolongs f. Finally, the choice $\bar{f} := f_\Sigma$ furnishes a map as desired.

(ii) If A is endowed with a unit element 1_A, besides the maps f_k introduced in point (i), we must consider the linear map

$$f_0 : \mathbb{K} \to A$$
$$\alpha \mapsto \alpha \cdot 1_A$$

and we derive – as above – the existence of a (unique) linear map

$$f_\Sigma : \bigoplus_{k \geq 0} \underbrace{V \otimes \cdots \otimes V}_{k \text{ times}} = \mathscr{T}(V) \to A$$

verifying $f_\Sigma(v_1 \otimes \cdots \otimes v_k) = f(v_1) \star \cdots \star f(v_k)$ (for $k \in \mathbb{N}$) and $f_\Sigma(1_\mathbb{K}) = 1_A$. The choice $\bar{f} := f_\Sigma$ again ends the proof (furnishing this time a UAA morphism).

(iii) Let us consider the inclusion map

$$\iota : V \hookrightarrow \mathscr{T}(V).$$

From our hypothesis, there exists one and only one UAA morphism ι^φ such that the diagram

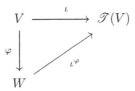

commutes. On the other hand, part (ii) of the theorem ensures that there exists a unique $\bar{\varphi} : \mathscr{T}(V) \to W$ such that the diagram

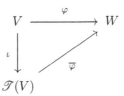

commutes. As usual, we will show that ι^φ and $\bar{\varphi}$ are inverse to each other. This is easily seen by checking that $\iota^\varphi \circ \bar{\varphi}$ and $\bar{\varphi} \circ \iota^\varphi$ close the following two trivial diagrams:

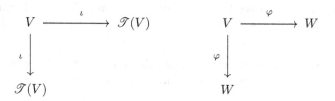

which uniquely admit $\mathrm{Id}_{\mathscr{T}(V)}$ and Id_W, respectively, as closing linear maps.

Let us now prove that φ is injective. Given $v, w \in V$ such that $\varphi(v) = \varphi(w)$, we have

$$\overline{\varphi}(v) = \overline{\varphi}(\iota(v)) = \varphi(v) = \varphi(w) = \overline{\varphi}(\iota(w)) = \overline{\varphi}(w),$$

and the result follows from the injectivity of $\overline{\varphi}$.

Furthermore, let us prove that W is generated as an algebra by the set $\{1_{\mathbb{K}}\} \cup \varphi(V)$. Indeed, let us first remark that $\overline{\varphi} \circ \iota^\varphi$ is the identity element of W, so that, for every $w \in W$, we have

$$w = \overline{\varphi}(\iota^\varphi(w)). \tag{7.2}$$

Since $\iota^\varphi(w) \in \mathscr{T}(V)$, we have

$$\iota^\varphi(w) = c \cdot 1_{\mathbb{K}} + \sum_{k=1}^{N} v_1^k \otimes \cdots \otimes v_k^k,$$

where $c \in \mathbb{K}$ and all the v_i^k are in V. Hence, recalling that $\overline{\varphi}$ is a UAA morphism prolonging φ, we deduce from (7.2) (denoting by $\circledast$ the algebra operation on W)

$$w = \overline{\varphi}\left(c \cdot 1_{\mathbb{K}} + \sum_{k=1}^{N} v_1^k \otimes \cdots \otimes v_k^k\right)$$

$$= c \cdot \overline{\varphi}(1_{\mathbb{K}}) + \sum_{k=1}^{N} \overline{\varphi}(v_1^k) \circledast \cdots \circledast \overline{\varphi}(v_k^k)$$

$$= c \cdot 1_W + \sum_{k=1}^{N} \varphi(v_1^k) \circledast \cdots \circledast \varphi(v_k^k).$$

Clearly the above far right-hand side is an element of the algebra generated by $\{1_W\} \cup \varphi(V)$. The arbitrariness of $w \in W$ completes the argument.

Finally, in order to prove the canonical isomorphism $W \simeq \mathscr{T}(\varphi(V))$, it suffices to show that, if $\iota : \varphi(V) \to W$ is the set inclusion, the couple (W, ι) has the universal property of the tensor algebra $\mathscr{T}(\varphi(V))$. To this aim, let A be any UA algebra and let $f : \varphi(V) \to A$ be any linear map. We need to show the existence of a unique UAA morphism $f^\iota : W \to A$ such that $f^\iota \circ \iota \equiv f$ on $\varphi(V)$, i.e.,

$$(\bigstar) \qquad\qquad (f^\iota \circ \iota)(\varphi(v)) = f(\varphi(v)), \quad \forall v \in V.$$

The uniqueness of such a UAA morphism f^ι follows from the fact that $\{1_W\} \cup \varphi(V)$ generates W, as an algebra. Let us turn to its existence: We consider the linear map $f \circ \varphi : V \to A$. By the first part of (iii), we know that $(f \circ \varphi)^\varphi : W \to A$ is a UAA morphism such that $(f \circ \varphi)^\varphi(\varphi(v)) = (f \circ \varphi)(v)$ for every $v \in V$. If we set $f^\iota := (f \circ \varphi)^\varphi$ then we are done with $(\bigstar)$. □

7.4 Proofs of Sect. 2.3.1

Proof (of Theorem 2.58, page 94). In order to stress the contribution of each of the hypotheses (H1), (H2), (H3), (H4) on page 94, we will split the statement 2.58 into two new statements (Theorems 7.5 and 7.7 below). The proof of Theorem 2.58 follows from these results.

We need first of all the following (not standard) definition:

Definition 7.4. The couple (A, Ω) is a *semi topological algebra* if $(A, +, *)$ is an associative algebra and Ω is a topology on A such that the maps

$$A \times A \ni (a, b) \mapsto a + b, a * b \in A, \quad \mathbb{K} \times A \ni (k, a) \mapsto k\, a \in A$$

are continuous in the associated product topologies, $\mathbb{K}$ being equipped with the discrete topology. We remark that a semi topological algebra is a topological algebra (see footnote on page 94) iff it is Hausdorff.

Theorem 7.5. *Let $(A, *)$ be an associative algebra and let $\{\Omega_k\}_{k \in \mathbb{N}}$ be a family of subsets of A verifying conditions* (H1), (H2), (H3) *on page 94. Then the family*

$$\mathcal{B} := \emptyset \cup \left\{ a + \Omega_k \right\}_{a \in A,\ k \in \mathbb{N}} \tag{7.3}$$

is a basis for a topology Ω on A endowing A with the structure of a semi topological algebra. Even more, the topology Ω is induced by the semimetric[3] $d : A \times A \to [0, \infty)$ defined as follows (posing $\exp(-\infty) := 0$)

$$d(x, y) := \exp(-\nu(x - y)), \quad \text{for all } x, y \in A, \tag{7.4}$$

---

[3]A semimetric d is defined in the same way as a metric, except that the proprty "$x = y \Leftrightarrow d(x, y) = 0$" is replaced by "$x = y \Rightarrow d(x, y) = 0$".

where $\nu : A \to \mathbb{N} \cup \{0, \infty\}$ is defined by $\nu(z) := \sup \{n \geq 1 \mid z \in \Omega_n\}$, i.e.,

$$\nu(z) := \begin{cases} \text{if } z \neq 0, & \max \{n \geq 1 \mid z \in \Omega_n\} \\ \text{if } z = 0, & \infty. \end{cases} \tag{7.5}$$

The triangle inequality for d holds in the stronger form:

$$d(x, y) \leq \max\{d(x, z), d(z, y)\}, \quad \text{for every } x, y, z \in A. \tag{7.6}$$

Finally, we have

$$d(x, y) = 0 \quad \Longleftrightarrow \quad x - y \in \bigcap_{n \in \mathbb{N}} \Omega_n = \overline{\{0\}}^{\Omega}. \tag{7.7}$$

Proof. This proof is a little laborious, so, for the Reader's convenience, we will spilt it into several steps.

Step 1. Let us check that the family $\mathcal{B}$ has the properties of a basis for a topology Ω. It is clear that the union of the sets in $\mathcal{B}$ covers A. So it suffices to prove that, given $a, b \in A$, and $h, k \in \mathbb{N}$, the following fact holds:

for every $x \in (a + \Omega_k) \cap (b + \Omega_h)$ there exists $B \in \mathcal{B}$

such that
$$x \in B \subseteq (a + \Omega_k) \cap (b + \Omega_h).$$

Now, x can be written in this two ways:

$$x = a + \omega_k = b + \omega_h,$$

where $\omega_k \in \Omega_k$ and $\omega_h \in \Omega_h$. Let us define $j := \max\{h, k\}$, and let us choose $B := x + \Omega_j$. Then B contains x. Also, for every $\omega_j \in \Omega_j$, we have

$$x + \omega_j = a + \omega_k + \omega_j \in a + \Omega_k,$$

thanks to properties (H2), (H3). Analogously, we have $x \in b + \Omega_h$, and this concludes the proof that $\mathcal{B}$ is a basis for a topology (which is necessarily unique: its open sets are the unions of the subfamilies of $\mathcal{B}$).

Step 2. We will now show that (A, Ω) is a semi topological algebra, i.e., the operations

$$A \times A \to A \qquad\qquad \mathbb{K} \times A \to A \qquad\qquad A \times A \to A$$

$$(a, b) \mapsto a + b, \qquad\qquad (k, a) \mapsto k \cdot a, \qquad\qquad (a, b) \mapsto a * b,$$

are continuous in the associated topologies (we fix the topology Ω on A, the discrete topology on $\mathbb{K}$, and the product topology on each of the Cartesian products involved above).

Let us start with the sum. We need to check that, given an open set ϑ of A, its inverse image via the operation $+$ is an open set too. Let us define

$$\Delta := \{(x, y) \in A \mid x + y \in \vartheta\}.$$

If $(x_0, y_0) \in \Delta$, then there certainly exist $a \in A$, $h \in \mathbb{N}$ such that

$$x_0 + y_0 \in a + \Omega_h \subseteq \vartheta.$$

If we set

$$\vartheta_1 := x_0 + \Omega_h, \quad \vartheta_2 := y_0 + \Omega_h,$$

we have, by property (H1),

$$\vartheta_1 + \vartheta_2 = x_0 + y_0 + \Omega_h \subseteq a + \Omega_h \subseteq \vartheta.$$

In this way we have found an open set $\vartheta_1 \times \vartheta_2$ such that

$$(x_0, y_0) \in \vartheta_1 \times \vartheta_2 \subseteq \Delta,$$

so we have proved that Δ is open in the topology Ω.

As for multiplication by a scalar element, let ϑ be an open set of A. Given $(k_0, x_0) \in \Delta := \{(k, a) \in \mathbb{K} \times A \mid k \cdot a \in \vartheta\}$, we look for an open subset of Δ containing (k_0, x_0). As above, there certainly exist $a \in A$, $h \in \mathbb{N}$ such that

$$k_0 \cdot x_0 \in a + \Omega_h \subseteq \vartheta.$$

So it suffices to choose $\vartheta_1 \times \vartheta_2$, where

$$\vartheta_1 := \{k_0\}, \quad \vartheta_2 := x_0 + \Omega_h.$$

Indeed, with this choice we have (Ω_k being an ideal)

$$\vartheta_1 \cdot \vartheta_2 = k_0 \cdot (x_0 + \Omega_h) = k_0 \cdot x_0 + \Omega_h \subseteq a + \Omega_h \subseteq \vartheta.$$

Finally, the continuity of the algebra operation $*$ can be proved as in the case of the sum, and this verification is left to the Reader.

Step 3. Let us now check that the function d defined in the statement is a semimetric, i.e., it satisfies:

(SM.1)   For every $x, y \in A$, $d(x, y) \geq 0$ and $d(x, x) = 0$.
(SM.2)   For every $x, y \in A$, $d(x, y) = d(y, x)$.
(SM.3)   For every $x, y, z \in A$, $d(x, y) \leq d(x, z) + d(z, y)$.

(SM.1): The equality $d(x, x) = 0$ for every $x \in A$ follows immediately from $\nu(0) = +\infty$.

(SM.2): Let $z \in A$. Then thanks to property (H1)

$$z \in \Omega_n \iff -z \in \Omega_n.$$

From the definition of ν, we thus have

$$\nu(z) = \nu(-z) \quad \forall z \in A,$$

which immediately leads to $d(x, y) = d(y, x)$, for every $x, y \in A$.

(SM.3): As for the triangle inequality, an even stronger property holds for d: For every $x, y, z \in A$, we have

$$d(x, y) \leq \max\{d(x, z), d(z, y)\} \qquad \big(\leq d(x, z) + d(z, y)\big).$$

In other words, (A, d) is an ultrametric space. To prove this, let us assume that, for example, $\max\{d(x, z), d(z, y)\} = d(x, z)$. This means that

$$(\bigstar) \qquad \nu(x - z) \leq \nu(z - y).$$

Let us write $x - y$ as

$$x - y = (x - z) + (z - y).$$

Observing that $x - z \in \Omega_{\nu(x-z)}$, while $z - y \in \Omega_{\nu(z-y)} \subseteq \Omega_{\nu(x-z)}$, the property $(\bigstar)$ implies that $x - y$ belongs to $\Omega_{\nu(x-z)}$, whence $\nu(x - y) \geq \nu(x - z)$. We can thus conclude

$$d(x, y) \leq d(x, z) = \max\{d(x, z), d(z, y)\}.$$

Step 4. For the sake of completeness (the fact being well-known for metric spaces), let us now see why *a semimetric d induces a topology by means of its open balls* (and the empty set). For every $x \in A$, and for every positive $r \in \mathbb{R}$, let us define

$$B_d(x, r) := \{y \in A \mid d(x, r) < r\}.$$

We claim that the family

$$\mathcal{B}_d := \emptyset \cup \{B_d(x, r)\}_{x \in A, r > 0}$$

is a basis for a topology, say Ω_d. It is trivial that, for every $a \in A$, there exists a set in $\mathcal{B}_d$ containing a. Let us consider the intersection of two generic sets in $\mathcal{B}_d$. Let $a \in B_d(x, r) \cap B_d(y, s)$, and let us define

$$\rho := \min\{r - d(x, a), s - d(y, a)\} > 0.$$

We claim that

$$B_d(a, \rho) \subseteq B_d(x, r) \cap B_d(y, s).$$

Indeed, for every $b \in B_d(a, \rho)$, we have

$$d(b, x) \leq d(b, a) + d(a, x) < \rho + d(a, x)$$
$$\leq r - d(x, a) + d(a, x) = r,$$

so $B_d(a, \rho) \subseteq B_d(x, r)$. The inclusion $B_d(a, \rho) \subseteq B_d(y, s)$ can be shown analogously, and the proof is complete.

Step 5. We now check that the two couples (A, Ω) and (A, Ω_d) coincide as topological spaces. This requires the following

Lemma 7.6. *The function ν defined in (7.5) verifies*

$$\nu(a + b) \geq \min\{\nu(a), \nu(b)\}.$$

Proof. Of course $a \in \Omega_{\nu(a)}$, $b \in \Omega_{\nu(b)}$, so that

$$a + b \in \Omega_{\nu(a)} + \Omega_{\nu(b)} \subseteq \Omega_{\min\{\nu(a), \nu(b)\}}.$$

By the definition of ν, this means that $\nu(a + b) \geq \min\{\nu(a), \nu(b)\}$.   □

Let us now compare Ω and Ω_d. First, let us show that every element of the basis $\mathcal{B}_d$ of the topology Ω_d is an open set in the topology Ω.
  With the notation introduced in Step 4, let $y \in B_d(x, r)$. We have to prove:

$$\exists\, h \in \mathbb{N} : \quad y + \Omega_h \subseteq B_d(x, r). \tag{7.8}$$

We consider two different cases:

- If $d(x, y) = 0$, then $\nu(x - y) = \infty$, i.e.,

$$x - y \in \bigcap_{h \in \mathbb{N}} \Omega_h.$$

So, for every $h \in \mathbb{N}$ and every $w_h \in \Omega_h$, we have

$$y + w_h = x + \underbrace{(y - x)}_{\in \Omega_h} + \underbrace{w_h}_{\in \Omega_h} \in x + \Omega_h.$$

We can thus write, for some $\widetilde{\omega}_h \in \Omega_h$,

$$y + \omega_h = x + \widetilde{\omega}_h,$$

so we can obtain

$$d(y + \omega_h, x) = d(x + \widetilde{\omega}_h, x) = \exp(-\nu(x + \widetilde{\omega}_h - x))$$
$$= \exp(-\nu(\widetilde{\omega}_h)) \leq \exp(-h).$$

Finally, it suffices to choose h large enough to have $e^{-h} < r$ and (7.8) follows.

• If $d(x, y) \neq 0$, let us argue by contradiction and suppose that (7.8) does not hold, i.e.,

$$\forall h \in \mathbb{N}\ \exists \omega_h \in \Omega_h \quad \text{such that} \quad y + \omega_h \notin B_d(x, r).$$

Then the following chain of equivalences holds:

$$d(y + \omega_h, x) \geq r \Longleftrightarrow \exp(-\nu(y + \omega_h - x)) \geq r \Longleftrightarrow$$
$$- \nu(y + \omega_h - x) \geq \ln(r) \Longleftrightarrow \nu(\omega_h + y - x) \leq -\ln(r). \tag{7.9}$$

By Lemma 7.6,

$$\nu(\omega_h + y - x) \geq \min\{\nu(\omega_h), \nu(y - x)\} \geq \min\{h, \nu(y - x)\},$$

so that (by exploiting the far right-hand side of (7.9))

$$\min\{h, \nu(y - x)\} \leq \ln(r).$$

Allowing h tend to ∞, this shows that

$$d(x, y) = \exp(\nu(x - y)) \geq r,$$

which contradicts our assumption.

The last part of Step 5 consists of showing the reverse inclusion: every element of the basis $\mathcal{B}$ of the topology Ω is an open set of the topology Ω_d. To this end, let us consider the set $a + \Omega_h$. It suffices to prove:

$$\forall \omega_h \in \Omega_h,\ \exists r > 0 \quad \text{such that} \quad B_d(a + \omega_h, r) \subseteq a + \Omega_h. \tag{7.10}$$

We argue by contradiction, supposing that there exists $\omega_h \in \Omega_h$ such that, for all $r > 0$ it holds that $B_d(a + \omega_h, r) \not\subseteq a + \Omega_h$. This is equivalent to the existence of $\omega_h \in \Omega_h$ such that

$$\forall r > 0 \ \exists z \in B_d(a + \omega_h, r) \quad \text{such that} \quad z \notin a + \Omega_h.$$

So we have $d(z, a + \omega_h) < r$, but $z - a \notin \Omega_h$. Since

$$z - a = z - (a + \omega_h) + \omega_h,$$

and $\omega_h \in \Omega_h$, then $z - (a + \omega_h) \notin \Omega_h$ (otherwise $z - a$ would belong to Ω_h too, which is not true). We have thus derived $z - (a + \omega_h) \notin \Omega_h$. This yields

$$\nu(z - (a + \omega_h)) < h. \tag{7.11}$$

Indeed, if we had $\nu(z - (a + \omega_h)) \geq h$, the definition of ν would produce

$$z - (a + \omega_h) \in \Omega_{\nu(z-(a+\omega_h))} \subseteq \Omega_h.$$

Now, we have already remarked that $d(z, a + \omega_h) < r$, which is equivalent to

$$\exp(-\nu(z - (a + \omega_h))) < r,$$

and this is equivalent, in its turn, to

$$\nu((z - (a + \omega_h)) > -\ln(r). \tag{7.12}$$

Summing up:

$$h \overset{(7.11)}{>} \nu(z - (a + \omega_h)) \overset{(7.12)}{>} -\ln(r).$$

Hence $h > -\ln(r)$. Thanks to the arbitrariness of $r > 0$, letting $r \to 0^+$, we get $h = \infty$, which is absurd.

Step 6. We are left to prove that (7.7) holds. First of all, unraveling the definitions of d and ν, we recognize that

$$d(x, y) = 0 \ \Leftrightarrow \ \nu(x - y) = \infty \ \Leftrightarrow \ x - y \in \bigcap_{h \in \mathbb{N}} \Omega_h.$$

This gives the first part of (7.7). Finally, we have to prove

$$\bigcap_{n \in \mathbb{N}} \Omega_n = \overline{\{0\}}^{\Omega}.$$

- Let us prove the inclusion $\bigcap_{n \in \mathbb{N}} \Omega_n \subseteq \overline{\{0\}}^{\Omega}$.

  Given $\omega \in \bigcap_{n \in \mathbb{N}} \Omega_n$, let us assume, by contradiction, that $\omega \notin \overline{\{0\}}^{\Omega}$. Let

  $$\mathcal{F} := \{F \text{ is a closed subset of A w.r.t. } \Omega \text{ such that } 0 \in F\}.$$

As $\overline{\{0\}}^{\,\Omega} = \bigcap_{F\in\mathcal{F}} F$, there must exist a closed set F containing the origin but not ω:

$$(\star) \qquad \omega \in A\setminus F, \quad 0 \notin A\setminus F.$$

Since $A\setminus F$ is open, there exists $h\in\mathbb{N}$ such that $\omega + \Omega_h \subseteq A\setminus F$. We claim that $\omega \notin \Omega_h$: this contradicts our assumption and completes the proof. Indeed, if ω were in Ω_h, we would have

$$0 \in \Omega_h = \omega + \Omega_h \subseteq A\setminus F,$$

which is not possible, in view of $(\star)$.

- Let us now turn to check the second inclusion $\bigcap_{n\in\mathbb{N}} \Omega_n \supseteq \overline{\{0\}}^{\,\Omega}$. Again, we will argue by contradiction: given $\omega \in \overline{\{0\}}^{\,\Omega}$, let us assume that $\omega \notin \bigcap_{n\in\mathbb{N}} \Omega_n$. Let us choose

$$h \in \mathbb{N} \text{ such that } \omega \notin \Omega_h.$$

Let us define

$$\widetilde{F} := A \setminus (\omega + \Omega_h).$$

We have $0 \in \widetilde{F}$, otherwise $\omega \in \Omega_h$ would hold. Thus $\widetilde{F}$ is a closed set containing the origin. We now show that $\omega \notin \widetilde{F}$. This will give a contradiction with $\omega \in \overline{\{0\}}^{\,\Omega} = \bigcap_{F\in\mathcal{F}} F$. But $\omega \notin \widetilde{F}$ is immediately proved: if ω were in $\widetilde{F}$,

$$\widetilde{F} \ni \omega = \omega + 0 \in \omega + \Omega_h = A\setminus\widetilde{F},$$

which is clearly a contradiction.

The proof is now complete. $\square$

Theorem 7.7. *Let $(A, *)$ be an associative algebra and let $\{\Omega_k\}_{k\in\mathbb{N}}$ be a family of subsets of A verifying conditions* (H1), (H2), (H3), (H4) *on page 94.*

Then, following the notation of Theorem 7.5, (A, Ω) is a topological algebra. Furthermore, the semimetric d in (7.4) is a metric inducing the topology Ω and it endows A with the structure of an ultrametric space (see (7.6)).

Proof. In view of Theorem 7.5, all we have to do is to show that, as long as property (H4) holds, A is a *Hausdorff* space.

We remark that, in view of (7.7), the semimetric d introduced in (7.4) is a metric if and only if $\bigcap_{n\in\mathbb{N}} \Omega_n = \{0\}$, which is precisely condition (H4). So it suffices to prove that *a semimetric space is Hausdorff if and only if it is a metric space.* This is proved as follows.

Let (A, d) be a metric space. Let us consider $a, b \in A$ with $a \neq b$. Their distance $d(a, b)$ must then be positive. Let us define

$$r := \tfrac{1}{3} d(a, b).$$

We claim that

$$B_d(a, r) \cap B_d(b, r) = \emptyset.$$

Indeed, if $\xi \in B_d(a, r) \cap B_d(b, r)$, we would have

$$d(a, b) \leq d(a, \xi) + d(\xi, b) < 2r = \tfrac{2}{3} d(a, b),$$

which is a contradiction. Thus we have found two disjoint neighborhoods of a and b, whence (A, d) is Hausdorff.

Suppose now that (A, d) is a Hausdorff semimetric space. Let us assume on the contrary that there exist $a, b \in A$ such that $a \neq b$ and $d(a, b) = 0$. By the Hausdorff condition, there exist two disjoint d-balls $B_d(a, \varepsilon)$ and $B_d(b, \varepsilon)$. Given $\xi \in B_d(a, \varepsilon)$, the following fact holds:

$$d(\xi, b) \leq d(\xi, a) + d(a, b) = d(\xi, a) < \varepsilon \Rightarrow \xi \in B_d(b, \varepsilon).$$

This contradicts the fact that the intersection $B_d(a, \varepsilon) \cap B_d(b, \varepsilon)$ is empty. $\square$

Proof (of Remark 2.61, page 95). We let $x, y, z, \xi, \eta \in A$, $k \in \mathbb{K} \setminus \{0\}$.

(1) By the definition of d, we have

$$d(x + z, y + z) = \exp\left(-\sup\{n \geq 0 \,|\, (x + z) - (y + z) \in \Omega_n\}\right)$$
$$= \exp\left(-\sup\{n \geq 0 \,|\, x - y \in \Omega_n\}\right) = d(x, y).$$

(2) Again, by the definition of d, we have

$$d(k\,x, k\,y) = \exp\left(-\sup\{n \geq 0 \,|\, k\,x - k\,y \in \Omega_n\}\right)$$
$$= \exp\left(-\sup\{n \geq 0 \,|\, x - y \in \Omega_n\}\right) = d(x, y).$$

Indeed, in the second equality we used the fact that the sets Ω_n are vector spaces (being ideals) and the fact that $k \neq 0$.

(3) Let us recall that

$$\nu(x * y, \xi * \eta) = \sup\{n \in \mathbb{N} \,|\, x * y - \xi * \eta \in \Omega_n\}.$$

The proof is an easy computation:

$$x * y - \xi * \eta = x * y - \xi * \eta + \xi * y - \xi * y = (x - \xi) * y + \xi * (y - \eta).$$

The definition of ν ensures that the far right-hand side belongs to

$$\Omega_{\nu(x-\xi)} * y + \xi * \Omega_{\nu(y-\eta)} \overset{(H1)}{=} \Omega_{\nu(x-\xi)} + \Omega_{\nu(y-\eta)} \overset{(H2)}{\subseteq} \Omega_{\min\{\nu(\xi-x),\nu(\eta-y)\}}.$$

Summing up, we have proved that $x * y - \xi * \eta \in \Omega_{\min\{\nu(\xi-x),\nu(\eta-y)\}}$: this means that $\nu(x * y, \xi * \eta) \geq \min\{\nu(x - \xi), \nu(y - \eta)\}$. So we have

$$\begin{aligned}
d(x * y, \xi * \eta) &= \exp(-\nu(x * y, \xi * \eta)) \\
&\leq \exp\left(-\min\{\nu(x - \xi), \nu(y - \eta)\}\right) \\
&= \exp\left(\max\{-\nu(x - \xi), -\nu(y - \eta)\}\right) \\
&= \max\left\{\exp(-\nu(x - \xi)), \exp(-\nu(y - \eta))\right\} \\
&= \max\{d(x, \xi), d(y, \eta)\},
\end{aligned}$$

as we claimed.   □

Proof (of Proposition 2.65, page 97). (a) follows from Theorem 2.58; (b) follows from Remark 2.61; (c) follows from Remark 2.62. We are left to prove (d), (e).

(d) Let $z = (z_j)_{j \geq 0} \in A$. Since $z \in \Omega_j$ iff $z_0 = \cdots = z_{j-1} = 0$, it is easily seen that

$$\begin{aligned}
\nu(z) &= \begin{cases} \max\{j \geq 0 \mid z_0 = \cdots = z_{j-1} = 0\}, & \text{if } z \neq 0, \\ \infty, & \text{if } z = 0 \end{cases} \\
&= \begin{cases} \min\{j \geq 0 \mid z_j \neq 0\}, & \text{if } z \neq 0, \\ \infty, & \text{if } z = 0. \end{cases}
\end{aligned} \tag{7.13}$$

Then (2.72) follows from $d(z) = \exp(-\nu(z))$ (with the usual convention $\exp(-\infty) := 0$).

(e) With the notation of the assertion, it is obviously not restrictive to suppose that $\beta = 0$. We have $\lim_{n \to \infty} b_n = 0$ iff $\lim_{n \to \infty} d(b_n) = 0$, that is, (by $d(z) = \exp(-\nu(z))$) $\lim_{n \to \infty} \nu(b_n) = \infty$, or equivalently (by definition of limit!):

$$(\bigstar) \qquad \forall J \geq 0 \; \exists N_J \in \mathbb{N} : \quad n \geq N_J \text{ implies } \nu(b_n) \geq J + 1.$$

Now, by the first equality in (7.13) and the notation $b_n = (a_j^{(n)})_{j \geq 0}$, we see that $(\bigstar)$ is equivalent to

$$(2\bigstar) \quad \forall J \geq 0 \; \exists N_J \in \mathbb{N} : \quad n \geq N_J \text{ implies } a_j^{(n)} = 0 \text{ for } j = 0, \ldots, J.$$

This is (2.73) under our non-restrictive assumption $\beta = (a_j)_{j \geq 0} = 0$.   □

7.5 Proofs of Sect. 2.3.2

Proof (of Theorem 2.67, page 98). If (Y_1, δ_1) and (Y_2, δ_2) are isometric completions of (X, d), there exist two metric spaces $Y_{1,0}$ and $Y_{2,0}$, respectively subspaces of Y_1 and Y_2, which are dense in the corresponding spaces and such that $(Y_{1,0}, \delta_1)$ and $(Y_{2,0}, \delta_2)$ are both isometric (in the sense of metric spaces) to (X, d). Let us call α and β the two isometries:

$$\alpha : X \to Y_{1,0} \quad \text{and} \quad \beta : X \to Y_{2,0}.$$

Notice that, for every $\eta, \eta' \in Y_{1,0}$ it holds that

$$\delta_2\left((\beta \circ \alpha^{-1})(\eta), (\beta \circ \alpha^{-1})(\eta')\right) = d\left(\alpha^{-1}(\eta), \alpha^{-1}(\eta')\right) = \delta_1(\eta, \eta'). \quad (7.14)$$

Given $y \in Y_1$, there exists a sequence $\{\eta_n\}_n$ in $Y_{1,0}$ which tends to y in the metric δ_1. We claim that the following actually defines a function:

$$\gamma : Y_1 \longrightarrow Y_2$$

$$y \mapsto \gamma(y) = \lim_{n \to \infty} (\beta \circ \alpha^{-1})(\eta_n).$$

Let us show that γ is well defined. First, we show that the sequence

$$\{(\beta \circ \alpha^{-1})(\eta_n)\}_n$$

admits a limit in Y_2. To this end, (Y_2, δ_2) being complete, it suffices to show that this sequence is Cauchy. In fact, given $n, m \in \mathbb{N}$,

$$\delta_2\left((\beta \circ \alpha^{-1})(\eta_n), (\beta \circ \alpha^{-1})(\eta_m)\right) \overset{(7.14)}{=} \delta_1(\eta_n, \eta_m).$$

Now, for every $\epsilon > 0$, there exists $n_\epsilon \in \mathbb{N}$ such that the last term is smaller than ϵ as long as $n, m > n_\epsilon$, so the same is true of the first term.

Secondly, we claim that the limit defining γ does not depend on the choice of $\{\eta_n\}_n$. Given another sequence $\{\eta'_n\}_n$ in $Y_{1,0}$ which tends to y, we have

$$\delta_2\left((\beta \circ \alpha^{-1})(\eta_n), (\beta \circ \alpha^{-1})(\eta'_n)\right) \overset{(7.14)}{=} \delta_1(\eta_n, \eta'_n).$$

Now, the above right-hand side vanishes (as $\eta_n, \eta'_n \to y$), so the same is true of the left-hand side: this shows that $(\beta \circ \alpha^{-1})(\eta_n)$ and $(\beta \circ \alpha^{-1})(\eta'_n)$ (which are indeed both convergent as already argued) tend to the same limit.

Finally, γ is an isometry. If $y, y' \in Y_1$, let us consider two sequences η_n, η'_n in $Y_{1,0}$ converging respectively to y, y'. By the definition of γ (and by the continuity properties of any distance of a metric space), we have:

$$\delta_2\left(\gamma(y), \gamma(y')\right) = \delta_2\left(\lim_{n\to\infty}(\beta\circ\alpha^{-1})(\eta_n), \lim_{n\to\infty}(\beta\circ\alpha^{-1})(\eta'_n)\right)$$

$$= \lim_{n\to\infty}\delta_2\left((\beta\circ\alpha^{-1})(\eta_n), (\beta\circ\alpha^{-1})(\eta'_n)\right)$$

$$\overset{(7.14)}{=} \lim_{n\to\infty}\delta_1(\eta_n, \eta'_n) = \delta_1(y, y').$$

This concludes the proof.
   We remark that

$$\gamma|_{Y_{1,0}} \equiv \beta\circ\alpha^{-1},$$

and, roughly speaking, γ is the prolongation by continuity of $\beta\circ\alpha^{-1}$. Indeed, if $y \in Y_{1,0}$, the constant sequence $\eta_n := y$ does the job for defining $\gamma(y)$.   □

Proof (of Theorem 2.68, page 98). The proof is split in many steps.

   I. First of all, $\sim$ is obviously an equivalence relation on $\mathcal{C}$ (by the axioms of a metric for d) and the function $\tilde{d}$ introduced in the statement of the theorem is well defined. That is, the limit in (2.75) exists and does not depend on the choice of the representative sequence.
   I.i. We claim that the limit in (2.75) actually exists. To this aim ($\mathbb{R}$ being complete!) it suffices to show that the sequence $\{d(x_n, y_n)\}_n$ is Cauchy, for every $(x_n)_n, (y_n)_n$ in $\mathcal{C}$. Indeed, let $(x_n)_n$ and $(y_n)_n$ be two Cauchy sequences in X, whence

$$\forall\, \varepsilon > 0 \quad \exists\, n_\varepsilon \in \mathbb{N}: \qquad d(x_n, x_m),\ d(y_n, y_m) < \varepsilon \quad \forall\, n, m \geq n_\varepsilon.$$

We thus have (by a repeated application of the triangle inequality)

$$|d(x_m, y_m) - d(x_n, y_n)| \leq |d(x_m, y_m) - d(x_n, y_m)| + |d(x_n, y_m) - d(x_n, y_n)|$$

$$\leq d(x_m, x_n) + d(y_m, y_n) < 2\,\varepsilon,$$

provided that $n, m \geq n_\varepsilon$. The first claim is proved.
   I.ii. We then claim that the limit in (2.75) does not depend on the choices of the representative sequences. Indeed, let $(x_n)_n \sim (x'_n)_n$ and $(y_n)_n \sim (y'_n)_n$. By definition of $\sim$ we have

$$\forall\, \varepsilon > 0 \quad \exists\, n_\varepsilon \in \mathbb{N}: \qquad d(x_n, x'_n),\ d(y_n, y'_n) < \varepsilon \quad \forall\, n \geq n_\varepsilon.$$

We thus deduce that, for $n \geq n_\varepsilon$,

$$|d(x_n, y_n) - d(x'_n, y'_n)| \leq |d(x_n, y_n) - d(x'_n, y_n)| + |d(x'_n, y_n) - d(x'_n, y'_n)|$$

$$\leq d(x_n, x'_n) + d(y_n, y'_n) < 2\,\varepsilon.$$

This proves that $\lim_{n\to\infty}(d(x_n, y_n) - d(x'_n, y'_n)) = 0$, so that (since the limits of $d(x_n, y_n)$ and $d(x'_n, y'_n)$, as $n \to \infty$, do exist thanks to part I.i)

$$\lim_{n\to\infty} d(x_n, y_n) = \lim_{n\to\infty} d(x'_n, y'_n),$$

and the second claim is proved.

If we consider the map α in (2.76), the well posed definition of $\tilde{d}$ now gives, for every $x, y \in X$,

$$\tilde{d}(\alpha(x), \alpha(y)) = \lim_{n\to\infty} d(x, y) = d(x, y). \tag{7.15}$$

II. We now prove that $\tilde{d}$ is actually a metric on $\tilde{X}$. If $(x_n)_n$ is an element of $\mathcal{C}$, we agree to denote by $\tilde{x}$ the associated element on the quotient $\tilde{X} = \mathcal{C}/\sim$, namely $\tilde{x} = [(x_n)_n]_\sim$ (analogously for $\tilde{y}$, $\tilde{z}$ and so on).

Let $\tilde{x}, \tilde{y} \in \tilde{X}$. We obviously have $\tilde{d}(\tilde{x}, \tilde{y}) \geq 0$ and $\tilde{d}(\tilde{x}, \tilde{x}) = 0$. On the other hand, if $\tilde{d}(\tilde{x}, \tilde{y}) = 0$, this means that $\lim_{n\to\infty} d(x_n, y_n) = 0$ so that $(x_n)_n \sim (y_n)_n$, by the definition of $\sim$, whence $\tilde{x} = \tilde{y}$. Finally, the symmetry and the triangle inequality for $\tilde{d}$ are obvious consequences of the same properties of d: for example the triangle inequality follows from this simple argument (all limits exist by part I.i):

$$\tilde{d}(\tilde{x}, \tilde{y}) = \lim_{n\to\infty} d(x_n, y_n) \leq \lim_{n\to\infty} (d(x_n, z_n) + d(z_n, y_n))$$

$$= \lim_{n\to\infty} d(x_n, z_n) + \lim_{n\to\infty} d(z_n, y_n) = \tilde{d}(\tilde{x}, \tilde{z}) + \tilde{d}(\tilde{z}, \tilde{y}).$$

Hence we now know that $(\tilde{X}, \tilde{d})$ is a metric space and identity (7.15) then means that α is an isometry of metric spaces, from X onto $X_0 := \alpha(X)$ (as a subspace of $\tilde{X}$).

III. Let us now show that $(\tilde{X}, \tilde{d})$ is an isometric completion of (X, d), that is, the two conditions of Definition 2.66 on page 98 are verified.

III.i. We first claim that X_0 is dense in $(\tilde{X}, \tilde{d})$. Let $\tilde{x} = [(x_n)_n]_\sim$ be an arbitrary element of $\tilde{X}$. We prove that

$$\tilde{x} = \lim_{N\to\infty} \alpha(x_N) \quad \text{in } (\tilde{X}, \tilde{d}), \tag{7.16}$$

which justifies our claim, since $\alpha(x_N) \in X_0$ for every $N \in \mathbb{N}$. Obviously, (7.16) is equivalent to

$$\lim_{N\to\infty} \tilde{d}(\tilde{x}, \alpha(x_N)) = 0.$$

Note that this is a double-limit problem, for (by definition of $\tilde{d}$ and of α)

$$\lim_{N\to\infty} \tilde{d}(\tilde{x}, \alpha(x_N)) = \lim_{N\to\infty} \left(\lim_{n\to\infty} d(x_n, x_N)\right).$$

In view of this last fact, (7.16) turns out to be a straightforward consequence of the fact that $(x_n)_n$ is a Cauchy sequence.

III.ii. Furthermore, we claim that the metric space $(\tilde{X}, \tilde{d})$ is complete. This is indeed the main task of the proof. To this end, let $\{\tilde{x}_p\}_{p\in\mathbb{N}}$ be a Cauchy sequence in $(\tilde{X}, \tilde{d})$. This means that

$$\forall\, \varepsilon > 0 \quad \exists\, \mu(\varepsilon) \in \mathbb{N}: \qquad \tilde{d}(\tilde{x}_p, \tilde{x}_q) < \varepsilon \quad \forall\, p, q \geq \mu(\varepsilon). \tag{7.17}$$

We can choose a representative sequence for any $\tilde{x}_p$, that is, there exists a double sequence $x_{p,n} \in X$ such that

$$\tilde{x}_p = [(x_{p,n})_n]_\sim, \quad \text{for every } p \in \mathbb{N}.$$

By definition of $\tilde{X} = \mathcal{C}/\!\sim$, for every fixed $p \in \mathbb{N}$, the sequence $(x_{p,n})_n$ is Cauchy in X, hence

$$\forall\, p \in \mathbb{N}, \ \forall\, \varepsilon > 0 \quad \exists\, n(p, \varepsilon) \in \mathbb{N}: \qquad d(x_{p,n}, x_{p,m}) < \varepsilon \quad \forall\, n, m \geq n(p, \varepsilon). \tag{7.18}$$

We set

$$\nu(p) := n(p, 1/p), \quad \text{for every } p \in \mathbb{N}. \tag{7.19}$$

We aim to prove that, setting

$$\xi_p := x_{p,\nu(p)} \quad \text{for any } p \in \mathbb{N},$$

the "diagonal" sequence $(\xi_p)_p$ is in $\mathcal{C}$, that is, it is Cauchy in (X, d). This requires some work.

Unraveling (7.18), the definition of $\nu(p)$ gives:

$$\forall\, p \in \mathbb{N} \quad \exists\, \nu(p) \in \mathbb{N}: \qquad d(x_{p,n}, x_{p,m}) < \tfrac{1}{p} \quad \forall\, n, m \geq \nu(p). \tag{7.20}$$

Let $\varepsilon > 0$ be fixed henceforth. Let $p, q \in \mathbb{N}$ be arbitrarily fixed and such that (see also (7.17) for the choice of μ)

$$p, q > \max\{3/\varepsilon, \mu(\varepsilon/3), \mu(\varepsilon)\}. \tag{7.21}$$

Having

$$\tilde{d}(\tilde{x}_p, \tilde{x}_q) = \lim_{n\to\infty} d(x_{p,n}, x_{q,n}),$$

the fact that p, q are greater than $\mu(\varepsilon/3)$ ensures that, by (7.17),

$$\exists\,\bar{n} = \bar{n}(\varepsilon, p, q) \in \mathbb{N}: \quad d(x_{p,n}, x_{q,n}) < \varepsilon/3 \quad \forall\, n \geq \bar{n}. \qquad (7.22)$$

We then fix $n^* \in \mathbb{N}$ (depending on ε, p, q) such that

$$n^* \geq \max\{\bar{n}, \nu(p), \nu(q)\}. \qquad (7.23)$$

With all the above choices, we have the following chain of inequalities:

$$
\begin{aligned}
d(\xi_p, \xi_q) &= d(x_{p,\nu(p)}, x_{q,\nu(q)}) \\
&\leq d(x_{p,\nu(p)}, x_{p,n^*}) + d(x_{p,n^*}, x_{q,n^*}) + d(x_{q,n^*}, x_{q,\nu(q)}) \\
&\overset{(7.21)}{\leq} \tfrac{1}{p} + \varepsilon/3 + \tfrac{1}{q} < \varepsilon.
\end{aligned}
$$

To derive the first inequality, we used the following facts:

- For the first and third summands we used (7.20), as $n^* > \nu(p), \nu(q)$ in view of (7.23).
- For the second summand we used (7.22), as $n^* \geq \bar{n}$ again by (7.23).

This estimate, together with the arbitrariness of p, q as in (7.21), proves that $(\xi_p)_p$ is a Cauchy sequence.

We are thus entitled to consider $\widetilde{\xi} := [(\xi_p)_p]_\sim$ in $\widetilde{X}$. We aim to show that

$$\lim_{n \to \infty} \widetilde{x}_n = \widetilde{\xi} \quad \text{in } (\widetilde{X}, \widetilde{d}).$$

We actually prove that $\lim_{n \to \infty} \widetilde{d}(\widetilde{x}_n, \widetilde{\xi}) = 0$, which is a double limit problem, for this means that

$$\lim_{n \to \infty} \left(\lim_{p \to \infty} d(x_{n,p}, x_{p,\nu(p)}) \right) = 0. \qquad (7.24)$$

Let $\varepsilon > 0$ be fixed. Since, as we proved above, $(x_{p,\nu(p)})_p$ is Cauchy, we infer

$$\exists\, j(\varepsilon) \in \mathbb{N}: \quad d(x_{n,\nu(n)}, x_{m,\nu(m)}) < \varepsilon/2 \quad \forall\, n, m \geq j(\varepsilon). \qquad (7.25)$$

Let us set, with this choice of $j(\varepsilon)$,

$$\sigma(\varepsilon) := \max\{2/\varepsilon, j(\varepsilon)\}, \qquad \overline{\sigma}(\varepsilon, n) := \max\{j(\varepsilon), \nu(n)\}. \qquad (7.26)$$

We now take any $p, n \in \mathbb{N}$ such that

$$n \geq \sigma(\varepsilon) \quad \text{and} \quad p \geq \overline{\sigma}(\varepsilon, n). \qquad (7.27)$$

Then one has

$$d(x_{n,p}, x_{p,\nu(p)}) \le d(x_{n,p}, x_{n,\nu(n)}) + d(x_{n,\nu(n)}, x_{p,\nu(p)})$$

(here we use (7.25), as $n, p \ge j(\varepsilon)$ thanks to (7.27) and

the fact that $\sigma(\varepsilon), \overline{\sigma}(\varepsilon, n) \ge j(\varepsilon)$, see (7.26))

$$\le d(x_{n,p}, x_{n,\nu(n)}) + \varepsilon/2 \le 1/n + \varepsilon/2$$

(here we used (7.20), as $p, \nu(n) \ge \nu(n)$)

since $p \ge \overline{\sigma}(\varepsilon, n) \ge \nu(n)$, again in view of (7.26))

$$\le \varepsilon \quad \text{(as } n \ge \sigma(\varepsilon) \ge 2/\varepsilon, \text{ by (7.27) and (7.26)).}$$

Let us finally fix $n \in \mathbb{N}$ such that $n \ge \sigma(\varepsilon)$. Then for every $p \ge \overline{\sigma}(\varepsilon, n)$ we have proved that

$$d(x_{n,p}, x_{p,\nu(p)}) \le \varepsilon.$$

Letting $p \to \infty$, we get

$$\lim_{p \to \infty} d(x_{n,p}, x_{p,\nu(p)}) \le \varepsilon.$$

Since this holds for every $n \ge \sigma(\varepsilon)$, then (7.24) follows. The proof is now complete.   □

Lemma 7.8. *Let X_1, X_2 be topological spaces, and let us assume that X_1 satisfies the first axiom of countability. A map $f : X_1 \to X_2$ is continuous at a point $\tau \in X_1$ iff it is sequence-continuous*[4] *in τ.*

Proof. The standard proof is left to the Reader.   □

Proof (of Theorem 2.69, page 99). Let $(\widetilde{A}, \widetilde{d})$ be the isometric completion of (A, d) as in Theorem 2.68. We will show that $\widetilde{A}$ has a natural structure of topological UA algebra. First of all, the structure of vector space is preserved, via the operations

$$\left[(x_n)_n \right]_\sim \widetilde{+} \left[(y_n)_n \right]_\sim := \left[(x_n + y_n)_n \right]_\sim,$$

$$k \left[(x_n)_n \right]_\sim := \left[(k\, x_n)_n \right]_\sim, \quad k \in \mathbb{K}. \tag{7.28}$$

Their well-posedness is a simple verification, as we now show. Given two Cauchy sequences $(x_n)_n, (\bar{x}_n)_n$ in A representing via $\sim$ the same element of $\widetilde{A}$, and given two Cauchy sequences $(y_n)_n, (\bar{y}_n)_n$ in A representing via $\sim$ the

---

[4]We say that f is *sequence-continuous* in τ if, for every sequence $\{\tau_n\}_n$ in X_1 converging to τ, it holds that $\lim_{n \to \infty} f(\tau_n) = f(\tau)$ in X_2.

same element of $\widetilde{A}$, we have (by applying twice Remark 2.61-1)

$$\begin{aligned}
d(\bar{x}_n + \bar{y}_n, x_n + y_n) &= d(\bar{x}_n - x_n, y_n - \bar{y}_n) \\
&\leq d(\bar{x}_n - x_n, 0) + d(0, y_n - \bar{y}_n).
\end{aligned} \tag{7.29}$$

Now, as $(x_n)_n \sim (\bar{x}_n)_n$, that is

$$0 = \lim_{n \to \infty} d(\bar{x}_n, x_n) = \lim_{n \to \infty} d(\bar{x}_n - x_n, 0),$$

there certainly exists for every $p \in \mathbb{N}$ a $\mu(p) \in \mathbb{N}$ such that

$$\bar{x}_n - x_n \in \Omega_p,$$

as long as $n > \mu(p)$. The same holds for $\bar{y}_n - y_n$, so that (by the estimate in (7.29)) the distance $d(\bar{x}_n + \bar{y}_n, x_n + y_n)$ tends to 0 as $n \to \infty$.

An analogous argument holds for multiplication by an element of $\mathbb{K}$.

Let us now turn to consider the algebra operation

$$\big[(x_n)_n\big]_\sim \widetilde{*} \big[(y_n)_n\big]_\sim := \big[(x_n * y_n)_n\big]_\sim.$$

We claim that the operation $\widetilde{*}$ is well-posed, that is, for every pair of A-valued Cauchy sequences $(x_n)_n$ and $(y_n)_n$, the sequence $(x_n * y_n)_n$ is Cauchy too; secondly, we shall see that the definition of $\widetilde{*}$ does not depend on the choice of the representatives.

Indeed, to begin with, we show that $d(x_n * y_n, x_m * y_m)$ tends to zero as $n, m \to \infty$. This follows immediately from Remark (2.61)-3 which yields

$$d(x_n * y_n, x_m * y_m) \leq \max \big\{ d(x_n, x_m), d(y_n, y_m) \big\}.$$

The above right-hand side vanishes as $n, m \to \infty$ since $(x_n)_n$ and $(y_n)_n$ are Cauchy and we are done.

Let us now turn to check why the operation $\widetilde{*}$ is well posed on equivalence classes. As above, let us consider two Cauchy sequences $(x_n)_n, (\bar{x}_n)_n$ in A representing via $\sim$ the same element of $\widetilde{A}$, and two Cauchy sequences $(y_n)_n, (\bar{y}_n)_n$ in A representing via $\sim$ the same element of $\widetilde{A}$. We then have (again by Remark (2.61)-3)

$$d(x_n * y_n, \bar{x}_n * \bar{y}_n) \leq \max \big\{ d(x_n, \bar{x}_n), d(y_n, \bar{y}_n) \big\} \xrightarrow[n \to \infty]{} 0,$$

for $(x_n)_n \sim (\bar{x}_n)_n$ and $(y_n)_n \sim (\bar{y}_n)_n$.

Furthermore, $\widetilde{*}$ is bilinear and associative. These properties follow immediately from the definition of $\widetilde{*}$ and, respectively, the bilinearity and the associativity of $*$.

As for the existence of a unit in $\widetilde{A}$, it is clear that the element

$$1_{\widetilde{A}} := [(1_A)_n]_\sim$$

is such an element.

Finally, $\widetilde{A}$ is a *topological* algebra, that is, the maps

$$\widetilde{A} \times \widetilde{A} \to \widetilde{A} \qquad\qquad \mathbb{K} \times \widetilde{A} \to \widetilde{A} \qquad\qquad \widetilde{A} \times \widetilde{A} \to \widetilde{A}$$

$$(a, b) \mapsto a \widetilde{+} b, \qquad\qquad (k, a) \mapsto k\, a, \qquad\qquad (a, b) \mapsto a \widetilde{*} b,$$

are continuous with respect to the associated topologies. To show this, we shall make use of Lemma 7.8. In fact, as a metric space, $\widetilde{A}$ satisfies the first axiom of countability,[5] so it will suffice to prove that each of the three maps is sequence continuous. Let us start with the sum. Given the $\widetilde{A}$-valued sequences $\{\widetilde{x}_n\}_n, \{\widetilde{y}_n\}_n$ with limits respectively $\widetilde{x}$ and $\widetilde{y}$, we need to prove that

$$\lim_{n \to \infty} (\widetilde{x}_n \widetilde{+} \widetilde{y}_n) = \widetilde{x} \widetilde{+} \widetilde{y}.$$

Recalling that every element in $\widetilde{A}$ is represented by a Cauchy sequence in A, let us set the following notation:

$$\widetilde{x}_n = [(x_{n,p})_p]_\sim \qquad \widetilde{x} = [(x_p)_p]_\sim \qquad \widetilde{y}_n = [(y_{n,p})_p]_\sim \qquad \widetilde{y} = [(y_p)_p]_\sim.$$

Let us evaluate $\widetilde{d}(\widetilde{x}_n \widetilde{+} \widetilde{y}_n, \widetilde{x} \widetilde{+} \widetilde{y})$: unraveling the definitions, we have

$$\begin{aligned}
\widetilde{d}(\widetilde{x}_n \widetilde{+} \widetilde{y}_n, \widetilde{x} \widetilde{+} \widetilde{y}) &= \lim_{p \to \infty} d(x_{n,p} + y_{n,p}, x_p + y_p) \\
&= \lim_{p \to \infty} d(x_{n,p} + y_{n,p} - x_p, y_p) \\
&= \lim_{p \to \infty} d(x_{n,p} - x_p, y_p - y_{n,p}) \\
&\leq \lim_{p \to \infty} (d(x_{n,p} - x_p, 0) + d(0, y_p - y_{n,p})) \\
&= \lim_{p \to \infty} (d(x_{n,p}, x_p) + d(y_p, y_{n,p})) \\
&= \widetilde{d}(\widetilde{x}_n, \widetilde{x}) \widetilde{+} \widetilde{d}(\widetilde{y}_n, \widetilde{y}).
\end{aligned}$$

Now, the last term of this chain of equalities and inequalities tends to zero as n tends to ∞ by construction.

---

[5] The family $\{B_{\widetilde{d}}(t, \frac{1}{n})\}_{n \in \mathbb{N}}$ is a basis of neighborhoods of $t \in \widetilde{A}$.

Let us now look at the case of the multiplication by an element of the field. The case $k = 0$ is trivial, so we can assume $k \neq 0$. With the above notation, we have (see Remark 2.61-2)

$$\widetilde{d}(k\,\widetilde{x}_n, k\,\widetilde{x}) = \lim_{p \to \infty} d(k\,x_{p,n}, k\,x_p) = \lim_{p \to \infty} d(x_{p,n}, x_p) = \widetilde{d}(\widetilde{x}_n, \widetilde{x}),$$

and the continuity of the multiplication by a scalar follows.

Finally, let us check that the algebra operation $\widetilde{*}$ is sequence continuous as well. Using the same notations as above, we have

$$\widetilde{d}(\widetilde{x}_n \widetilde{*} \widetilde{y}_n, \widetilde{x} \widetilde{*} \widetilde{y}) = \lim_{p \to \infty} d(x_{n,p} * y_{n,p}, x_p * y_p)$$

$$\leq \lim_{p \to \infty} \max\{d(x_{n,p}, x_p), d(y_{n,p}, y_p)\}$$

$$= \max\{\widetilde{d}(\widetilde{x}_n, \widetilde{x}), \widetilde{d}(\widetilde{y}_n, \widetilde{y})\},$$

where the inequality is a consequence of Remark 2.61-3. The last term tends to zero as n tends to ∞, so the operation $\widetilde{*}$ is actually sequence continuous.

The last step of this proof is to show that the isometry defined in (2.76),

$$\alpha : A \to A_0 \subseteq \widetilde{A}, \quad a \mapsto [(a_n)_n]_\sim \quad \text{with } a_n = a \text{ for every } n \in \mathbb{N}$$

is, in this setting, an isomorphism of UA algebras, but this is easily checked:

- α is linear: for every $k, j \in \mathbb{K}$, $a, b \in A$,

$$\alpha(k\,a + j\,b) = [(k\,a + j\,b)_n]_\sim = k\,[(a)_n]_\sim + j\,[(b)_n]_\sim = k\,\alpha(a) + j\,\alpha(b);$$

- α is an algebra morphism:

$$\alpha(a * b) = [(a * b)_n]_\sim = [(a)_n]_\sim \widetilde{*} [(b)_n]_\sim = \alpha(a)\,\widetilde{*}\,\alpha(b).$$

This concludes the proof. $\qquad\qquad\qquad\qquad\qquad\qquad\qquad\qquad\qquad\square$

Proof (of Remark 2.70, page 99). Let the notation in Remark 2.70 hold. By our assumptions on A and B it follows directly that B is an isometric completion of A and that the inclusion $\iota : A \hookrightarrow B$ is both an isomorphism of metric spaces onto $\iota(A)$ and a UAA isomorphism onto $\iota(A)$.

By the "uniqueness" of metric completions in Proposition 2.67, we have an isometry of metric spaces $\gamma : (B, \delta) \to (\widetilde{A}, d)$ (see the proof of Proposition 2.67, page 417) as follows:

$$\gamma : B \quad \longrightarrow \qquad\qquad \tilde{A}$$

$$y \quad \mapsto \quad \lim_{n\to\infty} (\alpha \circ \iota^{-1})(\eta_n),$$

where, given $y \in B$, $\{\eta_n\}_n$ is any sequence in A tending to y. We simply need to check that γ is not only an isometry of metric spaces, *but also a UAA isomorphism*. Given $a, b \in B$, let us consider two A-valued sequences $\{a_n\}_n$, $\{b_n\}_n$, tending respectively to a and b. Thanks to the continuity of $\star$ we have $A \ni a_n \star b_n \to a \star b$, so that (by the definition of γ) we infer

$$\begin{aligned}
\gamma(a \star b) &= \lim_{n\to\infty} (\alpha \circ \iota^{-1})(a_n \star b_n) \\
&= \lim_{n\to\infty} \alpha \left(\iota^{-1}(a_n) * \iota^{-1}(b_n) \right) \\
&= \lim_{n\to\infty} \left(\alpha \circ \iota^{-1}(a_n) \right) \tilde{*} \left(\alpha \circ \iota^{-1}(a_n) \right) \\
&= \gamma(a) \tilde{*} \gamma(b).
\end{aligned}$$

As for the linearity of γ, it is a consequence of the linearity of ι and α. Given $j, k \in \mathbb{K}$ (and a_n, b_n as above), we have

$$\begin{aligned}
\gamma(k\,a + j\,b) &= \lim_{n\to\infty} \alpha \circ \iota^{-1}(k\,a_n + j\,b_n) \\
&= \lim_{n\to\infty} \left(k(\alpha \circ \iota^{-1})(a_n) + j(\alpha \circ \iota^{-1})(b_n) \right) \\
&= k\,\gamma(a) + j\,\gamma(b).
\end{aligned}$$

This ends the proof.   □

Proof (of Theorem 2.72, page 100). To begin with, we prove that $\{\tilde{\Omega}_k\}_k$ is a topologically admissible family in $(B, \star)$:

(H1.) $\tilde{\Omega}_k = \varphi(\Omega_k)$ is an ideal of B, since Ω_k is an ideal of A (and φ is a UAA morphism).

(H2.) $\tilde{\Omega}_1 = \varphi(\Omega_1) = B$, for $\Omega_1 = A$ and φ is onto; $\tilde{\Omega}_k \supseteq \tilde{\Omega}_{k+1}$, for the same property holds for the sets Ω_k.

(H3.) We have, for every $h, k \in \mathbb{N}$,

$$\tilde{\Omega}_h \star \tilde{\Omega}_k = \varphi(\Omega_h * \Omega_k) \subseteq \varphi(\Omega_{h+k}) = \tilde{\Omega}_{h+k}.$$

(H4.) $\bigcap_{k\in\mathbb{N}} \tilde{\Omega}_k = \bigcap_{k\in\mathbb{N}} \varphi(\Omega_k) = \varphi\left(\bigcap_{k\in\mathbb{N}} \Omega_k \right) = \varphi(\{0\}) = 0$. Here the injectivity of φ has been exploited.

Let us consider the function $\widetilde{\varphi}$ mentioned on page 100:

$$\widetilde{\varphi} : \widetilde{A} \to \widetilde{B} \text{ defined by } \quad \widetilde{\varphi}([(a_n)_n]_\sim) := [(\varphi(a_n))_n]_\sim. \tag{7.30}$$

We have the following facts:

- $\widetilde{\varphi}$ is well posed. Let $(a_n)_n$ be a Cauchy sequence in A. For every $p \in \mathbb{N}$, there exists $n(p) \in \mathbb{N}$ such that $a_n - a_m \in \Omega_p$ as long as $n, m \geq n(p)$. Thus

$$\varphi(a_n) - \varphi(a_m) = \varphi(a_n - a_m) \in \varphi(\Omega_p) = \widetilde{\Omega}_p,$$

  as long as $n, m \geq n(p)$. So $(\varphi(a_n))_n$ is a Cauchy sequence in B.
- $\widetilde{\varphi}$ "prolongs" φ. Let $a \in A$, and let α_A and α_B be functions constructed as in (2.76) respectively for A and B. Then we have

$$\widetilde{\varphi}(\alpha_A(a)) = \widetilde{\varphi}[(a)_n]_\sim = [(\varphi(a))_n]_\sim = \alpha_B(\varphi(a)).$$

- $\widetilde{\varphi}$ is obviously a linear map.
- $\widetilde{\varphi}$ is a UAA morphism. Let us first check that $\widetilde{\varphi}$ is unital:

$$\widetilde{\varphi}(1_{\widetilde{A}}) = \widetilde{\varphi}[(1_A)_n]_\sim = [(\varphi(1_A))_n]_\sim = [1_B]_\sim = 1_{\widetilde{B}}.$$

Secondly, $\widetilde{\varphi}$ preserves the algebra operation:

$$\widetilde{\varphi}([(x_n)_n]_\sim \, \widetilde{*} \, [(y_n)]_\sim) \overset{(2.77)}{=} \widetilde{\varphi}[(x_n * y_n)_n]_\sim \overset{(2.79)}{=} [(\varphi(x_n * y_n))_n]_\sim$$

$$= [(\varphi(x_n) \star \varphi(y_n))_n]_\sim \overset{(2.77)}{=} [(\varphi(x_n))_n]_\sim \, \widetilde{\star} \, [(\varphi(y_n))_n]_\sim$$

$$\overset{(2.79)}{=} \widetilde{\varphi}([(x_n)_n]_\sim) \, \widetilde{\star} \, \widetilde{\varphi}([(y_n)_n]_\sim).$$

- $\widetilde{\varphi}$ preserves the associated metrics. Applying (2.75) and (2.79), we have

$$\widetilde{\delta}\Big(\widetilde{\varphi}([(x_n)_n]_\sim), \widetilde{\varphi}([(y_n)_n]_\sim)\Big) = \lim_{n \to \infty} \delta(\varphi(x_n), \varphi(y_n))$$

$$\overset{(\star)}{=} \lim_{n \to \infty} d(x_n, y_n) = \widetilde{d}([(x_n)_n]_\sim, [(y_n)_n]_\sim).$$

The equality $(\star)$ is derived as follows: thanks to (2.65) and (2.66),

$$\delta(\varphi(x_n), \varphi(y_n)) = \exp\big(-\sup\{\, n \geq 1 \mid \varphi(x_n - y_n) \in \widetilde{\Omega}_n\}\big)$$

$$= \exp\big(-\sup\{\, n \geq 1 \mid x_n - y_n \in \Omega_n\}\big) = d(x_n, y_n).$$

- $\widetilde{\varphi}$ is clearly surjective. Indeed, let $(y_n)_n$ be a Cauchy sequence in B. Then, as φ^{-1} is a UAA isomorphism with $\varphi^{-1}(\widetilde{\Omega}_n) = \Omega_n$, the sequence

$(\varphi^{-1}(y_n))_n$ is Cauchy in A, and we have

$$[(y_n)_n]_\sim = \widetilde{\varphi}\big([(\varphi^{-1}(y_n))_n]_\sim\big).$$

As for its injectivity, let us assume that $(a_n)_n$ and $(a'_n)_n$ are two A-valued Cauchy sequences satisfying

$$\widetilde{\varphi}([(a_n)_n]_\sim) = \widetilde{\varphi}([(a'_n)_n]_\sim).$$

Then we have

$$[(\varphi(a_n))_n]_\sim = [(\varphi(a'_n))_n]_\sim \Leftrightarrow$$
$$\varphi(a_n) = \varphi(a'_n) + \varepsilon_n, \quad \text{with } \varepsilon_n \to 0 \text{ in } B.$$

From the injectivity and linearity of φ, we have

$$a_n = a'_n + \varphi^{-1}(\varepsilon_n). \tag{7.31}$$

Now, by definition of the metric on B we have

$$\forall p \quad \exists n(p) \in \mathbb{N}: \qquad \varepsilon_n \in \widetilde{\Omega}_p \quad \forall n \geq n(p).$$

By applying φ^{-1}, we obtain

$$\forall p \quad \exists n(p) \in \mathbb{N}: \qquad \varphi^{-1}(\varepsilon_n) \in \Omega_p \quad \forall n \geq n(p).$$

This means that $\varphi^{-1}(\varepsilon_n) \to 0$ in A, and so, by exploiting (7.31), we derive

$$[(a_n)_n]_\sim = [(a'_n)_n]_\sim.$$

So the two metric spaces are isomorphic and the proof is complete. $\qquad\square$

7.6 Proofs of Sect. 2.3.3

Proof (of Theorem 2.75, page 102). First of all, we observe that the topology induced by $\{\Omega_k\}_{k\geq 0}$ on A coincides with the topology induced on A by the topology of $\widehat{A}$, which in its turn is induced by the family $\{\widehat{\Omega}_k\}_{k\geq 0}$: this is an immediate consequence of (2.85).

More is true: we claim that the metric $\widehat{d}$ coincides with d on $A \times A$. Indeed, if $a, b \in A$, we have

$$\widehat{d}(a,b) = \exp\big(-\sup\{n \geq 0 \mid a - b \in \widehat{\Omega}\}\big)$$
$$\overset{(2.85)}{=} \exp\big(-\sup\{n \geq 0 \mid a - b \in \Omega\}\big) = d(a,b).$$

So, as a consequence of Remark 2.70 on page 99, all that we have to show is:

(i) The metric space $\widehat{A}$, with the metric induced by $\{\widehat{\Omega}_k\}_k$, is complete.

(ii) A is dense in $\widehat{A}$ with the topology induced by $\{\widehat{\Omega}_k\}_k$.

Here are the proofs:

(i) Let $\{w_k\}_{k \in \mathbb{N}}$ be a Cauchy sequence in $\widehat{A}$ denoted by

$$w_k := (u_0^k, u_1^k, \ldots), \qquad k \in \mathbb{N}.$$

As $\{w_k\}_k$ is Cauchy, we have

$$\forall\, h \in \mathbb{N} \cup \{0\}, \ \exists\, k(h) \in \mathbb{N} : \quad w_n - w_m \in \widehat{\Omega}_{h+1}, \quad \forall\, n, m \geq k(h),$$

$$\text{that is, } (u_0^n - u_0^m, u_1^n - u_1^m, \ldots) \in \widehat{\Omega}_{h+1}, \quad \forall\, n, m \geq k(h),$$

$$\text{that is, } u_j^n - u_j^m = 0 \quad \forall\, j = 0, \ldots, h, \quad \forall\, n, m \geq k(h).$$

It is not restrictive to suppose that

$$k(h+1) \geq k(h), \quad \text{for every } h \in \mathbb{N} \cup \{0\}.$$

Let us define

$$w := (u_0^{k(0)}, u_1^{k(1)}, \ldots u_h^{k(h)}, \ldots) \in \widehat{A}.$$

We claim that w_k tends to w in $\widehat{A}$.

To this end, let us see what happens as h ranges over $\mathbb{N}$:

- If $h = 0$, we have

$$\forall\, n \geq k(0), \quad u_0^n = u_0^{k(0)} \quad \text{and} \quad w_n = (u_0^{k(0)}, u_1^n, u_2^n, \ldots).$$

- If $h = 1$,

$$\forall\, n \geq k(1), \quad \begin{array}{c} u_0^n = u_0^{k(1)} = u_0^{k(0)}, \ u_1^n = u_1^{k(1)}, \\ w_n = (u_0^{k(0)}, u_1^{k(1)}, u_2^n, \ldots). \end{array}$$

- For $h > 1$,

$$\forall\, n \geq k(h), \quad w_n = (u_0^{k(0)}, u_1^{k(1)}, \ldots u_h^{k(h)}, u_{h+1}^n, \ldots).$$

So, for every $h \geq 0$, we have

$$w_n - w = (\underbrace{0, 0, \ldots, 0}_{h+1 \text{ times}}, *, *, \ldots) \in \widehat{\Omega}_{h+1},$$

that is, w_k tends to w in the $\{\widehat{\Omega}_p\}_p$ topology, as claimed.

(ii) Let $u = (u_0, u_1, u_2, \ldots) \in \widehat{A}$. We claim that the A-valued sequence $\{w_k\}_k$ defined by

$$w_k := (u_0, u_1, \ldots, u_k, 0, 0, \ldots) \quad \forall\, k \in \mathbb{N} \cup \{0\}$$

tends to u in the $\{\widehat{\Omega}_p\}_p$ topology, whence the desired density of A in $\widehat{A}$ will follow. We have that

$$u - w_k = (0, \ldots, 0, u_{k+1}, u_{k+2}, \ldots) \in \widehat{\Omega}_{k+1}.$$

Then the following holds:

$$\forall\, h \in \mathbb{N}, \quad u - w_k \in \widehat{\Omega}_{k+1} \subseteq \widehat{\Omega}_h \text{ as long as } k \geq h - 1.$$

This means that

$$\forall\, h \in \mathbb{N}, \quad \exists k(h) := h - 1 : \quad u - w_k \in \widehat{\Omega}_h, \quad \forall\, k \geq k(h),$$

which is precisely $\lim_{k \to \infty} w_k = w$ in the given topology. $\qquad \square$

Proof (of Lemma 2.79, page 103). We have to prove that φ is uniformly continuous. Since (by Theorem 2.75) the restriction to $A \times A$ of the metric induced on $\widehat{A}$ by the sets $\widehat{\Omega}_k^A$ coincides with the the metric induced on A by the sets Ω_k^A, we have to prove that

$$\forall \varepsilon > 0 \quad \exists \delta_\varepsilon > 0 : \quad (x, y \in A, \; d_A(x, y) < \delta_\varepsilon) \;\Rightarrow\; d_B(\varphi(x), \varphi(y)) < \varepsilon. \tag{7.32}$$

Here we have denoted by d_A and by d_B the metrics induced respectively on A and on B by the sets Ω_k^A and Ω_k^B.

Let $\varepsilon > 0$ be fixed. If the sequence $\{k_n\}_n$ is as in the hypothesis (2.88), we have $\lim_{n \to \infty} k_n = \infty$. Hence there exists $n(\varepsilon) \in \mathbb{N}$ such that

$$k_n > \ln(1/\varepsilon) \text{ for every } n \geq n(\varepsilon). \tag{7.33}$$

We set $\delta_\varepsilon := \exp(-n(\varepsilon))$. We claim that this choice of δ_ε gives (7.32). Indeed, if $x, y \in A$ are such that $d_A(x, y) < \delta_\varepsilon$, by (2.68) we infer the existence of $n_0 > \ln(1/\delta_\varepsilon) = n(\varepsilon)$ such that $x - y \in \Omega_{n_0}^A$. This latter fact, together with the second part of (2.88) and the linearity of φ, yields

$$\varphi(x) - \varphi(y) = \varphi(x - y) \in \varphi(\Omega_{n_0}^A) \subseteq \Omega_{k_{n_0}}^B.$$

As a consequence, by also exploiting the very definition (2.65)–(2.66) of d_B, we infer (recall that $n_0 > n(\varepsilon)$)

$$d_B(\varphi(x), \varphi(y)) \le \exp(-k_{n_0}) \overset{(7.33)}{<} \exp(-\ln(1/\varepsilon)) = \varepsilon.$$

This proves (7.32) with the claimed choice of δ_ε.

Since φ is uniformly continuous, a general result on metric spaces (see Lemma 7.9 below) ensures that there exists a continuous function $\widehat{\varphi} : \widehat{A} \to \widehat{B}$ prolonging φ. Obviously, $\widehat{\varphi}$ is linear since φ is.

Finally, the theorem is completely proved if we show that φ is a UAA morphism, provided φ is. First we have $\widehat{\varphi}(1_A) = \varphi(1_A) = 1_B$. Moreover, suppose $a, b \in \widehat{A}$ and choose sequences $\{a_k\}_k$, $\{b_k\}_k$ in A such that $\lim_{k\to\infty} a_k = a$ and $\lim_{k\to\infty} b_k = b$ w.r.t. the metric of $\widehat{A}$ (such sequences do exist since A is dense in $\widehat{A}$). Then for every $a, b \in A$ we have (using the same symbol $*$ for the operations on A, B and the same symbol $\widehat{*}$ for the corresponding operations on $\widehat{A}, \widehat{B}$):

$$\widehat{\varphi}(a \,\widehat{*}\, b) \overset{(1)}{=} \widehat{\varphi}\left(\lim_{k\to\infty} a_k \,\widehat{*}\, b_k\right) \overset{(2)}{=} \lim_{k\to\infty} \widehat{\varphi}(a_k \,\widehat{*}\, b_k) \overset{(3)}{=} \lim_{k\to\infty} \varphi(a_k * b_k)$$

$$\overset{(4)}{=} \lim_{k\to\infty} \varphi(a_k) * \varphi(b_k) \overset{(5)}{=} \lim_{k\to\infty} \widehat{\varphi}(a_k) \,\widehat{*}\, \widehat{\varphi}(b_k)$$

$$\overset{(6)}{=} \widehat{\varphi}\left(\lim_{k\to\infty} a_k\right) \widehat{*}\, \widehat{\varphi}\left(\lim_{k\to\infty} b_k\right) \overset{(7)}{=} \widehat{\varphi}(a) \,\widehat{*}\, \widehat{\varphi}(b).$$

Here we used the following facts:

1. $(\widehat{A}, \widehat{*})$ is a topological algebra (and $a_k \to a$, $b_k \to b$ as $k \to \infty$).
2. $\widehat{\varphi}$ in continuous.
3. We invoked (2.83) (being $a_k, b_k \in A$) and $\widehat{\varphi} \equiv \varphi$ on A.
4. φ is a UAA morphism.
5. An analogue of (2.83) relatively to B (being $\varphi(a_k), \varphi(b_k) \in B$) and again $\widehat{\varphi} \equiv \varphi$ on A.
6. $(\widehat{B}, \widehat{*})$ is a topological algebra and $\widehat{\varphi}$ is continuous.
7. $\lim_{k\to\infty} a_k = a$ and $\lim_{k\to\infty} b_k = b$ in $\widehat{A}$.

This completes the proof.   □

Here we used the following result of Analysis.

Lemma 7.9. Let (X, d_X) and (Y, d_Y) be metric spaces. Suppose also that Y is complete. Let $A \subseteq X$, $B \subseteq Y$ and let $f : A \to B$ be uniformly continuous.
  Then there exists a unique continuous function $\overline{f} : \overline{A} \to \overline{B}$ prolonging f. Moreover, $\overline{f}$ is uniformly continuous.

Proof. The uniform continuity of f means that

$$\forall \, \varepsilon > 0 \ \ \exists \, \delta_\varepsilon > 0 : \quad (a, a' \in A, \ d_X(a, a') < \delta_\varepsilon) \Rightarrow d_Y(f(a), f(a')) < \varepsilon.$$
$$(7.34)$$

If $\overline{f}$ is as in the assertion and $A \ni a_n \xrightarrow[n\to\infty]{} a \in \overline{A}$, then one must have

$$\overline{f}(a) = \overline{f}\left(\lim_{n\to\infty} a_n\right) = \lim_{n\to\infty} \overline{f}(a_n) = \lim_{n\to\infty} f(a_n),$$

which proves the uniqueness of $\overline{f}$. It also suggests how to define $\overline{f} : \overline{A} \to \overline{B}$. Indeed, if $\alpha \in \overline{A}$, we choose any sequence $\{a_n\}_n$ in A such that $a_n \xrightarrow[n\to\infty]{} \alpha$ in X and we set

$$\overline{f}(\alpha) := \lim_{n\to\infty} f(a_n).$$

First of all, we have to show that this limit exists in Y and that the definition of $\overline{f}(\alpha)$ does not depend on the particular sequence $\{a_n\}_n$ as above:

Existence of the limit: Since Y is complete, it suffices to show that $\{f(a_n)\}_n$ is a Cauchy sequence in Y. Since $\{a_n\}_n$ is convergent, it is also a Cauchy sequence, hence

$$\forall\, \sigma > 0 \quad \exists\, n_\sigma \in \mathbb{N} : \quad (n, m \in \mathbb{N}, \ n, m \geq n_\sigma) \Rightarrow d_X(a_n, a_m) < \sigma. \tag{7.35}$$

Now let $\varepsilon > 0$ be fixed and let δ_ε be as in (7.34). By invoking (7.35) with $\sigma := \delta_\varepsilon$ we infer the existence of $n_{\delta_\varepsilon} =: N_\varepsilon$ such that $d_X(a_n, a_m) < \delta_\varepsilon$ whenever $n, m \geq N_\varepsilon$. Hence, thanks to (7.34) we get

$$d_Y(f(a_n), f(a_m)) < \varepsilon \quad \forall\, n, m \geq N_\varepsilon.$$

The arbitrariness of ε proves that $\{f(a_n)\}_n$ is a Cauchy sequence in Y, hence it is convergent.

Independence of the sequence: Let $\{a_n\}_n$ and $\{a'_n\}_n$ be sequences in A both converging to $\alpha \in \overline{A}$. We have to show that

$$\lim_{n\to\infty} f(a_n) = \lim_{n\to\infty} f(a'_n),$$

i.e., that the d_Y-distance of these two limits is null. By the continuity of the d_Y-distance on its arguments, this is equivalent to proving

$$\lim_{n\to\infty} d_Y(f(a'_n), f(a_n)) = 0. \tag{7.36}$$

(Indeed, recall that we already know that the limits of $f(a_n)$, $f(a'_n)$, as $n \to \infty$, do exist in Y.) Let $\varepsilon > 0$ be fixed and let δ_ε be as in (7.34). As $\lim_{n\to\infty} d_X(a'_n, a_n) = 0$, there exists $n_\varepsilon \in \mathbb{N}$ such that $d_X(a'_n, a_n) < \delta_\varepsilon$ if $n \geq n_\varepsilon$. As a consequence, by (7.34) we get

$$d_Y(f(a'_n), f(a_n)) < \varepsilon \quad \forall\, n \geq n_\varepsilon.$$

This, together with the arbitrariness of $\varepsilon > 0$, proves (7.36).

With the above definition of $\overline{f}$, it is obvious that $\overline{f}$ prolongs f and that $\overline{f}(\overline{A}) \subseteq \overline{B}$. To end the proof, we are left to show that $\overline{f}$ is uniformly continuous. Let $\alpha, \alpha' \in \overline{A}$ and choose sequences $\{a_n\}_n$ and $\{a'_n\}_n$ in A converging to α and α' respectively. We thus have

$$\forall \varepsilon_1 > 0 \; \exists n_1(\varepsilon_1) \in \mathbb{N} : \quad n \geq n_1(\varepsilon_1) \Rightarrow d_X(a_n, \alpha), d_X(a'_n, \alpha') < \varepsilon_1. \quad (7.37)$$

By the definition of $\overline{f}$, we have that $f(a_n)$ and $f(a'_n)$ converge in Y to $\overline{f}(\alpha)$ and $\overline{f}(\alpha')$, respectively. Hence, this yields

$$\forall \varepsilon_2 > 0 \; \exists n_2(\varepsilon_2) \in \mathbb{N} : \quad n \geq n_2(\varepsilon_2) \Rightarrow \left(\begin{array}{c} d_Y(f(a_n), \overline{f}(\alpha)) < \varepsilon_2 \\ d_Y(f(a'_n), \overline{f}(\alpha')) < \varepsilon_2. \end{array} \right) \quad (7.38)$$

Now, let $\varepsilon > 0$ be fixed and let δ_ε be as in (7.34). Let $\alpha, \alpha' \in \overline{A}$ be such that $d_X(\alpha, \alpha') < \frac{1}{3} \delta_e$. We claim that this implies $d_Y(\overline{f}(\alpha), \overline{f}(\alpha')) < 3\varepsilon$, which proves that $\overline{f}$ is uniformly continuous, ending the proof.

Indeed, let us fix any $n \in \mathbb{N}$ such that

$$n \geq \max\{n_2(\varepsilon), n_1(\delta_\varepsilon/3)\} \quad (7.39)$$

(where n_1 and n_2 are as in (7.37) and (7.38)). Then, if $d_X(\alpha, \alpha') < \frac{1}{3} \delta_e$, we get

$$d_X(a_n, a'_n) \leq d_X(a_n, \alpha) + d_X(\alpha, \alpha') + d_X(\alpha', a'_n) \leq = \tfrac{1}{3} \delta_e + \tfrac{1}{3} \delta_e + \tfrac{1}{3} \delta_e = \delta_\varepsilon,$$

since the first and the third summands do not exceed $\frac{1}{3} \delta_\varepsilon$, thanks to (7.37) (and the choice of n in (7.39)). Thus, from (7.34) we get $d_Y(f(a_n), f(a'_n)) < \varepsilon$, so that we finally infer

$$d_Y(\overline{f}(\alpha), \overline{f}(\alpha')) \leq d_Y(\overline{f}(\alpha), f(a_n)) + d_Y(f(a_n), f(a'_n)) + d_Y(f(a'_n), \overline{f}(\alpha'))$$
$$< \varepsilon + \varepsilon + \varepsilon = 3\varepsilon.$$

Indeed, note that the first and the third summands are less than ε in view of (7.38), due to $n \geq n_2(\varepsilon)$ (see (7.39)). This completes the proof. $\qquad \square$

Proof (of Theorem 2.82, page 106). We shall prove that the function

$$\widehat{\mathcal{T}}(V) \otimes \widehat{\mathcal{T}}(V) \ni (u_i)_i \otimes (v_j)_j \mapsto \left(u_i \otimes v_j \right)_{i,j} \in \widehat{\mathcal{T} \otimes \mathcal{T}}(V) \quad (7.40)$$

is indeed a UAA morphism. First of all, it is well-posed, because, for every pair $i, j \in \mathbb{N}$, $u_i \otimes v_j \in \mathcal{T}_{i,j}(V)$ and it clearly maps the identity $1 \otimes 1$ of $\mathcal{T} \otimes \mathcal{T}$ to $(1 \otimes 1, 0, 0, \ldots) \in \widehat{\mathcal{T} \otimes \mathcal{T}}$. Secondly, let us check that it preserves the algebra operations. In $\widehat{\mathcal{T}}(V) \otimes \widehat{\mathcal{T}}(V)$, we have the operation, say $\square$, obtained as described in Proposition 2.41 on page 81, starting from the $\cdot$ operation of $\widehat{\mathscr{T}}$: explicitly, this amounts to

$$((u_i)_i \otimes (v_j)_j) \boxdot ((\widetilde{u}_i)_i \otimes (\widetilde{v}_j)_j) = ((u_i)_i \cdot (\widetilde{u}_i)_i) \otimes ((v_j)_j \cdot (\widetilde{v}_j)_j)$$

$$= \left(\sum_{a+b=i} u_a \otimes \widetilde{u}_b \right)_i \otimes \left(\sum_{\alpha+\beta=j} v_\alpha \otimes \widetilde{v}_\beta \right)_j .$$

Via the identification in (7.40), the above far right-hand side is mapped to

$$\left(\left(\sum_{a+b=i} u_a \otimes \widetilde{u}_b \right) \otimes \left(\sum_{\alpha+\beta=j} v_\alpha \otimes \widetilde{v}_\beta \right) \right)_{i,j}$$

$$= \left(\sum_{\substack{a+b=i \\ \alpha+\beta=j}} (u_a \otimes \widetilde{u}_b) \otimes (v_\alpha \otimes \widetilde{v}_\beta) \right)_{i,j} .$$

On the other hand, again via the identification in (7.40), $(u_i)_i \otimes (v_j)_j$ and $(\widetilde{u}_i)_i \otimes (\widetilde{v}_j)_j$ are mapped respectively to

$$(u_i \otimes v_j)_{i,j} \quad \text{and} \quad (\widetilde{u}_i \otimes \widetilde{v}_j)_{i,j},$$

and the composition of these two elements in $\widehat{\mathscr{T} \otimes \mathscr{T}}$ is given by (see (2.93) on page 105, which immediately extends from $\mathscr{T} \otimes \mathscr{T}$ to $\widehat{\mathscr{T} \otimes \mathscr{T}}$)

$$(u_i \otimes v_j)_{i,j} \bullet (\widetilde{u}_i \otimes \widetilde{v}_j)_{i,j} = \left(\sum_{\substack{a+\alpha=i \\ b+\beta=j}} (u_a \otimes v_b) \bullet (\widetilde{u}_\alpha \otimes \widetilde{v}_\beta) \right)_{i,j \geq 0}$$

$$= \left(\sum_{\substack{a+\alpha=i \\ b+\beta=j}} (u_a \cdot \widetilde{u}_\alpha) \otimes (v_b \cdot \widetilde{v}_\beta) \right)_{i,j}$$

$$= \left(\sum_{\substack{a+\alpha=i \\ b+\beta=j}} (u_a \otimes \widetilde{u}_\alpha) \otimes (v_b \otimes \widetilde{v}_\beta) \right)_{i,j} ,$$

and this is exactly (with a different yet equivalent notation) what we got above. This proves that the given identification is in fact a UAA morphism. $\square$

Proof (of Remark 2.83, page 106). As $\{\alpha_k\}_k$ and $\{\beta_k\}_k$ are such that

$$\lim_{k \to \infty} \alpha_k = \alpha \quad \text{and} \quad \lim_{k \to \infty} \beta_k = \beta$$

in $\widehat{\mathscr{T}}(V)$, we know that

$$\forall\, n \in \mathbb{N},\ \exists\, k(n) \in \mathbb{N}:\qquad \alpha_k - \alpha,\, \beta_k - \beta \in \textstyle\prod_{i \geq n} \mathscr{T}_i(V),\quad \forall\, k \geq k(n).$$

Furthermore, it holds that

$$\alpha_k \otimes \beta_k - \alpha \otimes \beta = \alpha_k \otimes \beta_k - \alpha \otimes \beta_k + \alpha \otimes \beta_k - \alpha \otimes \beta$$
$$= (\alpha_k - \alpha) \otimes \beta_k + \alpha \otimes (\beta_k - \beta).$$

Now, the result follows from the fact that

$$x \in \textstyle\prod_{i \geq n} \mathscr{T}_i(V),\ y \in \widehat{\mathscr{T}}(V) \quad \Longrightarrow \quad x \otimes y,\, y \otimes x \in \textstyle\prod_{i+j \geq n} \mathscr{T}_{i,j}(V).$$

Let us prove this last claim: with the above meaning of x, y we have

$$x \otimes y = (0, \ldots, 0, x_{n+1}, x_{n+2}, \ldots) \otimes (y_0, y_1, \ldots)$$
$$\equiv (\underbrace{0, 0, \ldots, 0}_{n + 1 \text{ times}}, *, *, \ldots) \in \textstyle\prod_{i+j \geq n} \mathscr{T}_{i,j}(V),$$

where the identification in (2.82) has been used. □

7.7 Proofs of Sect. 2.4

Proof (of Theorem 2.92, page 110). (i). We split the proof in different steps.

(i.1) *Existence.* As f is linear and A is a UA algebra, recalling Theorem 2.38-(ii), we know that there exists one and only one UAA morphism

$$\overline{f} : \mathscr{T}(\mathfrak{g}) \to A$$

extending f. So let us define

$$f^{\mu} : \mathscr{U}(\mathfrak{g}) \to A$$
$$[t]_{\mathscr{J}} \mapsto \overline{f}(t).$$

In order to prove our statement, we will check that the so defined f^{μ} has all the required properties:

- f^{μ} is well posed. This is equivalent to prove that $\mathscr{J} \subseteq \ker(\overline{f})$. The generic element of $\mathscr{J}$ can be written as a linear combination of elements of this form: $t \cdot (x \otimes y - y \otimes x - [x, y]) \cdot t'$, where t, t' belong to $\mathscr{T}(\mathfrak{g})$, and x, y to $\mathfrak{g}$. As $\overline{f}$ is linear, let us just evaluate the images of such elements via $\overline{f}$:

$$\overline{f}\big(t \cdot (x \otimes y - y \otimes x - [x,y]) \cdot t'\big)$$

$$= \overline{f}(t) * \Big(\overline{f}(x) * \overline{f}(y) - \overline{f}(y) * \overline{f}(x) - \overline{f}([x,y])\Big) * \overline{f}(t')$$

$$= \overline{f}(t) * \underbrace{\big(f(x) * f(y) - f(y) * f(x) - f([x,y])\big)}_{= \,0 \text{ for } f \text{ is a Lie algebra morphism}} * \overline{f}(t') = 0.$$

- f^μ is linear:

$$f^\mu(a[t]_{\mathscr{J}} + b[t']_{\mathscr{J}}) = f^\mu([at + bt']_{\mathscr{J}}) = \overline{f}(at + bt')$$

$$= a\overline{f}(t) + b\overline{f}(t') = af^\mu([t]_{\mathscr{J}}) + bf^\mu([t']_{\mathscr{J}}).$$

- f^μ is an algebra morphism:

$$f^\mu([t]_{\mathscr{J}}[t']_{\mathscr{J}}) = f^\mu([t \cdot t']_{\mathscr{J}}) = \overline{f}(t \cdot t')$$

$$= \overline{f}(t) * \overline{f}(t') = f^\mu([t]_{\mathscr{J}}) * f^\mu([t']_{\mathscr{J}}).$$

- f^μ is unital:

$$f^\mu(1_{\mathscr{U}(\mathfrak{g})}) = f^\mu([1_{\mathbb{K}}]_{\mathscr{J}}) = \overline{f}(1_{\mathbb{K}}) = 1_A.$$

- $f^\mu \circ \mu = f$: for every $x \in \mathfrak{g}$ we have

$$(f^\mu \circ \mu)(x) = f^\mu([x]_{\mathscr{J}}) = \overline{f}(x) = f(x).$$

(i.2) *Uniqueness.*

Suppose that g^μ verifies all the properties required in the statement. Let us define

$$g : \mathscr{T}(\mathfrak{g}) \to A$$

$$t \mapsto g^\mu([t]_{\mathscr{J}}).$$

For every $x \in \mathfrak{g}$, we have $g(x) = g^\mu([x]_{\mathscr{J}}) = f(x)$, so g extends f. Moreover g is clearly linear. Furthermore, g is an associative algebra morphism:

$$g(v_1 \otimes \ldots \otimes v_k) = g^\mu([v_1 \otimes \ldots \otimes v_k]_{\mathscr{J}}) = g^\mu([v_1]_{\mathscr{J}} \cdots [v_k]_{\mathscr{J}})$$

$$= g^\mu([v_1]_{\mathscr{J}}) * \cdots * g^\mu([v_k]_{\mathscr{J}}) = g(v_1) * \cdots * g(v_k).$$

Thanks to Theorem 2.38-(ii), we can conclude that $g \equiv \overline{f}$, where $\overline{f}$ is the unique UAA morphism closing the diagram

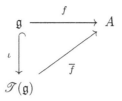

so that

$$g^\mu([t]_\mathscr{J}) = g(t) = \overline{f}(t) = f^\mu([t]_\mathscr{J}), \qquad \forall\, t \in \mathscr{T}(\mathfrak{g}),$$

i.e., $g^\mu = f^\mu$, and the proof is complete.

(ii) The proof of the first part of (ii) is standard. The second part is deduced as follows. Remark 2.89 implies that the set $\{1_U\} \cup \varphi(\mathfrak{g})$ is a set of algebra generators for U. Indeed, since $\varphi^\mu : \mathscr{U}(\mathfrak{g}) \to U$ is a UAA isomorphism (by the first part) and since $\{\pi(1)\} \cup \mu(\mathfrak{g})$ is a set of algebra generators for $\mathscr{U}(\mathfrak{g})$ then

$$\varphi^\mu\big(\{\pi(1)\} \cup \mu(\mathfrak{g})\big) = \{\varphi^\mu(1_{\mathscr{U}(\mathfrak{g})})\} \cup \varphi^\mu(\mu(\mathfrak{g})) = \{1_U\} \cup \varphi(\mathfrak{g})$$

is a set of algebra generators for U.

To prove that $U \simeq \mathscr{U}(\varphi(\mathfrak{g}))$, it suffices to show that U has the universal property of $\mathscr{U}(\varphi(\mathfrak{g}))$: this can be done arguing exactly as in the proof (on page 404) of Theorem 2.38 as we show in what follows.

Let A be any UA algebra and let $f : \varphi(\mathfrak{g}) \to A$ be any Lie algebra morphism. Let $\iota : \varphi(\mathfrak{g}) \hookrightarrow U$ be the inclusion map. We need to show the existence of a unique UAA morphism $f^\iota : U \to A$ such that $f^\iota \circ \iota \equiv f$ on $\varphi(\mathfrak{g})$, i.e.,

$$(\bigstar) \qquad\qquad (f^\iota \circ \iota)(\varphi(x)) = f(\varphi(x)), \quad \forall\, x \in \mathfrak{g}.$$

The uniqueness of such a UAA morphism f^ι follows from the fact that $\{1_U\} \cup \varphi(\mathfrak{g})$ generates U, as an algebra. Let us turn to its existence: We consider the linear map $f \circ \varphi : \mathfrak{g} \to A$. We know that $(f \circ \varphi)^\varphi : U \to A$ is a UAA morphism such that $(f \circ \varphi)^\varphi(\varphi(x)) = (f \circ \varphi)(x)$ for every $x \in \mathfrak{g}$. If we set $f^\iota := (f \circ \varphi)^\varphi$ then we are trivially done with $(\bigstar)$. $\qquad\square$

Proof (of Proposition 2.93, page 111). Let $\iota : \mathcal{L}(V) \to \mathscr{T}(V)$ be the inclusion. It is enough to show that the pair $(\mathscr{T}(V), \iota)$ has the universal property characterizing (up to canonical isomorphism) the enveloping algebra of $\mathcal{L}(V)$ (and then to apply Theorem 2.92-(ii)). To this end, we have to prove that, for every UA algebra A and every Lie algebra morphism $\alpha : \mathcal{L}(V) \to A$, there exists a unique UAA morphism $\alpha^\iota : \mathscr{T}(V) \to A$ such that

$$(\alpha^\iota \circ \iota)(t) = \alpha(t) \quad \text{for every } t \in \mathcal{L}(V). \tag{7.41}$$

To this aim, let us consider the linear map $\alpha|_V : V \to A$. From Theorem 2.38-(ii), there exists a UAA morphism $\overline{\alpha|_V} : \mathcal{T}(V) \to A$ prolonging $\alpha|_V$. We set $\alpha^\iota := \overline{\alpha|_V}$, and we prove (7.41). Let us denote by $[\cdot,\cdot]_\otimes$ and $[\cdot,\cdot]_A$ the commutators on $\mathcal{T}(V)$ and A respectively. Since $\mathcal{L}(V)$ is Lie-generated by V (see Proposition 2.47) it suffices to prove (7.41) when $t = [v_1 \cdots [v_{n-1}, v_n]_\otimes \cdots]_\otimes$, for any $n \in \mathbb{N}$ and $v_1, \ldots, v_n \in V$. We have

$$(\alpha^\iota \circ \iota)(t) = (\alpha^\iota \circ \iota)\big([v_1 \cdots [v_{n-1}, v_n]_\otimes \cdots]_\otimes\big)$$
$$= \big([\alpha^\iota(v_1) \cdots [\alpha^\iota(v_{n-1}), \alpha^\iota(v_n)]_A \cdots]_A\big)$$
$$= [\alpha(v_1) \cdots [\alpha(v_{n-1}), \alpha(v_n)]_A \cdots]_A = \alpha\big([v_1 \cdots [v_{n-1}, v_n]_\otimes \cdots]_\otimes\big) = \alpha(t).$$

In the first equality we used the fact that, α^ι being a UAA morphism, it is also a Lie algebra morphism of the associated commutator-algebras; in the second equality we used $\alpha^\iota = \overline{\alpha|_V}$ and the latter map coincides with α on V; in the third equality we used the fact that $\alpha : \mathcal{L}(V) \to A$ is a Lie algebra morphism and the fact that the Lie algebra structure on $\mathcal{L}(V)$ is the one induced by the commutator of $\mathcal{T}(V)$. The uniqueness of a UAA morphism $\alpha^\iota : \mathcal{T}(V) \to A$ satisfying (7.41) is granted by the fact that (7.41) defines α^ι on V uniquely, so that α^ι is uniquely defined throughout, since $\mathcal{T}(V)$ is generated by $\{1_\mathbb{K}\} \cup V$, as an algebra.

Proof (of Theorem 2.94, page 111). Recall the projection map

$$\phi : \mathcal{T}(\mathfrak{g}) \to \mathrm{Sym}(\mathfrak{g}), \quad t \mapsto [t]_{\mathcal{H}}$$

introduced in (10.2) on page 501 and consider its restriction to $\mathfrak{g}$. Let $\mathcal{X} := \{x_i\}_{i \in \mathcal{J}}$ be a fixed basis of $\mathfrak{g}$, such that $\mathcal{J}$ is totally ordered by $\preccurlyeq$, as in the statement of the PBW Theorem. For the sake of notational convenience, we set henceforth

$$x_{i_0} := 1_\mathbb{K} \qquad \text{(where } i_0 \notin \mathcal{J}).$$

Thanks to Theorem 10.20 on page 512 (and the fact that $\phi(x_{i_0}) = \phi(1_\mathbb{K})$ is the unit of $\mathrm{Sym}(\mathfrak{g})$), the set

$$\mathcal{A} := \big\{\phi(x_{i_0}) * \phi(x_{i_1}) * \cdots * \phi(x_{i_p}) \,\big|\, p \geq 0, \; i_1, \ldots, i_p \in \mathcal{J}, \; i_1 \preccurlyeq \ldots \preccurlyeq i_p\big\}$$

is a linear basis for the symmetric algebra $\mathrm{Sym}(\mathfrak{g})$. For any given $p \in \mathbb{N} \cup \{0\}$, let us denote by $S_p(\mathfrak{g})$ the subspace of $\mathrm{Sym}(\mathfrak{g})$ spanned by the symmetric powers of degree $\leq p$, that is,

$$S_p(\mathfrak{g}) := \bigoplus_{n=0}^p \mathrm{Sym}_n(\mathfrak{g}).$$

Let us remark that, for every $p \geq 0$, the subsystem of $\mathcal{A}$ given by

$$\mathcal{A}_p := \left\{ \phi(x_{i_0}) * \phi(x_{i_1}) * \cdots * \phi(x_{i_p}) \mid i_1, \ldots, i_p \in \mathcal{I},\ i_1 \preccurlyeq \ldots \preccurlyeq i_p \right\}$$

is a basis of $\mathrm{Sym}_p(\mathfrak{g})$ (thanks to the same cited theorem).

We now have to introduce a notation "$\le$", useful for the proof which follows.

Definition 7.10. Let x be an element of the basis $\mathcal{X} = \{x_i\}_{i \in \mathcal{I}}$ of $\mathfrak{g}$ and let s be an element of the basis $\mathcal{A}$ of $\mathrm{Sym}(\mathfrak{g})$. We allow the writing

$$\phi(x) \le s$$

if and only if one of the following conditions is satisfied:

- $s = \phi(x_{i_0})$.
- $x = x_i$ and $s = \phi(x_{i_0}) * \phi(x_{i_1}) * \cdots * \phi(x_{i_n})$ with $i \preccurlyeq i_1$.

With the above definition at hands, we are able to prove the next result.[6]

Lemma 7.11. *With the above notation, there exists a Lie algebra morphism*

$$\rho : \mathfrak{g} \longrightarrow \mathrm{End}(\mathrm{Sym}(\mathfrak{g}))$$

such that, for every $s \in \mathcal{A}$ and for every $x \in \mathcal{X}$ one has the following properties:

(1) $s \in \mathcal{A}_p$ implies $\rho(x)(s) - \phi(x) * s \in S_p(\mathfrak{g})$.
(2) $\phi(x) \le s$ implies $\rho(x)(s) = \phi(x) * s$.

Proof. I. We define $\rho(x)(s)$ for every $x \in \mathfrak{g}$ and then, one step at a time, for $s \in \mathrm{Sym}_0(\mathfrak{g})$, $s \in \mathrm{Sym}_1(\mathfrak{g})$, and so on.

(I.0). Let $s \in \mathcal{A}_0$, that is, $s = \phi(1_{\mathbb{K}})$. We set

$$\rho(x)(\phi(1_{\mathbb{K}})) := \phi(x), \quad \text{when } x \in \mathcal{X}. \tag{7.42}$$

By extending this definition *linearly* both in x and in s, we can unambiguously define $\rho(x)(s)$ for every $x \in \mathfrak{g}$ and $s \in \mathrm{Sym}_0(\mathfrak{g})$. Note that, by (7.42),

$$\rho(x)(s) \in S_1(\mathfrak{g}), \quad \text{for every } x \in \mathfrak{g} \text{ and every } s \in \mathrm{Sym}_0(\mathfrak{g}). \tag{7.43}$$

Moreover, when $s = \phi(1_{\mathbb{K}})$, (7.42) ensures the validity of statement (2) – and consequently of statement (1) – of the lemma.

(I.1). Let $s \in \mathcal{A}_1$, $s = \phi(x_i)$ for a fixed $i \in \mathcal{I}$. We begin by defining $\rho(x)(s)$ for $x = x_j \in \mathcal{X}$ for some $j \in \mathcal{I}$. We set

---

[6]As usual, if V is a vector space, we denote by $\mathrm{End}(V)$ the vector space of the endomorphisms of V, which is a UA algebra with the composition of maps, and is – in its turn – a Lie algebra with the associated commutator. Moreover, if V, W are vector spaces (on the same field), we denote by $\mathrm{Hom}(V, W)$ the vector space of the linear maps $\varphi : V \to W$.

$$\rho(x_j)(\phi(x_i)) := \begin{cases} (j \preccurlyeq i) & \phi(x_j) * \phi(x_i) \\ (i \preccurlyeq j) & \phi(x_i) * \phi(x_j) + \rho([x_j, x_i])(\phi(1_{\mathbb{K}})). \end{cases} \tag{7.44}$$

Note that this defines $\rho(x_j)(\phi(x_i))$ as an element of $S_2(\mathfrak{g})$ (recall indeed (7.43)). As a consequence, we can use bi-linearity to define $\rho(x)(s)$ for any $x \in \mathfrak{g}$ and any $s \in \mathrm{Sym}_1(\mathfrak{g})$; furthermore, by means of the above step (I.0) and by gluing together $\mathrm{Sym}_0(\mathfrak{g})$ and $\mathrm{Sym}_1(\mathfrak{g})$, we unambiguously define $\rho(x)(s)$ for every $x \in \mathfrak{g}$ and every $s \in S_1(\mathfrak{g})$ in such a way that

$$\rho(x)(s) \in S_2(\mathfrak{g}), \quad \text{for every } x \in \mathfrak{g} \text{ and every } s \in S_1(\mathfrak{g}). \tag{7.45}$$

We now show that this choice of ρ fulfills (1) and (2) of the lemma (plus a morphism-like property). Indeed, as for (1) we have:

$$\rho(x_j)(\phi(x_i)) - \phi(x_j) * \phi(x_i) = \begin{cases} (j \preccurlyeq i) & 0, \\ (i \preccurlyeq j) & \rho([x_j, x_i])(\phi(1_{\mathbb{K}})), \end{cases}$$

and the far right-hand belongs to $S_1(\mathfrak{g})$, in view of (7.43). Next, as for (2) we have $\phi(x) \leq s$ if and only if (recall that $x = x_j$ and $s = \phi(x_i)$) $j \preccurlyeq i$ so that (2) is trivially verified by the very (7.44).

We have an extra property of ρ, namely:

$$\rho(y)(\rho(z)(t)) - \rho(z)(\rho(y)(t)) = \rho([y, z])(t) \quad \left(\begin{array}{l} \text{for all } y, z \in \mathfrak{g} \\ \text{for all } t \in S_0(\mathfrak{g}) \end{array} \right). \tag{7.46}$$

To prove (7.46), we note that (by linearity on all the arguments), we can restrict to prove it when $y, z \in \mathcal{X}$ and $t \in \mathcal{A}_0$. So we can suppose that $y = x_j$, $z = x_i$ and (by the skew-symmetric rôles of y, z in (7.46)) we can additionally suppose that $j \preccurlyeq i$. So we have (recall that $*$ is Abelian!)

$$\rho(x_j)\big(\rho(x_i)(\phi(1_{\mathbb{K}}))\big) - \rho(x_i)\big(\rho(x_j)(\phi(1_{\mathbb{K}}))\big)$$

$$\overset{(7.42)}{=} \rho(x_j)(\phi(x_i)) - \rho(x_i)(\phi(x_j))$$

$$\overset{(7.44)}{=} \phi(x_j) * \phi(x_i) - \big(\phi(x_j) * \phi(x_i) + \rho([x_i, x_j])(\phi(1_{\mathbb{K}}))\big)$$

$$= -\rho([x_i, x_j])(\phi(1_{\mathbb{K}})) = \rho([x_j, x_i])(\phi(1_{\mathbb{K}})).$$

(I.p). Inductively we suppose that, for a given $p \geq 1$, we have defined $\rho(x)(s)$ for $x \in \mathfrak{g}$ and $s \in S_p(\mathfrak{g})$, in such a way that it depends linearly on x and on s and that the following properties hold:

$$\rho(x)(s) \in S_{j+1}(\mathfrak{g}), \quad \left(\begin{array}{l} \text{for every } x \in \mathfrak{g}, s \in S_j(\mathfrak{g}) \\ \text{and every } j = 0, 1, \ldots, p \end{array} \right); \tag{7.47}$$

$$\rho(x)(s) - \phi(x) * s \in S_j(\mathfrak{g}), \quad \left(\begin{array}{l} \text{for every } x \in \mathfrak{g},\, s \in \mathcal{A}_j \\ \text{and every } j = 0, 1, \ldots, p \end{array} \right); \tag{7.48}$$

$$\phi(x) \leq s \text{ implies } \rho(x)(s) = \phi(x) * s, \quad \left(\begin{array}{l} \text{for every } x \in \mathfrak{X},\, s \in \mathcal{A}_j(\mathfrak{g}) \\ \text{and every } j = 0, 1, \ldots, p \end{array} \right); \tag{7.49}$$

$$\rho(y)(\rho(z)(t)) - \rho(z)(\rho(y)(t)) = \rho([y, z])(t) \quad \left(\begin{array}{l} \text{for every } y, z \in \mathfrak{g} \\ \text{and every } t \in S_{p-1}(\mathfrak{g}) \end{array} \right). \tag{7.50}$$

(I.$p+1$). We now show how to define $\rho(x)(s)$ for $x \in \mathfrak{g}$ and $s \in S_{p+1}(\mathfrak{g})$ satisfying all the above properties up to the step $p + 1$, with the sole hypothesis that the statements in (I.p) above do hold.

It suffices to define $\rho(x)(s)$ for $x \in \mathfrak{g}$ and $s \in \mathrm{Sym}_{p+1}(\mathfrak{g})$; in turns, by eventually defining $\rho(x)(s)$ with a linearity argument, it suffices to take $x \in \mathfrak{X}$ and $s \in \mathcal{A}_{p+1}$. Hence, let $x = x_i \in \mathfrak{X}$ and $s = \phi(x_{i_1}) * \cdots * \phi(x_{i_{p+1}})$ be fixed, for $i, i_1, \ldots, i_{p+1} \in \mathcal{I}$ with $i_1 \preccurlyeq \cdots \preccurlyeq i_{p+1}$. We distinguish two cases, depending on whether $\phi(x)$ is greater than or less than s.

– If $\phi(x) \leq \phi(x_{i_1}) * \cdots * \phi(x_{i_{p+1}})$, condition (7.49) for the $(p + 1)$-th case immediately forces us to set

$$\rho(x)(\phi(x_{i_1}) * \cdots * \phi(x_{i_p})) := \phi(x) * \phi(x_{i_1}) * \cdots * \phi(x_{i_{p+1}}). \tag{7.51}$$

Condition (7.49) (up to $p+1$) is completely fulfilled. Moreover, in the present case $\phi(x) \leq s$, (7.48) and (7.47) are verified as well (all up to the case $p + 1$).

– Let us now consider the case $\phi(x) \not\leq \phi(x_{i_1}) * \cdots * \phi(x_{i_{p+1}})$, that is, $i_1 \preccurlyeq i$ and $i_1 \neq i$. We have $s = s' * \bar{s}$, where

$$s' := \phi(x_{i_1}), \qquad \bar{s} := \phi(x_{i_2}) * \cdots * \phi(x_{i_{p+1}}) \in \mathcal{A}_p.$$

Thanks to the inductive hypothesis (7.48), we have that $\rho(x)(\bar{s}) - \phi(x) * \bar{s}$ belongs to $S_p(\mathfrak{g})$, and its image via $\rho(x_{i_1})$ makes sense. We can thus define for general x and s

$$\rho(x)(s) := \rho([x, x_{i_1}])(\bar{s}) + \rho(x_{i_1})(\rho(x)(\bar{s}) - \phi(x) * \bar{s}) + \phi(x) * s. \tag{7.52}$$

This formula for ρ satisfies (7.47) (indeed, the first two summands belong to $S_{p+1} \subseteq S_{p+2}$, whilst the last belongs to S_{p+2}). As a consequence, (7.47) is completely fulfilled. Furthermore (for the same reasons) we have

$$\rho(x)(s) - \phi(x) * s \stackrel{(7.52)}{=} \rho([x, x_{i_1}])(\bar{s}) + \rho(x_{i_1})(\rho(x)(\bar{s}) - \phi(x) * \bar{s}) \in S_{p+1}(\mathfrak{g}),$$

so condition (7.48) is completely verified.

We are left to verify (7.50) when $p - 1$ is replaced by p. To begin with, notice that $s' \leq \phi(x) * \bar{s}$ (recalling that $s' = \phi(x_{i_1})$, $\phi(x) = \phi(x_i)$ and $i_1 \preccurlyeq i$), so that we have

$$\rho(x_{i_1})(\phi(x) * \bar{s}) \overset{(7.51)}{=} \phi(x_{i_1}) * \phi(x) * \bar{s} = s' * \phi(x) * \bar{s}$$
$$= \phi(x) * s' * \bar{s} = \phi(x) * s.$$

Hence equation 7.52 may now be rewritten as[7]

$$\rho(x)(\phi(x_{i_1}) * \bar{s}) = \rho([x, x_{i_1}])(\bar{s}) + \rho(x_{i_1})(\rho(x)(\bar{s})). \qquad (7.53)$$

In its turn, noticing that $\phi(x_{i_1}) \leq \bar{s}$ (indeed $\bar{s} = \phi(x_{i_2}) * \cdots$, with $i_1 \preccurlyeq i_2$) and hence applying (7.51), the identity (7.53) can be rewritten as

$$\rho(x)(\rho(x_{i_1})(\bar{s})) - \rho(x_{i_1})(\rho(x)(\bar{s})) = \rho([x, x_{i_1}])(\bar{s}).$$

The last equation ensures that the morphism condition (7.50) is verified for all $t \in \mathcal{A}_p$ and all $y, z \in \mathcal{X}$, as long as the following is satisfied:

$$\phi(z) \leq t \quad \& \quad \phi(z) < \phi(y).$$

So all we have to do in order to complete the proof is to check that (7.50) is verified in all the other cases.

First of all, we remark that, if (7.50) holds for some $y, z \in \mathfrak{g}$, then it also holds when y and z are interchanged. So we obtain automatically that (7.50) is also verified in the case

$$\phi(y) \leq t \quad \& \quad \phi(y) < \phi(z).$$

Secondly, the case $y = z$ is trivial. Consequently, all the cases when $\phi(z) \leq t$ as well as the cases when $\phi(y) \leq t$ hold true. [Indeed, notice that when $x_i, x_j \in \mathcal{X}$, condition $\phi(x_i) \leq \phi(x_j)$ is equivalent to $i \preccurlyeq j$ so that – via ϕ – the symbol $\leq$ defines a total ordering on the elements of $\mathcal{X}$.]

As a consequence, the very last case to consider is

$$\phi(y) \not\leq t \quad \& \quad \phi(z) \not\leq t.$$

As above, let us set $t := t' * \bar{t}$, with

$$t' := \phi(x_{i_1}) \qquad \bar{t} := \phi(x_{i_2}) * \cdots * \phi(x_{i_p}).$$

---

[7]We observe that $\rho(x)(s)$ cannot be defined straightaway as in (7.53), because the summand $\rho(x_{i_1})(\rho(x)(\bar{s}))$ is not, a priori, well-defined.

Then necessarily $t' \leq \phi(y)$, $t' \leq \phi(z)$, and we have

$$\rho(z)(t) \overset{(7.51)}{=} \rho(z)(\rho(x_{i_1})(\bar{t})) \overset{(7.50)}{=} \rho(x_{i_1})(\rho(z)(\bar{t})) + \rho([z, x_{i_1}])(\bar{t})$$

$$\text{(we add and subtract } \rho(x_{i_1})(\phi(z) * \bar{t}))$$

$$= \rho(x_{i_1})(\phi(z) * \bar{t}) + \rho(x_{i_1})(\rho(z)(\bar{t}) - \phi(z) * \bar{t}) + \rho([z, x_{i_1}])(\bar{t}).$$

Applying the map $\rho(y)$ to both sides, we obtain

$$\rho(y)(\rho(z)(t)) = \rho(y)(\rho(x_{i_1})(\phi(z) * \bar{t}))$$
$$+ \rho(y)(\rho(x_{i_1})(\rho(z)(\bar{t}) - \phi(z) * \bar{t})) + \rho(y)(\rho([z, x_{i_1}])(\bar{t})).$$

As $t' \leq \phi(z) * \bar{t}$ and $t' \leq \phi(y)$, the morphism condition is verified for the first term on the right side. Thanks to the inductive hypothesis (7.50), the morphism condition can be applied to the last two terms as well. We get

$$\rho(y)(\rho(z)(t)) = \rho(x_{i_1})(\rho(y)(\phi(z) * \bar{t})) + \rho([y, x_{i_1}])(\phi(z) * \bar{t})$$
$$+ \rho(x_{i_1})\left(\rho(y)(\rho(z)(\bar{t}) - \phi(z) * \bar{t})\right)$$
$$+ \rho([y, x_{i_1}])(\rho(z)(\bar{t}) - \phi(z) * \bar{t})$$
$$+ \rho([z, x_{i_1}])(\rho(y)(\bar{t})) + \rho([y, [z, x_{i_1}]])(\bar{t})$$
$$= \rho(x_{i_1})(\rho(y)(\rho(z)(\bar{t}))) + \rho([y, x_{i_1}])(\rho(z)(\bar{t}))$$
$$+ \rho([z, x_{i_1}])(\rho(y)(\bar{t})) + \rho([y, [z, x_{i_1}]])(\bar{t}). \tag{7.54}$$

An analogous result is found upon exchange of y and z:

$$\rho(z)(\rho(y)(t)) = \rho(x_{i_1})(\rho(z)(\rho(y)(\bar{t}))) + \rho([z, x_{i_1}])(\rho(y)(\bar{t}))$$
$$+ \rho([y, x_{i_1}])(\rho(z)(\bar{t})) + \rho([z, [y, x_{i_1}]])(\bar{t}). \tag{7.55}$$

Let us subtract equation (7.55) from equation (7.54). We obtain

$$\rho(y)(\rho(z)(t)) - \rho(z)(\rho(y)(t)) = \rho(x_{i_1})\left(\rho(y)(\rho(z)(\bar{t})) - \rho(z)(\rho(y)(\bar{t}))\right)$$
$$+ \rho([y, [z, x_{i_1}]])(\bar{t}) - \rho([z, [y, x_{i_1}]])(\bar{t}). \tag{7.56}$$

Now, using the Jacobi identity to rewrite the last two terms, and applying the inductive hypothesis twice, (7.56) reduces to

$$\rho(y)(\rho(z)(t)) - \rho(z)(\rho(y)(t)) = \rho(x_{i_1})(\rho([y, z])(\bar{t})) - \rho([x_{i_1}, [z, z]])(\bar{t})$$
$$= \rho([y, z])(\rho(x_{i_1})(\bar{t})) \overset{(7.51)}{=} \rho([y, z])(t),$$

which is precisely the morphism condition we set out to prove.

(II). *End of the Lemma.* We notice that we have defined $\rho(x)(\cdot)$ on every S_p in such a way that the definition on S_{p+1} agrees with that on S_p. This defines $\rho(x)(\cdot)$ unambiguously – as a linear map – on the whole of $\mathrm{Sym}(\mathfrak{g})$. Moreover, this also defines ρ as a linear map on $\mathfrak{g}$ with values in $\mathrm{End}(\mathrm{Sym}(\mathfrak{g}))$. Finally, condition (7.48) gives (1) in the assertion of the Lemma, condition (7.49) gives (2), and condition (7.50) ensures that ρ is actually a Lie algebra morphism. This ends the proof of the lemma.
□

Let us turn back to the proof of Theorem 2.94.

First of all we aim to show that the set

$$1,\quad X_{i_1}\cdots X_{i_n},\qquad where\quad n\in\mathbb{N},\ i_1,\ldots,i_n\in\mathfrak{I},\ i_1\preccurlyeq\ldots\preccurlyeq i_n.\qquad(7.57)$$

spans $\mathscr{U}(\mathfrak{g})$ as a vector space. We already know from Remark 2.89 that the set $\{1_{\mathbb{K}}\}\cup\mu(\mathfrak{g})$ generates $\mathscr{U}(\mathfrak{g})$ as an algebra. This means that any element $g\in\mathscr{U}(\mathfrak{g})$ may be written as a linear combination of products of the form $\mu(g_1)\cdots\mu(g_k)$, for some $k\in\mathbb{N}$ and for some g_i in $\mathfrak{g}$. In order to prove that the cited system spans $\mathscr{U}(\mathfrak{g})$, it is not restrictive to consider the case when the above g_i are elements of the basis $\{x_i\}_{i\in\mathfrak{I}}$ of $\mathfrak{g}$, and that g is of the form $g=\mu(x_{i_1})\cdots\mu(x_{i_k})$.

Let us proceed by induction on k. When $k=2$, if $i_1\preceq i_2$, we are done. If not, we have

$$g=\mu(x_{i_1})\mu(x_{i_2})=\mu([x_{i_1},x_{i_2}])+\mu(x_{i_2})\mu(x_{i_1}),$$

and, as $[x_{i_1},x_{i_2}]$ belongs to $\mathfrak{g}$, again we are done.

Let us now prove the inductive step. For k generic, we have

$$g=\mu(x_{i_1})\cdots\mu(x_{i_k})=X_{i_1}\cdots X_{i_k}.$$

The same argument as for the case $k=2$ shows how to interchange two consecutive X_i, modulo an element which is the product of $k-1$ factors (to which we are entitled to apply the inductive hypothesis). This shows at once how to reduce g to a linear combination of elements as in (7.57).

We are only left to prove the *linear independence* of the elements of the form (7.57), which is the hard task in the proof of the PBW Theorem. This motivates the use of the above Lemma 7.11, from where we inherit the notations.

By this lemma, we know the existence of a Lie algebra morphism $\rho:\mathfrak{g}\to\mathrm{End}(\mathrm{Sym}(\mathfrak{g}))$, so that we can apply the characteristic property of the Universal Enveloping Algebra (see Theorem 2.92, page 110) and infer the existence of a unique UAA morphism ρ^{μ} closing the following diagram:

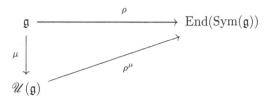

To prove the linear independence of the system in (7.57) it will suffice to show that the image via ρ^μ of any formally non-trivial linear combination of the elements in (7.57) is different from 0. But this is easily done: thanks to condition (2) of Lemma 7.11, we have

$$\rho^\mu(X_{i_1}\cdots X_{i_n})(\phi(1_\mathbb{K})) = \rho^\mu(\mu(x_{i_1})\cdots\mu(x_{i_n}))(\phi(1_\mathbb{K}))$$

$$= \Big(\rho^\mu(\mu(x_{i_1}))\circ\cdots\circ\rho^\mu(\mu(x_{i_n}))\Big)(\phi(1_\mathbb{K})) = \Big(\rho(x_{i_1})\circ\cdots\circ\rho(x_{i_n})\Big)(\phi(1_\mathbb{K}))$$

$$= \phi(x_{i_1})\ast\cdots\ast\phi(x_{i_n}).$$

So the image of a linear combination of several terms, each of the form $X_{i_1}\cdots X_{i_n}$, evaluated in $\phi(1_\mathbb{K})$, is a linear combination of elements of the given basis of $\mathrm{Sym}(\mathfrak{g})$ and thus is different from 0 unless its coefficients are all 0.

[We explicitly remark that the fact that ρ verifies condition (1) of Lemma 7.11 is apparently not used in the proof of Theorem 2.94. Nevertheless, it has been essential in the construction of the morphism ρ.]   □

7.8 Miscellanea of Proofs

Proof (of Proposition 4.12, page 203).

(i) The linearity of ∂_t is obvious. Moreover, if $p = (a_j)_j$ and $q = (b_j)_j$ are arbitrary elements of $A[[t]]$, we have

$$\partial_t(p\ast q) = \partial_t\left(\sum_{j\geq 0}\Big(\textstyle\sum_{i=0}^j a_i\star b_{j-i}\Big)t^j\right)$$

$$= \sum_{j\geq 0}(j+1)\Big(\textstyle\sum_{i=0}^{j+1} a_i\star b_{j+1-i}\Big)t^j.$$

On the other hand,

$$(\partial_t p) * q + p * \partial_t(q) = \left(\sum_{j\geq 0}(j+1)\,a_{j+1}\,t^j\right) * \left(\sum_{j\geq 0} b_j\,t^j\right)$$

$$+ \left(\sum_{j\geq 0} a_j\,t^j\right) * \left(\sum_{j\geq 0}(j+1)\,b_{j+1}\,t^j\right)$$

$$= \sum_{j\geq 0}\left\{\sum_{i=0}^{j}\Big((i+1)a_{i+1} \star b_{j-i}\right.$$

$$\left. + a_i \star (j-i+1)b_{j-i+1}\Big)\right\} t^j.$$

The above sum over i in curly braces is equal to

$$\sum_{h=1}^{j+1} h\,a_h \star b_{j-h+1} + \sum_{i=0}^{j} a_i \star (j-i+1)b_{j-i+1}$$

$$= (j+1)\,a_{j+1} \star b_0 + \sum_{h=1}^{j} h\,a_h \star b_{j-h+1}$$

$$+ (j+1)\,a_0 \star b_{j+1} + \sum_{i=1}^{j} a_i \star (j-i+1)b_{j-i+1}$$

$$= (j+1)\,a_{j+1} \star b_0 + (j+1)\,a_0 \star b_{j+1} + \sum_{i=1}^{j}(j+1)\,a_i \star b_{j-i+1}$$

$$= (j+1)\sum_{i=0}^{j+1} a_i \star b_{j-i+1}.$$

Hence $\partial_t(p * q) = (\partial_t p) * q + p * \partial_t(q)$.

(ii) Now that we know that ∂_t is a derivation of $A[[t]]$, (4.66) follows from the following general fact: Let D be a derivation of the unital associative algebra $(B, \circledast)$. Then for every $b \in B$ and $m \in \mathbb{N}$ we have

$$D(b^{\circledast\, m}) = \sum_{k=0}^{m-1} b^{\circledast\, k} \circledast (Db) \circledast b^{\circledast\, m-k-1}. \qquad (7.58)$$

We prove (7.58) by induction on m. The case $m = 1$ is trivial. Assuming (7.58) to hold for $1, 2, \ldots, m$, we prove it for $m+1$:

$$D(b^{\circledast\, m+1}) = D(b^{\circledast\, m} \circledast b) = D(b^{\circledast\, m}) \circledast b + b^{\circledast\, m} \circledast Db$$

(by the inductive hypothesis, using bilinearity and associativity of $\circledast$)

$$= \sum_{k=0}^{m-1} b^{\circledast\, k} \circledast (Db) \circledast b^{\circledast\, m-k} + b^{\circledast\, m} \circledast Db$$

$$= \sum_{k=0}^{m} b^{\circledast\, k} \circledast (Db) \circledast b^{\circledast\, m-k}.$$

(iii) By the very Definition 4.11 of ∂_t, it follows that $\partial_t(\widehat{U}_{N+1}) \subseteq \widehat{U}_N$, for every $N \in \mathbb{N}$ (where $\widehat{U}_N$ is as in (4.59)). Since the sets $\widehat{U}_N$ form a neighborhood of 0 and $\partial_t 0 = 0$, this proves that ∂_t is continuous at 0. By the linearity of ∂_t this is enough to prove (iii) (also recalling that $A[[t]]$ is a topological algebra).

(iv) For every $a \in A$, we have

$$\partial_t \exp(a\,t) \overset{(4.62)}{=} \partial_t \left(\sum_{k=0}^{\infty} \frac{a^{\star k}}{k!} t^k \right)$$

$$\overset{(4.65)}{=} \sum_{k=0}^{\infty} (k+1) \frac{a^{\star k+1}}{(k+1)!} t^k = \sum_{k=0}^{\infty} \frac{a^{\star k+1}}{k!} t^k$$

$$((4.57) \text{ ensures that } a^{\star k+1} \text{ equals } a * a^{\star k} \text{ and } a^{\star k} * a,$$

$$\text{and Remark 4.10-(3) can be used to derive that } a\,t^k = t^k * a)$$

$$= \begin{cases} \sum_{k=0}^{\infty} (a * a^{\star k} * t^k)/k! \\ \sum_{k=0}^{\infty} (a^{\star k} * t^k * a)/k! \end{cases}$$

$$= \quad \text{(by the continuity of } *) \quad \begin{cases} a * \exp(a\,t) \\ \exp(a\,t) * a. \end{cases}$$

This ends the proof. $\square$

Proof (of Remark 4.16, page 208). With the notation in Remark 4.13, we consider the map $ev_1 : A[t] \to A$. It may seem natural, in view of a possible application of this map to (4.76), to try to prolong ev_1 to a continuous map from $A[[t]]$ to $\widehat{A}$. Unfortunately, this is not possible.[8] Hence, we must first restrict ev_1 to a narrower domain. To this end, let

$$W := \left\{ p \in A[t] \;\middle|\; p = \sum_{k \geq 0} a_k\, t^k \quad \text{with } a_k \in \mathscr{T}_k(\mathbb{Q}\langle x, y \rangle) \text{ for } k \geq 0 \right\};$$

(7.59)

$$\widehat{W} := \left\{ p \in A[[t]] \;\middle|\; p = \sum_{k \geq 0} a_k\, t^k \quad \text{with } a_k \in \mathscr{T}_k(\mathbb{Q}\langle x, y \rangle) \text{ for } k \geq 0 \right\}.$$

We remark that $W, \widehat{W}$ are subalgebras of $A[t], A[[t]]$ respectively. Indeed, if $p = \sum_{k \geq 0} a_k\, t^k, q = \sum_{k \geq 0} b_k\, t^k$ with $a_k, b_k \in \mathscr{T}_k(\mathbb{Q}\langle x, y \rangle)$, then

---

[8]For example, the sequence t^{*N} vanishes, as $N \to \infty$, in $A[t]$, but the sequence $ev_1(t^{*N}) = 1$ does not.

$$p * q = \sum_{k \geq 0} \left(\underbrace{\sum_{i+j=k} \underbrace{a_i \cdot b_j}_{\in \mathscr{T}_{i+j}}}_{\in \mathscr{T}_k} \right) * t^k.$$

Moreover we have

$$p \in \widehat{W} \cap \widehat{U}_1 \quad \Rightarrow \quad \exp(p) \in \widehat{W}. \tag{7.60}$$

Indeed, if p is as above, an explicit calculation furnishes

$$\exp(p) = \sum_{h=0}^{\infty} \frac{1}{h!} \left(\sum_{k \geq 0} a_k \, t^k \right)^{*h} = \sum_{j=0}^{\infty} \left(\sum_{h=0}^{j} \frac{1}{h!} \sum_{\alpha \in \mathbb{N}^h : |\alpha| = j} a_{\alpha_1} \cdots a_{\alpha_h} \right) t^j,$$

and the sum in parentheses belongs to $\mathscr{T}_j(\mathbb{Q}\langle x, y \rangle)$, since $a_{\alpha_1} \cdots a_{\alpha_h} \in \mathscr{T}_{|\alpha|}$. We claim that the map

$$\Theta := \mathrm{ev}_1|_{\widehat{W}} : \widehat{W} \longrightarrow \widehat{\mathscr{T}}(\mathbb{Q}\langle x, y \rangle), \qquad \sum_{k \geq 0} a_k \, t^k \mapsto \sum_{k \geq 0} a_k$$

has the following properties:

- *It is well posed.* This follows immediately from the definition of $\widehat{W}$ in (7.59).
- *It is a UAA morphism.* Θ is obviously linear. Moreover, one has

$$\Theta\left(\left(\textstyle\sum_{k \geq 0} a_k \, t^k \right) * \left(\textstyle\sum_{k \geq 0} b_k \, t^k \right) \right) = \sum_{k \geq 0} \left(\textstyle\sum_{i+j=k} a_i \cdot b_j \right)$$

$$= \left(\textstyle\sum_{k \geq 0} a_k \right) \widehat{\cdot} \left(\textstyle\sum_{k \geq 0} b_k \right) = \Theta\left(\textstyle\sum_{k \geq 0} a_k \, t^k \right) \widehat{\cdot} \, \Theta\left(\textstyle\sum_{k \geq 0} b_k \, t^k \right).$$

- *It is continuous.* This follows immediately from the linearity of Θ, together with the following fact (see also (4.59)):

$$\Theta(\widehat{U}_N) \subseteq \textstyle\prod_{k \geq N} \mathscr{T}_k(\mathbb{Q}\langle x, y \rangle).$$

- *It commutes with the exponential map*, i.e., it has the property:

$$\Theta(\exp(p)) = \mathrm{Exp}(\Theta(p)), \quad \forall \, p \in \widehat{W} \cap \widehat{U}_1. \tag{7.61}$$

Here Exp on the right-hand side is the corresponding exponential map in $\widehat{\mathscr{T}}(\mathbb{Q}\langle x, y \rangle)$. The proof of this fact comes from the following identities:

$$\Theta(\exp(p)) = \Theta\left(\lim_{N\to\infty} \sum_{n=0}^{N} \frac{p^{*\,n}}{n!}\right) \quad \text{(by the continuity and linearity of } \Theta\text{)}$$

$$= \lim_{N\to\infty} \sum_{n=0}^{N} \Theta(p^{*\,n})/n! \quad \text{(as } \Theta \text{ is a UAA morphism)}$$

$$= \lim_{N\to\infty} \sum_{n=0}^{N} \left(\Theta(p)\right)^{\widehat{\ }\,n}/n! = \mathrm{Exp}(\Theta(p)).$$

Now take the identity in (4.76) and note that we are entitled to apply Θ to both sides, thanks to (7.60), thanks to the fact that

$$x\,t, \ y\,t, \ \textstyle\sum_{j=1}^{\infty} Z_j(x, y)\,t^j \quad \text{all belong to } \ \widehat{W} \cap \widehat{U}_1.$$

We thus get the identity in $\widehat{\mathscr{T}}(\mathbb{Q}\langle x, y\rangle)$:

$$\Theta\left(\exp\left(\sum_{j=1}^{\infty} Z_j(x, y)\,t^j\right)\right) = \Theta\left(\exp(x\,t) * \exp(y\,t)\right). \tag{7.62}$$

We finally recognize that this gives (4.9). Indeed, on the one hand we have

$$\Theta\left(\exp\left(\sum_{j=1}^{\infty} Z_j(x, y)\,t^j\right)\right) \overset{(7.61)}{=} \mathrm{Exp}\left(\Theta\left(\sum_{j=1}^{\infty} Z_j(x, y)\,t^j\right)\right)$$

$$\text{(by definition of } \Theta \text{ and } Z_j(x, y) \in \mathscr{T}_j(\mathbb{Q}\langle x, y\rangle))$$

$$= \mathrm{Exp}\left(\sum_{j=1}^{\infty} Z_j(x, y)\right) \overset{(4.75)}{=} \mathbf{e}^{\sum_{j=1}^{\infty} \mathbf{F}_j^{\mathscr{T}(\mathbb{Q}\langle x, y\rangle)}(x, y)} = \mathbf{e}^{\mathbf{F}(x, y)}.$$

On the other hand, since Θ is a UAA morphism, we have

$$\Theta\left(\exp(x\,t) * \exp(y\,t)\right) = \Theta(\exp(x\,t)) \widehat{\ } \Theta(\exp(y\,t))$$

$$\overset{(7.61)}{=} \mathrm{Exp}(\Theta(x\,t)) \widehat{\ } \mathrm{Exp}(\Theta(y\,t)) = \mathbf{e}^x \widehat{\ } \mathbf{e}^y.$$

This gives (4.9), as we claimed. $\qquad\square$

Proof (of Theorem 4.30, page 240). [This lemma may be proved by invoking very general properties of derivations of free UA algebras. Since we had no purposes in developing this general theory elsewhere in this Book, it is our concern in the present context to prove this lemma by explicit arguments only.] We set $\mathscr{T}_k := \mathscr{T}_k(\mathbb{K}\langle X\rangle)$ for every $k \geq 0$. Analogously for the notations $\mathscr{T}$ and $\widehat{\mathscr{T}}$. We begin by defining D inductively on the spaces $\mathscr{T}_k$. For $k = 0$, we define $D \equiv 0$ on $\mathscr{T}_0 = \mathbb{K}$. For $k = 1$ we define D on $\mathscr{T}_1 = \mathbb{K}\langle X\rangle$ as the unique linear map satisfying (4.141). This is well defined as X is a linear basis of $\mathbb{K}\langle X \rangle$. To define D on $\mathscr{T}_k$, we first recall that, by Proposition 2.35 on page 77, the system

$$\{x_1 \cdots x_k \mid x_1, \ldots, x_k \in X\}$$

is a linear basis of $\mathscr{T}_k$. Inductively, once D has been defined on $\mathscr{T}_0, \ldots, \mathscr{T}_k$ we define D on $\mathscr{T}_{k+1}$ as the unique linear map such that

$$D(x_1 \cdots x_{k+1}) = D(x_1)\, x_2 \cdots x_{k+1} + x_1\, D(x_2 \cdots x_{k+1}),$$

for any choice of $x_1, \ldots, x_{k+1}$ in X. This defines unambiguously $D : \mathscr{T} \to \widehat{\mathscr{T}}$. Note that the definition of D also gives

$$D\big(\bigoplus_{k \geq N} \mathscr{T}_k\big) \subseteq \prod_{k \geq N-1} \mathscr{T}_k.$$

Hence, by arguing as in Lemma 2.79 on page 103, D can be extended by continuity in a unique way to a linear map, still denoted by D, defined on $\widehat{\mathscr{T}}$. All that is left to prove is that D is a derivation of $\widehat{\mathscr{T}}$.

As D is linear and continuous, it suffices to prove that

$$D(u \cdot v) = D(u) \cdot v + u \cdot D(v)$$

when u, v are elementary products of elements of X. Equivalently, all we need to prove is that

$$D(x_1 \cdots x_n) = D(x_1 \cdots x_i)\, x_{i+1} \cdots x_n + x_1 \cdots x_i\, D(x_{i+1} \cdots x_n), \qquad (7.63)$$

for every $x_1, \ldots, x_n \in X$, every $n \geq 2$ and every i such that $1 \leq i < n$. We prove this by induction on n. The case $n = 2$ comes from the very definition of D. We now prove the assertion in the n-th step, supposing it to hold for the previous steps. Let $1 \leq i < n$. If $i = 1$, (7.63) is trivially true from the definition of D. We can suppose then that $1 < i < n$. Then we have

$$D(x_1 \cdots x_n) = D(x_1)\, x_2 \cdots x_n + x_1\, D(x_2 \cdots x_n)$$

(by the induction hypothesis)

$$= D(x_1)\, x_2 \cdots x_n + x_1\, \big(D(x_2 \cdots x_i)\, x_{i+1} \cdots x_n + x_2 \cdots x_i\, D(x_{i+1} \cdots x_n)\big).$$

On the other hand,

$$D(x_1 \cdots x_i)\, x_{i+1} \cdots x_n + x_1 \cdots x_i\, D(x_{i+1} \cdots x_n)$$

$$= \big(D(x_1)\, x_2 \cdots x_i + x_1\, D(x_2 \cdots x_i)\big)\, x_{i+1} \cdots x_n$$

$$\quad + x_1 \cdots x_i\, D(x_{i+1} \cdots x_n)$$

$$= D(x_1)\, x_2 \cdots x_n + x_1\, D(x_2 \cdots x_i)\, x_{i+1} \cdots x_n + x_1 \cdots x_i\, D(x_{i+1} \cdots x_n).$$

This is precisely what we got from the former computation, and the proof is complete. □

Proof (of Theorem 4.40, page 253). The assertion is trivial when $k = 1$. We then fix $k \geq 2$, $a_1, \ldots, a_k \in A$ and we prove (4.179) by induction on $n \geq 1$. The case $n = 1$ is equivalent to

$$D(a_1 * \cdots * a_k) = D(a_1) * a_2 * \cdots * a_k + a_1 * D(a_2) * \cdots * a_k +$$
$$+ \cdots + a_1 * a_2 \cdots * a_{k-1} * D(a_k), \tag{7.64}$$

which easily follows (this time by induction on k) from the very definition of a derivation. We next prove (4.179) in the $(n+1)$-th case, supposing it has been verified at the first and n-th steps:

$$D^{n+1}(a_1 * \cdots * a_k) = D\big(D^n(a_1 * \cdots * a_k)\big)$$

(by the inductive hypothesis)

$$= D\bigg(\sum_{\substack{0 \leq i_1,\ldots,i_k \leq n \\ i_1 + \cdots + i_k = n}} \frac{n!}{i_1! \cdots i_k!} D^{i_1} a_1 * \cdots * D^{i_k} a_k \bigg)$$

(using linearity of D and (4.179) in the case $n = 1$)

$$= \sum_{\substack{0 \leq i_1,\ldots,i_k \leq n \\ i_1 + \cdots + i_k = n}} \frac{n!}{i_1! \cdots i_k!} D^{i_1+1} a_1 * D^{i_2} a_2 * \cdots * D^{i_k} a_k + \cdots$$

$$\cdots + \sum_{\substack{0 \leq i_1,\ldots,i_k \leq n \\ i_1 + \cdots + i_k = n}} \frac{n!}{i_1! \cdots i_k!} D^{i_1} a_1 * \cdots * D^{i_{k-1}} a_{k-1} * D^{i_k+1} a_k$$

$$= \sum_{\substack{j_1 \geq 1,\, j_2,\ldots,j_k \geq 0 \\ j_1 + \cdots + j_k = n+1}} \frac{n!\, D^{j_1} a_1 * D^{j_2} a_2 * \cdots * D^{j_k} a_k}{(j_1 - 1)!\, j_2! \cdots j_k!} + \cdots$$

$$\cdots + \sum_{\substack{j_1,\ldots,j_{k-1} \geq 0,\, j_k \geq 1 \\ j_1 + \cdots + j_k = n+1}} \frac{n!\, D^{j_1} a_1 * \cdots * D^{j_{k-1}} a_{k-1} * D^{j_k} a_k}{j_1! \cdots j_{k-1}!\, (j_k - 1)!}.$$

We now apply the obvious equality

$$\frac{j}{j!} = \begin{cases} 0, & \text{if } j = 0, \\ \frac{1}{(j-1)!}, & \text{if } j \geq 1 \end{cases}$$

to each of the k summands (for $j = j_i$ in the i-th summand, $i = 1, \ldots, k$). As a consequence, we get (with respect to the previous sum we are here adding k summands which are all vanishing, so that we have equality)

$$D^{n+1}(a_1 * \cdots * a_k) = \sum_{\substack{j_1,j_2,\ldots,j_k \geq 0 \\ j_1+\cdots+j_k=n+1}} \frac{n!\, j_1}{j_1! \cdots j_k!} D^{j_1} a_1 * \cdots * D^{j_k} a_k + \cdots$$

$$\cdots + \sum_{\substack{j_1,\ldots,j_{k-1},j_k \geq 0 \\ j_1+\cdots+j_k=n+1}} \frac{n!\, j_k}{j_1! \cdots j_k!} D^{j_1} a_1 * \cdots * D^{j_k} a_k$$

$$= \sum_{\substack{j_1,\ldots,j_k \geq 0 \\ j_1+\cdots+j_k=n+1}} \frac{n!(n+1)}{j_1! \cdots j_k!} D^{j_1} a_1 * \cdots * D^{j_k} a_k,$$

which proves (4.179) in the case $n+1$.   □

Proof (of Theorem 4.42, page 256). Let us consider the unique UAA morphism

$$\theta : \mathcal{T}(V) \longrightarrow \mathrm{End}(\mathcal{T}_+(V)) \quad \text{such that}$$

$$\begin{cases} \theta(1) = \mathrm{Id}_{\mathcal{T}_+(V)}, \quad \theta(v_1) = \mathrm{ad}\,(v_1), \\ \theta(v_1 \otimes \cdots \otimes v_k) = \mathrm{ad}\,(v_1) \circ \cdots \circ \mathrm{ad}\,(v_k), \end{cases} \tag{7.65}$$

for every $v_1, \ldots, v_k \in V$ and every $k \geq 2$. The existence of θ is a consequence of the universal property of $\mathcal{T}(V)$ in Theorem 2.38-(ii). Here, given $v \in V$, we are considering the usual adjoint map $\mathrm{ad}\,(v)$ as an endomorphism of $\mathcal{T}_+(V)$:

$$\mathrm{ad}\,(v) : \mathcal{T}_+(V) \to \mathcal{T}_+(V), \quad w \mapsto \mathrm{ad}\,(v)(w) = v \cdot w - w \cdot v;$$

note that this really is an endomorphism of $\mathcal{T}_+(V)$, since $\mathcal{T} \cdot \mathcal{T}_+ \subseteq \mathcal{T}_+$ and $\mathcal{T}_+ \cdot \mathcal{T} \subseteq \mathcal{T}_+$.

For every $t = (t_k)_k \in \widehat{\mathcal{T}}(V)$, where $t_k \in \mathcal{T}_k(V)$ for every $k \geq 0$, we claim that the following formula defines an endomorphism of $\widehat{\mathcal{T}}_+(V)$, which we denote $\widehat{\theta}(t)$:

$$\widehat{\mathcal{T}}_+(V) \ni \tau = (\tau_k)_k \mapsto \widehat{\theta}(t)(\tau) := \sum_{h,k=0}^{\infty} \theta(t_h)(\tau_k), \tag{7.66}$$

where $\tau_k \in \mathcal{T}_k(V)$ for every $k \geq 0$ and $\tau_0 = 0$.

Well posedness of $\widehat{\theta}(t)$: We need to show that the double series in the far right-hand side of (7.66) is convergent in $\widehat{\mathcal{T}}_+(V)$. This derives from the following computation: for every $H, K, P, Q \in \mathbb{N}$ we have

$$\sum_{h=H}^{H+P} \sum_{k=K}^{K+Q} \theta(t_h)(\tau_k) \in \bigoplus_{j=K+H}^{\infty} \mathcal{T}_j(V),$$

so that

$$\sum_{h=H}^{H+P} \sum_{k=K}^{K+Q} \theta(t_h)(\tau_k) \xrightarrow{H,K \to \infty} 0 \quad \text{in } \widehat{\mathscr{T}}(V),$$

uniformly in $P, Q \geq 0$. Furthermore the cited double series converges to an element of $\widehat{\mathscr{T}}_+$ since $\tau_0 = 0$ and $\theta(t_h)(\tau_k) \in \mathscr{T}_{h+k}(V)$.

$\widehat{\theta}(t)$ *belongs to* $\mathrm{End}(\widehat{\mathscr{T}}_+(V))$:   The linearity of $\tau \mapsto \widehat{\theta}(t)(\tau)$ is a simple consequence of the linearity of any map $\theta(t_h)$ (and of the convergence of the double series in (7.66)).

We have so far defined a well posed map

$$\widehat{\theta} : \widehat{\mathscr{T}}(V) \longrightarrow \mathrm{End}(\widehat{\mathscr{T}}_+(V)), \quad t \mapsto \widehat{\theta}(t).$$

We further need to prove that this map has the following properties:

$\widehat{\theta}$ *is linear:*   This is a consequence of the linearity of the map $t \mapsto \theta(t)$ (and again of the convergence of the double series in (7.66)).

$\widehat{\theta}$ *is unital:*   We have $1_{\widehat{\mathscr{T}}} = (t_h)_h$ with $t_0 = 1_{\mathbb{K}}$ and $t_h = 0$ for every $h \geq 1$. As a consequence (recalling that $\theta(1_{\mathbb{K}})$ is the identity of $\widehat{\mathscr{T}}_+$) the definition of $\widehat{\theta}$ gives

$$\widehat{\theta}(1_{\widehat{\mathscr{T}}})(\tau) = \sum_{h,k=0}^{\infty} \theta(t_h)(\tau_k) = \sum_{k=0}^{\infty} \theta(t_0)(\tau_k) = \sum_{k=0}^{\infty} \tau_k = \tau.$$

$\widehat{\theta}$ *is an algebra morphism:*   Let us take $t = (t_k)_k$ and $t' = (t'_k)_k$ in $\widehat{\mathscr{T}}$ (where $t_k, t'_k \in \mathscr{T}_k$ for every $k \geq 0$). We need to show that

$$\widehat{\theta}(t \cdot t') = \widehat{\theta}(t) \circ \widehat{\theta}(t').$$

To this aim, we prove the equality of these maps on an arbitrary element $\tau = (\tau_k)_k$ of $\widehat{\mathscr{T}}_+$ (where $\tau_0 = 0$ and $\tau_k \in \mathscr{T}_k$ for every $k \geq 0$):

$$\widehat{\theta}(t \cdot t')(\tau) = \widehat{\theta}\Big(\big(\textstyle\sum_{i+j=h} t_i \cdot t'_j \big)_h \Big)(\tau) \overset{(7.66)}{=} \sum_{h,k=0}^{\infty} \theta\big(\textstyle\sum_{i+j=h} t_i \cdot t'_j \big)(\tau_k)$$

(recall that θ is a UAA morphism)

$$= \sum_{h,k=0}^{\infty} \Big(\sum_{i+j=h} \theta(t_i) \circ \theta(t'_j) \Big)(\tau_k) = \sum_{h,k=0}^{\infty} \sum_{i+j=h} \theta(t_i)\big(\theta(t'_j)(\tau_k)\big)$$

$$= \sum_{k=0}^{\infty} \sum_{h=0}^{\infty} \sum_{i+j=h} \theta(t_i)\big(\theta(t'_j)(\tau_k)\big) = \sum_{a,b,c=0}^{\infty} \theta(t_a)\big(\theta(t'_b)(\tau_c)\big).$$

On the other hand we have

$$\widehat{\theta}(t) \circ \widehat{\theta}(t')(\tau) = \widehat{\theta}(t)\big(\widehat{\theta}(t')(\tau)\big) \stackrel{(7.66)}{=} \widehat{\theta}(t)\Big(\sum_{h,k=0}^{\infty} \theta(t'_h)(\tau_k)\Big)$$

$$= \widehat{\theta}(t)\Big(\sum_{r=0}^{\infty} \sum_{h+k=r} \theta(t'_h)(\tau_k)\Big)$$

(note that $\theta(t'_h)(\tau_k) \in \mathscr{T}_{h+k}(V)$)

$$\stackrel{(7.66)}{=} \sum_{s,r=0}^{\infty} \theta(t_s)\big(\sum_{h+k=r} \theta(t'_h)(\tau_k)\big) = \sum_{s,r=0}^{\infty} \sum_{h+k=r} \theta(t_s)\big(\theta(t'_h)(\tau_k)\big)$$

$$= \sum_{s=0}^{\infty}\sum_{r=0}^{\infty} \sum_{h+k=r} \theta(t_s)\big(\theta(t'_h)(\tau_k)\big) = \sum_{a,b,c=0}^{\infty} \theta(t_a)\big(\theta(t'_b)(\tau_c)\big).$$

$\widehat{\theta}$ *satisfies* (4.187):   We already know that $\widehat{\theta}$ is unital. Let now $v_1 \in V = \mathscr{T}_1(V)$. For any $\tau = (\tau_k)_k \in \widehat{\mathscr{T}}_+$ we have

$$\widehat{\theta}(v_1)(\tau) = \sum_{k=0}^{\infty} \theta(v_1)(\tau_k) = \sum_{k=0}^{\infty} \text{ad}\,(v_1)(\tau_k)$$

$$= \text{ad}\,(v_1)\big(\sum_{k=0}^{\infty} \tau_k\big) = \text{ad}\,(v_1)(\tau).$$

The verification that $\widehat{\theta}(v_1 \otimes \cdots \otimes v_k) = \text{ad}\,(v_1) \circ \cdots \circ \text{ad}\,(v_k)$ follows analogously, by using the fact that $v_1 \otimes \cdots \otimes v_k \in \mathscr{T}_k(V)$ and the fact that $\theta(v_1 \otimes \cdots \otimes v_k) = \text{ad}\,(v_1) \circ \cdots \circ \text{ad}\,(v_k)$.

$\widehat{\theta}(t)$ *is continuous, for every* t: This follows immediately from the fact that, for every $N \in \mathbb{N} \cup \{0\}$ and every $t \in \widehat{\mathscr{T}}$ we have:

$$\tau \in \prod_{k \geq N} \mathscr{T}_k(V) \quad \Rightarrow \quad \widehat{\theta}(t)(\tau) \in \prod_{k \geq N} \mathscr{T}_k(V), \qquad (7.67)$$

which is a simple consequence of (7.66).

$\widehat{\theta}$ *satisfies* (4.188):   First we prove that the series in the right-hand side of (4.188) converges in $\widehat{\mathscr{T}}_+$, provided that $\sum_k \gamma_k$ converges in $\widehat{\mathscr{T}}$. This is a consequence of the following fact: *for every fixed* $\tau \in \widehat{\mathscr{T}}_+$, *we have*

$$\lim_{k \to \infty} \gamma_k = 0 \text{ in } \widehat{\mathscr{T}} \quad \Rightarrow \quad \lim_{k \to \infty} \widehat{\theta}(\gamma_k)(\tau) = 0 \text{ in } \widehat{\mathscr{T}}_+.$$

This follows directly from the fact that, for every $N \in \mathbb{N} \cup \{0\}$ and every $\tau \in \widehat{\mathscr{T}}_+$ we have:

$$t \in \prod_{k \geq N} \mathscr{T}_k(V) \quad \Rightarrow \quad \widehat{\theta}(t)(\tau) \in \prod_{k \geq N+1} \mathscr{T}_k(V), \qquad (7.68)$$

which is a simple consequence of (7.66). For every fixed $k \geq 0$, we have

$$\gamma_k = (\gamma_{k,i})_i \quad \text{with } \gamma_{k,i} \in \mathscr{T}_i(V) \text{ for every } i \geq 0.$$

This obviously gives

$$t := \sum_{k=0}^{\infty} \gamma_k = \left(\sum_{k=0}^{\infty} \gamma_{k,i}\right)_i,$$

with $\sum_{k=0}^{\infty} \gamma_{k,i} \in \mathscr{T}_i(V)$ for every $i \geq 0$, each of these sums being convergent (in view of $\lim_k \gamma_k = 0$). Let also $\tau = (\tau_k)_k \in \widehat{\mathscr{T}_+}$, with $\tau_k \in \mathscr{T}_k$ for every $k \geq 0$ and $\tau_0 = 0$. Then we have the following computation

$$\sum_{k=0}^{\infty} \widehat{\theta}(\gamma_k)(\tau) = \sum_{k=0}^{\infty} \left(\sum_{i,j \geq 0} \widehat{\theta}(\gamma_{k,i})(\tau_j)\right) = \sum_{k=0}^{\infty} \left(\sum_{i=0}^{\infty} \widehat{\theta}(\gamma_{k,i})\left(\sum_{j=0}^{\infty} \tau_j\right)\right)$$

$$= \sum_{k=0}^{\infty} \left(\sum_{i=0}^{\infty} \widehat{\theta}(\gamma_{k,i})(\tau)\right) = \sum_{i=0}^{\infty} \left(\sum_{k=0}^{\infty} \widehat{\theta}(\gamma_{k,i})(\tau)\right) = \sum_{i=0}^{\infty} \widehat{\theta}\left(\sum_{k=0}^{\infty} \gamma_{k,i}\right)(\tau)$$

$$= \sum_{i=0}^{\infty} \widehat{\theta}(t_i)(\tau) = \widehat{\theta}(t)(\tau).$$

In the second equality we used the fact that the double series over i, j equals the iterated series $\sum_i \sum_j$ (as was discovered early in the proof), together with the continuity of $\widehat{\theta}(\gamma_{k,i})$; in the fourth equality we used an analogous argument on convergent double series, based on the fact that

$$\widehat{\theta}(\gamma_{k,i})(\tau) \in \prod_{j \geq i+1} \mathscr{T}_j(V).$$

In the fifth equality we used the fact that, for every fixed i, the sum $\sum_{k=0}^{\infty} \widehat{\theta}(\gamma_{k,i})(\tau)$ is finite and $\widehat{\theta}$ is linear.

$\widehat{\theta}$ satisfies (4.189): First we prove that

$$\theta(\ell)(\tau) = \operatorname{ad}(\ell)(\tau), \quad \text{for every } \ell \in \mathcal{L}(V) \text{ and } \tau \in \mathscr{T}_+(V). \qquad (7.69)$$

Let $\ell \in \mathcal{L}(V)$. Then (7.69) follows if we show that $\theta(\ell)$ and $\operatorname{ad}(\ell)$ coincide as endomorphisms of $\mathscr{T}_+(V)$. In its turn, this is equivalent to the identity of $\widetilde{\theta} := \theta|_{\mathcal{L}(V)}$ and the map

$$\widetilde{A} := \operatorname{ad}|_{\mathcal{L}(V)} : \mathcal{L}(V) \to \operatorname{End}(\mathscr{T}_+(V)), \quad \xi \mapsto \operatorname{ad}(\xi).$$

It is easily seen that $\widetilde{A}$ is a Lie algebra morphism (use the Jacobi identity!). Also $\widetilde{\theta}$ is an LA morphism, since θ is a UAA morphism and $\mathcal{L}(V)$ is a Lie algebra. Hence, the equality of $\widetilde{A}$ and $\widetilde{\theta}$ follows if we prove that they are equal on a system of Lie generators for $\mathcal{L}(V)$, namely on V: for every $v \in V$ we have in fact $\widetilde{\theta}(v) = \theta(v) = \mathrm{ad}\,(v)$, by (7.65).

Now, we turn to prove (4.189). Given $\ell \in \overline{\mathcal{L}(V)}$, we have $\ell = \sum_{k=1}^{\infty} \ell_k$, with $\ell_k \in \mathcal{L}_k(V)$, for every $k \geq 1$. Then we obtain

$$\widehat{\theta}(\ell)(\tau) = \sum_{h,k=1}^{\infty} \theta(\ell_h)(\tau_k) \stackrel{(7.69)}{=} \sum_{h,k=1}^{\infty} \mathrm{ad}\,(\ell_h)(\tau_k) = \sum_{h,k=1}^{\infty} [\ell_h, \tau_k]$$

$$= \Big[\sum_{h=1}^{\infty} \ell_h, \sum_{k=1}^{\infty} \tau_k\Big] = [\ell, \tau] = \mathrm{ad}\,(\ell)(\tau).$$

This ends the proof of the lemma.   □

Proof (of Lemma 6.4, page 378). We furnish a more explicit proof of (6.8), with the collateral aim to exhibit the single homogeneous components in $f_{i,j}$ (see (7.71) below). We fix a, b, i, j as in (6.8) and we set $I := \{1, 2, \ldots, i+j\}$. I^k denotes the k-fold Cartesian product of I with itself ($k \in \mathbb{N}$) and its elements are denoted by $r = (r_1, \ldots, r_k)$. We have the formal expansion

$$\exp(Z^{\mathfrak{g}}(a,b)) = \sum_{k \geq 0} \frac{1}{k!} \Big(\sum_{r \geq 1} Z_r^{\mathfrak{g}}(a,b)\Big)^{\otimes k}$$

$$= \sum_{k=0}^{i+j} \frac{1}{k!} \sum_{r \in I^k} Z_{r_1}^{\mathfrak{g}} \otimes \cdots \otimes Z_{r_k}^{\mathfrak{g}}$$

$$+ \left\{\begin{array}{l} \text{summands not occurring} \\ \text{in the definition of } f_{i,j}(a,b) \end{array}\right\} =: A + \{B\}.$$

We define in I^k the equivalence relation $\sim$ by setting $r \sim \rho$ iff there exists a permutation $\sigma \in \mathfrak{S}_k$ such that $(\rho_1, \ldots, \rho_k) = (r_{\sigma(1)}, \ldots, r_{\sigma(k)})$. Thus, we have

$$A = \sum_{k=0}^{i+j} \frac{1}{k!} \sum_{r \in I_+^k} \Big(\sum_{\rho \sim r} Z_{\rho_1}^{\mathfrak{g}} \otimes \cdots \otimes Z_{\rho_k}^{\mathfrak{g}}\Big) \quad \text{where } I_+^k = \{r \in I^k : r_1 \leq \cdots \leq r_k\}.$$

Now, a combinatorial argument proves that, for every $r \in I_+^k$ there exists $\alpha(r) \in \mathbb{N}$ such that

$$\sum_{\sigma \in \mathfrak{S}_k} Z_{r_{\sigma(1)}}^{\mathfrak{g}} \otimes \cdots \otimes Z_{r_{\sigma(k)}}^{\mathfrak{g}} = \alpha(r) \sum_{\rho \sim r} Z_{\rho_1}^{\mathfrak{g}} \otimes \cdots \otimes Z_{\rho_k}^{\mathfrak{g}}. \qquad (7.70)$$

By inserting (7.70) in the expression of A, we get

$$A = \sum_{k=0}^{i+j} \sum_{r \in I_+^k} \alpha(r,k) \left(\sum_{\sigma \in \mathfrak{S}_k} Z_{r_{\sigma(1)}}^{\mathfrak{g}} \otimes \cdots \otimes Z_{r_{\sigma(k)}}^{\mathfrak{g}} \right), \quad \alpha(r,k) := (k!\,\alpha(r))^{-1}.$$

Next, we reorder $Z_r^{\mathfrak{g}}$ in (6.1) as $Z_r^{\mathfrak{g}} = \sum_{s=0}^{r} q_s^{(r)}(a,b)$, where

$$q_s^{(r)}(a,b) := \sum_{n=1}^{r} c_n \sum_{\substack{(h,k)\in N_n, \\ |k|=r-s,\, |h|=s}} \mathbf{c}(h,k) \left[a^{h_1} b^{k_1} \cdots a^{h_n} b^{k_n} \right]_{\mathfrak{g}}.$$

Hence, using different notations for the dummy index s, according to the fixed $\sigma \in \mathfrak{S}_k$, we get

$$A = \sum_{k=0}^{i+j} \sum_{r \in I_+^k} \alpha(r,k) \sum_{\sigma \in \mathfrak{S}_k} \left(\sum_{s_{\sigma(1)}=0}^{r_{\sigma(1)}} q_{s_{\sigma(1)}}^{(r_{\sigma(1)})}(a,b) \right) \otimes \cdots \otimes \left(\sum_{s_{\sigma(k)}=0}^{r_{\sigma(k)}} q_{s_{\sigma(k)}}^{(r_{\sigma(k)})}(a,b) \right).$$

By the definition of $f_{i,j}(a,b)$, we thus derive

$$f_{i,j}(a,b) = \sum_{k=0}^{i+j} \sum_{r \in I_+^k} \alpha(r,k) \sum_{\sigma \in \mathfrak{S}_k}$$

$$\times \sum_{\substack{1 \le s_{\sigma(1)} \le r_{\sigma(1)},\ldots,1 \le s_{\sigma(k)} \le r_{\sigma(k)} \\ s_{\sigma(1)}+\cdots+s_{\sigma(k)}=i,\ r_{\sigma(1)}+\cdots+r_{\sigma(k)}=i+j}} q_{s_{\sigma(1)}}^{(r_{\sigma(1)})}(a,b) \otimes \cdots \otimes q_{s_{\sigma(k)}}^{(r_{\sigma(k)})}(a,b).$$

Since any σ in the inner sum is a permutation of $\{1,2,\ldots,k\}$, this sum is over

$$|s| = s_1 + \cdots + s_k = i, \qquad 1 \le s_1 \le r_1, \ldots, 1 \le s_k \le r_k.$$
$$|r| = r_1 + \cdots + r_k = i+j,$$

As a consequence

$$f_{i,j}(a,b) = \sum_{k=0}^{i+j} \sum_{\substack{r \in I_+^k \\ |r|=i+j}} \alpha(r,k) \sum_{\substack{1 \le s_1 \le r_1,\ldots,1 \le s_k \le r_k \\ |s|=i}}$$

$$\times \left\{ \sum_{\sigma \in \mathfrak{S}_k} q_{s_{\sigma(1)}}^{(r_{\sigma(1)})}(a,b) \otimes \cdots \otimes q_{s_{\sigma(k)}}^{(r_{\sigma(k)})}(a,b) \right\}. \tag{7.71}$$

Evidently, this proves that $f_{i,j}(a,b)$ is thus the sum of elements in $\mathscr{TS}(\mathfrak{g})$ (those in curly braces) and the proof is complete. $\qquad \square$

Chapter 8
Construction of Free Lie Algebras

THE aim of this chapter is twofold. On the one hand (Sect. 8.1), we complete the missing proof from Chap. 2 concerning the existence of a free Lie algebra $\mathrm{Lie}(X)$ related to a set X. This proof relies on the *direct construction* of $\mathrm{Lie}(X)$ as a quotient of the free non-associative algebra $\mathrm{Lib}(X)$. Furthermore, we prove that $\mathrm{Lie}(X)$ is isomorphic to $\mathcal{L}(\mathbb{K}\langle X\rangle)$, and the latter provides a free Lie algebra *over* X.

On the other hand (Sect. 8.2), we turn to construct a very important class of Lie algebras, the free *nilpotent* Lie algebras generated by a set. Applications of this class of algebras to Lie group theory and to Analysis can be found in [62, 63, 150, 163, 172] (see also [21]).

8.1 Construction of Free Lie Algebras Continued

Let the notation of Sect. 2.2 on page 87 apply. Our main aim is to prove Theorem 2.54 on page 91, whose statement we reproduce here, for convenience of reading.

Theorem 8.1. *Let X be any set and, with the notation in* (2.56) *and* (2.57), *let us consider the map*

$$\varphi : X \to \mathrm{Lie}(X), \quad x \mapsto \pi(x), \tag{8.1}$$

that is,[1] $\varphi \equiv \pi|_X$. *Then:*

1. *The couple* $(\mathrm{Lie}(X), \varphi)$ *is a free Lie algebra related to X (see Definition 2.50 on page 88).*

---

[1] More precisely, the map φ is the composition

$$X \xrightarrow{\ \iota\ } M(X) \xrightarrow{\ \chi\ } \mathrm{Lib}(X) \xrightarrow{\ \pi\ } \mathrm{Lie}(X).$$

Via the identification $X \equiv \chi(X) \xrightarrow{\ \iota\ } \mathrm{Lib}(X)$ we can write $\varphi \equiv \pi|_X$.

A. Bonfiglioli and R. Fulci, *Topics in Noncommutative Algebra*, Lecture Notes in Mathematics 2034, DOI 10.1007/978-3-642-22597-0_8, © Springer-Verlag Berlin Heidelberg 2012

2. The set $\{\varphi(x)\}_{x \in X}$ is *independent* in $\mathrm{Lie}(X)$, whence φ is injective.
3. The set $\varphi(X)$ *Lie-generates* $\mathrm{Lie}(X)$, that is, the smallest Lie subalgebra of $\mathrm{Lie}(X)$ containing $\varphi(X)$ coincides with $\mathrm{Lie}(X)$.

The proof of this theorem may be derived by collecting together various results in Bourbaki [27, Chapitre II, §2], the only source (to the best of our knowledge) defining free Lie algebras as "explicit" quotients of the algebra of the free magma. Unfortunately, in the proofs of the above statements, Bourbaki makes use of some prerequisites (from another Bourbaki book [25]) concerned with quotients and ideals in the *associative* setting (whereas $\mathfrak{a}$ and $\mathrm{Lib}(X)$ are not associative). Hence, we have felt the need to produce all the details of Theorem 8.1, by providing also the relevant *non-associative* prerequisites (see Lemma 8.2 below).

Proof. We split the proof of the statement into its three parts.

(1). According to Definition 2.50 on page 88, we have to prove that, for every Lie algebra $\mathfrak{g}$ and every map $f : X \to \mathfrak{g}$, there exists a unique Lie algebra morphism $f^{\varphi} : \mathrm{Lie}(X) \to \mathfrak{g}$, such that the following fact holds

$$f^{\varphi}(\varphi(x)) = f(x) \quad \text{for every } x \in X, \tag{8.2}$$

thus making the following a commutative diagram:

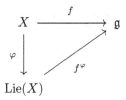

–Existence. Let $\mathfrak{g}$ and f be as above. We temporarily equip $\mathfrak{g}$ only with its algebra structure, i.e., of a magma endowed with the bilinear operation

$$\mathfrak{g} \times \mathfrak{g} \ni (a, b) \mapsto [a, b]_{\mathfrak{g}}.$$

By part (i) of Theorem 2.27 on page 70, there exists a unique algebra morphism $f^{\chi} : \mathrm{Lib}(X) \to \mathfrak{g}$ with the following property

$$f^{\chi}(\chi(x)) = f(x), \quad \text{for every } x \in X, \tag{8.3}$$

where $\chi|_X : X \to \mathrm{Lib}(X)$ is the composition of maps

$$X \overset{\iota}{\hookrightarrow} M(X) \overset{\chi}{\longrightarrow} \mathbb{K}\langle M(X)\rangle = \mathrm{Lib}(X).$$

For brevity, we set $h := f^{\chi}$. Note that the fact that h is a magma morphism ensures that

$$h(u * v) = [h(u), h(v)]_{\mathfrak{g}}, \quad \text{for every } u, v \in \mathrm{Lib}(X). \tag{8.4}$$

As $\mathfrak{g}$ is a Lie algebra, one has (see the notation in (2.55))

$$h(Q(a)) = h(J(a,b,c)) = 0, \quad \text{for every } a,b,c \in \text{Lib}(X). \tag{8.5}$$

Indeed, for $a,b,c \in \text{Lib}(X)$, we have

$$h(Q(a)) = h(a*a) \overset{(8.4)}{=} [h(a),h(a)]_{\mathfrak{g}} = 0,$$

$$h(J(a,b,c)) = h(a*(b*c) + b*(c*a) + c*(a*b))$$

$$\overset{(8.4)}{=} [h(a),[h(b),h(c)]_{\mathfrak{g}}]_{\mathfrak{g}} + [h(b),[h(c),h(a)]_{\mathfrak{g}}]_{\mathfrak{g}}$$

$$+ [h(c),[h(a),h(b)]_{\mathfrak{g}}]_{\mathfrak{g}} = 0.$$

Let $\mathfrak{a}$ denote, as usual, the magma ideal generated by the set

$$A := \{Q(a), J(a,b,c) \mid a,b,c \in \text{Lib}(X)\}.$$

We claim that

$$\mathfrak{a} \subseteq \ker(h). \tag{8.6}$$

Thanks to Lemma 8.2 below (see also the notation therein: we let P_n denote the corresponding sets related to A) and the fact that h is a magma morphism, (8.5) yields $h(P_n) = \{0\}$ for every $n \geq 0$, whence $h(\mathfrak{a}) = \{0\}$. This proves (8.6).

Furthermore, recalling that $\text{Lie}(X) = \text{Lib}(X)/\mathfrak{a}$, this ensures the well-posedness of the map (also recall that $\pi(t) = [t]_{\mathfrak{a}}$)

$$f^{\varphi} : \text{Lie}(X) \to \mathfrak{g}, \qquad f^{\varphi}(\pi(t)) := h(t), \quad \text{for every } t \in \text{Lib}(X).$$

Obviously f^{φ} is linear. We claim that f^{φ} is a Lie algebra morphism satisfying (8.2). Indeed, to begin with, we have for every $t,t' \in \text{Lib}(X)$,

$$f^{\varphi}\left([\pi(t),\pi(t')]\right) \overset{(2.58)}{=} f^{\varphi}\left(\pi(t*t')\right) = h(t*t') \overset{(8.4)}{=} [h(t),h(t')]_{\mathfrak{g}}$$

$$= \left[f^{\varphi}(\pi(t)), f^{\varphi}(\pi(t'))\right]_{\mathfrak{g}}, \quad \text{whence } f^{\varphi} \text{ is an LA morphism.}$$

As for (8.2), for every $x \in X$ we have

$$f^{\varphi}(\varphi(x)) \overset{(8.1)}{=} f^{\varphi}\left(\pi(\chi(x))\right) = h(\chi(x)) = f^{\chi}(\chi(x)) \overset{(8.3)}{=} f(x).$$

−*Uniqueness.* Let $\alpha : \text{Lie}(X) \to \mathfrak{g}$ be another LA morphism such that $\alpha(\varphi(x)) = f(x)$, for every $x \in X$. The map

$$\alpha \circ \pi : \text{Lib}(X) \to \mathfrak{g}, \quad t \mapsto \alpha(\pi(t))$$

is an algebra morphism, because $\pi : \mathrm{Lib}(X) \to \mathrm{Lie}(X)$ is an algebra morphism (see Proposition 2.53 on page 90) and the same is true of $\alpha :$ $\mathrm{Lie}(X) \to \mathfrak{g}$ (recall that an LA morphism is nothing but an algebra morphism, when the Lie algebras are thought of simply as algebras with their bracket operations!). The same argument ensures that $f^{\varphi} \circ \pi : \mathrm{Lib}(X) \to \mathfrak{g}$ is an algebra morphism. Furthermore, $\alpha \circ \pi$ coincides with $f^{\varphi} \circ \pi$ on X, for it holds that

$$\alpha(\pi(x)) \overset{(8.1)}{=} \alpha(\varphi(x)) = f(x) \overset{(8.2)}{=} f^{\varphi}(\varphi(x)) \overset{(8.1)}{=} f^{\varphi}(\pi(x)),$$

for every $x \in X$. These identities may be rewritten as

$$(\alpha \circ \pi)(\chi(x)) = (f^{\varphi} \circ \pi)(\chi(x)) = f(x), \quad \forall \, x \in X.$$

By the universal property of the free algebra $\mathrm{Lib}(X)$ (see Theorem 2.27 on page 70) the morphisms $\alpha \circ \pi$ and $f^{\varphi} \circ \pi$ coincide throughout $\mathrm{Lib}(X)$. Since $\pi : \mathrm{Lib}(X) \to \mathrm{Lie}(X)$ is surjective, this proves that α and f^{φ} coincide on the whole of $\mathrm{Lie}(X)$. This ends the proof of (1) of Theorem 8.1.

(2). Let $x_1, \ldots, x_p$ be p different elements in X and let $\lambda_1, \ldots, \lambda_p \in \mathbb{K}$ be such that

$$\lambda_1 \, \varphi(x_1) + \cdots + \lambda_p \, \varphi(x_p) = 0. \tag{8.7}$$

Fix any $i \in \{1, \ldots, p\}$. Finally, let us take $\mathfrak{g} = \mathbb{K}$ and $f : X \to \mathfrak{g}$ defined by $f(x_i) = 1$ and 0 otherwise. By part (1) of the proof, there exists a Lie algebra morphism $f^{\varphi} : \mathrm{Lie}(X) \to \mathfrak{g}$ such that $f^{\varphi}(\varphi(x)) = f(x)$, for every $x \in X$. By applying f^{φ} to (8.7), we get $\lambda_i = 0$ and the arbitrariness of i proves (2).

(3). Let L be the smallest Lie subalgebra of $\mathrm{Lie}(X)$ containing $\varphi(X)$. The map $\varphi_L : X \to L$ defined by $\varphi_L(x) := \varphi(x)$, for all $x \in X$ is well-defined. It is easily seen that the couple (L, φ_L) has the same universal property of $(\mathrm{Lie}(X), \varphi)$ as in Definition 2.50. By the "uniqueness" property of free Lie algebras as in Proposition 2.51-(1), it follows that L and $\mathrm{Lie}(X)$ are isomorphic Lie algebras and the isomorphism is the inclusion $L \hookrightarrow \mathrm{Lie}(X)$. Indeed, the inclusion $\iota : L \hookrightarrow \mathrm{Lie}(X)$ is the only LA morphism that closes the following diagram:

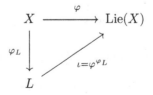

Hence (by the surjectivity of any isomorphism!) $L = \mathrm{Lie}(X)$.   □

Lemma 8.2. *Let* $(M, \cdot)$ *be an algebra (not necessarily associative) and let* $B \subseteq M$ *be any subset. Denote by* $\mathfrak{b}$ *the magma ideal in* M *generated by* B *(see Definition 2.52 on page 89). Define inductively* $P_0 := B$ *and, for any* $n \in \mathbb{N}$, $n \geq 1$,

$$P_n := \{p \cdot m \mid p \in P_{n-1},\ m \in M\} \cup \{m \cdot p \mid p \in P_{n-1},\ m \in M\}. \qquad (8.8)$$

Then we have (denoting by $\biguplus$ *the sum of vector subspaces of* M)

$$\mathfrak{b} = \biguplus\nolimits_{n \geq 0} \operatorname{span}(P_n). \qquad (8.9)$$

Proof. Let us denote by $\overline{\mathfrak{b}}$ the set on the right-hand side of (8.9). One has $\mathfrak{b} = \bigcap S$, where the sets S run over the magma ideals in M containing B. Obviously (by our definition of magma ideal), any such S contains $\overline{\mathfrak{b}}$, so that $\mathfrak{b} = \bigcap S \supseteq \overline{\mathfrak{b}}$. Vice versa, $\mathfrak{b} \subseteq \overline{\mathfrak{b}}$ will follow if we show that $\overline{\mathfrak{b}}$ is one of the sets S intervening in $\bigcap S = \mathfrak{b}$. Since $B = P_0 \subseteq \overline{\mathfrak{b}}$, we are left to prove that $\overline{\mathfrak{b}}$ is a magma ideal. Any element of $\overline{\mathfrak{b}}$ is of the form $v = \sum_n k_n\, p_n$, with $k_n \in \mathbb{K}$ and $p_n \in P_n$ (the sum being finite). This gives

$$m \cdot v = \sum\nolimits_n k_n\, m \cdot p_n, \quad \text{and} \quad v \cdot m = \sum\nolimits_n k_n\, p_n \cdot m,$$

and both of these elements belong to $\overline{\mathfrak{b}}$, as $m \cdot p_n, p_n \cdot m \in P_{n+1}$, in view of the very definition of P_{n+1}. $\qquad \square$

8.1.1 Free Lie Algebras over a Set

The rest of this section is devoted to constructing a free Lie algebra *over* X. The arguments presented here are inspired by those of Cartier in [33].

We denote by $\mathcal{L}(\mathbb{K}\langle X \rangle)$ the smallest Lie subalgebra of $\mathscr{T}(\mathbb{K}\langle X \rangle)$ containing X. As $\operatorname{Lie}(X)$ is a free Lie algebra generated by X, there exists a unique Lie algebra morphism

$$f : \operatorname{Lie}(X) \to \mathcal{L}(\mathbb{K}\langle X \rangle) \quad \text{such that } f(\varphi(x)) = x, \text{ for every } x \in X. \quad (8.10)$$

Our main task here is to show that f is an isomorphism, *without using* PBW.

Lemma 8.3. *We have the grading* $\operatorname{Lie}(X) = \bigoplus_{n=1}^{\infty} B_n$, *where* $B_1 = \operatorname{span}(\varphi(X))$, *and, for any* $n \geq 2$,

$$B_n = [B_1, B_{n-1}] = \operatorname{span}\Big\{ [\varphi(x), y] \ : \ x \in X,\ y \in B_{n-1} \Big\}.$$

Proof. It is easy to prove that $\mathrm{Lib}(X) = \bigoplus_{n \in \mathbb{N}} \mathrm{Lib}_n(X)$ defines a grading, where $\mathrm{Lib}_n(X)$ is the span of the non-associative words in $M(X)$ with n "letters" (see (2.18) on page 69 for the precise definition of $\mathrm{Lib}_n(X)$).

On the other hand, a simple argument shows that the magma ideal $\mathfrak{a}$ (introduced in the proof of Theorem 8.1) is also the magma ideal generated by the following elements

$$w.w, \ (w + w').(w + w') - w.w - w'.w',$$
$$w.(w'.w'') + w'.(w''.w) + w''.(w.w'),$$
$$(8.11)$$

with $w, w', w'' \in M(X)$. Indeed, let us set $B(w, w') := w.w' + w'.w$ for any $w, w' \in M(X)$. It holds

$$Q(w + w') - Q(w) - Q(w') = (w + w') * (w + w') - w.w - w'.w'$$
$$= w.w + w.w' + w'.w + w'.w' - w.w - w'.w' = w.w' + w'.w = B(w, w').$$

Moreover, for every $w_1, \ldots, w_n \in M(X)$ and every $\lambda_1, \ldots, \lambda_n \in \mathbb{K}$

$$Q(\lambda_1 w_1 + \cdots + \lambda_n w_n) = (\lambda_1 w_1 + \cdots + \lambda_n w_n) * (\lambda_1 w_1 + \cdots + \lambda_n w_n)$$

$$= \sum_{i=1}^{n} \lambda_i^2 w_i.w_i + \sum_{1 \le i < j \le n} \lambda_i \lambda_j (w_i.w_j + w_j.w_i)$$

$$= \sum_{i=1}^{n} \lambda_i^2 Q(w_i, w_i) + \sum_{1 \le i < j \le n} \lambda_i \lambda_j B(w_i, w_j).$$

All these identities prove that

$$\mathrm{span}\{Q(a) \,|\, a \in \mathrm{Lib}(X)\} = \mathrm{span}\Big\{Q(w), B(w, w') \,\big|\, w, w' \in M(X)\Big\}. \quad (8.12)$$

Moreover, if $a = \sum_i \lambda_i w_i$, $b = \sum_j \lambda'_j w'_j$ $c = \sum_k \lambda''_k w''_k$ (where $\lambda_i, \lambda'_j, \lambda''_k$ are scalars, w_i, w'_j, w''_k are in $M(X)$ and the sums are all finite) then

$$J(a, b, c) = \sum_{i,j,k} \lambda_i \lambda'_j \lambda''_k J(w_i, w'_j, w''_k).$$

This proves that

$$\mathrm{span}\{J(a, b, c) \,|\, a, b, c \in \mathrm{Lib}(X)\} = \mathrm{span}\Big\{J(w, w', w'') \,\big|\, w, w', w'' \in M(X)\Big\}.$$
$$(8.13)$$

Gathering (8.12) and (8.13) together, we deduce that $\mathfrak{a}$ equals the magma ideal generated by the elements in (8.11).

As a consequence of the form of the elements of $\mathrm{Lib}(X)$ in (8.11), it is easily seen that one has a grading $\mathfrak{a} = \bigoplus_{n \in \mathbb{N}} \mathfrak{a}_n$, with $\mathfrak{a}_n \subseteq \mathrm{Lib}_n(X)$, for every $n \in \mathbb{N}$. This gives (use Lemma 8.4 below)

$$\mathrm{Lie}(X) = \mathrm{Lib}(X)/\mathfrak{a} = \Big(\bigoplus_{n \in \mathbb{N}} \mathrm{Lib}_n(X) \Big) \Big/ \Big(\bigoplus_{n \in \mathbb{N}} \mathfrak{a}_n \Big)$$

$$= \bigoplus_{n \in \mathbb{N}} (\mathrm{Lib}_n(X)/\mathfrak{a}) \quad \Big(\text{also isomorphic to } \bigoplus_{n \in \mathbb{N}} (\mathrm{Lib}_n(X)/\mathfrak{a}_n) \Big).$$

As a consequence, we have $\mathrm{Lie}(X) = \bigoplus_{n=1}^{\infty} C_n$, where $C_n = \mathrm{Lib}_n(X)/\mathfrak{a}$ is nothing but the span of the higher-order brackets of degree n of the elements of $\varphi(X)$ (where the bracketing is taken in any arbitrary order). In turns, thanks to Theorem 2.15 on page 60, we have $C_n = B_n$, where B_n is the span of the degree n *right-nested* brackets of the elements of $\varphi(X)$, namely $[\varphi(x_1) \cdots [\varphi(x_{n-1}), \varphi(x_n)] \cdots]$, for $x_1, \ldots, x_n \in X$. It is a simple proof to check that $[B_n, B_m] \subseteq B_{n+m}$, for every $n, m \in \mathbb{N}$ (it suffices to argue as in the derivation of (2.11), page 60). $\qquad \square$

Here we have used the following result:

Lemma 8.4. *Let V, V_i, W_i (for $i \in \mathcal{I}$) be vector spaces such that $W_i \subseteq V_i \subseteq V$ for every $i \in \mathcal{I}$ and such that $V = \bigoplus_{i \in \mathcal{I}} V_i$. Set $W := \bigoplus_{i \in \mathcal{I}} W_i$ (the sum being direct in view of the other hypotheses). Then we have*

$$V/W = \bigoplus_{i \in \mathcal{I}} V_i/W.$$

Proof. For the sake of brevity, we prove the assertion for $\mathcal{I} = \{1, 2\}$. We then have $V = V_1 \oplus V_2$, $W = W_1 \oplus W_2$ and $W_1 \subseteq V_1$, $W_2 \subseteq V_2$. Any $v \in V$ uniquely determines v_1, v_2 such that $v = v_1 + v_2$ with $v_i \in V_i$ ($i = 1, 2$). The linear structure of V/W gives

$$[v]_W = [v_1]_W + [v_2]_W.$$

This proves $V/W = V_1/W + V_2/W$. We claim that the sum is direct. If

$$[v_1]_W + [v_2]_W = [v_1']_W + [v_2']_W \quad (\text{with } v_i, v_i' \in V_i, \, i = 1, 2),$$

we have $v_1 + v_2 - (v_1' + v_2') \in W$ so that there exist $w_1 \in W_1$, $w_2 \in W_2$ such that $v_1 + v_2 - (v_1' + v_2') = w_1 + w_2$ or equivalently

$$\underbrace{(v_1 - v_1')}_{\in V_1} + \underbrace{(v_2 - v_2')}_{\in V_2} = \underbrace{w_1}_{\in W_1 \subseteq V_1} + \underbrace{w_2}_{\in W_2 \subseteq V_2}.$$

By uniqueness of the decomposition in the direct sum $V_1 \oplus V_2$, this gives $v_i - v_i' = w_i$ $(i = 1, 2)$ whence, since $w_i \in W$,

$$[v_1]_W = [v_1']_W, \quad [v_2]_W = [v_2']_W.$$

This proves that the sum $V_1/W + V_2/W$ is direct.   □

We recall that f is the unique Lie algebra morphism

$$f : \mathrm{Lie}(X) \to \mathcal{L}(\mathbb{K}\langle X \rangle) \quad \text{such that}$$
$$f(\varphi(x)) = x \quad \text{for every } x \in X. \tag{8.14}$$

Thanks to Lemma 8.3, the following map is well-posed:

$$\delta : \mathrm{Lie}(X) \to \mathrm{Lie}(X),$$
$$\delta(\textstyle\sum_n b_n) := \sum_n n\, b_n \qquad (\text{where } b_n \in B_n \text{ for every } n \in \mathbb{N}). \tag{8.15}$$

In the remainder of the section, for any vector space V we denote by $\mathrm{End}(V)$ the set of the endomorphisms of V. Clearly, $\mathrm{End}(V)$ is a UA algebra, when equipped with the usual composition of maps. By part (2) of Theorem 2.40 on page 79, there exists a unique UAA morphism

$$\theta : \mathscr{T}(\mathbb{K}\langle X \rangle) \to \mathrm{End}(\mathrm{Lie}(X)), \quad \text{such that}$$
$$\theta(x) = \mathrm{ad}\,(\varphi(x)), \quad \text{for every } x \in X. \tag{8.16}$$

(As usual, $\varphi(x) = \pi(\chi(x)) = [\chi(x)]_a$ for every $x \in X$.) Finally, it is straightforwardly proved that there exists a unique linear map

$$g : \mathscr{T}(\mathbb{K}\langle X \rangle) \to \mathrm{Lie}(X) \quad \text{with}$$
$$\begin{cases} g(1_\mathbb{K}) = 0, \quad g(x_1) = \varphi(x_1), \\ g(x_1 \otimes \cdots \otimes x_k) = [\varphi(x_1) \cdots [\varphi(x_{k-1}), \varphi(x_k)] \cdots], \end{cases} \tag{8.17}$$

for every $x_1, \ldots, x_k \in X$ and every $k \geq 2$.

All the above maps are related by the following result, proved by Cartier in [33] (only when X is finite, but the argument generalizes straightaway): the proof is closely related to that of the Dynkin, Specht, Wever Lemma 3.26 on page 145.

Lemma 8.5. *With all the above notation, we have:*

1. *δ is a derivation of the Lie algebra $\mathrm{Lie}(X)$ and $\delta \circ \varphi \equiv \varphi$ on X.*
2. *$g(x \cdot y) = \theta(x)(g(y))$, for every $x \in \mathscr{T}(\mathbb{K}\langle X \rangle)$, $y \in \mathscr{T}_+(\mathbb{K}\langle X \rangle)$.*
3. *$\theta \circ f \equiv \mathrm{ad}$ on $\mathrm{Lie}(X)$, that is, $\theta(f(\xi))(\eta) = [\xi, \eta]$, for every $\xi, \eta \in \mathrm{Lie}(X)$.*
4. *$g \circ f \equiv \delta$ on $\mathrm{Lie}(X)$.*

See the diagram below:

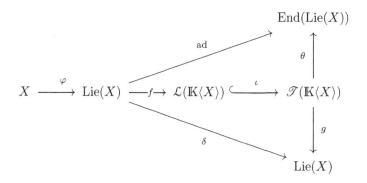

Proof. We split the proof in four steps.

(1): A simple verification: Take elements $t, t' \in \text{Lie}(X) = \bigoplus_{n \geq 1} B_n$ (see the notation in Lemma 8.3), with $t_n, t'_n \in B_n$ for every $n \geq 1$ (and t_n, t'_n are equal to 0 for n large enough). Then we have

$$\delta([t, t']) = \delta\left(\left(\sum_{i+j=n} [t_i, t'_j] \right)_{n \geq 1} \right) = \left(n \sum_{i+j=n} [t_i, t'_j] \right)_{n \geq 1}$$

$$= \left(\sum_{i+j=n} (i+j) [t_i, t'_j] \right)_{n \geq 1} = \left(\sum_{i+j=n} [i \, t_i, t'_j] \right)_n + \left(\sum_{i+j=n} [t_i, j \, t'_j] \right)_n$$

$$= [\delta(t), t'] + [t, \delta(t')].$$

Next we prove that

$$\delta(\varphi(x)) = \varphi(x), \quad \text{for every } x \in X. \tag{8.18}$$

Indeed, from the definition of B_n in Lemma 8.3, we have $\varphi(X) \subset B_1$ so that $\delta(\varphi(x)) = 1 \cdot \varphi(x)$ (in view of (8.15)) and (8.18) holds.

(2): If $x = k \in \mathscr{T}_0(V)$, (2) is trivially true, indeed we have $g(k \cdot y) = g(k \, y) = k \, g(y)$ (since g is linear) and (recalling that θ is a UAA morphism, hence unital)

$$\theta(k)(g(y)) = k \, \text{Id}_{\text{Lie}(X)}(g(y)) = k \, g(y).$$

Thus we are left to prove (2) when both x, y belong to $\mathscr{T}_+(\mathbb{K}\langle X \rangle)$; moreover, by linearity, we can assume without loss of generality that $x = v_1 \otimes \cdots \otimes v_k$ and $y = w_1 \otimes \cdots \otimes w_h$ with $h, k \geq 1$ and where the vs and ws are elements of X:

$$g(x \cdot y) = g\big(v_1 \otimes \cdots \otimes v_k \otimes w_1 \otimes \cdots \otimes w_h\big)$$

$$\overset{(8.17)}{=} [\varphi(v_1), \ldots [\varphi(v_k), [\varphi(w_1), \ldots [\varphi(w_{h-1}), \varphi(w_h)] \ldots]] \ldots]$$

$$= \mathrm{ad}\,(\varphi(v_1)) \circ \cdots \circ \mathrm{ad}\,(\varphi(v_k))\big([\varphi(w_1), \ldots [\varphi(w_{h-1}), \varphi(w_h)] \ldots]\big)$$

(by (8.16) and (8.17), the cases $h = 1$ and $h > 1$ being analogous)

$$= \theta(v_1 \otimes \cdots \otimes v_k)(g(w_1 \otimes \cdots \otimes w_h)) = \theta(x)(g(y)).$$

(3): Let $\xi \in \mathrm{Lie}(X)$. Then (4) follows if we show that $\theta(f(\xi))$ and $\mathrm{ad}\,(\xi)$ coincide (note that they are both endomorphisms of $\mathrm{Lie}(X)$). In its turn, this is equivalent to equality between $\theta \circ f$ and the map

$$\mathrm{ad} : \mathrm{Lie}(X) \to \mathrm{End}(\mathrm{Lie}(X)), \quad \xi \mapsto \mathrm{ad}\,(\xi).$$

Now recall that ad is a Lie algebra morphism (see Lemma 3.28 on page 150: this is essentially the meaning of the Jacobi identity!). Also $\theta \circ f = \theta \circ \iota \circ f$ is an LA morphism: indeed f is an LA morphism by construction, ι is an LA morphism trivially, θ is a (commutator-) LA morphism since it is a UAA morphism. Hence, equality of ad and $\theta \circ f$ follows if we prove that they are equal on a system of Lie generators for $\mathrm{Lie}(X)$, for example on $\varphi(X)$ (recall Theorem 8.1-3): for every $x \in X$ we in fact have

$$(\theta \circ f)(\varphi(x)) \overset{(8.14)}{=} \theta(x) \overset{(8.16)}{=} \mathrm{ad}\,(\varphi(x)).$$

(4): We claim that $g \circ f$ is a derivation of $\mathrm{Lie}(X)$: for $\xi, \eta \in \mathrm{Lie}(X)$ one has

$$(g \circ f)[\xi, \eta] \overset{(\mathrm{i})}{=} g([f(\xi), f(\eta)]) \overset{(\mathrm{ii})}{=} g\big(f(\xi) \cdot f(\eta) - f(\eta) \cdot f(\xi)\big)$$

$$\overset{(\mathrm{iii})}{=} \theta(f(\xi))(g(f(\eta))) - \theta(f(\eta))(g(f(\xi)))$$

$$\overset{(\mathrm{iv})}{=} [\xi, g(f(\eta))] - [\eta, g(f(\xi))] = [\xi, g(f(\eta))] + [g(f(\xi)), \eta].$$

Here we applied the following:

(i): f is a Lie algebra morphism.
(ii): The bracket of $\mathcal{L}(\mathbb{K}\langle X \rangle)$ is the commutator of $(\mathscr{T}(\mathbb{K}\langle X \rangle), \cdot)$.
(iii): We invoke part (2) of the proof together with $\mathcal{L}(\mathbb{K}\langle X \rangle) \subset \mathscr{T}_+(\mathbb{K}\langle X \rangle)$.
(iv): We invoke part (3) of the proof.

Moreover, for every $x \in X$ one has

$$(g \circ f)(\varphi(x)) \overset{(8.14)}{=} g(x) \overset{(8.17)}{=} \varphi(x) \overset{(8.18)}{=} \delta(\varphi(x)).$$

Since $g \circ f$ and δ are derivations of $\mathrm{Lie}(X)$ coinciding on $\varphi(X)$ (which is a system of Lie generators for $\mathrm{Lie}(X)$ by Theorem 8.1-3), (4) follows.   □

Corollary 8.6. *Let* $\mathbb{K}$ *be a field of characteristic zero, as usual. If* f *is as in* (8.14), f *is an isomorphism of Lie algebras and* $\mathcal{L}(\mathbb{K}\langle X\rangle)$ *is a free Lie algebra over* X.

Proof. From (4) in Lemma 8.5 and the injectivity of δ, we derive the injectivity of f: Indeed, if $\ell, \ell' \in \mathrm{Lie}(X)$ are such that $f(\ell) = f(\ell')$ then

$$\delta(\ell) = (g \circ f)(\ell) = g(f(\ell)) = g(f(\ell')) = (g \circ f)(\ell') = \delta(\ell').$$

Writing ℓ, ℓ' as $(\ell_n)_n$ and $(\ell'_n)_n$ respectively (with $\ell_n, \ell'_n \in B_n$ for every $n \in \mathbb{N}$), the identity $\delta(\ell) = \delta(\ell')$ is equivalent to

$$n\, \ell_n = n\, \ell'_n \quad \text{for every } n \in \mathbb{N}.$$

Since $\mathbb{K}$ has characteristic zero, this is possible iff $\ell_n = \ell'_n$ for every n, that is, $\ell = \ell'$. This proves the injectivity of f.

As for the surjectivity of f, it suffices to show that the set of (linear) generators for $\mathcal{L}(\mathbb{K}\langle X\rangle)$

$$x, \quad [x_n, \cdots [x_2, x_1]\cdots] \qquad (n \geq 2, \quad x, x_1, \ldots, x_n \in X)$$

belongs to the image of f. This is a consequence of (8.14) and of the fact that f is an LA morphism, gathered together to produce the following identity:

$$f\big([\varphi(x_n), \cdots [\varphi(x_2), \varphi(x_1)]\cdots]\big) = [f(\varphi(x_n)), \cdots [f(\varphi(x_2)), f(\varphi(x_1))]\cdots]$$

$$= [x_n, \cdots [x_2, x_1]\cdots].$$

This ends the proof.   □

8.2   Free Nilpotent Lie Algebra Generated by a Set

The aim of this section is to introduce the definition of the free nilpotent Lie algebra generated by a set. We begin with the relevant definition. We recall that, given a set X and a field $\mathbb{K}$, we denote by $\mathbb{K}\langle X\rangle$ the free vector space over X (see Definition 2.3 on page 51). We also recall that, given a vector space V, we denote by $\mathcal{L}(V)$ the free Lie algebra generated by V, according to Definition 2.46 on page 85.

Definition 8.7 (Free Nilpotent Lie Algebra Generated by a Set). Let X be a nonempty set, let $r \geq 1$ be a fixed integer and let $\mathbb{K}$ be any field. We denote by $\mathfrak{N}_r(X)$ the vector space (over $\mathbb{K}$) obtained as the quotient of $\mathcal{L}(\mathbb{K}\langle X\rangle)$ by its subspace $\mathfrak{R}_{r+1}(X) := \bigoplus_{k \geq r+1} \mathcal{L}_k(\mathbb{K}\langle X\rangle)$, that is,

$$\mathfrak{N}_r(X) := \mathcal{L}(\mathbb{K}\langle X\rangle)/\mathfrak{R}_{r+1}(X), \quad \text{where}$$

$$\mathfrak{R}_{r+1}(X) = \text{span}\Big\{ [x_1 \cdots [x_{k-1}, x_k] \cdots] \,\Big|\, x_1, \ldots, x_k \in X, \ k \geq r+1 \Big\}. \tag{8.19}$$

We call $\mathfrak{N}_r(X)$ the *free Lie algebra generated by X nilpotent of step r*.

In the sequel, retaining X, r and $\mathbb{K}$ fixed as above, we write for short

$$V := \mathbb{K}\langle X\rangle, \quad \mathfrak{R} := \mathfrak{R}_{r+1}(X),$$

and we also denote by

$$\pi : \mathcal{L}(\mathbb{K}\langle X\rangle) \to \mathfrak{N}_r(X), \qquad \ell \mapsto [\ell]_{\mathfrak{R}} \tag{8.20}$$

the associated projection. Note that *the restriction of π to X* (thought of as a subset of $\mathbb{K}\langle X\rangle = \mathcal{L}_1(\mathbb{K}\langle X\rangle) \subset \mathcal{L}(V)$) *is injective*: indeed if $x, x' \in X$ and if one has $\pi(x) = \pi(x')$, that is, $x - x' \in \mathfrak{R}$, then we have

$$\mathcal{L}_1(V) \ni x - x' \in \mathfrak{R} = \bigoplus_{k \geq r+1} \mathcal{L}_k(V),$$

and this is possible if and only if $x - x' = 0$ (as $r + 1 \geq 2$). So, on occasion, by identifying X with $\pi(X)$, we shall think of X as a subset of $\mathfrak{N}_r(X)$. Here we have some properties of $\mathfrak{N}_r(X)$.

Proposition 8.8. $\mathfrak{N}_r(X)$ *is a Lie algebra, nilpotent of step $\leq r$. Furthermore, $\mathfrak{N}_r(X)$ is isomorphic (as a Lie algebra) to*

$$\bigoplus_{k=1}^{r} \mathcal{L}_k(\mathbb{K}\langle X\rangle),$$

equipped with the Lie bracket

$$\Big[\textstyle\sum_{i=1}^{r} \ell_i, \sum_{j=1}^{r} \ell'_j\Big]_r := \sum_{i+j \leq r} [\ell_i, \ell'_j], \tag{8.21}$$

$$\ell_k, \ell'_k \in \mathcal{L}_k(\mathbb{K}\langle X\rangle) \quad \forall\, k = 1, \ldots, r.$$

Proof. By means of (2.51), stating that $[\mathcal{L}_i(V), \mathcal{L}_j(V)] \subseteq \mathcal{L}_{i+j}(V)$ (for every $i, j \geq 1$), we have

$$[\ell, r] \in \mathfrak{R}, \quad \text{for every } \ell \in \mathcal{L}(V) \text{ and every } r \in \mathfrak{R}.$$

In other words $\mathfrak{R}$ is an *ideal* of $\mathcal{L}(V)$. Then it is immediately seen that $\mathfrak{N}_r(X)$ is a Lie algebra, when equipped with the Lie bracket

$$[\pi(\ell), \pi(\ell')] := \pi([\ell, \ell']), \quad \ell, \ell' \in \mathcal{L}(V), \tag{8.22}$$

the bracket in the above right-hand side denoting the usual Lie bracket of $\mathcal{L}(V)$ (inherited from the commutator of the tensor algebra $\mathscr{T}(V)$). This also proves that π is a Lie algebra morphism.

We next have to prove that $\mathfrak{N}_r(X)$ is nilpotent of step $\leq r$. Indeed, for every $k \geq r$ and every $\ell, \ldots, \ell_{k+1} \in \mathcal{L}(V)$ we have

$$[\pi(\ell_1)\ldots[\pi(\ell_k),\pi(\ell_{k+1})]] = \pi([\ell_1 \ldots [\ell_k, \ell_{k+1}]]) = 0,$$

since $[\ell_1 \ldots [\ell_k, \ell_{k+1}]] \in \mathfrak{R}$.

Finally we note that, as $\mathcal{L}(V) = \bigoplus_{k=1}^{\infty} \mathcal{L}_k(V)$, the restriction of π to $\bigoplus_{k=1}^{r} \mathcal{L}_k(V)$ is an isomorphism of vector spaces. More explicitly, this map (call it α say) is given by

$$\alpha : \bigoplus_{k=1}^{r} \mathcal{L}_k(V) \longrightarrow \mathfrak{N}_r(X), \quad \textstyle\sum_{k=1}^{r} \ell_k \mapsto \pi\left(\sum_{k=1}^{r} \ell_k\right) \tag{8.23}$$

(where $\ell_k \in \mathcal{L}_k(V)$ for all $k = 1, \ldots, r$).

This map becomes a Lie algebra morphism when $\bigoplus_{k=1}^{r} \mathcal{L}_k(V)$ is endowed with the following operation (here $\ell_k, \ell_k' \in \mathcal{L}_k(V)$ for all $k = 1, \ldots, r$):

$$\alpha^{-1}\left(\left[\alpha\left(\textstyle\sum_{i=1}^{r} \ell_i\right), \alpha\left(\textstyle\sum_{j=1}^{r} \ell_j'\right)\right]\right)$$
$$= \alpha^{-1}\left(\pi\left(\textstyle\sum_{i+j\leq r}[\ell_i, \ell_j'] + \sum_{i+j\geq r}[\ell_i, \ell_j']\right)\right)$$
$$= \alpha^{-1}\left(\pi\left(\textstyle\sum_{i+j\leq r}[\ell_i, \ell_j']\right)\right) = \alpha^{-1}\left(\alpha\left(\textstyle\sum_{i+j\leq r}[\ell_i, \ell_j']\right)\right)$$
$$= \textstyle\sum_{i+j\leq r}[\ell_i, \ell_j'],$$

which is precisely the operation in (8.21). This ends the proof. $\qquad\square$

Remark 8.9. The last statement of Proposition 8.8 says roughly that $\mathfrak{N}_r(X)$ can be obtained from the Lie algebra $\mathcal{L}(\mathbb{K}\langle X\rangle)$ simply by setting to zero all the brackets of height $\geq r + 1$.

Moreover, by looking at the proof of Proposition 8.8, we deduce that any element of $\mathfrak{N}_r(X)$ is a linear combination of elements of the form

$$[\pi(x_1) \cdots [\pi(x_{k-1}), \pi(x_k)] \cdots] \quad \text{with} \quad x_1, \ldots, x_k \in X, \ k \leq r.$$

In particular (see also (8.22)) this proves that $\pi(X)$ generates $\mathfrak{N}_r(X)$ as a Lie algebra (actually, brackets of height $\leq r$ suffice!).

The most important property of $\mathfrak{N}_r(X)$ is the following one, according to which we shall call $\mathfrak{N}_r(X)$ – by full right – *the free nilpotent Lie algebra of step r generated by* X.

Theorem 8.10 (Universal Property of $\mathfrak{N}_r(X)$). *Let X be a set, let $r \in \mathbb{N}$ and suppose $\mathbb{K}$ is a field. Let also $\mathfrak{N}_r(X)$ be as in (8.19). Then we have the following properties:*

(i) *For every Lie algebra $\mathfrak{n}$ (over $\mathbb{K}$), nilpotent of step less than or equal to r, and for every map $f : X \to \mathfrak{n}$, there exists a unique Lie algebra morphism $f^\pi :$ $\mathfrak{N}_r(X) \to \mathfrak{n}$ prolonging f, or – more precisely – with the following property*

$$f^\pi(\pi(x)) = f(x) \quad \text{for every } x \in X, \tag{8.24}$$

thus making the following a commutative diagram:

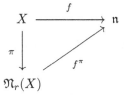

(ii) *Conversely, suppose N, φ are respectively a Lie algebra nilpotent of step $\leq r$ and a map $\varphi : X \to N$ with the following property: For every Lie algebra $\mathfrak{n}$ nilpotent of step $\leq r$ and every map $f : X \to \mathfrak{n}$, there exists a unique Lie algebra morphism $f^\varphi : N \to \mathfrak{n}$ such that*

$$f^\varphi(\varphi(x)) = f(x) \quad \text{for every } x \in X, \tag{8.25}$$

thus making the following a commutative diagram:

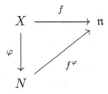

Then N is canonically isomorphic, as a Lie algebra, to $\mathfrak{N}_r(X)$, the isomorphism being $\varphi^\pi : \mathfrak{N}_r(X) \to N$ (see the notation in (i) above) and its inverse being $\pi^\varphi : N \to \mathfrak{N}_r(X)$. Furthermore, φ is injective. Actually it holds that $\varphi = \varphi^\pi \circ (\pi|_X)$. Finally we have $N \simeq \mathfrak{N}_r(\varphi(X))$, canonically.

Proof. (i). By Theorem 2.56 on page 91, there exists a Lie algebra morphism $F : \mathcal{L}(\mathbb{K}\langle X \rangle) \to \mathfrak{n}$ prolonging f. We set

$$f^\pi : \mathfrak{N}_r(X) \to \mathfrak{n}, \quad f^\pi(\pi(\ell)) := F(\ell) \quad (\ell \in \mathcal{L}(V)). \tag{8.26}$$

We claim that the definition is well posed, that is, $\mathfrak{R} \subseteq \ker(F)$. Indeed, any element of $\mathfrak{R}$ is a linear combination of right-nested brackets of X with heights $\geq r + 1$, that is, of elements of the form:

$$\ell = [x_1 \cdots [x_{k-1}, x_k] \cdots], \quad \text{with} \quad x_1, \ldots, x_k \in X, \ k \geq r+1.$$

Then we have (recall that F is an LA morphism)

$$F(\ell) = [F(x_1) \cdots [F(x_{k-1}), F(x_k)]_\mathfrak{n} \cdots]_\mathfrak{n} = 0,$$

the last equality following from the fact that $\mathfrak{n}$ is nilpotent of step $\leq r$ and $k \geq r+1$. We next prove that f^π is an LA morphism satisfying (8.24). This latter fact is obvious, for we have (for every $x \in X$)

$$f^\pi(\pi(x)) \overset{(8.26)}{=} F(x) = f(x),$$

the last equality following from the fact that F prolongs f. As for f^π being an LA morphism, we have (for every $\ell, \ell' \in \mathcal{L}(V)$)

$$f^\pi([\pi(\ell), \pi(\ell')]) \overset{(8.22)}{=} f^\pi\left(\pi([\ell, \ell'])\right) \overset{(8.26)}{=} F([\ell, \ell'])$$

$$= [F(\ell), F(\ell')]_\mathfrak{n} \overset{(8.26)}{=} \left[f^\pi(\pi(\ell)), f^\pi(\pi(\ell'))\right]_\mathfrak{n}.$$

Finally, we have to focuss on the uniqueness part. By Remark 8.9, we know that any element of $\mathfrak{N}_r(X)$ is a linear combination of elements of the form

$$n = [\pi(x_1) \cdots [\pi(x_{k-1}), \pi(x_k)] \cdots], \quad \text{with} \quad x_1, \ldots, x_k \in X, \ k \leq r.$$

An LA morphism from $\mathfrak{N}_r(X)$ to X satisfying (8.24) necessarily maps the above n into $[f(x_1) \cdots [f(x_{k-1}), f(x_k)]_\mathfrak{n} \cdots]_\mathfrak{n}$. This proves the uniqueness of such an LA morphism.

(ii). Part (ii) of the theorem follows by arguing as in the proof of Theorem 2.6 (see page 393). We recall the scheme of the proof. We have the commutative diagrams (recalling that $\mathfrak{N}_r(X)$ and N are nilpotent Lie algebras of step not exceeding r; see also Proposition 8.8)

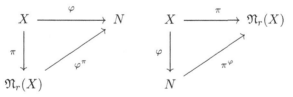

Obviously, the following are commutative diagrams too

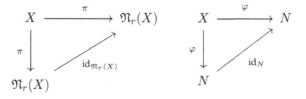

Now, the maps $\pi^\varphi \circ \varphi^\pi : \mathfrak{N}_r(X) \to \mathfrak{N}_r(X)$, $\varphi^\pi \circ \pi^\varphi : N \to N$ are LA morphisms such that (thanks to the first couple of commutative diagrams)

$$(\pi^\varphi \circ \varphi^\pi)(\pi(x)) = \pi(x) \quad \forall\, x \in X, \qquad (\varphi^\pi \circ \pi^\varphi)(\varphi(x)) = \varphi(x) \quad \forall\, x \in X.$$

Hence, by the uniqueness of the LA morphisms in the "diagonal" arrows in the last couple of commutative diagrams above, we have

$$\pi^\varphi \circ \varphi^\pi \equiv \mathrm{id}_{\mathfrak{N}_r(X)}, \qquad \varphi^\pi \circ \pi^\varphi \equiv \mathrm{id}_N.$$

From the first commutative diagram in the first couple of diagrams, we infer that $\varphi \equiv \varphi^\pi \circ (\pi|_X)$, whence (since $\pi|_X$ is injective and φ^π is bijective) φ is injective. Finally, in order to prove that $N \simeq \mathfrak{N}_r(\varphi(X))$, due to what we have alre ady proved of part (ii), it suffices to show that N and the map $\iota : \varphi(X) \hookrightarrow N$ have the following property: For every Lie algebra $\mathfrak{n}$ nilpotent of step $\leq r$ and every map $f : \varphi(X) \to \mathfrak{n}$, there exists a unique Lie algebra morphism $f^\iota : N \to \mathfrak{n}$ such that

$$f^\iota(\iota(t)) = f(t) \quad \text{for every } t \in \varphi(X).$$

Now, this follows by taking $f^\iota := (f \circ \varphi)^\varphi$.   □

Example 8.11. Let $s \in \mathbb{N} \cup \{0\}$ be fixed. Consider the following polynomial vector fields in $\mathbb{R}^2$ (here vector fields are thought of as linear differential operators of order 1)

$$X = \partial_x, \qquad Y_s = \frac{x^s}{s!}\, \partial_y.$$

If $s = 0$ we have $[X, Y_s] = 0$. If $s \geq 1$, it is immediately seen that (with the usual bracket of vector fields) one has

$$\underbrace{[X \ldots [X, Y_s]]}_{k \text{ times}} = \begin{cases} \dfrac{x^{s-k}}{(s-k)!}\, \partial_y, & \text{if } 1 \leq k \leq s, \\ 0, & \text{if } k \geq s+1, \end{cases}$$

whereas all right-nested brackets with more than one Y_s vanish identically. This proves that the Lie algebra generated by $\{X, Y_s\}$

$$\mathfrak{n}_s := \mathrm{Lie}\{X, Y_s\}$$

(thought of as a subalgebra of the algebra of the smooth vector fields in $\mathbb{R}^2$) is nilpotent of step $s + 1$. The arbitrariness of $s \in \mathbb{N} \cup \{0\}$ proves the existence of nilpotent Lie algebras (Lie-generated by two elements) of any step of nilpotency.

Corollary 8.12. *Let X be a set with at least two distinct elements. Then the step of nilpotency of $\mathfrak{N}_r(X)$ is precisely r.*

Proof. Let $r \in \mathbb{N}$. We know from Proposition 8.8 that $\mathfrak{N}_r(X)$ is nilpotent of step less than or equal to r. Suppose now that X has two distinct elements, say x_1, x_2. We have to prove that there exists at least a Lie bracket of elements of $\mathfrak{N}_r(X)$ of height r which is not vanishing.

To this end, consider the Lie algebra $\mathfrak{n}_{r-1} = \mathrm{Lie}\{X, Y_{r-1}\}$ resulting by taking $s := r - 1$ in the above Example 8.11. We know that $\mathfrak{n}_{r-1}$ is nilpotent of step $s + 1 = r$ and we have

$$\underbrace{[X \ldots [X, Y_{r-1}]]}_{r-1 \text{ times}} = \partial_y \neq 0. \tag{8.27}$$

Consider the map

$$f : X \to \mathfrak{n}_{r-1}, \quad f(x) := \begin{cases} X, & \text{if } x = x_1, \\ Y_{r-1}, & \text{if } x = x_2, \\ 0, & \text{otherwise.} \end{cases} \tag{8.28}$$

Then by Theorem 8.10-(i) there exists a unique Lie algebra morphism $f^\pi : \mathfrak{N}_r(X) \to \mathfrak{n}_{r-1}$ satisfying

$$f^\pi(\pi(x)) = f(x) \quad \text{for every } x \in X. \tag{8.29}$$

Now let us consider the element of $\mathfrak{N}_r(X)$ defined by

$$n := \underbrace{[\pi(x_1) \ldots [\pi(x_1)}_{r-1 \text{ times}}, \pi(x_2)]].$$

Then, exploiting the fact that f^π is an LA morphism, we have

$$f^\pi(n) = f^\pi\big([\pi(x_1) \ldots [\pi(x_1), \pi(x_2)]]\big)$$
$$= [f^\pi(\pi(x_1)) \ldots [f^\pi(\pi(x_1)), f^\pi(\pi(x_2))]]$$
$$\overset{(8.29)}{=} [f(x_1) \ldots [f(x_1), f(x_2)]] \overset{(8.28)}{=} [X \ldots [X, Y_{r-1}]] \overset{(8.27)}{\neq} 0.$$

This proves that $f^\pi(n) \neq 0$, whence $n \neq 0$. But now we note that n is a Lie bracket of height r of elements of $\mathfrak{N}_r(X)$. This proves that the step of nilpotency of $\mathfrak{N}_r(X)$ is precisely r. $\qquad\qquad \square$

Theorem 8.13 (Stratification of Free Nilpotent Lie Algebras). *Let* $r \in \mathbb{N}$, $r \geq 2$ *and let* X *be a nonempty set. Then* $\mathfrak{N}_r(X)$ *admits a stratification, that is, a direct sum decomposition*

$$\mathfrak{N}_r(X) = \bigoplus_{k=1}^{r} V_k, \quad \text{where} \quad [V_1, V_k] = \begin{cases} V_{k+1}, & \text{for every } k = 1, \ldots, r-1, \\ \{0\}, & \text{for } k = r. \end{cases}$$
$$\tag{8.30}$$

Proof. Arguing up to the isomorphism α in (8.23) of the proof of Proposition 8.8, we can suppose that

$$\mathfrak{N}_r(X) = \bigoplus_{k=1}^{r} \mathcal{L}_k(\mathbb{K}\langle X \rangle),$$

and that the Lie bracket of $\mathfrak{N}_r(X)$ is given by $[\cdot, \cdot]_r$, introduced in (8.21). We claim that the collection of vector spaces defined by

$$V_k := \mathcal{L}_k(\mathbb{K}\langle X \rangle), \quad \text{for every } k = 1, \ldots, r, \tag{8.31}$$

satisfies (8.30). To begin with, let $k \in \{1, \ldots, r - 1\}$ and let us set, for short, $V = \mathbb{K}\langle X \rangle$. We have

$$[V_1, V_k]_r = [\mathcal{L}_1(V), \mathcal{L}_k(V)]_r = [\mathcal{L}_1(V), \mathcal{L}_k(V)] \subseteq \mathcal{L}_{1+k}(V) = V_{k+1}.$$

Indeed, the first and last equalities follow from the definition of V_k; the second equality is a direct consequence of the definition of $[\cdot, \cdot]_r$ in (8.21) (and the fact that $1 + k \leq r$); the "$\subseteq$" sign follows from (2.51) on page 85. In order to prove that the equality $[V_1, V_k]_r = V_{k+1}$ actually holds, we note that (in view of (2.49) on page 85) an element of $V_{k+1} = \mathcal{L}_{k+1}(\mathbb{K}\langle X \rangle)$ is a linear combination of nested brackets of the form

$$[\underbrace{x_1}_{\in X \subseteq V_1} , \underbrace{[x_2 \ldots [x_k, x_{k+1}] \ldots]}_{\in \mathcal{L}_{k+1}(\mathbb{K}\langle X \rangle) = V_k}]], \qquad x_1, \ldots, x_{k+1} \in X,$$

which is *de visu* an element of $[V_1, V_k] = [V_1, V_k]_r$.

Finally, we are left to prove that $[V_1, V_r]_r = \{0\}$. This is a simple consequence of the definition of $[\cdot, \cdot]_r$ in (8.21): indeed, an arbitrary element of $V_1 = \mathcal{L}_1(V)$ can be written as

$$\sum_{i=1}^{r} \ell_i, \quad \text{with } \ell_1 \in \mathcal{L}_1(V) \text{ and } \ell_2 = \cdots = \ell_r = 0,$$

whilst an arbitrary element of $V_r = \mathcal{L}_r(V)$ can be written as

$$\sum_{j=1}^{r} \ell'_j, \quad \text{with } \ell'_r \in \mathcal{L}_r(V) \text{ and } \ell'_1 = \cdots = \ell'_{r-1} = 0,$$

so that the generic element of $[V_1, V_r]_r$ turns out to be

$$\left[\sum_{i=1}^{r} \ell_i, \sum_{j=1}^{r} \ell'_j \right]_r \overset{(8.21)}{=} \sum_{i+j \leq r} [\ell_i, \ell'_j] = \sum_{1+j \leq r} [\ell_1, \ell'_j] = 0.$$

This completes the proof. $\square$

It is convenient, for future reference, to introduce a separate notation for $\mathfrak{N}_r(X)$ when $X = \{x_1, \ldots, x_m\}$ is finite. (As usual, all linear structures are referred to the same fixed field $\mathbb{K}$.)

Definition 8.14. Let $m \in \mathbb{N}$, $m \geq 2$ and let $r \in \mathbb{N}$ be fixed. We say that $\mathfrak{f}_{m,r}$ is the *free Lie algebra with m generators $x_1, \ldots, x_m$ and nilpotent of step r* if:

(i) $\mathfrak{f}_{m,r}$ is a Lie algebra generated by its m distinct elements $x_1, \ldots, x_m$ i.e.,

$$x_1, \ldots, x_m \in \mathfrak{f}_{m,r} \qquad \text{and} \quad \mathfrak{f}_{m,r} = \mathrm{Lie}\{x_1, \ldots, x_m\};$$
$$\mathrm{card}(\{x_1, \ldots, x_m\}) = m$$

(ii) $\mathfrak{f}_{m,r}$ is nilpotent of step r;

(iii) For every Lie algebra $\mathfrak{n}$ nilpotent of step not exceeding r and for every map $f : \{x_1, \ldots, x_m\} \to \mathfrak{n}$, there exists a unique LA morphism $\overline{f}$ from $\mathfrak{f}_{m,r}$ to $\mathfrak{n}$ which extends f.

Taking into account Definition 8.7, we simply have

$$\mathfrak{f}_{m,r} = \mathfrak{N}_r(\{x_1, \ldots, x_m\})$$

(where any x_i has been identified to $\pi(x_i)$). The existence of $\mathfrak{f}_{m,r}$ follows from Theorem 8.10 and Corollary 8.12 (indeed, note that $\{x_1, \ldots, x_m\}$ contains at least two elements, as $m \geq 2$).

Remark 8.15. Note that $\mathfrak{f}_{m,r}$ is finite dimensional, for every $m \geq 2$ and every $r \geq 1$. Indeed, we have

$$\mathfrak{f}_{m,r} = \bigoplus_{k=1}^{r} \mathcal{L}_k(\mathbb{K}\langle x_1, \ldots, x_m \rangle),$$

and $\mathcal{L}_k(\mathbb{K}\langle x_1, \ldots, x_m \rangle)$ is spanned by right-nested brackets of the form

$$[x_{i_1} \ldots [x_{i_{k-1}}, x_{i_k}] \ldots], \quad \text{where } i_1, \ldots, i_k \in \{1, \ldots, m\},$$

and there are only finitely many of these right-nested brackets. Hence

$$H(m, r) := \dim(\mathfrak{f}_{m,r}), \qquad m, r \in \mathbb{N}, \ m \geq 2 \qquad (8.32)$$

defines a finite positive integer. For example, it is easily seen that

$$H(m, 2) = \frac{m(m+1)}{2}.$$

Chapter 9
Formal Power Series in One Indeterminate

THE aim of this chapter is to collect some prerequisites on formal power series in one indeterminate, needed in this Book. One of the main aims is to furnish a *purely algebraic* proof of the fact that, by substituting into each other – in any order – the two series

$$\sum_{n=1}^{\infty} \frac{x^n}{n!} \quad \text{and} \quad \sum_{n=1}^{\infty} \frac{(-1)^{n+1}\, x^n}{n},$$

one obtains the result x.

As far as Calculus is concerned, this may appear as a trivial fact, since the above are the Maclaurin series of $e^x - 1$ and $\ln(1+x)$, respectively, which are *de facto* inverse functions to each other. But if we pause a moment to think of how many results of differential calculus (though basic) we would invoke in formalizing this reasoning (and in making it as self-contained as possible), we should agree that a more algebraic proof must be hidden behind the fact that the coefficients $1/n!$ and $(-1)^{n+1}/n$ actually combine together in such a well-behaved way.

The main aim of this chapter is to give such an algebraic approach to the study of these series and to the fact that they are inverse to one another. We explicitly remark how very stimulating is it to observe that the study of the composition of the above series can be faced both with arguments of Analysis and with arguments of Algebra. This is undoubtedly a common link with the possible multiplex approaches to the CBHD formula, as we presented them in the first Part of this Book.

In reaching our goal, we shall need some basic facts concerning formal power series in one indeterminate. The interested Reader is referred to, e.g., Henrinci [82], for a comprehensive treatise on the subject.

A. Bonfiglioli and R. Fulci, *Topics in Noncommutative Algebra*, Lecture Notes in Mathematics 2034, DOI 10.1007/978-3-642-22597-0_9, © Springer-Verlag Berlin Heidelberg 2012

9.1 Operations on Formal Power Series in One Indeterminate

In the sequel, $\mathbb{K}$ *will denote a fixed field of characteristic zero*. We denote by $\mathbb{K}[t]$ and $\mathbb{K}[[t]]$ respectively, the algebra of the polynomials in the indeterminate t over $\mathbb{K}$ and the algebra of the formal power series in the indeterminate t over $\mathbb{K}$ (see e.g., Sect. 4.3.1 page 199).

For the elements of $\mathbb{K}[t]$ and $\mathbb{K}[[t]]$, we shall indifferently use the sequence-type notation

$$(a_n)_n \quad (\text{where } a_n \in \mathbb{K} \text{ for every } n \in \mathbb{N} \cup \{0\}),$$

or the series-type notation (identified with the former)

$$\sum_{n=0}^{\infty} a_n \, t^n \quad (\text{where } a_n \in \mathbb{K} \text{ for every } n \in \mathbb{N} \cup \{0\})$$

The following notation will apply as well

$$(a_0, a_1, \ldots, a_n, \ldots).$$

Thus, the only difference between $\mathbb{K}[t]$ and $\mathbb{K}[[t]]$ is that, for an element of the former, the elements a_n are null for n large enough. Obviously, $\mathbb{K}[t]$ is a subspace of $\mathbb{K}[[t]]$.

Note that, according to the results of Sect. 2.3, $\mathbb{K}[[t]]$ is not only an algebraic object, *but it has topological and metric properties too*, being the (isometric) completion of $\mathbb{K}[t]$: A basis of neighborhoods of the origin of $\mathbb{K}[[t]]$ is given by the well-known sets $\widehat{U}_N = \{ \sum_{n=N}^{\infty} k_n \, t^n \mid k_n \in \mathbb{K}, \ \forall \, n \geq N \}$, whilst the metric of $\mathbb{K}[[t]]$ is given by

$$d\left(\sum_{n=0}^{\infty} a_n \, t^n, \sum_{n=0}^{\infty} b_n \, t^n \right) = \begin{cases} e^{-\min\{n \,:\, a_n \neq b_n\}}, & \text{if } (a_n)_n \neq (b_n)_n, \\ 0, & \text{if } (a_n)_n = (b_n)_n. \end{cases}$$

9.1.1 The Cauchy Product of Formal Power Series

We recall that the algebraic operation of multiplication on $\mathbb{K}[[t]]$ is given by the product

$$(a_n)_n \cdot (b_n)_n = \left(\sum_{i=0}^{n} a_i \, b_{n-i} \right)_n,$$

that is, equivalently, it is given by the usual *Cauchy product* of series

$$\sum_{n=0}^{\infty} a_n \, t^n \cdot \sum_{n=0}^{\infty} b_n \, t^n = \sum_{n=0}^{\infty} \left(\sum_{i=0}^{n} a_i \, b_{n-i} \right) t^n. \tag{9.1}$$

Obviously, $\mathbb{K}[t]$ is a subalgebra of $\mathbb{K}[[t]]$. Note also that $\mathbb{K}[[t]]$ (whence $\mathbb{K}[t]$) is a *commutative algebra*, for $\cdot$ is evidently Abelian. The identity of $(\mathbb{K}[[t]], \cdot)$ is

$$1 = (1, 0, 0, \cdots).$$

Proposition 9.1. *An element* $(a_n)_n$ *of* $\mathbb{K}[[t]]$ *admits an inverse with respect to* $\cdot$ *if and only if* $a_0 \neq 0$. *In this case, its inverse is*

$$(b_n)_n \quad \text{where} \quad b_n = \begin{cases} a_0^{-1}, & \text{if } n = 0, \\ -a_0^{-1} \sum_{i=0}^{n-1} b_i \, a_{n-i}, & \text{if } n \geq 1. \end{cases} \tag{9.2}$$

Proof. Let $a := (a_n)_n \in \mathbb{K}[[t]]$ be fixed. Since $(\mathbb{K}[[t]], \cdot)$ is Abelian, we can restrict to study the left-invertibility of a. We have $(b_n)_n \cdot (a_n)_n = 1$ iff

$$(\bigstar) \qquad \begin{cases} b_0 \, a_0 = 1, \\ b_n \, a_0 + \sum_{i=0}^{n-1} b_i \, a_{n-i} = 0, & \text{if } n \geq 1. \end{cases}$$

Thus the condition $a_0 \neq 0$ is necessary for the left-invertibility of a; and it is sufficient, for if $a_0 \neq 0$, then $(\bigstar)$ is uniquely satisfied by $(b_n)_n$ as in (9.2). $\square$

When f is invertible in $\mathbb{K}[[t]]$ with respect to $\cdot$, we denote its inverse by $\frac{1}{f}$ (preserving the notation f^{-1} for something else, see (9.11) below).

9.1.2 Substitution of Formal Power Series

An important "operation" on formal power series is the following one. We recall that we denote by $\mathbb{K}[[t]]_+$ the following subspace of $\mathbb{K}[[t]]$:

$$\mathbb{K}[[t]]_+ := \big\{ (b_n)_n \in \mathbb{K}[[t]] : b_0 = 0 \big\}. \tag{9.3}$$

We recognize that

$$\mathbb{K}[[t]]_+ = \widehat{U}_1 = \big\{ \sum_{n=1}^{\infty} b_n \, t^n \mid b_n \in \mathbb{K}, \ \forall \, n \geq 1 \big\}.$$

Definition 9.2 (Substitution of Formal Power Series). Given $f = (a_n)_n \in \mathbb{K}[[t]]$ and $g = (b_n)_n \in \mathbb{K}[[t]]_+$, we set

$$f \circ g := \left(a_0, a_1 \, b_1, a_1 \, b_2 + a_2 \, b_1^2, \ldots, \sum_{k=1}^{n} a_k \sum_{i_1 + \cdots + i_k = n} b_{i_1} \cdots b_{i_k}, \ldots \right). \tag{9.4}$$

[There are other possible equivalent ways of writing the coefficients of $f \circ g$, which – in Calculus – are related to the iterated derivatives of a composition of functions; one such a way is given by the so-called *Faà di Bruno formula*.[1]]

We thus get a function

$$\circ : \mathbb{K}[[t]] \times \mathbb{K}[[t]]_+ \to \mathbb{K}[[t]], \quad (f, g) \mapsto f \circ g.$$

Note that if $f, g \in \mathbb{K}[[t]]_+$, then $f \circ g \in \mathbb{K}[[t]]_+$. In other words, $\circ$ *is a binary operation on* $\mathbb{K}[[t]]_+$ (we shall soon prove that $(\mathbb{K}[[t]]_+, \circ)$ is indeed a monoid, see Theorem 9.4 below).

Since the elements of $\mathbb{K}[[t]]$ are not functions, the $\circ$ operation must not be confused with the ordinary composition of functions. Nonetheless, the notation is motivated by the following important result.

Theorem 9.3. *Let $\circ$ be the map in Definition 9.2. Then for every $f \in \mathbb{K}[[t]]$ with $f = \sum_{k=0}^{\infty} a_k \, t^k$, and every $g \in \mathbb{K}[[t]]_+$, we have*

$$f \circ g = \sum_{k=0}^{\infty} a_k \, g^{\cdot k}, \tag{9.5}$$

the series on the right-hand side being interpreted as a series (hence, a limit) in the metric space $(\mathbb{K}[[t]], d)$.

In other words, we have

$$\sum_{n=0}^{\infty} a_n \, t^n \circ \sum_{n=1}^{\infty} b_n \, t^n = \sum_{k=0}^{\infty} a_k \left(\sum_{i=1}^{\infty} b_i \, t^i \right)^{\cdot k} = \lim_{K \to \infty} \sum_{k=0}^{K} a_k \left(\sum_{i=1}^{\infty} b_i \, t^i \right)^{\cdot k},$$

the limit being performed in the metric space $(\mathbb{K}[[t]], d)$.

Proof. First we prove that the series on the right-hand side of (9.5) is convergent. To this aim, by Remark 2.76-3, page 102 (recall that $\mathbb{K}[[t]] = \widehat{\mathbb{K}[t]}$ and that $\mathbb{K}[t]$ is a graded algebra), it is sufficient to prove that

$$\lim_{k \to \infty} g^{\cdot k} = 0 \quad \text{in } (\mathbb{K}[[t]], d). \tag{9.6}$$

This is easily seen: as $g \in \mathbb{K}[[t]]_+ = \widehat{U}_1$, we have $g^{\cdot k} \in \widehat{U}_k$ and (9.6) follows from the fact that $\{\widehat{U}_k\}_{k \in \mathbb{N}}$ is a basis of neighborhoods of the origin.

---

[1]According to this formula, if f, g are smooth functions of $x \in \mathbb{R}$ (and $f \circ g$ is well posed) it holds that

$$\frac{d^n}{d\,x^n}(f(g(x))) = \sum_{\pi \in P} f^{(|\pi|)}(g(x)) \cdot \prod_{B \in \pi} g^{(|B|)}(x),$$

where P is the set of all the partitions of $\{1, \ldots, n\}$ and $|\cdot|$ denotes cardinality.

We are left to show the equality in (9.5). We have, for any $k \in \mathbb{N}$,

$$g^{\cdot k} = \left(\lim_{I \to \infty} \sum_{i=1}^{I} b_i \, t^i \right)^{\cdot k} \qquad \text{(recall that } (\mathbb{K}[[t]], \cdot) \text{ is a topological algebra)}$$

$$= \lim_{I \to \infty} \left(\sum_{i=1}^{I} b_i \, t^i \right)^{\cdot k} = \lim_{I \to \infty} \sum_{1 \le i_1, \ldots, i_k \le I} b_{i_1} \cdots b_{i_k} \, t^{i_1 + \cdots + i_k}$$

$$= \lim_{I \to \infty} \left\{ \left(\sum_{i_1 + \cdots + i_k \le I} + \sum_{\substack{1 \le i_1, \ldots, i_k \le I \\ i_1 + \cdots + i_k \ge I+1}} \right) b_{i_1} \cdots b_{i_k} \, t^{i_1 + \cdots + i_k} \right\}$$

$$\overset{(\star)}{=} \lim_{I \to \infty} \sum_{n=1}^{I} \sum_{i_1 + \cdots + i_k = n} b_{i_1} \cdots b_{i_k} \, t^n = \sum_{n=1}^{\infty} \left(\sum_{i_1 + \cdots + i_k = n} b_{i_1} \cdots b_{i_k} \right) t^n.$$

As for the starred equality, we used the fact that

$$\sum_{\substack{1 \le i_1, \ldots, i_k \le I \\ i_1 + \cdots + i_k \ge I+1}} b_{i_1} \cdots b_{i_k} \, t^{i_1 + \cdots + i_k}$$

vanishes as $I \to \infty$ (indeed this is an element of $\widehat{U}_{I+1}$).

We thus obtained the expected formula

$$\left(\sum_{i=1}^{\infty} b_i \, t^i \right)^{\cdot k} = \sum_{n=1}^{\infty} \left(\sum_{i_1 + \cdots + i_k = n} b_{i_1} \cdots b_{i_k} \right) t^n \qquad (k \in \mathbb{N}). \qquad (9.7)$$

This gives, by a simple reordering argument,

$$\sum_{k=0}^{\infty} a_k \, g^{\cdot k} = a_0 + \sum_{k=1}^{\infty} a_k \left(\sum_{n=1}^{\infty} \sum_{i_1 + \cdots + i_k = n} b_{i_1} \cdots b_{i_k} \, t^n \right)$$

$$= a_0 + \sum_{n=1}^{\infty} \left(\sum_{k=1}^{n} a_k \sum_{i_1 + \cdots + i_k = n} b_{i_1} \cdots b_{i_k} \right) t^n.$$

Hence, (9.5) follows by the definition of $f \circ g$ in (9.4). $\qquad \square$

Theorem 9.4. *Let $\circ$ be the map in Definition 9.2. Then $(\mathbb{K}[[t]]_+, \circ)$ is a monoid, i.e., $\circ$ is an associative operation on $\mathbb{K}[[t]]$ endowed with the unit $t = (0, 1, 0, 0, \ldots)$. Moreover, the set of the invertible elements of $\mathbb{K}[[t]]_+$ with respect to $\circ$ is*

$$\{ (a_n)_n \in \mathbb{K}[[t]] \mid a_0 = 0 \text{ and } a_1 \ne 0 \}.$$

An element of $\mathbb{K}[[t]]_+$ is invertible with respect to $\circ$ iff it has a right inverse or iff it has a left inverse.

Proof. We already know that $\circ$ is binary on $\mathbb{K}[[t]]_1$. To avoid confusion, we temporarily set $e := (0, 1, 0, \ldots)$. The fact that

$$f \circ e = f = e \circ f \quad \text{for every } f \in \mathbb{K}[[t]]_+$$

follows directly from (9.5). The main task is to prove the associativity of $\circ$. To this end, let $f, g, h \in \mathbb{K}[[t]]_+$, with $f = \sum_{k=1}^{\infty} a_k\, t^k$ and $g = \sum_{i=1}^{\infty} b_i\, t^i$. We have

$$(f \circ g) \circ h \overset{(9.5)}{=} \sum_{n=1}^{\infty} (f \circ g)_n\, h^{\cdot n} \overset{(9.4)}{=} \sum_{n=1}^{\infty} \left(\sum_{k=1}^{n} a_k \sum_{i_1+\cdots+i_k=n} b_{i_1} \cdots b_{i_k} \right) h^{\cdot n}.$$

On the other hand, it holds that

$$f \circ (g \circ h) \overset{(9.5)}{=} \sum_{k=1}^{\infty} a_k\, (g \circ h)^{\cdot k} \overset{(9.5)}{=} \sum_{k=1}^{\infty} a_k \left(\sum_{i=1}^{\infty} b_i\, h^{\cdot i} \right)^{\cdot k}$$

$$= \sum_{k=1}^{\infty} a_k \sum_{i_1,\ldots,i_k \geq 1} b_{i_1} \cdots b_{i_k}\, h^{\cdot i_1+\cdots+i_k}$$

(by a simple reordering argument)

$$= \sum_{n=1}^{\infty} \left(\sum_{k=1}^{n} a_k \sum_{i_1+\cdots+i_k=n} b_{i_1} \cdots b_{i_k} \right) h^{\cdot n}.$$

This proves that $(f \circ g) \circ h = f \circ (g \circ h)$, and the associativity follows.

To end the proof, we have to characterize the elements of $\mathbb{K}[[t]]$ which are invertible w.r.t. $\circ$. Let $f, g \in \mathbb{K}[[t]]_+$, with $f = \sum_{k=1}^{\infty} a_k\, t^k$ and $g = \sum_{i=1}^{\infty} b_i\, t^i$. Then $f \circ g = e$ if and only if (see (9.4))

$$\begin{cases} a_1\, b_1 = 1, \\ \sum_{k=1}^{n} a_k \sum_{i_1+\cdots+i_k=n} b_{i_1} \cdots b_{i_k} = 0, \quad \text{if } n \geq 2. \end{cases} \tag{9.8}$$

It is then clear that the condition $a_1 \neq 0$ is necessary for the right and the left invertibility of f. We show that it is also sufficient.

Left invertibility. For fixed g as above, with $b_1 \neq 0$, (9.8) can be uniquely solved for $f = (a_n)_n$ by the recursion formula

$$\begin{cases} a_1 = b_1^{-1}, \\ a_n = -b_1^{-n} \sum_{k=1}^{n-1} a_k \sum_{i_1+\cdots+i_k=n} b_{i_1} \cdots b_{i_k}, \quad \text{if } n \geq 2. \end{cases} \tag{9.9}$$

[Note that on the right-hand side of (9.9) only $a_1, \ldots, a_{n-1}$ are involved.] Hence *the condition $b_1 \neq 0$ is equivalent to the left-invertibility of $g = (b_n)_n$.*

Right invertibility. Fixed f as above, with $a_1 \neq 0$, (9.8) can be uniquely solved for $g = (b_n)_n$ by the recursion formula

$$
\begin{cases}
b_1 = a_1^{-1}, \\
b_n = -a_1^{-1} \sum_{k=2}^{n} a_k \sum_{i_1 + \cdots + i_k = n} b_{i_1} \cdots b_{i_k}, & \text{if } n \geq 2.
\end{cases} \tag{9.10}
$$

[Note that in the right-hand side of (9.10) only $b_1, \ldots, b_{n-1}$ are involved.] Hence *the condition $a_1 \neq 0$ is equivalent to the right-invertibility of $f = (a_n)_n$.*

Summing up, since $(\mathbb{K}[[t]]_+, \circ)$ is a monoid, the condition $a_1 \neq 0$ is equivalent to the invertibility of $f = (a_n)_n$ (see Lemma 9.5 below). □

Here we used the following simple lemma.

Lemma 9.5. *Let $(A, *)$ be a monoid (i.e., $*$ is an associative, binary operation on the set A endowed with a unit). Then an element of A is invertible iff it is equipped with a left and a right inverse; in this case the right and the left inverses coincide.*

Proof. Let l and r be, respectively, a left and a right $*$-inverse for $x \in A$, i.e., $l * x = 1 = x * r$. We are left to show that $l = r$. This is easily seen: $l = l * 1 = l * (x * r) = (l * x) * r = 1 * r = r$. □

Remark 9.6. Note that the above lemma ensures that, for fixed $(a_n)_n$ with $a_0 = 0$ and $a_1 \neq 0$, the sequence $(b_n)_n$ inductively defined by

$$
(\bigstar 1) \qquad
\begin{cases}
b_0 = 0, \quad b_1 = a_1^{-1}, \\
b_n = -a_1^{-1} \sum_{k=2}^{n} a_k \sum_{i_1 + \cdots + i_k = n} b_{i_1} \cdots b_{i_k}, & \text{if } n \geq 2,
\end{cases}
$$

coincides with the sequence inductively defined by

$$
(\bigstar 2) \qquad
\begin{cases}
b_0 = 0, \quad b_1 = a_1^{-1}, \\
b_n = -a_1^{-n} \sum_{k=1}^{n-1} b_k \sum_{i_1 + \cdots + i_k = n} a_{i_1} \cdots a_{i_k}, & \text{if } n \geq 2.
\end{cases}
$$

Indeed, $(b_n)_n$ in $(\bigstar 1)$ is the right inverse of $(a_n)_n$ in $(\mathbb{K}[[t]], \circ)$, whereas $(b_n)_n$ in $(\bigstar 2)$ is its left inverse (and these must necessarily coincide!). We remark that a direct proof of the equivalence of $(\bigstar 1)$ and $(\bigstar 2)$ (without, say, invoking Lemma 9.5) seems not so easy. □

When $f \in \mathbb{K}[[t]]_+$ is invertible with respect to $\circ$, we denote its inverse by f^{-1}. Thus we have

$$
f \circ f^{-1} = t, \qquad f^{-1} \circ f = t. \tag{9.11}
$$

An equivalent way to restate Theorem 9.4 is the following one:

Theorem 9.7. *Let $\circ$ be the map in Definition 9.2. Let $\mathfrak{F}(\mathbb{K}[[t]]_+)$ denote the set of the functions defined on $\mathbb{K}[[t]]_+$ with values in $\mathbb{K}[[t]]_+$ itself.*

For every $f \in \mathbb{K}[[t]]_+$, we can define an element $\Lambda(f) \in \mathfrak{F}(\mathbb{K}[[t]]_+)$ as follows:

$$\Lambda(f) : \mathbb{K}[[t]]_+ \longrightarrow \mathbb{K}[[t]]_+, \quad \Lambda(f)(g) := f \circ g. \qquad (9.12)$$

Then Λ defines a monoid-morphism of $(\mathbb{K}[[t]]_+, \circ)$ to $\mathfrak{F}(\mathbb{K}[[t]]_+)$ (the latter being equipped with the operation of composition of functions). In other words, we have

$$\Lambda(t) = \mathrm{Id}_{\mathbb{K}[[t]]_+}, \quad \Lambda(f \circ g) = \Lambda(f) \circ \Lambda(g), \quad \forall\, f, g \in \mathbb{K}[[t]]_+. \qquad (9.13)$$

Here, the second $\circ$ symbol denotes the ordinary composition of functions, whereas the first one is the operation defined in (9.4). Moreover, for every $f = (a_n)_n \in \mathbb{K}[[t]]_+$ with $a_1 \neq 0$, we have

$$\Lambda(f^{-1}) = \big(\Lambda(f)\big)^{-1}. \qquad (9.14)$$

Here, $(\cdot)^{-1}$ in the left-hand side denotes the inversion on $(\mathbb{K}[[t]]_+, \circ)$, whereas $(\cdot)^{-1}$ in the right-hand side denotes the inverse of a function.

Proof. Since t is the identity of $(\mathbb{K}[[t]]_+, \circ)$, it holds that $\Lambda(t)(g) = g \circ t = g$, for every $g \in \mathbb{K}[[t]]_+$, that is, $\Lambda(t) = \mathrm{Id}_{\mathbb{K}[[t]]_+}$. Moreover, the second identity of (9.13) is clearly equivalent to

$$(f \circ g) \circ h = f \circ (g \circ h), \quad \forall\, f, g, h \in \mathbb{K}[[t]]_+,$$

which is the property of associativity of $\circ$ on $\mathbb{K}[[t]]_+$, proved in Theorem 9.4. This proves that $\Lambda : \mathbb{K}[[t]]_+ \to \mathfrak{F}(\mathbb{K}[[t]]_+)$ is a monoid morphism. In particular, (9.14) follows.[2] This ends the proof. $\qquad \square$

9.1.3 The Derivation Operator on Formal Power Series

Another important operation on formal power series is the following one, resemblant to the well-known derivative of differentiable functions. We set

---

[2]Indeed, if $(A, \circledast)$ and $(B, \odot)$ are monoids and $\varphi : A \to B$ is a monoid morphism, for every $\circledast$-invertible element $a \in A$, it holds that

$$1_B = \varphi(1_A) = \varphi(a^{\circledast\, -1} \circledast a) = \varphi(a^{\circledast\, -1}) \odot \varphi(a).$$

This proves that $\varphi(a)$ is $\odot$-invertible in B and

$$\varphi(a)^{\odot\, -1} = \varphi(a^{\circledast\, -1}).$$

$$\partial_t : \mathbb{K}[[t]] \longrightarrow \mathbb{K}[[t]]$$

$$(a_n)_n \mapsto (a_1, 2\, a_2, 3\, a_3, \ldots, (n+1)\, a_{n+1}, \ldots). \tag{9.15}$$

We recognize the usual operator of "derivation with respect to t":

$$\partial_t \left(\sum_{n=0}^{\infty} a_n * t^n \right) = \sum_{n=1}^{\infty} n\, a_n * t^{n-1} = \sum_{k=0}^{\infty} (k+1)\, a_{k+1} * t^k.$$

This is exactly the operation introduced, in a more general setting, in Definition 4.11, page 203. In the sequel, the notation

$$f' := \partial_t f, \quad f \in \mathbb{K}[[t]]$$

will apply as well. The well-behaved properties of ∂_t with respect to the operations introduced so far on $\mathbb{K}[[t]]$ are summarized in the following theorem (the Reader will recognize the analogues with results from Calculus).

Theorem 9.8. *Let ∂_t be the operator on $\mathbb{K}[[t]]$ defined in (9.15). Then the following results hold.*

(a) ∂_t *is a derivation of the algebra $(\mathbb{K}[[t]], \cdot)$.*
(b) ∂_t *is continuous, w.r.t. the usual topology on $\mathbb{K}[[t]]$.*
(c) *For every $f \in \mathbb{K}[[t]]$ and every $g \in \mathbb{K}[[t]]_+$, we have*

$$(f \circ g)' = (f' \circ g) \cdot g'. \tag{9.16}$$

(d) *For every $f \in \mathbb{K}[[t]]_+$ invertible with respect to $\circ$, then f' has a reciprocal with respect to $\cdot$ and we have*

$$(f^{-1})' \circ f = \frac{1}{f'}. \tag{9.17}$$

Proof. (a) and (b). These follow, respectively, from Proposition 4.12 (page 203), parts (i) and (iii).
  (c). If $f \in \mathbb{K}[[t]]$ and $g \in \mathbb{K}[[t]]_+$, we have

$$(f \circ g)' \overset{(9.5)}{=} \partial_t \sum_{k=0}^{\infty} a_k\, g^{\cdot\, k} \quad \text{(from the continuity of } \partial_t, \text{ see part (b))}$$

$$= \sum_{k=1}^{\infty} a_k\, \partial_t(g^{\cdot\, k}) \quad \left(\begin{array}{c} \partial_t \text{ is a derivation, see part (a)} \\ \text{and } (\mathbb{K}[[t]], \cdot) \text{ is Abelian} \end{array} \right)$$

$$= \sum_{k=1}^{\infty} a_k\, k\, g^{\cdot\, k-1} \cdot g' = (f' \circ g) \cdot g'.$$

(d). Set $f = (a_n)_n$. Since f is invertible w.r.t. $\circ$, by Theorem 9.4 we have $a_1 \neq 0$. Hence the zero-degree component of f' (i.e., a_1) is non-vanishing, whence, by Proposition 9.1, f' has a reciprocal. We have $f^{-1} \circ f = t$. We now apply ∂_t to this identity, getting (by part (b) of this theorem)

$$1 = \partial_t(t) = \partial_t(f^{-1} \circ f) \stackrel{(9.16)}{=} ((f^{-1})' \circ f) \cdot f'.$$

Then (9.17) follows by definition of reciprocal.   □

9.1.4   The Relation Between the exp and the log Series

The main result of this section is contained in the following Theorem 9.9. First we fix some notation: we set

$$E := \left(0, 1, \ldots, \tfrac{1}{n!}, \ldots\right) = \sum_{n=1}^{\infty} \tfrac{1}{n!} t^n;$$

(9.18)

$$L := \left(0, 1, \ldots, \tfrac{(-1)^{n+1}}{n}, \ldots\right) = \sum_{n=1}^{\infty} \tfrac{(-1)^{n+1}}{n} t^n.$$

[Note that these are the Maclaurin series expansions of the functions $e^x - 1$ and $\ln(1 + x)$, respectively.] We remark that $E, L \in \mathbb{K}[[t]]_+$ and both are invertible w.r.t. $\circ$ for their degree-one coefficient is non-vanishing.

Theorem 9.9. *With the notation in (9.18), E and L are inverse to each other with respect to $\circ$, that is,*

$$E \circ L = t = L \circ E.$$

(9.19)

Proof. Roughly speaking, we shall steal an idea from ODE's: we shall show that L and E^{-1} solve the same "Cauchy problem", whence they coincide.

The proof is split in several steps, some having an independent interest.

I. We have

$$L' = \frac{1}{(1, 1, 0, 0, \ldots, 0, \ldots)}.$$

(9.20)

[Roughly speaking, this is $(\ln(1 + x))' = 1/(1 + x)$.] By definition of L, we have

$$L' = \left(1, -\tfrac{1}{2} 2, \ldots, \tfrac{(-1)^{n+2}}{n+1} (n + 1), \ldots\right) = (1, -1, 1, \ldots, (-1)^n, \ldots).$$

We claim that the above right-hand side is precisely the reciprocal of $(b_n)_n := (1, 1, 0, 0, \ldots)$. This follows from this computation:

$$(1, -1, \ldots, (-1)^n, \ldots) \cdot (1, 1, 0, 0, \ldots) = \left(\sum_{j=0}^{n} (-1)^{n-j} b_j \right)_n$$

(by construction, $b_0 = b_1 = 1$ and $b_j = 0$ if $j \geq 2$)

$$= \left(b_0, -b_0 + b_1, \ldots, (-1)^n b_0 + (-1)^{n-1} b_1, \ldots \right) = (1, 0, 0, \ldots, 0, \ldots).$$

II. Let $f, g \in \mathbb{K}[[t]]$ have the same zero-degree components and suppose they satisfy $f' = g'$. Then $f = g$.

[Roughly speaking, this is the uniqueness of the solution of a Cauchy problem.] Let us set $f = (a_n)_n$ and $g = (b_n)_n$. By hypothesis, we have $a_0 = b_0$. Also, by the hypothesis $f' = g'$, we have $(n+1)a_{n+1} = (n+1)b_{n+1}$ for every $n \geq 0$. Since $\mathbb{K}$ has characteristic zero, this is equivalent to $a_{n+1} = b_{n+1}$ for every $n \geq 0$. Summing up, all the coefficients of f and g coincide, that is, $f = g$.

III. Let $h \in \mathbb{K}[[t]]_+$. Then we have

$$(f \cdot g) \circ h = (f \circ h) \cdot (g \circ h), \quad \forall \, f, g \in \mathbb{K}[[t]]. \tag{9.21}$$

Let us set $f = \sum_{i=0}^{\infty} a_i t^i$ and $g = \sum_{j=0}^{\infty} b_j t^j$. We have

$$(f \circ h) \cdot (g \circ h) \overset{(9.5)}{=} \left(\sum_{i=0}^{\infty} a_i h^{\cdot i} \right) \cdot \left(\sum_{j=0}^{\infty} b_j h^{\cdot j} \right)$$

(recall $(\mathbb{K}[[t]], \cdot)$ is a topological algebra and the above series converge)

$$= \sum_{i,j \geq 0} a_i b_j h^{\cdot i+j} = \sum_{n=0}^{\infty} \left(\sum_{i+j=n} a_i b_j \right) t^n \overset{(9.5)}{=} (f \cdot g) \circ h.$$

IV. Let $f = (a_n)_n$ and $g = (b_n)_n$ with $a_0 \neq 0$ and $b_0 = 0$. Then $f \circ g$ has a reciprocal with respect to $\cdot$ and

$$\frac{1}{f} \circ g = \frac{1}{f \circ g}. \tag{9.22}$$

Since $a_0 \neq 0$, f has a reciprocal $\frac{1}{f}$; since the zero-degree coefficient of $f \circ g$ coincides with that of f, then $f \circ g$ also has a reciprocal. We have $\frac{1}{f} \cdot f = 1$. Due to $g \in \mathbb{K}[[t]]$, we can apply (9.21), thus getting

$$1 = 1 \circ g = \left(\tfrac{1}{f} \cdot f \right) \circ g = \left(\tfrac{1}{f} \circ g \right) \cdot (f \circ g),$$

which is precisely (9.22).

We are now in a position to complete the proof. We have

$$E' = E + 1. \tag{9.23}$$

[Roughly speaking, this is $(e^x - 1)' = e^x = (e^x - 1) + 1$.] Indeed this follows from:

$$E' = \left(1, 1, \ldots, (n+1)\,\tfrac{1}{(n+1)!}, \ldots\right) = (1, 0, \ldots) + \left(0, 1, \ldots, \tfrac{1}{n!}, \ldots\right) = 1 + E.$$

Since the component of degree 1 of E is non-vanishing, E admits a $\circ$-inverse. We claim that

$$(E^{-1})' = \frac{1}{(1,1,0,0,0,\ldots)}. \tag{9.24}$$

We show that this easily completes the proof. Indeed, we have

$$(E^{-1})' \overset{(9.24)}{=} \frac{1}{(1,1,0,0,0,\ldots)} \overset{(9.20)}{=} L';$$

moreover the zero-degree components of E^{-1} and of L coincide (see e.g., (9.9) for computation of the zero-degree component of E^{-1}); by part II of the proof, these facts together imply that E^{-1} and L do coincide, that is, L is the inverse of E, so that (9.19) follows.

We are thus left with the proof of the claimed (9.24). We have

$$(E^{-1})' = (E^{-1})' \circ (0, 1, 0, 0, \ldots) = (E^{-1})' \circ (E \circ E^{-1})$$

$$= \left((E^{-1})' \circ E\right) \circ E^{-1} \overset{(9.16)}{=} \frac{1}{E'} \circ E^{-1} \overset{(9.22)}{=} \frac{1}{E' \circ E^{-1}}$$

$$\overset{(9.23)}{=} \frac{1}{(E+1) \circ E^{-1}} = \frac{1}{E \circ E^{-1} + 1 \circ E^{-1}}$$

$$= \frac{1}{(0, 1, 0, \ldots) + (1, 0, 0, \ldots)} = \frac{1}{(1, 1, 0, 0, 0, \ldots)}.$$

This proves (9.24) and the proof is complete.   □

Remark 9.10. From the identities in (9.19), we can obviously deduce a family of relations among the coefficients of the series E and L. Namely, if we let $E = (b_n)_n$ and $L = (c_n)_n$, i.e.,

$$b_n := \frac{1}{n!}, \qquad c_n := \frac{(-1)^{n+1}}{n} \qquad \forall\, n \in \mathbb{N}, \tag{9.25}$$

the identities in (9.19) (and the definition of the coefficients of the substitution, see (9.4)) are equivalent to the following ones

$$\begin{cases} b_1\, c_1 = 1 \\ \displaystyle\sum_{k=1}^{n} b_k \sum_{i_1+\cdots+i_k=n} c_{i_1}\cdots c_{i_k} = 0, \quad n \geq 2; \end{cases}$$

and
$$\begin{cases} c_1\, b_1 = 1 \\ \displaystyle\sum_{k=1}^{n} c_k \sum_{i_1+\cdots+i_k=n} b_{i_1}\cdots b_{i_k} = 0, \quad n \geq 2. \end{cases} \tag{9.26}$$

In the following arguments, we show how to derive identities in any associative algebra, starting from identities in $\mathbb{K}[[t]]$.

First, we observe that we have the isomorphism

$$\mathbb{K}[[t]] \simeq \widehat{\mathscr{T}}(\mathbb{K}\langle x\rangle), \tag{9.27}$$

both as UA algebras and as topological spaces, via the same isomorphism. Indeed, it is easily seen that the map

$$\varphi : \mathbb{K}[[t]] \longrightarrow \widehat{\mathscr{T}}(\mathbb{K}\langle x\rangle)$$

$$\sum_{n=0}^{\infty} a_n\, t^n \mapsto \left(a_0, a_1\, x, a_2\, x \otimes x, \ldots, a_n\, x^{\otimes n}, \ldots \right) \tag{9.28}$$

is both a UAA isomorphism and a homeomorphism (when domain and codomain are endowed with the usual topologies). By restricting φ to $\mathbb{K}[t]$ we obtain another remarkable isomorphism of UA algebras:

$$\mathbb{K}[t] \simeq \mathscr{T}(\mathbb{K}\langle x\rangle). \tag{9.29}$$

Lemma 9.11. *Let $\{a_n\}_{n\geq 1}$ and $\{b_n\}_{n\geq 1}$ be any pair of sequences in $\mathbb{K}$. For every $M, N \geq 1$ there exists a formal power series*

$$\mathcal{R}_{M,N} \in \widehat{U}_{\min\{N+1,M+1\}}, \tag{9.30}$$

such that the following identity holds in $\mathbb{K}[[t]]$:

$$\sum_{m=1}^{\infty} a_m \left(\sum_{n=1}^{\infty} b_n\, t^n \right)^m = \sum_{m=1}^{M} a_m \left(\sum_{n=1}^{N} b_n\, t^n \right)^m + \mathcal{R}_{M,N}. \tag{9.31}$$

Proof. With the above notation, Newton's binomial formula gives

$$\left(\sum_{n=1}^{\infty} b_n t^n\right)^m = \left(\sum_{n=1}^{N}\cdots + \sum_{n=N+1}^{\infty}\cdots\right)^m =: (A+B)^m$$

$$= \left(\sum_{n=1}^{N} b_n t^n\right)^m + \underbrace{\sum_{h=0}^{m-1}\binom{m}{h} A^h \cdot B^{m-h}}_{=: G_{m,N} \in \widehat{U}_{m+N}}.$$

Indeed (with clear meanings), $A \in \widehat{U}_1$, $B \in \widehat{U}_{N+1}$, so that

$$G_{m,N} \in \sum_{h=0}^{m-1} \widehat{U}_{h+(N+1)(m-h)} \subseteq \widehat{U}_{\min_{h\in[0,m-1]} h+(N+1)(m-h)} \subseteq \widehat{U}_{m+N}.$$

This gives

$$\sum_{m=1}^{\infty} a_m\left(\sum_{n=1}^{\infty} b_n t^n\right)^m = \sum_{m=1}^{M}\cdots + \sum_{m=M+1}^{\infty}\cdots \in \sum_{m=1}^{M} a_m\left(\sum_{n=1}^{N} b_n t^n\right)^m$$

$$+ \underbrace{\sum_{m=1}^{M} a_m\, G_{m,N}}_{\in \widehat{U}_{N+1}} + \widehat{U}_{M+1},$$

and (9.30)–(9.31) follow from $\widehat{U}_{N+1} + \widehat{U}_{M+1} \subseteq \widehat{U}_{\min\{N+1,M+1\}}$.   □

Proposition 9.12. *Let $\{a_n\}_{n\geq 1}$ and $\{b_n\}_{n\geq 1}$ be any pair of sequences in $\mathbb{K}$. For every $n \in \mathbb{N}$, let*

$$c_n := \sum_{k=1}^{n} a_k \sum_{i_1+\cdots+i_k=n} b_{i_1}\cdots b_{i_k} \qquad (n \geq 1),$$

that is (according to Definition 9.2), c_n is the n-th coefficient of the formal power series obtained by substitution of $\sum_{n=1}^{\infty} b_n t^n$ in $\sum_{n=1}^{\infty} a_n t^n$.

Then for every $M, N \geq 1$ there exists a polynomial $\mathcal{R}_{M,N} \in \mathbb{K}[t]$ of the form

$$\mathcal{R}_{M,N} = \sum_{n=\min\{N,M\}+1}^{NM} r_n t^n, \qquad \text{(for suitable scalars r_n)}, \qquad (9.32)$$

such that the following identity holds in $\mathbb{K}[t]$:

$$\sum_{m=1}^{M} a_m\left(\sum_{n=1}^{N} b_n t^n\right)^m = \sum_{n=1}^{\min\{N,M\}} c_n t^n + \mathcal{R}_{M,N}. \qquad (9.33)$$

Proof. This follows by a straightforward expansion of the right-hand side of (9.33) and by the definition of c_n (or by the aid of Lemma 9.11). □

Now consider the following two facts:

- any associative algebra can be embedded in a UA algebra (see Remark 5.6, page 273);
- $\mathbb{K}[t]$ is isomorphic to $\mathscr{T}(\mathbb{K}\langle x\rangle)$ (see (9.29)) whence it inherits the universal property of the free associative algebra $\text{Libas}(\{x\}) \simeq \mathscr{T}(\mathbb{K}\langle x\rangle)$ (see Theorems 2.28 and 2.40).

As a consequence, given an associative algebra $(A, *)$ over $\mathbb{K}$ and an element $z \in A$, we derive the existence of a unique associative algebra homomorphism

$$\Phi_z : \mathbb{K}[t]_+ \to A, \qquad \text{such that} \quad \Phi_z(t) = z. \tag{9.34}$$

[Recall that $\mathbb{K}[t]_+ = \{\sum_{n=1}^{N} a_n t^n \mid N \in \mathbb{N}, a_1, \ldots, a_N \in \mathbb{K}\}$.]

Then by a "substitution" argument, as a corollary of Proposition 9.12, we straightforwardly obtain the following result.

Theorem 9.13. *Let $\{a_n\}_{n\geq 1}$, $\{b_n\}_{n\geq 1}$ and $\{c_n\}_{n\geq 1}$ be as in Proposition 9.12.*
Then for every $M, N \geq 1$ there exists a polynomial $\mathcal{R}_{M,N} \in \mathbb{K}[t]$ of the form (9.32) such that, for every associative algebra $(A, *)$ over $\mathbb{K}$ and every $z \in A$,

$$\sum_{m=1}^{M} a_m \left(\sum_{n=1}^{N} b_n z^{*n} \right)^{*m} = \sum_{n=1}^{\min\{N,M\}} c_n z^{*n} + \Phi_z(\mathcal{R}_{M,N}). \tag{9.35}$$

Here Φ_z is the associative algebra homomorphism in (9.34). In particular, the "remainder term" $\Phi_z(\mathcal{R}_{M,N})$ is a $\mathbb{K}$-linear combination in A of powers z^{*n} with $n \in \{\min\{N, M\} + 1, \ldots, N M\}$.

Remark 9.14. As a very particular case of Theorem 9.13, if $a_n = \frac{1}{n!}$ and $b_n = \frac{(-1)^n}{n}$ or if $a_n = \frac{(-1)^n}{n}$ and $b_n = \frac{1}{n!}$, we get the following identity (in view of the relations (9.26) existing between the coefficients a_n and b_n):

$$\sum_{m=1}^{M} a_m \left(\sum_{n=1}^{N} b_n z^{*n} \right)^{*m} = z + \left(\begin{array}{c} \text{a } \mathbb{Q}\text{-linear combination of powers } z^{*n} \\ \text{with } n \in \{\min\{N, M\} + 1, \ldots, N M\} \end{array} \right). \tag{9.36}$$

This holds true on every associative algebra $(A, *)$ over a field of characteristic zero, and for every $z \in A$. For example, A may be the algebra of smooth vector fields on some open set $\Omega \subseteq \mathbb{R}^N$ with the operation of composition (here, we are referring to a smooth vector field as a linear partial differential operator of first order with smooth coefficients).

9.2   Bernoulli Numbers

The aim of this section is to recall the definition of the so-called *Bernoulli numbers* B_n and to collect some useful identities involving them. These numbers intervene in many arguments concerning with the CBHD Theorem. For example, we invoked the B_n in Sect. 4.5, when giving another "short" proof of the CBHD Theorem.

Our approach in this section will be different from that in the preceding section, in that we shall make use of elementary differential calculus (in $\mathbb{R}$ or in $\mathbb{C}$). This will allow us to streamline our arguments and to furnish a useful tool in handling with formal power series identities, frequent in the literature. We will take the opportunity to establish a result (see Lemma 9.17 below) allowing us to fill the link between Calculus and the algebraic aspects of formal power series.

We begin with the central definition.

Definition 9.15 (Bernoulli Numbers). Let us define, inductively, a sequence of rational numbers $\{B_n\}_n$ by the following recursion formula:

$$B_0 := 1, \qquad B_n := -n! \sum_{k=0}^{n-1} \frac{B_k}{k!\,(n+1-k)!} \qquad (n \geq 1). \qquad (9.37)$$

The B_n are referred to[3] as the *Bernoulli numbers*.

For example, the first few B_n are:

n	0	1	2	4	6	8
B_n	1	$-\frac{1}{2}$	$\frac{1}{6}$	$-\frac{1}{30}$	$\frac{1}{42}$	$-\frac{1}{30}$

n	10	12	14	16	18	20
B_n	$\frac{5}{66}$	$-\frac{691}{2730}$	$\frac{7}{6}$	$-\frac{3617}{510}$	$\frac{43867}{798}$	$-\frac{174611}{330}$

while $B_{2k+1} = 0$ for every $k \geq 1$.

[Note that the fact that $B_3, B_5, \ldots$ vanish is not obvious from the definition (9.37): it will be proved below.]

---

[3]Some authors use alternative notations and definitions for the Bernoulli numbers, see e.g. [176, §1.1].

There are plenty of interesting relations involving the Bernoulli numbers B_n. We here confine ourselves in proving those occurred in this Book (namely, (4.111a)–(4.111c) and (4.112) in Chap. 4). For a comprehensive study of Bernoulli numbers (and the so-called *Bernoulli polynomials*) in the theory of special functions, the interested Reader is referred to, e.g., Wang and Guo [176, §1.1].

Let us consider the real function

$$g : \mathbb{R} \to \mathbb{R}, \quad g(x) := \begin{cases} \dfrac{x}{e^x - 1}, & \text{if } x \neq 0, \\ 1, & \text{if } x = 0. \end{cases} \tag{9.38}$$

We shall write $g(x) = \frac{e^x - 1}{x}$ even when $x = 0$, with the obvious meaning. It is easily seen that g is real analytic: however, the Maclaurin series of g converges to g only on the interval $(-2\pi, 2\pi)$. Obviously, this depends on the fact that, among the non-removable singularities of the complex function $\frac{z}{e^z - 1}$, the ones closest to the origin are $\pm 2\pi i$. We set

$$\varphi_n := \left(\frac{d}{dx} \right)^n \Big|_0 g(x), \quad \text{whence} \quad g(x) = \sum_{n=0}^{\infty} \frac{\varphi_n}{n!} x^n \quad \text{if } |x| < 2\pi.$$

[We shall soon discover that the constants φ_n are the same as the Bernoulli numbers!]

Thanks to Lemma 9.17 below, ensuring that Maclaurin series behave under multiplication like formal power series, we have the following identities (valid for $|x| < 2\pi$):

$$1 = \frac{e^x - 1}{x} g(x) \overset{(9.39)}{=} \sum_{j=0}^{\infty} \frac{t^j}{(j+1)!} \cdot \sum_{k=0}^{\infty} \frac{\varphi_k}{k!} x^k$$

$$= \sum_{n=0}^{\infty} x^n \left(\sum_{k=0}^{n} \frac{\varphi_k}{k! \, (n+1-k)!} \right).$$

By equating the coefficients of x^n from the left-/right-hand sides, we obtain

$$\begin{cases} 1 = \varphi_0, \\ 0 = \displaystyle\sum_{k=0}^{n} \frac{\varphi_k}{k! \, (n+1-k)!} = \frac{\varphi_n}{n!} + \sum_{k=0}^{n-1} \frac{\varphi_k}{k! \, (n+1-k)!}, \quad (n \geq 1). \end{cases}$$

This recursion formula is precisely the recursion formula (9.37) defining the Bernoulli numbers, so that

$$\varphi_n = B_n, \quad \text{for all } n \in \mathbb{N} \cup \{0\}.$$

We have thus derived the following facts on the Bernoulli numbers:

$$B_n := \left(\frac{d}{dx}\right)^n\Big|_0 \frac{x}{e^x - 1}, \quad \text{and} \quad \frac{x}{e^x - 1} = \sum_{n=0}^{\infty} \frac{B_n}{n!} x^n \quad \text{if } |x| < 2\pi. \quad (9.39)$$

Let now $x \in \mathbb{R} \setminus \{0\}$. We have

$$g(x) + \frac{1}{2} x = \frac{x}{e^x - 1} + \frac{1}{2} x = \frac{e^x + 1}{e^x - 1} \frac{x}{2} = \frac{e^{x/2} + e^{-x/2}}{e^{x/2} - e^{-x/2}} \frac{x}{2} = \frac{\cosh(x/2)}{\sinh(x/2)} \frac{x}{2}.$$

If we set

$$k : \mathbb{R} \to \mathbb{R}, \quad k(x) := \begin{cases} \cosh(x/2) \dfrac{x/2}{\sinh(x/2)}, & \text{if } x \neq 0, \\ 1, & \text{if } x = 0, \end{cases} \quad (9.40)$$

we have proved that

$$g(x) + \frac{x}{2} = k(x), \quad \text{for all } x \in \mathbb{R} \quad (9.41)$$

(the value $x = 0$ being recovered by passing to the limit $x \to 0$ in the above computations). Thanks to the second identity in (9.39), we derive from (9.41) the Maclaurin expansion for k (recall that $B_1 = -1/2$):

$$k(x) = 1 + \sum_{n=2}^{\infty} \frac{B_n}{n!} x^n \quad \text{if } |x| < 2\pi.$$

Now, from its very definition (9.40), we recognize that k *is an even function*, so that its derivatives at 0 of odd orders do vanish. The above expansion of k thus ensures that

$$B_{2k+1} = 0 \quad \text{for every } k \geq 1, \quad (9.42)$$

so that the Maclaurin expansions of g and k are actually

$$g(x) = -\frac{x}{2} + \sum_{n=0}^{\infty} \frac{B_{2n}}{(2n)!} x^{2n}, \quad k(x) = \sum_{n=0}^{\infty} \frac{B_{2n}}{(2n)!} x^{2n} \quad \text{if } |x| < 2\pi. \quad (9.43)$$

Remark 9.16. In the above calculations, we have proved that $g(x) = -x/2 + k(x)$, where k is an even function. Thus, $g(-x) = x/2 + k(x)$. This gives the identity $g(-x) = x + k(x)$, that is,

$$\frac{-x}{e^{-x} - 1} = x + \frac{x}{e^x - 1}, \quad (9.44)$$

which is, at the same time, an identity for $x \in \mathbb{R}$, an identity for complex x such that $|x| < 2\pi$, and an identity between the corresponding formal power series.

In the above computations, we used the following simple result.

Lemma 9.17. *Let f be a real-valued C^∞ function defined on an open real-interval containing 0. We write*

$$f \sim \sum_{k=0}^\infty a_k t^k$$

to mean that the above formal power series is the Maclaurin series expansion of f. In other words, this means that

$$a_k = \frac{f^{(k)}(0)}{k!} \quad \text{for all } k \geq 0, \qquad \text{where} \quad f^{(k)}(0) := \frac{d^k f}{d x^k}(0).$$

Let $\varepsilon > 0$ be fixed. Let $f, g : (-\varepsilon, \varepsilon) \to \mathbb{R}$ be of class C^∞ and let $f \sim \sum_{k=0} a_k t^k$ and $g \sim \sum_{k=0} b_k t^k$. Then

$$f \cdot g \sim \left(\sum_{k=0} a_k t^k \right) \cdot \left(\sum_{k=0} b_k t^k \right). \tag{9.45}$$

The $\cdot$ symbol in the left-hand side denotes the (point-wise) product of functions, whereas the same symbol in the right-hand side is the (Cauchy) product in $\mathbb{R}[[t]]$ in (9.1).

Another way to state the above result is the following: Given $\varepsilon > 0$, the map

$$C^\infty((-\varepsilon, \varepsilon), \mathbb{R}) \longrightarrow \mathbb{R}[[t]], \quad f \mapsto \sum_{k=0}^\infty \frac{f^{(k)}(0)}{k!} t^k$$

is a UAA morphism (when the usual corresponding products are considered). [Obviously, analogous results hold replacing $\mathbb{R}$ with $\mathbb{C}$, C^∞ with C^ω and $(-\varepsilon, \varepsilon)$ with the complex disc about 0 with radius ε.]

Proof. In view of the definition of $\sim$ and of the Cauchy product in $\mathbb{K}[[t]]$, (9.45) is equivalent to

$$\frac{1}{k!} \frac{d^k (f g)}{d x^k}(0) = \sum_{n=0}^k \frac{f^{(n)}(0)}{n!} \frac{g^{(k-n)}(0)}{(k-n)!},$$

but this immediately[4] follows from the fact that $f \mapsto f'$ is a derivation of $C^\infty((-\varepsilon, \varepsilon), \mathbb{R})$. □

---

[4] Indeed, if D is a derivation of an associative algebra $(A, *)$, we have

For the sake of completeness, we furnish the analogue of Lemma 9.17 in the case of composition of functions (and of formal power series). This result is much more delicate than the case of the product, since *the derivative ∂_t is not a derivation of the algebra* $(\mathbb{K}[[t]]_+, \circ)$.

Theorem 9.18. *Let $\sim$ have the same meaning as in Lemma 9.17. Let $\varepsilon > 0$ be fixed. Let $f, g : (-\varepsilon, \varepsilon) \to \mathbb{R}$ be of class C^∞ and let $f \sim \sum_{k=0} a_k t^k$ and $g \sim \sum_{k=0} b_k t^k$. Suppose that $b_0 = 0$.*
  Then the composite of functions $f \circ g$ is well posed on a neighborhood of 0 and

$$f \circ g \sim \left(\sum_{k=0} a_k t^k \right) \circ \left(\sum_{k=1} b_k t^k \right). \tag{9.46}$$

Here, the $\circ$ symbol in the left-hand side denotes the ordinary composition of functions, whereas that in the right-hand side is the operation of substitution of formal power series as in Definition 9.2.

Note that (9.46) is equivalent to the following *formula for the iterated derivative of the composite of functions*

$$\frac{d^n (f \circ g)}{d x^n}(0) = n! \sum_{k=1}^n \frac{f^{(k)}(0)}{k!} \sum_{i_1 + \cdots + i_k = n} \frac{g^{(i_1)}(0) \cdots g^{(i_k)}(0)}{i_1! \cdots i_k!}. \tag{9.47}$$

Proof. Since $g(0) = b_0 = 0$, by continuity there exists a small $\delta_\varepsilon > 0$ such that $\delta_\varepsilon < \varepsilon$ and $g(x) \in (-\varepsilon, \varepsilon)$ whenever $|x| < \delta_\varepsilon$, so that $f \circ g$ is well posed (and obviously C^∞) on $(-\delta_\varepsilon, \delta_\varepsilon)$. Let $n \in \mathbb{N}$ be fixed. By Taylor's formula with the Peano remainder, we have

$$f(y) = \sum_{k=0}^n \frac{f^{(k)}(0)}{k!} y^k + \mathcal{O}(y^{n+1}), \qquad \text{as } y \to 0,$$

$$g(x) = \sum_{i=1}^n \frac{g^{(i)}(0)}{i!} x^i + \mathcal{O}(x^{n+1}), \qquad \text{as } x \to 0.$$

Since $g(x) = \mathcal{O}(x)$ as $x \to 0$ (recall that $g(0) = 0$), from the substitution $y = g(x)$ in the above expansions, we easily get

$$f(g(x)) = \sum_{k=0}^n \frac{f^{(k)}(0)}{k!} \left(\sum_{i=1}^n \frac{g^{(i)}(0)}{i!} x^i \right)^k + \mathcal{O}(x^{n+1}), \qquad \text{as } x \to 0.$$

$$D^k(a * b) = \sum_{n=0}^k \binom{k}{n} (D^n a) * (D^{k-n} b), \qquad \forall\, a, b \in A, \ \forall\, k \in \mathbb{N},$$

as a simple inductive argument shows.

Moreover, we notice that the polynomial in the above right-hand side equals

$$f(0) + \sum_{k=1}^{n} \frac{f^{(k)}(0)}{k!} \sum_{1 \le i_1, \dots, i_n \le n} \frac{g^{(i_1)}(0) \cdots g^{(i_k)}(0)}{i_1! \cdots i_k!} x^{i_1 + \cdots + i_k}$$

$$= f(0) + \sum_{h=1}^{n} x^h \left(\sum_{k=1}^{h} \frac{f^{(k)}(0)}{k!} \sum_{i_1 + \cdots + i_k = h} \frac{g^{(i_1)}(0) \cdots g^{(i_k)}(0)}{i_1! \cdots i_k!} \right) + \mathcal{O}(x^{n+1}),$$

as $x \to 0$. Since a smooth function which is a $\mathcal{O}_{x \to 0}(x^{n+1})$ as $x \to 0$ has vanishing n-th order derivative at 0, collecting the above expansions we have

$$\frac{\mathrm{d}^n (f \circ g)}{\mathrm{d} x^n}(0) = \frac{\mathrm{d}^n}{\mathrm{d} x^n}\Big|_0 \left\{ \sum_{h=1}^{n} x^h \left(\sum_{k=1}^{h} \frac{f^{(k)}(0)}{k!} \sum_{i_1 + \cdots + i_k = h} \frac{g^{(i_1)}(0) \cdots g^{(i_k)}(0)}{i_1! \cdots i_k!} \right) \right\}$$

$$= n! \sum_{k=1}^{n} \frac{f^{(k)}(0)}{k!} \sum_{i_1 + \cdots + i_k = n} \frac{g^{(i_1)}(0) \cdots g^{(i_k)}(0)}{i_1! \cdots i_k!}.$$

This is (9.47), which is equivalent to (9.46), by definition of $\sim$ and of the composite of formal power series. □

Chapter 10
Symmetric Algebra

I**N** this chapter, we recall the basic facts we needed about the so-called symmetric algebra (of a vector space), which we used in Chap. 6 in exhibiting the relationship between the CBHD Theorem and the PBW Theorem.

Throughout, V will denote a fixed vector space over the field $\mathbb{K}$. Moreover, $\mathbb{K}$ *is supposed to have characteristic zero.* This hypothesis will be crucial in Theorems 10.9, 10.17, 10.19 and in Proposition 10.16 below, whereas – as the Reader will certainly realize – all other definitions and results hold without this restriction.

To lighten the reading, the chapter is split in two parts: the main results (Sect. 10.1) and the proofs of these results (Sect. 10.2).

10.1 The Symmetric Algebra and the Symmetric Tensor Space

Definition 10.1 (Symmetric Algebra of a Vector Space). Let V be a vector space and let $\mathscr{T}(V)$ be its tensor algebra. We denote by $\mathscr{H}(V)$ the two-sided ideal of $\mathscr{T}(V)$ generated by the elements of the form $x \otimes y - y \otimes x$, where $x, y \in V$. Then the quotient algebra

$$\mathrm{Sym}(V) := \mathscr{T}(V)/\mathscr{H}(V) \tag{10.1}$$

is called *the symmetric algebra of* V. Throughout this section, the map

$$\phi : \mathscr{T}(V) \to \mathrm{Sym}(V), \quad t \mapsto [t]_{\mathscr{H}} \tag{10.2}$$

denotes the corresponding projection. Moreover, the induced algebra operation on $\mathrm{Sym}(V)$ is denoted by $*$, namely

$$* : \mathrm{Sym}(V) \times \mathrm{Sym}(V) \to \mathrm{Sym}(V), \quad \phi(t) * \phi(t') = \phi(t \cdot t') \quad \forall\, t, t' \in \mathscr{T}(V).$$

A. Bonfiglioli and R. Fulci, *Topics in Noncommutative Algebra*, Lecture Notes in Mathematics 2034, DOI 10.1007/978-3-642-22597-0_10, © Springer-Verlag Berlin Heidelberg 2012

Remark 10.2. 1. *The set* $\{\phi(1_{\mathbb{K}})\}\cup\phi(V)$ *is a set of algebra generators for* $\mathrm{Sym}(V)$.
   This follows from the fact that $\{1_{\mathbb{K}}\} \cup V$ is a set of algebra generators for
   $\mathscr{T}(V)$ and the fact that $\phi : V \to \mathrm{Sym}(V)$ is a UAA morphism.
2. *The map* $\phi|_V : V \to \mathrm{Sym}(V)$ *is injective.* Indeed, if $v \in V$ is such that
   $\phi(v) = 0$, we have $v \in \mathscr{H}(V)$, whence $v = 0$ since $\mathscr{H}(V) \subset \bigoplus_{n\geq 2} \mathscr{T}_n(V)$.

Proposition 10.3. *With the notation of Definition 10.1, the algebra* $(\mathrm{Sym}(V), *)$
is an Abelian UA algebra.

Proof. We have to prove that $s * s' = s' * s$ for every $s, s' \in \mathrm{Sym}(V)$. This is
equivalent to $t \cdot t' - t' \cdot t \in \mathscr{H}(V)$, for every $t, t' \in \mathscr{T}(V)$. Obviously, we can
restrict the proof to the case when t, t' are elementary tensors. In general,
given $u_1, \ldots, u_k \in V$, we have

$$u_1 \otimes \cdots u_i \otimes u_{i+1} \cdots \otimes u_k = u_1 \otimes \cdots (u_i \otimes u_{i+1} - u_{i+1} \otimes u_i) \cdots \otimes u_k +$$

$$+ u_1 \otimes \cdots u_{i+1} \otimes u_i \cdots \otimes u_k$$

$$\equiv u_1 \otimes \cdots u_{i+1} \otimes u_i \cdots \otimes u_k \quad \mathrm{mod}\ \mathscr{H}(V).$$

This ensures that the ϕ-images of two elementary tensors with an interchanged pair of consecutive factors do coincide in $\mathrm{Sym}(V)$. An inductive
argument then shows that

$$(v_1 \otimes \cdots \otimes v_n) \otimes (w_1 \otimes \cdots \otimes w_m) \equiv (w_1 \otimes \cdots \otimes w_m) \otimes (v_1 \otimes \cdots \otimes v_n) \ \mathrm{mod}\ \mathscr{H}(V),$$

for any choice of vectors v_i, w_j in V and n, m in $\mathbb{N}$. This ends the proof. $\square$

The ideal $\mathscr{H}(V)$ will frequently be written simply as $\mathscr{H}$. By the very
definition of $\mathscr{H}$, the elements of $\mathscr{H}$ are linear combinations of tensors of
the form

$$t \cdot (x \otimes y - y \otimes x) \cdot t', \quad \text{with } t, t' \in \mathscr{T}(V) \text{ and } x, y \in V.$$

Hence $\mathscr{H}$ is spanned by tensors of the form

$$v_1 \otimes \cdots \otimes v_n \otimes (x \otimes y - y \otimes x) \otimes w_1 \otimes \cdots \otimes w_m, \tag{10.3}$$

where $n, m \in \mathbb{N} \cup \{0\}$, $x, y \in V$ and the vectors v_i and w_j belong to V (we
set $v_1 \otimes \cdots \otimes v_n = 1$ when $n = 0$ and analogously for $w_1 \otimes \cdots \otimes w_m$).
   Note that we have $\mathscr{H} \subset \bigoplus_{n\geq 2} \mathscr{T}_n(V)$. Since $\mathscr{H}$ is generated by homoge-
nous tensors, it is easily seen that $\mathscr{H}$ admits the grading

$$\mathscr{H}(V) = \bigoplus_{n\geq 2} \mathscr{H}_n(V), \quad \text{where } \mathscr{H}_n(V) := \mathscr{T}_n(V) \cap \mathscr{H}(V). \tag{10.4}$$

Indeed, it holds that $\mathscr{H}_n \cdot \mathscr{H}_m \subseteq \mathscr{H}_{n+m}$, for every $n, m \geq 2$ (the shorthand
$\mathscr{H}_n := \mathscr{H}_n(V)$ applies).

Throughout this section, given $n \in \mathbb{N}$, we denote by $\mathfrak{S}_n$ the group of permutations of the set $\{1, 2, \ldots, n\}$ (i.e., the set of the bijections of the set $\{1, 2, \ldots, n\}$). The group operation on $\mathfrak{S}_n$ is the composition of functions. With this notation at hand, we see that in the proof of Proposition 10.3 we have shown that

$$v_1 \otimes \cdots \otimes v_n - v_{\sigma(1)} \otimes \cdots \otimes v_{\sigma(n)} \in \mathcal{H}(V),$$

$$\text{for any choice of } n \in \mathbb{N}, \sigma \in \mathfrak{S}_n \text{ and } v_1, \ldots, v_n \in V. \tag{10.5}$$

Proposition 10.4. *With the above notation, for every $n \in \mathbb{N}, n \geq 2$,*

$$\mathcal{H}_n(V) = \mathrm{span}\Big\{ v_1 \otimes \cdots \otimes v_n - v_{\sigma(1)} \otimes \cdots \otimes v_{\sigma(n)} \;\Big|\; \sigma \in \mathfrak{S}_n, \, v_1, \ldots, v_n \in V \Big\}. \tag{10.6}$$

Proof. The inclusion $\mathcal{H}_n \supseteq \mathrm{span}\{\cdots\}$ in (10.6) follows from (10.5). The reverse inclusion can be easily argued by exploiting the form (10.3) of a system of generators for $\mathcal{H}(V)$. Indeed, (10.3) shows that an element of $\mathcal{H}_n$ is a linear combination of tensors like

$$v_1 \otimes \cdots \otimes v_k \otimes \big(v_{k+1} \otimes v_{k+2} - v_{k+2} \otimes v_{k+1} \big) \otimes v_{k+3} \otimes \cdots \otimes v_n.$$

In its turn, the latter can be rewritten as $v_1 \otimes \cdots \otimes v_n - v_{\sigma(1)} \otimes \cdots \otimes v_{\sigma(n)}$ with

$$\sigma(i) = \begin{cases} i, & \text{if } i \in \{1, \ldots, k-2\} \cup \{k+3, \ldots, n\}, \\ k+2, & \text{if } i = k+1, \\ k+1, & \text{if } i = k+2. \end{cases}$$

This ends the proof. Indeed, this proves something more, namely that *in (10.6) we can replace $\mathfrak{S}_n$ by the set of transpositions of $\{1, \ldots, n\}$ or even by the set of the transpositions that exchange two consecutive integers.* $\square$

Gathering together the gradings $\mathcal{T} = \bigoplus_{n \geq 0} \mathcal{T}_n$ and $\mathcal{H} = \bigoplus_{n \geq 0} \mathcal{H}_n$ (here we have set $\mathcal{H}_0 := \{0\} =: \mathcal{H}_1$) and the fact that $\mathcal{H}_n \subset \mathcal{T}_n$ for every $n \geq 0$, we easily obtain the decomposition

$$\mathrm{Sym}(V) = \bigoplus_{n \geq 0} \mathrm{Sym}_n(V), \quad \text{where } \mathrm{Sym}_n(V) := \mathcal{T}_n(V)/\mathcal{H}(V), \tag{10.7}$$

which is also a grading, for it holds that

$$\mathrm{Sym}_n(V) * \mathrm{Sym}_m(V) \subseteq \mathrm{Sym}_{n+m}(V), \quad \text{for all } n, m \geq 0. \tag{10.8}$$

The linear set $\mathrm{Sym}_n(V)$ is called *the n-th symmetric power of V*. We have $\mathrm{Sym}_0(V) = \phi(\mathbb{K}) \simeq \mathbb{K}$, $\mathrm{Sym}_1(V) = \phi(V) \simeq V$, and, more important,

$$\mathrm{Sym}_n(V) = \mathscr{T}_n(V)/\mathscr{H}(V) \simeq \mathscr{T}_n(V)/\mathscr{H}_n(V), \quad \text{for all } n \geq 0. \qquad (10.9)$$

Indeed, the map

$$\mathscr{T}_n(V)/\mathscr{H}(V) \ni [t]_{\mathscr{H}} \mapsto [t]_{\mathscr{H}_n} \in \mathscr{T}_n(V)/\mathscr{H}_n(V)$$

(since $t \in \mathscr{T}_n$) is well posed and it is an isomorphism of vector spaces.[1]

Remark 10.5. Since $\mathscr{T}_n$ is spanned by $\{v_1 \otimes \cdots \otimes v_n \,|\, v_1, \ldots, v_n \in V\}$, $\mathrm{Sym}_n(V) = \phi(\mathscr{T}_n(V))$ is spanned by $\{\phi(v_1) * \cdots * \phi(v_n) \,|\, v_1, \ldots, v_n \in V\}$ (recall that $\phi : \mathscr{T}(V) \to \mathrm{Sym}(V)$ is a UAA morphism).

The linear set $\mathrm{Sym}_n(V)$ has a characterizing property, which we state in Proposition 10.6 below. First we recall that, given sets U, V and $n \in \mathbb{N}$, a map $\varphi : V^n \to U$ (here V^n is the n-fold Cartesian product of V with itself) is called *symmetric* if

$$\varphi(v_{\sigma(1)}, \ldots, v_{\sigma(n)}) = \varphi(v_1, \ldots, v_n), \quad \forall\, \sigma \in \mathfrak{S}_n, \ \forall\, v_1, \ldots, v_n \in V.$$

Proposition 10.6. *Let V be a vector space. If ϕ is as in (10.2), we set*

$$\varphi : V^n \to \mathrm{Sym}_n(V), \quad \varphi(v_1, \ldots, v_n) := \phi(v_1) * \cdots * \phi(v_n).$$

Then φ is n-linear and symmetric. Moreover, for every vector space U and every symmetric n-linear map $\beta : V^n \to U$, there exists a unique linear map $\beta^\varphi : \mathrm{Sym}_n(V) \to U$ such that

$$\beta^\varphi(\varphi(v)) = \beta(v) \quad \text{for every } v \in V^n, \qquad (10.10)$$

thus making the following diagram commute:

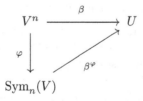

Proof. Note that the above map φ equals

$$\varphi(v_1, \ldots, v_n) = \phi(v_1 \otimes \cdots \otimes v_n) = [v_1 \otimes \cdots \otimes v_n]_{\mathscr{H}}, \qquad \forall\, v_1, \ldots, v_n \in V.$$

---

[1] Indeed, if $t, t' \in \mathscr{T}_n$ and $[t]_{\mathscr{H}} = [t']_{\mathscr{H}}$ then $t - t' \in \mathscr{T}_n \cap \mathscr{H} = \mathscr{H}_n$ so that $[t]_{\mathscr{H}_n} = [t']_{\mathscr{H}_n}$. Also, the map is injective for, if $[t]_{\mathscr{H}_n} = [t']_{\mathscr{H}_n}$ then $t - t' \in \mathscr{H}_n \subset \mathscr{H}$, so that $[t]_{\mathscr{H}} = [t']_{\mathscr{H}}$. Finally, the map is obviously linear and onto.

For the proof of this proposition, see page 514. □

Also the symmetric algebra $\mathrm{Sym}(V)$ has a universal property:

Theorem 10.7 (Universal Property of $\mathrm{Sym}(V)$). *Let V be a vector space and let $\mathrm{Sym}(V)$ and ϕ be as in Definition 10.1.*

(i) *For every Abelian UA algebra A and every linear map $f : V \to A$, there exists a unique UAA morphism $f^\phi : \mathrm{Sym}(V) \to A$ such that*

$$f^\phi(\phi(v)) = f(v) \quad \text{for every } v \in V, \tag{10.11}$$

thus making the following diagram commute:

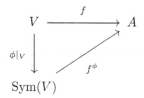

(ii) *Conversely, suppose W, φ are respectively an Abelian UA algebra and a linear map $\varphi : V \to W$ with the following property: For every Abelian UA algebra A and every linear map $f : V \to A$, there exists a unique UAA morphism $f^\varphi : W \to A$ such that*

$$f^\varphi(\varphi(v)) = f(v) \quad \text{for every } v \in V, \tag{10.12}$$

thus making the following diagram commute:

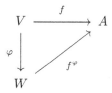

Then W and $\mathrm{Sym}(V)$ are canonically isomorphic, as UA algebras, the isomorphism being (see the notation in (ii) above) $\varphi^\phi : \mathrm{Sym}(V) \to W$ and its inverse being $\phi^\varphi : W \to \mathrm{Sym}(V)$. Furthermore, φ is injective and W is generated, as an algebra, by the set $\{1_W\} \cup \varphi(V)$. Actually it holds that $\varphi = \varphi^\phi \circ (\phi|_V)$. Finally we have $W \simeq \mathrm{Sym}(\varphi(V))$, canonically.

Proof. See page 515. □

Remark 10.8. More generally, the following fact holds: *For every UA algebra $(A, \circledast)$ (not necessarily Abelian) and every linear map $f : V \to A$ such that*

$$f(x) \circledast f(y) - f(y) \circledast f(x) = 0, \quad \forall\, x, y \in V,$$

there exists a unique UAA morphism $f^\phi : \mathrm{Sym}(V) \to A$ such that (10.11) holds.

We have already used the following result back in Chap. 6.

Theorem 10.9. *Let $n \in \mathbb{N}$. Suppose $\mathbb{K}$ is of characteristic zero. Then, with the notation in Definition 10.1, the set $\mathrm{Sym}_n(V) = \mathscr{T}_n(V)/\mathscr{H}(V)$ is spanned by the elements of the form $(\phi(v))^{*n}$, with $v \in V$. More explicitly, we have*

$$\mathrm{Sym}_n(V) = \mathrm{span}\{\phi(v^{\otimes n}) \mid v \in V\} = \mathrm{span}\{w^{*n} \mid w \in \phi(V)\}. \tag{10.13}$$

More generally, this holds provided that $n! \cdot 1_\mathbb{K}$ is invertible in $\mathbb{K}$.

Proof. We give two proofs of this fact: One is short but indirect; the other "constructive", but it involves Lemma 10.10 below (whose proof is a bit tricky). See page 515. $\square$

The following lemma is useful:

Lemma 10.10. *Let $(A, *)$ be a ring. Then for every $n \in \mathbb{N}$ and $x_1, \ldots, x_n \in A$, we have (using "card" to denote set-cardinality)*

$$\sum_{\sigma \in \mathfrak{S}_n} x_{\sigma(1)} * \cdots * x_{\sigma(n)} = (-1)^n \sum_{H \subseteq \{1,\ldots,n\}} (-1)^{\mathrm{card}(H)} \left(\sum_{i \in H} x_i\right)^{*n}. \tag{10.14}$$

(When $H = \emptyset$, we have set $\sum_{i \in H} x_i = 0$.) In particular, if A is Abelian we get

$$n!\, x_1 * \cdots * x_n = (-1)^n \sum_{H \subseteq \{1,\ldots,n\}} (-1)^{\mathrm{card}(H)} \left(\sum_{i \in H} x_i\right)^{*n}. \tag{10.15}$$

Proof. See page 516. $\square$

We next turn to realizing $\mathrm{Sym}(V)$ as a subset (which is not a subalgebra, though) of $\mathscr{T}(V)$. The fact that $\mathbb{K}$ has null characteristic allows us to carry out the following construction. Let $n \in \mathbb{N}$ be fixed. We consider on V^n (the n-fold Cartesian product of V with itself) the map

$$\beta : V^n \to \mathscr{T}_n(V), \quad \beta(v_1, \ldots, v_n) := \frac{1}{n!} \sum_{\sigma \in \mathfrak{S}_n} v_{\sigma(1)} \otimes \cdots \otimes v_{\sigma(n)}.$$

We claim that β is *n-linear and symmetric*. For instance, leaving the verification of the n-linearity to the Reader, we show that β is symmetric: Indeed, for every fixed $\sigma \in \mathfrak{S}_n$ and every $w_1, \ldots, w_n \in V$ we have

$$\beta(w_{\sigma(1)}, \ldots, w_{\sigma(n)}) = \quad (\text{set } v_i := w_{\sigma(i)} \text{ for every } i = 1, \ldots, n)$$

$$= \frac{1}{n!} \sum_{\tau \in \mathfrak{S}_n} v_{\tau(1)} \otimes \cdots \otimes v_{\tau(n)} = \frac{1}{n!} \sum_{\tau \in \mathfrak{S}_n} w_{\sigma(\tau(1))} \otimes \cdots \otimes w_{\sigma(\tau(n))}$$

$$(10.16)$$

$$(\text{obviously } \tau \in \mathfrak{S}_n \text{ iff } \tau = \sigma^{-1} \circ \rho \text{ with } \rho \in \mathfrak{S}_n)$$

$$= \frac{1}{n!} \sum_{\rho \in \mathfrak{S}_n} w_{\rho(1)} \otimes \cdots \otimes w_{\rho(n)} = \beta(w_1, \ldots, w_n).$$

By Proposition 10.6, there exists a unique linear map

$$S_n : \mathrm{Sym}_n(V) \to \mathscr{T}_n(V) \tag{10.17}$$

such that $S_n(\phi(v_1 \otimes \cdots \otimes v_n)) = \beta(v_1, \ldots, v_n)$, i.e., such that

$$S_n([v_1 \otimes \cdots \otimes v_n]_{\mathscr{H}}) = \frac{1}{n!} \sum_{\sigma \in \mathfrak{S}_n} v_{\sigma(1)} \otimes \cdots \otimes v_{\sigma(n)}, \quad \forall\, v_1, \ldots, v_n \in V.$$

$$(10.18)$$

Note that we have (see the computation in (10.16))

$$S_n([v_{\sigma(1)} \otimes \cdots \otimes v_{\sigma(n)}]_{\mathscr{H}}) = S_n([v_1 \otimes \cdots \otimes v_n]_{\mathscr{H}}),$$

$$\text{for every } \sigma \in \mathfrak{S}_n \text{ and every } v_1, \ldots, v_n \in V. \tag{10.19}$$

Let now $\sigma \in \mathfrak{S}_n$ be fixed. We consider the function from V^n to $\mathscr{T}_n(V)$ mapping $(v_1, \ldots, v_n) \in V^n$ to $v_{\sigma(1)} \otimes \cdots \otimes v_{\sigma(n)}$. Obviously, this map is n-linear, so that (by the universal property of the tensor product) there exists a unique linear map

$$R_\sigma : \mathscr{T}_n(V) \to \mathscr{T}_n(V)$$

such that

$$R_\sigma(v_1 \otimes \cdots \otimes v_n) = v_{\sigma(1)} \otimes \cdots \otimes v_{\sigma(n)}, \quad \forall\, v_1, \ldots, v_n \in V. \tag{10.20}$$

Note that we have

$$R_\sigma \circ R_\tau = R_{\sigma \circ \tau}, \quad (R_\sigma)^{-1} = R_{\sigma^{-1}}, \quad \text{for every } \sigma, \tau \in \mathfrak{S}_n, \tag{10.21}$$

as can be seen by comparing the action of these functions on elementary tensors (which span $\mathscr{T}_n$). Since any R_σ is an endomorphism of $\mathscr{T}_n$ (actually, an automorphism), the formula

$$Q_n := \tfrac{1}{n!} \sum_{\sigma \in \mathfrak{S}_n} R_\sigma$$

defines in its turn an endomorphism of $\mathcal{T}_n(V)$. More explicitly

$$Q_n : \mathcal{T}_n(V) \to \mathcal{T}_n(V)$$

is the (unique) linear map such that

$$Q_n(v_1 \otimes \cdots \otimes v_n) = \frac{1}{n!} \sum_{\sigma \in \mathfrak{S}_n} v_{\sigma(1)} \otimes \cdots \otimes v_{\sigma(n)}, \quad \forall \, v_1, \ldots, v_n \in V. \quad (10.22)$$

We call Q_n *the symmetrizing operator* and we say that $Q_n(v)$ is *the symmetrization* of $v \in \mathcal{T}_n(V)$. Note that we have

$$Q_n(v_{\sigma(1)} \otimes \cdots \otimes v_{\sigma(n)}) = Q_n(v_1 \otimes \cdots \otimes v_n),$$

$$(10.23)$$

for every $\sigma \in \mathfrak{S}_n$ and every $v_1, \ldots, v_n \in V$.

Remark 10.11. For every $v \in \mathcal{T}_n(V)$, we have $v - Q_n(v) \in \mathcal{H}_n(V)$. Indeed, it suffices to prove this for a set of generators of $\mathcal{T}_n$, namely the elementary tensors. To this aim, let $v = v_1 \otimes \cdots \otimes v_n$, with $v_1, \ldots, v_n \in V$. In view of Proposition 10.4, we have

$$v_1 \otimes \cdots \otimes v_n - R_\sigma(v_1 \otimes \cdots \otimes v_n) \in \mathcal{H}_n(V), \quad \forall \, \sigma \in \mathfrak{S}_n.$$

Summing up over all σ in $\mathfrak{S}_n$ we obtain (recall that $\mathrm{card}(\mathfrak{S}_n) = n!$)

$$n! \, v_1 \otimes \cdots \otimes v_n - \sum_{\sigma \in \mathfrak{S}_n} R_\sigma(v_1 \otimes \cdots \otimes v_n) \in \mathcal{H}_n(V),$$

which can be rewritten as $n! \, (v - Q_n(v)) \in \mathcal{H}_n(V)$. The fact that $n! \neq 0$ (recall that $\mathbb{K}$ has characteristic zero) proves that $v - Q_n(v) \in \mathcal{H}_n(V)$, as claimed. $\qquad\qquad\square$

The link between Q_n in (10.22) and S_n in (10.18) is (see the proof of Proposition 10.6, page 514):

$$S_n([v]_{\mathcal{H}(V)}) = Q_n(v), \quad \text{for every } v \in \mathcal{T}_n(V). \quad (10.24)$$

This means that the following is a commutative diagram:

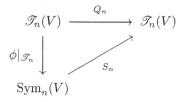

We give the following definition:

Definition 10.12 (Symmetric Tensor Space of Order n). Let $n \in \mathbb{N}$ and consider the notation in (10.20). We set

$$\mathscr{S}_n(V) := \{ v \in \mathscr{T}_n(V) \mid v = Q_n(v) \}, \qquad (10.25)$$

and we call it the *space of the symmetric tensors of order n* (on V). We also set $\mathscr{S}_0(V) := \mathbb{K}$.

Obviously, $\mathscr{S}_n(V)$ (sometimes shortened to $\mathscr{S}_n$) is a linear subspace of $\mathscr{T}_n(V)$.

Remark 10.13. With the notation of Definition 10.12, we have

$$\mathscr{S}_n(V) = \{ v \in \mathscr{T}_n(V) \mid v = R_\sigma(v) \ \forall \sigma \in \mathfrak{S}_n \}. \qquad (10.26)$$

Indeed, the inclusion $\mathscr{S}_n(V) \supseteq \{\cdots\}$ in (10.25) is obvious (for $\mathrm{card}(\mathfrak{S}_n)=n!$). Conversely, if $v \in \mathscr{S}_n(V)$ we have, for any $\tau \in \mathfrak{S}_n$,

$$R_\tau(v) = R_\tau(Q_n(v)) = \tfrac{1}{n!} \sum_{\sigma \in \mathfrak{S}_n} R_\tau(R_\sigma(v))$$

$$\overset{(10.21)}{=} \tfrac{1}{n!} \sum_{\sigma \in \mathfrak{S}_n} R_{\tau \circ \sigma}(v) = \tfrac{1}{n!} \sum_{\rho \in \mathfrak{S}_n} R_\rho(v) = Q_n(v) = v.$$

Here we used the fact that $\tau \circ \mathfrak{S}_n = \mathfrak{S}_n$.

Remark 10.14. From the very definitions (10.25) of $\mathrm{Sym}_n(V)$ and (10.22) of Q_n (together with the fact that elementary tensors span $\mathscr{T}_n$), we infer

$$\mathscr{S}_n(V) = \mathrm{span} \left\{ \sum_{\sigma \in \mathfrak{S}_n} v_{\sigma(1)} \otimes \cdots \otimes v_{\sigma(n)} \;\middle|\; v_1, \ldots, v_n \in V \right\}. \qquad (10.27)$$

Yet another characterization of $\mathrm{Sym}_n(V)$:

Remark 10.15. By making use of (10.27) and of (10.14) in Lemma 10.10, we obtain

$$\mathscr{S}_n(V) = \mathrm{span}\{ v^{\otimes n} \mid v \in V \}. \qquad (10.28)$$

Indeed, it suffices to argue as in the proof of Theorem 10.9 (page 515), thus obtaining an explicit representation of the symmetrization of $v_1 \otimes \cdots \otimes v_n$ in terms of n-powers, as follows:

$$\sum_{\sigma \in \mathfrak{S}_n} v_{\sigma(1)} \otimes \cdots \otimes v_{\sigma(n)} = \sum_{H \subseteq \{1,\ldots,n\}} (-1)^{n+\mathrm{card}(H)} \left(\sum_{i \in H} v_i\right)^{\otimes n} \quad (10.29)$$

$$Q_n(v_1 \otimes \cdots \otimes v_n) = \sum_{H \subseteq \{1,\ldots,n\}} \frac{(-1)^{n+\mathrm{card}(H)}}{n!} \left(\sum_{i \in H} v_i\right)^{\otimes n}. \quad (10.30)$$

We summarize the characterizations of $\mathscr{S}_n(V)$ found in Remarks 10.13, 10.14 and 10.15 in the following proposition.

Proposition 10.16 (Characterizations of $\mathscr{S}_n(V)$). *Let $\mathbb{K}$ be of characteristic zero. Let $n \in \mathbb{N}$ and let $\mathscr{S}_n(V)$ be as in Definition 10.12. Then we have:*

$$\mathscr{S}_n(V) = \mathrm{span}\{v^{\otimes n} \mid v \in V\}. \quad (10.31a)$$

$$= \{v \in \mathscr{T}_n(V) \mid v = Q_n(v)\} \quad (10.31b)$$

$$= \{v \in \mathscr{T}_n(V) \mid v = R_\sigma(v) \; \forall \sigma \in \mathfrak{S}_n\} \quad (10.31c)$$

$$= \mathrm{span}\left\{\sum_{\sigma \in \mathfrak{S}_n} v_{\sigma(1)} \otimes \cdots \otimes v_{\sigma(n)} \mid v_1, \ldots, v_n \in V\right\}. \quad (10.31d)$$

The following important theorem holds, providing a representation of the space $\mathrm{Sym}_n(V)$ as a subset (rather than a quotient) of $\mathscr{T}_n(V)$.

Theorem 10.17 (The Isomorphism of Vector Spaces $\mathscr{S}_n(V) \simeq \mathrm{Sym}_n(V)$). *Let $\mathbb{K}$ be of characteristic zero. Let $n \in \mathbb{N}$ and let all the above notation apply. Then*

$$\mathscr{S}_n(V) = S_n(\mathrm{Sym}_n(V)),$$

and $S_n : \mathrm{Sym}_n(V) \to \mathscr{S}_n(V)$ is an isomorphism of vector spaces. The inverse isomorphism $S_n^{-1} : \mathscr{S}_n(V) \to \mathrm{Sym}_n(V)$ coincides with the restriction $\phi|_{\mathscr{S}_n(V)}$ of the natural projection, i.e.,

$$S_n^{-1}(s) = \phi(s) = [s]_{\mathscr{H}_n(V)}, \quad \text{for every } s \in \mathscr{S}_n(V). \quad (10.32)$$

Finally, the map $Q_n : \mathscr{T}_n(V) \to \mathscr{T}_n(V)$ in (10.22) is a projector of $\mathscr{T}_n(V)$ onto $\mathscr{S}_n(V)$ with kernel $\mathscr{H}_n(V)$ so that $\mathscr{T}_n(V) = \mathscr{S}_n(V) \oplus \mathscr{H}_n(V)$.

Proof. See page 518. □

So, the linear map $Q_n : \mathscr{T}_n(V) \to \mathscr{T}_n(V)$ has the following properties:

1. $Q_n(\mathscr{T}_n(V)) = \mathscr{S}_n(V)$ so that $Q_n : \mathscr{T}_n(V) \to \mathscr{S}_n(V)$ is onto.
2. $Q_n|_{\mathscr{S}_n(V)}$ is the identity of $\mathscr{S}_n(V)$ and $Q_n^2 = Q_n$ on $\mathscr{T}_n(V)$.
3. $\ker(Q_n) = \mathscr{H}_n(V)$ and $\mathscr{T}_n(V) = \mathscr{S}_n(V) \oplus \mathscr{H}_n(V)$.

By Theorem 10.17, the arrows in the following diagram are all isomorphisms of vector spaces (and the diagram is commutative):

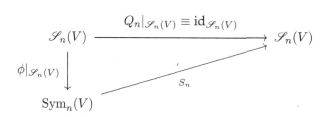

The following is a very natural definition.

Definition 10.18 (The Symmetric Tensor Space $\mathscr{S}(V)$). Let V be a vector space. For $n \in \mathbb{N} \cup \{0\}$, let $\mathscr{S}_n(V) \subseteq \mathscr{T}_n(V)$ be as in Definition 10.12. Then we set

$$\mathscr{S}(V) := \bigoplus_{n \geq 0} \mathscr{S}_n(V), \tag{10.33}$$

and we call $\mathscr{S}(V)$ the *symmetric tensor space of V*. Note that $\mathscr{S}(V)$ is a vector subspace of the tensor algebra $\mathscr{T}(V)$ of V.

For the sake of convenience, we introduce the maps

$$S_0 : \mathrm{Sym}_0(V) \to \mathscr{T}_0(V) \qquad Q_0 : \mathscr{T}_0(V) \to \mathscr{T}_0(V)$$
$$[k]_{\mathscr{H}} \mapsto k \qquad\text{and}\qquad k \mapsto k$$

(k being an arbitrary element of $\mathbb{K}$). We "glue" together the maps $\{S_n\}_{n \geq 0}$ to define the (unique) linear map

$$S : \mathrm{Sym}(V) \to \mathscr{T}(V), \quad \text{such that } S|_{\mathrm{Sym}_n(V)} \equiv S_n, \text{ for every } n \geq 0. \tag{10.34}$$

Analogously, we "glue" together the maps $\{Q_n\}_{n \geq 0}$ to define the (unique) linear map

$$Q : \mathscr{T}(V) \to \mathscr{T}(V), \quad \text{such that } Q|_{\mathscr{T}_n(V)} \equiv Q_n, \text{ for every } n \geq 0. \tag{10.35}$$

We call Q the *symmetrizing operator of $\mathscr{T}(V)$*.

Theorem 10.19 (The Isomorphism of Vector Spaces $\mathscr{S}(V) \simeq \mathrm{Sym}(V)$. *Let $\mathbb{K}$ be of characteristic zero. Then the set $\mathscr{S}(V) \subset \mathscr{T}(V)$ is isomorphic, as a vector space, to $\mathrm{Sym}(V)$ (the symmetric algebra of V).*

Indeed, the map $S : \mathrm{Sym}(V) \to \mathscr{S}(V)$ is an isomorphism of vector spaces and its inverse is the restriction of the natural projection: $S^{-1} = \phi|_{\mathscr{S}(V)}$.

Finally, the map $Q : \mathscr{T}(V) \to \mathscr{T}(V)$ is a projector of $\mathscr{T}(V)$ onto $\mathscr{S}(V)$ with kernel $\mathscr{H}(V)$, so that $\mathscr{T}(V) = \mathscr{S}(V) \oplus \mathscr{H}(V)$. More precisely, we have:

  i. $Q(\mathscr{T}(V)) = \mathscr{S}(V)$ so that $Q : \mathscr{T}(V) \to \mathscr{S}(V)$ is onto;
 ii. $Q|_{\mathscr{S}(V)}$ is the identity of $\mathscr{S}(V)$ and $Q^2 = Q$ on $\mathscr{T}(V)$.
iii. $\ker(Q) = \mathscr{H}(V)$ and $\mathscr{T}(V) = \mathscr{S}(V) \oplus \mathscr{H}(V)$.

Proof. This immediately follows from Theorem 10.17 and the very definitions of S and Q.   □

By Theorem 10.19, the arrows in the following diagram are all isomorphisms of vector spaces (and the diagram is commutative):

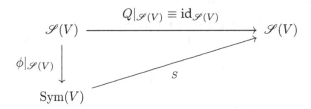

10.1.1   Basis for the Symmetric Algebra

The rest of the section is devoted to construct bases for the symmetric algebra $\mathrm{Sym}(V)$ and the symmetric tensor space $\mathscr{S}(V)$.

Throughout the section, V denotes a vector space and $\mathcal{B} = \{b_i\}_{i \in \mathfrak{I}}$ denotes a fixed indexed basis for V; moreover we assume that $\mathfrak{I}$ (or, equivalently, $\mathcal{B}$) is ordered by the relation $\preccurlyeq$. (Recall that any nonempty set can be ordered.[2]) As usual, V is identified to $\mathscr{T}_1(V) \hookrightarrow \mathscr{T}(V)$. Finally, recall that we are assuming that the underlying field $\mathbb{K}$ has characteristic zero.

Our main aim here is to prove the following theorem.

Theorem 10.20 (Basis for the Symmetric Algebra $\mathrm{Sym}(V)$). *Let $\mathcal{B} = \{b_i\}_{i \in \mathfrak{I}}$ be a basis for V and let $\mathfrak{I}$ be totally ordered by the relation $\preccurlyeq$. As usual, let ϕ and $*$ be as in Definition 10.1. For brevity, we set*

$$B_i := \phi(b_i), \quad \textit{for every } i \in \mathfrak{I}. \tag{10.36}$$

---

[2]Actually, any nonempty set A can be *well* ordered, that is, A can be endowed with a total ordering $\preccurlyeq$ (a relation on A which is reflexive, antisymmetric and transitive and such that, for any pair $a, b \in A$, it holds that $a \preccurlyeq b$ or $b \preccurlyeq a$) such that every nonempty subset B of A possesses a smallest element (i.e., there exists $b \in B$ such that $b \preccurlyeq x$ for every $x \in B$). See [108, Theorem 4.1-Appendix 2, page 892], where Zorn's Lemma is required. We shall not make explicit use of the *well* ordering.

Then we have the following results.

1. *For every $n \in \mathbb{N}$, the system*

$$\mathcal{A}_n := \left\{ B_{i_1} * \cdots * B_{i_n} \,\middle|\, i_1, \ldots, i_n \in \mathfrak{I}, \; i_1 \preccurlyeq \cdots \preccurlyeq i_n \right\}$$

is a basis for $\mathrm{Sym}_n(V)$.
2. *The set $\mathcal{A} := \{\phi(1_{\mathbb{K}})\} \cup \bigcup_{n \in \mathbb{N}} \mathcal{A}_n$, i.e., the system*

$$\mathcal{A} = \{\phi(1_{\mathbb{K}})\} \cup \left\{ B_{i_1} * \cdots * B_{i_n} \,\middle|\, n \in \mathbb{N}, \; i_1, \ldots, i_n \in \mathfrak{I}, \; i_1 \preccurlyeq \cdots \preccurlyeq i_n \right\}$$

is a basis for $\mathrm{Sym}(V)$.

Proof. See page 519.   □

Equivalently, the set $\mathcal{A}_n$ can be rewritten as the set of vectors

$$\phi\big(b_{i_1} \otimes \cdots \otimes b_{i_n}\big) = \big[b_{i_1} \otimes \cdots \otimes b_{i_n}\big]_{\mathscr{H}(V)} = \big[b_{i_1} \otimes \cdots \otimes b_{i_n}\big]_{\mathscr{H}_n(V)},$$

as $i_1, \ldots, i_n$ run through $\mathfrak{I}$ with $i_1 \preccurlyeq \cdots \preccurlyeq i_n$.

Theorem 10.21 (Basis for the Symmetric Tensor Space $\mathscr{S}(V)$). *Let $\mathcal{B} = \{b_i\}_{i \in \mathfrak{I}}$ be a basis for V and let $\mathfrak{I}$ be totally ordered by the relation $\preccurlyeq$. Then we have the following facts.*

1. *For every $n \in \mathbb{N}$, the system*

$$\mathcal{C}_n := \left\{ Q_n\big(b_{i_1} \otimes \cdots \otimes b_{i_n}\big) \,\middle|\, i_1, \ldots, i_n \in \mathfrak{I}, \; i_1 \preccurlyeq \cdots \preccurlyeq i_n \right\}$$

is a basis for $\mathscr{S}_n(V)$.
2. *The set $\mathcal{C} := \{1_{\mathbb{K}}\} \cup \bigcup_{n \in \mathbb{N}} \mathcal{C}_n$, i.e., the system*

$$\mathcal{C} = \{1_{\mathbb{K}}\} \cup \left\{ Q_n\big(b_{i_1} \otimes \cdots \otimes b_{i_n}\big) \,\middle|\, n \in \mathbb{N}, \; i_1, \ldots, i_n \in \mathfrak{I}, \; i_1 \preccurlyeq \cdots \preccurlyeq i_n \right\}$$

is a basis for $\mathscr{S}(V)$.

Proof. This is a straightforward corollary of Theorem 10.20, together with the fact that (as stated in Theorem 10.17) $S_n : \mathrm{Sym}_n(V) \to \mathscr{S}_n(V)$ is an isomorphism of vector spaces and it holds that

$$S_n(B_{i_1} * \cdots * B_{i_n}) = S_n\big(\phi(b_{i_1}) * \cdots * \phi(b_{i_n})\big) = S_n\big(\phi\big(b_{i_1} \otimes \cdots \otimes b_{i_n}\big)\big)$$

$$= S_n\big([b_{i_1} \otimes \cdots \otimes b_{i_n}]_{\mathscr{H}}\big) = Q_n\big(b_{i_1} \otimes \cdots \otimes b_{i_n}\big),$$

thanks to (10.24).   □

Equivalently (see (10.22)), the set $\mathcal{C}_n$ can be rewritten as the set of the vectors

$$\frac{1}{n!} \sum_{\sigma \in \mathfrak{S}_n} b_{i_{\sigma(1)}} \otimes \cdots \otimes b_{i_{\sigma(n)}},$$

as $i_1, \ldots, i_n$ run through $\mathcal{J}$ with $i_1 \preccurlyeq \cdots \preccurlyeq i_n$.

10.2   Proofs of Sect. 10.1

Proof (of Proposition 10.6, page 504). First we prove that $\varphi(V^n) \subseteq \mathrm{Sym}_n(V)$. By the very definition of $*$, we have

$$\varphi(v_1, \ldots, v_n) = \phi(v_1) * \cdots * \phi(v_n) = \phi(v_1 \otimes \cdots \otimes v_n)$$
$$= [v_1 \otimes \cdots \otimes v_n]_{\mathcal{H}} \in \mathrm{Sym}_n(V).$$

The n-linearity of φ is a consequence of the bilinearity of $*$ and the linearity of ϕ. The symmetry of φ is a straightforward consequence of the fact that $\mathrm{Sym}(V)$ is Abelian or, equivalently, of identity (10.6).

Let now U be a vector space and let $\beta : V^n \to U$ be a symmetric n-linear map. By the universal property of the tensor product (see Theorem 2.30), there exists a linear map $\overline{\beta} : \mathcal{T}_n(V) \to U$ such that

$$\overline{\beta}(v_1 \otimes \cdots \otimes v_n) = \beta(v_1, \ldots, v_n), \text{ for every } v_1, \ldots, v_n \in V. \tag{10.37}$$

We set

$$\beta^\varphi : \mathrm{Sym}_n(V) \to U, \quad \beta^\varphi([t]_{\mathcal{H}}) := \overline{\beta}(t), \quad \text{for every } t \in \mathcal{T}_n.$$

This map has the following properties:

1. *It is well-posed.* This follows from $\overline{\beta}|_{\mathcal{H}_n} \equiv 0$, which derives from the computation

$$\overline{\beta}\big(v_1 \otimes \cdots \otimes v_n - v_{\sigma(1)} \otimes \cdots \otimes v_{\sigma(n)}\big)$$
$$= \beta(v_1, \ldots, v_n) - \beta(v_{\sigma(1)}, \ldots, v_{\sigma(n)}) = 0,$$
$$\text{for every } \sigma \in \mathfrak{S}_n \text{ and every } v_1, \ldots, v_n \in V,$$

where the linearity of $\overline{\beta}$, (10.37) and the symmetry of β have been used.
2. *It is linear.* This follows from the linearity of $\overline{\beta}$.
3. *It satisfies (10.10).* Indeed, one has

$$\beta^\varphi(\varphi(v_1, \ldots, v_n)) = \beta^\varphi(\phi(v_1) * \cdots * \phi(v_n))$$
$$= \beta^\varphi(\phi(v_1 \otimes \cdots \otimes v_n)) = \beta^\varphi([v_1 \otimes \cdots \otimes v_n]_{\mathcal{H}})$$
$$= \overline{\beta}(v_1 \otimes \cdots \otimes v_n) = \beta(v_1, \ldots, v_n).$$

Finally, suppose $\gamma : \mathrm{Sym}_n(V) \to V$ is linear and it satisfies (10.10). Since this is equivalent to $\gamma([v_1 \otimes \cdots \otimes v_n]_{\mathscr{H}}) = \beta(v_1, \ldots, v_n)$, we see that γ is pre-assigned on a set of generators for $\mathscr{T}_n/\mathscr{H} = \mathrm{Sym}_n(V)$. Hence, β^φ is the unique linear map satisfying (10.10). This ends the proof. □

Proof (of Theorem 10.7 and Remark 10.8, page 505). Let $(A, \circledast)$ be an Abelian UA algebra and $f : V \to A$ a linear map. By the universal property of the tensor algebra, there exists a (unique) UAA morphism $\overline{f} : \mathscr{T}(V) \to A$ prolonging f. We set

$$f^\phi : \mathrm{Sym}(V) \to A, \quad f^\phi([t]_{\mathscr{H}}) := \overline{f}(t) \quad \text{for every } t \in \mathscr{T}(V).$$

The map f^ϕ has the following properties:

1. *It is well-posed.* This follows from $\overline{\beta}|_{\mathscr{H}} \equiv 0$, which derives from this computation (here $x, y \in V$ and $t, t' \in \mathscr{T}(V)$):

$$\overline{f}\big(t \cdot (x \otimes y - y \otimes x) \cdot t'\big) = \quad (\overline{f} \text{ is a UAA morphism})$$
$$\overline{f}(t) \circledast (f(x) \circledast f(y) - f(y) \circledast f(x)) \circledast \overline{f}(t') = 0,$$

   since $f(x) \circledast f(y) - f(y) \circledast f(x) = 0$, for $(A, \circledast)$ is Abelian.
2. *It is linear.* This follows from the linearity of $\overline{f}$.
3. *It satisfies (10.11).* Indeed, for every $v \in V$, one has $f^\phi(\phi(v)) = f^\phi([v]_{\mathscr{H}}) = \overline{f}(v) = f(v)$, for $\overline{f}$ prolongs f.

The computation in (1) above also proves that Remark 10.8 holds. The uniqueness of a UAA morphism satisfying (10.11) is a consequence of Remark 10.2-1. Part (ii) of the Theorem is standard. □

Proof (of Theorem 10.9, page 506). Let us set $P_n := \mathrm{span}\{\phi(v^{\otimes n}) \,|\, v \in V\}$. The inclusion $\mathrm{Sym}_n(V) \supseteq P_n$ in (10.13) is obvious. We give two proofs of the reverse inclusion:

First Proof. We argue by contradiction, supposing $P_n \subsetneqq \mathrm{Sym}_n(V)$. Since $\mathrm{Sym}_n(V) = \mathrm{span}\{\phi(v_1 \otimes \cdots \otimes v_n) \,|\, v_1, \ldots, v_n \in V\}$, this means that there exists $w \in \mathrm{Sym}_n(V) \backslash P_n$, with $w = \phi(v_1 \otimes \cdots \otimes v_n)$, for suitable $v_1, \ldots, v_n \in V$. Let $\Xi : \mathrm{Sym}_n(V) \to \mathbb{Q}$ be a linear map such that $\Xi(w) = 1$ and $\Xi \equiv 0$ on P_n. For every $\lambda_1, \ldots, \lambda_n \in \mathbb{Q}$, we have

$$\phi\big((\lambda_1 v_1 + \cdots + \lambda_n v_n)^{\otimes n}\big) \overset{(1)}{=} \big(\lambda_1 \phi(v_1) + \cdots + \lambda_n \phi(v_n)\big)^{*n}$$

$$\overset{(2)}{=} \sum_{k_1 + \cdots + k_n = n} \frac{n!}{k_1! \cdots k_n!} (\phi(v_1))^{*k_1} * \cdots * (\phi(v_n))^{*k_n} \lambda_1^{k_1} \cdots \lambda_n^{k_n}. \tag{10.38}$$

Here we applied the following facts:

(1) $\phi : \mathscr{T}(V) \to \mathrm{Sym}(V)$ is a UAA morphism.
(2) $(\mathrm{Sym}(V), *)$ is Abelian (together with the ordinary multinomial theorem).

The far left-hand side of (10.38) belongs to P_n; hence, applying Ξ to (10.38) we get

$$0 = \sum_{k_1+\cdots+k_n=n} \frac{n!}{k_1!\cdots k_n!} \Xi\left((\phi(v_1))^{*k_1} * \cdots * (\phi(v_n))^{*k_n}\right) \lambda_1^{k_1}\cdots\lambda_n^{k_n}$$

$$= \lambda_1\cdots\lambda_n \sum_{k_1=\cdots=k_n=1} n!\,\Xi\underbrace{(\phi(v_1)*\cdots*\phi(v_n))}_{=\phi(v_1\otimes\cdots\otimes v_n)=w} + \sum_{\substack{k_1+\cdots+k_n=n \\ (k_1,\ldots,k_n)\neq(1,\ldots,1)}} [\cdots]$$

$$= n!\,\lambda_1\cdots\lambda_n + \sum_{\substack{k_1+\cdots+k_n=n \\ (k_1,\ldots,k_n)\neq(1,\ldots,1)}} a(k_1,\ldots,k_n)\,\lambda_1^{k_1}\cdots\lambda_n^{k_n}$$

$$=: n!\,\lambda_1\cdots\lambda_n + H(\lambda_1,\ldots,\lambda_n),$$

for suitable $a(k_1,\ldots,k_n) \in \mathbb{Q}$. Since $\lambda_1,\ldots,\lambda_n \in \mathbb{Q}$ are arbitrary, by a continuity argument we get the polynomial identity

$$0 = n!\,\lambda_1\cdots\lambda_n + H(\lambda_1,\ldots,\lambda_n), \qquad \forall\,\lambda_1,\ldots,\lambda_n \in \mathbb{R}.$$

Applying to this identity the differential operator $\partial/(\partial_{\lambda_1}\cdots\partial_{\lambda_n})$, then setting $(\lambda_1,\ldots,\lambda_n) = 0$, and by observing that, in the sum defining H one at least among $k_1,\ldots,k_n$ is ≥ 2, we get $0 = n!$, in contradiction with the invertibility of $n!$.

Second Proof. By Remark 10.5, we have

$$\mathrm{Sym}_n(V) = \mathrm{span}\{\phi(v_1)*\cdots*\phi(v_n)\,|\,v_1,\ldots,v_n \in V\}.$$

Since $(\mathrm{Sym}(V),*)$ is Abelian, part two of Theorem 10.10 (and the hypothesis of $n!$ being invertible in $\mathbb{K}$) gives

$$\phi(v_1)*\cdots*\phi(v_n) = \frac{(-1)^n}{n!} \sum_{H\subseteq\{1,\ldots,n\}} (-1)^{\mathrm{card}(H)} \left(\sum_{i\in H}\phi(v_i)\right)^{*n}$$

$$= \frac{(-1)^n}{n!} \sum_{H\subseteq\{1,\ldots,n\}} (-1)^{\mathrm{card}(H)} \phi\left(\left(\sum_{i\in H}v_i\right)^{\otimes n}\right).$$

Since the above far right-hand side is evidently an element of P_n, we have $\mathrm{Sym}_n(V) \subseteq P_n$ and the proof is complete. $\square$

Proof (of Theorem 10.10, page 506). We prove (10.14), since (10.15) is a consequence of it (as $\mathrm{card}(\mathfrak{S}_n) = n!$).

   Let $\mathfrak{I} = \{1,\ldots,n\}$ and let $C = \{0,1\}^{\mathfrak{I}}$, that is, C is the set of the mappings of $\mathfrak{I}$ into $\{0,1\}$. If $\mathfrak{P}(\mathfrak{I})$ denotes the powerset of $\mathfrak{I}$, the mapping

$$\mathfrak{P}(\mathfrak{I}) \ni H \mapsto \chi_H \in C$$

is a bijection.[3] Note that we have

$$
\left.\begin{aligned}
\operatorname{card}(H) &= \chi_H(1) + \cdots + \chi_H(n) \\
\sum_{i\in H} x_i &= \sum_{i\in\{1,\ldots,n\}} \chi_H(i)\, x_i
\end{aligned}\right\} \qquad \text{for every } H \in \mathfrak{P}(\mathfrak{I}).
$$

Consequently, the sum on the right-hand side of (10.14) equals

$$
\sum_{H\subseteq\{1,\ldots,n\}} (-1)^{\operatorname{card}(H)} \left(\sum_{i\in H} x_i \right)^{*n}
$$

$$
= \sum_{a\in C} (-1)^{a(1)+\cdots+a(n)} \left(\sum_{i\in\{1,\ldots,n\}} a(i)\, x_i \right)^{*n}
$$

$$
= \sum_{a\in C} (-1)^{a(1)+\cdots+a(n)} \sum_{i_1,\ldots,i_n\in\{1,\ldots,n\}} a(i_1)\cdots a(i_n)\, x_{i_1} * \cdots * x_{i_n}
$$

(by interchanging the two sums)

$$
= \sum_{i_1,\ldots,i_n\in\{1,\ldots,n\}} c(i_1,\ldots,i_n)\, x_{i_1} * \cdots * x_{i_n} =: (\star),
$$

where

$$
c(i_1,\ldots,i_n) := \sum_{a\in C} (-1)^{a(1)+\cdots+a(n)} a(i_1)\cdots a(i_n). \tag{10.39}
$$

We split the Cartesian product $\mathfrak{I}^n$ in two parts: the set (say A) of the n-tuples $(i_1,\ldots,i_n)$ which are a permutation of $\mathfrak{I} = \{1,\ldots,n\}$, and its complementary set $\mathfrak{I}^n \setminus A$ (say B). Hence

$$
(\star) = \sum_{(i_1,\ldots,i_n)\in\mathfrak{I}^n} c(i_1,\ldots,i_n)\, x_{i_1} * \cdots * x_{i_n}
$$

$$
= \left\{ \sum_{(i_1,\ldots,i_n)\in A} + \sum_{(i_1,\ldots,i_n)\in B} \right\} c(i_1,\ldots,i_n)\, x_{i_1} * \cdots * x_{i_n}.
$$

We claim that

$$
\sum_{(i_1,\ldots,i_n)\in A} c(i_1,\ldots,i_n)\, x_{i_1} * \cdots * x_{i_n} = (-1)^n \sum_{\sigma\in\mathfrak{S}_n} x_{\sigma(1)} * \cdots * x_{\sigma(n)},
$$

$$
\tag{10.40}
$$

$$
\sum_{(i_1,\ldots,i_n)\in B} c(i_1,\ldots,i_n)\, x_{i_1} * \cdots * x_{i_n} = 0. \tag{10.41}
$$

---

[3] As usual, χ_H denotes the characteristic function of H (on $\mathfrak{I}$), that is, $\chi_H(i) = 1$ iff $i \in H$ and $\chi_H(i) = 0$ iff $i \in \mathfrak{I} \setminus H$; when $H = \emptyset$, this means $\chi_H \equiv 0$.

Note that (10.40)–(10.41) complete the proof of the theorem. We now turn to prove these claimed equalities:

(10.40): By the definition of A, $(i_1, \ldots, i_n) \in A$ iff there exists $\sigma \in \mathfrak{S}_n$ such that $i_1 = \sigma(1), \ldots, i_n = \sigma(n)$, so that

$$\sum_{(i_1,\ldots,i_n) \in A} c(i_1, \ldots, i_n)\, x_{i_1} * \cdots * x_{i_n}$$

$$= \sum_{\sigma \in \mathfrak{S}_n} c(\sigma(1), \ldots, \sigma(n))\, x_{\sigma(1)} * \cdots * x_{\sigma(n)}$$

$$= (-1)^n \sum_{\sigma \in \mathfrak{S}_n} x_{\sigma(1)} * \cdots * x_{\sigma(n)}.$$

Indeed, in the sum (10.39) for $c(\sigma(1), \ldots, \sigma(n))$, the only non-vanishing contribution is given by $a \equiv 1$ (this being the permutation $\{\sigma(1), \ldots, \sigma(n)\} = \{1, \ldots, n\}$), so that $c(\sigma(1), \ldots, \sigma(n)) = (-1)^n$ for every $\sigma \in \mathfrak{S}_n$.

(10.41): We prove something more, namely $c(i_1, \ldots, i_n) = 0$ for every $(i_1, \ldots, i_n) \in B$. To this end, let us fix $(i_1, \ldots, i_n) \in B$ and let us note that, by the definition of B, $(i_1, \ldots, i_n)$ is not a permutation of $(1, \ldots, n)$ so that there exists at least one $j \in \{1, \ldots, n\}$ distinct from every $i_1, \ldots, i_n$. Next we split C in two parts: the set, say C', of those $a \in C$ such that $a(j) = 0$ and the set, say C'', of those $a \in C$ such that $a(j) = 1$. Now, the map

$$C' \ni a \mapsto \widehat{a} := a + \chi_{\{j\}} \in C''$$

is a bijection. Hence

$$c(i_1, \ldots, i_n) = \left\{ \sum_{a \in C'} + \sum_{a \in C''} \right\} (-1)^{a(1)+\cdots+a(n)} a(i_1) \cdots a(i_n)$$

$$= \sum_{a \in C'} \left\{ (-1)^{a(1)+\cdots+a(n)} a(i_1) \cdots a(i_n) + (-1)^{\widehat{a}(1)+\cdots+\widehat{a}(n)} \widehat{a}(i_1) \cdots \widehat{a}(i_n) \right\}$$

$$\left(\text{note that } \widehat{a}(1) + \cdots + \widehat{a}(n) = a(1) + \cdots + a(n) + 1 \text{ and } \widehat{a}(i_1) \cdots \widehat{a}(i_n) \right.$$

$$\left. \text{equals } a(i_1) \cdots a(i_n) \text{ because } j \text{ is distinct from every } i_1, \ldots, i_n \right)$$

$$= \sum_{a \in C'} (-1)^{a(1)+\cdots+a(n)} \{1 + (-1)\}\, a(i_1) \cdots a(i_n) = 0.$$

This completes the proof. $\square$

Proof (of Theorem 10.17, page 510). We claim that the following facts hold for the map $Q_n : \mathcal{T}_n(V) \to \mathcal{T}_n(V)$ in (10.22):

1. $Q_n(\mathcal{T}_n(V)) = \mathcal{S}_n(V)$, whence $Q_n : \mathcal{T}_n(V) \to \mathcal{S}_n(V)$ is onto.
2. $Q_n^2 = Q_n$ on $\mathcal{T}_n(V)$.
3. $\ker(Q_n) = \mathcal{H}_n(V)$.

We argue as follows:

1. The first claim follows from (10.22) and from (10.27) (by recalling that $\mathscr{T}_n(V)$ is spanned by elementary tensors).
2. If $v \in \mathscr{T}_n$, by (1) we have $Q_n(v) \in \mathscr{S}_n(V)$. By definition (10.25) of $\mathscr{S}_n(V)$, Q_n leaves unchanged all elements of $\mathscr{S}_n(V)$, whence $Q_n(Q_n(v)) = Q_n(v)$ and this proves the second claim.
3. The inclusion $\mathscr{H}_n(V) \subseteq \ker(Q_n)$ is a consequence of (10.6) together with (10.23). Conversely, let $v \in \ker(Q_n)$. By Remark 10.11, we have $v - Q_n(v) \in \mathscr{H}_n(V)$, whence $v = v - 0 = v - Q_n(v)$ belongs to $\mathscr{H}_n(V)$. The inclusion $\ker(Q_n) \subseteq \mathscr{H}_n(V)$ is thus proved too.

As a consequence, for any $v \in \mathscr{T}_n(V)$, we can write

$$v = \underbrace{v - Q_n(v)}_{\in \mathscr{H}_n(V)} + \underbrace{Q_n(v)}_{\in \mathscr{S}_n(V)}, \quad \text{whence } \mathscr{T}_n(V) = \mathscr{H}_n(V) + \mathscr{S}_n(V).$$

The fact that $\mathscr{T}_n(V) = \mathscr{H}_n(V) \oplus \mathscr{S}_n(V)$ follows from $\mathscr{H}_n(V) \cap \mathscr{S}_n(V) = \{0\}$. Indeed, if $v \in \mathscr{H}_n(V) \cap \mathscr{S}_n(V)$ then

$$0 = Q_n(v) = v.$$

The first equality is a consequence of $v \in \mathscr{H}_n(V) = \ker(Q_n)$, the second equality is a consequence of $v \in \mathscr{S}(V)$ and the fact that $Q_n|_{\mathscr{S}(V)}$ is the identity (see (10.25)). By Proposition 2.2-(ii), the map

$$\mathscr{T}_n(V)/\ker(Q_n) \to Q_n(\mathscr{T}_n(V)), \quad [t]_{\ker(Q_n)} \mapsto Q_n(t) \quad (\text{for } t \in \mathscr{T}_n(V))$$

is an isomorphism of vector spaces. Actually, this is precisely the map $S_n :$ $\mathrm{Sym}_n(V) \to \mathscr{S}_n(V)$ introduced in (10.17), for $\mathscr{H}_n(V) = \ker(Q_n)$, $\mathrm{Sym}_n(V) = \mathscr{T}_n(V)/\mathscr{H}_n(V)$ (see (10.9)), $Q_n(\mathscr{T}_n(V)) = \mathscr{S}_n(V)$ and thanks to the fact that $S_n([t]_{\mathscr{H}_n(V)}) = Q_n(t)$ for every $t \in \mathscr{T}_n(V)$ (see (10.24)).

   We are left to prove (10.32). Since we have proved that $S_n : \mathrm{Sym}_n(V) \to \mathscr{S}_n(V)$ is invertible, in order to find its inverse it suffices to exhibit the right inverse. Then (10.32) will follow if we show that $S_n(\phi(s)) = s$ for every $s \in \mathscr{S}_n(V)$: this is a consequence of the following equalities, valid for any $s \in \mathscr{S}_n(V)$,

$$S_n(\phi(s)) = S_n([s]_{\mathscr{H}}) \overset{(10.24)}{=} Q_n(s) \overset{(10.25)}{=} s.$$

The theorem is thus completely proved. $\square$

Proof (of Theorem 10.20, page 512). Part (2) of the assertion follows from part (1), by recalling that (see (10.7)) $\mathrm{Sym}(V) = \bigoplus_{n=0}^{\infty} \mathrm{Sym}_n(V)$ (and $\mathrm{Sym}_0 = \phi(\mathbb{K})$). Hence, we turn to prove that, for every $n \in \mathbb{N}$, $\mathcal{A}_n$ is a basis for $\mathrm{Sym}_n(V)$.

- A_n generates $\mathrm{Sym}_n(V)$. This follows from the following facts:

$$\mathrm{Sym}_n(V) \overset{(10.7)}{=} \mathscr{T}_n(V)/\mathscr{H}(V) \overset{(10.2)}{=} \phi(\mathscr{T}_n(V)) = \quad \text{(by Proposition 2.35-1)}$$

$$= \phi\Big(\mathrm{span}\big\{b_{i_1} \otimes \cdots \otimes b_{i_n} \,\big|\, i_1, \ldots, i_n \in \mathfrak{I}\big\}\Big)$$

$$= \mathrm{span}\big\{\phi(b_{i_1} \otimes \cdots \otimes b_{i_n}) \,\big|\, i_1, \ldots, i_n \in \mathfrak{I}\big\}$$

$$\overset{(10.36)}{=} \mathrm{span}\big\{B_{i_1} * \cdots * B_{i_n} \,\big|\, i_1, \ldots, i_n \in \mathfrak{I}\big\}.$$

Now, since $(\mathrm{Sym}(V), *)$ is an Abelian algebra, any product $B_{i_1} * \cdots * B_{i_n}$ is equal to an analogous product with $i_1 \preccurlyeq \cdots \preccurlyeq i_n$, and this completes the proof of the fact that A_n generates $\mathrm{Sym}_n(V)$.

- A_n is linearly independent. Let $p \in \mathbb{N}$, and, for every $k \in \{1, \ldots, p\}$, let $\lambda_k \in \mathbb{K}$, $i_1^k, \ldots, i_n^k \in \mathfrak{I}$ with $i_1^k \preccurlyeq \ldots \preccurlyeq i_n^k$ and such that the n-tuples

$$(i_1^1, \ldots, i_n^1), \ldots, (i_1^p, \ldots, i_n^p)$$

are pairwise distinct. It is easily seen that this implies that

$(\star)$: *no two of these n-tuples are related to one another*
by a permutation of their entries.

Now we suppose that

$$0 = \sum_{k=1}^p \lambda_k \, B_{i_1^k} * \cdots * B_{i_n^k}.$$

Applying to both sides of this equality the linear map S_n, we get

$$0 = \sum_{k=1}^p \lambda_k \left(\frac{1}{n!} \sum_{\sigma_k \in \mathfrak{S}_n} b_{i_{\sigma_k(1)}^k} \otimes \cdots \otimes b_{i_{\sigma_k(n)}^k}\right),$$

by recalling that $B_{i_1^k} * \cdots * B_{i_n^k} = \big[b_{i_1^k} \otimes \cdots \otimes b_{i_n^k}\big]_{\mathscr{H}(V)}$ and by invoking (10.24). Note that, thanks to the remark in $(\star)$, if $h \neq k$, then the n-tuples

$$\big(i_{\sigma_h(1)}^h, \ldots, i_{\sigma_h(n)}^h\big), \quad \big(i_{\sigma_k(1)}^k, \ldots, i_{\sigma_k(n)}^k\big)$$

are necessarily distinct: Otherwise, since $\sigma_h, \sigma_k \in \mathfrak{S}_n$, the n-tuples $(i_1^h, \ldots, i_n^h)$ and $(i_1^k, \ldots, i_n^k)$ would be permutations of the same n-tuple, contradicting $(\star)$. By Proposition 2.35-(1) (applied to the basis $\{b_i\}_{i \in \mathfrak{I}}$) this implies that

$(2\star)$: $$\frac{\lambda_k}{n!} \sum_{\sigma \in \mathfrak{S}_n} b_{i_{\sigma(1)}^k} \otimes \cdots \otimes b_{i_{\sigma(n)}^k} = 0, \quad \text{for every } k = 1, \ldots, p.$$

Now we remark a crucial (combinatorial) fact: the summands in the above sum over $\mathfrak{S}_n$ can be *grouped together* in such a way that

$$\sum_{\sigma \in \mathfrak{S}_n} b_{i_{\sigma(1)}^k} \otimes \cdots \otimes b_{i_{\sigma(n)}^k} = \sum_{j=1}^{N} c_j \, b_{\alpha_1^j} \otimes \cdots \otimes b_{\alpha_n^j},$$

where the coefficients c_j are *positive integers* and the n-tuples

$$\left(\alpha_1^1, \ldots, \alpha_n^1\right), \ldots, \left(\alpha_1^N, \ldots, \alpha_n^N\right)$$

are *pairwise distinct*. As a consequence, once again by invoking Proposition 2.35-(1), (2$\star$) can hold if and only if $\lambda_k = 0$, for every $k \in \{1, \ldots, p\}$. This completes the proof of the fact that $\mathcal{A}_n$ is a linearly independent set.   □

Appendix A
List of the Basic Notation

Algebraic Structures

$\mathbb{K}\langle S \rangle$ 51
$\prod_{i \in \mathfrak{I}} V_i$ 53
$\bigoplus_{i \in \mathfrak{I}} V_i$ 53
$[U, V]$ 59
$\mathrm{Lie}\{U\}$ 59
$M_n(X)$, $M(X)$ 63
$\coprod_{n \in \mathbb{N}}$ 63
$\mathrm{Mo}(X)$ 65
M_{alg} 67
$\mathrm{Lib}(X)$ 69
$\mathrm{Libas}(X)$ 69
$\mathrm{Lib}_n(X)$ 69
$\mathrm{Libas}_n(X)$ 70
$U \otimes V$ 73
$\mathscr{T}_k(V)$ 75
$\mathscr{T}(V)$ 75
$U_k(V)$ 76
$\mathscr{T}_+(V)$ 76
$\mathscr{T}_{i,j}(V)$ 81
$\mathscr{T}(V) \otimes \mathscr{T}(V)$ 82
$K_k(V)$ 82
$W_k(V)$ 82
$(\mathscr{T} \otimes \mathscr{T})_+(V)$ 82
K 84
$\mathcal{L}(V)$ 85
$\mathcal{L}_n(V)$ 85
$\mathrm{Lie}(X)$ 90
$\mathfrak{a}$ 90

A. Bonfiglioli and R. Fulci, *Topics in Noncommutative Algebra*, Lecture Notes
in Mathematics 2034, DOI 10.1007/978-3-642-22597-0,
© Springer-Verlag Berlin Heidelberg 2012

$\mathcal{L}(\mathbb{K}\langle X\rangle)$ 91

$\tilde{X}$ 98

Ω_k 101

$\widehat{A}$ 101

$\widehat{\Omega}_k$ 101

Ω_k^A 103

U_k 104

W_k 104

$\widehat{\mathscr{T}}(V)$ 104

$\widehat{\mathscr{T}\otimes\mathscr{T}}(V)$ 104

$\widehat{U}_k$ 104

$\widehat{W}_k$ 104

$\mathbb{K}\langle x_1,\ldots,x_n\rangle$ 106

$\mathbb{K}[x]$, $\mathbb{K}[[x]]$ 107

$\mathscr{J}(\mathfrak{g})$, $\mathscr{J}$ 108

$\mathscr{U}(\mathfrak{g})$, $\mathscr{U}$ 108

$\mathbb{K}$ 117

$\widehat{A}_+$ 117

$\overline{\mathcal{L}(V)}$ 125

$\Gamma(V)$ 142

$A[t]$ 200

$A_k[t]$ 200

$A[[t]]$ 201

$\mathbf{H}$ 258

H_N 268

$\mathcal{H}$ 209

A_1 273

A^n 275

$(A,*,\|\cdot\|)$ 292

$(\mathfrak{g},[\cdot,\cdot]_\mathfrak{g},\|\cdot\|)$ 292

$\mathfrak{n}_n$, $\mathfrak{n}_{n+1}$ 320

$c_{i,j}^k$ 332

$\mathfrak{N}_r(X)$ 469

$\mathfrak{R}_r$, $\mathfrak{R}_{r+1}$ 469

$\mathfrak{f}_{m,r}$ 477

$\mathbb{K}[[t]]_+$ 481

E, L 488

$\mathrm{Sym}(V)$ 501

$\mathscr{H}(V)$ 501

$\mathscr{H}_n(V)$ 503

$\mathrm{Sym}_n(V)$ 503

$\mathscr{S}_n(V)$ 509

$\mathscr{S}(V)$ 511

Operations

$[\cdot,\cdot]_*, [\cdot,\cdot]_A$ 61
$w.w'$ 63
$*$ 67
$\otimes$ 73
$u \cdot v$ 75
$\bullet$ 81, 81
$\bigotimes$ 83
$[\cdot,\cdot]_\otimes$ 85
π 90
$[\cdot,\cdot]$ 90
$d(x,y)$ 94
ν 94
$\tilde{\delta}$ 98
$\sim$ 98
$[\cdot]_\sim$ 98
$+, \tilde{*}$ 99
$\widehat{*}$ 101
$t \cdot t', t \bullet t'$ 105
$t \widehat{\cdot} t', t \widehat{\bullet} t'$ 105
$\widehat{\bullet}$ 123
$\eta_N(u,v)$ 131
$\mathbf{F}_j^A(u,v)$ 179
$(k_1,a_1) \star (k_2,a_2)$ 274
$D_{(h,k)}^{\mathfrak{g}}(a,b)$ 278
$\eta_N^{\mathfrak{g}}(a,b)$ 279
$Z_j^{\mathfrak{g}}(a,b)$ 279
$\diamond_{\mathbf{n}}$ 320
$\Xi_n(a,b)$ 339
$[\cdot,\cdot]_r$ 470
$\circ$ 481
$*$ 501

Maps

χ 51
F_Σ 54
$\ell(w)$ 63, 65
φ 91
$\Phi_{a,b}$ 107
$\Phi_{a,b,c}$ 107

π 108

μ 108

j 111

$\preccurlyeq$ 111

Exp 119

Log 119

$\mathrm{Exp}_{\curvearrowright}, \mathrm{Log}_{\curvearrowright}$ 122

$\mathrm{Exp}_{\otimes}, \mathrm{Log}_{\otimes}$ 122

$\mathrm{Exp}_{\widehat{\bullet}}, \mathrm{Log}_{\widehat{\bullet}}$ 122

$\mathrm{Exp}_{\bullet}, \mathrm{Log}_{\bullet}$ 122

η_N 131

δ 133

$\widehat{\delta}$ 136

P 145

ϱ 147

P^* 148

ad 150

$\widehat{P}$ 154

$D^*_{(h,k)}$ 156

$D_{(h,k)}$ 157

$\mathbf{F}^A_j$ 179

$\mathbf{F}^*_j$ 179

$\mathbf{F}^{\widehat{\mathscr{T}}(V)}_j$ 179

$e^z, \mathbf{log}$ 181

$\mathbf{F}$ 181

exp, log 202

∂_t 203, 487

ev_a 204

$(\mathrm{ad}\, u)^{\circ h}$ 205

L_a, R_b 205

f_E 209

$S(\partial/\partial y)$ 240

g 254

D 254

$\widehat{g}, \widehat{D}$ 255

d 255

$\widehat{\theta}$ 256

π_N 268

$\mathcal{R}_N, \mathcal{R}_{N+1}$ 273

$\mathcal{R}^*_N, \mathcal{R}^*_{N+1}$ 275, 310

$D^{\mathfrak{g}}_{(h,k)}$ 278

$\eta^{\mathfrak{g}}_N$ 279

$Z^{\mathfrak{g}}_j$ 279

$\| \cdot \|_\varepsilon$ 280

$F(z)$ 302
$\mathcal{R}_N^{\mathfrak{g}}, \mathcal{R}_{N+1}^{\mathfrak{g}}$ 311
Ξ_n 339
f, δ, θ, g 466
$\Lambda(f)$ 486
ϕ 501
S_n 507
R_σ 507
Q_n 508
S, Q 511

Notation for the CBHD Theorem

$\left[u^{h_1} v^{k_1} \cdots u^{h_n} v^{k_n}\right]_{\mathfrak{g}}$ 125
$\left[u^{h_1} v^{k_1} \cdots u^{h_n} v^{k_n}\right]_{\otimes}$ 125
$Z_j(u, v)$ 126
$\blacklozenge$ 126
$|h|, h!$ 127
$\mathcal{N}_n$ 127
c_n 127
$\mathbf{c}(h, k)$ 127
$\diamond$ 128
$Z(u, v)$ 183
K_j 223, 228
B_n 224, 494
$H(x, y)$ 234
H_j^x 234
$Z_j^{\mathfrak{g}}(u, v), Z_j^*(u, v)$ 228, 279
$D_{(h,k)}^{\mathfrak{g}}(a, b)$ 278
$\eta_N^{\mathfrak{g}}(a, b)$ 279
D, Q 285, 296
$\mathcal{Q}$ 288
D_δ 297
$\widehat{D}_\rho$ 301
$\widehat{D}$ 302
γ_n 303
$\check{Q}$ 312
$\diamond_{\mathfrak{n}}$ 320
$\mathbf{D}, \mathbf{Q}$ 337
$\mathbf{E}$ 339
$\mathbf{E}_1, \mathbf{E}_2$ 339
$\mathbf{E}_0$ 341

References

1. Abbaspour, H., Moskowitz, M.: *Basic Lie Theory*, World Scientific, Hackensack, NJ, 2007
2. Abe, E.: *Hopf Algebras*, Cambridge Tracts in Mathematics, **74**. Cambridge University Press, 1980
3. Achilles, R., Bonfiglioli, A.: *The early proofs of the theorem of Campbell, Baker, Hausdorff and Dynkin*, preprint (2011)
4. Alekseev, A., Meinrenken, E.: *On the Kashiwara-Vergne Conjecture*, Invent. Math., **164**, 615–634 (2006)
5. Baker, H.F.: *On the exponential theorem for a simply transitive continuous group, and the calculation of the finite equations from the constants of structure*, Lond. M. S. Proc., **34**, 91–127 (1901)
6. Baker, H.F.: *Further applications of matrix notation to integration problems*, Lond. M. S. Proc., 34, 347–360 (1902)
7. Baker, H.F.: *On the calculation of the finite equations of a continuous group*, London M. S. Proc., **35**, 332–333 (1903)
8. Baker, H.F.: *Alternants and continuous groups*, Lond. M. S. Proc., **(2) 3**, 24–47 (1905)
9. Belinfante, J.G.F., Kolman, B.: *A Survey of Lie Groups and Lie Algebras with Applications and Computational Methods*, SIAM Monograph, 1972
10. Bialynicki-Birula, I., Mielnik, B., Plebański, J.: *Explicit solution of the continuous Baker-Campbell-Hausdorff problem and a new expression for the phase operator*, Ann. Phys., **51**, 187–200 (1969)
11. Biermann, K.-R.: *Die Mathematik und ihre Dozenten an der Berliner Universität 1810-1933. Stationen auf dem Wege eines mathematischen Zentrums von Weltgeltung*, 2nd improved ed., Akademie-Verlag, Berlin, 1988
12. Birkhoff, G.: *Continuous groups and linear spaces*, Mat. Sb., **1**, 635-642 (1936)
13. Birkhoff, G.: *Analytic groups*, Trans. Amer. Math. Soc., **43**, 61-101 (1938)
14. Blanes, S., Casas, F.: *On the convergence and optimization of the Baker-Campbell-Hausdorff formula*, Linear Algebra Appl., **378**, 135–158 (2004)
15. Blanes, S., Casas, F., Oteo, J.A., Ros, J.: *Magnus and Fer expansions for matrix differential equations: The convergence problem*, J. Phys. A: Math. Gen., **22**, 259–268 (1998)
16. Blanes, S., Casas, F., Oteo, J.A., Ros, J.: *The Magnus expansion and some of its applications*, Phys. Rep., **470**, 151–238 (2009)
17. Bonfiglioli, A.: *An ODE's version of the formula of Baker, Campbell, Dynkin and Hausdorff and the construction of Lie groups with prescribed Lie algebra*, Mediterr. J. Math., **7**, 387–414 (2010)

18. Bonfiglioli, A.: *The contribution of Ernesto Pascal to the so-called Campbell-Hausdorff formula*, in preparation (2011)
19. Bonfiglioli, A., Fulci, R.: *A new proof of the existence of free Lie algebras and an application*, to appear in ISRN Geometry (2011)
20. Bonfiglioli, A., Lanconelli, E.: *Lie groups constructed from Hörmander operators. Fundamental solutions and applications to Kolmogorov-Fokker-Planck equations*, to appear in Communications on Pure and Applied Analysis (2011)
21. Bonfiglioli, A., Lanconelli, E., Uguzzoni, F.: *Stratified Lie Groups and Potential Theory for their sub-Laplacians*, Springer Monographs in Mathematics, New York, NY, Springer 2007
22. Bonfiglioli, A., Lanconelli, E.: *On left invariant Hörmander operators in* $\mathbb{R}^N$. *Applications to Kolmogorov-Fokker-Planck equations*, Journal of Mathematical Sciences, **171**, 22–33 (2010)
23. Bose, A. *Dynkin's method of computing the terms of the Baker-Campbell-Hausdorff series*, J. Math. Phys., **30**, 2035–2037 (1989)
24. Boseck, H., Czichowski, G., Rudolph, K.-P.: *Analysis on topological groups - general Lie theory*, Teubner-Texte zur Mathematik, **37**, Leipzig: BSB B. G. Teubner Verlagsgesellschaft, 1981
25. Bourbaki, N.: *Éléments de Mathématique. Fasc. XXVI: Groupes et Algèbres de Lie. Chap. I: Algèbres de Lie*, Actualités scientifiques et industrielles, 1285. Paris: Hermann, 1960.
26. Bourbaki, N.: *Éléments de Mathématique. Fasc. XIII: Algèbre 1. Chap. I, II, III*, Actualités scientifiques et industrielles, 1285. Paris: Hermann, 1970.
27. Bourbaki, N.: *Éléments de Mathématique. Fasc. XXXVII: Groupes et Algèbres de Lie. Chap. II: Algèbres de Lie libres. Chap. III: Groupes de Lie*, Actualités scientifiques et industrielles, 1349. Paris: Hermann, 1972.
28. Campbell, J.E.: *On a law of combination of operators bearing on the theory of continuous transformation groups*, Lond. M. S. Proc., **28**, 381-390 (1897)
29. Campbell, J.E.: *Note on the theory of continuous groups*, American M. S. Bull., **4**, 407–408 (1897)
30. Campbell, J.E.: *On a law of combination of operators (second paper)*, Lond. M. S. Proc., **29**, 14–32 (1898)
31. Campbell, J.E.: *Introductory treatise on Lie's theory of finite continuous transformation groups*, Oxford: At the Clarendon Press, 1903
32. Cartier, P.: *Théorie des algèbres de Lie. Topologie des groupes de Lie.*, vol. **1**, École Normale Supérieure, Séminaire "Sophus Lie" 1954/55, Paris: Secrétariat mathématique, 1955
33. Cartier, P.: *Demonstration algébrique de la formule de Hausdorff.*, Bull. Soc. Math. France, **84**, 241–249 (1956)
34. Cartier, P.: *A primer of Hopf algebras*, in: Frontiers in number theory, physics, and geometry II (Cartier, P. editor et al.), Papers from the meeting, Les Houches, France (March 9–21, 2003), Springer, Berlin, p. 537–615 (2007)
35. Casas, F.: *Sufficient conditions for the convergence of the Magnus expansion*, J. Phys. A: Math. Theor., **40**, 15001–15017 (2007)
36. Casas, F., Murua, A.: *An efficient algorithm for computing the Baker-Campbell-Hausdorff series and some of its applications*, J. Math. Phys., **50**, 033513-1–033513-23 (2009)
37. Chen, K.-T.: *Integration of paths, geometric invariants and a generalized Baker-Hausdorff formula*, Annals of Mathematics, **65**, 163–178 (1957)
38. Chevalley, C.: *Theory of Lie Groups*, Princeton University Press, London, 1946
39. Christ, M., Nagel, A., Stein, E.M., Wainger, S.: *Singular and maximal Radon transforms: Analysis and geometry*, Ann. of Math., **150**, 489–577 (1999)
40. Citti, G., Manfredini, M.: *Implicit function theorem in Carnot-Carathéodory spaces*, Commun. Contemp. Math., **8**, 657–680 (2006)
41. Cohen, A.: *An introduction to the Lie theory of one parameter groups*, GE Stechert & Co.: New York 1931

42. Czyż, J.: *On Lie supergroups and superbundles defined via the Baker-Campbell- Hausdorff formula*, J. Geom. Phys., **6**, 595–626 (1989)

43. Czyż, J.: *Paradoxes of Measures and Dimensions Originating in Felix Hausdorff's Ideas*, World Scientific Publishing Co. Inc., 1994.

44. Czichowski, G.: *Hausdorff und die Exponentialformel in der Lie-Theorie*, Sem. Sophus Lie, **2**, 85–93 (1992)

45. Dăscălescu, S., Năstăsescu, C., Raianu, S.: *Hopf Algebras. An Introduction*, Pure and Applied Mathematics, **235**, Marcel Dekker, New York, 2001

46. Day, J., So, W., Thompson, R.C.: *Some properties of the Campbell-Baker-Hausdorff series*, Linear and Multilinear Algebra, **29**, 207–224 (1991)

47. Dixmier, J.: *L'application exponentielle dans les groupes de Lie résolubles*, Bull. Soc. Math. France, **85**, 113–121 (1957)

48. Djoković, D. Ž.: *An elementary proof of then Baker-Campbell-Hausdorff-Dynkin formula*, Math. Z., **143**, 209–211 (1975)

49. Djoković, D. Ž., Hofmann, K.H.: *The surjectivity question for the exponential function of real Lie groups: A status report*, J. Lie Theory, **7**, 171–199 (1997)

50. Douady A., Lazard, M.: *Espaces fibrés en algèbres de Lie et en groupes*, Invent. Math., **1**, 133-151 (1966)

51. Dragt, A.J., Finn, J.M.: *Lie series and invariant functions for analytic symplectic maps*, J. Math. Phys., **17**, 2215–2217 (1976)

52. Duistermaat J.J., Kolk J.A.C.: *Lie Groups*, Universitext. Berlin: Springer-Verlag, 2000.

53. Duleba, I.: *On a computationally simple form of the generalized Campbell-Baker-Hausdorff-Dynkin formula*, Systems Control Lett., **34**, 191–202 (1998)

54. Dynkin, E.B.: *Calculation of the coefficients in the Campbell-Hausdorff formula* (Russian), Dokl. Akad. Nauk SSSR (N.S.), **57**, 323–326 (1947)

55. Dynkin, E.B.: *On the representation by means of commutators of the series* $\log(e^x e^y)$ *for noncommutative x and y* (Russian), Mat. Sbornik (N.S.), **25** (67), 155–162 (1949)

56. Dynkin, E.B.: *Normed Lie algebras and analytic groups*, Uspehi Matem. Nauk (N.S.), **5**, n.1(35), 135–186 (1950)

57. Dynkin, E.B.: *Selected papers of E.B. Dynkin with commentary*, Edited by A.A. Yushkevich, G.M. Seitz and A.L. Onishchik, American Mathematical Society: Providence, RI, 2000

58. Eggert, A.:, *Extending the Campbell-Hausdorff multiplication*, Geom. Dedicata, **46**, 35–45 (1993)

59. Eichler, M.: *A new proof of the Baker-Campbell-Hausdorff formula*, J. Math. Soc. Japan, **20**, 23–25 (1968)

60. Eisenhart, L.P.: *Continuous Groups of Transformations*, Princeton, University Press, 1933. Reprinted by Dover Publications: New York, 1961

61. Eriksen, E.: *Properties of higher-order commutator products and the Baker-Hausdorff formula*, J. Math. Phys., **9**, 790–796 (1968)

62. Folland, G.B.: *Subelliptic estimates and function spaces on nilpotent Lie groups*, Ark. Mat., **13**, 161–207 (1975)

63. Folland, G.B., Stein, E.M.: *Hardy spaces on homogeneous groups*, Mathematical Notes, **28**, Princeton University Press, Princeton; University of Tokyo Press, Tokyo, 1982

64. Friedrichs, K.O.: *Mathematical aspects of the quantum theory of fields. V Fields modified by linear homogeneous forces*, Comm. Pure Appl. Math., **6**, 1–72 (1953)

65. Gilmore, R.: *Baker-Campbell-Hausdorff formulas*, J. Math. Phys., **15**, 2090–2092 (1974)

66. Glöckner, H.: *Algebras whose groups of units are Lie groups*, Stud. Math., **153**, 147–177 (2002)

67. Glöckner, H.: *Infinite-dimensional Lie groups without completeness restrictions*, in: Geometry and Analysis on Finite and Infinite-dimensional Lie Groups (eds. A. Strasburger, W. Wojtynski, J. Hilgert and K.-H. Neeb), Banach Center Publ., **55**, 43–59 (2002)

68. Glöckner, H.: *Lie group structures on quotient groups and universal complexifications for infinite-dimensional Lie groups*, J. Funct. Anal., **194**, 347–409 (2002)

69. Glöckner, H., Neeb, K.-H.: *Banach-Lie quotients, enlargibility, and universal complexifications*, J. Reine Angew. Math., **560**, 1–28 (2003)

70. Godement R.: *Introduction à la théorie des groupes de Lie. Tome 2*, Publications Mathématiques de l'Université Paris VII, Université de Paris VII, U.E.R. de Mathématiques, Paris, 1982

71. Goldberg, K.: *The formal power series for* log $e^x e^y$, Duke Math. J., **23**, 13–21 (1956)

72. Gorbatsevich, V.V., Onishchik, A.L., Vinberg, E.B.: *Foundations of Lie Theory and Lie Transformation Groups*, Springer: New York, 1997

73. Gordina, M.: *Hilbert–Schmidt groups as infinite-dimensional Lie groups and their Riemannian geometry*, J. Funct. Anal., **227**, 245–272 (2005)

74. Grivel, P.-P.: *Une histoire du théorème de Poincaré-Birkhoff-Witt*, Expo. Math., **22**, 145–184 (2004)

75. Gromov, M.: *Carnot-Carathéodory spaces seen from within*, Sub-Riemannian Geometry, Progr. Math., **144**, Birkhäuser, Basel, 1996

76. Hairer, E., Lubich, Ch., Wanner, G.: *Geometric Numerical Integration. Structure-Preserving Algorithms for Ordinary Differential Equations*, Springer Series in Computational Mathematics 31, Springer-Verlag:Berlin, 2006

77. Hall, B.C.: *Lie Groups, Lie Algebras, and Representations: an Elementary Introduction*, Graduate Texts in Mathematics, Springer-Verlag: New York, 2003

78. Hausdorff, F.: *Die symbolische Exponentialformel in der Gruppentheorie*, Leipz. Ber., **58**, 19–48 (1906)

79. Hausner, M., Schwartz, J.T.: *Lie Groups. Lie Algebras*, Notes on Mathematics and Its Applications, Gordon and Breach: New York-London-Paris, 1968

80. Hazewinkel, Michiel (ed.): *Encyclopaedia of mathematics. Volume 5. An updated and annotated translation of the Soviet "Mathematical Encyclopaedia"*, Dordrecht etc.: Kluwer Academic Publishers, **9**, 1990

81. Helgason, S.: *Differential Geometry, Lie Groups, and Symmetric Spaces*, Pure and Applied Mathematics, Academic Press: New York, 1978

82. Henrici, P.: *Applied and Computational Complex analysis. Vol. 1: Power Series, Integration, Conformal Mapping, Location of Zeros*, Pure and Applied Mathematics, **15**, John Wiley&Sons, Wiley-Interscience Publ., New York, 1974

83. Hilgert, J., Hofmann, K.H.: *On Sophus Lie's fundamental theorem*, J. Funct. Anal., **67**, 293–319 (1986)

84. Hilgert, J., Neeb, K.-H.: *Lie-Gruppen und Lie-Algebren*, Vieweg, Braunschweig, 1991

85. Hochschild, G.P.: *The structure of Lie groups*, Holden-Day Inc., San Francisco, 1965

86. Hofmann, K.H.: *Die Formel von Campbell, Hausdorff und Dynkin und die Definition Liescher Gruppen*, Theory of sets and topology, 251–264; VEB Deutsch Verlag Wissensch., Berlin, 1972

87. Hofmann, K.H.: *Théorie directe des groupes de Lie. I, II, III, IV*, Séminaire Dubreil, Algèbre, tome **27**, n.1 (1973/74), Exp. 1 (1–24), Exp. 2 (1–16), Exp. 3 (1–39), Exp. 4 (1–15); 1975

88. Hofmann, K.H., Djoković, D. Ž.: *The exponential map in real Lie algebras*, Journal of Lie theory, **7**, 177–199 (1997)

89. Hofmann, K.H., Michaelis, W.J.: *Commuting exponentials in a Lie group*, Math. Proc. Camb. Phil. Soc., **141**, 317–338 (2006)

90. Hofmann, K.H., Morris, S.A.: *Sophus Lie's third fundamental theorem and the adjoint functor theorem*, J. Group Theory, **8**, 115–133 (2005)

91. Hofmann, K.H., Morris, S.A.: *The Structure of Compact Groups. A primer for the student – a handbook for the expert*, 2nd revised edition, de Gruyter Studies in Mathematics **25**, Walter de Gruyter: Berlin, 2006 (1st edition: 1998)

92. Hofmann, K.H., Morris, S.A.: *The Lie theory of connected pro-Lie goups. A structure theory for pro-Lie algebras, pro-Lie groups, and connected locally compact groups*, EMS Tracts in Mathematics **2**, Zürich: European Mathematical Society (2007)

93. Hofmann, K.H., Neeb, K.-H.: *Pro-Lie groups which are infinite-dimensional Lie groups* Math. Proc. Camb. Philos. Soc., **146**, 351–378 (2009)

94. Hörmander, L.: *Hypoelliptic second order differential equations*, Acta Math., **119**, 147–171 (1967)

95. Humphreys, J.E.: *Introduction to Lie algebras and representation theory*, Graduate Texts in Mathematics, **9**, New York-Heidelberg: Springer-Verlag, 1972

96. Ince, E.L.: *Ordinary Differential Equations*, Longmans, Green & Co.: London, 1927 (reprinted by Dover Publications Inc.: New York 1956)

97. Iserles, A., Munthe-Kaas, H.Z., Nørsett, S.P., Zanna, A.: *Lie-group methods*, Acta Numer., **9**, 215–365 (2000)

98. Iserles, A., Nørsett, S.P.: *On the solution of linear differential equations in Lie groups*, Phil. Trans. R. Soc. A, **357**, 983–1019 (1999)

99. Jacobson, N.: *Lie algebras*, Interscience Tracts in Pure and Applied Mathematics, **10**, New York-London: Interscience Publ., John Wiley & Sons, 1962

100. Kashiwara, M., Vergne, M.: *The Campbell-Hausdorff formula and invariant hyperfunctions*, Invent. Math., **47**, 249–272 (1978)

101. Kawski, M.: *The combinatorics of nonlinear controllability and noncommuting flows*, In: Mathematical Control Theory, ICTP Lect. Notes 8, 223–311 (2002)

102. Klarsfeld, S., Oteo, J.A.: *Recursive generation of higher-order terms in the Magnus expansion*, Phys. Rev. A, **39**, 3270–3273 (1989)

103. Kolsrud, M.: *Maximal reductions in the Baker-Hausdorff formula*, J. Math. Phys., **34**, 270–286 (1993)

104. Kumar, K.: *On expanding the exponential*, J. Math. Phys., **6**, 1928–1934 (1965)

105. Kurlin, V.: *The Baker-Campbell-Hausdorff formula in the free metabelian Lie algebra*, J. Lie Theory, **17**, 525–538 (2007)

106. Klarsfeld, S., Oteo, J.A.: *The Baker-Campbell-Hausdorff formula and the convergence of Magnus expansion*, J. Phys. A: Math. Gen., **22**, 4565–4572 (1989)

107. Kobayashi, H., Hatano, N., Suzuki, M.: *Goldberg's theorem and the Baker-Campbell-Hausdorff formula*, Phys. A, **250**, 535–548 (1998)

108. Lang, S.: *Algebra*, Graduate Texts in Mathematics, **211**, New York, NY: Springer-Verlag, 2002

109. Macdonald, I.G.: *On the Baker-Campbell-Hausdorff series*, Aust. Math. Soc. Gaz., **8**, 69–95 (1981)

110. Magnani, V.: *Lipschitz continuity, Aleksandrov theorem and characterizations for H-convex functions*, Math. Ann., **334**, 199–233 (2006)

111. Magnus, W.: *A connection between the Baker-Hausdorff formula and a problem of Burnside*, Ann. of Math., **52**, 111–126 (1950)

112. Magnus, W.: *On the exponential solution of differential equations for a linear operator*, Comm. Pure Appl. Math., **7**, 649–673 (1954)

113. Magnus, W., Karrass, A., Solitar, D.: *Combinatorial Group Theory*, Interscience, New York, 1966

114. McLachlan, R.I., Quispel, R.: *Splitting methods*, Acta Numerica, **11**, 341–434 (2002)

115. Mérigot, M.: *Domaine de convergence de la série de Campbell-Hausdorff*, multigraphie interne, Nice University (1974)

116. Michel, J.: *Bases des algèbres de Lie libre; application à l'étude de la formule de Campbell-Hausdorff*, Thèse 3ème cycle, Math., Univ. Paris-Sud (Orsay), 1974

117. Michel, J.: *Calculs dans les algèbres de Lie libre: la série de Hausdorff et le problème de Burnside*, Astérisque, **38**, 139–148 (1976)

118. Michel, J.: *Bases des algèbres de Lie et série de Hausdorff*, Séminaire Dubreil. Algèbre, **27** n.1 (1973-1974), exp. n.6, 1–9 (1974)

119. Mielnik, B., Plebański, J.: *Combinatorial approach to Baker-Campbell-Hausdorff exponents*, Ann. Inst. Henri Poincaré, **12**, 215–254 (1970)

120. Milnor, J.W., Moore, J.C.: *On the structure of Hopf algebras*, Ann. Math., **81**, 211–264 (1965)

121. Moan, P.C.: *Efficient approximation of Sturm-Liouville problems using Lie-group methods*, Technical Report 1998/NA11, Department of Applied Mathematics and Theoretical Physics, University of Cambridge, England, 1998

122. Moan, P.C., Niesen, J.: *Convergence of the Magnus series*, Found. Comput. Math., **8**, 291–301 (2008)

123. Moan, P.C., Oteo, J.A.: *Convergence of the exponential Lie series*, J. Math. Phys., **42**, 501–508 (2001)

124. Montanari, A., Morbidelli, D.: *Nonsmooth Hörmander vector fields and their control balls*, Trans. Amer. Math. Soc. (to appear), http://arxiv.org/abs/0812.2369

125. Montgomery, S.: *Hopf algebras and their actions on rings*, Regional Conference Series in Mathematics, **82**, American Math. Society, Providence, 1993

126. Montgomery, D., Zippin, L.: *Topological transformation groups*, Interscience Publishers: New York, London 1955

127. Morbidelli, D.: *Fractional Sobolev norms and structure of Carnot-Carathèodory balls for Hörmander vector fields*, Studia Math., **139**, 213–242 (2000)

128. Murray, F.J.: *Perturbation theory and Lie algebras*, J. Math. Phys. **3**, 451–468 (1962)

129. Nagel, A., Stein, E.M., Wainger, S.: *Balls and metrics defined by vector fields I, basic properties*, Acta Math., **155**, 103–147 (1985)

130. Neeb, K.-H.: *Towards a Lie theory of locally convex groups*, Jpn. J. Math. (3) **1**, 291–468 (2006)

131. Newman, M., So, W., Thompson, R.C.: *Convergence domains for the Campbell-Baker-Hausdorff formula*, Linear Multilinear Algebra, **24**, 301–310 (1989)

132. Newman, M., Thompson, R. C.: *Numerical values of Goldberg's coefficients in the series for* $\log(e^x e^y)$, Math. Comp., **48**, 265–271 (1987)

133. Omori, H.: *Infinite-Dimensional Lie Groups*, Translations of Math. Monographs **158**, Amer. Math. Soc., 1997

134. Oteo, J.A.: *The Baker-Campbell-Hausdorff formula and nested commutator identities*, J. Math. Phys., **32**, 419–424 (1991)

135. Pascal, E.: *Sopra alcune indentitá fra i simboli operativi rappresentanti trasformazioni infinitesime*, Lomb. Ist. Rend. (2), **34**, 1062–1079 (1901)

136. Pascal, E.: *Sulla formola del prodotto di due trasformazioni finite e sulla dimostrazione del cosidetto secondo teorema fondamentale di Lie nella teoria dei gruppi*, Lomb. Ist. Rend. (2) **34**, 1118–1130 (1901)

137. Pascal, E.: *Sopra i numeri bernoulliani*, Lomb. Ist. Rend. (2), **35**, 377–389 (1902)

138. Pascal, E.: *Del terzo teorema di Lie sull'esistenza dei gruppi di data struttura*, Lomb. Ist. Rend. (2), **35**, 419–431 (1902)

139. Pascal, E.: *Altre ricerche sulla formola del prodotto di due trasformazioni finite e sul gruppo parametrico di un dato*, Lomb. Ist. Rend. (2), **35**, 555–567 (1902)

140. Pascal, E.: *I Gruppi Continui di Trasformazioni (Parte generale della teoria)*, Manuali Hoepli, Nr. 327 bis 328; Milano: Hoepli, **11**, 1903

141. Poincaré, H.: *Sur les groupes continus*, C. R. **128**, 1065–1069 (1899)

142. Poincaré, H.: *Sur les groupes continus*, Cambr. Trans., **18**, 220–255 (1900)

143. Poincaré, H.: *Quelques remarques sur les groupes continus*, Rend. Circ. Mat. Palermo, **15**, 321–368 (1901)

144. Reutenauer, C.: *Free Lie Algebras*, London Mathematical Society Monographs (New Series), **7**, Oxford: Clarendon Press, 1993

145. Reinsch, M.W.: *A simple expression for the terms in the Baker-Campbell-Hausdorff series*, J. Math. Phys., **41**, 2434–2442 (2000)

146. Richtmyer, R.D., Greenspan, S.: *Expansion of the Campbell-Baker-Hausdorff formula by computer* Comm. Pure Appl. Math., **18**, 107–108 (1965)

147. Robart, Th.: *Sur l'intégrabilité des sous-algèbres de Lie en dimension infinie*, Can. J. Math., **49**, 820–839 (1997)

148. Robart, Th.: *On Milnor's regularity and the path-functor for the class of infinite dimensional Lie algebras of CBH type*, in Algebras, Groups and Geometries, **21**, (2004)
149. Rossmann, W.: *Lie Groups. An Introduction Through Linear Groups*, Oxford Graduate Texts in Mathematics, **5**, Oxford University Press, 2002
150. Rothschild, L.P., Stein, E.M.: *Hypoelliptic differential operators and nilpotent groups*, Acta Math., **137**, 247–320 (1976)
151. Sagle, A.A., Walde, R.E.: *Introduction to Lie groups and Lie algebras*, Pure and Applied Mathematics, **51**, New York-London, Academic Press, 1973.
152. Schmid, R.: *Infinite-dimensional Lie groups and algebras in Mathematical Physics*, Advances in Mathematical Physics, **2010** (2010) doi:10.1155/2010/280362
153. Schmid, W.: *Poincaré and Lie groups*, Bull. Amer. Math. Soc. (New Series), **6**, 175–186 (1982)
154. Schur, F.: *Neue Begründung der Theorie der endlichen Transformationsgruppen*, Math. Ann., **35**, 161–197 (1890)
155. Schur, F.: *Beweis für die Darstellbarkeit der infinitesimalen Transformationen aller transitiven endlichen Gruppen durch Quotienten beständig convergenter Potenzreihen*, Leipz. Ber., **42**, 1–7 (1890)
156. Schur, F.: *Zur Theorie der endlichen Transformationsgruppen*, Math. Ann., **38**, 263–286 (1891)
157. Schur, F.: *Ueber den analytischen Charakter der eine endliche continuirliche Transformationsgruppe darstellenden Functionen*, Math. Ann., **41**, 509–538 (1893)
158. Sepanski, M.R.: *Compact Lie Groups*, Graduate Texts in Mathematics, **235**, Springer, New York, 2007
159. Serre, J.P.: *Lie algebras and Lie groups*, 1964 Lectures given at Harvard University; 1st ed.: W.A. Benjamin, Inc., New York, 1965; 2nd ed.: Lecture Notes in Mathematics, **1500**, Berlin: Springer-Verlag, 1992.
160. Strichartz, R.S.: *The Campbell-Baker-Hausdorff-Dynkin formula and solutions of differential equations*, J. Funct Anal., **72**, 320–345 (1987)
161. Specht, W.: *Die linearen Beziehungen zwischen höheren Kommutatoren*, Math. Z., **51**, 367–376 (1948)
162. Suzuki, M.: *On the convergence of exponential operators–the Zassenhaus formula, BCH formula and systematic approximants*, Commun. Math. Phys., **57**, 193–200 (1977)
163. Stein, E.M.: *Harmonic Analysis: Real-Variable Methods, Orthogonality, and Oscillatory Integrals*, Princeton Mathematical Series **43**, Princeton, NJ: Princeton University Press, 1993
164. Steinberg, S.: *Applications of the Lie algebraic formulas of Baker, Campbell, Hausdorff, and Zassenhaus to the calculation of explicit solutions of partial differential equations*, J. Differential Equations, **26**, 404–434 (1977)
165. Sweedler, M.E.: *Hopf Algebras*, W.A Benjamin Inc., New York, 1969
166. Thompson, R. C.: *Cyclic relations and the Goldberg coefficients in the Campbell-Baker-Hausdorff formula*, Proc. Amer. Math. Soc., **86**, 12–14 (1982)
167. Thompson, R. C.: *Convergence proof for Goldberg's exponential series*, Linear Algebra Appl., **121**, 3–7 (1989)
168. Ton-That, T., Tran, T.D.: *Poincaré's proof of the so-called Birkhoff-Witt theorem*, Rev. Histoire Math., **5**, 249–284 (1999)
169. Tu, L.W.: *Une courte démonstration de la formule de Campbell-Hausdorff*, J. Lie Theory, **14**, 501–508 (2004)
170. Van Est, W.T., Korthagen, J.: *Non-enlargible Lie algebras*, Nederl. Akad. Wetensch. Proc. Ser. A 67, Indag. Math., **26**, 15–31 (1964)
171. Varadarajan, V.S.: *Lie groups, Lie algebras, and their representations*, reprint of the 1974 edition, Graduate Texts in Mathematics **102**, New York: Springer-Verlag, 1984.
172. Varopoulos, N.T., Saloff-Coste, L., Coulhon, T.: *Analysis and geometry on groups*, Cambridge University Press, Cambridge Tracts in Mathematics **100**, Cambridge 1992
173. Vasilescu, F.-H.: *Normed Lie algebras*, Can. J. Math., **24**, 580–591 (1972)

174. Veldkamp, F.D.: *A note on the Campbell-Hausdorff formula*, J. Algebra, **62**, 477–478 (1980)
175. Vinokurov, V.A.: *The logarithm of the solution of a linear differential equation, Hausdorff's formula, and conservation laws*, Soviet Math. Dokl., **44**, 200–205 (1992). Translated from **319**, no.5 (1991)
176. Wang, Z.X., Guo, D.R.: *Special Functions*, World Scientific Publishing Co. Inc., Singapore, 1989.
177. Wei, J.: *Note on the global validity of the Baker-Hausdorff and Magnus theorems*, J. Mathematical Phys., **4**, 1337–1341 (1963)
178. Weiss, G.H., Maradudin, A.A.: *The Baker Hausdorff formula and a problem in Crystal Physics*, J. Math. Phys., **3**, 771–777 (1962)
179. Wever, F.: *Operatoren in Lieschen Ringen*, J. Reine Angew. Math., **187**, 44–55 (1949)
180. Wichmann, E.H.: *Note on the algebraic aspect of the integration of a system of ordinary linear differential equations*, J. Math. Phys., **2**, 876–880 (1961)
181. Wilcox, R.M.: *Exponential operators and parameter differentiation in quantum physics*, J. Math. Phys., **8**, 962–982 (1967)
182. Wojtyński, W.: *Quasinilpotent Banach-Lie algebras are Baker-Campbell-Hausdorff*, J. Funct. Anal., **153**, 405–413 (1998)
183. Yosida, K.: *On the exponential-formula in the metrical complete ring*, Proc. Imp. Acad. Tokyo, **13**, 301–304 (1937); Collect. Papers Fac. Sci. Osaka Univ. A **5**, Nr. 40, (1937)

Index

♦ operation, 126
◇ operation, 128

Abel's Lemma, 356
algebra, 56
   associative, 56
   Banach, 292
   derivation (of an), 57
   filtered, 58
   free (non-associative), 69
   generators, 57
   graded, 58
   Hopf, 168
   Lie, 56
   morphism, 57
   normed, 292
   of a magma, 68
   quotient, 58
   tensor, 75
   topological, 94
   UA, 56
   unital associative, 56
analytic function (on a Banach space), 347
associative algebra, 56
associativity
   Banach-Lie algebra, 313

Banach algebra, 292
Banach-Lie algebra, 293
basis
   of $\mathcal{T}(V) \otimes \mathcal{T}(V)$, 83
   of the symmetric algebra, 512
   of the symmetric tensor space, 513
   of the tensor algebra, 77
   of the tensor product, 74

Bernoulli numbers, 223, 494
bialgebra, 165
bracket, 56
   nested, 59

Campbell, Baker, Hausdorff Theorem, 141
Campbell, Baker, Hausdorff, Dynkin
      Theorem, 125
CBHD Theorem, 125
   commutative case, 121
   for formal power series, 181
coalgebra, 163
commutator, 56, 61
   nested, 59
commutator-algebra, 61
convention, 62, 83, 85, 104, 117, 125, 292

derivation, 57
   (with respect to a morphism), 231
derivative
   formal power series, 486
   polynomial, 203
Dynkin's Theorem, 151
Dynkin, Specht, Wever Lemma, 145

elementary tensor, 73
evaluation map, 204
exponential, 119
external direct sum, 53

filtered
   algebra, 58
filtration, 58

formal power series
   on a graded algebra, 101
   in one indeterminate, 480
   of an endomorphism, 209
   substitution, 481
free
   (non-associative) algebra, 69
   Lie algebra generated by a vector space,
      85
   Lie algebra related to a set, 88
   magma, 63
   monoid, 66
   nilpotent Lie algebra generated by a
      set, 469
   UA algebra, 69
   vector space, 51
free (non-associative) algebra, 69
free Lie algebra generated by a vector
     space, 85
free Lie algebra related to a set, 88
free magma, 63
free monoid, 66
free nilpotent Lie algebra generated by a
     set, 469
   stratification, 475
free UA algebra, 69
free vector space, 51
   universal property (of the), 52
Friedrichs's Theorem, 133, 137

generators
   algebra, 57
   Lie algebra, 57
   magma, 57
   monoid, 57
Goldberg presentation, 360
graded algebra, 58
   metric (related to a), 97
grading, 58
grouplike element, 167

Hausdorff group, 143
Hopf algebra, 168

isometric completion, 98
   of a UA algebra, 99

Jacobi identity, 56

LA morphism, 57
left-nested, 59
Lie algebra, 56
   Banach, 293
   commutator, 61
   free, related to a set, 88
   generated by a vector space, 85
   generators, 57
   morphism, 57
   normed, 292
   related to an associative algebra, 61
Lie bracket, 56
Lie subalgebra generated by a set, 59
logarithm, 119

magma, 56
   free, 63
   generators, 57
   ideal, 89
   morphism, 57
   unital, 56
Magnus group, 118
metric
   related to a graded algebra, 97
monoid, 56
   free, 66
   generators, 57
   morphism, 57
morphism
   algebra, 57
   LA, 57
   Lie algebra, 57
   magma, 57
   monoid, 57
   UAA, 57
   unital associative algebra, 57

nilpotency, 324, 474
norm
   compatible, 292, 293
   normally convergent (series of functions),
     278
normed algebra, 292
normed Lie algebra, 292

operation
   $\blacklozenge$, 126
   $\diamond$, 128
operator $S\,(\partial/\partial y)$, 240

PBW, 111
Poincaré-Birkhoff-Witt, 111
polynomial over a UA algebra, 199
power series
    formal, 101
primitive element, 167
product space, 53

quotient algebra, 58

right-nested, 59

semigroup, 56
stratification, 475
structure constants, 332
substitution (formal power series), 481
symmetric algebra, 501
symmetric tensor space, 509
symmetrizing operator, 508

tensor algebra, 75
    basis (of the), 77
tensor product, 73
    basis (of the), 74
    of algebras, 81
theorem
    $\mathcal{L}(\mathbb{K}\langle X \rangle) \simeq \mathrm{Lie}(X)$, 91
    $\mathcal{T}(\mathbb{K}\langle X \rangle) \simeq \mathrm{Libas}(X)$, 79
    $\widehat{\mathcal{T}} \otimes \widehat{\mathcal{T}}$ is a subalgebra of $\widehat{\mathcal{T} \otimes \mathcal{T}}$, 106
    associativity, 313
    associativity (nilpotent case), 320
    Campbell, Baker, Hausdorff, 141
    Cartier, 258
    CBHD, 125
    completion of a metric space, 98
    conjugation by an exponential, 205
    convergence, 285, 296
    convergence (improved), 301
    Djoković, 208
    double limits, 318
    Dynkin, 151
    Dynkin, Specht, Wever, 145
    Eichler, 188
    finite associativity, 310
    free Lie algebra, 91
    Friedrichs, 133, 137
    fundamental estimate, 281, 296
    Hausdorff group, 143
    nested brackets, 60
    on the completion of metric spaces, 98

on the formal power series, 102
PBW, 111
prolongation, 103
rate of convergence, 298
real-analiticity, 288
Reutenauer, 248
    third fundamental (of Lie), 328
    Varadarajan, 227
third fundamental theorem of Lie, 328
topological algebra, 94
topologically admissible family, 94
topology induced by an admissible family,
    95

UA algebra
    free, 69
UAA morphism, 57
ultrametric
    inequality, 94
    space, 94
uniformly convergent (series of function),
    278
unital associative algebra, 56
    morphism, 57
unital magma, 56
universal enveloping algebra, 108
universal property
    $\mathcal{L}(\mathbb{K}\langle x, y, z \rangle)$, 107
    $\mathcal{L}(\mathbb{K}\langle x, y \rangle)$, 107
    $\mathcal{T}(\mathbb{K}\langle x, y, z \rangle)$, 107
    $\mathcal{T}(\mathbb{K}\langle x, y \rangle)$, 107
    algebra of a magma, 68
    algebra of a monoid, 68
    external direct sum, 54
    free algebra, 70
    free Lie algebra generated by a vector
        space, 86
    free magma, 64
    free monoid, 66
    free nilpotent Lie algebra, 472
    free UA algebra, 71
    free vector space, 52
    symmetric algebra, 505
    tensor algebra, 77
    tensor product, 74
    universal enveloping algebra, 110

von Neumann, 119

word, 65

LECTURE NOTES IN MATHEMATICS   Springer

Edited by J.-M. Morel, B. Teissier; P.K. Maini

Editorial Policy (for the publication of monographs)

1. Lecture Notes aim to report new developments in all areas of mathematics and their applications - quickly, informally and at a high level. Mathematical texts analysing new developments in modelling and numerical simulation are welcome.

   Monograph manuscripts should be reasonably self-contained and rounded off. Thus they may, and often will, present not only results of the author but also related work by other people. They may be based on specialised lecture courses. Furthermore, the manuscripts should provide sufficient motivation, examples and applications. This clearly distinguishes Lecture Notes from journal articles or technical reports which normally are very concise. Articles intended for a journal but too long to be accepted by most journals, usually do not have this "lecture notes" character. For similar reasons it is unusual for doctoral theses to be accepted for the Lecture Notes series, though habilitation theses may be appropriate.

2. Manuscripts should be submitted either online at www.editorialmanager.com/lnm to Springer's mathematics editorial in Heidelberg, or to one of the series editors. In general, manuscripts will be sent out to 2 external referees for evaluation. If a decision cannot yet be reached on the basis of the first 2 reports, further referees may be contacted: The author will be informed of this. A final decision to publish can be made only on the basis of the complete manuscript, however a refereeing process leading to a preliminary decision can be based on a pre-final or incomplete manuscript. The strict minimum amount of material that will be considered should include a detailed outline describing the planned contents of each chapter, a bibliography and several sample chapters.

   Authors should be aware that incomplete or insufficiently close to final manuscripts almost always result in longer refereeing times and nevertheless unclear referees' recommendations, making further refereeing of a final draft necessary.

   Authors should also be aware that parallel submission of their manuscript to another publisher while under consideration for LNM will in general lead to immediate rejection.

3. Manuscripts should in general be submitted in English. Final manuscripts should contain at least 100 pages of mathematical text and should always include

   - a table of contents;
   - an informative introduction, with adequate motivation and perhaps some historical remarks: it should be accessible to a reader not intimately familiar with the topic treated;
   - a subject index: as a rule this is genuinely helpful for the reader.

   For evaluation purposes, manuscripts may be submitted in print or electronic form (print form is still preferred by most referees), in the latter case preferably as pdf- or zipped psfiles. Lecture Notes volumes are, as a rule, printed digitally from the authors' files. To ensure best results, authors are asked to use the LaTeX2e style files available from Springer's web-server at:

   ftp://ftp.springer.de/pub/tex/latex/svmonot1/ (for monographs) and
   ftp://ftp.springer.de/pub/tex/latex/svmultt1/ (for summer schools/tutorials).

Additional technical instructions, if necessary, are available on request from lnm@springer.com.

4. Careful preparation of the manuscripts will help keep production time short besides ensuring satisfactory appearance of the finished book in print and online. After acceptance of the manuscript authors will be asked to prepare the final LaTeX source files and also the corresponding dvi-, pdf- or zipped ps-file. The LaTeX source files are essential for producing the full-text online version of the book (see http://www.springerlink. com/openurl.asp?genre=journal&issn=0075-8434 for the existing online volumes of LNM). The actual production of a Lecture Notes volume takes approximately 12 weeks.

5. Authors receive a total of 50 free copies of their volume, but no royalties. They are entitled to a discount of 33.3 % on the price of Springer books purchased for their personal use, if ordering directly from Springer.

6. Commitment to publish is made by letter of intent rather than by signing a formal contract. Springer-Verlag secures the copyright for each volume. Authors are free to reuse material contained in their LNM volumes in later publications: a brief written (or e-mail) request for formal permission is sufficient.

Addresses:
Professor J.-M. Morel, CMLA,
École Normale Supérieure de Cachan,
61 Avenue du Président Wilson, 94235 Cachan Cedex, France
E-mail: morel@cmla.ens-cachan.fr

Professor B. Teissier, Institut Mathématique de Jussieu,
UMR 7586 du CNRS, Équipe "Géométrie et Dynamique",
175 rue du Chevaleret
75013 Paris, France
E-mail: teissier@math.jussieu.fr

For the "Mathematical Biosciences Subseries" of LNM:

Professor P. K. Maini, Center for Mathematical Biology,
Mathematical Institute, 24-29 St Giles,
Oxford OX1 3LP, UK
E-mail : maini@maths.ox.ac.uk

Springer, Mathematics Editorial, Tiergartenstr. 17,
69121 Heidelberg, Germany,
Tel.: +49 (6221) 487-8259

Fax: +49 (6221) 4876-8259
E-mail: lnm@springer.com